SCHAUM'S SOLVED PROBLEMS SERIES

1000 SOLVED PROBLEMS IN

HEAT TRANSFER

by

Donald R. Pitts, Ph.D.
The University of Tennessee

Leighton E. Sissom, Ph.D.
Tennessee Technological University

McGRAW-HILL, INC.
New York St. Louis San Francisco Auckland Bogotá Caracas Hamburg Lisbon London Madrid Mexico Milan Montreal New Delhi Paris San Juan São Paulo Singapore Sydney Tokyo Toronto

Donald R. Pitts, Ph.D., *Professor and Head, Mechanical and Aerospace Engineering, University of Tennessee, Knoxville.*
Dr. Pitts has more than 20 years of teaching experience in the thermal-fluid sciences at Tennessee Technological University, Clemson University, and the University of Tennessee.

Leighton E. Sissom, Ph.D., P.E., *President, Leighton E. Sissom and Associates, Cookeville, Tennessee.*
Dr. Sissom has 30 years of university experience in engineering teaching, research, and administration at Tennessee Technological University; his field of technical specialization is thermal-fluid sciences. He has held several national offices in the American Society of Mechanical Engineers and the American Society for Engineering Education.

Project supervision was done by The Total Book.

Library of Congress Cataloging-in-Publication Data

Pitts, Donald R.
1000 solved problems in heat transfer / Donald R. Pitts, Leighton E. Sissom.
p. cm.
ISBN 0-07-050204-8
1. Heat – Transmission – Problems, exercises, etc. I. Sissom, Leighton E. II. Title. III. Title: One thousand solved problems in heat transfer. IV. Title: Schaum's 1000 solved problems in heat transfer.
QC320.34.P58 1991
621.402′2′076 – dc20 89-12878
CIP

1 2 3 4 5 6 7 8 9 0 SHP/SHP 9 4 3 2 1 0

ISBN 0-07-050204-8

CONTENTS

TO THE STUDENT

This book contains 1010 completely solved problems in heat transfer. Most of the problems pertain to engineering heat transfer as typically encountered in junior-year and senior-year study of that topic in many of the engineering disciplines, specifically aerospace, chemical, electrical, industrial, mechanical, and nuclear engineering. You will find that a knowledge of the mathematics of differential equations, including partial differential equations, is required for certain of the solutions. The order of topics is that followed in several widely used heat transfer textbooks: conduction, convection, phase change, applications to heat exchangers, and radiation. Within each of these topics, the order of the material is important. In convection, for example, a traversal of Chapter 4 should precede any attempt to work the problems in Chapters 5, 6, and 7. In conduction, Chapter 1 should be mastered prior to attempting Chapters 2 and 3. Chapters 8 (Boiling and Condensation) and 10 (Radiation) are relatively independent of the rest of the book, whereas Chapter 9 (Heat Exchangers) depends on a knowledge of conductive and convective heat transfer. Chapter 10 (Radiation) is more self-contained than any of the others.

Numerical problems are about equally divided between the SI and the British Engineering system of units. Parallel problems in the two systems are identified by a superscript **D** on the problem numbers.

Any number appearing in a solution is "given" in the problem statement, taken from an accompanying illustration or table, or previously computed in the problem or a cited companion problem. It also may be a conversion factor or a property or factor obtainable from a cited table or illustration in an appendix. Unusual or infrequent symbols are defined in the problems where they are used.

All the problems in this book have been checked, but it is possible that some mistakes have gone unnoticed. The authors would appreciate being notified of any error that you find, for correction in future printings. We wish you Godspeed in your study of heat transfer and hope you find the problems in this book helpful.

CHAPTER 1

One-Dimensional, Steady-State Conduction

1.1 FOURIER'S LAW; THERMAL CONDUCTIVITY

1.1[D] To effect a bond between two metal plates, 1 and 6 in. thick, heat is uniformly applied through the thinner plate by a radiant heat source. The bonding epoxy must be held at 120 °F for a short time. When the heat source is adjusted to have a steady value of 96 Btu/h · in^2, a thermocouple installed on the side of the thinner plate next to the source gives a temperature of 160 °F. What is the thermal conductivity of the 1-in. metal plate? (See Fig. 1-1.)

▌ The heat-conduction equation for this one-dimensional case may be written

$$\frac{q}{A} = k\,\frac{T_{\text{hot}} - T_{\text{cold}}}{\Delta x}$$

where Δx is the plate thickness. All quantities are known except k, which can be easily found:

$$k = \frac{q}{A}\,\frac{\Delta x}{\Delta T} = \left(96\ \frac{\text{Btu}}{\text{h}\cdot\text{in}^2}\right)\left[\frac{1\ \text{in.}}{(160-120)\ \text{°F}}\right]\left(\frac{12\ \text{in.}}{1\ \text{ft}}\right) = 28.8\ \text{Btu/h}\cdot\text{ft}\cdot\text{°F}$$

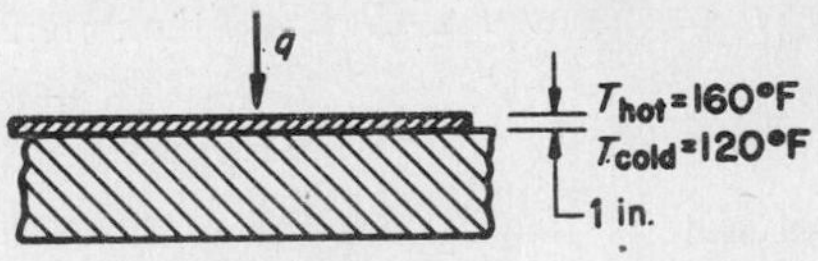

Fig. 1-1

1.2[D] Determine the thermal conductivity k for the thin plate of Fig. 1-1 for the following parameters:

$$\Delta x = 2.54\ \text{cm} \qquad T_{\text{hot}} = 71.1\ \text{°C} = 344.3\ \text{K}$$
$$q/A = 43\,600\ \text{J/s}\cdot\text{m}^2 \qquad T_{\text{cold}} = 48.9\ \text{°C} = 322\ \text{K}$$

▌ The thermal conductivity is given by

$$k = \frac{q}{A}\,\frac{\Delta x}{\Delta T} = \left(43\,600\ \frac{\text{J}}{\text{s}\cdot\text{m}^2}\right)\left(\frac{0.0254\ \text{m}}{344.3\ \text{K} - 322\ \text{K}}\right) = 49.66\ \text{J/s}\cdot\text{m}\cdot\text{K} = 49.66\ \text{W/m}\cdot\text{K}$$

since a joule is a watt-second. *Note*: Except for round-off of significant figures, the parameters of this problem are the converted values from Problem 1.1. By direct conversion, using the conversion factor given in Table A-2 (Appendix A), the thermal conductivity is

$$k = \left(28.8\ \frac{\text{Btu}}{\text{h}\cdot\text{ft}\cdot\text{°F}}\right)(1.7295771) = 49.81\ \text{J/m}\cdot\text{s}\cdot\text{K}$$

This differs only slightly from the above value but serves to illustrate the significance of significant figures.

1.3[D] Determine the steady-state heat transfer rate per unit area through a 1.5-in.-thick homogeneous slab with its two faces maintained at uniform temperatures of 100 °F and 70 °F. The thermal conductivity of the material is 0.11 Btu/h · ft · °F.

▌ The physical problem is shown in Fig. 1-2. For the steady state, the heat transfer rate is given by

$$\frac{q}{A} = -k\left(\frac{T_2 - T_1}{x_2 - x_1}\right) = -\left(\frac{0.11\ \text{Btu}}{\text{h}\cdot\text{ft}\cdot\text{°F}}\right)\left[\left(\frac{70-100}{1.5/12}\right)\frac{\text{°F}}{\text{ft}}\right] = +26.40\ \text{Btu/h}\cdot\text{ft}^2$$

1.4[D] What is the heat transfer rate through the slab of Fig. 1-2 if its thermal conductivity is 0.19 J/m · s · K and its thickness is 3.81 cm? The surface temperatures are 37.8 °C and 21.1 °C.

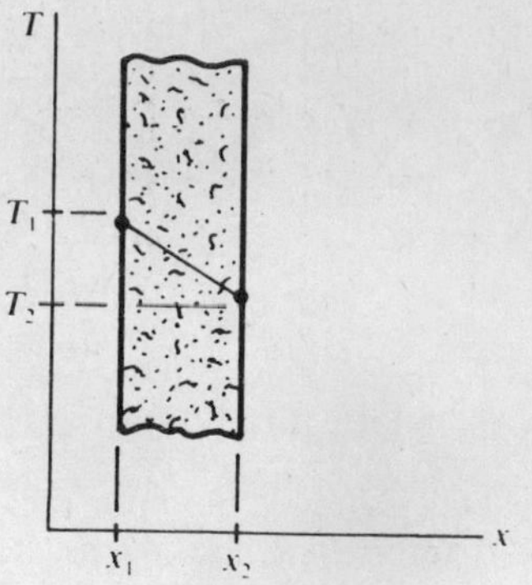

Fig. 1-2

The heat transfer rate is given by

$$\frac{q}{A} = -k\left(\frac{T_2 - T_1}{x_2 - x_1}\right) = -\left(0.19 \frac{\text{J}}{\text{m}\cdot\text{s}\cdot{}^\circ\text{C}}\right)\left[\frac{(21.1 - 37.8)\ {}^\circ\text{C}}{0.0381\ \text{m}}\right] = 83.28 \frac{\text{J}}{\text{s}\cdot\text{m}^2} = 83.28\ \text{W/m}^2$$

1.5 Estimate the value of thermal conductivity, in W/m · K, for steam at 850 K and $13.789 \times 10^6\ \text{N/m}^2$ (13.789 MPa).

In order to use Table B-5 (Appendix B), it will be convenient to convert from SI temperature and pressure units to British engineering units. Using Table A-1,

$$T\ (^\circ\text{R}) = \tfrac{9}{5}[T(\text{K})] = 1530$$

$$p = (13.789 \times 10^6\ \text{N/m}^2)\left(\frac{1\ \text{atm}}{1.01325 \times 10^5\ \text{N/m}^2}\right) = 136.1\ \text{atm}$$

From Table B-5, the critical state is $p_c = 218.3$ atm, $T_c = 1165.3$ °R. Thus

$$P_r = \frac{136.1}{218.3} \approx 0.623 \qquad T_r = \frac{1530}{1165.3} \approx 1.31$$

From Figure B-3, $k/k_1 \approx 1.14$; from Table B-4, $k_1 \approx 0.0368$ Btu/h · ft · °F. With the aid of Table A-2,

$$k \approx (1.14)\left(0.0368 \frac{\text{Btu}}{\text{h}\cdot\text{ft}\cdot{}^\circ\text{F}}\right)\left(\frac{1.72957\ \text{J/m}\cdot\text{s}\cdot\text{K}}{1\ \text{Btu/h}\cdot\text{ft}\cdot{}^\circ\text{F}}\right) = 0.0725\ \text{J/m}\cdot\text{s}\cdot\text{K} = 0.0725\ \text{W/m}\cdot\text{K}$$

1.6 Determine k in British engineering units (Btu/h · ft · °F) for nitrogen gas at 80 °F and 2000 psia. (For gases other than air or steam, Figure B-4 may be used as an approximation for k/k_1.)

From Table B-5, the critical state is $p_c = 33.5$ atm, $T_c = 226.9$ °R. Thus

$$P_r = \frac{(2000/14.7)\ \text{atm}}{33.5\ \text{atm}} = 4.06 \qquad T_r = \frac{539.7\ {}^\circ\text{R}}{226.9\ {}^\circ\text{R}} = 2.379$$

From Figure B-4, $k/k_1 \approx 1.20$. From Table B-4, $k_1 = 0.01514$ Btu/h · ft · °F, so

$$k = (0.01514)(1.20) = 0.0182\ \text{Btu/h}\cdot\text{ft}\cdot{}^\circ\text{F}$$

1.7 Verify the following conversion factors: (*a*) 1 Btu/ft² · h = 3.1525 W/m²; (*b*) 1 Btu/h = 0.292875 W; (*c*) 1 Btu/h · ft · °F = 1.729 W/m · K.

(*a*)

$$\left(1 \frac{\text{Btu}}{\text{ft}^2\cdot\text{h}}\right)\left(\frac{1054.35\ \text{J}}{1\ \text{Btu}}\right)\left(\frac{1\ \text{ft}^2}{(0.3048\ \text{m})^2}\right)\left(\frac{1\ \text{h}}{3600\ \text{s}}\right) = 3.1525\ \text{J/s}\cdot\text{m}^2 = 3.1525\ \text{W/m}^2$$

(*b*)

$$\left(1 \frac{\text{Btu}}{\text{h}}\right)\left(\frac{1054.35\ \text{J}}{1\ \text{Btu}}\right)\left(\frac{1\ \text{h}}{3600\ \text{s}}\right) = 0.292875\ \text{W}$$

(*c*)

$$\left(1 \frac{\text{Btu}}{\text{ft}\cdot\text{h}\cdot{}^\circ\text{F}}\right)\left(\frac{1054.35\ \text{J}}{1\ \text{Btu}}\right)\left(\frac{1\ \text{ft}}{0.3048\ \text{m}}\right)\left(\frac{1\ \text{h}}{3600\ \text{s}}\right)\left(\frac{9\ {}^\circ\text{F}}{5\ \text{K}}\right) = 1.72958\ \text{W/m}\cdot\text{K}$$

1.8^D A plane wall 0.5 ft thick, of a homogeneous material with $k = 0.25$ Btu/h · ft · °F, has steady and uniform surface temperatures $T_1 = 70$ °F and $T_2 = 160$ °F (see Fig. 1-3). Determine the heat transfer rate in the positive x-direction per square foot of surface area.

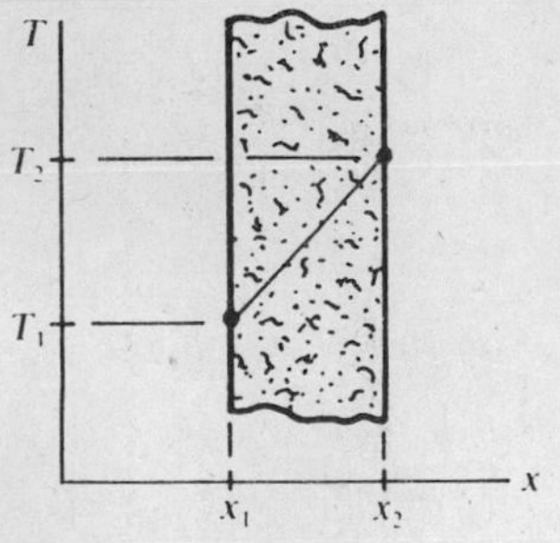

Fig. 1-3

▌ The heat transfer rate is given by

$$\frac{q}{A} = -k\left(\frac{T_2 - T_1}{x_2 - x_1}\right) = -\left(0.25 \frac{\text{Btu}}{\text{h}\cdot\text{ft}\cdot{}^\circ\text{F}}\right)\left[\frac{(160 - 70)\ {}^\circ\text{F}}{0.5\ \text{ft}}\right] = -45\ \text{Btu/h}\cdot\text{ft}^2$$

The minus sign indicates that the heat is transferred in the negative x-direction (i.e., from x_2 to x_1), which is known from the physical problem since $T_2 > T_1$.

1.9^D What is the heat transfer rate for Fig. 1-3 if the material has a thermal conductivity of 0.432 W/m·K and is 15.24 cm thick? $T_1 = 21.1\ {}^\circ\text{C}$ and $T_2 = 71.1\ {}^\circ\text{C}$.

▌ The heat transfer is given by

$$\frac{q}{A} = -k\left(\frac{T_2 - T_1}{x_2 - x_1}\right) = -\left(0.432 \frac{\text{W}}{\text{m}\cdot\text{K}}\right)\left[\left(\frac{71.1 - 21.1}{0.1524}\right)\frac{\text{K}}{\text{m}}\right] = -141.73\ \text{W/m}^2$$

1.10^D The inside temperature of a house is to be maintained such that the inside surface temperature of the windows is 70 °F. How much heat is transferred by conduction through a 4- by 8-ft picture window if the outside glass surface temperature is (*a*) 92 °F? (*b*) 32 °F? Assume $\frac{1}{8}$-in. glass.

▌
$$q = -kA\frac{\Delta T}{\Delta x} = -kA\left(\frac{T_o - T_i}{x_o - x_i}\right)$$

From Table B-2, k of glass ≈ 0.45 Btu/h·ft·°F. Assume the glass surface is the same temperature as air. Let $+x$ be the direction from the inside to the outside of the room.

(*a*)
$$q = -\left(0.45 \frac{\text{Btu}}{\text{h}\cdot\text{ft}\cdot{}^\circ\text{F}}\right)[(4\ \text{ft}) \times (8\ \text{ft})]\left[\frac{(92 - 70)\ {}^\circ\text{F}}{\left(\frac{1}{8}\ \text{in.}\right)\left(\frac{1\ \text{ft}}{12\ \text{in.}}\right)}\right] = -3.05 \times 10^4\ \text{Btu/h}\quad \text{(into room)}$$

(*b*)
$$q = -\left(0.45 \frac{\text{Btu}}{\text{h}\cdot\text{ft}\cdot{}^\circ\text{F}}\right)(32\ \text{ft}^2)\left[\frac{(32 - 70)\ {}^\circ\text{F}}{\frac{1}{96}\ \text{ft}}\right] = +5.23 \times 10^4\ \text{Btu/h}\quad \text{(out of room)}$$

1.11^D What is the heat transfer rate through a 1.2 × 2.5 m, 0.3-cm-thick, plate glass window ($k = 0.78$ W/m·K), for inside and outside temperatures of 21 °C and 33 °C, respectively?

▌
$$q = -kA\frac{\Delta T}{\Delta x} = -\left(0.78 \frac{\text{W}}{\text{m}\cdot{}^\circ\text{C}}\right)[(1.2)(2.5)\ \text{m}^2]\left[\left(\frac{33 - 21}{0.003}\right)\frac{{}^\circ\text{C}}{\text{m}}\right]$$
$$= -9360\ \text{W} = -9.36\ \text{kW}$$

The minus sign indicates transfer from outside to inside.

1.2 WALLS AND SHELLS

1.12 Determine an expression for the steady-state heat transfer rate for the two-material plane composite wall shown in Fig. 1-4(*a*).

▌ For a steady state, the one-dimensional heat flux q is the same at each interface:

$$q = q_{12} = q_{23} = q_3$$

With interface temperatures T_1, T_2, and T_3 as shown, the heat transfer rate through material a is

$$q = -k_a A\frac{T_2 - T_1}{L_a} \tag{1}$$

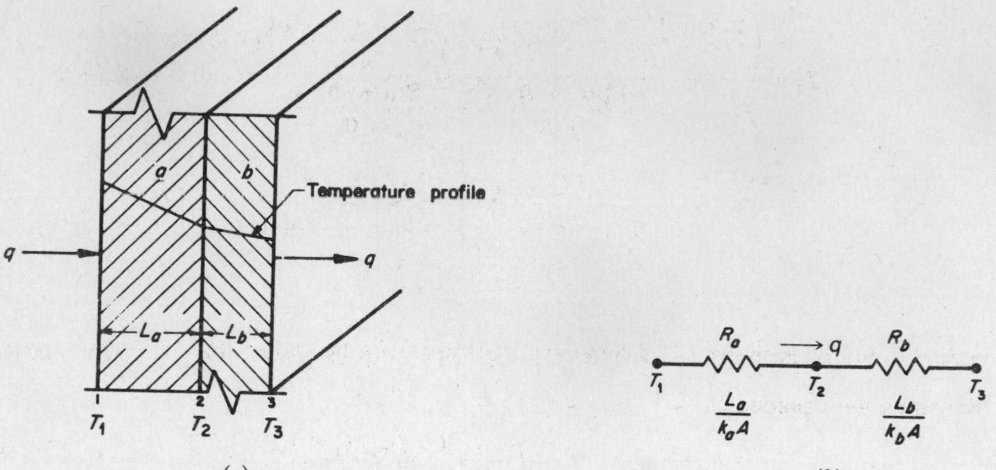

Fig. 1-4

and through material b is

$$q = -k_b A \frac{T_3 - T_2}{L_b} \tag{2}$$

Alternatively,

$$T_1 - T_2 = q \frac{L_a}{k_a A} \quad \text{and} \quad T_2 - T_3 = q \frac{L_b}{k_b A}$$

whence, by addition,

$$T_1 - T_3 = q\left(\frac{L_a}{k_a A} + \frac{L_b}{k_b A}\right)$$

and this yields the desired expression,

$$q = \frac{T_1 - T_3}{(L_a/k_a A) + (L_b/k_b A)} \tag{3}$$

If each $L_i/k_i A$ is interpreted as the *thermal resistance* of a single thickness of conductive material, and if temperature is interpreted as *thermal potential* or *voltage*, then (*3*) becomes Ohm's law for the electric network shown in Fig. 1-4(*b*).

1.13[D] For steady one-dimensional conduction, find the temperature interface T_2 in Fig. 1-5 ($k_a = 10$ Btu/h · ft · °F, $k_b = 1$ Btu/h · ft · °F), when L_b is (*a*) 1 in.; (*b*) 2 in.

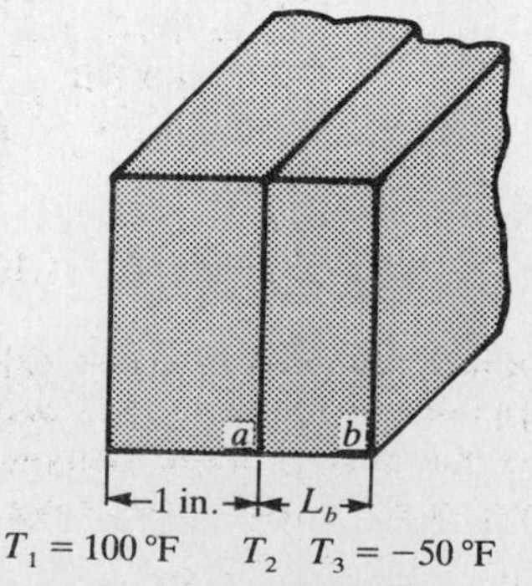

Fig. 1-5

Eliminating q between (*1*) and (*2*) of Problem 1.12 and solving for T_2, one obtains

$$T_2 = \frac{(k_a/L_a)T_1 + (k_b/L_b)T_3}{(k_a/L_a) + (k_b/L_b)} \tag{1}$$

(*a*)
$$T_2 = \frac{\left(\dfrac{10\ \text{Btu/h}\cdot\text{ft}\cdot°\text{F}}{\frac{1}{12}\ \text{ft}}\right)(100\ °\text{F}) + \left(\dfrac{1\ \text{Btu/h}\cdot\text{ft}\cdot°\text{F}}{\frac{1}{12}\ \text{ft}}\right)(-50\ °\text{F})}{\dfrac{10\ \text{Btu/h}\cdot\text{ft}\cdot°\text{F}}{\frac{1}{12}\ \text{ft}} + \dfrac{1\ \text{Btu/h}\cdot\text{ft}\cdot°\text{F}}{\frac{1}{12}\ \text{ft}}} = 86.36\ °\text{F}$$

(*b*)
$$T_2 = \frac{\left(\dfrac{10}{\frac{1}{12}}\right)(100) + \left(\dfrac{1}{\frac{1}{6}}\right)(-50)}{\left(\dfrac{10}{\frac{1}{12}}\right) + \left(\dfrac{1}{\frac{1}{6}}\right)} = 92.86\ °\text{F}$$

1.14 Show that in (*1*) of Problem 1.13 a different length unit may be chosen for the L_i than occurs in the k_i.

▌ Rewriting the equation as

$$T_2 = \frac{(k_a/k_b)(L_b/L_a)T_1 + T_3}{(k_a/k_b)(L_b/L_a) + 1}$$

we see that the length units of the k's and of the L's cancel separately.

1.15[D] Referring to Fig. 1-4(*a*), what is the interface temperature T_2 for the composite wall when $L_a = 3$ picas, $L_b = 5$ picas, $T_1 = 40\ °\text{C}$, $T_3 = -45\ °\text{C}$, $k_a = 17\ \text{W/m}\cdot\text{K}$, and $k_b = 1.8\ \text{W/m}\cdot\text{K}$. [1 pica = 4.23 mm.]

▌ By (*1*) of Problem 1.13, and the result of Problem 1.14,

$$T_2 = \frac{(\frac{17}{3})(40) + (1.8/5)(-45)}{(\frac{17}{3}) + (1.8/5)} = 34.9\ °\text{C}$$

1.16 An industrial furnace wall is constructed of 0.7-ft-thick fireclay brick having $k = 0.6\ \text{Btu/h}\cdot\text{ft}\cdot°\text{F}$. This is covered on the outer surface with a 0.1-ft-thick layer of insulating material having $k = 0.04\ \text{Btu/h}\cdot\text{ft}\cdot°\text{F}$. The innermost surface is at 1800 °F and the outermost is at 100 °F. Calculate the steady-state heat transfer rate per square foot.

▌ With the brick denoted a and the insulation b [see Fig. 1-4(*a*)], we have from (*3*) of Problem 1.12:

$$\frac{q}{A} = \frac{T_1 - T_3}{\dfrac{\Delta x_a}{k_a} + \dfrac{\Delta x_b}{k_b}} = \frac{(1800 - 100)\ °\text{F}}{\left(\dfrac{0.7}{0.6} + \dfrac{0.1}{0.04}\right)\dfrac{\text{ft}}{\text{Btu/h}\cdot\text{ft}\cdot°\text{F}}} = 464\ \text{Btu/h}\cdot\text{ft}^2 \qquad (1)$$

1.17 A frequently encountered engineering problem is the determination of the thickness of insulation that will result in a specified heat flux. If the maximum allowable heat transfer rate for the furnace of Problem 1.16 is 300 Btu/h · ft², the brick wall is unchanged, and the same insulation material is to be used, how thick must the insulation be?

▌ Substitute in (*1*) of Problem 1.16:

$$300\ \frac{\text{Btu}}{\text{h}\cdot\text{ft}^2} = \frac{1700\ °\text{F}}{\left(\dfrac{0.7\ \text{ft}}{0.6} + \dfrac{\Delta x_b}{0.04}\right)\dfrac{\text{h}\cdot\text{ft}\cdot°\text{F}}{\text{Btu}}}$$

Now solve for Δx_b:

$$\Delta x_b = [(0.04)(\tfrac{1700}{300} - 1.166)]\ \text{ft} \approx 0.18\ \text{ft}$$

1.18 The ceilings of many American homes consist of a $\frac{5}{8}$-in.-thick sheet of Celotex board supported by joists, with the space between the joists filled with loose rock wool insulation (density = 4 lbm/ft³); see Fig. 1-6(*a*). Neglecting the effect of the wooden joists, determine the heat transfer rate per unit area for a ceiling lower surface temperature of 85 °F and a rock wool upper surface temperature of 45 °F.

▌ From Table B-2, the thermal conductivity of this rock wool at 65 °F average temperature is $k_i \approx 0.0192\ \text{Btu/h}\cdot\text{ft}\cdot°\text{F}$, and that of the Celotex board is $k_c = 0.028\ \text{Btu/h}\cdot\text{ft}\cdot°\text{F}$. (The latter value is at 90 °F, which should be acceptable for this problem.) By the electrical analogy [Fig. 1-6(*b*)],

$$\frac{q}{A} = \frac{T_1 - T_3}{R_c + R_i}$$

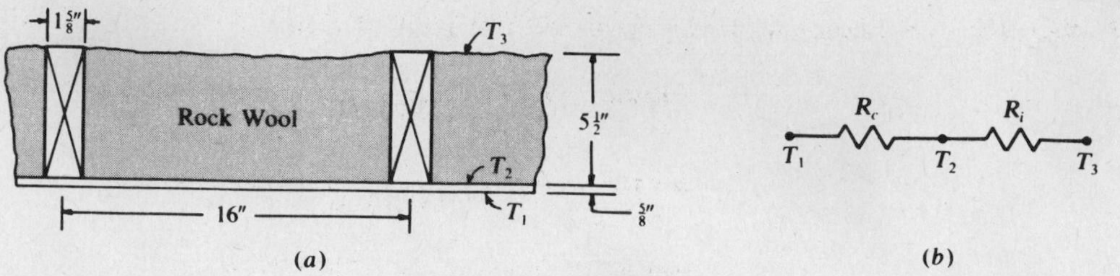

Fig. 1-6

The thermal resistances per unit area of the Celotex and the rock wool are

$$R_c = \frac{(0.625/12)\ \text{ft}}{0.028\ \text{Btu/h}\cdot\text{ft}\cdot{}^\circ\text{F}} = 1.86\ \text{h}\cdot\text{ft}^2\cdot{}^\circ\text{F/Btu} \qquad R_f = \frac{(5.5/12)\ \text{ft}}{0.0192\ \text{Btu/h}\cdot\text{ft}\cdot{}^\circ\text{F}} = 23.87\ \text{h}\cdot\text{ft}^2\cdot{}^\circ\text{F/Btu}$$

Hence
$$\frac{q}{A} = \frac{(85-45)\ {}^\circ\text{F}}{(1.86+23.87)\ \text{h}\cdot\text{ft}^2\cdot{}^\circ\text{F/Btu}} = 1.55\ \text{Btu/h}\cdot\text{ft}^2$$

1.19 A composite three-layered wall is formed of a 0.5-cm-thick aluminum plate, a 0.25-cm-thick layer of sheet asbestos, and a 2.0-cm-thick layer of rock wool (density = 64 kg/m^3); the asbestos is the center layer. The outer aluminum surface is at 500 °C, and the outer rock wool surface is at 50 °C. Determine the heat flow per unit area.

▌ From Table B-1 at 500 °C, $k_{al} = 268.08$ W/m · K; from Table B-2 at 51 °C, $k_{asb} = 0.1660$ W/m · K; from Table B-2 at 93 °C, $k_{rw} = 0.0548$ W/m · K. Note that the asbestos sheet average temperature is certainly greater than 51 °C, but this is the only k-value available in Appendix B. Also, the outer two thermal conductivities were taken at reasonable temperatures for this problem. The heat flow is given by

$$\frac{q}{A} = \frac{(500-50)\ {}^\circ\text{C}}{\left(\dfrac{0.5\times10^{-2}}{268.08} + \dfrac{0.25\times10^{-2}}{0.1660} + \dfrac{2.0\times10^{-2}}{0.0548}\right)\dfrac{\text{m}}{\text{W/m}\cdot{}^\circ\text{C}}} = 1184.08\ \text{W/m}^2$$

1.20 Repeat Problem 1.19 for a two-layer composite wall consisting of the asbestos sheet and the rock wool, with the same overall temperature difference.

▌ Using the thermal conductivities found in Problem 1.19, we get

$$\frac{q}{A} = \frac{(500-50)\ \text{K}}{\left(\dfrac{0.25\times10^{-2}}{0.1660} + \dfrac{2.0\times10^{-2}}{0.0548}\right)\dfrac{\text{m}}{\text{W/m}\cdot\text{K}}} = 1184.14\ \text{W/m}^2$$

Clearly, the thermal resistance of the aluminum sheet is negligibly small.

1.21 A load-bearing masonry wall consists of a 4-in. brick outer face with $\frac{3}{8}$-in. mortar joints, an 8-in. concrete wall, and a $\frac{5}{8}$-in. insulating board on the inside. (See Fig. 1-7.) The outer and inner temperatures are 10 °F and 70 °F, respectively. Determine the heat flux.

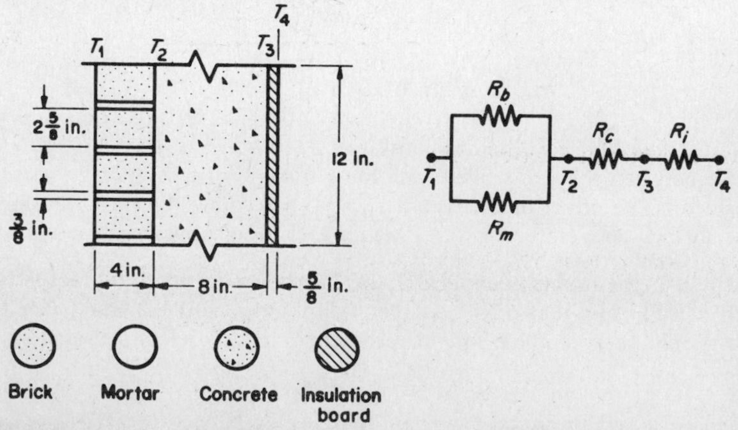

Fig. 1-7

Resistances are calculated for unit height and unit width of wall.

$$\textit{Brick} \qquad R_b = \frac{\frac{4}{12}}{k_b(4)(2\frac{5}{8})/12} = 1.002\ \text{h}\cdot{}^\circ\text{F/Btu}$$

$$\textit{Mortar} \qquad R_m = \frac{\frac{4}{12}}{k_m(4)(\frac{3}{8})/12} = 6.06\ \text{h}\cdot{}^\circ\text{F/Btu}$$

$$\textit{Concrete} \qquad R_c = \frac{\frac{8}{12}}{k_c(1)} = 1.234\ \text{h}\cdot{}^\circ\text{F/Btu}$$

$$\textit{Insulating board} \qquad R_i = \frac{(\frac{5}{8})(\frac{1}{12})}{k_i(1)} = 0.578\ \text{h}\cdot{}^\circ\text{F/Btu}$$

where physical properties are taken from Appendix B, except $k_m = 0.44$ Btu/h · ft · °F. Then, by analogy with dc electric network theory,

$$q = \frac{T_1 - T_4}{\Sigma R} = \frac{T_1 - T_4}{[R_b R_m/(R_b + R_m)] + R_c + R_i} = \frac{10 - 70}{[(1.002)(6.06)]/(1.002 + 6.06)] + 1.234 + 0.578} = -22.5\ \text{Btu/h}$$

The minus sign indicates that the direction of the heat flux is from the inside to the outside, since we were solving for the heat flux from outer to inner surfaces. This simplified method of treating a composite wall assumes one-dimensional heat transfer, which is exactly true only in the simple case where all materials have equal thermal conductivities.

1.22 For most materials, thermal conductivity varies with temperature; often one assumes a linear relationship $k = k_0(1 + bT)$. For this linear relationship, evaluate the thermal conductivity at the mean temperature of the material.

By linearity, the conductivity at the mean is the mean of the extreme conductivities. Hence, writing $k_m \equiv k[(T_1 + T_2)/2]$, $k_1 \equiv k(T_1)$, $k_2 \equiv k(T_2)$, we have

$$k_m = \frac{k_1 + k_2}{2}$$

1.23 Obtain an analytical expression for the temperature distribution $T(x)$ in the plane wall of Fig. 1-8 having uniform surface temperatures T_1 and T_2 at x_1 and x_2, respectively, and a thermal conductivity which varies linearly with temperature: $k = k_0(1 + bT)$.

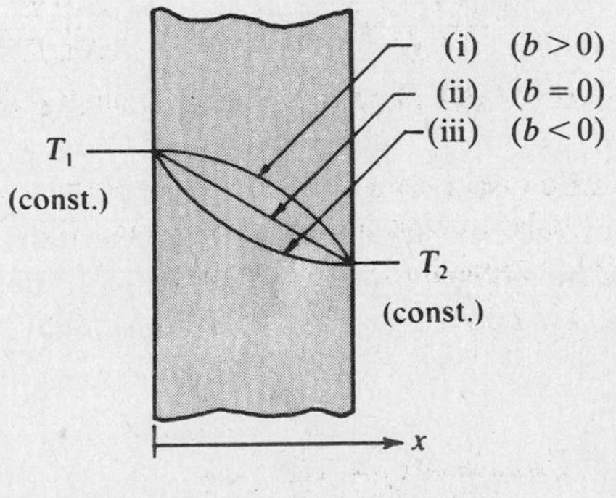

Fig. 1-8

Separating variables in Fourier's law,

$$q = -kA\,\frac{dT}{dx}$$

and integrating from x_1 to arbitrary x yields

$$\frac{q}{A}\int_{x_1}^{x} dx = -k_0 \int_{T_1}^{T} (1 + bT)\,dT$$

$$\frac{q}{A}(x - x_1) = -k_0\left[\left(T + \frac{b}{2}T^2\right) - \left(T_1 + \frac{b}{2}T_1^2\right)\right]$$

or, completing the squares on the right,

$$-\frac{2q}{bAk_0}(x-x_1)=\left(T+\frac{1}{b}\right)^2-\left(T_1+\frac{1}{b}\right)^2 \tag{1}$$

Since the curve (*1*) must pass through the point (x_2, T_2),

$$\frac{2q}{bAk_0}(x_2-x_1)-\left(T_2+\frac{1}{b}\right)^2-\left(T_1+\frac{1}{b}\right)^2=2\left(T_m+\frac{1}{b}\right)(T_2-T_1) \tag{2}$$

with $T_m=(T_1+T_2)/2$. Divide (*1*) by (*2*) and solve for T, obtaining

$$T=-\frac{1}{b}+\left[\left(T_1+\frac{1}{b}\right)^2+2\left(T_m+\frac{1}{b}\right)\left(\frac{T_2-T_1}{x_2-x_1}\right)(x-x_1)\right]^{1/2} \qquad (b\neq 0) \tag{3}$$

If $b=0$, Fourier's law gives directly

$$T=T_1+\frac{T_2-T_1}{x_2-x_1}(x-x_1) \qquad (b=0) \tag{4}$$

1.24 Refer to Problem 1.23. Verify that for $b>0$ the temperature profile is concave downward, as drawn in Fig. 1-8.

I Differentiate Fourier's law,

$$-\frac{q}{Ak_0}=(1+bT)\frac{dT}{dx}$$

with respect to x:

$$0=b\left(\frac{dT}{dx}\right)^2+(1+bT)\frac{d^2T}{dx^2}$$

or

$$\frac{d^2T}{dx^2}=-\left(\frac{b}{1+bT}\right)\left(\frac{dT}{dx}\right)^2<0$$

The second derivative is negative, as was to be shown. Note that the result is independent of the boundary conditions.

1.25 What is the thermal conductivity, in W/m · K, of diatomaceous earth at 250 °C?

I Interpolating between the values given in Table B-2,

$$\frac{k_{250}-0.039}{0.046-0.039}=\frac{250-204}{316-204}$$

from which

$$k_{250}=0.042\ \text{Btu/h}\cdot\text{ft}\cdot{}^\circ\text{F}=0.073\ \text{W/m}\cdot\text{K}$$

1.26 Determine an expression for the heat loss from an insulated thick-walled pipe, as shown in Fig. 1-9. The inside pipe radius is r_1, the outside pipe radius (and inside insulation radius) is r_2, and the outside insulation radius is r_3. The temperatures corresponding to the three radii are T_1, T_2, and T_3, respectively.

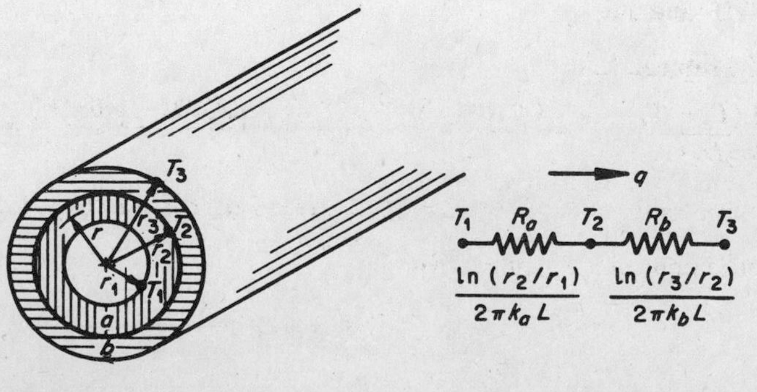

Fig. 1-9

For a pipe of length L, the area for radial heat flow in the system is $A = 2\pi rL$, and substitution in Fourier's law yields

$$q = -(2\pi rL)k\,\frac{dT}{dr} \tag{1}$$

For the pipe section (material a) only, the boundary conditions are $T = T_1$, at $r = r_1$; $T = T_2$, at $r = r_2$. Thus, integration of (1) gives

$$q = \frac{2\pi k_a L(T_1 - T_2)}{\ln (r_2/r_1)} \tag{2}$$

It follows that the thermal resistance of material a is

$$R_a = \frac{\ln (r_2/r_1)}{2\pi k_a L} \tag{3}$$

In a completely analogous manner, the thermal resistance of the insulation material is

$$R_b = \frac{\ln (r_3/r_2)}{2\pi k_b L} \tag{4}$$

For the resistances (2) and (3), the electrical analogy gives

$$q = \frac{2\pi L(T_1 - T_3)}{\dfrac{\ln (r_2/r_1)}{k_a} + \dfrac{\ln (r_3/r_2)}{k_b}} \tag{5}$$

1.27 A 3-in.-OD steel pipe is covered with a $\frac{1}{2}$-in. layer of asbestos ($\rho = 36\ \text{lbm/ft}^3$) that is covered in turn with a 2-in. layer of glass wool ($\rho = 4\ \text{lbm/ft}^3$). Determine (*a*) the steady-state heat transfer rate per lineal foot and (*b*) the interfacial temperature between the asbestos and the glass wool if the pipe outer surface temperature is 400 °F and the glass wool outer temperature is 100 °F.

From Table B-2 at 392 °F and 200 °F, for asbestos and glass wool, respectively,

$$k_{\text{asb}} = 0.120\ \frac{\text{Btu}}{\text{h}\cdot\text{ft}\cdot{}^\circ\text{F}} \qquad k_{\text{gw}} = 0.0317\ \frac{\text{Btu}}{\text{h}\cdot\text{ft}\cdot{}^\circ\text{F}}$$

These temperatures should be reasonably close to the average values for the two materials.

(*a*) By (5) of Problem 1.26,

$$\frac{q}{L} = \frac{2\pi[(400 - 100)\ {}^\circ\text{F}]}{\dfrac{\ln (2.0/1.5)}{k_{\text{asb}}} + \dfrac{\ln (4.0/2.0)}{k_{\text{gw}}}} = \frac{2\pi(300\ {}^\circ\text{F})}{\dfrac{0.288}{0.120\ \text{Btu/h}\cdot\text{ft}\cdot{}^\circ\text{F}} + \dfrac{0.693}{0.0317\ \text{Btu/h}\cdot\text{ft}\cdot{}^\circ\text{F}}} = 77.69\ \text{Btu/h}\cdot\text{ft}$$

(*b*) Since the heat transfer rate per foot is now known, the single-layer equation, (2) of Problem 1.26, can be used to determine the interfacial temperature. Thus, considering the glass wool layer,

$$T_2 - (100\ {}^\circ\text{F}) = \left[\frac{77.69}{2\pi(0.0317)}\ln\frac{4.0}{2.0}\right]\ {}^\circ\text{F} \qquad \text{or} \qquad T_2 = 370.37\ {}^\circ\text{F}$$

We could have used instead the asbestos layer to find T_2, since q/L is the same for either layer in steady state. Note that the average temperature of the glass wool is about 235 °F, which is reasonably close to the temperature at which the thermal conductivity was chosen.

1.28 Determine the steady-state heat transfer rate q from a 20-ft-long cylinder having inside and outside radii of 9 and 10 ft. The thermal conductivity is 1.0 Btu/h · ft · °F, and the inside and outside temperatures are, respectively, 400 °F and 100 °F.

Using (2) of Problem 1.26,

$$q = \frac{-k2\pi L(T_2 - T_1)}{\ln (r_2/r_1)} = \frac{-(1\ \text{Btu/h}\cdot\text{ft}\cdot{}^\circ\text{F})(2\pi)(20\ \text{ft})[(100 - 400)\ {}^\circ\text{F}]}{\ln \frac{10}{9}} = 358\,357\ \text{Btu/h} \qquad \text{(heat out)}$$

1.29 Verify that the large radii in Problem 1.28 allow the cylinder to be treated as a flat plate.

The area of the equivalent plate is given by

$$A = 2\pi\left(\frac{r_1 + r_2}{2}\right)L = 2\pi\left(\frac{10 + 9}{2}\right)(20) = 380\pi\ \text{ft}^2$$

Then,

$$q = -kA\frac{T_2 - T_1}{r_2 - r_1} = -(1\ \text{Btu/h}\cdot\text{ft}\cdot{}^\circ\text{F})(380\pi\ \text{ft}^2)\left[\frac{(100-400)\ {}^\circ\text{F}}{(10-9)\ \text{ft}}\right] = 358\,142\ \text{Btu/h} \qquad \text{(heat out)}$$

The error in q is only about 0.06%.

1.30 A hollow cylinder having inner and outer radii r_1 and r_2, respectively, is subjected to a steady heat transfer resulting in constant surface temperatures T_1 and T_2 at r_1 and r_2. If the thermal conductivity can be expressed as $k = k_0(1 + b\theta)$, where $\theta = T - T_{\text{ref}}$, obtain an expression for the heat transfer rate per unit length of the cylinder.

▌ In terms of θ, Fourier's law is

$$q = -kA_r\frac{d\theta}{dr}$$

where A_r is the area normal to r. Substituting $k = k_0(1 + b\theta)$ and $A_r = 2\pi rL$, where L is cylinder length, results in

$$q = -k_0(1 + b\theta)(2\pi rL)\frac{d\theta}{dr} \qquad \text{or} \qquad \frac{q}{L}\frac{dr}{r} = -2\pi k_0(1 + b\theta)\,d\theta$$

In the steady state, q/L is constant. Hence, integrating through the cylinder wall yields, after rearrangement,

$$\frac{q}{L} = -2\pi k_0\left[1 + \frac{b}{2}(\theta_2 + \theta_1)\right]\frac{\theta_2 - \theta_1}{\ln(r_2/r_1)} \qquad (1)$$

But the mean thermal conductivity is given by

$$k_m = k_0\left(1 + b\,\frac{\theta_2 + \theta_1}{2}\right)$$

so that (*1*) simplifies to

$$\frac{q}{L} = -2\pi k_m\frac{\theta_2 - \theta_1}{\ln(r_2/r_1)} = -2\pi k_m\frac{T_2 - T_1}{\ln(r_2/r_1)} \qquad (2)$$

1.31 In a single experiment with a 2-cm-thick sheet of pure copper having one face maintained at 500 °C and the other at 300 °C, the measured heat flux per unit area is 3.633 MW/m² (1 MW = 10^6 W). A reported value of k for this material at 150 °C is 371.9 W/m · K. Represent k in the form $k = k_{\text{ref}}(1 + b\theta)$ (see Problem 1.30), and determine b.

▌ For the experiment,

$$\frac{q}{A} = 3.633 \times 10^6\ \frac{\text{W}}{m^2} = k_m\left|\frac{\Delta T}{\Delta x}\right| = k_m\left(\frac{200\ \text{K}}{2\times 10^{-2}\ \text{m}}\right) \qquad \text{or} \qquad k_m = 363.3\ \text{W/m}\cdot\text{K}$$

Using $T_{\text{ref}} = 150$ °C and $k_{\text{ref}} = 371.9$ W/m · K, we have the fitting

$$363.3\ \frac{\text{W}}{\text{m}\cdot\text{K}} = \left(371.9\ \frac{\text{W}}{\text{m}\cdot\text{K}}\right)\left[1 + b\,\frac{(500-150)+(300-150)}{2}\ \text{K}\right]$$

which gives $b = -9.25 \times 10^{-5}\ \text{K}^{-1}$.

1.32 Test the accuracy of the linear formula

$$k = (371.9)[1 - (9.25\times 10^{-5})(T - 150)] \quad \text{W/m}\cdot\text{K}$$

for T in °C, found in Problem 1.31.

▌ The accuracy of this expression may be checked by comparison with tabulated values in Appendix B. At 300 °C,

$$k = (371.9)[1 - (9.25\times 10^{-5})(300 - 150)] = 366.74\ \text{W/m}\cdot\text{K} = 212.04\ \text{Btu/h}\cdot\text{ft}\cdot{}^\circ\text{F}$$

At 500 °C,

$$k = (371.9)[1 - (9.25\times 10^{-5})(500 - 150)] = 359.86\ \text{W/m}\cdot\text{K} = 208.06\ \text{Btu/h}\cdot\text{ft}\cdot{}^\circ\text{F}$$

These values are in reasonable agreement with the tabulated data.

1.33 A thick-walled copper cylinder has an inside radius of 1 cm and an outside radius of 1.8 cm. The inner and outer surfaces are held at 305 °C and 295 °C, respectively. Assume that k varies linearly with temperature, with T_{ref}, k_{ref}, and b the same as in Problem 1.31. Determine the rate of heat loss per unit length.

▌ From Problem 1.31, with $T_m = 300$ °C,

$$k_m = (371.9)[1 - (9.25 \times 10^{-5})(150)] = 366.74 \text{ W/m}\cdot\text{K}$$

Hence, by (2) of Problem 1.30,

$$\frac{q}{L} = -2\pi\left(366.74 \frac{\text{W}}{\text{m}\cdot\text{K}}\right)\frac{(295 - 305)\text{ K}}{\ln(1.8/1)} = 39.203 \text{ kW/m}$$

1.34 An industrial oven wall is made up of 9 in. of fireclay brick (inside), 4 in. of kaolin insulating brick, and 8 in. of masonry brick (outside). The inner and outer surface temperatures, T_1 and T_4, are 400 °F and 70 °F, respectively. Neglecting the resistance of the mortar joints, determine the temperatures T_2 and T_3 at the intermediate surfaces. (See Fig. 1-10.)

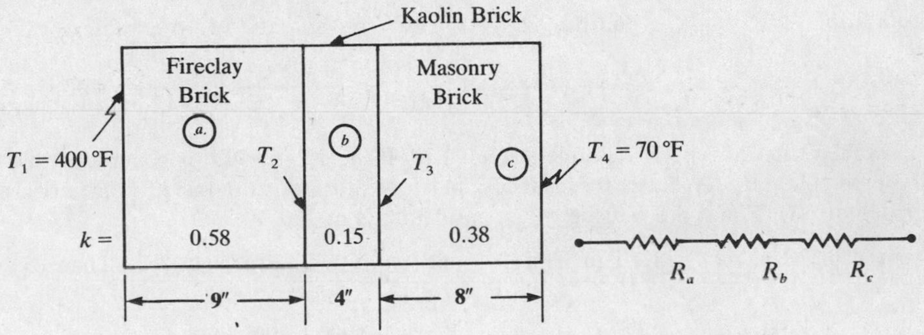

Fig. 1-10

▌
$$q = \frac{T_1 - T_4}{R_a + R_b + R_c} = \frac{T_1 - T_2}{R_a} = \frac{T_1 - T_3}{R_a + R_b} \tag{1}$$

From the data,

$$AR_a = \frac{L_a}{k_a} = \frac{\frac{9}{12}\text{ ft}}{0.58 \text{ Btu/h}\cdot\text{ft}\cdot°\text{F}} = 1.293 \text{ h}\cdot\text{ft}^2\cdot°\text{F/Btu}$$

$$AR_b = 2.222 \text{ h}\cdot\text{ft}^2\cdot°\text{F/Btu}$$

$$AR_c = 1.754 \text{ h}\cdot\text{ft}^2\cdot°\text{F/Btu}$$

Therefore, solving the equations (1),

$$T_2 = T_1 - \frac{R_a}{R_a + R_b + R_c}(T_1 - T_4) = 400 - \frac{1.293}{5.269}(330) = 319 \text{ °F}$$

$$T_3 = T_1 - \frac{R_a + R_b}{R_a + R_b + R_c}(T_1 - T_4) = 400 - \frac{3.515}{5.269}(330) = 180 \text{ °F}$$

1.35 Determine expressions for the heat transfer rate and thermal resistance in the spherical shell $r_1 \le r \le r_2$.

▌ The area at radius r is $4\pi r^2$, which may be substituted into Fourier's law, giving

$$q = -k(4\pi r^2)\frac{dT}{dr}$$

Separating variables and integrating, we get

$$q\int_{r_1}^{r_2}\frac{dr}{r^2} = -4\pi k\int_{T_1}^{T_2} dT$$

$$q\left(\frac{1}{r_1} - \frac{1}{r_2}\right) = 4\pi k(T_1 - T_2) \tag{1}$$

$$q = \frac{4\pi k(T_1 - T_2)}{(1/r_1) - (1/r_2)}$$

From (1) we read off the thermal resistance as

$$R = \frac{(1/r_1) - (1/r_2)}{4\pi k} \tag{2}$$

As with the plane wall and cylinder, these results may also be extended to multilayer cases.

1.36[D] The inside surface of a spherical iron shell, of inner radius 5 in. and outer radius 6 in., is at a uniform temperature of 100 °F. The entire sphere is immersed in boiling water at 212 °F. Assuming that the outer surface is at the same temperature as the water, what is the heat transfer rate?

From Table B-1, $k_{\text{iron}} = 36.6\ \text{Btu/h}\cdot\text{ft}\cdot\text{°F}$; ($1$) of Problem 1.35 now gives

$$q = \frac{4\pi k(T_1 - T_2)}{(1/r_1) - (1/r_2)} = \frac{4\pi(36.6\ \text{Btu/h}\cdot\text{ft}\cdot\text{°F})[(100 - 212)\ \text{°F}]}{[(\frac{12}{5}) - (\frac{12}{6})]\ \text{ft}^{-1}} = -12\,871.49\ \text{Btu/h} \qquad \text{inward}$$

1.37[D] Find the inside temperature of a spherical iron shell immersed in boiling water at 100 °C, if the inner radius is 12 cm, the outer radius is 15 cm, and the heat transfer rate into the sphere is 3.77 kW.

From Table B-1, $k_{\text{iron}} = (36.6)(1.7296) = 63.30\ \text{W/m}\cdot\text{K}$, and ($1$) of Problem 1.35 yields

$$T_1 = T_2 + \frac{q}{4\pi k}\left(\frac{1}{r_1} - \frac{1}{r_2}\right) = (100\ \text{°C}) + \frac{(-3770\ \text{W})}{4\pi(63.30\ \text{W/m}\cdot\text{°C})}\left(\frac{1}{0.012\ \text{m}} - \frac{1}{0.015\ \text{m}}\right) = 21.04\ \text{°C}$$

1.38 A thick-walled tube of type-347 stainless steel, 1-in.-ID and 2-in.-OD, is covered with a 2-in. layer of molded pipe covering. What is the heat loss per foot of tube if the inside wall temperature of the pipe is maintained at 900 °F and the outside of the insulation is at 100 °F?

From Table B-1, $k_s = 11.0\ \text{Btu/h}\cdot\text{ft}\cdot\text{°F}$ and $k_i = 0.051\ \text{Btu/h}\cdot\text{ft}\cdot\text{°F}$. Then ($5$) of Problem 1.26 gives

$$\frac{q}{L} = \frac{2\pi(T_1 - T_3)}{\dfrac{\ln(r_2/r_1)}{k_s} + \dfrac{\ln(r_3/r_2)}{k_i}} = \frac{2\pi(900 - 100)}{\dfrac{\ln(1/0.5)}{11.0} + \dfrac{\ln(3/1)}{0.051}} = 232.39\ \text{Btu/h}\cdot\text{ft}$$

1.39 (*a*) A 4-in.-OD mild-steel pipe carrying chilled water is to be covered with a 1-in. thickness of asbestos and a 1-in. thickness of rock wool; see Fig. 1-11. The pipe surface is at 35 °F, and the outer insulation surface is at 80 °F. To achieve the optimum insulating effect, which insulation should be placed next to the pipe? (*b*) For a single layer of insulation of thickness τ, express the diameter D of the pipe in terms of τ, L, k, q, and ΔT.

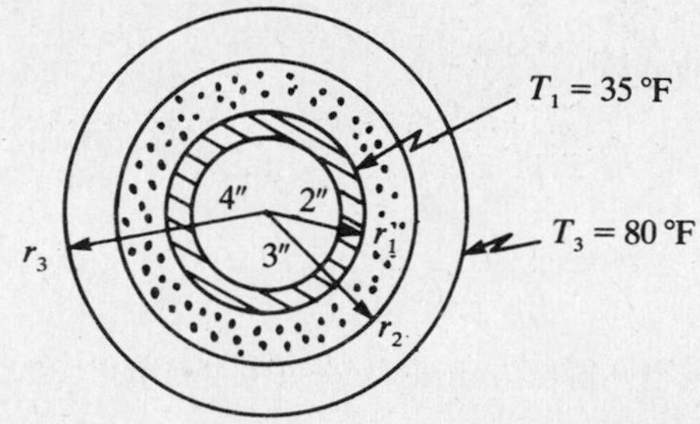

Fig. 1-11

From Table B-2, $k_{\text{asb}} = 0.087\ \text{Btu/h}\cdot\text{ft}\cdot\text{°F}$ and $k_{\text{rw}} = 0.017\ \text{Btu/h}\cdot\text{ft}\cdot\text{°F}$.

(*a*) The basic equation $q = \Delta T/\Sigma R_{\text{th}}$ implies that for minimum q, ΣR_{th} must be a maximum. For each cylindrical layer, Problem 1.26 gives

$$R_{\text{th}} = \frac{\ln(r_o/r_i)}{2\pi L}$$

Case I. Assume asbestos next to pipe. For 1 ft of pipe,

$$\Sigma R_{\text{th}} = \frac{\ln\frac{3}{2}}{2\pi(1)(0.087)} + \frac{\ln\frac{4}{3}}{2\pi(1)(0.017)} = 3.4326\ \text{h}\cdot\text{°F/Btu}$$

Case II. Assume rock wool next to pipe. For 1 ft of pipe,

$$\Sigma R_{\text{th}} = \frac{\ln \frac{3}{2}}{2\pi(1)(0.017)} + \frac{\ln \frac{4}{3}}{2\pi(1)(0.087)} = 4.321 \text{ h}\cdot{}^\circ\text{F/Btu}$$

Thus, rock wool should be placed next to the pipe.

(*b*) By (2) of Problem 1.26,

$$q = \frac{2\pi k L(\Delta T)}{\ln\left(\frac{D/2+\tau}{D/2}\right)}$$

Solving for D,

$$D = \frac{2\tau}{\exp\left[2\pi k L(\Delta T)/q\right] - 1}$$

1.40 What thickness of rock-wool insulation is needed to guarantee that the temperature of the outer surface of a kitchen oven will not exceed 120 °F? The maximum oven temperature, maintained by the thermostatic control, is 500 °F; the maximum steady-state electric energy input is 4.4 kW, and the oven is 2 by 2 by 2 ft.

▌ The surface area of the oven is $6(2 \text{ ft})(2 \text{ ft}) = 24 \text{ ft}^2$; hence,

$$\frac{q}{A} = \frac{(4400 \text{ W})(3.4129 \text{ Btu/h}\cdot\text{W})}{24 \text{ ft}^2} = 625.7 \text{ Btu/h}\cdot\text{ft}^2$$

By Fourier's law, $q/A = -k(\Delta T/\Delta x)$, or

$$\Delta x = \frac{-k\,\Delta T}{q/A} = \frac{-(0.017 \text{ Btu/h}\cdot\text{ft}\cdot{}^\circ\text{F})[(120-500)\ {}^\circ\text{F}]}{625.7 \text{ Btu/h}\cdot\text{ft}^2} = 0.0103 \text{ ft} = 0.1239 \text{ in.}$$

1.41 A 10-in.-OD mild-steel pipe of $\frac{1}{2}$-in. wall thickness is covered by two layers of insulating materials. This pipe is used to convey heated air in a test arrangement and is not a permanent installation. (See Fig. 1-12.) (*a*) Determine the heat transfer rate per lineal foot of pipe. (*b*) Determine T_3. (*c*) If the insulating materials could be interchanged, i.e., if the 85% magnesia could be placed next to the pipe, what would be the heat transfer per lineal foot? (The magnesia is not suitable for long-term use above 600 °F.) (*d*) Compare the results of parts (*a*) and (*c*).

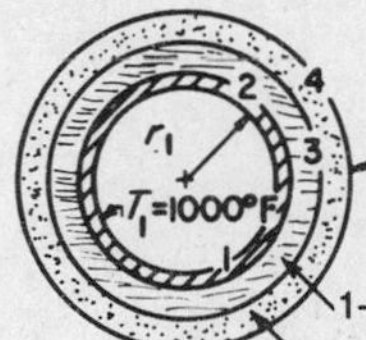

Fig. 1-12

▌ $k_{12} = 22 \text{ Btu/h}\cdot\text{ft}\cdot{}^\circ\text{F}$ $\quad k_{23} = 0.051 \text{ Btu/h}\cdot\text{ft}\cdot{}^\circ\text{F}$ $\quad k_{34} = 0.032 \text{ Btu/h}\cdot\text{ft}\cdot{}^\circ\text{F}$

(*a*) By Problem 1.26,

$$\frac{q}{L} = \frac{2\pi(T_1 - T_4)}{\frac{\ln(r_2/r_1)}{k_{12}} + \frac{\ln(r_3/r_2)}{k_{23}} + \frac{\ln(r_4/r_3)}{k_{34}}} = \frac{2\pi(1000-32)}{\frac{\ln(5/4.5)}{22} + \frac{\ln(6/5)}{0.051} + \frac{\ln(7/6)}{0.032}} = 724.5 \text{ Btu/h}\cdot\text{ft}$$

(*b*)

$$724.5 = \frac{2\pi(T_1 - T_3)}{\frac{\ln(r_2/r_1)}{k_{12}} + \frac{\ln(r_3/r_2)}{k_{23}}}$$

$$T_3 = T_1 - \frac{724.5}{2\pi}\left[\frac{\ln(r_2/r_1)}{k_{12}} + \frac{\ln(r_3/r_2)}{k_{23}}\right]$$

$$= 1000 - (115.31)(0.00478 + 3.58) = 586.61\ {}^\circ\text{F}$$

(c) $$\frac{q}{L}=\frac{2\pi(T_1-T_4)}{\dfrac{\ln(r_2/r_1)}{k_{12}}+\dfrac{\ln(r_3/r_2)}{k_{34}}+\dfrac{\ln(r_4/r_3)}{k_{23}}}=\frac{1936\pi}{\dfrac{\ln(5/4.5)}{22}+\dfrac{\ln(6/5)}{0.032}+\dfrac{\ln(7/6)}{0.051}}=697.09\ \text{Btu/h}\cdot\text{ft}$$

(d) No significant change, since the resistance of the 85% magnesia and the high-temperature insulation are approximately the same.

1.42 Derive an expression for the total heat transfer rate from the tank, of thermal conductivity k, shown in Fig. 1-13.

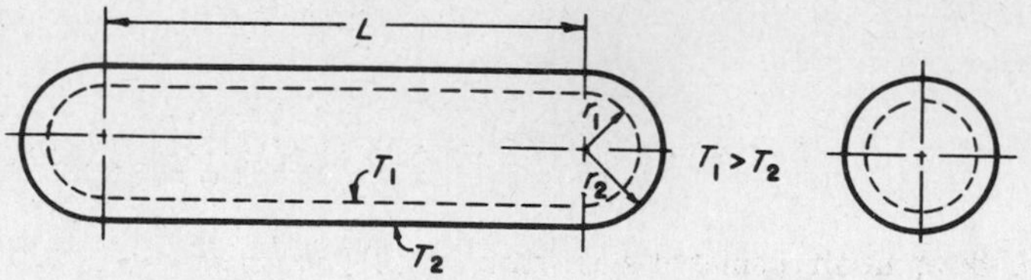

Fig. 1-13

The total heat transfer rate will be the sum of that for a cylinder (Problem 1.26) and for a sphere (Problem 1.35):

$$q=\frac{2\pi kL(T_1-T_2)}{\ln(r_2/r_1)}+\frac{4\pi k(T_1-T_2)}{(1/r_1)-(1/r_2)}=2\pi k(T_1-T_2)\left[\frac{L}{\ln(r_2/r_1)}+\frac{2}{(1/r_1)-(1/r_2)}\right]$$

1.43 The annular space between two thin concentric spherical shells having radii of 4 and 6 in. is filled with bulk powdered insulating material. What wattage is required from an electric resistance heater located in the center of the smaller sphere in order to maintain a temperature difference between the two spherical shells equal to 40 °F? Assume that the average thermal conductivity of the insulating material is 0.04 Btu/h · ft · °F.

By (1) of Problem 1.35,

$$q=\frac{4\pi k(T_1-T_2)}{(1/r_1)-(1/r_2)}=\frac{4\pi(0.04)(40)}{(\frac{12}{4})-(\frac{12}{6})}=20.1\ \text{Btu/h}$$

Thus, $P=(20.1\ \text{Btu/h})/(3.4129\ \text{Btu/h}\cdot\text{W})=5.89\ \text{W}$.

1.44 Two materials are in perfect thermal contact. The steady-state temperature distributions are as shown in Fig. 1-14. If the thermal conductivity of the 3-in.-thick material is $k_{12}=10$ Btu/h · ft · °F, what is the thermal conductivity k_{23} of the 5-in.-thick material?

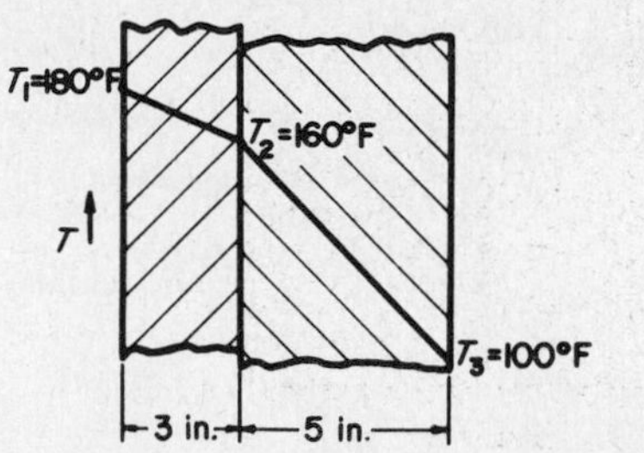

Fig. 1-14

Fourier's law gives

$$\frac{q}{A}=-10\left(\frac{160-180}{3}\right)=-k_{23}\left(\frac{100-160}{5}\right)$$

whence $k_{23}=\frac{50}{9}=5.\dot{5}$ Btu/h · ft · °F.

1.45 In Problem 1.16 what is the steady-state interfacial temperature T_2 between the brick and the insulation?

Working from the brick side of the furnace, the heat transfer rate is given by

$$\frac{q}{A}=\frac{T_1-T_2}{\Delta x_a/k_a}$$

Thus,

$$T_2 = T_1 - \left(\frac{q}{A}\right)\left(\frac{\Delta x_a}{k_a}\right) = 1800 - (464)\left(\frac{0.7}{0.6}\right) = 1259\ °F$$

1.46 For the ceiling of Problem 1.18, what is the temperature at the center of the rock wool insulation ($2\frac{3}{4}$ in. below upper surface)?

▮ Using the upper surface as a reference, the heat transfer rate is given by

$$\frac{q}{A} = \frac{T - T_3}{\Delta x / k_{rw}}$$

whence

$$T = T_3 + \left(\frac{q}{A}\right)\left(\frac{\Delta x}{k_{rw}}\right) = 45 + (1.55)\left(\frac{2.75/12}{0.0192}\right) = 63.5\ °F$$

1.47 Steam at 120 °C flows in an insulated steel pipe. The pipe inner radius is 10 cm and the outer radius is 11 cm. This is covered with a 3-cm-thick layer of asbestos having a density of 577 kg/m^3. The outer asbestos surface is at 45 °C. Using mild-steel thermal-conductivity data at 100 °C and asbestos thermal-conductivity data at 70 °C, determine the heat transfer rate from the steam per meter of pipe length.

▮ For the asbestos,

$$\rho = \frac{577\ \text{kg/m}^3}{16.02\ \dfrac{\text{kg/m}^3}{\text{lbm/ft}^3}} = 36\ \text{lbm/ft}^3$$

Interpolating in Table B-2 to get the thermal conductivity for the asbestos,

$$k_b = [0.087 + (0.7)(0.111 - 0.087)](1.7296) = 0.180\ \text{W/m}\cdot°\text{C}$$

Then, by (5) of Problem 1.26,

$$\frac{q}{L} = \frac{2\pi(T_1 - T_3)}{\dfrac{\ln(r_2/r_1)}{k_a} + \dfrac{\ln(r_3/r_2)}{k_b}} = \frac{2\pi(120 - 45)}{\dfrac{\ln\frac{11}{10}}{(26)(1.7296)} + \dfrac{\ln\frac{14}{11}}{0.180}} = 351\ \text{W/m}$$

1.48 Find the heat transfer rate through the composite wall shown in Fig. 1-15. Assume a width (into the paper) of 12 in.

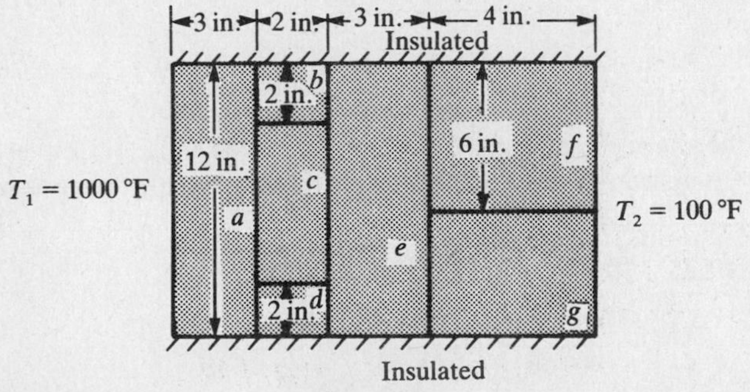

Fig. 1-15

material	thermal conductivity k, Btu/h·ft·°F
a	100
b	40
c	10
d	60
e	30
f	40
g	20

▮ $$q = \frac{\Delta T}{\Sigma R_{th}} \qquad \text{where} \qquad R_{th} = \frac{\Delta x}{kA}$$

The electrical analog for the wall is given in Fig. 1-16, where the thermal resistances are evaluated as

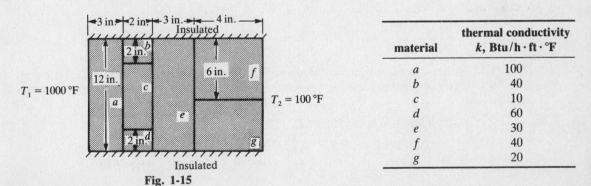

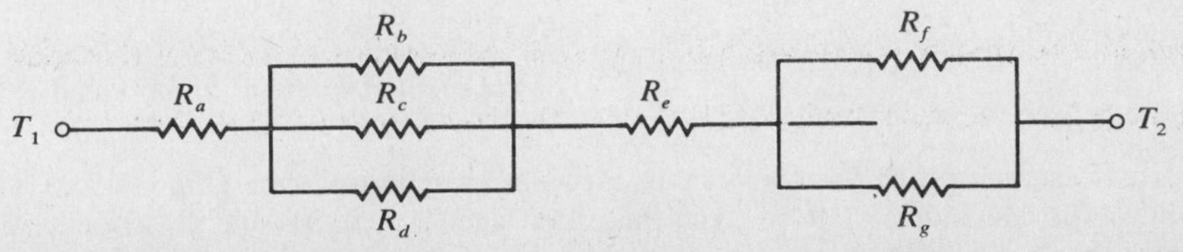

Fig. 1-16

$$R_a = \frac{\frac{3}{12}\text{ ft}}{(100\text{ Btu/h}\cdot\text{ft}\cdot{}^\circ\text{F})(1\text{ ft}^2)} = 0.00250\text{ h}\cdot{}^\circ\text{F/Btu}$$

$$R_b = \frac{\frac{2}{12}}{(40)(\frac{2}{12})} = \frac{1}{40}\,\frac{\text{h}\cdot{}^\circ\text{F}}{\text{Btu}} \qquad R_e = \frac{\frac{3}{12}}{(30)(1)} = 0.00833\text{ h}\cdot{}^\circ\text{F/Btu}$$

$$R_c = \frac{\frac{2}{12}}{(10)(\frac{8}{12})} = \frac{1}{40}\,\frac{\text{h}\cdot{}^\circ\text{F}}{\text{Btu}} \qquad R_f = \frac{\frac{4}{12}}{(40)(\frac{6}{12})} = \frac{1}{60}\,\frac{\text{h}\cdot{}^\circ\text{F}}{\text{Btu}}$$

$$R_d = \frac{\frac{2}{12}}{(60)(\frac{2}{12})} = \frac{1}{60}\,\frac{\text{h}\cdot{}^\circ\text{F}}{\text{Btu}} \qquad R_g = \frac{\frac{4}{12}}{(20)(\frac{6}{12})} = \frac{1}{30}\,\frac{\text{h}\cdot{}^\circ\text{F}}{\text{Btu}}$$

For the parallel circuits:

$$\frac{1}{R_1} = \frac{1}{R_b} + \frac{1}{R_c} + \frac{1}{R_d} = 40 + 40 + 60 = 140 \qquad \text{or} \qquad R_1 = \frac{1}{140} = 0.00714\text{ h}\cdot{}^\circ\text{F/Btu}$$

$$\frac{1}{R_2} = \frac{1}{R_f} + \frac{1}{R_g} = 60 + 30 = 90 \qquad \text{or} \qquad R_2 = \frac{1}{90} = 0.0111\text{ h}\cdot{}^\circ\text{F/Btu}$$

Therefore

$$\Sigma R_{\text{th}} = R_a + R_1 + R_e + R_2 = 0.00250 + 0.00714 + 0.00833 + 0.0111 = 0.02907\text{ h}\cdot{}^\circ\text{F/Btu}$$

whence $q = 900/0.02907 = 30\,960$ Btu/h.

1.49 A spherical shell, of radii r_1 and r_2, is made of a material with thermal conductivity $k = k_0T^2$. Derive an expression for the heat transfer rate, if the surfaces are held at temperatures T_1 and T_2, respectively.

▌ Fourier's law takes the form

$$q = -(k_0T^2)(4\pi r^2)\frac{dT}{dr}$$

Separating variables and integrating,

$$-\frac{q}{4\pi k_0}\int_{r_1}^{r_2}\frac{dr}{r^2} = \int_{T_1}^{T_2} T^2\,dT$$

$$-\frac{q}{4\pi k_0}\left(\frac{1}{r_1} - \frac{1}{r_2}\right) = \frac{1}{3}(T_2^3 - T_1^3)$$

$$q = \frac{4\pi k_0(T_1^3 - T_2^3)}{3\left(\frac{1}{r_1} - \frac{1}{r_2}\right)}$$

1.50 Problem 1.32 presents an expression for the thermal conductivity of copper, in SI units. Convert this expression to British engineering units. Compare k from the resulting expression with the four values in Table B-1.

▌

$$k = (371.9\text{ W/m}\cdot\text{K})\{1 - (9.25\times10^{-5}\,{}^\circ\text{C}^{-1})[T({}^\circ\text{C}) - 150\,{}^\circ\text{C}]\}$$

$$= \left(\frac{371.9\text{ W/m}\cdot\text{K}}{1.7296\,\dfrac{\text{W/m}\cdot\text{K}}{\text{Btu/h}\cdot\text{ft}\cdot{}^\circ\text{F}}}\right)\left\{1 - \frac{9.25\times10^{-5}\,{}^\circ\text{C}^{-1}}{(9\,{}^\circ\text{F})/(5\,{}^\circ\text{C})}[T({}^\circ\text{F}) - 302\,{}^\circ\text{F}]\right\}$$

$$= (215.02)[1 - (5.13\times10^{-5})(T - 302)]\quad \text{Btu/h}\cdot\text{ft}\cdot{}^\circ\text{F}$$

with T now in °F. At the four given temperatures,

$$\left.\begin{aligned} k_{32^\circ\text{F}} &= 218.0\text{ Btu/h}\cdot\text{ft}\cdot{}^\circ\text{F} \\ k_{212^\circ\text{F}} &= 216.01 \end{aligned}\right\}\ \textit{deviate from tabulated values}$$

$$\left.\begin{aligned} k_{572^\circ\text{F}} &= 212.04 \\ k_{932^\circ\text{F}} &= 208.06 \end{aligned}\right\}\ \textit{approximately equal to tabulated values}$$

1.51 A deep-sea research probe has the form of a two-layer spherical shell (Fig. 1-17). The inner layer is mild steel with an inside radius of 10 in., and the outer layer is type-304 stainless steel. Each layer is 1 in.

thick, and the two layers are in perfect thermal contact. The electronic gear inside the probe will give off energy, resulting in a heat transfer rate of approximately 5000 Btu/h · ft² (based on outside surface area) when the unit is surrounded by 40 °F seawater. The inner surface wall should be under 125 °F, for safe operation of the electronic equipment. Estimate the inner surface temperature under these conditions, assuming the outer surface to be at the water temperature.

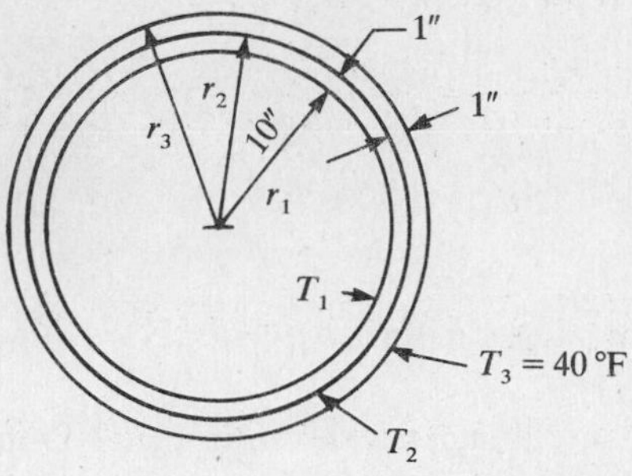

Fig. 1-17

▮ From Table B-1, $k_{ss} = 8.0$ Btu/h · ft · °F and $k_{ms} = 26.5$ Btu/h · ft · °F. The thermal resistances are, by (*2*) of Problem 1.35,

$$R_{ms} = \frac{(1/r_1) - (1/r_2)}{4\pi k_{ms}} \qquad R_{ss} = \frac{(1/r_2) - (1/r_3)}{4\pi k_{ss}}$$

Therefore

$$\Sigma R_{th} = \frac{1}{4\pi}\left[\frac{(1/r_1) - (1/r_2)}{k_{ms}} + \frac{(1/r_2) - (1/r_3)}{k_{ss}}\right]$$

$$= \frac{1}{4\pi}\left[\frac{[(\frac{12}{10}) - (\frac{12}{11})]\ \text{ft}^{-1}}{26.5\ \text{Btu/h}\cdot\text{ft}\cdot{}^\circ\text{F}} + \frac{[(\frac{12}{11}) - (\frac{12}{12})]\ \text{ft}^{-1}}{8\ \text{Btu/h}\cdot\text{ft}\cdot{}^\circ\text{F}}\right] = \frac{1}{4\pi}(15.4 \times 10^{-3})\ \text{h}\cdot{}^\circ\text{F/Btu}$$

and (*1*) of Problem 1.35 becomes

$$q = (4\pi r_3^2)(5000\ \text{Btu/h}\cdot\text{ft}^2) = \frac{T_1 - T_3}{\Sigma R_{th}}$$

Substitution of known values and solution for T_1 gives $T_1 = 117$ °F.

1.52 Estimate the conductive heat transfer rate in the insulated copper wire between the two liquid surfaces for two thermocouples respectively located in boiling water and in an ice bath (Fig. 1-18). The wire length between the surfaces is 14 in., and the wire is AWG No. 28 (0.0126 in. diameter) and is pure copper.

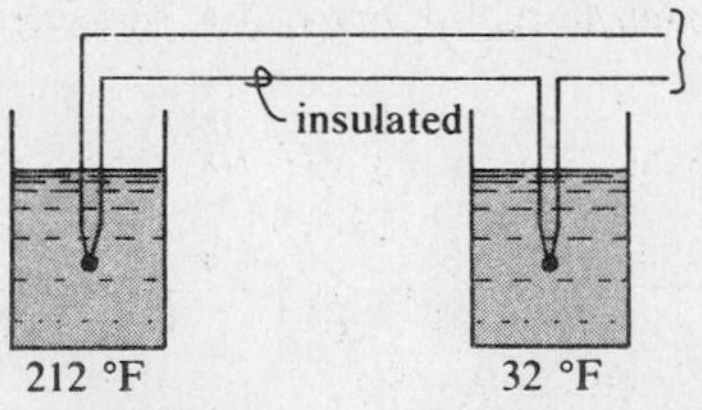

Fig. 1-18

▮ $$A = \frac{\pi d^2}{4} = \frac{\pi}{4}\left(\frac{0.0126}{12}\ \text{ft}\right)^2 = 1.04 \times 10^{-5}\ \text{ft}^2 \qquad T_m = \frac{212 + 32}{2} = 122\ {}^\circ\text{F}$$

Interpolating in Table B-1,

$$k_m = 224 - \left(\frac{122 - 32}{212 - 32}\right)(224 - 218) = 221\ \text{Btu/h}\cdot\text{ft}\cdot{}^\circ\text{F}$$

Hence,

$$q = -k_m A \frac{\Delta T}{\Delta x} = -(221)(1.04 \times 10^{-5})\left(\frac{32 - 212}{\frac{14}{12}}\right) = 0.355\ \text{Btu/h}$$

1.53 Approximate $k(T)$ for 1% mild carbon steel by a linear expression over the temperature range 0 °C to 300 °C.

Noting that the data in Table B-1 are linear, we may use them to get the reference value and the slope in the linear relation

$$k = k_0(1 + bT) \equiv k_0 + aT$$

At 32 °F, $26.5 = k_0 + a32$; at 572 °F, $25 = k_0 + a572$. Solving simultaneously, one obtains

$$a = -2.78 \times 10^{-3}\ \text{°F}^{-1} \qquad k_0 = 26.41\ \text{Btu/h} \cdot \text{ft} \cdot \text{°F}$$

Therefore,

$$k = (26.41)(1 - 1.05 \times 10^{-4}T)\ \text{Btu/h} \cdot \text{ft} \cdot \text{°F} \qquad (T \text{ in °F})$$

or

$$k = (45.84)(1 - 1.89 \times 10^{-4}T)\ \text{W/m} \cdot \text{K} \qquad (T \text{ in °C})$$

1.54 If the spaces of the rubber insulating wall shown in Fig. 1-19 are filled with dry air at an average temperature of 32 °F, determine the thermal resistance of the wall and compare it with that of a solid rubber wall.

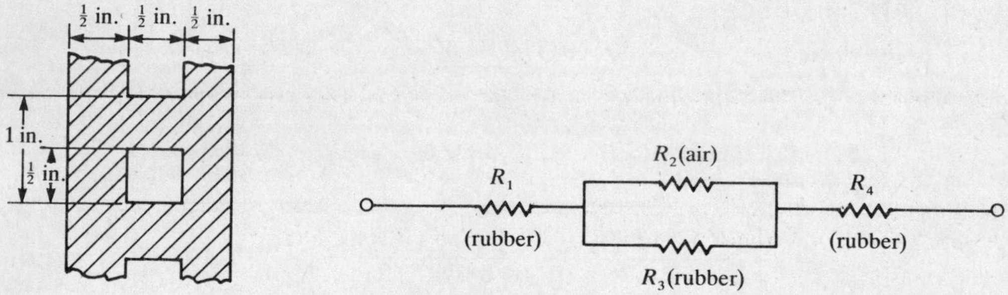

Fig. 1-19

From Table B-2, $k(\text{rubber}) = 0.087\ \text{Btu/h} \cdot \text{ft} \cdot \text{°F}$; from Table B-4, $k(\text{air}) = 0.014\ \text{Btu/h} \cdot \text{ft} \cdot \text{°F}$. Thus, in the electrical analogy,

$$R_1 = R_4 = \frac{\frac{1}{24}\ \text{ft}}{(1\ \text{ft})[(\frac{1}{12})\ \text{ft}](0.087\ \text{Btu/h} \cdot \text{ft} \cdot \text{°F})} = 5.75\ \text{h} \cdot \text{°F/Btu}$$

$$R_2 = \frac{\frac{1}{24}}{(1)(\frac{1}{24})(0.014)} = 71.43\ \text{h} \cdot \text{°F/Btu}$$

$$R_3 = \frac{\frac{1}{24}}{(1)(\frac{1}{24})(0.087)} = 11.49\ \text{h} \cdot \text{°F/Btu}$$

The resistance of the parallel combination is given by

$$\frac{1}{R_p} = \frac{1}{71.43} + \frac{1}{11.49} \qquad \text{or} \qquad R_p = 9.90\ \text{h} \cdot \text{°F/Btu}$$

Thus, for the composite wall, $R_{\text{th}} = 5.75 + 9.90 + 5.75 = 21.4\ \text{h} \cdot \text{°F/Btu}$.
For the solid wall (per foot of depth, 1 in. high),

$$R_{\text{th}} = \frac{\frac{3}{24}}{(1)(\frac{1}{12})(0.087)} = 17.24\ \text{h} \cdot \text{°F/Btu}$$

Problems 1.55–1.57 show how equations for steady, one-dimensional heat transfer may be obtained by specializing the unsteady, three-dimensional equation.

1.55 Derive the general conduction equation using the nomenclature of Fig. 1-20.

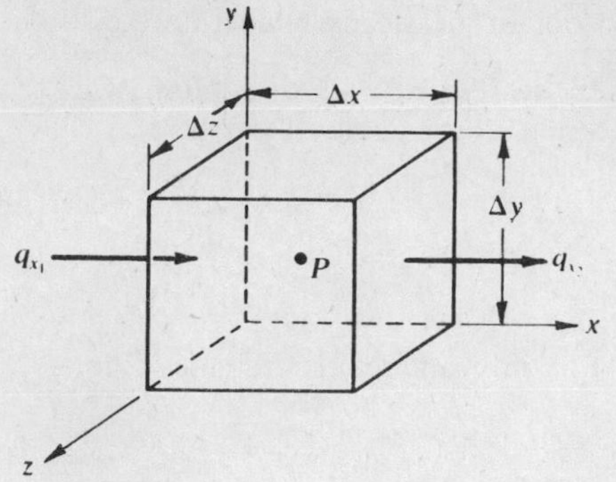

Fig. 1-20

For the control volume in Fig. 1-20, the first law of thermodynamics may be expressed as

(Rate of heat transfer in) + (rate of work in) + (rate of other energy conversion)
= (rate of heat transfer out) + (rate of work out) + (rate of internal energy storage) (1)

For an incompressible substance the net work done on the control volume is converted to internal energy. Denoting the rate of energy conversion (from work, chemical reaction, etc.) as q''', (1) becomes

$$q_{x_1} + q_{y_1} + q_{z_1} + q'''(\Delta x\,\Delta y\,\Delta z) = q_{x_2} + q_{y_2} + q_{z_2} + \frac{\partial U}{\partial t} \tag{2}$$

Examine the heat transfer terms in (2). In the x-direction, the two terms may be grouped to form

$$q_{x_1} - q_{x_2} = (\Delta y\,\Delta z)\left[\left(k\,\frac{\partial T}{\partial x}\right)_{x_1} - \left(k\,\frac{\partial T}{\partial x}\right)_{x_2}\right] \tag{3}$$

by application of Fourier's law. Notice that k may be temperature-dependent and hence spatially dependent. By a Taylor's series expansion about the center point P,

$$\left(k\,\frac{\partial T}{\partial x}\right)_{x_1} = \left(k\,\frac{\partial T}{\partial x}\right)_P + \left(-\frac{\Delta x}{2}\right)\frac{\partial}{\partial x}\left(k\,\frac{\partial T}{\partial x}\right)\bigg|_P + \cdots$$

$$\left(k\,\frac{\partial T}{\partial x}\right)_{x_2} = \left(k\,\frac{\partial T}{\partial x}\right)_P + \left(\frac{\Delta x}{2}\right)\frac{\partial}{\partial x}\left(k\,\frac{\partial T}{\partial x}\right)\bigg|_P + \cdots$$

so that (3) becomes (omitting the subscript P)

$$q_{x_1} - q_{x_2} = (\Delta y\,\Delta z)\left[(\Delta x)\,\frac{\partial}{\partial x}\left(k\,\frac{\partial T}{\partial x}\right) + \cdots\right] \tag{4}$$

Similarly,

$$q_{y_1} - q_{y_2} = (\Delta x\,\Delta z)\left[(\Delta y)\,\frac{\partial}{\partial y}\left(k\,\frac{\partial T}{\partial y}\right) + \cdots\right] \tag{5}$$

$$q_{z_1} - q_{z_2} = (\Delta x\,\Delta y)\left[(\Delta z)\,\frac{\partial}{\partial z}\left(k\,\frac{\partial T}{\partial z}\right) + \cdots\right] \tag{6}$$

Finally, the internal energy storage per unit volume and per unit temperature is the product of density and specific heat, so that

$$\frac{\partial U}{\partial t} = \rho c(\Delta x\,\Delta y\,\Delta z)\,\frac{\partial T}{\partial t} \tag{7}$$

Substituting (4) through (7) in (2), dividing by the volume element $\Delta x\,\Delta y\,\Delta z$, and taking the limit as Δx, Δy, and Δz simultaneously approach zero yields the general conduction equation

$$\frac{\partial}{\partial x}\left(k\,\frac{\partial T}{\partial x}\right) + \frac{\partial}{\partial y}\left(k\,\frac{\partial T}{\partial y}\right) + \frac{\partial}{\partial z}\left(k\,\frac{\partial T}{\partial z}\right) + q''' = \rho c\,\frac{\partial T}{\partial t} \tag{8}$$

In the usual case, k, ρ, and c are constants and $q''' \equiv 0$, yielding the familiar

$$\nabla^2 T = \frac{1}{\alpha}\,\frac{\partial T}{\partial t} \qquad \left(\alpha \equiv \frac{k}{\rho c}\right) \tag{9}$$

1.56 From (8) of Problem 1.55 obtain the steady temperature distribution across a large plane wall.

For steady-state transfer $(\partial T/\partial t = 0)$ in a bi-infinite wall $(\partial T/\partial y = \partial T/\partial z = 0)$ of constant thermal conductivity, the general conduction equation becomes

$$\frac{d^2T}{dx^2} = 0$$

whence T is linear in x.

1.57 (*a*) Write (9) of Problem 1.55 in cylindrical coordinates (Fig. 1-21). (*b*) From (*a*) retrieve (2) of Problem 1.26.

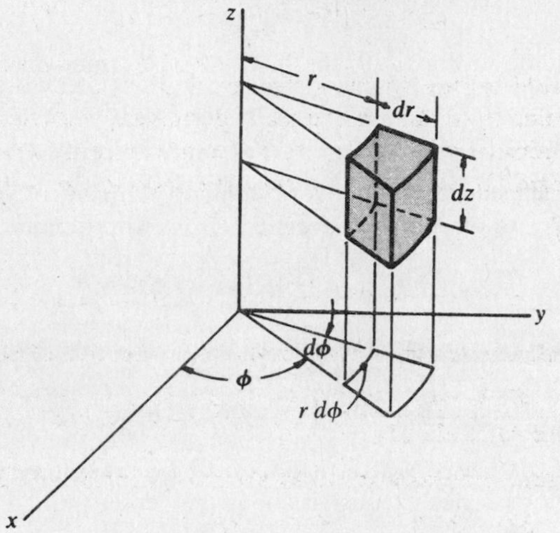

Fig. 1-21

(*a*)

$$\frac{\partial^2 T}{\partial r^2} + \frac{1}{r}\frac{\partial T}{\partial r} + \frac{1}{r^2}\frac{\partial^2 T}{\partial \phi^2} + \frac{\partial^2 T}{\partial z^2} = \frac{1}{\alpha}\frac{\partial T}{\partial t}$$

(*b*) For steady-state heat transfer through a long cylindrical shell, the equation of (*a*) reduces to

$$\frac{d^2T}{dr^2} + \frac{1}{r}\frac{dT}{dr} = 0 \qquad \text{or} \qquad \frac{1}{r}\frac{d}{dr}\left(r\frac{dT}{dr}\right) = 0$$

A first integration gives

$$r\frac{dT}{dr} = B \qquad \text{or} \qquad dT = B\frac{dr}{r}$$

and a second integration yields $T = B \ln r + c$. The boundary conditions, $T(r_1) = T_1$ and $T(r_2) = T_2$, determine

$$B = \frac{T_1 - T_2}{\ln (r_1/r_2)}$$

Fourier's law becomes

$$q = -k(2\pi rL)\frac{dT}{dr} = -k(2\pi L)B = 2\pi kL\left[\frac{T_1 - T_2}{\ln (r_2/r_1)}\right]$$

which is the required equation.

1.58 A cold-storage room has walls constructed of a 4-in. layer of corkboard contained between double wooden (fir) walls, each $\frac{1}{2}$ in. thick. (*a*) Find the rate of heat removal, in Btu/h · ft^2, if the wall temperature is 10 °F inside the room and 70 °F outside the room. (*b*) Compute the temperature at the interface between the outer wall and the corkboard.

(*a*) From Table B-2, $k_{\text{cork}} = 0.025$ Btu/h · ft · °F and $k_{\text{fir}} = 0.063$ Btu/h · ft · °F. Thermal resistances $(A = 1 \text{ ft}^2)$ are then

$$R_{\text{fir}} = \frac{(\Delta x)_{\text{fir}}}{k_{\text{fir}}A} = \frac{(0.5/12)\text{ ft}}{(0.063\text{ Btu/h}\cdot\text{ft}\cdot{}^\circ\text{F})(1\text{ ft}^2)} = 0.66\text{ h}\cdot{}^\circ\text{F/Btu}$$

$$R_{\text{cork}} = \frac{(\Delta x)_{\text{cork}}}{k_{\text{cork}}A} = \frac{\frac{4}{12}}{(0.025)(1)} = 13.33\text{ h}\cdot{}^\circ\text{F/Btu}$$

The heat transfer rate is then given by

$$q = \frac{T_1 - T_2}{\Sigma R} = \frac{(10-70)\ {}^\circ\text{F}}{(0.66 + 13.33 + 0.66)\text{ h}\cdot{}^\circ\text{F/Btu}} = -4.10\text{ Btu/h}$$

This is heat *into* the cold-storage room; thus, a heat pump or refrigerator must remove 4.10 Btu from the room each hour. **(*b*)** Denote the inside and outside corkboard temperatures T_{ci} and T_{co}. Then, across the outer fir wall,

$$q = -4.10 = \frac{T_{co} - 70}{0.66} \qquad \text{or} \qquad T_{co} = 67.3\ {}^\circ\text{F}$$

1.59 The outer wall of the cold-storage room of Problem 1.58 is exposed to air which is at a temperature of 70 °F and has a dew point of 60 °F. If the air is permitted to diffuse freely through the walls of the wood and cork, its moisture content will start to condense at the point where the wall temperature is 60 °F and it will freeze at 32 °F. Find the zones of moisture and frost in the wall.

▌ From Problem 1.58,

$$-4.10 = \frac{10 - T_{ci}}{0.66} \qquad \text{or} \qquad T_{ci} = 12.71\ {}^\circ\text{F}$$

Let Δx_1 denote the thickness of cork between the cross sections at 60 °F and $T_{co} = 67.3$ °F; then

$$-q = \frac{67.3 - 60}{(\Delta x_1)/(0.025)(1)} = \frac{67.3 - 12.71}{13.33} \qquad \text{whence} \qquad \Delta x_1 = 0.045\text{ ft} = 0.535\text{ in.}$$

Let Δx_2 denote the thickness of cork between the cross sections at 32 °F and $T_{ci} = 12.71$ °F; then

$$-q = \frac{32 - 12.71}{(\Delta x_2)/(0.025)(1)} = \frac{67.3 - 12.71}{13.33} \qquad \text{whence} \qquad \Delta x_2 = 0.123\text{ ft} = 1.474\text{ in.}$$

This treatment assumes that the thermal conductivity is not affected by moisture.

1.60 To prevent diffusion of air and moisture within cold-storage insulation, its surfaces are usually sealed with some impervious material. Assuming that the corkboard of Problems 1.58 and 1.59 is treated in this way, though the wood is still pervious, find the minimum thickness of cork that will prevent condensation of moisture at the outer cork-wood interface. All the other conditions of Problems 1.58 and 1.59 hold.

▌ If the outer surface of the cork is at 60 °F, then the temperature drop through either layer of wood is 10 °F (both layers are the same). Then $\Delta t_{\text{cork}} = 40$ °F. Letting Δx be the required cork thickness,

$$-q = \frac{40}{(\Delta x)/(0.025)(1)} = \frac{70 - 10}{0.66 + [(\Delta x)/(0.025)(1)] + 0.66}$$

which yields $\Delta x = 0.066\text{ ft} = 0.792\text{ in.}$

1.61 A furnace is to be built with a layer of firebrick ($k = 0.85$ Btu/h · ft · °F) on the inside. This is covered with an 8.5-in. layer of insulating brick ($k = 0.09$ Btu/h · ft · °F) and then by 6 in. of building brick ($k =$ 0.50 Btu/hr · ft · °F) on the exterior. The interior of the furnace is at 2100 °F, and the exterior is at 150 °F. **(*a*)** Determine the thickness of firebrick necessary to keep the temperature of the insulating brick below 1700 °F. **(*b*)** Calculate the inside temperature T_3 of the building brick.

▌ For a 1-ft² area,

$$\text{Resistance of building brick} = \frac{\frac{6}{12}}{(0.50)(1)} = 1\text{ h}\cdot{}^\circ\text{F/Btu}$$

$$\text{Resistance of insulating brick} = \frac{8.5/12}{(0.09)(1)} = 7.87\text{ h}\cdot{}^\circ\text{F/Btu}$$

whence

$$q = \frac{1700 - 150}{7.87 + 1} = 175\text{ Btu/h}$$

(*a*) Let thickness of firebrick be Δx (in.). Then

$$175 = \frac{2100 - 1700}{(\Delta x/12)/(0.85)(1)} \qquad \text{or} \qquad \Delta x = 23.3 \text{ in.}$$

(*b*)
$$T_3 = 150 + \left(\frac{1}{1 + 7.87}\right)(1700 - 150) = 325\ °\text{F}$$

1.62 A "thermopane" window, 10 ft by 4 ft, consists of two layers of glass, each $\frac{1}{4}$ in. thick, separated by a layer of dry, stagnant air, also $\frac{1}{4}$ in. thick. If the temperature drop through the composite system is 30 °F, (*a*) find the heat loss through the window. (*b*) If the window replaces a single glass pane, $\frac{3}{8}$ in. thick, at an increase in cost of \$120, find the number of days of operation to pay for the window. Coal with a heating value of 13 200 Btu/lbm costs \$20 per ton; the furnace efficiency can be assumed to be 50 percent. The thermal conductivity of the window glass is 0.5 Btu/h · ft · °F.

▮ (*a*)
$$\text{Thermal resistance of glass layer} = \frac{\frac{1}{48}}{(0.5)(40)} = 0.00104\ \text{h} \cdot °\text{F/Btu}$$

$$\text{Thermal resistance of air} = \frac{\frac{1}{48}}{(0.015)(40)} = 0.0347\ \text{h} \cdot °\text{F/Btu}$$

Hence
$$q = \frac{30}{(2)(0.00104) + 0.0347} = 815\ \text{Btu/h}$$

(*b*)
$$\text{Thermal resistance of single pane} = \frac{\frac{3}{8}/12}{(0.5)(40)} = 0.00156\ \text{h} \cdot °\text{F/Btu}$$

which gives $q = 30/0.00156 = 19\,250$ Btu/h.

$$\text{Extra fuel needed} = \frac{(19\,250 - 815)(24)}{(13\,200)(0.50)} = 67\ \text{lbm/day}$$

$$\text{Payout time} = \frac{\$120}{(\frac{67}{2000}\ \text{ton/day})(\$20/\text{ton})} = 180\ \text{days}$$

1.63 A brewery fermentation tank 37 ft in diameter is situated in a room which has a temperature of 60 °F. The tank is constructed of $\frac{3}{8}$-in. welded-steel plate with a $\frac{1}{2}$-in. glass lining. The temperature at the interface between the glass and the contents of the tank is known to be 120 °F. Assuming that the thermal resistance offered by the air film on the outside of the tank equals the combined resistances of the glass and steel, calculate the temperature at the glass-steel interface and the temperature at the steel-air interface. The thermal conductivity of the glass is 0.45 Btu/h · ft · °F.

▮ Treat the tank as a flat plate (see Problems 1.28 and 1.29).

$$R_{\text{glass}} = \frac{\frac{1}{24}}{(0.45)(1)} = 0.0926\ \text{h} \cdot °\text{F/Btu}$$

$$R_{\text{steel}} = \frac{\frac{3}{8}/12}{(26)(1)} = 0.0012\ \text{h} \cdot °\text{F/Btu}$$

Hence $R_{\text{air}} = 0.0926 + 0.0012 = 0.0938$ h · °F/Btu and $\Sigma R = 2R = 0.1876$ h · °F/Btu. Since temperature drop is proportional to resistance,

$$\Delta t_{\text{air}} = \left(\frac{0.0938}{0.1876}\right)(120 - 60) = 30\ °\text{F} \qquad \Delta t_{\text{steel}} = \left(\frac{0.0012}{0.1876}\right)(120 - 60) = 0.4\ °\text{F}$$

from which the air-steel interface is at 90 °F and the steel-glass interface is at 90.4 °F.

1.64 A cylindrical exhaust duct has a constant inside temperature of 600 °F. It is insulated on the exterior with a 4-in. layer of rock wool which has an outer surface temperature of 100 °F. The duct has an ID of 3.5 in. and $\frac{1}{4}$-in.-thick walls of a ceramic material with a thermal conductivity of 0.88 Btu/h · ft · °F. Find the rate of heat loss from the duct per lineal foot. The thermal conductivity of rock wool is represented by the equation $k = 0.025 + (5 \times 10^{-5})T$, where T is in °F and k is in Btu/h · ft · °F.

▮ The system is shown in cross section in Fig. 1-22. The theory of Problem 1.30 applies. For the ceramic, $b = 0$ and $k_m = k_{\text{ref}} = 0.88$ Btu/h · ft · °F; thus, (*2*) of Problem 1.30 gives

$$\frac{q}{L} = 2\pi(0.88)\frac{600 - T_i}{\ln(2/1.75)} \qquad (1)$$

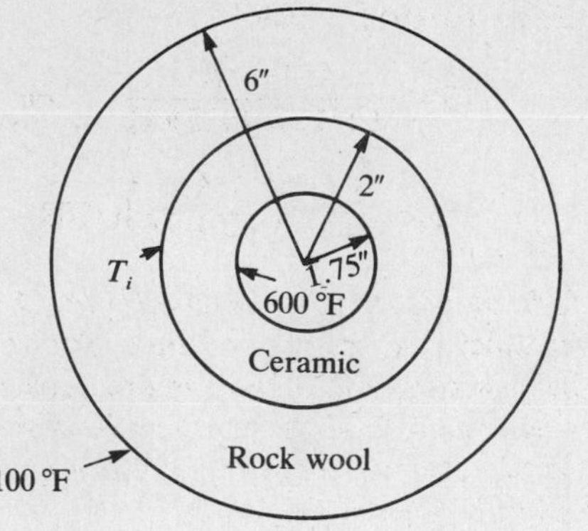

Fig. 1-22

For the rock wool, with $T_{\text{ref}} = 0\ ^\circ\text{F}$,

$$k_m = 0.025 + (5 \times 10^{-5})\left(\frac{T_i + 100}{2}\right)$$

so that

$$\frac{q}{L} = 2\pi\left[0.025 + (5 \times 10^{-5})\left(\frac{T_i + 100}{2}\right)\right]\left(\frac{T_i - 100}{\ln \frac{6}{2}}\right) \qquad (2)$$

Eliminating T_i between (*1*) and (*2*), one obtains a quadratic in q/L, with solution

$$\frac{q}{L} = 120\ \text{Btu/h} \cdot \text{ft}$$

1.65 For constant internal heat generation per unit volume q'''—e.g., electric resistance elements, nuclear-fission elements, chemically reacting systems—obtain the steady-state temperature distribution corresponding to the asymmetrical boundary conditions of Fig. 1-23(*a*). Also find the steady-state heat transfer rate within the wall.

▌ Assuming the plane wall is large, the problem is one-dimensional in space; the governing equation, (*8*) of Problem 1.55, reduces to

$$\frac{d^2T}{dx^2} + \frac{q'''}{k} = 0$$

Integrating twice, we get

$$T = -\frac{q'''}{2k}x^2 + C_1x + C_2 = -\frac{q'''}{2k}(x + D_1)^2 + D_2 \qquad (1)$$

By (*1*), the temperature profile is a section of a parabola having vertex $(-D_1, D_2)$. The parabola is concave downward or upward according as $q''' > 0$ or $q''' < 0$. The two constants are evaluated from the boundary conditions:

$$T_1 = -\frac{q'''}{2k}D_1^2 + D_2$$

$$T_2 = -\frac{q'''}{2k}(2L + D_1)^2 + D_2$$

Solving simultaneously,

$$D_1 = -L + \frac{k(T_1 - T_2)}{2Lq'''}$$

$$D_2 = \frac{q'''L^2}{2k} + \frac{T_1 + T_2}{2} + \frac{k(T_1 - T_2)^2}{8L^2q'''} \qquad (2)$$

The required heat transfer is given by Fourier's law as

$$q = -kA\frac{dT}{dx} = Aq'''(x + D_1)$$

$$= q'''A(x - L) + \frac{kA(T_1 - T_2)}{2L} \qquad (3)$$

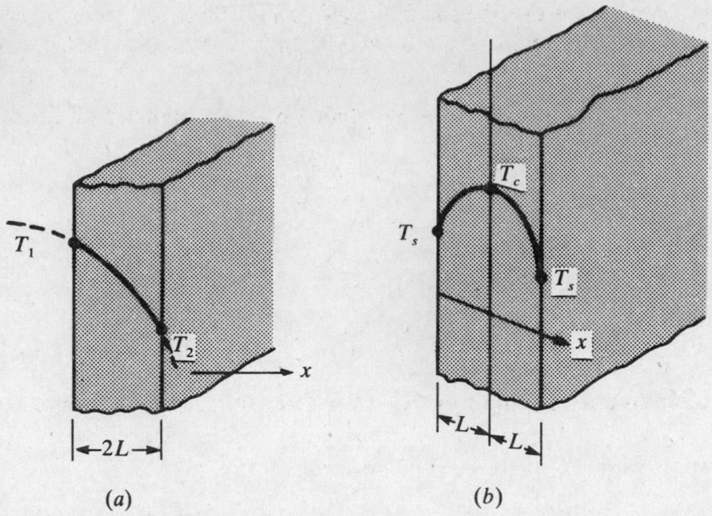

Fig. 1-23

1.66 Solve Problem 1.65 for the symmetrical boundary conditions of Fig. 1-23(*b*).

▎ In Problem 1.64 set $T_1 = T_2 = T_s$, giving

$$D_1 = -L \qquad D_2 = \frac{q'''L^2}{2k} + T_s = T_c \tag{2}$$

Hence, the temperature distribution is given by

$$T = T_c - \frac{q'''}{2k}(x - L)^2 \tag{1}$$

and the heat transfer rate by

$$q = q'''A(x - L) \tag{3}$$

1.67 Obtain (*3*) of Problem 1.65 directly from a superposition argument.

▎ The total heat q is the sum of q_{int}, the heat generated internally, and $q_{\Delta T}$, the heat flowing in response to the temperature gradient $T_1 - T_2$ imposed across the wall. Now, by considering a rectangular box with ends on the center plane and on the plane through station x, it is easy to see that

$$q_{\text{int}} = q''' \times (\text{volume of box}) = q'''A(x - L)$$

(The sources q''' within the mirror-image box send their heat to the mirror-image station $2L - x$.) Further, we know that

$$q_{\Delta T} = \frac{T_1 - T_2}{R_{\text{th}}} = \frac{T_1 - T_2}{2L/kA}$$

This completes the argument.

1.68 Consider a plate with uniform heat generation q''', as shown in Fig. 1-23(*a*). For $k = 200\ \text{W/m} \cdot \text{K}$, $q''' = 40\ \text{MW/m}^3$, $T_1 = 160\ °\text{C}$ (at $x = 0$), $T_2 = 100°\text{C}$ (at $x = 2L$), and a plate thickness of 20 mm, determine q/A at (*a*) $x = 0$; (*b*) $x = 2L$; (*c*) $x = L$.

▎ Use (*3*) of Problem 1.65.

(*a*)
$$\left.\frac{q}{A}\right|_0 = (40 \times 10^6)(0 - 0.010) + \frac{200(160 - 100)}{0.020}$$
$$= +2 \times 10^5\ \text{W/m}^2 = +0.2\ \text{MW/m}^2 \quad \text{(rightward)}$$

(*b*)
$$\left.\frac{q}{A}\right|_{2L} = (40 \times 10^6)(0.010) + (6 \times 10^5) = 1\ \text{MW/m}^2$$

(*c*)
$$\left.\frac{q}{A}\right|_L = \frac{0.2 + 1}{2} = 0.6\ \text{MW/m}^2$$

1.69 Find the steady temperature distribution in a long solid cylinder of radius r_s and thermal conductivity k, if the surface is maintained at temperature T_s and there is uniform internal heat generation q''' per unit volume.

The easiest route is that suggested by Problem 1.67. Per unit length of cylinder, Fourier's law gives

$$q(r) = \pi r^2 q''' = -k(2\pi r)\frac{dT}{dr}$$

Integrating,

$$\int_T^{T_s} dT = -\frac{q'''}{2k}\int_r^{r_s} r\,dr$$

$$T - T_s = \frac{q'''}{4k}(r_s^2 - r^2) \qquad (1)$$

Denoting the centerline temperature by T_c, we can rewrite (1) in the convenient nondimensional form

$$\frac{T - T_s}{T_c - T_s} = 1 - \left(\frac{r}{r_s}\right)^2 \qquad (2)$$

It is apparent from (1) or (2) that the temperature strictly decreases with increasing r.

1.70 An AWG No. 10 stainless-steel wire, 1 ft long, is used as an electric resistance heater in a laboratory experiment. The measured voltage drop across the wire is 20 V, and the measured current is 40 A. The wire surface temperature, measured with an attached thermocouple, is 600 °F. Find the maximum temperature in the wire.

In the notation of Problem 1.69, we are asked to find T_c. Since the given power input,

$$(20\text{ V})(40\text{ A}) = 800\text{ W} = 2730\text{ Btu/h}$$

appears as heat within the 1-ft-long wire,

$$q''' = \frac{\text{heat}}{\text{volume}} = \frac{2730\text{ Btu/h}}{\pi\left(\frac{0.051}{12}\text{ ft}\right)^2(1\text{ ft})} = 48.1 \times 10^6\text{ Btu/h}\cdot\text{ft}^3$$

Using $k = 10\text{ Btu/h}\cdot\text{ft}\cdot°\text{F}$ and $r = 0$ in (1) of Problem 1.69,

$$T_c = T_s + \frac{q''' r_s^2}{4k} = 600 + \frac{(4.81\times10^7)(18.07\times10^{-6})}{(4)(10)} = 622\text{ °F}$$

[We could have avoided much arithmetic by noting that q''' is inversely proportional to r_s^2.]

1.71[D] A 200-A electric current flows through a stainless-steel wire 0.1 in. in diameter. The resistivity of the wire is $7\times10^{-9}\ \Omega\cdot\text{m}$, and its length is 10 ft. If the outer surface temperature of the wire is 500 °F, what is its center temperature? Assume a thermal conductivity $k = 11\text{ Btu/h}\cdot\text{ft}\cdot°\text{F}$.

To avoid unnecessary computation (see Problem 1.70), solve the problem algebraically. Denoting resistivity and electric resistance by ρ and R, respectively, one has

$$q''' = \frac{\text{power input}}{\text{volume}} = \frac{I^2R}{\pi r_s^2 L} = \frac{I^2(\rho L/\pi r_s^2)}{\pi r_s^2 L} = \frac{\rho I^2}{\pi^2 r_s^4}$$

Hence, with $d_s \equiv 2r_s$ and noting that $1\text{ W} = 1\text{ A}^2\cdot\Omega$,

$$T_c = T_s + \frac{q''' r_s^2}{4k} = T_s + \frac{\rho}{k}\left(\frac{I}{\pi d_s}\right)^2$$

$$= (500\text{ °F}) + \left[\frac{(7\times10^{-9}\ \Omega\cdot\text{m})/(0.3048\text{ m/ft})}{11\text{ Btu/h}\cdot\text{ft}\cdot°\text{F}}\right]\left[\frac{200\text{ A}}{\pi(\frac{1}{120}\text{ ft})}\right]^2\left(\frac{3600\text{ Btu/h}}{1054.4\text{ A}^2\cdot\Omega}\right)$$

$$= 541.5\text{ °F}$$

1.72[D] An electric resistance-heater wire is 2 mm in diameter. The electric resistivity is $8\times10^{-7}\ \Omega\cdot\text{m}$, and the thermal conductivity is 19.0 W/m·K. For a steady current of 150 A passing through the wire, determine the temperature rise from the surface to the centerline, in °C.

▮ By the formula of Problem 1.71,

$$T_c - T_s = \frac{\rho}{k}\left(\frac{I}{\pi d_s}\right)^2 = \frac{8\times10^{-7}}{19.0}\left[\frac{150}{\pi(2\times10^{-3})}\right]^2$$

$$= 24.0\ \text{K} = 24.0\ ^\circ\text{C}$$

1.73 A wall has a freshly plastered layer which is 0.5 in. thick. If q''' due to the "curing" of the plaster is approximately constant at 5000 Btu/h · ft³, the outer surface is insulated (no heat transfer), and the inner surface is held at 90 °F, determine the steady-state temperature of the outer surface. Assume that k of the fresh plaster is 0.5 Btu/h · ft · °F (which is higher than the values listed in Appendix B due to increased moisture content).

▮ Imagine the plaster layer reflected in its outer surface, producing a 1-in.-thick layer with both surfaces at 90 °F. The condition $q = 0$ at the center plane automatically obtains [see (*3*) of Problem 1.66]. Thus, we are looking for T_c of the 1-in. layer, which is given by (*2*) of Problem 1.66 as

$$T_c = \frac{q'''L^2}{2k} + T_s = \frac{5000(0.5/12)^2}{2(0.5)} + 90 = 98.68\ ^\circ\text{F}$$

1.74 Often heat flux is primarily due to convective transport (by a moving fluid) at the boundary of a system. This mechanism of heat removal (or addition) at boundaries is described by *Newton's law of cooling*,

$$q = hA\,\Delta T \qquad (1)$$

where $h \equiv$ convective heat transfer coefficient
$A \equiv$ area perpendicular to direction of heat flux
$\Delta T \equiv$ temperature drop across boundary

What is the relation between convection and conduction?

▮ Combining Newton's and Fourier's equations, we get

$$\frac{q}{A} = -k\left.\frac{dT}{dx}\right|_{\text{boundary}} = h\,\Delta T \qquad (2)$$

It is important to note a basic difference between convection and Fourier's equation: k is a property of the conducting medium, whereas h is a function of many variables, including velocity and various fluid properties.

1.75 A fluid of low electrical conductivity at 200 °F is heated by a very long, immersed iron bar, 1 in. thick by 5 in. wide; the 1-in. surfaces are insulated so that fluid is in contact with the 5-in.-wide surfaces only. Heat is generated uniformly in the bar at a rate of 100 000 Btu/h · ft³ by passing an electric current through it. Determine the surface conductance required to maintain the temperature of the bar below 400 °F.

▮ For the highest allowable mid-plane temperature, the surface temperature of the bar is given by (*2*) of Problem 1.66 $(2L = 1\ \text{in.})$ as

$$T_s = T_c - \frac{q'''L^2}{2k} = 400 - \frac{(100\,000)(\frac{1}{24})^2}{2(36.6)} = 397.63\ ^\circ\text{F}$$

where k has been estimated from Table B-1. Then, by Newton's law of cooling at one surface,

$$\bar{h} = \frac{q/A}{\Delta T} = \frac{q'''L}{\Delta T} = \frac{(100\,000)(\frac{1}{24})}{397.63 - 200} = 21.12\ \text{Btu/h}\cdot\text{ft}^2\cdot{}^\circ\text{F}$$

1.76 When boundary temperatures are unknown, it is often convenient to work with an overall heat transfer coefficient U which incorporates the convective heat transfer coefficient h and the thermal conductivity of the medium: one defines U by the equation

$$q = UA(\Delta T)_{\text{overall}} \qquad (1)$$

Evaluate U for the transfer of heat from fluid i to fluid o through planar medium a (Fig. 1-24).

▮ Applying (*1*) of Problem 1.74 at the two surfaces yields

$$\frac{q}{A} = \bar{h}_i(T_i - T_1) = \bar{h}_o(T_2 - T_o)$$

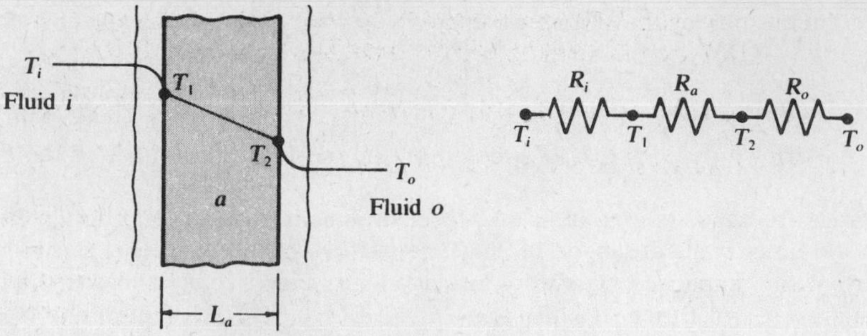

Fig. 1-24

or

$$q = \frac{T_i - T_1}{1/\bar{h}_i A} = \frac{T_2 - T_o}{1/\bar{h}_o A} \tag{2}$$

where the overbar on $\bar{h}$ denotes an average value for the entire surface. In agreement with the electrical analogy, $1/\bar{h}A$ can be thought of as a thermal resistance due to the convective boundary. Thus, the electrical analog to this problem is that of three resistances in series. Here, $R_a = L_a/k_a A$ is the conductive resistance due to homogeneous material a. Since the conductive heat flow within the solid must exactly equal the convective heat flow at the boundaries, we have

$$q = \frac{T_i - T_o}{(1/\bar{h}_i A) + (L_a/k_a A) + (1/\bar{h}_o A)} \tag{3}$$

Comparison of (*3*) with the definition (*1*) yields

$$U = \frac{1}{1/\bar{h}_i + L_a/k_a + 1/\bar{h}_o} \tag{4}$$

The generalization of (*4*) to any geometry is

$$U = \frac{1}{\bar{A}\,\Sigma R_{\text{th}}} \tag{5}$$

where $\bar{A}$ is a representative transverse (to heat flux) area for the system.

1.77 Derive an expression for the steady-state, overall heat transfer coefficient U for the cylindrical system shown in Fig. 1-25.

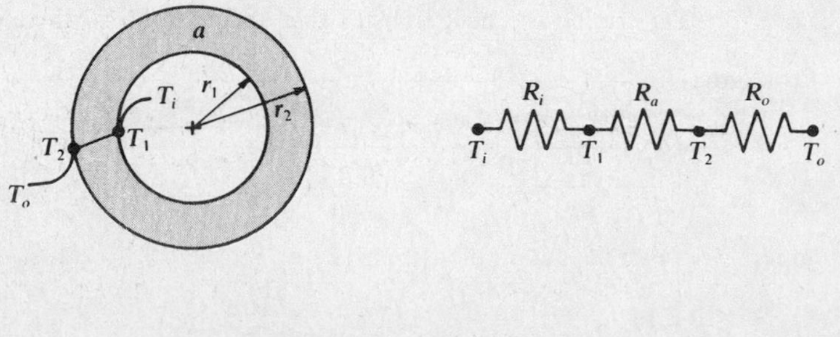

Fig. 1-25

▮ The thermal resistances are (from Problems 1.76 and 1.26):

$$R_1 = \text{inside convective } R_{\text{th}} = \frac{1}{2\pi r_1 L \bar{h}_i}$$

$$R_a = \text{conductive } R_{\text{th}} \text{ due to material } a = \frac{\ln (r_2/r_1)}{2\pi k_a L}$$

$$R_o = \text{outside convective } R_{\text{th}} = \frac{1}{2\pi r_2 L \bar{h}_o}$$

and for the lateral area of the system we choose $\bar{A} = A_o = 2\pi r_2 L$. Then, by (5) of Problem 1.76,

$$U_o = \left[\frac{r_2}{r_i \bar{h}_i} + \frac{r_2 \ln (r_2/r_1)}{k_a} + \frac{1}{\bar{h}_o}\right]^{-1} \tag{1}$$

where the subscript o denotes that U is based on the outside surface area of the cylinder.

1.78 A food-storage freezer is operating in a 60 °F environment and maintaining a −20 °F inside temperature. The outer and inner walls are made of 26-gauge (0.0184-in.) sheet steel. Between the two layers of sheet steel is 2 in. of fine unpacked glass-wool insulation (see Fig. 1-26). The average convective heat transfer coefficients are $\bar{h}_o = 4\ \text{Btu/h}\cdot\text{ft}^2\cdot{}^\circ\text{F}$ and $\bar{h}_i = 2\ \text{Btu/h}\cdot\text{ft}^2\cdot{}^\circ\text{F}$. Determine the overall heat transfer coefficient U.

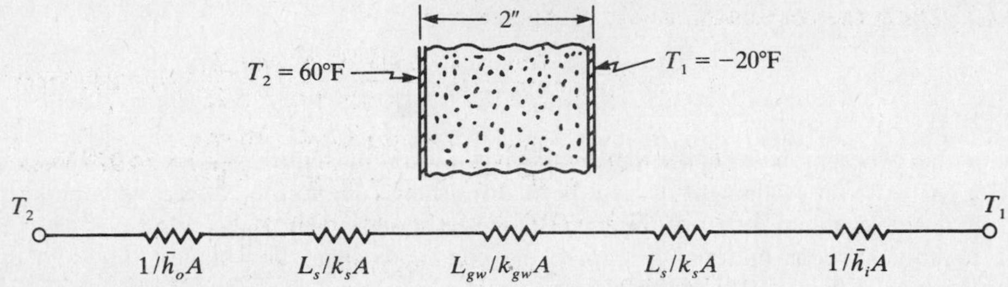

Fig. 1-26

▮ By (4) of Problem 1.76 (generalized),

$$U = \frac{1}{\dfrac{1}{\bar{h}_o} + \dfrac{L_s}{k_s} + \dfrac{L_{gw}}{k_{gw}} + \dfrac{L_s}{k_s} + \dfrac{1}{\bar{h}_i}}$$

Now, from Table B-2, $k_{gw} = 0.022\ \text{Btu/h}\cdot\text{ft}\cdot{}^\circ\text{F}$; and from Table B-1, at the mean temperature 20 °F,

$$\frac{L_s}{k_s} = \frac{0.0184}{12(26.5)} \approx 0$$

Therefore,

$$U \approx \frac{1}{\dfrac{1}{\bar{h}_o} + \dfrac{L_{gw}}{k_{gw}} + \dfrac{1}{\bar{h}_i}} = \frac{1}{\left(\dfrac{1}{4} + \dfrac{\frac{2}{12}}{0.022} + \dfrac{1}{2}\right)\dfrac{\text{h}\cdot\text{ft}^2\cdot{}^\circ\text{F}}{\text{Btu}}} = 0.1201\ \text{Btu/h}\cdot\text{ft}^2\cdot{}^\circ\text{F}$$

1.79 A 6-in.-thick concrete wall, having thermal conductivity $k = 0.50\ \text{Btu/h}\cdot\text{ft}\cdot{}^\circ\text{F}$, is exposed to air at 70 °F on the inside and air at 20 °F on the outside. The heat transfer coefficients are $\bar{h}_i = 2.0\ \text{Btu/h}\cdot\text{ft}^2\cdot{}^\circ\text{F}$ and $\bar{h}_o = 10\ \text{Btu/h}\cdot\text{ft}^2\cdot{}^\circ\text{F}$. Determine the heat transfer rate and the two surface temperatures of the wall.

▮ By (2) of Problem 1.76,

$$\frac{q}{A} = \frac{T_i - T_o}{\dfrac{1}{\bar{h}_i} + \dfrac{L_a}{k_a} + \dfrac{1}{\bar{h}_o}} = \frac{(70-20)\ {}^\circ\text{F}}{\left(\dfrac{1}{2} + \dfrac{0.5}{0.50} + \dfrac{1}{10}\right)\dfrac{\text{h}\cdot\text{ft}^2\cdot{}^\circ\text{F}}{\text{Btu}}} = 31.25\ \text{Btu/h}\cdot\text{ft}^2$$

The surface temperatures can be determined from (2) of Problem 1.76; thus

$$T_1 = T_i - \frac{q}{A}\frac{1}{\bar{h}_i} = (70\ {}^\circ\text{F}) - \left(31.25\ \frac{\text{Btu}}{\text{h}\cdot\text{ft}^2}\right)\left(\frac{\text{h}\cdot\text{ft}^2\cdot{}^\circ\text{F}}{2.0\ \text{Btu}}\right) = 54.375\ {}^\circ\text{F}$$

$$T_2 = T_o + \frac{q}{A}\frac{1}{\bar{h}_o} = (20\ {}^\circ\text{F}) + \left(31.25\ \frac{\text{Btu}}{\text{h}\cdot\text{ft}^2}\right)\left(\frac{\text{h}\cdot\text{ft}^2\cdot{}^\circ\text{F}}{10\ \text{Btu}}\right) = 23.125\ {}^\circ\text{F}$$

1.80 A 12-in.-thick brick outer wall is used in an office building in a southern city with no insulation or added internal finish. On a winter day the following temperatures were measured: inside air temperature, $T_i = 70\ {}^\circ\text{F}$; outside air temperature, $T_o = 15\ {}^\circ\text{F}$; inside surface temperature, $T_1 = 56\ {}^\circ\text{F}$; outside surface temperature, $T_2 = 19.5\ {}^\circ\text{F}$. Using $k = 0.76\ \text{Btu/h}\cdot\text{ft}\cdot{}^\circ\text{F}$ (from Table B-2), estimate the average values of the inner and outer heat transfer coefficients, $\bar{h}_i$ and $\bar{h}_0$.

The heat transfer rate per unit area may be determined by applying Fourier's law to the solid brick wall:

$$\frac{q}{A} = -k\frac{\Delta T}{\Delta x} = -\left(0.76\ \frac{\text{Btu}}{\text{h}\cdot\text{ft}\cdot{}^\circ\text{F}}\right)\frac{(19.5-56)\ {}^\circ\text{F}}{1\ \text{ft}} = 27.74\ \text{Btu/h}\cdot\text{ft}^2$$

Then

$$27.74\ \text{Btu/h}\cdot\text{ft}^2 = \bar{h}_i[(70-56)\ {}^\circ\text{F}] \qquad \text{or} \qquad \bar{h}_i = 1.981\ \text{Btu/h}\cdot\text{ft}^2\cdot{}^\circ\text{F}$$

and

$$27.74\ \text{Btu/h}\cdot\text{ft}^2 = \bar{h}_o[(19.5-15)\ {}^\circ\text{F}] \qquad \text{or} \qquad \bar{h}_o = 6.164\ \text{Btu/h}\cdot\text{ft}^2\cdot{}^\circ\text{F}$$

1.81 (*a*) Determine U for the situation of Problem 1.80, using SI units throughout. (*b*) From U and the overall temperature difference, determine q/A in W/m^2. Compare this with the value found in Problem 1.80.

(*a*) Using the conversion factors of Appendix A,

$$\bar{h}_i = \left(1.981\ \frac{\text{Btu}}{\text{h}\cdot\text{ft}^2\cdot{}^\circ\text{F}}\right)\left(\frac{3.1524\ \text{W/m}^2}{1\ \text{Btu/h}\cdot\text{ft}^2}\right)\left(\frac{9\ {}^\circ\text{K}}{5\ \text{K}}\right) = 11.241\ \text{W/m}^2\cdot\text{K}$$

$$\bar{h}_o = \left(6.164\ \frac{\text{Btu}}{\text{h}\cdot\text{ft}^2\cdot{}^\circ\text{F}}\right)\left(\frac{3.1524\ \text{W/m}^2}{1\ \text{Btu/h}\cdot\text{ft}^2}\right)\left(\frac{9\ {}^\circ\text{F}}{5\ \text{K}}\right) = 34.976\ \text{W/m}^2\cdot\text{K}$$

$$L_a = (12\ \text{in.})(0.0254\ \text{min/in.}) = 0.3048\ \text{m}$$

$$k_a = \left(0.76\ \frac{\text{Btu}}{\text{h}\cdot\text{ft}\cdot{}^\circ\text{F}}\right)\left(\frac{1.729577\ \text{W/m}\cdot\text{K}}{1\ \text{Btu/h}\cdot\text{ft}\cdot{}^\circ\text{F}}\right) = 1.314\ \text{W/m}\cdot\text{K}$$

Applying (*4*) of Problem 1.76,

$$U = \frac{1}{\left(\dfrac{1}{11.241} + \dfrac{0.3048}{1.314} + \dfrac{1}{34.976}\right)\dfrac{\text{m}^2\cdot\text{K}}{\text{W}}} = 2.861\ \text{W/m}^2\cdot\text{K}$$

(*b*) Since $(\Delta T)_{\text{overall}} = [(70-15)\ {}^\circ\text{F}](5\ \text{K}/9\ {}^\circ\text{K}) = 30.555\ \text{K}$, (*1*) of Problem 1.76 gives

$$\frac{q}{A} = (2.861\ \text{W/m}^2\cdot\text{K})(30.555\ \text{K}) = 87.421\ \text{W/m}^2$$

Converting to British engineering units,

$$\frac{q}{A} = (87.421\ \text{W/m}^2)\left(\frac{1\ \text{Btu/h}\cdot\text{ft}^2}{3.1524\ \text{W/m}^2}\right) = 27.73\ \text{Btu/h}\cdot\text{ft}^2$$

which agrees well with Problem 1.80.

1.82 Steam at 250 °F flows in an insulated pipe. The pipe is mild steel and has an inside radius of 2.0 in. and an outside radius of 2.25 in. The pipe is covered with a 1-in. layer of 85% magnesia. The inside heat transfer coefficient, $\bar{h}_i$, is 15 $\text{Btu/h}\cdot\text{ft}^2\cdot{}^\circ\text{F}$, and the outside coefficient, $\bar{h}_o$, is 2.2 $\text{Btu/h}\cdot\text{ft}^2\cdot{}^\circ\text{F}$. Determine the overall heat transfer coefficient U_o and the heat transfer rate from the steam per foot of pipe length, if the surrounding air temperature is 65 °F.

Use the two-material generalization of (*1*) of Problem 1.77. In terms of Fig. 1-27:

$$k_{\text{st}} = 26\ \text{Btu/h}\cdot\text{ft}\cdot{}^\circ\text{F}$$

$$r_1 = 2.0\ \text{in.} \qquad k_{\text{mag}} = 0.041\ \text{Btu/h}\cdot\text{ft}\cdot{}^\circ\text{F}$$

$$r_2 = 2.25\ \text{in.} \qquad \bar{h}_i = 15\ \text{Btu/h}\cdot\text{ft}^2\cdot{}^\circ\text{F}$$

$$r_3 = 3.25\ \text{in.} \qquad \bar{h}_o = 2.2\ \text{Btu/h}\cdot\text{ft}^2\cdot{}^\circ\text{F}$$

where thermal conductivity data are from Tables B-1 and B-2, at temperatures reasonably close to the expected average material temperatures. Thus

$$U_o = \left[\frac{3.25}{(2.0)(15)} + \frac{(3.25/12)\ln(2.25/2.0)}{26} + \frac{(3.25/12)\ln(3.25/2.25)}{0.041} + \frac{1}{2.2}\right]^{-1} = 0.3341\ \text{Btu/h}\cdot\text{ft}^2\cdot{}^\circ\text{F}$$

Clearly, the thermal resistance of the steel pipe wall is negligibly small in this problem. The heat transfer rate per unit length of the pipe is

$$\frac{q}{L} = U_o\frac{A_o}{L}(\Delta T)_{\text{overall}} = (0.3341\ \text{Btu/h}\cdot\text{ft}^2\cdot{}^\circ\text{F})(2\pi)\left(\frac{3.25}{12}\ \text{ft}\right)[(250-65)\ {}^\circ\text{F}] = 105.18\ \text{Btu/h}\cdot\text{ft}$$

1.83 In Problem 1.82 the thermal conductivity of the magnesia was taken at 200 °F. Determine the two surface temperatures of the magnesia, for the calculated q/L, and evaluate k_{mag} at the average temperature.

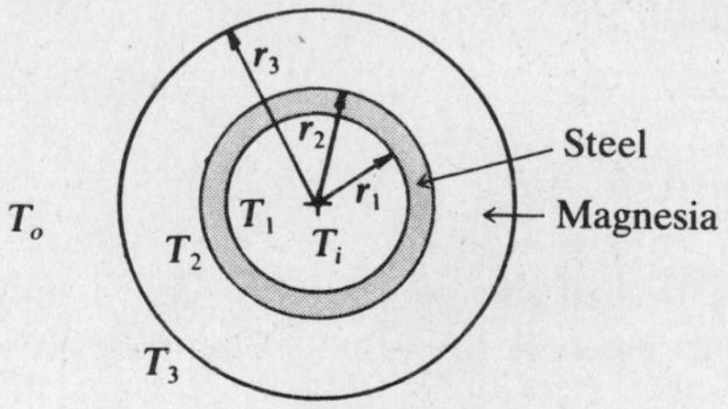

Fig. 1-27

▮ Refer to Fig. 1-27. Since $\bar{h}_o$ and $\bar{h}_i$ are specified together with T_o and T_i, we need to determine T_2 beginning with T_i and working from the inside out; and we should calculate T_3 beginning with T_o and working inward. From the steam to the inner wall of the steel pipe:

$$\frac{q}{L} = 2\pi r_1 \bar{h}_i (T_i - T_1)$$

$$T_1 = T_i - \left(\frac{q/L}{2\pi r_1 \bar{h}_i}\right) = 250 - \left[\frac{105.18}{2\pi(2.0/.12)(15)}\right] = 243.30\ °\text{F}$$

Through the steel pipe:

$$\frac{q}{L} = \frac{2\pi k_{st}(T_1 - T_2)}{\ln(r_2/r_1)}$$

$$T_2 = T_1 - \left[\frac{(q/L)\ln(r_2/r_1)}{2\pi k_{st}}\right] = 243.30 - \left[\frac{(105.18)\ln(2.25/2.0)}{2\pi(26)}\right] = 243.23\ °\text{F}$$

From the ambient air to the magnesia outer surface:

$$\frac{q}{L} = 2\pi r_3 \bar{h}_o (T_3 - T_o)$$

$$T_3 = 65 + \left[\frac{105.18}{2\pi(3.25/12)(2.2)}\right] = 93.09\ °\text{F}$$

Hence, the average temperature of the magnesia is

$$T_{avg} = \frac{243.23 + 93.09}{2} = 168.16\ °\text{F}$$

and by linear interpolation of the data of Table B-2, $k_{mag} \approx 0.0404$ Btu/h · ft · °F. However, it is questionable that Table B-2 is sufficiently accurate to justify recalculation of U_o and q/L for Problem 1.82.

1.84 A rectangular steel tank is filled with a liquid at 150 °F and exposed along the outside to air at 70 °F. The inner and outer convective heat transfer coefficients are $\bar{h}_i = 4.0$ Btu/h · ft² · °F and $\bar{h}_o = 1.5$ Btu/h · ft² · °F. The tank wall is ($\frac{1}{4}$)-in. mild steel ($k = 26$ Btu/h · ft · °F), and this is covered with a 1-in. layer of glass wool ($k = 0.024$ Btu/h · ft · °F). Determine (*a*) the overall heat transfer coefficient U and (*b*) the heat transfer rate per square foot.

▮ (*a*) By (4) of Problem 1.76 (extended to two material layers),

$$U = \left(\frac{1}{\bar{h}_i} + \frac{L_a}{k_a} + \frac{L_b}{k_b} + \frac{1}{\bar{h}_o}\right)^{-1} = \left(\frac{1}{4.0} + \frac{0.25/12}{26} + \frac{\frac{1}{12}}{0.024} + \frac{1}{1.5}\right)^{-1} = 0.2278\ \text{Btu/h}\cdot\text{ft}^2\cdot °\text{F}$$

Note that the thermal resistance due to the steel wall is negligible.

(*b*)
$$\frac{q}{A} = U\,\Delta T = (0.2278)(150 - 70) = 18.22\ \text{Btu/h}\cdot\text{ft}^2$$

1.85 Determine the optimum thickness of radial insulation, using the nomenclature of Fig. 1-28. Assume that $r - r_i$ must be kept small.

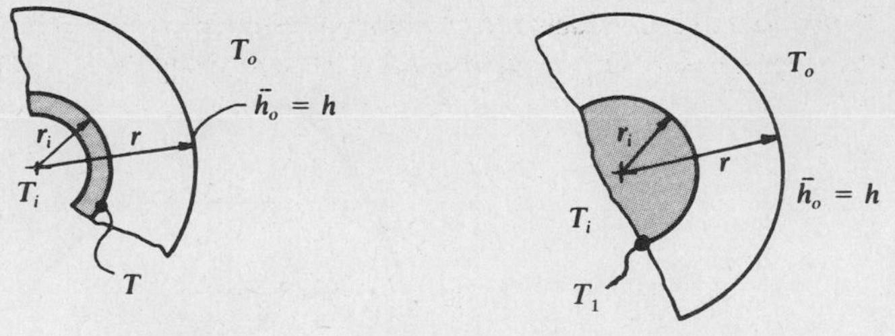

(*a*) Pipe System (*b*) Rod or Wire System

Fig. 1-28

▮ For a single layer of insulation material, with external convection only, Problem 1.77 gives

$$\frac{q}{L} = U_o \frac{A_o}{L} \Delta T = \frac{2\pi(T_1 - T_o)}{\dfrac{\ln(r/r_i)}{k} + \dfrac{1}{hr}} \tag{1}$$

In (*1*), T_1 can be replaced by T_i if the thermal resistance of the metal is negligible. As a function of r, q/L has no minimum, but it has a maximum [the denominator of (*1*) has a minimum] at

$$r = r_{crit} \equiv \frac{k}{h} \tag{2}$$

Thus, if $r_i < r_{crit}$, q/L *increases* with r (until $r = r_{crit}$). If $r_i > r_{crit}$, q/L steadily decreases as r increases.

1.86 Determine the critical radius, in millimeters, for an asbestos-covered pipe ($k_{asb} = 0.208$ W/m · K) if the external heat transfer coefficient is 1.5 Btu/h · ft² · °F.

▮ First, we need to convert h to SI units. Using Appendix A,

$$h = (1.5 \text{ Btu/h} \cdot \text{ft}^2 \cdot {}^\circ\text{F})\left(\frac{3.15248 \text{ W/m}^2}{1 \text{ Btu/h} \cdot \text{ft}^2}\right)\left(\frac{9\ {}^\circ\text{F}}{5 \text{ K}}\right) = 8.51 \text{ W/m}^2 \cdot \text{K}$$

Then, by (*2*) of Problem 1.85,

$$r_{crit} = \frac{k}{h} = \frac{0.208 \text{ W/m} \cdot \text{K}}{8.51 \text{ W/m}^2 \cdot \text{K}} = 0.0244 \text{ m} = 24.4 \text{ mm}$$

1.87 Plot q/L (Btu/h · ft) versus r (in.) for the situation of Problem 1.86, if $r_i = 0.5$ in., $T_i = 250$ °F, and $T_o = 70$ °F. Consider the range $r = r_i$ to $r = 1.5$ in.

▮ Converting units,

$$k = \left(0.208 \frac{\text{W}}{\text{m} \cdot \text{K}}\right)\left(\frac{1 \text{ Btu/h} \cdot \text{ft} \cdot {}^\circ\text{F}}{1.729577 \text{ W/m} \cdot \text{K}}\right)$$
$$= 0.120 \text{ Btu/h} \cdot \text{ft} \cdot {}^\circ\text{F}$$

By (*1*) of Problem 1.85, we have, in the specified units,

$$\frac{q}{L} = \frac{2\pi(250 - 70)}{\dfrac{\ln(r/0.5)}{0.120} + \dfrac{1}{(1.5)(r/12)}} = \frac{1130.97}{\dfrac{\ln(r/0.5)}{0.120} + \dfrac{8}{r}} \tag{1}$$

The graph of (*1*) is shown in Fig. 1-29. As calculated in Problem 1.85, the peak occurs at $r_{crit} \approx 1$ in., which value exceeds r_i.

1.88 In an attempt to reduce the energy loss from a ½-in.-OD metal hot-water line, a plumber decides to insulate the line with 0.47-in.-thick insulation having a thermal conductivity of 0.09 Btu/h · ft · °F. As a result of the high convective coefficient of the water and the low thermal resistance of the metal, the entire metal tube can be considered to remain at a uniform temperature of 180 °F. The line is surrounded by air at 80 °F, for which $\bar{h}_o = 1.5$ Btu/h · ft² · °F. In terms of percentage, how successful was the plumber in reducing the heat loss?

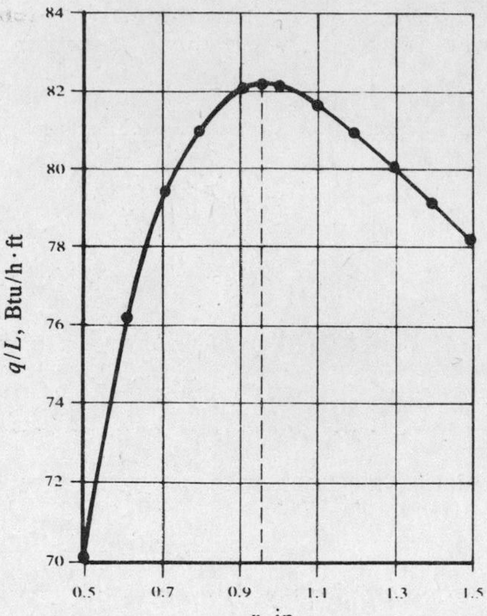

Fig. 1-29

In the notation of Fig. 1-28(*a*), equation (*1*) of Problem 1.85 gives for the bare pipe $(r = r_i)$:

$$\left(\frac{q}{L}\right)_{\text{bare}} = \frac{2\pi(T_1 - T_o)}{1/hr_i}$$

and for the insulated pipe:

$$\left(\frac{q}{L}\right)_{\text{ins}} = \frac{2\pi(T_1 - T_o)}{\dfrac{\ln(r/r_i)}{k} + \dfrac{1}{hr}}$$

Therefore,

$$\begin{aligned}\%\ \text{reduction} &= \frac{(q/L)_{\text{bare}} - (q/L)_{\text{ins}}}{(q/L)_{\text{bare}}}(100\%)\\ &= \left[1 - \frac{1}{(h/k)r_i \ln(r/r_i) + (r_i/r)}\right](100\%)\\ &= \left[1 - \frac{1}{(1.5/0.09)(\frac{1}{48})\ln(0.72/0.25) + (0.25/0.72)}\right](100\%)\\ &= -40\%\end{aligned}$$

Adding the insulation *increased* heat transfer by 40%.

1.89 A plane wall 5 in. thick generates heat internally at the rate of 10^4 Btu/h·ft³. One side of the wall is insulated, and the other side is exposed to an environment at 120 °F. The convective heat transfer coefficient between the wall and the environment is 100 Btu/h·ft²·°F. The thermal conductivity of the wall is 10 Btu/h·ft·°F. Find the two surface temperatures.

Use reflection, as in Problem 1.73. The faces of the 10-in. wall are at temperature T_s, where

$$q'''(AL) = hA(T_s - T_\infty)$$

$$T_s = \frac{q'''L}{h} + T_\infty = \frac{(10^4\ \text{Btu/h}\cdot\text{ft}^3)[(\frac{5}{12}\ \text{ft}]}{100\ \text{Btu/h}\cdot\text{ft}^2\cdot{}^\circ\text{F}} + (120\ {}^\circ\text{F}) = 161.67\ {}^\circ\text{F}$$

The insulated surface temperature is then

$$T_c = T_s + \frac{q'''}{2k}(L^2) = 161.67 + \left[\frac{10^4}{2(10)}\right]\left(\frac{5}{12}\right)^2 = 248.47\ {}^\circ\text{F}$$

1.90 A constant-temperature cold-water reservoir for a laboratory experiment consists of a 4-in.-ID type-304 stainless-steel pipe with ½-in. walls and a 2-in. layer of rock wool covered with a thin cotton cloth on the outside. The inner pipe surface in contact with an ice-water mixture is maintained at 32 °F. The outside convective coefficient is 2.0 Btu/h · ft^2 · °F, where the ambient temperature is 70 °F. Draw an electrical analog and determine the heat transfer rate per lineal foot from the room to the pipe. (See Fig. 1-30.)

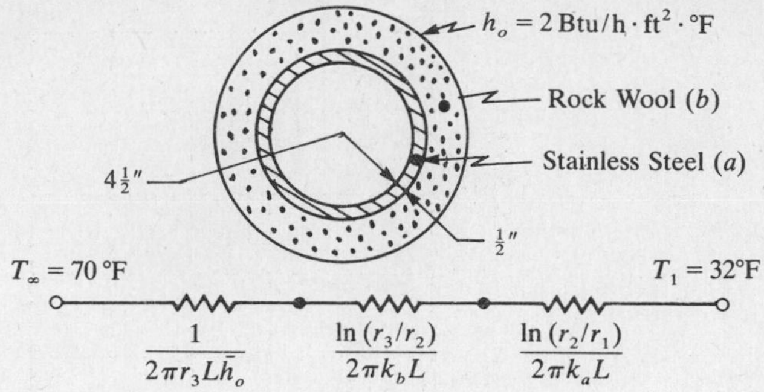

Fig. 1-30

▮ The three thermal resistances are displayed in Fig. 1-30. With $\bar{A} = A_0 = 2\pi r_3 L$, (5) of Problem 1.76 gives

$$U_o = \frac{1}{(2\pi r_3 L)\Sigma R_{th}} = \frac{1}{(1/\bar{h}_o) + [(r_3/k_b)\ln(r_3/r_2)] + [(r_3/k_a)\ln(r_2/r_1)]}$$

Substituting the data (k-values from Appendix B), we find $U_o = (1/13.48)$ Btu/h · ft^2 · °F. Hence,

$$\frac{q}{L} = U_o(2\pi r_3)(\Delta T) = \left(\frac{1}{13.48}\right)\left[2\pi\left(\frac{4.5}{12}\right)\right](32-70)$$
$$= -6.64 \text{ Btu/h} \cdot \text{ft} \quad \text{(into pipe)}$$

1.91 A $\frac{1}{8}$-in.-diameter copper heater wire 2 ft long is submerged in a 320 °F fluid. The voltage drop across the wire is 60 V, and the current measured is 120 A. Calculate the center temperature of the wire if $\bar{h}_o =$ 100 Btu/h · ft^2 · °F.

▮ Because of the extremely high heat transfer coefficient between the wire and the fluid, we can neglect any resistance and assume that the outer wire surface and fluid are at the same temperature. The problem is thus equivalent to Problem 1.70, so that (1 W = 1 V · A)

$$T_c = T_s + \frac{EI}{4\pi kL}$$
$$= (320\ °\text{F}) + \frac{(60\ \text{V})(120\ \text{A})(3.412\ \text{Btu/h} \cdot \text{V} \cdot \text{A})}{4\pi(216\ \text{Btu/h} \cdot \text{ft} \cdot °\text{F})(2\ \text{ft})}$$
$$= 324.51\ °\text{F}$$

1.92 A vapor-to-liquid heat-exchanger surface of 1000 in^2 face area is made of $\frac{3}{8}$-in. nickel ($k = 34$ Btu/h · ft · °F) with a $\frac{3}{64}$-in. plating of copper ($k = 218$ Btu/h · ft · °F) on the vapor side. The resistivity ($= L/k$) of the scale deposit on the vapor side is estimated to be 0.03 h · ft^2 · °F/Btu, and the vapor- and liquid-side convective coefficients are known to be 850 and 110 Btu/h · ft^2 · °F, respectively. The heated vapor is at 233 °F, and the liquid is at 155 °F. Find the (*a*) overall heat transfer coefficient U, (*b*) overall heat exchange, (*c*) temperature drop across the scale deposit, and (*d*) temperature at the copper-nickel interface. (See Fig. 1-31.)

▮ (*a*) Generalizing (4) of Problem 1.76,

$$U = \left[\frac{1}{850} + 0.03 + \frac{\frac{3}{64}}{(12)(218)} + \frac{\frac{3}{8}}{(12)(34)} + \frac{1}{110}\right]^{-1} = 24.27\ \text{Btu/h} \cdot \text{ft}^2 \cdot °\text{F}$$

(*b*)
$$q = UA(\Delta T) = (24.27\ \text{Btu/h} \cdot \text{ft}^2 \cdot °\text{F})(\tfrac{1000}{144}\ \text{ft}^2)[(233-155)\ °\text{F}] = 13\,146\ \text{Btu/h}$$

(*c*)
$$(\Delta T)_{\text{scale}} = qR_{\text{scale}} = q\left(\frac{L/k}{A}\right)_{\text{scale}}$$
$$= (13\,146)\left(\frac{0.03}{\frac{1000}{144}}\right) = 56.79\ °\text{F}$$

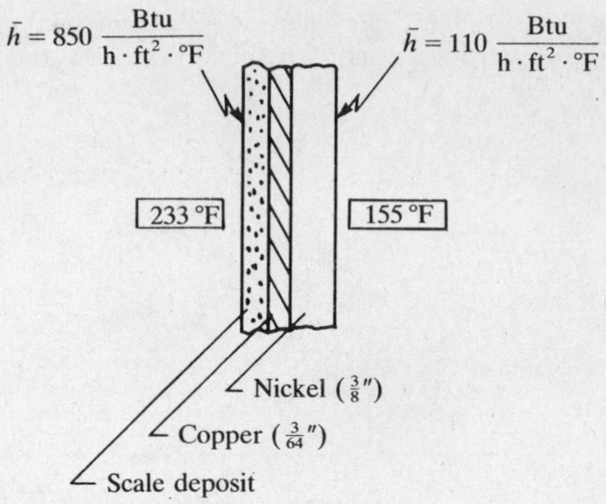

Fig. 1-31

(*d*) From Fourier's law applied to the nickel,

$$T_i = T_{\text{liq}} + \frac{L_{\text{Ni}}q}{k_{\text{Ni}}A} = 155 + \frac{[3/(8)(12)](13\,146)}{(34)(\frac{1000}{144})}$$

$$= 156.74\ °\text{F}$$

1.93 The outside walls of a house consist of 4-in. masonry brick, $\frac{1}{2}$-in. pine sheathing, 2- by 4-in. pine studs 16 in. on center, and $\frac{1}{2}$-in. Sheetrock ($k = 0.09$ Btu/h · ft · °F). Assume the thermal conductivity of the mortar joints to be equal to that of the brick. Find the overall heat transfer coefficient U if the outside heat transfer coefficient is 3.5 Btu/h · ft² · °F, the inside coefficient is 1.5 Btu/h · ft² · °F, and the air space between the studs has a heat transfer coefficient of 1.2 Btu/h · ft² · °F (for each internal surface). Calculate the heat transfer rate for a 1 ft depth per unit section (16 in.) of wall. Inside and outside air temperatures are 70 °F and 95 °F, respectively.

$$q = \frac{\Delta T}{\Sigma R_{\text{th}}}$$

$$k(\text{brick}) \equiv k_b = 0.38\ \text{Btu/h} \cdot \text{ft} \cdot °\text{F}$$
$$k(\text{plywood}) \equiv k_{\text{ply}} = 0.06\ \text{Btu/h} \cdot \text{ft} \cdot °\text{F}$$
$$k(\text{studs}) \equiv k_s = 0.14\ \text{Btu/h} \cdot \text{ft} \cdot °\text{F}$$
$$k(\text{Sheetrock}) \equiv k_{\text{ins}} = 0.09\ \text{Btu/h} \cdot \text{ft} \cdot °\text{F}$$

Per foot of depth,

$$R_b = \frac{\frac{4}{12}\ \text{ft}}{(1\ \text{ft})(\frac{16}{12}\ \text{ft})(0.38\ \text{Btu/h} \cdot \text{ft} \cdot °\text{F})} = 0.658\ \text{h} \cdot °\text{F/Btu}$$

$$R_{\text{ply}} = \frac{\frac{1}{2}/12}{(1)(\frac{16}{12})(0.06)} = 0.521\ \text{h} \cdot °\text{F/Btu}$$

$$R_s = \frac{\frac{4}{12}}{(1)(\frac{2}{12})(0.14)} = 14.28\ \text{h} \cdot °\text{F/Btu}$$

$$R_{\text{ins}} = \frac{\frac{1}{2}/12}{(1)(\frac{16}{12})(0.09)} = 0.347\ \text{h} \cdot °\text{F/Btu}$$

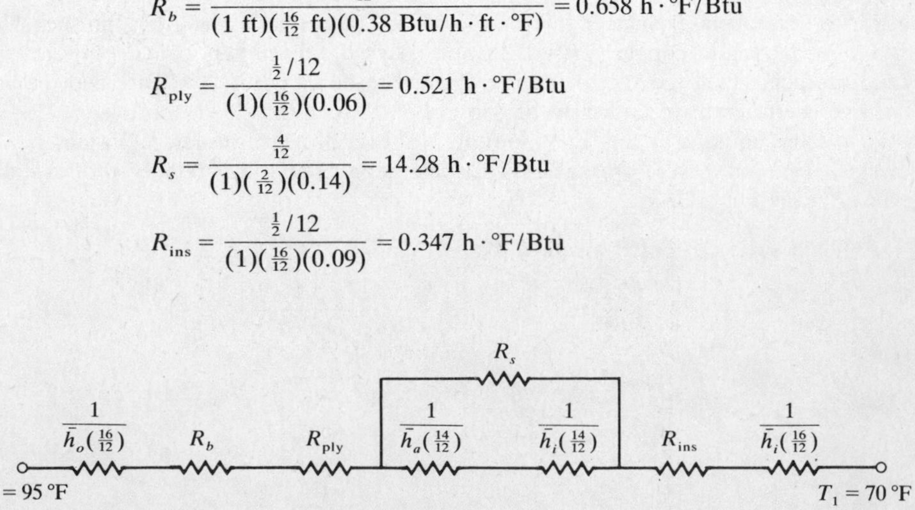

Fig. 1-32

The electrical analog (assuming convection only in air space) is shown in Fig. 1-32. Replacing the series-parallel combination by the equivalent

$$R_{\rm eq} = \frac{R_s\left[\dfrac{2}{\bar{h}_a(\frac{14}{12})}\right]}{R_s + \dfrac{2}{\bar{h}_a(\frac{14}{12})}} = 1.30 \text{ h}\cdot{}^\circ\text{F/Btu}$$

we obtain

$$q = \frac{\Delta T}{\dfrac{1}{\bar{h}_o(\frac{16}{12})} + R_b + R_{\rm ply} + R_{\rm eq} + R_{\rm ins} + \dfrac{1}{\bar{h}_i(\frac{16}{12})}}$$

$$= \frac{95-70}{\dfrac{1}{(3.5)(\frac{16}{12})} + 0.658 + 0.521 + 1.30 + 0.347 + \dfrac{1}{(1.5)(\frac{16}{12})}}$$

$$= 7.07 \text{ Btu/h}$$

1.94[D] Determine the critical radius, in inches, of insulation for asbestos felt, 40 laminations per inch, at 100 °F, if the external heat transfer coefficient is 1.2 Btu/h · ft^2 · °F.

▌ From Table B-2, $k = 0.033$ Btu/h · ft · °F; from (2) of Problem 1.85,

$$r_{\rm crit} = \frac{0.033}{1.2} = 0.028 \text{ ft} = 0.33 \text{ in.}$$

1.95[D] What is the critical radius, in millimeters, for 29.3-lbm/ft^3 asbestos at 0 °C for an external heat transfer coeficient of 5.0 W/m · K?

▌ From Table B-2, $k = (0.090)(1.7296) = 0.16$ W/m · K; thus,

$$r_{\rm crit} = \frac{0.16}{5.0} = 0.032 \text{ m} = 32 \text{ mm}$$

1.3 FINS

1.96 Fins are used to increase the effective surface area for convective heat transfer in heat exchangers, internal combustion engines, transistor configurations, etc. For the configuration shown in Fig. 1-33, find q, if (*a*) $L \to \infty$, with end at temperature of surrounding fluid; (*b*) L finite, with insulated end; and (*c*) L finite, with heat loss by convection at end.

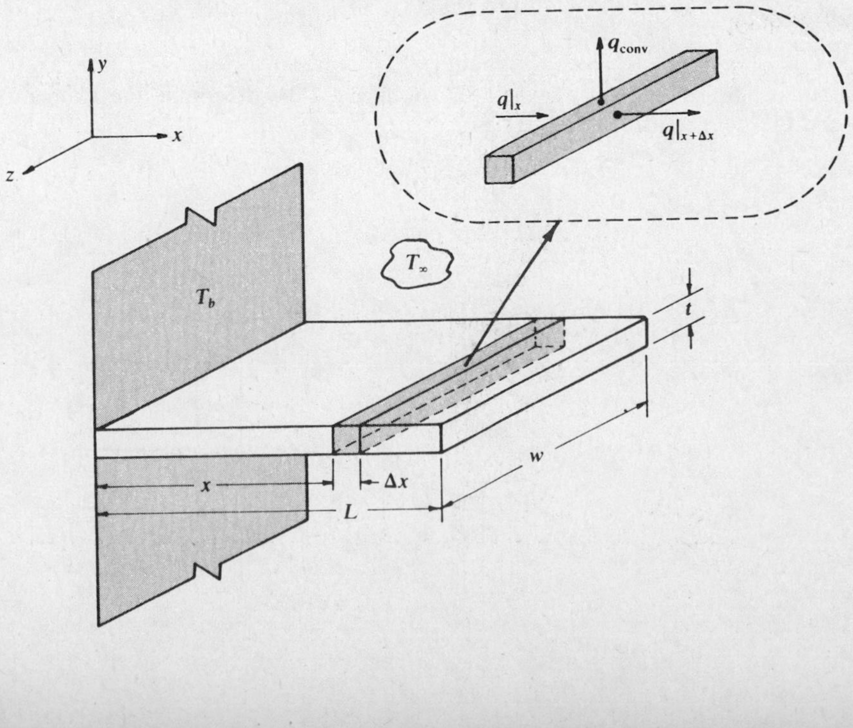

Fig. 1-33

The first law of thermodynamics, applied to an element of the fin of length Δx, gives, in the steady state,

$$(\text{Energy conducted in at } x) = (\text{energy conducted out at } x + \Delta x) + (\text{energy out by convection}) \qquad (1)$$

Assuming no temperature variation in the y- or z-directions, the three terms are respectively

$$q\Big|_x = -kA \frac{dT}{dx}\Big|_x \qquad q\Big|_{x+\Delta x} = -kA \frac{dT}{dx}\Big|_{x+\Delta x} \qquad q_{\text{conv}} = \bar{h}(P\,\Delta x)(T - T_\infty)$$

where P is the perimeter, $2(w + t)$. Substituting these three expressions in (1), dividing by Δx, and taking the limit as $\Delta x \to 0$ results in

$$\frac{d^2T}{dx^2} - \frac{\bar{h}P}{kA}(T - T_\infty) = 0 \qquad (2)$$

if the thermal conductivity k is constant. Letting $\theta \equiv T - T_\infty$ and $n \equiv \sqrt{\bar{h}P/kA}$, (2) becomes

$$\frac{d^2\theta}{dx^2} - n^2\theta = 0$$

which has the general solution

$$\theta(x) = C_1 \sinh nx + C_2 \cosh nx \qquad (3)$$

One boundary condition is $\theta(0) = T_b - T_\infty \equiv \theta_b$, which requires that $C_2 = \theta_b$. The heat transfer from the fin is most easily evaluated as

$$q = -k(wt)\frac{d\theta}{dx}\Big|_{x=0} = -kwtnC_1 \qquad (4)$$

(*a*) The second boundary condition is $\theta(+\infty) = 0$, which gives $C_1 = -\theta_b$. Hence, from (4), $q = kwtn\theta_b$.

(*b*) The second boundary condition is

$$\frac{d\theta}{dx}\Big|_{x=L} = 0 \qquad \text{whence} \qquad C_1 = -\theta_b \tanh nL$$

Hence, from (4), $q = kwtn\theta_b \tanh nL$.

(*c*) The second boundary condition is

$$-k\frac{d\theta}{dx}\Big|_{x=L} = \bar{h}_L\theta(L)$$

which gives

$$-C_1 = \frac{q}{kwtn} = \theta_b \frac{nk \sinh nL + \bar{h}_L \cosh nL}{\bar{h}_L \sinh nL + nk \cosh nL}$$

1.97 For the annular fin of uniform thickness shown in Fig. 1-34, determine the temperature distribution within the fin and the heat transfer into the base.

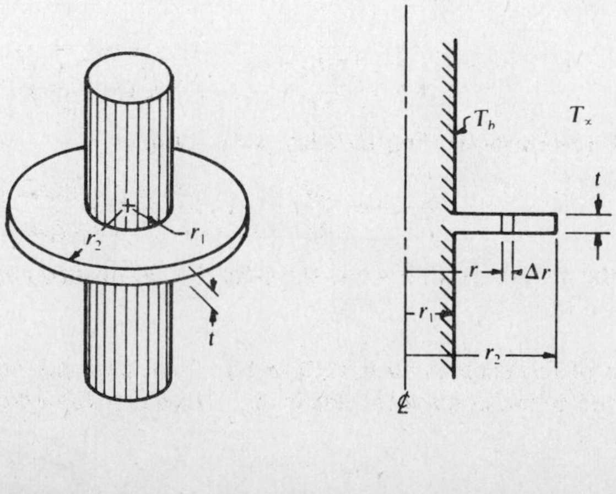

Fig. 1-34

For purely radial temperature variation, the first law of thermodynamics, applied to a cylindrical volume element, leads to

$$\frac{d^2\theta}{dr^2} + \frac{1}{r}\frac{d\theta}{dr} - \frac{2\bar{h}}{kt}\theta = 0 \tag{1}$$

This is a form of Bessel's differential equation of zero order, and it has the general solution

$$\theta = C_1 I_0(nr) + C_2 K_0(nr) \tag{2}$$

where $n \equiv \sqrt{2\bar{h}/kt}$
$I_0 \equiv$ modified Bessel function of the 1st kind
$K_0 \equiv$ modified Bessel function of the 2d kind

The constants C_1 and C_2 are determined by the boundary conditions, which are $\theta(r_1) = T_b - T_\infty \equiv \theta_b$ and, usually,

$$\left.\frac{d\theta}{dr}\right|_{r=r_2} = 0 \tag{3}$$

The assumption (3) of negligible heat loss from the end of the fin is realistic for $t \ll r_2 - r_1$.

With C_1 and C_2 evaluated [application of (3) requires the formulas $I_0'(x) = I_1(x)$, $K_0'(x) = -K_1(x)$], (2) becomes

$$\frac{\theta}{\theta_b} = \frac{I_0(nr)K_1(nr_2) + K_0(nr)I_1(nr_2)}{I_0(nr_1)K_1(nr_2) + K_0(nr_1)I_1(nr_2)} \tag{4}$$

Determining the heat loss from the fin by evaluating the conductive heat transfer rate into the base, we obtain

$$q = 2\pi kt\theta_b nr_1\left[\frac{K_1(nr_1)I_1(nr_2) - I_1(nr_1)K_1(nr_2)}{I_0(nr_1)K_1(nr_2) + K_0(nr_1)I_1(nr_2)}\right] \tag{5}$$

1.98 Calculate the rate of heat transfer into the base of the straight, isosceles-triangular fin shown in Fig. 1-35.

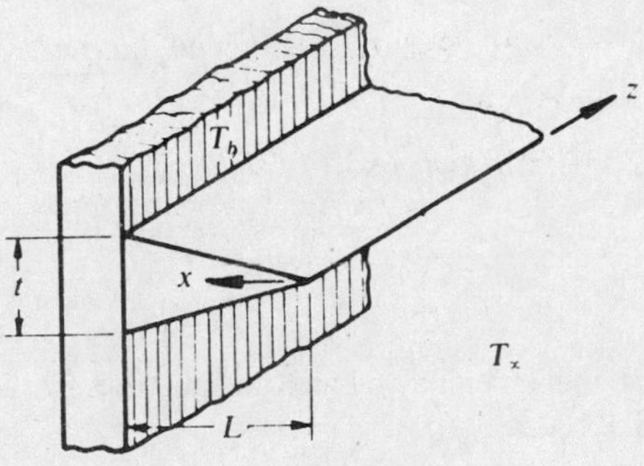

Fig. 1-35

For a very thin fin $(t/L \ll 1)$, the temperature distribution will be very nearly one-dimensional:

$$\frac{\theta}{\theta_b} = \frac{I_0(2px^{1/2})}{I_0(2pL^{1/2})} \tag{1}$$

where

$$p \equiv \sqrt{\frac{2f\bar{h}L}{kt}} \qquad f \equiv \sqrt{1 + \left(\frac{t}{2L}\right)^2} \approx 1$$

Then, for unit width (z-direction), Fourier's law and (1) give

$$q = +kt\left.\frac{d\theta}{dx}\right|_{x=L} = \left(\frac{kt\theta_b p}{L^{1/2}}\right)\left[\frac{I_1(2pL^{1/2})}{I_0(2pL^{1/2})}\right] \tag{2}$$

[We have reversed the sign in Fourier's law in order that q, as given by (2), represent flux *into the fin* at the base.]

1.99 Consider a tall stack of fins such as illustrated in Fig. 1-36; each fin has surface area A_f, and the area of the common base included between adjacent fins is A_b. Defining the *efficiency* of a fin as

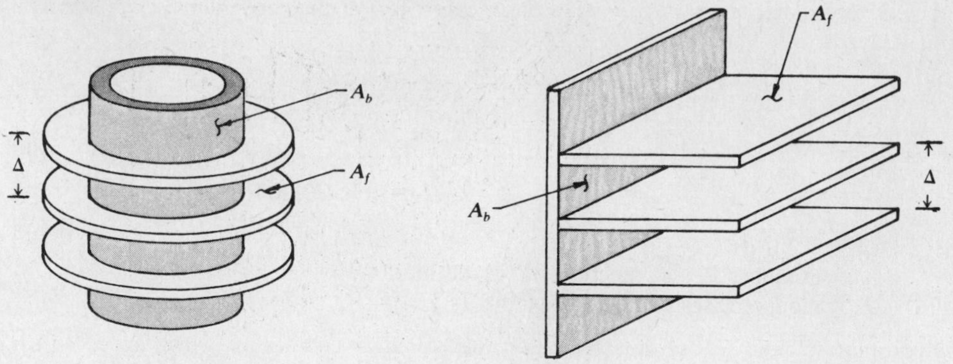

Fig. 1-36

$$\eta_f \equiv \frac{\text{actual heat transfer by fin}}{\text{heat transfer if entire fin were at base temperature}} \tag{1}$$

express the heat transfer rate per unit height, q', of the system.

I By (*1*), the heat transfer by a fin is $q_{\text{fin}} = \eta_f \bar{h} A_f \theta_b$; the heat transfer rate by an interfin section of the base is $q_{\text{base}} = \bar{h} A_b \theta_b$. Hence,

$$q' = \frac{q_{\text{fin}} + q_{\text{base}}}{\Delta} = \bar{h}\theta_b\left(\frac{A_b}{\Delta} + \eta_f \frac{A_f}{\Delta}\right) \tag{2}$$

(Note the assumption of a single $\bar{h}$ for all surfaces.)

1.100 Determine the thermal efficiency of a rectangular fin with no end heat loss.

I Problem 1.96(*b*) and the definition (*1*) of Problem 1.99 give, in the nomenclature of Fig. 1-33,

$$q_f = \frac{kwtn\theta_b \tanh nL}{\bar{h}(2Lt + 2Lw)\theta_b} \tag{1}$$

But, by definition, $n^2 = 2(w+t)\bar{h}/kwt$; so (*1*) becomes

$$\eta_f = \frac{\tanh nL}{nL} \tag{2}$$

1.101 It is physically obvious that the efficiency of any fin must be smaller than unity. Verify this for the rectangular fin of Problem 1.100.

I For all x,

$$\frac{\tanh x}{x} = \frac{\sinh x}{x \cosh x} = \frac{x + \dfrac{x^3}{3!} + \dfrac{x^5}{5!} + \cdots}{x + \dfrac{x^3}{2!} + \dfrac{x^5}{4!} + \cdots} < 1$$

because each term, after the first, in the numerator is smaller than the corresponding term in the denominator.

For small x, division of the power series yields

$$\frac{\tanh x}{x} \approx 1 - \frac{x^2}{3} \qquad (x \text{ small})$$

while for large x,

$$\frac{\tanh x}{x} \approx \frac{1}{x} \qquad (x \text{ large})$$

This behavior is apparent in Fig. 1-37, which graphs η_f for the rectangular fin against nL (with a scale factor $1/\sqrt{2}$, and with L replaced by $L_c \equiv L + t/2$ to account for tip loss). The figure also graphs the thermal efficiencies of the annular and triangular fins.

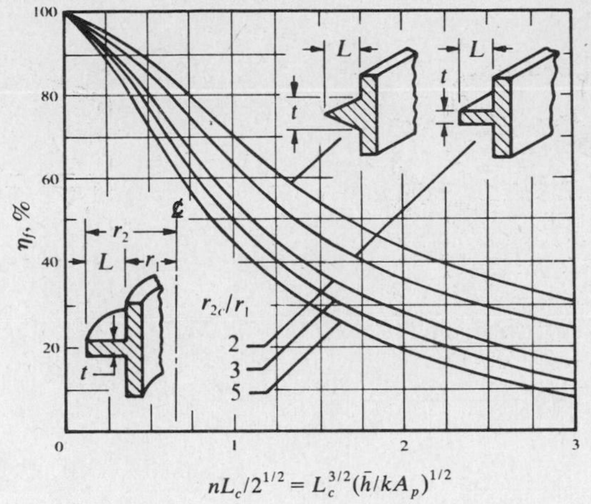

Fig. 1-37

1.102 Refer to Fig. 1-33. A rectangular fin, with dimensions $L = 2$ in. and $t = 0.05$ in. extends from a plane wall. The fin material is mild steel ($k = 26$ Btu/h · ft · °F), and the external heat transfer coefficient $\bar{h}$ may be taken as 10 Btu/h · ft^2 · °F. The surrounding air temperature is 80 °F, and the wall temperature is 300 °F. Calculate the heat loss rate per unit width of fin.

▌ Problem 1.96(*c*) applies; but since t/L and t/w are small in this case, the simpler result of Problem 1.96(*b*) may be used. With

$$n = \sqrt{\frac{\bar{h}2(w+t)}{kwt}} \approx \sqrt{\frac{2\bar{h}}{kt}} = \sqrt{\frac{2(10)}{(26)(0.05/12)}} = 13.6 \text{ ft}^{-1}$$

we have

$$\frac{q}{w} = ktn\theta_b \tanh nL = (26)\left(\frac{0.05}{12}\right)(13.6)(300 - 80) \tanh\left[(13.6)\left(\frac{2}{12}\right)\right]$$

$$= 316 \text{ Btu/h} \cdot \text{ft}$$

1.103 Annular, copper, 3-in.-OD fins are placed on a 1-in.-OD tube; each fin is 0.10 in. thick. The tube surface is at 450 °F, and the surrounding fluid temperature is 70 °F. Find the heat loss from each fin, if $\bar{h} =$ 240 Btu/h · ft^2 · °F.

▌ The easiest way to determine the heat flux from an annular fin is to use Fig. 1-37. The necessary parameters are

$$L_c = 1.50 - 0.5 + \frac{0.10}{2} = 1.05 \text{ in.} = 0.0875 \text{ ft}$$

$$\frac{r_{2c}}{r_1} = \frac{1.50 + (0.10/2)}{0.5} = 3.1$$

$$A_p = t(r_{2c} - r_1) = \frac{0.10(1.55 - 0.5)}{144} = 7.3 \times 10^{-4} \text{ ft}^2$$

$$L_c^{3/2}\left(\frac{\bar{h}}{kA_p}\right)^{1/2} = (0.0875)^{3/2}\left[\frac{240}{(215)(7.3 \times 10^{-4})}\right]^{1/2} = 1.01$$

Hence, $q_{\text{actual}} = \eta_f q_{450} = (0.49)(8560) = 4194$ Btu/h.

1.104 An aluminum cylindrical rod ($k = 132$ Btu/h · ft · °F), of diameter 0.375 in. and length 4 in., protrudes from a surface having a temperature of 200 °F. The rod is exposed to ambient air at 70 °F, and the heat transfer coefficient along the length and at the end is 1.5 Btu/h · ft^2 · °F. Determine the heat flux, (*a*) neglecting the heat transfer at the end and (*b*) accounting for the heat transfer at the end.

▌ The results of Problem 1.96, for the rectangular fin, apply to the cylindrical rod of radius r, if the mere replacements

$$P = 2(w + t) \rightarrow 2\pi r \qquad A = wt \rightarrow \pi r^2$$

are made. Thus,

$$n = \sqrt{\frac{2\bar{h}}{kr}} = \sqrt{\frac{2(1.5)}{132[0.375/2(12)]}} = 1.2060\ \text{ft}^{-1}$$

(*a*) By Problem 1.96(*b*),

$$q = k\pi r^2 n\theta_b \tanh nL$$
$$= (132)\pi\left(\frac{0.375}{24}\right)^2 (1.2060)(200-70)\tanh\left[(1.2060)\left(\frac{4}{12}\right)\right] = 6.058\ \text{Btu/h}$$

(*b*) By Problem 1.96(*c*), with $\bar{h}_L/nk = \bar{h}/nk = 0.00942$,

$$q = k\pi r^2 n\theta_b\left[\frac{\sinh nL + (\bar{h}/nk)\cosh nL}{\cosh nL + (\bar{h}/nk)\sinh nL}\right]$$
$$= (132)\pi\left(\frac{0.375}{24}\right)^2 (1.2060)(200-70)\left\{\frac{\sinh[(1.2060)(\frac{4}{12})] + 0.00942\cosh[(1.2060)(\frac{4}{12})]}{\cosh[(1.2060)(\frac{4}{12})] + 0.00942\sinh[(1.2060)(\frac{4}{12})]}\right\}$$
$$= 6.185\ \text{Btu/h}$$

1.105 A thin fin of length L has its two ends attached to two parallel walls which have temperatures T_1 and T_2 [Fig. 1-38(*a*)]. The fin loses heat by convection to the ambient air at T_∞. Obtain an analytical expression for the one-dimensional temperature distribution along the length of the fin.

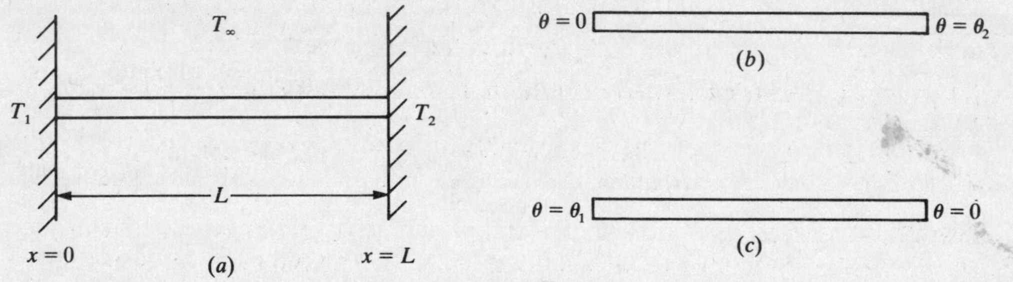

Fig. 1-38

From Problem 1.96 it is clear that the desired distribution (of $\theta \equiv T - T_\infty$) may be obtained as the superposition of the solutions to subproblems (*b*) and (*c*) of Fig. 1-38. Now, by (*3*) of Problem 1.96, the solution to (*b*) is given by the function

$$g(x;\theta_2) = \frac{\theta_2}{\sinh nL}\sinh nx$$

Therefore, the solution to (*c*) is given by

$$g(L-x;\theta_1) = \frac{\theta_1}{\sinh nL}\sinh n(L-x)$$

and so

$$\theta(x) = \frac{\theta_2 \sinh nx + \theta_1 \sinh n(L-x)}{\sinh nL} \qquad (1)$$

1.106 Find the heat transfer in Problem 1.105.

The total heat transfer rate is the sum of the two conductive heat transfer rates *into* the fin at its ends:

$$q = kA\left(\left.\frac{d\theta}{dx}\right|_{x=L} - \left.\frac{d\theta}{dx}\right|_{x=0}\right)$$

Differentiating (*1*) of Problem 1.105, we get

$$q = kAn\frac{\cosh nL - 1}{\sinh nL}(\theta_1 + \theta_2)$$

1.107 A simplified method of determining thermal conductivity is to use the temperature gradient in a small-diameter, long rod with one end attached to a high- (or low-) temperature source. What is the thermal conductivity of a $\frac{1}{2}$-in.-diameter rod extending from an oven into a 70 °F environment with an external heat transfer coefficient of 5.0 Btu/h · ft² · °F, if the temperatures detected by two thermocouples located 6 in. apart along the rod are 342 °F and 227 °F?

For an infinitely long rod, the temperature distribution is given by

$$\theta = \theta_b e^{-nx} \qquad \text{or} \qquad T - T_\infty = (T_b - T_\infty)e^{-nx} \tag{1}$$

where (Problem 1.104) $n^2 = 4\bar{h}/kd$. Writing $T(x_1) \equiv T_1$ and $T(x_1 + \Delta x) \equiv T_2$, we have from (1)

$$\frac{T_1 - T_\infty}{T_2 - T_\infty} = e^{n\,\Delta x} \qquad \text{or} \qquad \frac{342 - 70}{227 - 70} = e^{n/2}$$

Solving, $n = 1.098\ \text{ft}^{-1}$ and

$$k = \frac{4\bar{h}}{n^2 d} = \frac{4(5.0)}{(1.098)^2(\frac{1}{24})} = 398.34\ \text{Btu/h}\cdot\text{ft}\cdot{}^\circ\text{F}$$

1.108 The end of a very long cylindrical stainless-steel rod is attached to a heated wall, and its surface is in contact with a cold fluid. (*a*) If the rod diameter were doubled, by what percentage would the rate of heat removal increase? (*b*) If the rod were made of aluminum, by what percentage would the heat transfer rate change from that of the stainless steel?

In Problem 1.104(*a*), let $L \to \infty$, obtaining

$$q = k\pi r^2 n\theta_b \propto k^{1/2} d^{3/2}$$

(*a*) $$\%\ \text{increase} = (2^{3/2} - 1)(100\%) = 183\%$$

(*b*) Choosing k (stainless steel) = 9.4 Btu/h · ft · °F and k (aluminum) = 119 Btu/h · ft · °F,

$$\%\ \text{increase} = \frac{\sqrt{119} - \sqrt{9.4}}{\sqrt{9.4}}(100\%) = 256.53\%$$

1.109 In Problem 1.105, let $n = 3.0\ \text{m}^{-1}$, $\theta_1 = \theta_2 = 20\ {}^\circ\text{C}$, and $L = 1.0$ m. (*a*) Determine the temperature at $x = 0.4$ m, if the ambient temperature is 25 °C. (*b*) Obtain the same numerical result by exploiting the symmetry of the problem.

(*a*) Substitution of the data in (1) of Problem 1.105 yields $\theta(0.4) = 8.8873\ {}^\circ\text{C}$. Hence,

$$T(0.4) = 25 + 8.8873 = 33.89\ {}^\circ\text{C}$$

(*b*) As the problem is symmetrical about the midpoint of the rod, the insulated-end solution applies, with $L = 0.5$ m. Problem 1.96(*b*),

$$\theta(x) = -\theta_b \tanh nL \sinh nx + \theta_b \cosh nx = \theta_b \frac{\cosh n(L - x)}{\cosh nL}$$

and substitution of values yields $\theta(0.4) = 8.8874\ {}^\circ\text{C}$, as before.

1.110 A very long 1-cm-diameter copper rod ($k = 377$ W/m · K) is exposed to an environment at 22 °C. The base temperature of the rod is maintained at 150 °C. The heat transfer coefficient between the rod and the surrounding air is 11 W/m² · K. Determine the heat transfer rate from the rod to the surrounding air.

As in Problem 1.104(*a*),

$$n = \left(\frac{2\bar{h}}{kr}\right)^{1/2} = \left[\frac{2(11\ \text{W/m}^2\cdot\text{K})}{(377\ \text{W/m}\cdot\text{K})(0.005\ \text{m})}\right]^{1/2} = 3.416\ \text{m}^{-1}$$

and

$$q = kAn\theta_b = (377\ \text{W/m}\cdot\text{K})[\pi(0.005\ \text{m})^2](3.416\ \text{m}^{-1})[(150 - 22)\ \text{K}] = 12.948\ \text{W}$$

1.111 Repeat Problem 1.110 for finite lengths 2, 4, 8, . . . , 128 cm, assuming heat loss at the end, with $\bar{h}_L =$ 11 W/m² · K $= \bar{h}$.

▌ We apply the formula of Problem 1.104(*b*), with the following parameter values:

$$n = 3.416\ \text{m}^{-1} \qquad \text{(from Problem 1.110)}$$

$$\frac{\bar{h}}{nk} = \frac{11\ \text{W/m}^2\cdot\text{K}}{(3.416\ \text{m}^{-1})(3.77\ \text{W/m}\cdot\text{K})} = 0.00854$$

$$k\pi r^2 n\theta_b = 12.948\ \text{W} \qquad \text{(from Problem 1.110)}$$

For $L = 2$ cm,

$$nL = 0.06832 \qquad \sinh nL = 0.06837 \qquad \cosh nL = 1.00233$$

so that

$$q = (12.948\ \text{W})\left[\frac{0.06837 + (0.00854)(1.00233)}{1.00233 + (0.00854)(0.06837)}\right] = 0.993\ \text{W}$$

Repeating for 4, . . . , 128 cm, we obtain the results plotted in Fig. 1-39. This problem illustrates that when k is large there are significant differences between the finite-length and infinite-length cases.

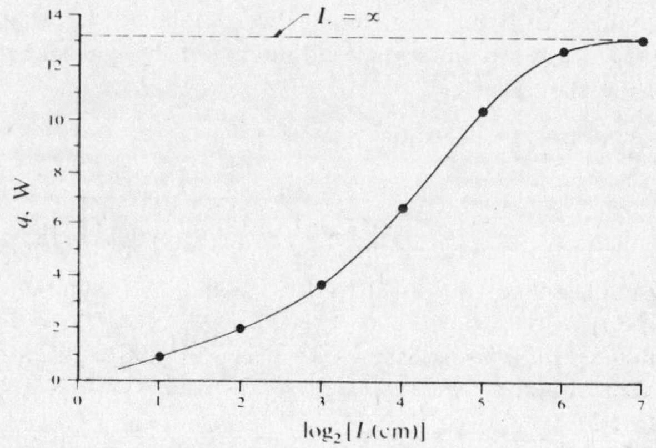

Fig. 1-39

1.112 An annular aluminum-alloy fin ($k = 90$ Btu/h · ft · °F) is mounted on a 1-in.-OD heated tube. The fin is of constant thickness $\frac{1}{64}$ in. and has outer radius 1.5 in. The tube wall temperature is 300 °F, the surrounding temperature is 70 °F, and the average convective heat transfer coefficient is 5 Btu/h · ft² · °F. Calculate the heat loss from the fin.

▌ Equation (5) of Problem 1.97 may be used, with the parameters

$$nr_1 = \left(\frac{2\bar{h}}{kt}\right)^{1/2} r_1 = \left[\frac{2(5)(12)}{90(\frac{1}{64})}\right]^{1/2}\left(\frac{0.5}{12}\right) = 0.3849 \qquad nr_2 = 3(nr_1) = 1.1547$$

Thus,

$$q = 2\pi(90)\left[\frac{1}{64(12)}\right](300 - 70)(0.3849)\left[\frac{K_1(0.385)I_1(1.155) - I_1(0.385)K_1(1.155)}{I_0(0.385)K_1(1.155) + K_0(0.385)I_1(1.155)}\right]$$

The modified Bessel functions are evaluated (approximately) from standard tables of higher functions, with the result

$$q = (65.18)\left[\frac{(2.2860)(0.6793) - (0.1961)(0.4667)}{(1.0374)(0.4667) + (1.1483)(0.6793)}\right] = 75.35\ \text{Btu/h}$$

1.113 Repeat Problem 1.112 using the fin-efficiency approach (see Problems 1.99–1.101) with **(*a*)** no length correction for the tip loss and **(*b*)** a length correction.

▌ (*a*)

$$\frac{nL}{2^{1/2}} = 2^{1/2}(nr_1) = (1.414)(0.3849) = 0.5443$$

From Fig. 1-37, $\eta_f \approx 0.75$, so that (see Problem 1.99)

$$q_{\text{fin}} = \eta_f \bar{h} A_f \theta_b \approx (0.75)(5)[(\tfrac{2}{144})\pi(1.5^2 - 0.5^2)](230) = 75.27\ \text{Btu/h}$$

which is very close to the previous solution.

(b)
$$\frac{nL_c}{2^{1/2}} = \left(\frac{nL}{2^{1/2}}\right) + \left(\frac{nt}{2^{3/2}}\right) = \left(\frac{nL}{2^{1/2}}\right)\left[1 + \left(\frac{1}{128}\right)\right]$$

The correction term is too small to make any difference in the reading of Fig. 1-37.

1.114 A 1.0-in.-OD tube is fitted with 2.0-in.-OD annular fins $\frac{3}{16}$ in. apart on centers. The fins are aluminum alloy $(k = 93\ \text{Btu/h}\cdot\text{ft}\cdot{}^\circ\text{F})$ and are of constant thickness 0.009 in. The external free convective heat transfer coefficient to the ambient air is 1.5 $\text{Btu/h}\cdot\text{ft}^2\cdot{}^\circ\text{F}$. For a tube wall temperature of 330 °F and an ambient temperature of 80 °F, determine the heat loss rate per foot of length of finned tube.

▮ Formula (2) of Problem 1.99 applies. Correcting for tip loss, we have

$$L_c = \frac{2.0 - 1.0}{2(12)} + \frac{0.009}{2(12)} = \frac{0.5045}{12}\ \text{ft}$$

$$L_c^{3/2}\left(\frac{\bar{h}}{kA_p}\right)^{1/2} = \left(\frac{0.5045}{12}\right)^{3/2}\left[\frac{1.5}{(93)(0.5045/12)(0.009/12)}\right]^{1/2} = 0.1950$$

$$\frac{r_{2c}}{r_1} = \frac{1.0045}{0.5} = 2.009$$

From Fig. 1-37, $\eta_f \approx 0.94$. Moreover,

$$A_b = 2\pi r_1 \Delta \qquad A_f = 2\pi(r_{2c}^2 - r_1^2)$$

Hence,

$$q' = 2\pi \bar{h}\theta_b r_1\left[1 + \eta_f\,\frac{(r_{2c}/r_1)^2 - 1}{\Delta/r_1}\right]$$

$$= 2\pi(1.5)(330 - 80)\left(\frac{0.5}{12}\right)\left[1 + (0.94)\,\frac{(2.009)^2 - 1}{\frac{3}{8}}\right] = 827.3\ \text{Btu/h}\cdot\text{ft}$$

1.115 Two long pieces of $\frac{1}{4}$-in.-square copper bar $(k = 207\ \text{Btu/h}\cdot\text{ft}\cdot{}^\circ\text{F})$ are to be silver-soldered together end to end. The surrounding air temperature is 80 °F, and the melting point of the solder is 1200 °F. If the heat transfer coefficient between the copper and the air is 3 $\text{Btu/h}\cdot\text{ft}^2\cdot{}^\circ\text{F}$, find the minimum energy input, in watts, to keep the soldered surface at 1200 °F.

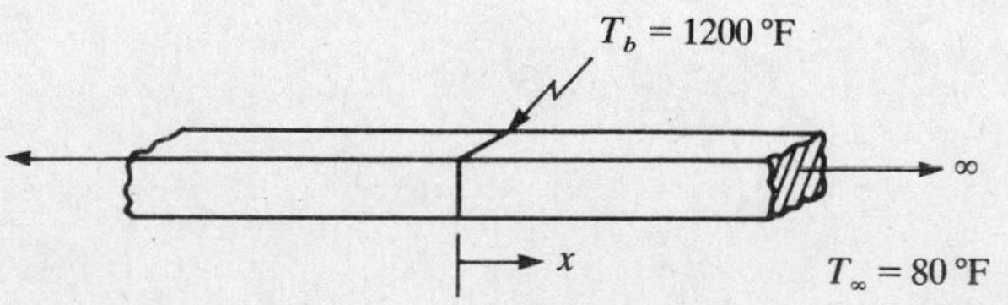

Fig. 1-40

▮ Choosing the junction of the two bars as the origin of x, we have an extended fin of constant cross section (Fig. 1-40). By symmetry, the total heat transfer (=energy input) is twice the heat transfer into $x > 0$. Thus, with $s \equiv$ side of square, .

$$q = 2\theta_b\sqrt{\bar{h}PkA} = 4s^{3/2}(\bar{h}k)^{1/2}(T_b - T_\infty)$$

$$= 4\left(\tfrac{1}{48}\right)^{3/2}[(3)(207)]^{1/2}(1200 - 80) = 335.78\ \text{Btu/h} = 98.38\ \text{W}$$

1.116 A long, 1-in.-diam. brass rod $(k = 60\ \text{Btu/h}\cdot\text{ft}\cdot{}^\circ\text{F})$ is heated by inserting part of it in a laboratory furnace. The major portion of the rod projects into ambient 80 °F air. During steady state, two temperatures 4 in. apart along the length of the rod are 312 °F and 215 °F. What is the effective external heat transfer coefficient $\bar{h}$?

▮ By Problem 1.107,

$$\frac{T_1 - T_\infty}{T_2 - T_\infty} = e^{n(\Delta x)}$$

or

$$n = \frac{1}{\Delta x}\ln\frac{T_1 - T_\infty}{T_2 - T_\infty} = (3\ \text{ft}^{-1})\ln\frac{312 - 80}{215 - 80} = 1.623\ \text{ft}^{-1}$$

Therefore

$$\bar{h} = \frac{k\,dn^2}{4} = \frac{(60)(\frac{1}{12})(1.623)^2}{4} = 3.3\ \text{Btu/h}\cdot\text{ft}^2\cdot{}^\circ\text{F}$$

1.117 A heat exchanger intended for cooling an electronics package in a space vehicle is to utilize rectangular aluminum fins ($k = 119$ Btu/h · ft · °F). The physical design dictates a fin length of 1.6 in., and weight and manufacturing considerations dictate the choice of either (*a*) 0.032-in. material spaced 0.25 in. apart or (*b*) 0.064 in. material spaced 0.50 in. apart. The design temperatures are 195 °F for the heat-exchanger wall and 68 °F for the ambient temperature. Assuming an external heat transfer coefficient $\bar{h}$ of 2.0 Btu/h · ft² · °F, select the better fin design, based on the calculated heat loss per unit surface area of the heat exchanger.

▮ Proceed as in Problem 1.102, but now make an end correction, so that

$$\frac{q}{w} = ktn\theta_b \tanh nL_c$$

with $L_c = L + (t/2)$ and $n^2 = 2\bar{h}/kt$.

(*a*) The data give $L_c = 0.1346$ ft and $n^2 = 12.6\ \text{ft}^{-2}$; thus,

$$\frac{q}{w} = (119)\left(\frac{0.032}{12}\right)(12.6)^{1/2}(195 - 68)\tanh[(12.6)^{1/2}(0.1346)] = 63.55\ \text{Btu/h}\cdot\text{ft}$$

(*b*) The data give $L_c = 0.136$ ft and $n^2 = 6.3\ \text{ft}^{-2}$; thus,

$$\frac{q}{w} = (119)\left(\frac{0.064}{12}\right)(6.3)^{1/2}(195 - 68)\tanh[(6.3)^{1/2}(0.136)] = 66.36\ \text{Btu/h}\cdot\text{ft}$$

Since two 0.032-in.-thick fins fit in the space of one 0.064-in.-thick fin, the former should be used.

1.118 Annular fins of mild steel ($k = 26$ Btu/h · ft · °F) are used on a 2-in.-OD steam line for heating highly viscous fluids in a railroad tank car. The fins are 6 in. OD, 0.102 in. thick, and $\frac{3}{4}$ in. apart. Assuming saturated steam at 50 psia in the pipe, that the pipe wall temperature is equal to that of the saturated steam, and that the external heat transfer coefficient is 50 Btu/h · ft² · °F, calculate the heat transfer rate per linear foot of piping for a fluid temperature of 60 °F.

▮ From steam tables, the pipe wall temperature is 281.01 °F ≈ 280 °F. The solution now follows that of Problem 1.114. The data determine the parameters in Fig. 1-37 as

$$L_c^{3/2}\left(\frac{\bar{h}}{kA_p}\right)^{1/2} = 2.58 \qquad \frac{r_{2c}}{r_1} = 3.06$$

whence $\eta_f \approx 0.18$. Then

$$q' = 2\pi\bar{h}\theta_b r_1\left[1 + \eta_f \frac{(r_{2c}/r_1)^2 - 1}{\Delta/r_1}\right]$$

$$= 2\pi(50)(280 - 60)\left(\frac{1.0}{12}\right)\left[1 + (0.18)\frac{(3.06)^2 - 1}{\frac{3}{4}}\right] = 17\,320\ \text{Btu/h}\cdot\text{ft}$$

1.119 To increase the heat dissipation from an air-cooled, flat cylinder wall the installation of fins is under consideration. The wall temperature is 1200 °F, and the heat transfer coefficient between the solid surface and the 120 °F ambient air is 15 Btu/h · ft² · °F. The two types under consideration are the rectangular fin and the triangular fin; either fin is to be 1 in. thick at the base, 4 in. long, and made from aluminum ($k =$ 155 Btu/h · ft · °F). Compare the effectiveness of the two fins based on the heat flow per unit weight.

▮ For the triangular fin, we have, in the notation of Problem 1.98,

$$f = \sqrt{1 + (\tfrac{1}{8})^2} = 1.007 \approx 1$$

$$p = \left(\frac{2f\bar{h}L}{kt}\right)^{1/2} = \left[\frac{2(15)(\frac{4}{12})}{(155)(\frac{1}{12})}\right]^{1/2} = 0.88\ \text{ft}^{-1/2}$$

Substitution in (2) of Problem 1.98 yields

$$q = \frac{(155)(\frac{1}{12})(1200 - 120)(0.88)}{(\frac{4}{12})^{1/2}}\ \frac{I_1[2(0.88)(\frac{4}{12})^{1/2}]}{I_0[2(0.88)(\frac{4}{12})^{1/2}]} = 9613\ \text{Btu/hr}\cdot\text{ft width}$$

The fin weight per foot of width is

$$\rho_w(\tfrac{1}{2}Lt) = \left(169\ \frac{\text{lbf}}{\text{ft}^3}\right)\left(\frac{1}{2}\right)\left(\frac{4\times 1}{144}\ \text{ft}^2\right) = 2.35\ \text{lbf/ft width}$$

Thus, heat flow per unit weight is $9613/2.35 = 4091$ Btu/h · lbf.

For the rectangular fin, with

$$n=\left(\frac{\bar{h}P}{kA}\right)^{1/2}=\left[\frac{(15)(\frac{26}{12})}{(155)(\frac{1}{12})}\right]^{1/2}=1.586\ \text{ft}^{-1}$$

$$nL=0.529$$

$$\frac{\bar{h}}{nk}=0.061$$

Problem 1.96(*c*) gives (assuming $\bar{h}_L=\bar{h}$)

$$q=kAn\theta_b\,\frac{\sinh nL\ +\ (\bar{h}/nk)\cosh nL}{\cosh nL\ +\ (\bar{h}/nk)\sinh nL}$$

$$=(155)\left(\frac{1}{12}\right)(1.586)\,\frac{\sinh 0.529+0.061\cosh 0.529}{\cosh 0.529+0.061\sinh 0.529}=11\,737\ \text{Btu/h}\cdot\text{ft width}$$

The fin weight per foot of width is twice that of the triangular fin, but the flux (11 737) is less than twice the flux $(2\times 9613=19\,226)$ of the triangular fin. Thus, the triangular fin is the more effective.

1.4 REVIEW

1.120 A 60-ft length of Nichrome IV ribbon, 0.051 in. thick and 0.50 in. wide, is used as a resistive heating element. The surface temperature of the ribbon is 1400 °F when the voltage drop is 110 V. Find the maximum temperature in the ribbon under these conditions. The product of electrical resistivity and thermal conductivity of Nichrome IV is $k\rho=25.7\times 10^{-6}\ \text{Btu}\cdot\Omega/\text{h}\cdot{}^\circ\text{F}$.

▌ The volume density of heat generation is given by

$$q'''=\frac{E^2/R}{LA}=\frac{E^2/(\rho L/A)}{LA}=\frac{1}{\rho}\left(\frac{E}{L}\right)^2$$

According to (*8*) of Problem 1.55, the steady-state temperature distribution obeys

$$\frac{d^2T}{dx^2}+\frac{q'''}{k}=0 \qquad (1)$$

under the approximation that heat is transferred only in the direction of the thickness of the ribbon (see Fig. 1-41). In (*1*), the constant term has the value

$$\frac{q'''}{k}=\frac{1}{k\rho}\left(\frac{E}{L}\right)^2$$

$$=\frac{1}{25.7\times 10^{-6}\ \text{Btu}\cdot\Omega/\text{h}\cdot{}^\circ\text{F}}\left(\frac{110\ \text{V}}{60\ \text{ft}}\right)^2\left(\frac{3.413\ \text{Btu/h}}{1\ \text{V}^2/\Omega}\right)$$

$$=4.47\times 10^5\ {}^\circ\text{F/ft}^2$$

Integration of (*1*), subject to the boundary conditions

$$\left.\frac{dT}{dx}\right|_{x=0}=0 \qquad T\left(\frac{0.051}{24\ \text{ft}}\right)=1400\ {}^\circ\text{F}$$

yields

$$T=1401-2.24\times 10^5x^2\quad {}^\circ\text{F} \qquad (2)$$

From (*2*) it is apparent that the maximum temperature is 1401 °F, at $x=0$.

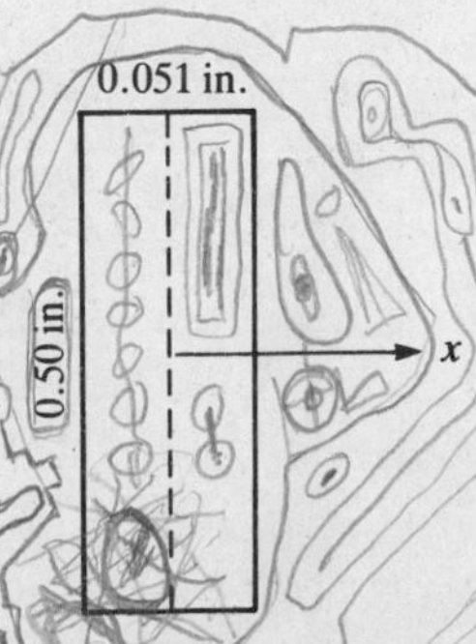

Fig. 1-41

1.121 A fuel element proposed for use in a nuclear power reactor consists of a 2-in.-diameter cylindrical core of fuel supported by a surrounding layer of 0.25-in.-thick aluminum cladding. The outside of the aluminum will be in contact with a heat transfer fluid which will keep the outer surface temperature of the cladding at 200 °F. The fuel in the core gives off heat at the rate of 5.8×10^7 Btu/h · ft³. Because of the adverse effect of temperature on the strength of the aluminum, the cladding temperature must not exceed 800 °F at any point. Is the proposed design satisfactory? For aluminum the thermal conductivity may be expressed as $k = K(1 - \alpha T)$ where $K = 118$ Btu/h · ft · °F and $\alpha = 4.95 \times 10^{-4}$ °F^{-1} in the temperature range 200 °F to 800 °F.

▮ A heat balance on an element of length L reads ($r_i \equiv$ radius of fuel core):

Heat generated in fuel = heat transferred through cladding

$$q'''\pi r_i^2 L = -k(2\pi r L)\frac{dT}{dr}$$

Integration gives ($r_o \equiv$ radius of cladding)

$$r_i^2 q''' \int_{r_i}^{r_o} \frac{dr}{r} = -2\int_{T_i}^{200} K(1 - \alpha T)\, dT$$

$$r_i^2 q''' \ln\frac{r_o}{r_i} = 2K\left[(T_i - 200) - \frac{\alpha}{2}(T_i^2 - 200^2)\right]$$

Substitution of numerical values produces the quadratic equation

$$2.48 \times 10^{-4} T_i^2 - T_i + 572 = 0$$

from which $T_i = 695$ °F, 3340 °F. The second solution is not valid, because the expression for k does not apply above 800 °F; hence the design is satisfactory.

1.122 A 2-in.-diameter sphere of radioactive material generates heat at 114 Btu/h. If the sphere is enclosed in a spherical shell of insulation ($k = 0.1$ Btu/h · ft · °F), what will be its temperature if the OD of the insulation is 2 ft? The outer surface of the insulation is at 100 °F. Assume the radioactive sphere to be at a uniform temperature.

▮ Apply (*1*) of Problem 1.35 to the insulation:

$$q = \frac{4\pi k(T_1 - T_2)}{(1/r_1) - (1/r_2)}$$

$$T_1 = T_2 + \frac{(1/r_1) - (1/r_2)}{4\pi k} q$$

$$= (100\ \text{°F}) + \frac{(12\ \text{ft}^{-1}) - (1\ \text{ft}^{-1})}{4\pi(0.1\ \text{Btu/h·ft·°F})}(114\ \text{Btu/h}) = 1098\ \text{°F}$$

1.123 The wall of a refrigerator consists of $\frac{1}{16}$ in. of sheet steel outside and $\frac{3}{8}$-in. plywood ($k = 0.06$ Btu/h · ft · °F) inside with a layer of glass-wool insulation between. The unit is designed to maintain an inside temperature of 10 °F while the temperature on the outside is 70 °F. What thickness of packed glass-wool insulation is needed to limit the heat transfer rate to 2 Btu/h · ft²?

▮ At the average temperature (70 + 10)/2 = 40 °F, Table B-1 gives

$$k_s \approx 26.5\ \text{Btu/h·ft·°F} \qquad k_{gw} \approx 0.017\ \text{Btu/h·ft·°F}$$

By the electrical analogy,

$$\frac{q}{A} = \frac{T_2 - T_1}{\dfrac{L_s}{k_s} + \dfrac{L_{gw}}{k_{gw}} + \dfrac{L_{ply.w}}{k_{ply.w}}}$$

$$10 = \frac{70 - 10}{\dfrac{(\frac{1}{16})(\frac{1}{12})}{26.5} + \dfrac{L_{gw}}{0.017} + \dfrac{(\frac{3}{8})(\frac{1}{12})}{0.06}}$$

Solving, $L_{gw} = 0.0931$ ft.

CHAPTER 2

Multidimensional, Steady-State Conduction

2.1 ANALYTICAL SOLUTION OF THE LAPLACE EQUATION; SUPERPOSITION

2.1 A very long (z-direction) rectangular bar has three of its lateral sides held at a fixed temperature T_0; the temperature distribution across the fourth side is sinusoidal (see Fig. 2-1). Find the temperature distribution within the bar, beginning with the Laplace equation (set $\partial T/\partial t = 0$ in (*9*) of Problem 1.54) and using the separation-of-variables technique.

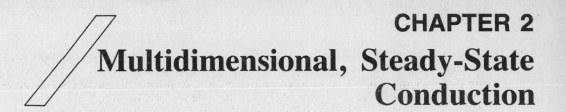

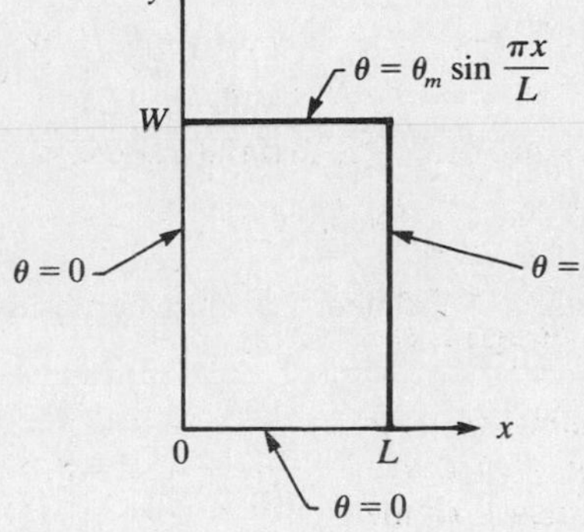

Fig. 2-1

I By using the shifted temperature $\theta \equiv T - T_0$ we may suppose the fixed temperature to be zero. Since there is no z-direction temperature gradient, the Laplace equation is

$$\frac{\partial^2\theta}{\partial x^2} + \frac{\partial^2\theta}{\partial y^2} = 0 \qquad (1)$$

subject to the boundary conditions

$$\text{(i)}\quad \theta(0, y) = 0 \qquad (0 < y < W)$$

$$\text{(ii)}\quad \theta(L, y) = 0 \qquad (0 < y < W)$$

$$\text{(iii)}\quad \theta(x, 0) = 0 \qquad (0 < x < L)$$

$$\text{(iv)}\quad \theta(x, W) = \theta_m \sin\frac{\pi x}{L} \qquad (0 < x < L)$$

Assume a solution of the form $\theta(x, y) = X(x)Y(y)$. When substituted into the Laplace equation this yields

$$-\frac{1}{X}\frac{d^2X}{dx^2} = \frac{1}{Y}\frac{d^2Y}{dy^2} \qquad (2)$$

The left side of (*2*), a function of x alone, can equal the right side, a function of y alone, only if both sides have a constant value, say $\lambda^2 > 0$. Then,

$$\frac{d^2X}{dx^2} + \lambda^2 X = 0 \qquad \frac{d^2Y}{dy^2} - \lambda^2 Y = 0$$

with general solutions

$$X = C_1 \cos\lambda x + C_2 \sin\lambda x \qquad Y = C_3 \cosh\lambda y + C_4 \sinh\lambda y$$

so that

$$\theta = (C_1 \cos\lambda x + C_2 \sin\lambda x)(C_3 \cosh\lambda y + C_4 \sinh\lambda y)$$

Now, applying the boundary conditions, (i) gives $C_1 = 0$ and (iii) gives $C_3 = 0$. Using these together with (ii) yields

$$0 = C_2 C_4 (\sin \lambda L)(\sinh \lambda y)$$

which requires that

$$\sin \lambda L = 0 \qquad \text{or} \qquad \lambda = \frac{n\pi}{L} \qquad (n \text{ a positive integer})$$

Because the original differential equation (*1*) is linear, the sum of any number of solutions constitutes a solution. Thus, θ can be written as the sum of an infinite series:

$$\theta = \sum_{n=1}^{\infty} C_n \sin \frac{n\pi x}{L} \sinh \frac{n\pi y}{L} \tag{3}$$

where the constants have been combined.

Finally, boundary condition (iv) gives

$$\theta_m \sin \frac{\pi x}{L} = \sum_{n=1}^{\infty} C_n \sin \frac{n\pi x}{L} \sinh \frac{n\pi W}{L} \qquad (0 < x < L)$$

which holds only if $C_2 = C_3 = C_4 = \cdots = 0$ and $C_1 = \theta_m / \sinh (\pi W/L)$. Therefore,

$$\theta = \theta_m \frac{\sinh (\pi y/L)}{\sinh (\pi W/L)} \sin \frac{\pi x}{L} \tag{4}$$

and $T = T_0 + \theta$.

2.2 Verify that the separation constant in Problem 2.1 must be positive.

▌ The equation for $X(x)$,

$$\frac{d^2X}{dx^2} + \lambda^2 X = 0$$

has general solutions (for nonpositive λ^2)

$$X = C_5 + C_6 x \qquad (\lambda^2 = 0) \qquad \text{and} \qquad X = C_7 \cosh (|\lambda|\, x) + C_8 \sinh (|\lambda|\, x) \qquad (\lambda^2 < 0)$$

In neither case could X fit a sine function along the edge $y = W$. More concisely, since the boundary function happens to be proportional to the x-direction eigenfunction corresponding to $\lambda^2 = (\pi/L)^2$, only that one (positive) value of the separation constant need be considered.

2.3 Problem 2.1 is changed so that the shifted temperature along $y = W$ is given by the arbitrary function $f(x)$. Find the temperature distribution within the bar.

▌ Everything in Problem 2.1 through (*3*) remains valid for the present problem. The new fourth boundary condition gives

$$f(x) = \sum_{n=1}^{\infty} C_n \sin \frac{n\pi x}{L} \sinh \frac{n\pi W}{L} \qquad (0 < x < L)$$

Thus, the quantities $C_n \sinh (n\pi W/L)$ must be the coefficients of the Fourier sine series for $f(x)$ in the interval $0 < x < L$. From the theory of Fourier series,

$$C_n \sinh \frac{n\pi W}{L} = \frac{2}{L} \int_0^L f(x) \sin \frac{n\pi x}{L}\, dx$$

and so

$$\theta = \frac{2}{L} \sum_{n=1}^{\infty} \left[\frac{1}{\sinh (n\pi W/L)} \int_0^L f(u) \sin \frac{n\pi u}{L}\, du \right] \sin \frac{n\pi x}{L} \sinh \frac{n\pi y}{L} \tag{1}$$

For the special case $f(x) = \theta_c = \text{constant}$, (*1*) reduces to

$$\theta = \frac{2\theta_c}{\pi} \sum_{n=1}^{\infty} \frac{(-1)^{n+1} + 1}{n \sinh (n\pi W/L)} \sin \frac{n\pi x}{L} \sinh \frac{n\pi y}{L}$$

$$= \frac{4\theta_c}{\pi} \sum_{j=1}^{\infty} \frac{1}{(2j-1) \sinh [(2j-1)\pi W/L]} \sin \frac{(2j-1)\pi x}{L} \sinh \frac{(2j-1)\pi y}{L} \tag{2}$$

Note that only odd-order terms appear in (*2*).

2.4 Determine the temperature at the point $(\frac{1}{2}, \frac{1}{2})$ in Fig. 2-2.

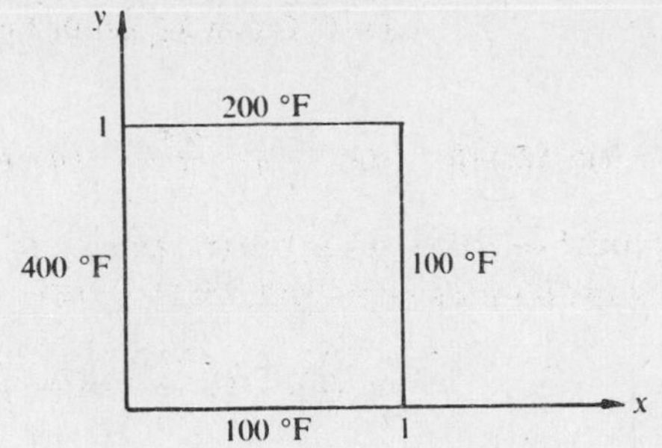

Fig. 2-2

▮ As was seen in Problem 1.104, a linear problem with more than one nonhomogeneous boundary condition can be resolved into a set of simpler problems each with the physical geometry of the original problem and each having only one nonhomogeneous boundary condition. The solutions to the simpler problems can be superposed (at the geometric point being considered) to yield the solution to the original problem.

In the present case two of the nonhomogeneous boundary conditions can be removed by defining $\theta = T - (110\ °\text{F})$. Then the resulting problem may be separated into the two subproblems of Fig. 2-3. Either by use of (2) of Problem 2.3 or by intuition, the solutions are $\theta_1(\frac{1}{2}, \frac{1}{2}) = 75\ °\text{F}$ and $\theta_2(\frac{1}{2}, \frac{1}{2}) = 25\ °\text{F}$. Consequently,

$$\theta(\tfrac{1}{2}, \tfrac{1}{2}) = 75 + 25 = 100\ °\text{F} \qquad \text{and} \qquad T(\tfrac{1}{2}, \tfrac{1}{2}) = \theta(\tfrac{1}{2}, \tfrac{1}{2}) + 100 = 200\ °\text{F}$$

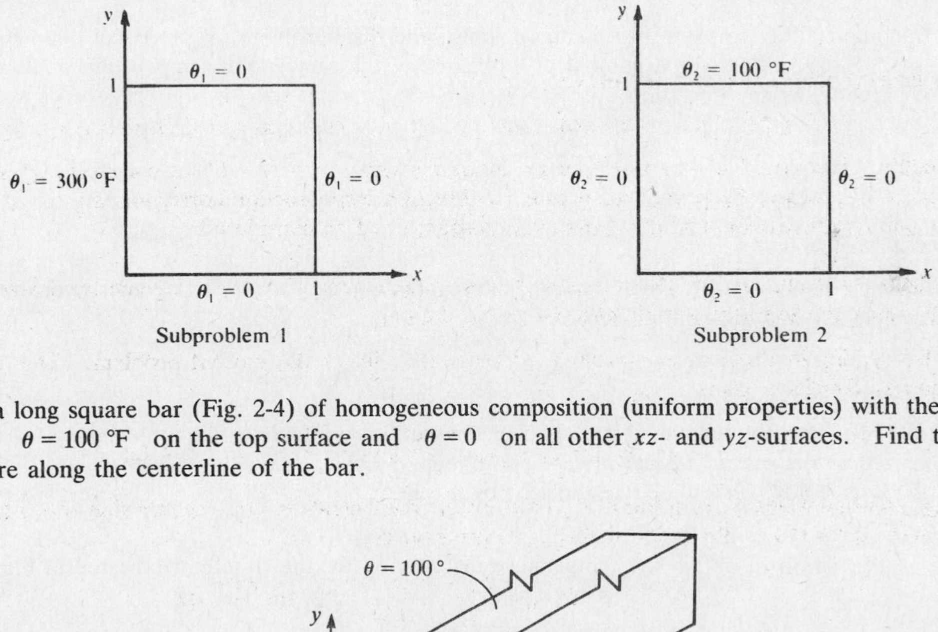

Fig. 2-3

2.5 Consider a long square bar (Fig. 2-4) of homogeneous composition (uniform properties) with the boundary conditions $\theta = 100\ °\text{F}$ on the top surface and $\theta = 0$ on all other xz- and yz-surfaces. Find the temperature along the centerline of the bar.

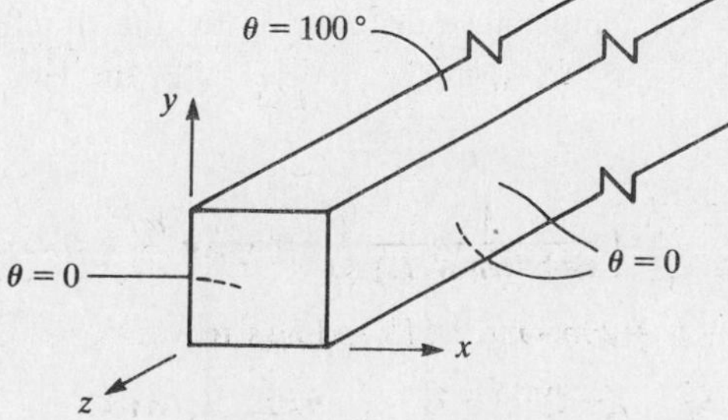

Fig. 2-4

▮ By (2) of Problem 2.3, with $L = W = 2x = 2y$,

$$\frac{\theta}{\theta_c} = \frac{2}{\pi} \sum_{n=1}^{\infty} \frac{(-1)^{n+1} + 1}{n} \left(\sin \frac{n\pi}{2}\right)\left(\frac{\sinh (n\pi/2)}{\sinh n\pi}\right)$$

Retaining only the first two odd-order terms, we find

$$\frac{\theta}{\theta_c} \approx \frac{2}{\pi}(0.398 - 0.00597) = 0.2498 \qquad \text{or} \qquad \theta \approx 24.98\ ^\circ\text{F}$$

2.6 Solve Problem 2.5 (exactly) by an appeal to symmetry and the superposition principle.

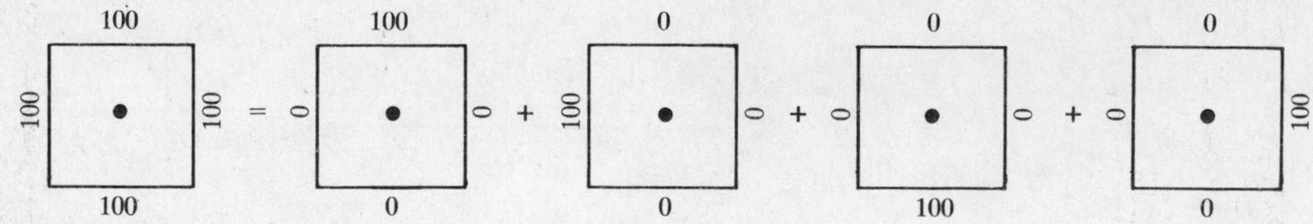

Fig. 2-5

▌ Consider the superposition symbolized in Fig. 2-5. The original problem, on the left, has the unique solution $\theta = 100$ throughout the square—in particular, $\theta = 100$ at the centerpoint. Let τ denote the centerpoint temperature in the first subproblem on the right, which is Problem 2.5. It is apparent from symmetry (the Laplacian operator ∇^2 is invariant under a rotation of the coordinate system) that the centerpoint temperature in all the subproblems is τ. Then, the superposition gives $100 = 4\tau$, or $\tau = 25$.

2.7 A very long, homogeneous bar has as cross section a regular n-sided polygon; side i is held at temperature T_i $(i = 1, 2, \ldots, n)$. Show that the temperature T_0 at the centerpoint is the arithmetic mean of the temperatures around the perimeter.

▌ Decompose the problem into n subproblems, where subproblem i replaces all boundary temperatures except T_i by zero. By the argument of Problem 2.6, the centerpoint temperature in subproblem i is T_i/n. Hence, by superposition,

$$T_0 = \frac{T_1}{n} + \frac{T_2}{n} + \cdots + \frac{T_n}{n} = \frac{T_1 + T_2 + \cdots + T_n}{n} \tag{1}$$

2.8 Generalize Problem 2.7.

▌ Letting $n \to \infty$ in a suitable fashion, the regular n-gon of Problem 2.7 goes over into a circle, and (*1*) of Problem 2.7 has the limiting form

$$T_0 = \frac{1}{2\pi}\int_0^{2\pi} T(\phi)\, d\phi$$

where $T(\phi)$ gives the angular distribution of temperature around the circumference. This result is recognized as the *mean-value theorem* for solutions of Laplace's equation in the plane. In three dimensions the theorem holds, with the circle replaced by a sphere.

2.9 Use superposition to obtain the temperature distribution for the situation indicated in Fig. 2-6.

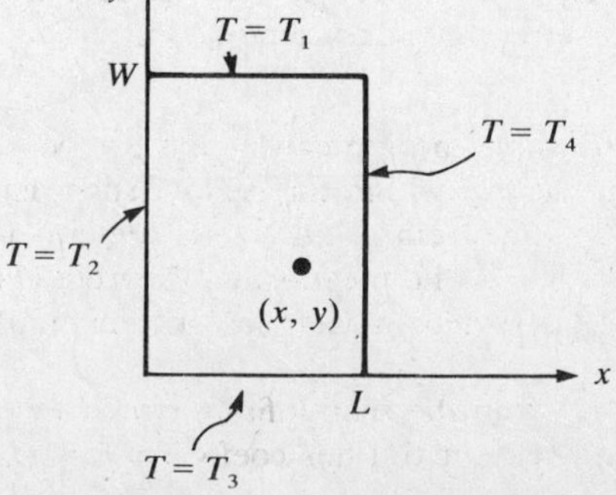

Fig. 2-6

Temporarily denote the function defined by (2) of Problem 2.3 as $F(x, y \mid \theta_c, L, W)$. Then the four subproblems corresponding to Fig. 2-6 have the respective solutions

$$F(x, y \mid T_1, L, W) \qquad F(y, L-x \mid T_2, W, L) \qquad F(L-x, W-y \mid T_3, L, W) \qquad F(W-y, x \mid T_4, W, L)$$

and so, by superposition,

$$\begin{aligned} T(x, y) = \frac{4}{\pi} \sum_{j=1}^{\infty} \frac{1}{2j-1} \Bigg\{ & \frac{T_1}{\sinh[(2j-1)\pi W/L]} \sin \frac{(2j-1)\pi x}{L} \sinh \frac{(2j-1)\pi y}{L} \\ & + \frac{T_2}{\sinh[(2j-1)\pi L/W]} \sin \frac{(2j-1)\pi y}{W} \sinh \frac{(2j-1)\pi(L-x)}{W} \\ & + \frac{T_3}{\sinh[(2j-1)\pi W/L]} \sin \frac{(2j-1)\pi(L-x)}{L} \sinh \frac{(2j-1)\pi(W-y)}{L} \\ & + \frac{T_4}{\sinh[(2j-1)\pi L/W]} \sin \frac{(2j-1)\pi(W-y)}{W} \sinh \frac{(2j-1)\pi x}{W} \Bigg\} \quad (1) \end{aligned}$$

Because (1) also holds in the reduced temperature θ, at least one of the boundary temperatures may be assumed to be zero.

2.10 Numerically test (1) of Problem 2.9 against the mean-value property of the square (Problem 2.7).

For $x = y = L/2 = W/2 \equiv a$, (1) reduces to

$$T(a, a) = (T_1 + T_2 + T_3 + T_4) \frac{2}{\pi} \sum_{j=1}^{\infty} \frac{(-1)^{j-1}}{(2j-1)\cosh(j-\frac{1}{2})\pi}$$

and the question is: Does the infinite series have the sum $\pi/8 \approx 0.3927$? Evaluating the first two terms,

$$\begin{aligned} \Sigma &= \frac{1}{\cosh \pi/2} - \frac{1}{3 \cosh 3\pi/2} + \cdots \\ &\approx \frac{1}{2.509} - \frac{1}{3(55.0)} + \cdots = 0.392^+ \end{aligned}$$

(On the money!)

2.11 For the square region of Fig. 2-2, determine the temperature at $(\frac{1}{4}, \frac{3}{4})$.

Equation (1) of Problem 2.9—in the dependent variable $\theta = T - 100$—gives, after trigonometric simplification,

$$\theta(\tfrac{1}{4}, \tfrac{3}{4}) = \frac{1600}{\pi} \sum_{j=1}^{\infty} \frac{\sinh[3(2j-1)\pi/4]}{(2j-1)\sinh(2j-1)\pi} \sin \frac{(2j-1)\pi}{4}$$

wherein the sine function assumes only the values $\pm 1/\sqrt{2}$. Retaining three terms of the series,

$$\begin{aligned} \theta(\tfrac{1}{4}, \tfrac{3}{4}) &\approx \frac{1600}{\pi\sqrt{2}} \left[\frac{5.22797}{(1)(11.5487)}(1) + \frac{587.241}{(3)(6195.82)}(1) + \frac{6.53704 \times 10^4}{(5)(3.3178 \times 10^6)}(-1) \right] \\ &\approx 173 \end{aligned}$$

whence $T(\frac{1}{4}, \frac{3}{4}) \approx 273$.

2.12 Given an arbitrary two-dimensional region, with an arbitrary temperature distribution prescribed along the boundary, show that the maximum boundary temperature is the maximum temperature for the whole region.

If, on the contrary, the absolute maximum temperature occurred at an interior point P, then P could be surrounded by a circle, lying entirely within the region, such that the temperature at each point of the circumference—and hence the mean temperature over the circumference—was less than the temperature at P. But this would be a violation of the mean-value theorem (Problem 2.8). Indeed, it follows that the temperature cannot even have a *relative* maximum (or minimum) inside the region.

2.13 Find the temperature distribution in the semi-infinite bar of Fig. 2-7. In terms of $\theta \equiv T - T_\infty$, the base temperature is θ_b; the convective heat transfer coefficient is assumed large.

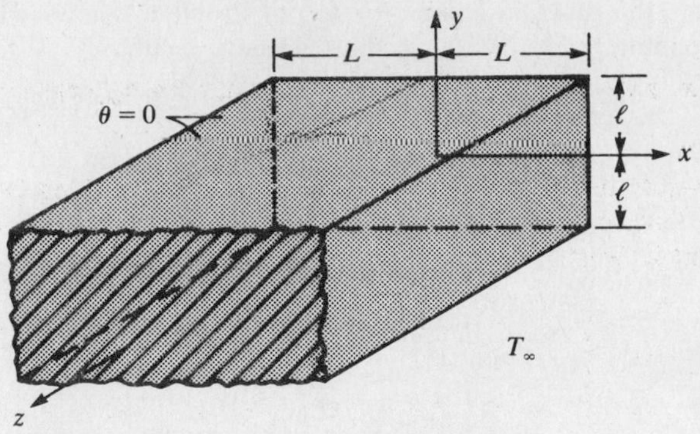

Fig. 2-7

▮ We must solve

$$\frac{\partial^2\theta}{\partial x^2} + \frac{\partial^2\theta}{\partial y^2} + \frac{\partial^2\theta}{\partial z^2} = 0$$

subject to the boundary conditions:

(i) θ even in x, with $\theta(L, y, z) = 0$

(ii) θ even in y, with $\theta(x, l, z) = 0$

(iii) $\theta(x, y, 0) = \theta_b$

(iv) $\lim_{z\to\infty} \theta(x, y, z) = 0$

Separation of variables leads to

$$-\frac{1}{X}\left(\frac{d^2X}{dx^2}\right) = \frac{1}{Y}\left(\frac{d^2Y}{dy^2}\right) + \frac{1}{Z}\left(\frac{d^2Z}{dz^2}\right) \equiv \lambda^2 \qquad (1)$$

The x-eigenfunctions, satisfying (i), are

$$X_n(x) = \cos \lambda_n x \qquad \text{where} \qquad \lambda_n = \frac{(n + \frac{1}{2})\pi}{L}$$

for $n = 0, 1, 2, \ldots$. Rearranging the second equality of (1),

$$-\frac{1}{Y}\left(\frac{d^2Y}{dy^2}\right) = \frac{1}{Z}\left(\frac{d^2Z}{dz^2}\right) - \lambda^2 \equiv \mu^2$$

which, together with (ii) and (iv), gives

$$Y_m(y) = \cos \mu_m y \qquad \text{where} \qquad \mu_m = \frac{(m + \frac{1}{2})\pi}{l}$$

for $m = 0, 1, 2, \ldots,$ and

$$Z_{mn}(z) = \exp\,[-(\lambda_n^2 + \mu_m^2)^{1/2} z]$$

Thus, by superposition,

$$\theta(x, y, z) = \sum_{n=0}^{\infty} \sum_{m=0}^{\infty} a_{mn} \cos \lambda_n x \cos \mu_m y \exp\,[-(\lambda_n^2 + \mu_m^2)^{1/2} z] \qquad (2)$$

Boundary condition (iii) requires

$$\theta_b = \sum_{n=0}^{\infty} \sum_{m=0}^{\infty} a_{mn} \cos \lambda_n x \cos \mu_m y \qquad (3)$$

(3) is seen to be the double Fourier cosine-series expansion of θ_b over the cross section of the rod. Applying orthogonality in the usual fashion, we find

$$a_{mn} = \theta_b \frac{\int_0^L \int_0^l \cos \lambda_n x \cos \mu_m y \, dx\, dy}{\int_0^L \int_0^l \cos^2 \lambda_n x \cos^2 \mu_m y \, dx\, dy} = \frac{4\theta_b(-1)^{n+m}}{(\lambda_n L)(\mu_m l)}$$

Finally, then,

$$\theta(x, y, z) = \frac{4\theta_b}{Ll} \sum_{n=0}^{\infty} \sum_{m=0}^{\infty} \frac{(-1)^{n+m}}{\lambda_n \mu_m} \cos \lambda_n x \cos \mu_m y \exp[-(\lambda_n^2 + \mu_m^2)^{1/2} z] \tag{4}$$

2.14 Determine the steady-state temperature distribution in an infinitely long two-dimensional strip of width w (Fig. 2-8). The base temperature distribution is $F(y)$, the ambient temperature is T_∞, and the convective heat transfer coefficient is large.

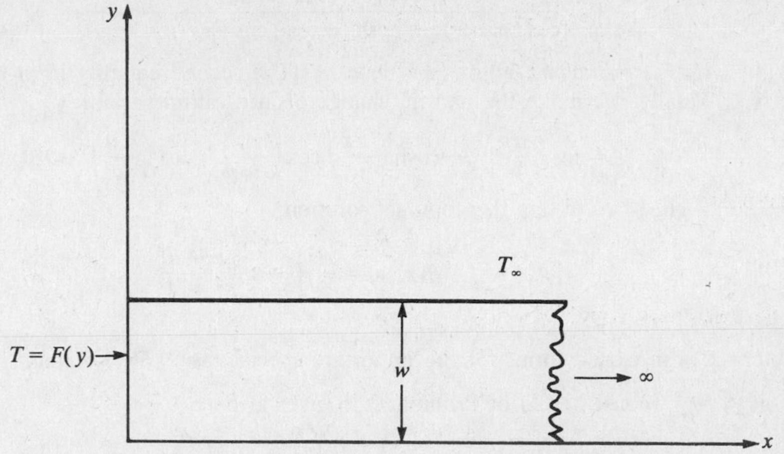

Fig. 2-8

Solving the two-dimensional Laplace equation for $\theta \equiv T - T_\infty$ by separation of variables (cf. Problem 2.13), we obtain the eigenfunction expansion

$$\theta(x, y) = \sum_{n=1}^{\infty} a_n e^{-\lambda_n x} \sin \lambda_n y \qquad \text{with} \qquad \lambda_n = \frac{n\pi}{w}$$

The nonhomogeneous boundary condition, $\theta(0, y) = F(y) - T_\infty \equiv f(y)$ for $0 < y < w$, yields the expression

$$f(y) = \sum_{n=1}^{\infty} a_n \sin \lambda_n y$$

which determines the a_n as the Fourier sine-series coefficients of the function $f(y)$:

$$a_n = \frac{2}{w} \int_0^w f(u) \sin \lambda_n u \, du$$

Hence

$$\theta(x, y) = \frac{2}{w} \sum_{n=1}^{\infty} \left[\int_0^w f(u) \sin \lambda_n u \, du \right] e^{-\lambda_n x} \sin \lambda_n y \tag{1}$$

2.15 Convert the series solution (*1*) of Problem 2.14 to an integral solution.

First rewrite (*1*) as $(\lambda_n = n\pi/w)$

$$\theta(x, y) = \frac{1}{w} \int_0^w f(u) \left\{ \sum_{n=1}^{\infty} 2\left(\sin \frac{n\pi}{w} u\right)\left(\sin \frac{n\pi}{w} y\right) e^{-n\pi x/w} \right\} du$$

We can sum the infinite series by applying the trigonometric identity

$$2 \sin \alpha \sin \beta = \cos(\alpha - \beta) - \cos(\alpha + \beta)$$

and going over to complex notation, wherein $\cos \phi = \mathbf{Re}\{e^{i\phi}\}$. Thus (postponing the taking of the real part until the end of the computation), using $\Sigma_{n=1}^{\infty} z^n = z/(1-z)$,

$$\begin{aligned} \sum_{n=1}^{\infty} &= \sum_{n=1}^{\infty} e^{n(\pi/w)[-x+i(u-y)]} - \sum_{n=1}^{\infty} e^{n(\pi/w)[-x+i(u+y)]} \\ &= \frac{e^{(\pi/w)[-x+i(u-y)]}}{1 - e^{(\pi/w)[-x+i(u-y)]}} - \frac{e^{(\pi/w)[-x+i(u+y)]}}{1 - e^{(\pi/w)[-x+i(u+y)]}} \\ &= \frac{1}{1 - e^{-\pi x/w} e^{i\pi(u-y)/w}} - (\text{same in } +y) \end{aligned}$$

Now, for real A and ψ,

$$\mathbf{Re}\left\{\frac{1}{1-Ae^{i\psi}}\right\} = \frac{1 - A\cos\psi}{1 - 2A\cos\psi + A^2}$$

Therefore, taking the real part of the complex sum and inserting it in the integral, one obtains

$$\theta(x, y) = \frac{1}{w}\int_0^w f(u)\left\{\frac{1 - e^{-\pi x/w}\cos[\pi(u-y)/w]}{1 - 2e^{-\pi x/w}\cos[\pi(u-y)/w] + e^{-2\pi x/w}} - (\text{same in } +y)\right\} du$$

$$= \frac{A}{w}\int_0^w \frac{f(u)\sin(\pi u/w)\,du}{[\cos(\pi u/w) - \cosh(\pi x/w)\cos(\pi y/w)]^2 + A^2}$$

in which $A = A(x, y) = \sinh(\pi x/w)\sin(\pi y/w) \geq 0$. [The second equality in (1) follows after considerable trigonometry.] Finally, we make the natural change of integration variable

$$\xi = \cos\left(\frac{\pi u}{w}\right) - \cosh\left(\frac{\pi x}{w}\right)\cos\left(\frac{\pi y}{w}\right) \equiv \cos\left(\frac{\pi u}{w}\right) - B(x, y)$$

and write $f(u) = g(\xi)$ to obtain the compact solution

$$\theta(x, y) = \frac{A}{\pi}\int_{-1-B}^{1-B} \frac{g(\xi)\,d\xi}{\xi^2 + A^2} \tag{2}$$

2.16 Solve Problem 2.14 in closed form, for the important special case $F(y) = \text{const.}$

▌ With $g(\xi) = \gamma = \text{const.}$, (2) of Problem 2.15 gives at once

$$\theta(x, y) = \frac{\gamma}{\pi}\left[\arctan\frac{1-B}{A} + \arctan\frac{1+B}{A}\right]$$

$$= \frac{\gamma}{\pi}\left[\arctan\frac{1 - \cosh(\pi x/w)\cos(\pi y/w)}{\sinh(\pi x/w)\sin(\pi y/w)} + \arctan\frac{1 + \cosh(\pi x/w)\cos(\pi y/w)}{\sinh(\pi x/w)\sin(\pi y/w)}\right]$$

Using the limiting values $\arctan(\pm\infty) = \pm\pi/2$, one easily verifies that $\theta(x, y)$ satisfies all four boundary conditions. It is noteworthy that only elementary calculus has been used to obtain this solution.

2.2 ELECTRICAL ANALOGY; CONDUCTIVE SHAPE FACTOR

2.17 Define the *conductive shape factor* for the two-dimensional region of Fig. 2-9.

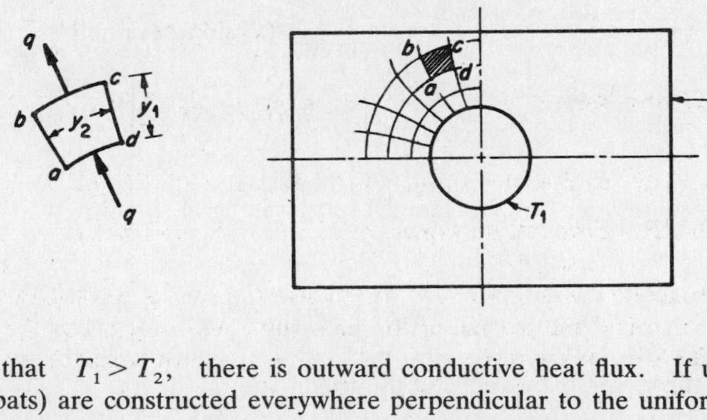

Fig. 2-9

▌ On the assumption that $T_1 > T_2$, there is outward conductive heat flux. If uniformly spaced lines (equipotentials or adiabats) are constructed everywhere perpendicular to the uniformly spaced isotherms, the result is a group of *heat-flow lanes*. In Fig. 2-9, there are six such lanes in each quadrant, and if we can determine the heat flow for each lane, we can calculate the total heat transfer. Applying Fourier's law to an element *a-b-c-d* of one of these passages yields for unit depth

$$q = \frac{ky_2(1)(T_{ad} - T_{bc})}{y_1}$$

If $y_1 = y_2$, the individual elements such as *a-b-c-d* become curvilinear squares. Note that $y_1 = (\overline{ab} + \overline{cd})/2$ and $y_2 = (\overline{ad} + \overline{bc})/2$. Since the isotherms are uniformly spaced, the drop in temperature potential across each square has the constant value

$$\Delta T = \frac{T_1 - T_2}{M}$$

where m is the number of squares in each flow lane. Then, if there are N such flow lanes for the entire configuration, we can express the total heat flux as

$$q = Nk\frac{T_1 - T_2}{M} \equiv Sk(T_1 - T_2)$$

where $S \equiv N/M$ is the conductive shape factor. The solution of a two-dimensional heat-conduction problem is reduced to the determination of S. Appendix C summarizes shape factors for some common geometrical configurations and thermal conditions of practical importance.

2.18 Develop a shape factor for a hollow sphere with inner and outer radii r_1 and r_2 and thermal conductivity k. The inner and outer temperatures, T_1 and T_2, respectively, may be assumed constant.

▮ From Problem 1.34,

$$q = \frac{4\pi(T_1 - T_2)}{(1/r_1) - (1/r_2)} \equiv Sk(T_1 - T_2)$$

so that $S = 4\pi r_1 r_2/(r_2 - r_1)$.

2.19 Part of a laboratory system consists of a long $\frac{3}{4}$-in.-OD copper tube embedded in the center of a square insulator (asbestos) with 3-in. sides. The purpose of the pipe is to deliver hot water at 30 psia to the system, but the system requires that the water must not be boiling. The energy loss per foot of pipe is measured to be 35 Btu/h · ft, and the outside surface temperature of the asbestos is 150 °F. Assuming the water and the tube are at the same temperature, does the water boil? (See Fig. 2-10.)

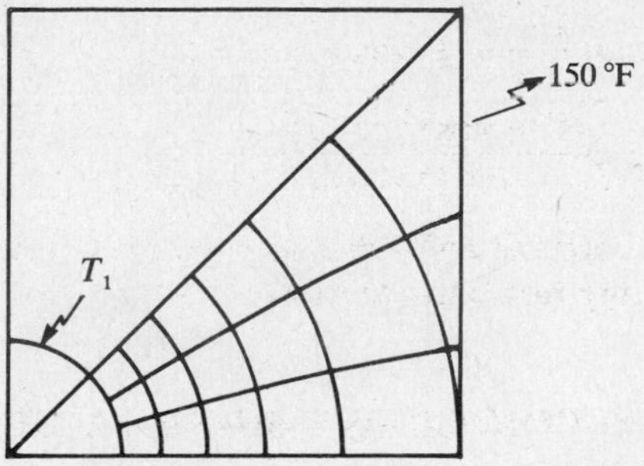

Fig. 2-10

▮ For one-eighth of the system (Fig. 2-10), we have, following Problem 2.17,

$$N = 3 \qquad M = \frac{5.1 + 5.3 + 5.7}{3} = \frac{16.1}{3}$$

giving $(S/L) = 8N/M = 4.464$. Then, with $k(\text{asbestos}) = 0.087$ Btu/h · ft · °F, we have $(q/L) = (S/L)k(\Delta T)$, or

$$35 \text{ Btu/h} \cdot \text{ft} = (4.464)(0.087 \text{ Btu/h} \cdot \text{ft} \cdot °\text{F})[(T_1 - 150)\ °\text{F}]$$

whence $T_1 = 240.2$ °F. From steam tables, at 30 psia, $T_{sat} = 250.33$ °F. Thus, the water is not boiling.

2.20 An underground concrete tunnel of 3- by 3-ft cross section has a 10-in.-OD steam main passing through its geometric center. Around the outside of the pipe and completely filling the tunnel is insulating material ($k = 0.05$ Btu/h · ft · °F). What is the rate of heat loss per foot from the pipe ($T_1 = 200$ °F) to the ground ($T_2 = 55$ °F)?

▮ The problem lends to freehand plotting (Fig. 2-11). For one-eighth of the region,

$$N = 5 \qquad M = \frac{8 + 8.2 + 8.4 + 9 + 9.5}{5} = 8.63$$

whence $(S/L) = 8(N/M) = 4.64$. Therefore,

$$\frac{q}{L} = \frac{S}{L}k(T_1 - T_2) = (4.64)(0.05 \text{ Btu/h} \cdot \text{ft} \cdot °\text{F})[(200 - 55)\ °\text{F}] = 33.6 \text{ Btu/h} \cdot \text{ft}$$

2.21 Consider an 8.0-in.-OD pipe with a 12.3-in.-thick insulation blanket (Fig. 2-12). By flux plotting determine the heat transfer rate per unit length if the inner surface of the insulation is at 300 °F, the outer surface is at 120 °F, and the thermal conductivity of the insulation is 0.35 Btu/h · ft · °F.

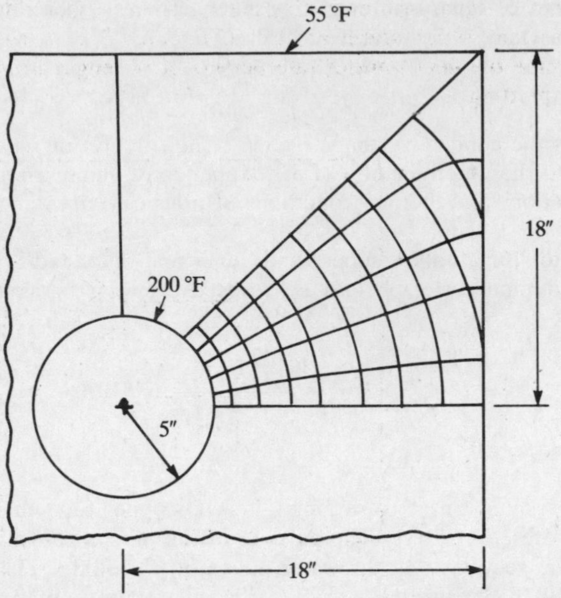

Fig. 2-11

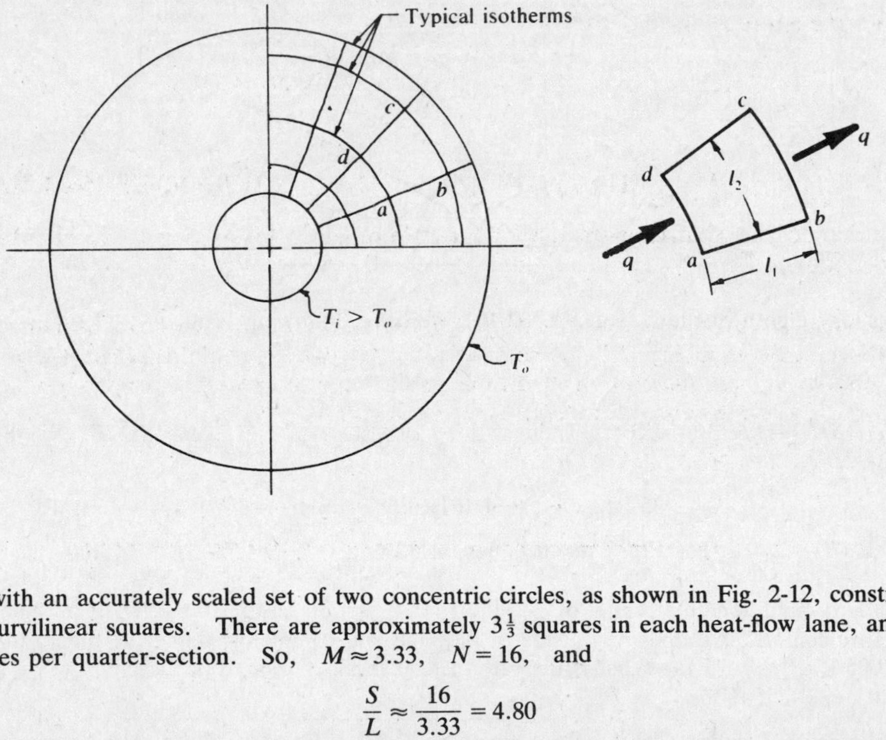

Fig. 2-12

▮ Starting with an accurately scaled set of two concentric circles, as shown in Fig. 2-12, construct a network of curvilinear squares. There are approximately $3\frac{1}{3}$ squares in each heat-flow lane, and there are four flow lanes per quarter-section. So, $M \approx 3.33$, $N = 16$, and

$$\frac{S}{L} \approx \frac{16}{3.33} = 4.80$$

Thus

$$\frac{q}{L} \approx 4.80\, k(T_i - T_o) = (4.80)(0.35\ \text{Btu/h}\cdot\text{ft}\cdot{}^\circ\text{F})[(300 - 120)\ {}^\circ\text{F}] = 302.4\ \text{Btu/h}\cdot\text{ft}$$

2.22 Check the result of Problem 2.21 against the analytical solution.

▮ By Problem 1.26, the exact solution is

$$\frac{q}{L} = \frac{2\pi k(T_i - T_o)}{\ln(r_o/r_i)} = (4.47)(0.35\ \text{Btu/h}\cdot\text{ft}\cdot{}^\circ\text{F})[(300 - 120)\ {}^\circ\text{F}] = 281.61\ \text{Btu/h}\cdot\text{ft}$$

The value of S/L, and hence that of q/L, is some 7% lower than that obtained by freehand plotting.

2.23 Consider a 4-in.-square block of fireclay having a 1-in.-square hole at the center, as shown in Fig. 2-13. If the inner and outer surface temperatures are 150 °C and 30 °C, respectively, and the thermal conductivity is 1.00 W/m · K, determine the heat transfer rate per meter of length from the inner surface to the outer surface, by freehand plotting.

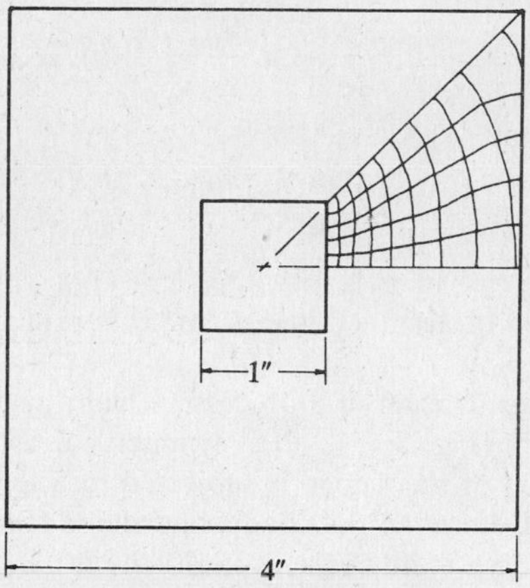

Fig. 2-13

From the freehand grid of curvilinear squares for one-eighth of the region,

$$M_{\text{avg}} \approx \frac{8.15 + 8.25 + 8.35 + 8.70 + 9.00}{5} \approx 8.5$$

and, for the entire block, $N = 8 \times 5 = 40$. Thus

$$\frac{S}{L} \approx \frac{40}{8.5} = 4.71$$

and

$$\frac{q}{L} = \frac{S}{L} k(T_i - T_o) \approx (4.71)(1.00\ \text{W/m} \cdot \text{K})[(150 - 30)\ \text{K}] = 564.7\ \text{W/m}$$

Notice that because S/L depends only on the ratio of linear dimensions, no conversion from inches to meters was required.

2.24 A 6-in.-thick wall of homogeneous material has 2.0-in.-OD tubes spaced at 6-in. intervals along the wall centerline, as shown in Fig. 2.14. The tubes carry hot water and are at constant wall temperature T_i. Both sides of the wall may be assumed to be at the same constant wall temperature, T_o. Determine by freehand plotting the conductive shape factor for one tube.

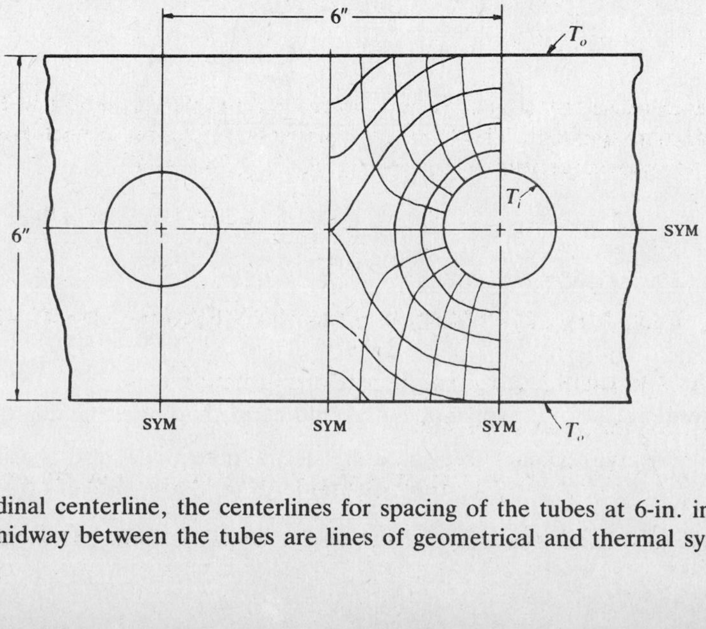

Fig. 2-14

The wall longitudinal centerline, the centerlines for spacing of the tubes at 6-in. intervals, and the centerlines located midway between the tubes are lines of geometrical and thermal symmetry, marked

SYM. Thus a freehand plot of isotherms and adiabatics for any quarter-section will suffice. Two such plots have been completed in Fig. 2-14. From the upper plot,

$$M_{avg} \approx \frac{3.9+4.1+4.7+5.2+5.9}{5} = 4.76$$

and S/L for the entire tube is

$$\frac{S}{L} \approx 4\left(\frac{N_{1/4}}{M_{avg}}\right) = 4\left(\frac{5}{4.76}\right) = 4.20$$

Likewise, from the lower plot,

$$M_{avg} \approx \frac{4.1+4.2+4.4+5.0+5.7}{5} = 4.68 \qquad \frac{S}{L} \approx 4\left(\frac{5}{4.68}\right) = 4.27$$

The agreement between the two plots is to within 2 percent. This, however, is more an indication of consistency in sketching than of accuracy; the accuracy is probably to within ±10 percent.

2.25 A long structural wedge-shaped beam in a processing plant has a cross section 4 in. high, 2 in. across the top, and 4 in. on the bottom (Fig. 2-15). It is symmetrical about a vertical centerline, and the sloping sides are insulated. The top of the wedge is subjected to a temperature of 1000 °F while the base is maintained at 400 °F. (*a*) Determine the conduction shape factor S by the potential-plotting technique. (*b*) If the thermal conductivity of the beam is 25 Btu/h · ft · °F, what is the heat transfer rate per lineal foot?

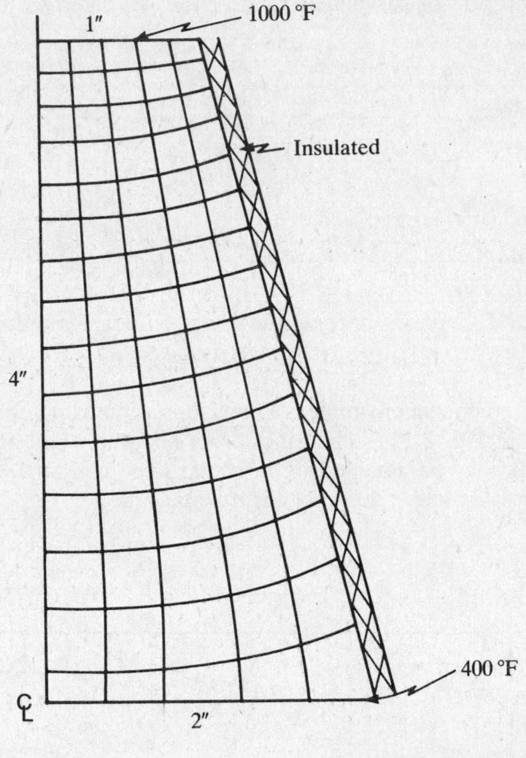

Fig. 2-15

▮ (*a*) $N = 5 \qquad M = \dfrac{14.4+14.5+14.6+14.7+14.9}{5} = 14.62$

so $S/L = N/M = 5/14.62 = 0.342$.

(*b*) $$\frac{q}{L} = \frac{S}{L} k\,\Delta T = (0.342)(25\ \text{Btu/h}\cdot\text{ft}\cdot{}^\circ\text{F})[(1000 - 400)\ {}^\circ\text{F}] = 51\,350\ \text{Btu/h}\cdot\text{ft}$$

2.26 Describe an alternative to the thermal-potential-plotting method of estimating the conductive shape factor.

▮ In a two-dimensional region of free space the electric potential V also obeys the Laplace equation:

$$\frac{\partial^2 V}{\partial x^2} + \frac{\partial^2 V}{\partial y^2} = 0$$

for steady-state conditions. Consequently, if the boundary conditions for V are similar to those for temperature and if the physical geometry of the problem is the same as for the thermal problem, then lines of constant electric potential are also lines of constant temperature. This analogy leads to a more accurate grid of curvilinear squares than obtainable by freehand plotting, and consequently to a somewhat better value of the conductive space factor.

2.27 A 4-in.-diam. steam line and a 2-in.-diam. chilled-water line for air conditioning are $D = 6$ in. apart on centers, in a large service trough packed with rock wool insulation, $k = 0.025$ Btu/h · ft · °F. The steam-line surface temperature is 280 °F, and the chilled-water-line surface temperature is 40 °F. Calculate the heat transfer rate to the water for 40 lineal feet of piping.

From Appendix C, the shape factor is

$$\frac{S}{L} = \frac{2\pi}{\cosh^{-1}[(D^2 - r_1^2 - r_2^2)/2r_1r_2]} = \frac{2\pi}{\cosh^{-1}[(6^2 - 2^2 - 1^2)/2(2)(1)]}$$

$$= \frac{2\pi}{\cosh^{-1} 7.75} = 2.295$$

[The relation $\cosh^{-1} x = \ln(x + \sqrt{x^2 - 1})$ is useful here.] Thus,

$$q = \left(\frac{S}{L} k\, \Delta T\right)L = (2.295)(0.025)(240)(40) = 551.0 \text{ Btu/h}$$

2.28 An electrical engineer who is studying heat transfer works Problem 2.27; it seems to him that he has seen the expression for S/L somewhere before. Has he?

Yes, he has. Recall that the *capicitance* of two electrically conductive bodies, bearing equal and opposite static charges $\pm Q$ and with a voltage difference $\Delta V > 0$ between them, is defined as

$$C \equiv \frac{Q}{\Delta V}$$

Now Q is proportional to the electric field strength (electric flux) at either conductor; thus we can write

$$\text{Electric flux} = k_{\text{elec}} C(\Delta V) \tag{1}$$

Comparing (*1*) and

$$\text{Heat flux} = kS(\Delta T) \tag{2}$$

we see that—since T and V correspond under the electrical analogy (Problem 2.26)—S and C also correspond; in fact, they are identical up to a constant factor. Thus the engineer has recognized the capacitance (per unit length) of a pair of long parallel conducting cylinders.

We see in the above another way of exploiting the electrical analogy to determine conductive shape factors.

2.29[D] A circular cylinder 1 in. in diameter and 1 ft long is used in a chemical experiment to measure the amount of heat given off by the reactants. After the reactants are mixed, the cylinder is placed vertically in a large block of material whose thermal conductivity is 25 Btu/h · ft · °F; the top of the cylinder is flush with the upper surface of the block. Upon reaching steady-state conditions, the temperature of the cylinder is 850 °F, and the temperature of the top of the block is 100 °F. What is the steady-state heat transfer rate?

From Appendix C,

$$S = \frac{2\pi L}{\ln(4L/d)} = \frac{2\pi(1 \text{ ft})}{\ln \dfrac{48 \text{ in.}}{1 \text{ in.}}} = 1.62 \text{ ft}$$

whence $$q = (1.62 \text{ ft})(25 \text{ Btu/h} \cdot \text{ft} \cdot {}^\circ\text{F})[(850 - 100)\ {}^\circ\text{F}] = 30\,375 \text{ Btu/h}$$

2.30 A long, 3-in.-OD sewer line is buried 3 ft below the surface of the earth ($k = 0.2$ Btu/h · ft · °F). If 10 Btu/h · ft of heat is given off from the line, how cold must the surface of the earth be for water to begin to freeze in the line, assuming negligible thermal resistance between the water and the pipe?

▌ From Appendix C,

$$\frac{S}{L} = \frac{2\pi}{\cosh^{-1}(z/r)} = \frac{2\pi}{\cosh^{-1}[3/(1.5/12)]} = 1.61$$

$$\frac{q}{L} = \frac{S}{L} k \, \Delta T$$

$$10 \text{ Btu/h} \cdot \text{ft} = (1.61)(0.2 \text{ Btu/h} \cdot \text{ft} \cdot °\text{F})[(32 \text{ °F}) - T_s]$$

$$T_s = 0.95 \text{ °F} \approx 1 \text{ °F}$$

2.31 Obtain the shape factor for Problem 2.30 from that for Problem 2.27.

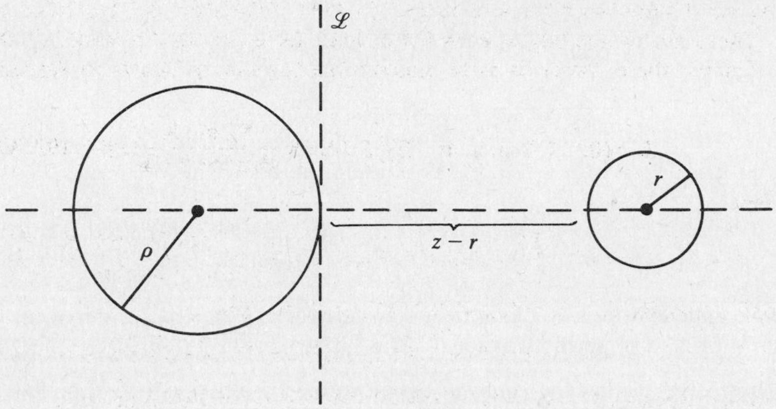

Fig. 2-16

▌ For the cross-sectional geometry of Fig. 2-16 the shape factor is given by $S/L = 2\pi/\cosh^{-1}\Theta$, where

$$\Theta = \frac{(z+\rho)^2 - \rho^2 - r^2}{2\rho r} = \frac{z}{r} + \frac{1}{\rho}\left(\frac{z^2 - r^2}{2r}\right)$$

Letting $\rho \to \infty$, the left-hand cylinder maps onto the fixed plane $\mathscr{L}$ and $\Theta \to z/r$. Thus we recover the shape factor for the geometry of Problem 2.30.

2.32^D Determine the heat transfer rate per unit length from a 2.0-in.-OD isothermal pipe located 5 in. below the surface of a thick cinder-concrete slab. The slab is very wide and very thick, resulting in the two-dimensional problem of Fig. 2.17.

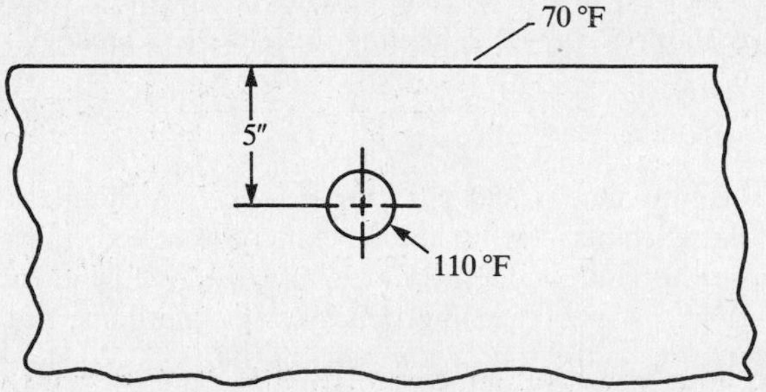

Fig. 2-17

▌ From Appendix C, with $D = 5$ in. and $r = 1.0$ in. $(D > 3r, \; L \gg r)$,

$$\frac{S}{L} = \frac{2\pi}{\ln(2D/r)} = \frac{2\pi}{\ln[2(5)/1]} = 2.73$$

From Table B-2, $k = 0.44 \text{ Btu/h} \cdot \text{ft} \cdot °\text{F}$ and thus

$$\frac{q}{L} = \frac{S}{L} k \, \Delta T = (2.73)(0.44 \text{ Btu/h} \cdot \text{ft} \cdot °\text{F})[(110 - 70) \text{ °F}] = 48.02 \text{ Btu/h} \cdot \text{ft}$$

2.33^D Repeat Problem 2.32 for a 4-cm-OD isothermal pipe at 40 °C located 15 cm below the upper surface of a thick cinder-concrete slab; the upper surface is at 25 °C.

▮ Using Appendix C, with $D = 15$ cm and $r = 2$ cm $(D > 3r,\ L \gg r)$,

$$\frac{S}{L} = \frac{2\pi}{\ln(2D/r)} = \frac{2\pi}{\ln(\frac{30}{2})} = 2.32$$

From Table B-2, $k = (0.44)(1.7296) = 0.76$ W/m · K; hence

$$\frac{q}{L} = \frac{S}{L} k\,\Delta T = (2.32)(0.76 \text{ W/m} \cdot \text{K})[(40-25)\text{ K}] = 26.48 \text{ W/m}$$

2.34 A cubical box having outside dimension 0.5 m is made of 50-mm-thick asbestos $(\rho = 36 \text{ lbm/ft}^3)$ sheets. If the inside surface temperature is 150 °C and the outside surface temperature is 50 °C, determine the rate of heat loss from the box.

▮ There are six 0.4-m-square surfaces [0.5 − 2(0.050)], which can be treated as one-dimensional conduction problems. There are twelve edges, each 0.4 m long, as shown in Appendix C; these are two-dimensional problems. Finally, there are eight three-dimensional corners, as shown in Appendix C. From Appendix B at 110 °C,

$$k = (0.111 \text{ Btu/h} \cdot \text{ft} \cdot {}^\circ\text{F})\left(1.729\,58\ \frac{\text{W/m} \cdot \text{K}}{\text{Btu/h} \cdot \text{ft} \cdot {}^\circ\text{F}}\right) = 0.192 \text{ W/m} \cdot \text{K}$$

Square surfaces

$$q_s = 6kA_s\frac{\Delta T}{\Delta n} = 6(0.192 \text{ W/m} \cdot \text{K})(0.4 \text{ m})^2\,\frac{(150-50)\text{ K}}{0.05 \text{ m}} = 368.64 \text{ W}$$

Edges From Appendix C, $S_c = (0.54)(0.4 \text{ m}) = 0.22$ m and

$$q_e = 12(S_e k\,\Delta T) = 12(0.22 \text{ m})(0.192 \text{ W/m} \cdot \text{K})(100 \text{ K}) = 50.69 \text{ W}$$

Corners From Appendix C, $S_c = (0.15)t = (0.15)(0.05 \text{ m}) = 0.0075$ m and

$$q_c = 8(S_c k\,\Delta T) = 8(0.0075 \text{ m})(0.192 \text{ W/m} \cdot \text{K})(100 \text{ K}) = 1.15 \text{ W}$$

The total heat transfer rate is then

$$q_{\text{total}} = q_s + q_e + q_c = 368.64 + 50.69 + 1.15 = 420.48 \text{ W}$$

say, 420 W, since clearly S_c and S_e are approximations.

2.35[D] The cavity of a volcano, considered to be at a uniform temperature, can be represented as a sphere with a 60-ft diameter. From sonar measurements, the center of the sphere is estimated to be about 220 ft below the surface of the earth $(k = 0.2 \text{ Btu/h} \cdot \text{ft} \cdot {}^\circ\text{F})$. The rate of heat transfer to the earth's surface is approximately 1 million Btu/h when the temperature of the earth's surface is 70 °F. Estimate the temperature of the volcanic cavity.

▮ From Appendix C,

$$S = \frac{4\pi r}{1 - \dfrac{r}{2D}} = \frac{4\pi(30 \text{ ft})}{1 - \dfrac{30 \text{ ft}}{2(220 \text{ ft})}} = 405 \text{ ft}$$

whence

$$T_{\text{cav}} = \frac{q}{Sk} + T_{\text{surf}} = \frac{10^6}{(405)(0.2)} + 70 = 12\,400\ {}^\circ\text{F}$$

2.36 Combustion products at an average temperature of 2000 °C flow through a 10-m-long, 20-cm-OD cylindrical duct which is shielded from the surroundings by a 60-cm by 60-cm square enclosure having a common centerline with the duct. The space between the duct and the enclosure is filled with magnesite. What is the heat transfer rate if the enclosure is kept at an average temperature of 20 °C?

▮ From Appendix C,

$$S = \frac{2\pi L}{\ln(0.54\ W/r)} = \frac{2\pi(10 \text{ m})}{\ln[(0.54)(60)/10]} = 53.45 \text{ m}$$

For the average insulation temperature $(2000 + 20)/2 = 1010$ °C, we get the thermal conductivity by interpolating in Table B-2:

$$k = 1.6 - \left(\frac{1010-650}{1205-650}\right)(1.6-1.1) = 1.28 \text{ Btu/h} \cdot \text{ft} \cdot {}^\circ\text{F} = 2.21 \text{ W/m} \cdot \text{K}$$

Hence $$q = Sk\,\Delta T = (53.45\text{ m})(2.21\text{ W/m}\cdot{}^\circ\text{C})[(2000-20)\ {}^\circ\text{C}] = 234\text{ kW}$$

2.37 A 60-cm-OD by 2-m-long drum of chemical waste is buried horizontally 1 m deep in coarse earth; the surface temperature is 20 °C. What is the heat transfer rate if the drum is at 40 °C?

▮ From Appendix C,

$$S = \frac{2\pi L}{\ln(2D/r)} = \frac{2\pi(2\text{ m})}{\ln[2(100)/30]} = 6.62\text{ m}$$

From Table B-2, $k = (0.30)(1.7296) = 0.52\text{ W/m}\cdot\text{K}$, and so

$$q = Sk\,\Delta T = (6.62\text{ m})(0.52\text{ W/m}\cdot{}^\circ\text{C})[(40-20)\ {}^\circ\text{C}] = 68.70\text{ W}$$

2.38^D A 3-m-ID service tunnel is at the earth temperature (≈15 °C). A 30-cm-OD pipe carries saturated steam (≈100 °C). What is the heat transfer rate per unit length from the steam line to the tunnel if the line is eccentrically located by 1 m?

▮ The shape factor for eccentric cylinders is given in Appendix C as

$$\frac{S}{L} = \frac{2\pi}{\cosh^{-1}\left(\dfrac{r_1^2 + r_2^2 - D^2}{2r_1r_2}\right)}$$

$$= \frac{2\pi}{\cosh^{-1}\left[\dfrac{(0.15)^2 + (1.50)^2 - (1.00)^2}{2(0.15)(1.50)}\right]} = \frac{2\pi}{\cosh^{-1} 2.83} = 8.38$$

Assuming stagnant air in the tunnel, with no convective effects, at about 25 °C, its thermal conductivity is given in Table B-4 as $k = (0.015)(1.7296) = 0.026\text{ W/m}\cdot\text{K}$. Therefore

$$\frac{q}{L} = \left(\frac{S}{L}\right)k\,\Delta T = (8.38)(0.026)(100-15) = 18.52\text{ W/m}$$

[In reality, convective effects often outweigh conduction.]

2.39 What is the heat transfer rate from a radioactive brick (5 cm × 10 cm × 20 cm) at 1000 °C buried 50 cm deep in coarse earth where the ambient temperature is 20 °C?

▮ From Appendix C,

$$S = 1.685L\left[\log\left(1+\frac{b}{a}\right)\right]^{-0.59}\left(\frac{b}{c}\right)^{-0.078} = (1.685)(20\text{ cm})\left[\log\left(1+\frac{50}{10}\right)\right]^{-0.59}\left(\frac{50}{5}\right)^{-0.078} = 32.84\text{ cm}$$

From Table B-2, $k = (0.30)(1.7296) = 0.52\text{ W/m}\cdot\text{K}$, so that

$$q = Sk\,\Delta T = (0.3284\text{ m})(0.52\text{ W/m}\cdot{}^\circ\text{C})[(1000-20)\ {}^\circ\text{C}] = 167.35\text{ W}$$

2.40^D A 30-cm-OD steam pipe is driven vertically 200 m into the earth ($k = 1.3\text{ W/m}\cdot\text{K}$) to recover oil. What is the heat loss rate if the pipe is at 160 °C and the earth at 18 °C?

▮ From Appendix C,

$$S = \frac{2\pi L}{\ln(2L/r)} = \frac{2\pi(200\text{ m})}{\ln(400/0.15)} = 123.31\text{ m}$$

whence $$q = (123.31\text{ m})(1.3\text{ W/m}\cdot{}^\circ\text{C})[(160-18)\ {}^\circ\text{C}] = 22.76\text{ kW}$$

2.41^D A 1-m-OD spherical container of radioactive waste generates heat at 15 kW. It is buried 5 m deep in earth ($k = 2.2\text{ W/m}\cdot\text{K}$), whose surface temperature is 16 °C. What is the temperature of the sphere?

▮ From Appendix C,

$$S = \frac{4\pi r}{1 - r/2D} = \frac{4\pi(0.5\text{ m})}{1 - 0.5/2(5)} = 6.61\text{ m}$$

whence $$T_{\text{sphere}} = \frac{q}{Sk} + T_{\text{surf}} = \frac{15\times10^3}{(6.61)(2.2)} + 16 = 1047\ {}^\circ\text{C}$$

2.42 The buildings of a university are heated by 4 miles of 20-cm-OD steam pipes buried 1 m below the surface of moist earth $(k = 2.0\ \text{W/m}\cdot\text{K})$. The pipe surface is at 140 °C, and the ground temperature is 15 °C. How much heat is lost to the ground?

▌ From Appendix C (1 mile = 1.609 km),

$$S = \frac{2\pi L}{\ln(2D/r)} = \frac{2\pi(6.436\ \text{km})}{\ln[2(100)/10]} = 13.5\ \text{km}$$

Hence $$q = (13.5\ \text{km})(2.0\ \text{W/m}\cdot{}^\circ\text{C})[(140 - 15)\ {}^\circ\text{C}] = 3380\ \text{kW}$$

2.43 How much fuel is wasted per heating season (110 days) in the university heating system of Problem 2.42 if heating oil with a heating value of 32 kW · h/gal is used? What is the cost if the oil is \$0.90/gal?

▌

$$\text{Fuel wasted} = \frac{(3380\ \text{kW})(24\ \text{h/day})(110\ \text{day})}{32\ \text{kW}\cdot\text{h/gal}} = 278\,850\ \text{gal}$$

$$\text{Cost} = (278\,850\ \text{gal})(0.90\ \$/\text{gal}) = \$250\,965$$

2.44 It is physically evident that a two- or three-dimensional region cannot support a steady temperature distribution if there is a net heat flux into the region from the surroundings. Verify this mathematically.

▌ For any three-dimensional function $F(x, y, z)$ defined on a three-dimensional region $\mathcal{R}$, the divergence theorem of calculus gives

$$\iiint_{\substack{\text{interior}\\ \text{of } \mathcal{R}}} \nabla^2 F\, dx\, dy\, dz = -\iint_{\substack{\text{surface}\\ \text{of } \mathcal{R}}} \frac{\partial F}{\partial n}\, dA \tag{1}$$

where $\partial/\partial m$ indicates differentiation along the inward normal to the surface element dA. In particular, if $F = T \equiv$ (steady-state temperature), then, because

$$\nabla^2 T = 0 \qquad \text{and} \qquad dq_{\text{in}} = -k(dA)\frac{\partial T}{\partial n}$$

(*1*) becomes $0 = q_{\text{in}}/k$, or $q_{\text{in}} = 0$. In other words, for a steady state in temperature to exist within the region, the heat influx must be exactly balanced by the heat efflux.

2.45 Check the solution of Problem 2.1 against the result of Problem 2.44.

▌ Assume unit depth in Fig. 2-1, and, for convenience, write $\Theta \equiv \theta_m/[\sinh(\pi W/L)]$, so that the solution (*4*) of Problem 2.1 becomes

$$\theta = \Theta \sin\frac{\pi x}{L} \sinh\frac{\pi y}{L}$$

Describing the rectangle in the counterclockwise sense, starting at the corner (0, 0), we have—with heat influx taken as positive—

$$q_{\text{bottom}} = -k\int_0^L \frac{\partial\theta}{\partial y}\bigg|_{y=0} dx = -\frac{k\pi\Theta}{L}\int_0^L \sin\frac{\pi x}{L}\, dx = -2k\Theta$$

$$q_{\text{right}} = -k\int_0^W \frac{\partial\theta}{\partial(-x)}\bigg|_{x=L} dy = -\frac{k\pi\Theta}{L}\int_0^W \sinh\frac{\pi y}{L}\, dy = -k\Theta\left(\cosh\frac{\pi W}{L} - 1\right)$$

$$q_{\text{top}} = -k\int_L^0 \frac{\partial\theta}{\partial(-y)}\bigg|_{y=W} (-dx) = -\frac{k\pi\Theta\cosh(\pi W/L)}{L}\int_L^0 \sin\frac{\pi x}{L}\, dx = +2k\Theta\cosh\frac{\pi W}{L}$$

$$q_{\text{left}} = -k\int_W^0 \frac{\partial\theta}{\partial x}\bigg|_{x=0} (-dy) = +\frac{k\pi\Theta}{L}\int_W^0 \sinh\frac{\pi y}{L}\, dy = q_{\text{right}}$$

Sure enough: $q_{\text{top}} = -(q_{\text{bottom}} + q_{\text{right}} + q_{\text{left}})$.

2.46 With reference to Problem 2.45, express the fractional heat losses from the bottom and sides as functions of the aspect ratio $\alpha \equiv L/W$.

▮
$$\frac{|q_{\text{bottom}}|}{q_{\text{top}}} = \frac{1}{\cosh(\pi/\alpha)}$$
$$\frac{|q_{\text{right}}|}{q_{\text{top}}} = \frac{|q_{\text{left}}|}{q_{\text{top}}} = \frac{1}{2}\left[1 - \frac{1}{\cosh(\pi/\alpha)}\right]$$

2.47 Evaluate the heat losses found in Problem 2.46 for $\alpha = 1, 3, 5, 10, 100$. What happens as $\alpha \to \infty$?

▮ Figure 2-18 presents the numerical results in diagrammatic form. For very large α we should expect the temperature to vary linearly through the very thin plate. This can be confirmed from the analytic solution, (*4*) of Problem 2.1, which may be rewritten as

$$\theta = \theta_{\text{top}} \frac{\sinh(\pi y/L)}{\sinh(\pi W/L)}$$

Holding L—and with it, θ_{top}—fixed, set $y = \lambda W$ $(0 < \lambda < 1)$ and let $W \to 0$:

$$\theta = \theta_{\text{top}} \frac{\sinh(\pi\lambda W/L)}{\sinh(\pi W/L)} \to \theta_{\text{top}} \frac{\pi\lambda W/L}{\pi W/L} = \lambda\theta_{\text{top}}$$

which is indeed a linear function of $\lambda = y/W$.

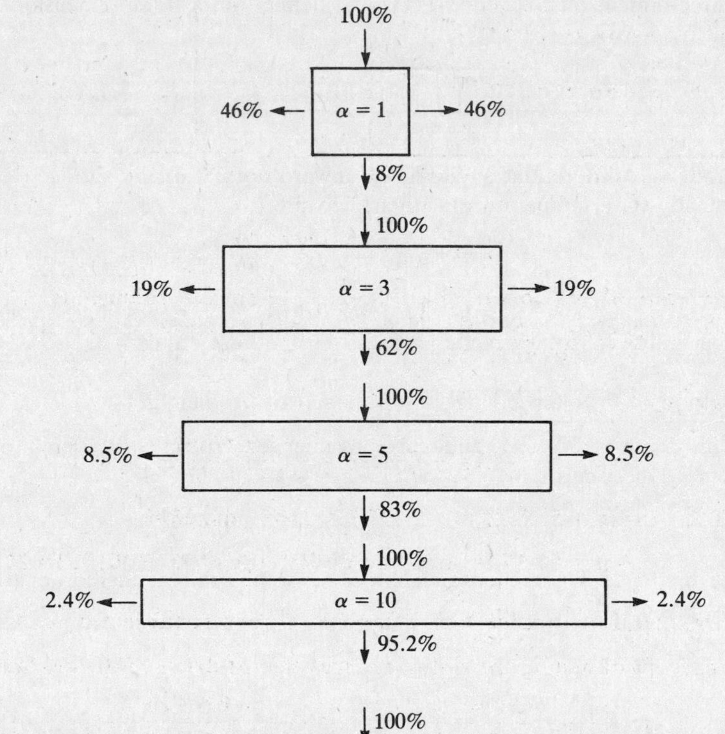

Fig. 2-18

2.48 Refer to Problems 2.46 and 2.47. For what aspect ratio does 90 percent of the heat escape through the bottom edge?

▮ Solving the equation

$$\frac{1}{\cosh(\pi/\alpha)} = 0.90 \qquad \text{or} \qquad \cosh\frac{\pi}{\alpha} = \frac{1}{0.90}$$

either by use of tables of the hyperbolic cosine or the logarithmic formula (Problem 2.27), one finds

$$\frac{\pi}{\alpha} = 0.4671 \qquad \alpha = 6.72$$

2.49 For $W = L = 1$, sketch isotherms of **(a)** the distribution (4) of Problem 2.1, with $\theta_m = 1$; **(b)** the distribution (2) of Problem 2.3, with $\theta_c = 1$. Contrast the two patterns.

▌ See Fig. 2-19. In (a), the boundary distribution is smooth; in particular, the reduced temperature θ is uniquely defined (as 0) at the upper corners. Here the isotherms are everywhere disjunct. But in (b) the boundary temperature is discontinuous at the upper corners. These become singular points of the solution, in which all isotherms are concurrent. Observe, however, that these boundary discontinuities are *not* propagated into the interior of the square. This is an important "smoothing" feature of the Laplace differential equation.

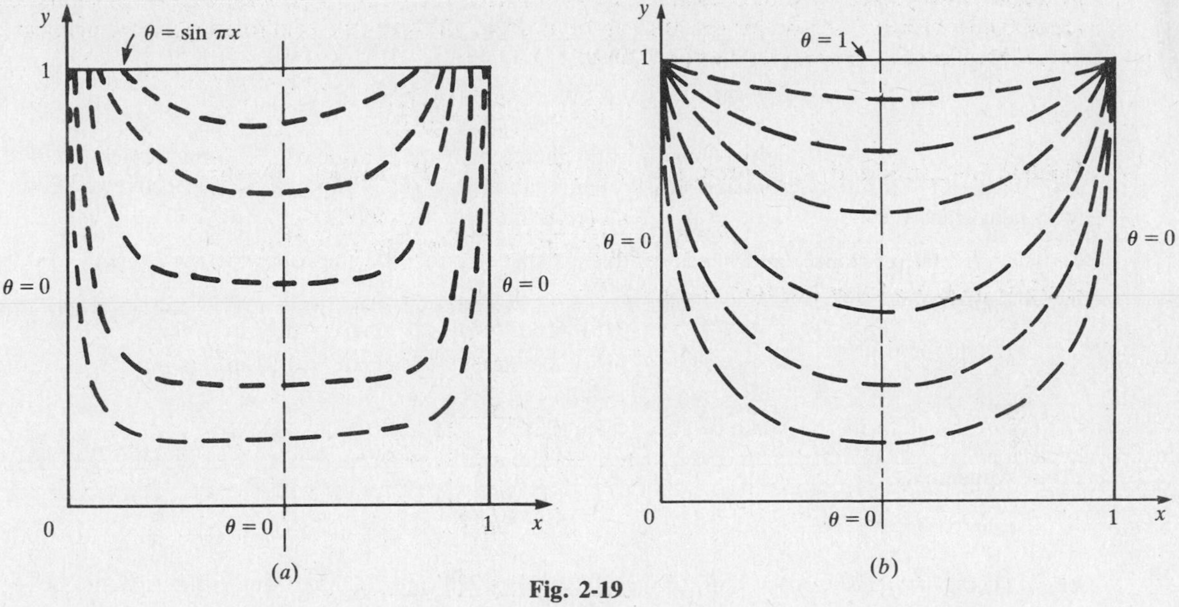

Fig. 2-19

2.50 Approximate the rate of heat loss from a 50-m-high, 25-cm-OD chimney to calm, dry outdoor air at 20 °C if its average wall temperature is 800 °C. Neglect convective effects on the outside of the stack. (This is not a good assumption; see Problem 2.51.)

▌ From Appendix C,

$$S = \frac{2\pi L}{\ln(2L/r)} = \frac{2\pi(50\text{ m})}{\ln[2(50)/0.125]} = 47\text{ m}$$

From Table B-3, $k = (0.345)(1.7296) = 0.597\text{ W/m}\cdot\text{K}$; therefore,

$$q = Sk\,\Delta T = (47\text{ m})(0.597\text{ W/m}\cdot{}^\circ\text{C})[(800-20)\ {}^\circ\text{C}] = 21.9\text{ kW}$$

2.51 Repeat Problem 2.50 if the outside convective heat transfer coefficient is $\bar{h}_0 = 15\text{ W/m}^2\cdot\text{K}$.

▌ Using the concept of adding thermal resistances (see Problem 1.75), the heat transfer is given by

$$q \approx \frac{\Delta T}{(1/kS) + (1/\bar{h}_o A_o)} = \frac{800-20}{[1/(0.597)(47)] + [1/(15)(50\pi/4)]} = 20.9\text{ kW}$$

Note that the convective effect is of the same order as that of conduction.

2.52 Approximate the average temperature of the combustion products in Problem 2.51 if the inside convective heat transfer coefficient is $\bar{h}_i = 80\text{ W/m}^2\cdot\text{K}$. Assume the wall thickness of the chimney is negligible; i.e., $A_o \approx A_i$.

▌ From the basic convection equation, $q = \bar{h}_i A_i (T_g - T_w)$, or

$$T_g = T_w + \frac{q}{\bar{h}_i A_i} = 800 + \frac{20\,900}{(80)(50\pi/4)} = 807\ {}^\circ\text{C}$$

2.53 A cylindrical 30-cm-diam. flue liner is centered in a 40-m-high, 60-cm-square chimney ($k = 1.34\text{ W/m}\cdot\text{K}$). The average temperature of the combustion products is 400 °C, and the ambient temperature is 15 °C.

Estimate the heat loss rate from the chimney if the inside and outside convective heat transfer coefficients are $\bar{h}_i = 90\ \text{W/m}^2\cdot\text{K}$ and $\bar{h}_o = 10\ \text{W/m}^2\cdot\text{K}$, respectively.

▌ From Appendix C,

$$S = \frac{2\pi L}{\ln(0.54W/r)} = \frac{2\pi(40\ \text{m})}{\ln[(0.54)(60)/15]} = 326.4\ \text{m}$$

Using the overall heat transfer coefficient concept (Problem 1.75),

$$q = \frac{\Delta T}{(1/\bar{h}_i A_i) + (1/kS) + (1/\bar{h}_o A_o)}$$

$$= \frac{400 - 15}{[1/(90)\pi(0.30)(40)] + [1/(1.34)(326.4)] + [1/(10)(2.40)(40)]}$$

$$= 107\,500\ \text{W} = 107.5\ \text{kW}$$

2.54 A furnace is 120 cm by 90 cm by 60 cm (outside dimensions) and is made of 15-cm-thick brick ($k = 1.43\ \text{W/m}\cdot\text{K}$). Inside and outside surface temperatures are 425 °C and 40 °C, respectively. Estimate the total heat transfer rate.

▌ The thermal resistance (and hence conductive shape factor) has the following components:

Plane walls (6)

$$A_1 = [120 - 2(15)][60 - 2(15)] = 2700\ \text{cm}^2 = 0.27\ \text{m}^2$$
$$A_2 = [120 - 2(15)][90 - 2(15)] = 5400\ \text{cm}^2 = 0.54\ \text{m}^2$$
$$A_3 = [90 - 2(15)][60 - 2(15)] = 1800\ \text{cm}^2 = 0.18\ \text{m}^2$$
$$A_{\text{walls}} = 2(A_1 + A_2 + A_3) = 1.98\ \text{m}^2$$

From Appendix C,

$$S_{\text{walls}} = \frac{A}{t} = \frac{1.98\ \text{m}^2}{0.15\ \text{m}} = 13.2\ \text{m}$$

Edges (12) $L = 4(120 - 30) + 4(90 - 30) + 4(60 - 30) = 720\ \text{cm} = 7.20\ \text{m}$, and from Appendix C,

$$S_{\text{edges}} = (0.54)L = (0.54)(7.20) = 3.89\ \text{m}$$

Corners (8) From Appendix C, $S = (0.15)t$ per corner; hence

$$S_{\text{corners}} = 8(0.15)(0.15\ \text{m}) = 0.18\ \text{m}$$

Consequently, $S_{\text{total}} = 13.2 + 3.89 + 0.18 = 17.27\ \text{m}$, and

$$q = kS(\Delta T) = (1.43\ \text{W/m}\cdot{}^\circ\text{C})(17.27\ \text{m})[(425 - 40)\ {}^\circ\text{C}] = 9.5\ \text{kW}$$

2.55 What is the maximum thickness of a plane wall of area A for conduction losses from the four corners to be less than $\frac{1}{2}$ percent of the overall losses?

▌ From Appendix C, $S_{\text{wall}} = A/t$ and $S_{\text{corners}} = 4(0.15t) = (0.60)t$; therefore,

$$\frac{q_{\text{corners}}}{q_{\text{wall}}} = \frac{kS_{\text{corners}}}{kS_{\text{wall}}} = \frac{(0.60)t}{A/t} \le 0.005$$

$$(0.60)t^2 \le (0.005)A$$

$$t^2 \le (0.00833)A$$

$$t \le (0.092)\sqrt{A}$$

2.56 Repeat Problem 2.53 for a 30-cm-square flue liner.

▌ From Appendix C, $S_{\text{wall}} = A/t$ and $S_{\text{edge}} = (0.54)L$; hence

$$S_{\text{walls}} = \frac{4(30\ \text{cm})(40\ \text{m})}{15\ \text{cm}} = 320\ \text{m} \qquad S_{\text{edges}} = 4(0.54)(0.15\ \text{m}) = 0.32\ \text{m} \qquad S_{\text{total}} = 320.32\ \text{m}$$

and

$$q = \frac{\Delta T}{(1/\bar{h}_i A_i) + (1/kS_{\text{total}}) + (1/\bar{h}_o A_o)}$$

$$= \frac{400 - 15}{[1/(90)4(0.30)(40)] + [1/(1.34)(320.32)] + [1/(10)4(0.60)(40)]}$$

$$= 106.9\ \text{kW}$$

2.57 What accounts for the difference between the results of Problems 2.53 and 2.56?

▮ The physical difference is in the mass of the wall through which conduction occurs, which is greater for the cylindrical flue liner (as shown by the shaded cross section in Fig. 2-20).

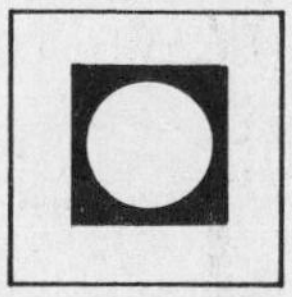

Fig. 2-20

2.58 A 1-m-diam. radioactive sphere generates heat at 4 kW. It is buried in wet earth $(k = 1.2\ \text{W/m}\cdot\text{K})$ with a surface temperature of 20 °C. How deep must it be buried to keep its surface temperature below 300 °C?

▮ Taking the shape factor S from Appendix C,

$$4000\ \text{W} = q = kS(\Delta T) = k\left[\frac{4\pi r}{1-(r/2D)}\right]\Delta T = (1.2)\left[\frac{4\pi(0.5)(300-20)}{1-(0.5/2D)}\right]$$

Solving, $D = 0.53$ m.

2.59 Repeat Problem 2.58 if the sphere generates 2.5 kW.

▮ One finds $D < 0$, which means that $T_{\text{sphere}} < 300$ °C at any positive depth (see Problem 2.60).

2.60 If the sphere of Problem 2.59 is buried at 0.53 m, what is its surface temperature?

▮

$$T_{\text{sphere}} = T_{\text{surface}} + \frac{q}{4\pi kr}\left(1-\frac{r}{2D}\right)$$

$$= 20 + \frac{2500}{4\pi(1.2)(0.5)}\left[1-\frac{0.5}{2(0.53)}\right] = 195.2\ ^\circ\text{C}$$

2.61 A 50-cm-OD pipeline for transporting crude oil is buried 2.0 m below the earth's surface $(k_e = 1.3\ \text{W/m}\cdot\text{K})$. If the surface temperature of the earth is 0 °C and the oil is at 120 °C, what is the rate of heat loss per unit length of pipe?

▮ From Appendix C,

$$\frac{S}{L} = \frac{2\pi}{\ln(2D/r)} = \frac{2\pi}{\ln[2(2)/0.25]} = 2.27$$

$$\frac{q}{L} = k_e\frac{S}{L}\Delta T = (1.3\ \text{W/m}\cdot\text{K})(2.27)[(120-0)\ \text{K}] = 354\ \text{W/m}$$

2.62 Repeat Problem 2.61 if the pipe is covered with 100-mm-thick cellular glass insulation $(k_i = 0.03\ \text{W/m}\cdot\text{K})$.

▮

$$\frac{q}{L} = \frac{\Delta T}{\dfrac{1}{k_e(S/L)} + \dfrac{1}{k_i\pi d_i}}$$

$$= \frac{120-0}{\dfrac{1}{(1.3)(2.27)} + \dfrac{1}{(0.03)\pi(0.70)}} = 7.75\ \text{W/m}$$

2.63 How deep must the pipe of Problem 2.61 be buried if it is insulated with the cellular glass insulation of Problem 2.62 and if 354 W/m must still be dissipated?

▮ The heat transfer equation at depth D is

$$\frac{q}{L} = \frac{\Delta T}{\dfrac{1}{k_e[2\pi/\ln(2D/r)]} + \dfrac{1}{k_i\pi d_i}}$$

Substituting the data and solving for D, one finds $D \approx 0$ m. Therefore, it is not necessary to bury the pipe at all. This illustrates the economics of digging a ditch vs. buying insulation.

2.64 Determine the ratio of heat transfer from a sphere of radius r buried in an infinite medium to a sphere of radius r at depth D in a semi-infinite medium.

▮ Using the shape factors from Appendix C,

$$\frac{q_{\text{inf}}}{q_{\text{semi}}} = \frac{4\pi r}{4\pi r/[1-(r/2D)]} = 1 - \frac{r}{2D}$$

2.65 At what ratio r/D is the heat transferred from a sphere buried in an infinite medium 99% of that from a sphere buried in a semi-infinite medium?

▮ From Problem 2.64, $r/D = 2(1-0.99) = 0.02$.

2.66 An electric heater 200 mm long and 5 mm in diameter is inserted in a hole normal to the surface of a large block of limestone whose surface is maintained at 20 °C. What is the temperature of the heater when its dissipation rate is 100 W?

▮ From Table B-2, $k_{100\,°\text{C}} = (0.73)(1.7296) = 1.26$ W/m · K, and from Appendix C,

$$S = \frac{2\pi L}{\ln(2L/r)} = \frac{2\pi(0.2\text{ m})}{\ln[2(200)/2.5]} = 0.25\text{ m}$$

Then
$$T_h = T_s + \frac{q}{kS} = 20 + \frac{100}{(1.26)(0.25)} = 337.5\text{ °C}$$

2.67[D] Hot water at 90 °C flows through a 30-mm-diam., thin-walled copper tube. It is enclosed by a 150-mm-diam. eccentric cylinder, with 20-mm eccentricity, which is maintained at 40 °C. The space between the cylinder is filled with dry sawdust. What is the heat transfer?

▮ From Table B-2, $k = 0.034(1.7296) = 0.059$ W/m · K. Using the shape factor from Appendix C for eccentric cylinders,

$$\frac{S}{L} = \frac{2\pi}{\cosh^{-1}\left(\dfrac{r_1^2 + r_2^2 - D^2}{2r_1r_2}\right)} = \frac{2\pi}{\cosh^{-1}\left[\dfrac{(15)^2 + (75)^2 - (20)^2}{2(15)(75)}\right]} = 4.1$$

and $q/L = (S/L)k\,\Delta T = (4.1)(0.059)(90-40) = 12.1$ W/m.

2.68 Comparing Problems 2.67 and 2.38 with Problem 2.27, we see that the shape factors for the "internal" and the "external" geometries are essentially the same, one formally arising from the other by a sign change in the argument of the $\cosh^{-1}$ function. How does this come about?

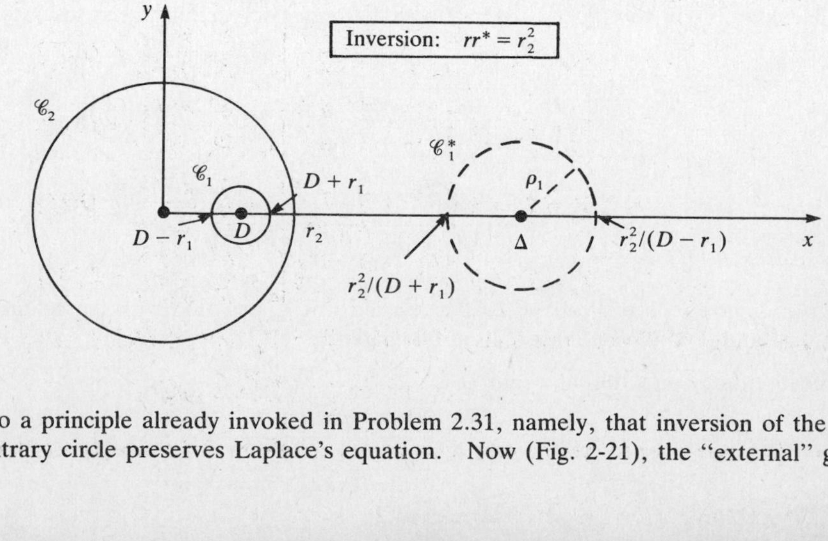

Fig. 2-21

▮ We appeal to a principle already invoked in Problem 2.31, namely, that inversion of the two-dimensional plane in an arbitrary circle preserves Laplace's equation. Now (Fig. 2-21), the "external" geometry ($\mathscr{C}_2$ and

$\mathscr{C}_1^*$) can be obtained from the "internal" geometry ($\mathscr{C}_1$ and $\mathscr{C}_2$) by an inversion of the plane in C_2. Under this mapping, provided $r_1 < D < r_2 - r_1$,

$$\Delta = \frac{1}{2}\left(\frac{r_2^2}{D+r_1} + \frac{r_2^2}{D-r_1}\right) = \frac{r_2^2 D}{D^2 - r_1^2}$$

$$\rho_1 = \frac{1}{2}\left(\frac{r_2^2}{D-r_1} - \frac{r_2^2}{D+r_1}\right) = \frac{r_2^2 r_1}{D^2 - r_1^2}$$

$$\rho_2 = r_2$$

so that (check it!)

$$\frac{\Delta^2 - \rho_1^2 - \rho_2^2}{2\rho_1\rho_2} = \frac{r_1^2 + r_2^2 - D^2}{2r_1 r_2}$$

2.69 Show that the result of Problem 2.31 is obtained as a special case of that of Problem 2.68.

▌ Let $r_1 \to D$ in Problem 2.68; then $\mathscr{C}_1^*$ becomes a vertical line $\mathscr{L}$ through the point $x = r_2^2/2D$, and the shape factor for the "external" geometry depends on the argument

$$\frac{D^2 + r_2^2 - D^2}{2Dr_2} = \frac{r_2}{2D} = \frac{x}{r_2}$$

2.70 A small cubical furnace, 1 m by 1 m by 1 m on the inside, is made of 10-cm-thick fireclay brick ($k = 1.06$ W/m · K). Determine the rate of heat loss if the inside is maintained at 600 °C and the outside at 150 °C.

▌ From Appendix C,

Per wall $\quad S = \dfrac{A}{t} = \dfrac{(1.0)(1.0)}{0.10} = 10 \text{ m}$

Per edge $\quad S = (0.54)L = (0.54)(1.0) = 0.54 \text{ m}$

Per corner $\quad S = (0.15)t = (0.15)(0.10) = 0.015 \text{ m}$

For the six walls, twelve edges, and eight corners, the total shape factor is

$$S = 6(10) + 12(0.54) + 8(0.015) = 66.60 \text{ m}$$

and

$$q = kS(\Delta T) = (1.06 \text{ W/m}\cdot{}^\circ\text{C})(66.60 \text{ m})[(600 - 150)\ {}^\circ\text{C}] = 31.8 \text{ kW}$$

2.3 FINITE-DIFFERENCE APPROXIMATIONS

2.71 Considering a general two-dimensional body (Fig. 2-22), determine an expression for the temperature of an interior nodal point n in terms of the temperatures of adjacent nodes.

▌ By Taylor's theorem,

$$T_1 = T_n - \delta \left.\frac{\partial T}{\partial x}\right|_n + \frac{\delta^2}{2!}\left.\frac{\partial^2 T}{\partial x^2}\right|_n - \frac{\delta^3}{3!}\left.\frac{\partial^3 T}{\partial x^3}\right|_n + O(\delta^4)$$

$$T_2 = T_n - \delta \left.\frac{\partial T}{\partial y}\right|_n + \frac{\delta^2}{2!}\left.\frac{\partial^2 T}{\partial y^2}\right|_n - \frac{\delta^3}{3!}\left.\frac{\partial^3 T}{\partial y^3}\right|_n + O(\delta^4)$$

$$T_3 = T_n + \delta \left.\frac{\partial T}{\partial x}\right|_n + \frac{\delta^2}{2!}\left.\frac{\partial^2 T}{\partial x^2}\right|_n + \frac{\delta^3}{3!}\left.\frac{\partial^3 T}{\partial x^3}\right|_n + O(\delta^4)$$

$$T_4 = T_n + \delta \left.\frac{\partial T}{\partial y}\right|_n + \frac{\delta^2}{2!}\left.\frac{\partial^2 T}{\partial y^2}\right|_n + \frac{\delta^3}{3!}\left.\frac{\partial^3 T}{\partial y^3}\right|_n + O(\delta^4)$$

Adding, and using $\nabla^2 T = 0$, one finds $T_1 + T_2 + T_3 + T_4 = 4T_n + O(\delta^4)$, or

$$T_n = \frac{T_1 + T_2 + T_3 + T_4}{4} + O(\delta^4) \tag{1}$$

Equation (*1*) is the discrete mean-value theorem for solutions of the two-dimensional Laplace equation.

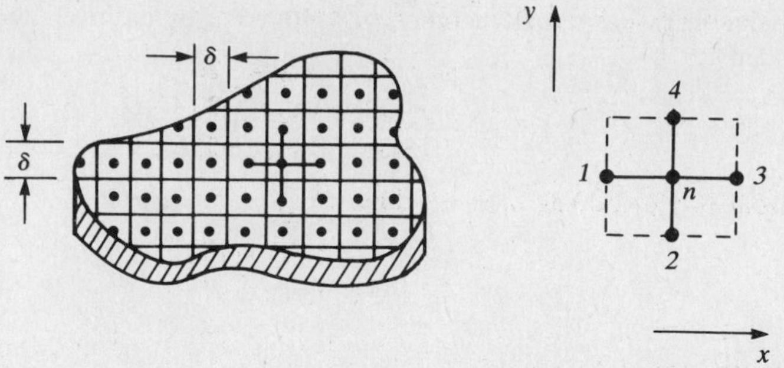

Fig. 2-22

Observe that

$$F_n = \frac{F_1 + F_2 + F_3 + F_4}{4} + O(\delta^2) \qquad (2)$$

for *any* two-dimensional function F; but (2) is inferior to (1) by two orders of magnitude.

2.72 Write the set of nodal temperature equations for a 6-in. square grid for the square chimney shown in Fig. 2.23. Assume the material to have uniform thermal conductivity, uniform inside temperature $T_i = 300$ °F, and uniform outside temperature $T_o = 100$ °F.

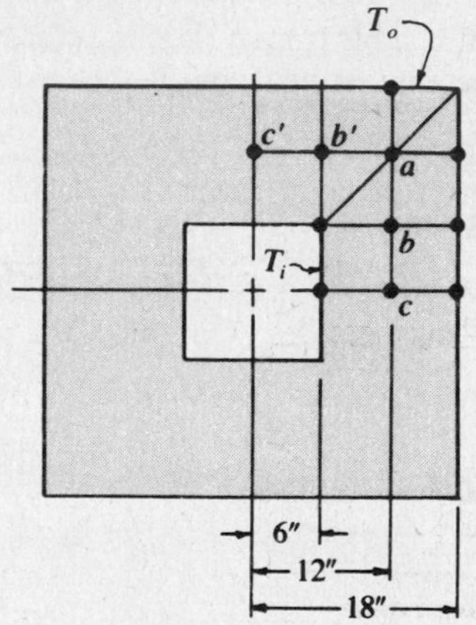

Fig. 2-23

In the indicated quadrant of the chimney, the only unknown nodal temperatures are T_a, T_b, and T_c since clearly $T'_b = T_b$, $T'_c = T_c$, etc. The nodal equations are, from Problem 2.71,

node *a*	$T'_b + T_b + 100 + 100 - 4T_a = 0$
node *b*	$300 + T_c + 100 + T_a - 4T_b = 0$
node *c*	$300 + T'_b + 100 + T_b - 4T_c = 0$

which rearrange into the system

$$2T_a - 9T_b \qquad = 100 \qquad (1a)$$

$$-T_a + 4T_b - T_c = 400 \qquad (2a)$$

$$-T_b + 2T_c = 200 \qquad (3a)$$

2.73 Solve the following system of equations, using *Gaussian elimination*:

$$x_1 + 2x_2 + 3x_3 = 20 \qquad (a)$$

$$x_1 - 3x_2 + x_3 = -3 \qquad (b)$$

$$2x_1 + x_2 + x_3 = 11 \qquad (c)$$

First, triangularize the given set of equations. This can always be accomplished by repeated application of three basic row operations: (i) multiplication of a row by a constant, (ii) addition of one row to another, (iii) interchange of two rows. Thus, eliminate x_1 from (b) and (c) by respectively adding to these equations -1 times (a) and -2 times (a); the result is

$$x_1 + 2x_2 + 3x_3 = 20 \qquad (a')$$
$$-5x_2 - 2x_3 = -23 \qquad (b')$$
$$-3x_2 - 5x_3 = -29 \qquad (c')$$

Now eliminate x_2 from (c') by adding to it $-\frac{3}{5}$ times (b'):

$$x_1 + 2x_2 + 3x_3 = 20$$
$$5x_2 + 2x_3 = 23$$
$$x_3 = 4$$

This is the triangularized set of equations.

Finally, "back substitute," beginning with the bottom equation and working upward, to obtain successively $x_2 = 3$ and $x_1 = 2$. Thus, the final solution is

$$x_1 = 2 \qquad x_2 = 3 \qquad x_3 = 4$$

2.74 Give a flow diagram for the solution of the matrix system

$$\begin{bmatrix} A(1,1) & A(1,2)\ldots A(1,N) \\ A(2,1) & A(2,2)\ldots A(2,N) \\ \cdots\cdots & \cdots\cdots\cdots\cdots \\ A(N,1) & A(N,2)\ldots A(N,N) \end{bmatrix} \begin{bmatrix} T(1) \\ T(2) \\ \cdots\cdots \\ T(N) \end{bmatrix} = \begin{bmatrix} B(1) \\ B(2) \\ \cdots\cdots \\ B(N) \end{bmatrix}$$

by Gaussian elimination.

I See Fig. 2-24.

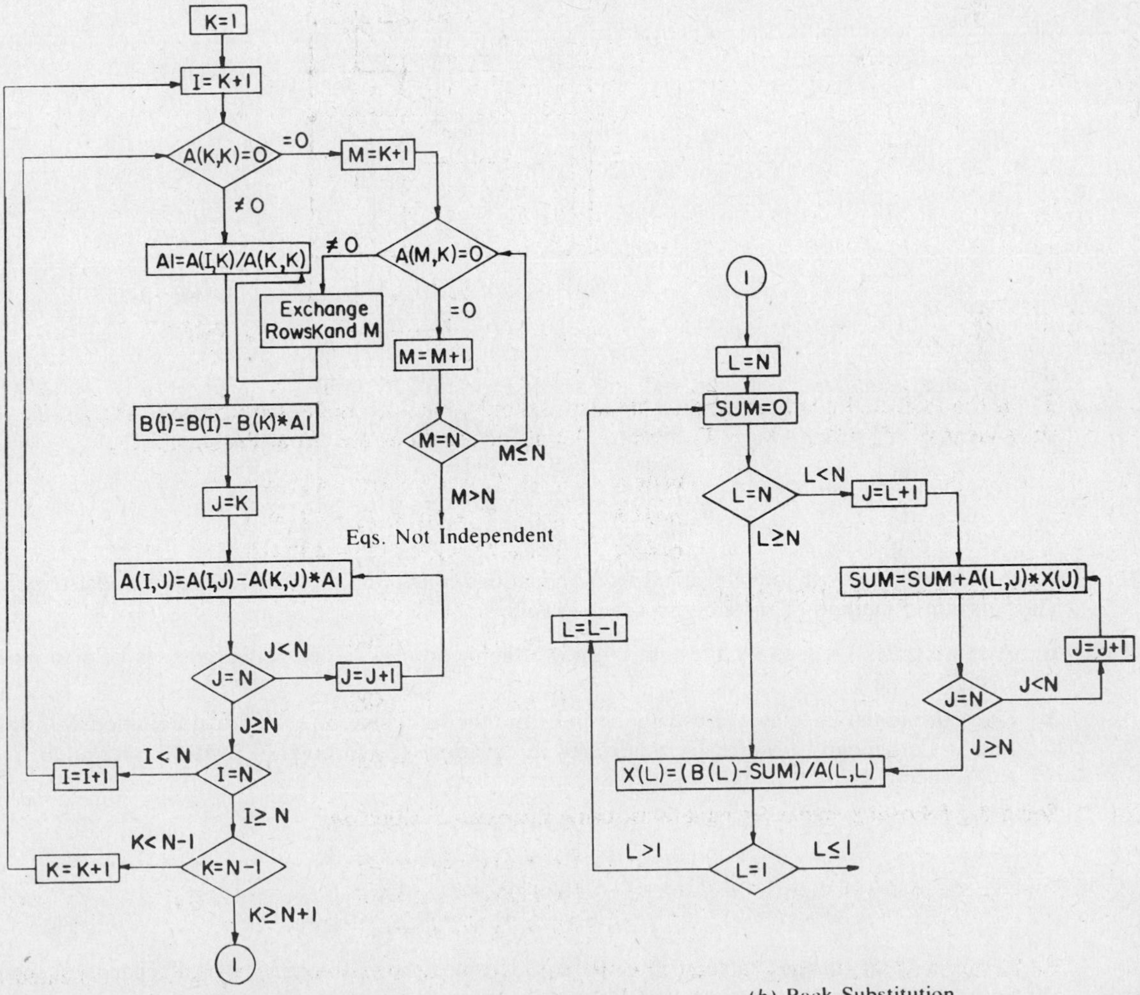

(a) Triangularization (b) Back Substitution

Fig. 2-24

2.75 Solve by Gaussian elimination the set of algebraic equations obtained in Problem 2.72.

▌ Multiplying ($1a$) by $\frac{1}{2}$ and adding to ($2a$) yields

$$2T_a - T_b = 100 \tag{1b}$$

$$+\tfrac{7}{2}T_b - T_c = 450 \tag{2b}$$

$$-T_b + 2T_c = 200 \tag{3b}$$

Multiplying ($2b$) by $\frac{2}{7}$ and adding to ($3b$) yields

$$2T_a - T_b = 100 \tag{1c}$$

$$\tfrac{7}{2}T_b - T_c = 450 \tag{2c}$$

$$+\tfrac{12}{7}T_c = \tfrac{2300}{7} \tag{3c}$$

Thus, by back substitution,

$$T_c = \tfrac{2300}{12} = 191.67\ \text{°F}$$
$$T_b = \tfrac{2}{7}(450 + 191.67) = 183.33\ \text{°F}$$
$$T_a = \tfrac{1}{2}(100 + 183.33) = 141.67\ \text{°F}$$

2.76 Use the *relaxation method* to determine the steady-state temperatures at the four interior nodal points of Fig. 2.25.

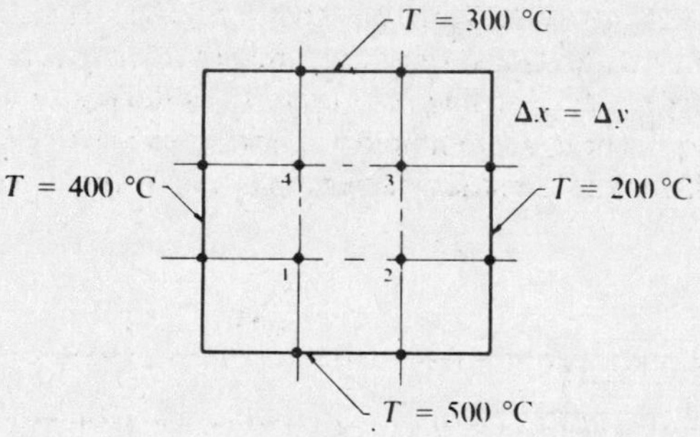

Fig. 2-25

▌ The nodal equations, obtained with the aid of Problem 2.71, are

node *1*	$400 + 500 + T_2 + T_4 - 4T_1 = 0$	(*1*)
node *2*	$500 + 200 + T_1 + T_3 - 4T_2 = 0$	(*2*)
node *3*	$200 + 300 + T_2 + T_4 - 4T_3 = 0$	(*3*)
node *4*	$300 + 400 + T_1 + T_3 - 4T_4 = 0$	(*4*)

which comprise a set of four linear algebraic equations containing the four unknown nodal temperatures. The relaxation method of solution proceeds as follows:

1. Assume (guess) values for the four unknown temperatures. Good initial guesses help to minimize the ensuing work.
2. Since the initial guesses will usually be in error, the right side of each nodal equation will differ from zero. Consequently, we replace the zeros in equations (*1*) through (*4*) with the *residuals* R_1, R_2, and R_4, respectively.

$$900 + T_2 + T_4 - 4T_1 = R_1 \tag{5}$$
$$700 + T_1 + T_3 - 4T_2 = R_1 \tag{6}$$
$$500 + T_2 + T_4 - 4T_3 = R_3 \tag{7}$$
$$700 + T_1 + T_3 - 4T_4 = R_4 \tag{8}$$

3. Set up a "unit change" table such as Table 2-1, which shows the effect of a 1° change of temperature at one node upon the residuals. The fact that a "block" (overall) unit change has the same effect upon all residuals is unusual, this being due to the overall problem symmetry.

TABLE 2-1

	ΔR_1	ΔR_2	ΔR_3	ΔR_4
$\Delta T_1 = +1$	−4	+1	0	+1
$\Delta T_2 = +1$	+1	−4	+1	0
$\Delta T_3 = +1$	0	+1	−4	+1
$\Delta T_4 = +1$	+1	0	+1	−4
Block change = +1	−2	−2	−2	−2

4. Calculate the initial residuals for the initially assumed temperatures using the "residual equations" (*5*) through (*8*).
5. Set up a relaxation table such as Table 2-2. Begin with the initially assumed temperatures and the resulting initial residuals. The left-hand column records the changes from the initially assumed temperature values. Notice that the procedure begins by "relaxing" the largest initial residual (or perhaps by making a block change, a technique useful when all residuals are of the same sign).

In the present problem, we should begin by reducing R_2 or R_3. Arbitrarily choose R_2 and proceed by overrelaxing slightly. Table 2-1 allows rapid calculation of the changes in the residuals without recourse to the equations. Notice that the +20° change in T_2 reduced the residuals at nodes *1* and *2* but unfortunately increased R_3.

The first row in Table 2-2 shows the new residuals and temperatures; the only temperature changed is underlined. Proceeding, we next relax the largest resulting residual, this being R_3. Following a temperature change of +25° at node 3, we see that $R_4 = 0$. This does not mean that we have obtained the correct temperature at node 4, but only that the set of as yet incorrect temperature values happens to satisfy equation (*4*) exactly. Proceeding, the largest residual is now R_2, which is reduced to 0 by a +5° change in T_2. This also reduces all remaining residuals to zero. A check is made by substituting the temperatures thus obtained into equations (*1*) through (*4*); this verifies the solution.

TABLE 2-2

	T_1	R_1	T_2	R_2	T_3	R_3	T_4	R_4
Initial values	400	−25	325	+75	275	+75	350	−25
$\Delta T_2 = +20$	400	−5	<u>345</u>	−5	275	+95	350	−25
$\Delta T_3 = +25$	400	−5	345	+20	<u>300</u>	−5	350	0
$\Delta T_2 = +5$	400	0	<u>350</u>	0	300	0	350	0
Check by equations		✓ 0		✓ 0		✓ 0		✓ 0
Solution	400		350		300		350	

2.77 Consider the block of insulation material shown in Fig. 2-26, with a square inside duct at 400 °F and an outside temperature of 100 °F. Find the temperature distribution by the relaxation method.

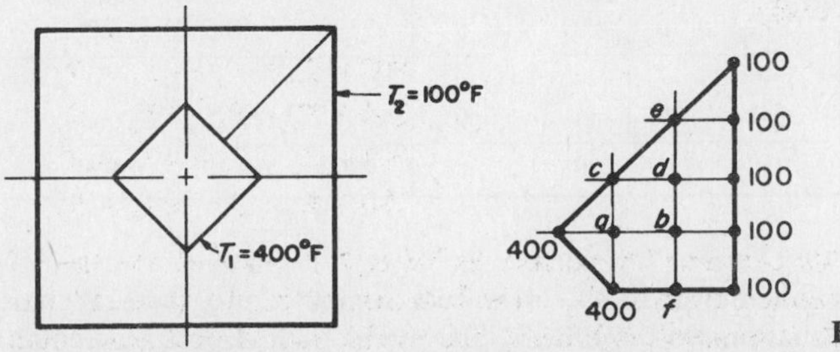

Fig. 2-26

I Apply the procedure of Problem 2.76 to the system

$$\begin{aligned}
&\textbf{node } a & 400 + 400 + T_b + T_c - 4T_a &= R_a\\
&\textbf{node } b & T_a + T_f + 100 + T_d - 4T_b &= R_b\\
&\textbf{node } c & T_a + T_d - 2T_c &= R_c\\
&\textbf{node } d & T_c + T_b + 100 + T_e - 4T_d &= R_d\\
&\textbf{node } e & T_d + 100 - 2T_e &= R_e\\
&\textbf{node } f & 200 + T_b + 50 - 2T_f &= R_f
\end{aligned}$$

Observe the three coefficients 2: because nodes c, e, and f are on lines of symmetry, only one-half of the heat flow through them can be attributed to the temperature value in the section of body shown. Using Table 2-3 to calculate changes in the residuals, relax the system as in Table 2-4. Here the computation was stopped when all residuals were reduced to an absolute value of 1.0 or less. This produces a maximum error of $\frac{1}{4}$ °F at an interior node and $\frac{1}{2}$ °F at a node on a line of symmetry.

TABLE 2-3

	R_a	R_b	R_c	R_d	R_e	R_f
$\Delta T_a = +1$	−4	+1	+1	0	0	0
$\Delta T_b = +1$	+1	−4	0	+1	0	+1
$\Delta T_c = +1$	+1	0	−2	+1	0	0
$\Delta T_d = +1$	0	+1	+1	−4	+1	0
$\Delta T_e = +1$	0	0	0	+1	−2	0
$\Delta T_f = +1$	0	+1	0	0	0	−2
Block = +1	−2	−1	0	−1	−1	−1

TABLE 2-4

	T_a	R_a	T_b	R_b	T_c	R_c	T_d	R_d	T_e	R_e	T_f	R_f
	330	−60	210	0	250	10	180	−10	150	−20	230	0
$\Delta T_a = -16$	314	4	210	−16	250	−6	180	−10	150	−20	230	0
$\Delta T_e = -11$	314	4	210	−16	250	−6	180	−21	139	2	230	0
$\Delta T_d = -6$	314	4	210	−22	250	−12	174	3	139	−4	230	0
$\Delta T_b = -6$	314	−2	204	2	250	−12	174	−3	139	−4	230	−6
$\Delta T_c = -6$	314	−8	204	2	244	0	174	−9	139	−4	230	−6
$\Delta T_f = -3$	314	−8	204	−1	244	0	174	−9	139	−4	227	0
Block = −1	313	−6	203	0	243	0	173	−8	138	−3	226	1
Check by equations		√ −6		√ 0		√ 0		√ −8		√ −3		√ 1
$\Delta T_d = -2$	313	−6	203	−2	243	−2	171	0	138	−5	226	1
$\Delta T_a = -2$	311	2	203	−4	243	−4	171	0	138	−5	226	1
$\Delta T_e = -3$	311	2	203	−4	243	−4	171	−3	135	1	226	1
$\Delta T_b = -1$	311	1	202	0	243	−4	171	−4	135	1	226	0
$\Delta T_c = -2$	311	−1	202	0	241	0	171	−6	135	1	226	0
$\Delta T_d = -1.5$	311	−1	202	−1.5	241	−1.5	169.5	0	135	−0.5	226	0
$\Delta T_c = -1.0$	311	−2	202	−1.5	240	0.5	169.5	−1	135	−0.5	226	0
Block = −1	310	0	201	−0.5	239	0.5	168.5	0	134	0.5	225	1
$\Delta T_f = +0.5$	310	0	201	0	239	0.5	168.5	0	134	0.5	225.5	0
Check by equations		√ 0		√ 0		√ 0.5		√ 0		√ 0.5		√ 0
Solution	310		201		239		168.5		134		225.5	

2.78 Using the flow diagram of Fig. 2-24, write a computer program for Gaussian elimination in Fortran IV. Allow for 10 equations in the DIMENSION and FORMAT statements.

Part 1. Triangularization

```
0060  DIMENSION A(10,10) , B(10) ,X(10)
0070  READ (5,1000) N
0080  N1 = N
0090  DO 100 I = 1 , N
0100  READ (5,1010) (A(I,J), J=1,N1)
0110  READ (5,1010) (B(I),I=1,N)
0120  WRITE (6,1020) N
0130  WRITE (6,1030)
0140  DO 200 I= 1,N
0200  WRITE (6,1040) (A(I,J),J=1,N1) , B(I)
0220  K = 1
0260  I = K+1
0270  IF (A(K,K).EQ.0) GO TO 410
0280  A1 = A(I,K)/A(K,K)
0290  B(I) = B(I)-B(K)*A1
0300  J = K
0310  A(I,J) = A(I,J)-A(K,J)*A1
0320  IF (J.GE.N) GO TO 350
0330  J = J+1
0340  GO TO 310
0350  IF (I.GE.N) GO TO 380
0360  I = I+1
0370  GO TO 270
0380  IF (K.GE.(N-1)) GO TO 550
0390  K = K+1
0400  GO TO 260
0410  M = K+1
0420  IF ((A(M,K)).NE.0) GO TO 460
0430  M = M+1
0440  IF (M.LE.N) GO TO 420
0450  WRITE (6,1050)
0455  GO TO 2000
0460  C1 = B(K)
0470  B(K) = B(M)
0480  B(M) = C1
0490  DO 520 J=1,N
0500  Z1=A(K,J)
0510  A(K,J) = A(M,J)
0520  A(M,J)=Z1
0530  GO TO 280
0550  WRITE (6,1060)
0560  DO 570 I=1,N
0570  WRITE (6,1040) (A(I,J),J=1,N1)
0575  WRITE (6,1070)
0580  WRITE (6,1040) (B(I),I=1,N)
```

Part 2. Back Substitution

```
0585  WRITE(6,1090)
0590  L=N
0600  SUM=0
0610  IF (L.LT.N) GO TO 700
0620  X(L)=(B(L)-SUM)/A(L,L
0640  IF (L.LE.1.0) GO TO 120
0650  L=L-1
0660  GO TO 600
0700  J=L+1
0710  SUM=SUM+A(L,J)*X(J)
0720  IF(J.GE.N) GO TO 620
0730  J=J+1
0740  GO TO 710
1000  FORMAT(I2)
1010  FORMAT(10 F8.3)
1020  FORMAT('1','THERE ARE',I3,2X,'EQUATIONS'//)
1030  FORMAT(' ','THE EQUATIONS ARE:'/)
1040  FORMAT('0',11F11.3)
1050  FORMAT('1','THE EQUATIONS ARE NOT INDEPENDENT')
1060  FORMAT('1',' MATRIX A TRIANGULARIZED'//)
1070  FORMAT('0',' MATRIX B TRIANGULARIZED'//)
1090  FORMAT('1',' THE EQUATION ROOTS ARE:'//)
1100  FORMAT('0',' ROOT #',I2,' = ',F10.3)
1200  DO 1210 L=1,N
1210  WRITE(6,1100) L, X(L)
2000  CONTINUE
      CALL EXIT
      END
```

Fig. 2-27

▌ Figure 2-27 gives one solution. It can easily be modified to handle more equations by appropriate renumbering in the DIMENSION statement and suitable FORMAT statement(s) changes. It is best not to heavily overdimension in the DIMENSION statement as this fixes computer storage requirements.

2.79 Refer to Fig. 2-28. (***a***) For the square grid shown, write the appropriate residual equations for the three interior nodes. (***b***) Set up a relaxation-pattern effect of unit change on the residuals. (***c***) Using a relaxation table, find T_1, T_2, and T_3. Reduce residuals to ± 10.

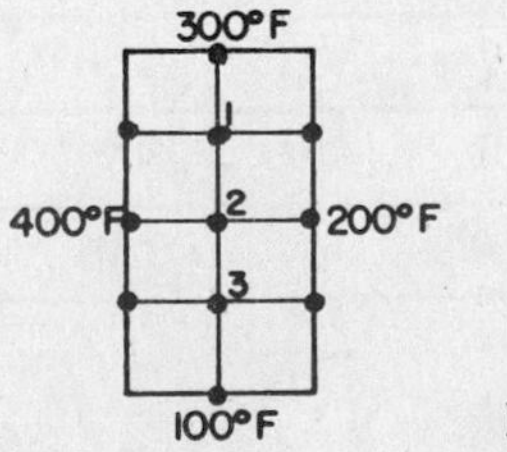

Fig. 2-28

▌ (*a*)

$$R_1 = 900 + T_2 - 4T_1$$
$$R_2 = 600 + T_1 + T_3 - 4T_2$$
$$R_3 = 700 + T_2 - 4T_3$$

(***b***) See Table 2-5.

(***c***) See Table 2-6.

TABLE 2-5

	ΔR_1	ΔR_2	ΔR_3
$\Delta T_1 = 1$	−4	1	0
$\Delta T_2 = 1$	1	−4	1
$\Delta T_2 = 1$	0	1	−4
Unit block	−3	−2	−3

TABLE 2-6

	T_1	R_1	T_2	R_2	T_3	R_3
Guess	300	−50	250	100	200	150
$\Delta T_3 = 40$				40/140	240	−160/(−10)
$\Delta T_2 = 40$		40/(−10)	290	−160/(−30)		40/30
$\Delta T_3 = 10$				10/(−10)	250	−40/10
Solution	300	−10	290	−10	250	−10

2.80 For Fig. 2-29 (square grid):

(*a*) Write the residual equations for points 1, 2, 3, and 4.

(*b*) Prepare an operation table for +1 change in temperature.

(*c*) Initially assuming $T_1 = T_2 = 300$ °F and $T_3 = T_4 = 200$ °F, determine T_1, T_2, T_3, and T_4 by relaxation.

(*d*) With the assumed temperatures of part (*c*), use block relaxation alone to solve for the nodal temperatures.

▮ (*a*)

$$R_1 = 600 + T_2 + T_3 - 4T_1$$
$$R_2 = 600 + T_1 + T_4 - 4T_2$$
$$R_3 = 200 + T_1 + T_4 - 4T_3$$
$$R_4 = 200 + T_2 + T_3 - 4T_4$$

(*b*) See Table 2-7.

(*c*) See Table 2-8.

(*d*) For block relaxation:

$$\frac{|\Sigma \text{ Residuals}|}{|\text{Total effect}|} = \frac{400}{8} = 50$$

	T_1	R_1	T_2	R_2	T_3	R_3	T_4	R_4
Starting	300	−100	300	−100	200	−100	200	−100
$\Delta T_i = -50$	250	0	250	0	150	0	150	0

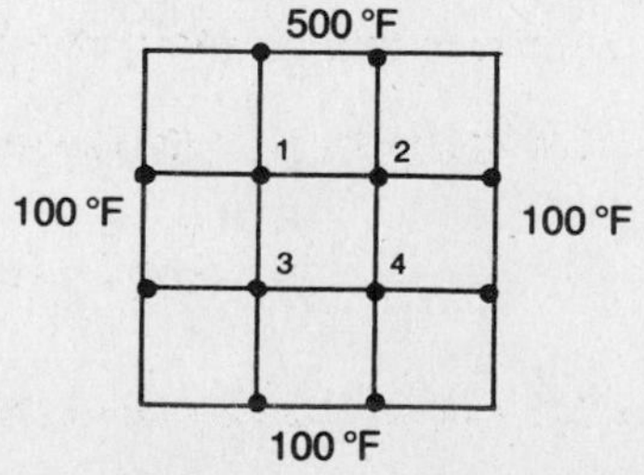

Fig. 2-29

TABLE 2-7

	ΔR_1	ΔR_2	ΔR_3	ΔR_4
$\Delta T_1 = 1$	−4	1	1	0
$\Delta T_2 = 1$	1	−4	0	1
$\Delta T_3 = 1$	1	0	−4	1
$\Delta T_4 = 1$	0	1	1	−4
Block	−2	−2	−2	−2

TABLE 2-8

	T_1	R_1	T_2	R_2	T_3	R_3	T_4	R_4
Starting	300	−100	300	−100	200	−100	200	−100
$\Delta T_1 = -25$	275	0		−125		−125		
$\Delta T_2 = -30$		−30	270	−5				−130
$\Delta T_4 = -40$				−45		−165	160	30
$\Delta T_3 = -40$		−70			160	−5		−10
$\Delta T_1 = -20$	255	10		−65		−25		
$\Delta T_2 = -10$		0	260	−25				−20
$\Delta T_3 = -5$		−5			155	−5		−25
$\Delta T_2 = -10$		−15	250	15				−35
$\Delta T_4 = -10$				5		−15	150	5
$\Delta T_3 = -5$		−20			150	5		0
Solution $\Delta T_1 = -5$	250	0	250	0	150	0	150	0

2.81 An extruded stainless-steel beam of the cross section shown in Fig. 2-30 is used as a structural member in a furnace where the outer surface temperature is 2000 °F. To prevent creep, cooling water flows through the beam. To determine the water flow rate necessary, the rate of heat flow must be estimated. The inside surface temperature is 70 °F. (*a*) Estimate the heat transfer rate by sketching the isotherms and heat-flow lines in the left section. (*b*) Assuming $T_d = 1700$ °F, $T_e = 1400$ °F, and $T_f = 1300$ °F, determine the residual for point *e*. (*c*) Express the rate of heat flow per unit length into the beam in terms of nodal temperatures on the 1-in. by 1-in. grid shown. (*d*) Determine the heat transfer rate by the relaxation process.

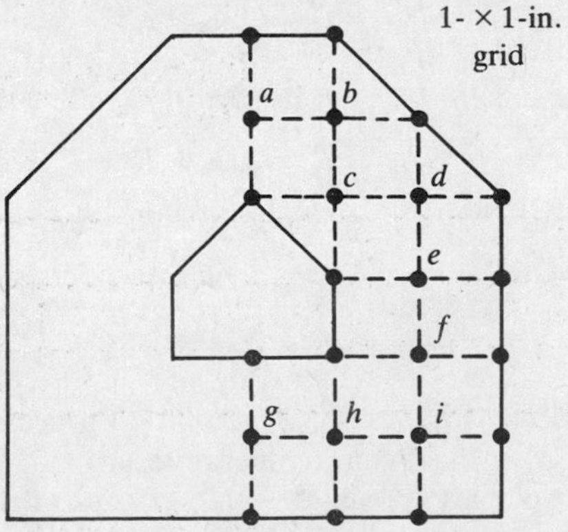

Fig. 2-30

▮ (*a*) From Table B-1, $k = 12.4$ Btu/h · ft · °F. By Problem 2.17 and Fig. 2-31,

$$\frac{S}{L} = 2\left(\frac{N}{M}\right) = 2\left(\frac{7}{2.6}\right) = 5.38$$

Hence $\dfrac{q}{L} \approx \dfrac{S}{L} k(T_1 - T_2) = (5.38)(12.4\text{ Btu/h}\cdot\text{ft}\cdot{}^\circ\text{F})[(2000 - 70)\ {}^\circ\text{F}] = 128\,750\text{ Btu/h}\cdot\text{ft}$

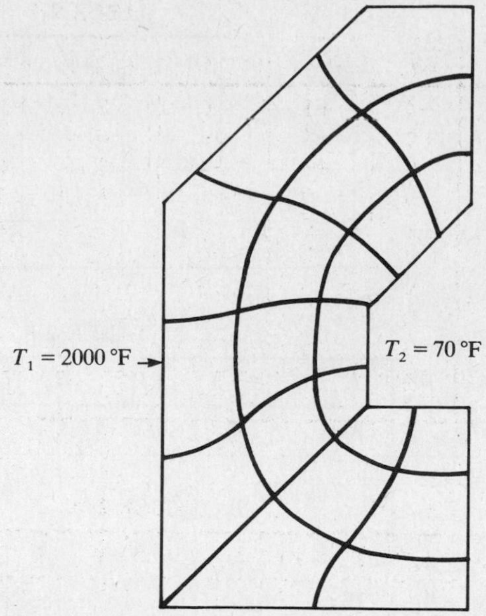

Fig. 2-31

(*b*) $$R_e = T_1 + T_2 + T_d + T_f - 4T_e = 2000 + 70 + 1700 + 1300 - 4(1400) = -530$$

(*c*) $$\frac{q}{L} = 2k \sum \frac{(\text{heat-flow path width})}{(\text{heat-flow path length})} (\Delta T)$$

$$= 2k\left[\frac{2000 - T_a}{2} + \frac{2000 - T_b}{2} + \frac{2000 - T_b}{2(\frac{3}{4})} + (2000 - T_b) + (2000 - T_d) + \frac{2000 - T_d}{2(\frac{3}{4})}\right.$$
$$\left. + \frac{2000 - T_d}{2} + (2000 - T_e) + (2000 - T_f) + 2(2000 - T_i) + (2000 - T_h) + \frac{2000 - T_g}{2}\right]$$

(*d*) For the residuals,

$$R_a = 2000 + 70 + 2T_b - 4T_a$$
$$R_b = 2000 + 2000 + T_c + T_a - 4T_b$$
$$R_c = 70 + 70 + T_b + T_d - 4T_c$$
$$R_d = 2000 + 2000 + T_e + T_c - 4T_d$$
$$R_e = 2000 + 70 + T_d + T_f - 4T_e$$
$$R_f = 2000 + 70 + T_e + T_i - 4T_f$$
$$R_g = 2000 + 70 + 2T_h - 4T_g$$
$$R_h = 2000 + 70 + T_g + T_i - 4T_h$$
$$R_i = 2000 + 2000 + T_h + T_f - 4T_i$$

Table 2-9 gives the unit-increase responses. A relaxation process is shown in Table 2-10. In the solution, the maximum $|R|$ is 10, giving a maximum temperature error of 2.5 °F. Substitution of the nodal temperatures in the formula of part (*c*) yields $q/L = 152\,023$ Btu/h · ft.

2.82 Addition of the nodal equations in Problem 2.76 or Problem 2.80 yields the result that the sum of the four nodal temperatures equals the sum of the four boundary temperatures. Is this result only approximate, like the nodal equations themselves?

▌ No; the result is exact for any four points that are symmetrically disposed about the center of the square. The proof is similar to that of Problem 2.7: One now superposes the original problem and the three derived problems obtained by rotating the set of boundary temperatures through 90°, 180°, and 270°.

TABLE 2-9

	ΔR_a	ΔR_b	ΔR_c	ΔR_d	ΔR_e	ΔR_f	ΔR_g	ΔR_h	ΔR_i
$\Delta T_a = +1$	−4	+1	0	0	0	0	0	0	0
$\Delta T_b = +1$	+2	−4	+1	0	0	0	0	0	0
$\Delta T_c = +1$	0	+1	−4	+1	0	0	0	0	0
$\Delta T_d = +1$	0	0	+1	−4	+1	0	0	0	0
$\Delta T_e = +1$	0	0	0	+1	−4	+1	0	0	0
$\Delta T_f = +1$	0	0	0	0	+1	−4	0	0	+1
$\Delta T_g = +1$	0	0	0	0	0	0	−4	+1	0
$\Delta T_h = +1$	0	0	0	0	0	0	+2	−4	+1
$\Delta T_i = +1$	0	0	0	0	0	+1	0	+1	−4
Block = +1	−2	−2	−2	−2	−2	−2	−2	−2	−2

TABLE 2-10

	T_a	R_a	T_b	R_b	T_c	R_c	T_d	R_d	T_e	R_e	T_f	R_f	T_g	R_g	T_h	R_h	T_i	R_i
Assume temp. and calculate *R*'s	1000	1070	1500	−200	800	−60	1500	−200	1000	570	1000	270	1000	70	1000	270	1200	1200
$\Delta T_i = +400$												670				670	1600	−400
$\Delta T_a = +300$	1300	−130		100														
$\Delta T_h = +200$														470	1200	−130		−200
$\Delta T_f = +200$										770	1200	−130						0
$\Delta T_e = +200$								0	1200	−30		70						
$\Delta T_g = +125$													1125	−30		−5		
$\Delta T_a = -50$	1250	70		50														
$\Delta T_c = -25$				25	775	40		−25										
$\Delta T_a = +25$	1275	−30		50														
$\Delta T_f = +15$										−15	1215	10						15
$\Delta T_b = +10$		−10	1510	10		50												
$\Delta T_c = +10$				20	785	10		−15										
$\Delta T_g = -10$													1115	10		−15		
$\Delta T_b = +5$		0	1515	0		15												
Check		√		√		√		√		√		√		√		√		√
residuals	1275	0	1515	0	785	15	1500	−15	1200	−15	1215	−10	1115	10	1200	−15	1600	15
$\Delta T_i = +5$												15				−10	1605	−5
$\Delta T_e = -5$								−20	1195	5		10						
$\Delta T_d = -5$						10	1495	0		0								
Check		√		√		√		√		√		√		√		√		√
residuals	1275	0	1515	0	785	10	1495	0	1195	0	1215	0	1115	10	1200	−10	1605	−5

2.83 Repeat Problem 2.72 using a 3-in. square grid of nodal points (Fig. 2-32).

▮ The nodal equations are

node *a*	$2T_b + 2T_f - 4T_a = 0$
node *b*	$300 + T_c + T_g + T_a - 4T_b = 0$
node *c*	$300 + T_d + T_h + T_b - 4T_c = 0$
node *d*	$300 + 2T_c + T_i - 4T_d = 0$
node *e*	$2T_f + 2T_k - 4T_e = 0$
node *f*	$T_a + T_g + T_l + T_e - 4T_f = 0$
node *g*	$T_b + T_h + T_m + T_f - 4T_g = 0$
node *h*	$T_c + T_i + T_n + T_g - 4T_h = 0$
node *i*	$T_d + 2T_h + T_o - 4T_i = 0$
node *j*	$2T_k + 2(100) - 4T_i = 0$

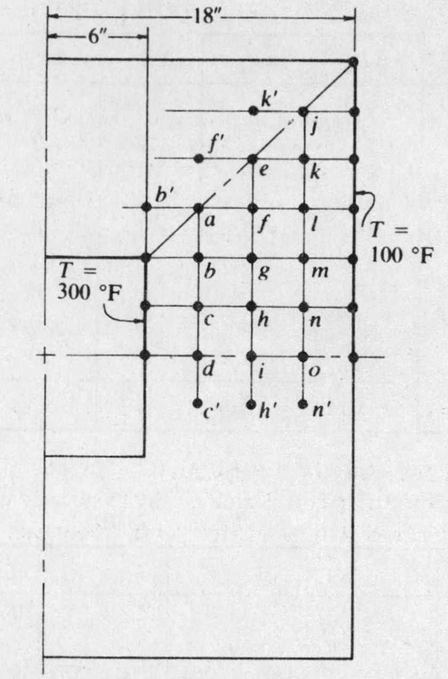

Fig. 2-32

node k	$T_e + T_l + 100 + T_j - 4T_k = 0$
node l	$T_f + T_m + 100 + T_k - 4T_l = 0$
node m	$T_g + T_n + 100 + T_l - 4T_m = 0$
node n	$T_h + T_o + 100 + T_m - 4T_n = 0$
node o	$T_i + 2T_n + 100 - 4T_o = 0$

which are displayed in matrix form as Table 2-11. The computer program of Problem 2.78 (suitably expanded in DIMENSION and FORMAT) yields the solution (where units are °F):

$T_a = 194.41$	$T_e = 140.34$	$T_i = 190.34$	$T_m = 137.58$
$T_b = 228.11$	$T_f = 160.71$	$T_j = 109.99$	$T_n = 142.23$
$T_c = 239.53$	$T_g = 178.51$	$T_k = 119.97$	$T_o = 143.70$
$T_d = 242.35$	$T_h = 187.65$	$T_l = 129.56$	

TABLE 2-11

a	*b*	*c*	*d*	*e*	*f*	*g*	*h*	*i*	*j*	*k*	*l*	*m*	*n*	*o*			
−4	2				2										T_a		0
1	−4	1				1									T_b		−300
	1	−4	1				1								T_c		−300
		2	−4					1							T_d		−300
				−4	2					2					T_e		0
1				1	−4	1					1				T_f		0
	1				1	−4	1					1			T_g		0
		1				1	−4	1					1		T_h	=	0
			1				2	−4						1	T_i		0
									−4	2					T_j		−200
				1					1	−4	1				T_k		−100
					1					1	−4	1			T_l		−100
						1					1	−4	1		T_m		−100
							1					1	−4	−1	T_n		−100
								1					2	−4	T_o		−100

2.84 Write a nodal equation for the boundary node n of Fig. 2-33, if convective heat transfer occurs at the boundary.

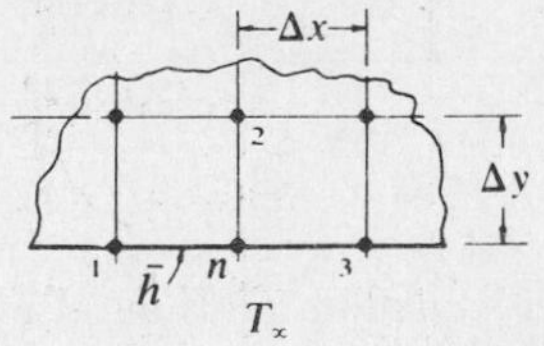

Fig. 2-33

▮ A steady-state energy balance gives

$$kL\left(\frac{\Delta y}{2}\right)\left(\frac{T_1 - T_n}{\Delta x}\right) + kL(\Delta x)\left(\frac{T_2 - T_n}{\Delta y}\right) + kL\left(\frac{\Delta y}{2}\right)\left(\frac{T_3 - T_n}{\Delta y}\right) + \bar{h}L(\Delta x)(T_\infty - T_n) = 0 \tag{1}$$

where L is the thickness of the body in the z-direction. Note that the effective horizontal conductance between nodes 1 and n, or that between 3 and n, involves only one-half the area that is associated with a horizontal conductance between a pair of adjacent interior nodal points. For a square grid with spacing δ, (*1*) simplifies to

$$\frac{1}{2}(T_1 + 2T_2 + T_3) + \frac{\bar{h}\delta}{k}(T_\infty) - \left(\frac{\bar{h}\delta}{k} + 2\right)T_n = 0 \tag{2}$$

2.85 Derive the temperature equation at a boundary corner node with one adjacent side insulated and one adjacent side subject to convective heat transfer, as shown in Fig. 2-34.

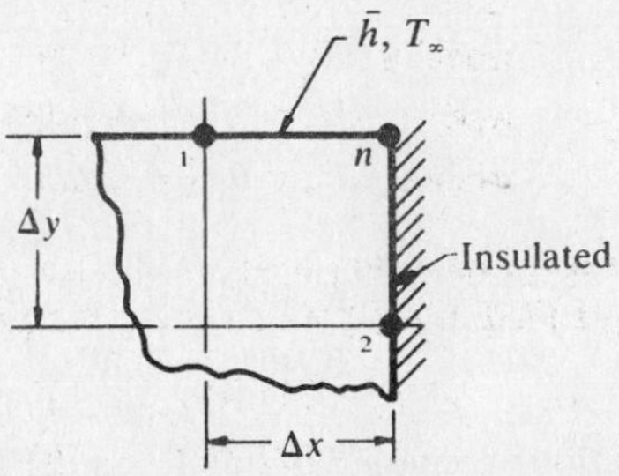

Fig. 2-34

▮ The rates of energy conducted between nodes 1 and n and between nodes 2 and n are, respectively,

$$kL\left(\frac{\Delta y}{2}\right)\left(\frac{T_1 - T_n}{\Delta x}\right) \quad \text{and} \quad kL\left(\frac{\Delta x}{2}\right)\left(\frac{T_2 - T_n}{\Delta y}\right)$$

where L is the depth perpendicular to the xy-plane. The rate of energy convected is

$$\bar{h}L\left(\frac{\Delta x}{2}\right)(T_\infty - T_n)$$

For steady state, the summation of energy transfer rates into node n must be zero, and thus

$$kL\left(\frac{\Delta y}{2}\right)\left(\frac{T_1 - T_n}{\Delta x}\right) + kL\left(\frac{\Delta x}{2}\right)\left(\frac{T_2 - T_n}{\Delta y}\right) + \bar{h}L\left(\frac{\Delta x}{2}\right)(T_\infty - T_n) = 0 \tag{1}$$

For a square grid $(\Delta x = \Delta y = \delta)$, this simplifies to

$$T_1 + T_2 + \frac{\bar{h}\delta}{k}(T_\infty) - \left(\frac{\bar{h}\delta}{k} + 2\right)T_n = 0 \tag{2}$$

2.86 Derive the temperature equation at a boundary corner node where both sides are insulated.

▮ Simply let $\bar{h} \to 0$ in the results of Problem 2.85. Note that (*2*) of Problem 2.85 becomes independent of the mesh size δ.

2.87 A cylindrical pin fin is attached to a 300 °F wall while its surface is exposed to a gas at 100 °F. The convective heat transfer coefficient is 20 Btu/h · ft² · °F. The fin is made of stainless steel with a thermal conductivity of 10 Btu/h · ft · °F. Set up the residual equations in $\theta = T - T_\infty$ (see Fig. 2-35).

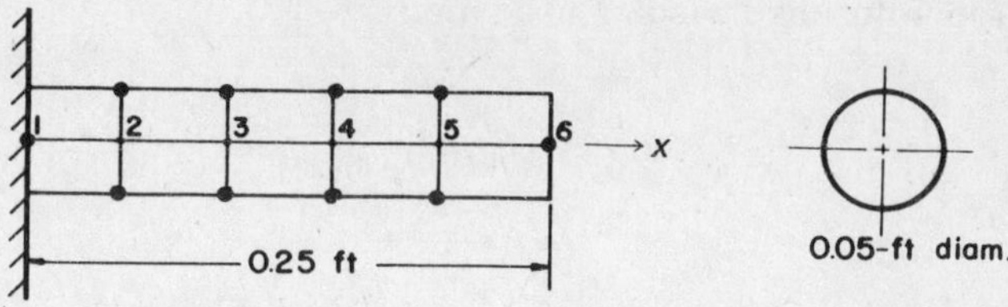

Fig. 2-35

▌ The same form of heat-balance equation holds at all interior nodes; e.g., at node *2*

$$0 = q_{1\text{-}2} + q_{3\text{-}2} + q_{\infty\text{-}2} = k\left(\frac{\pi D^2}{4}\right)\left(\frac{T_1 - T_2}{\Delta x}\right) + k\left(\frac{\pi D^2}{4}\right)\left(\frac{T_3 - T_2}{\Delta x}\right) + \bar{h}(\pi D\,\Delta x)(T_\infty - T_2)$$

$$R_2 = \theta_1 + \theta_3 - 2.4\theta_2$$

At the base node *1*,

$$R_1 = \frac{q_1}{k\pi D^2/(4\,\Delta x)} = \frac{4\,\Delta x}{k\pi D^2}\,q_b + \theta_2 - \left[1 + \frac{2\bar{h}(\Delta x)^2}{kD}\right]\theta_1$$

$$= 2.54q_b + \theta_2 - 240$$

and, at the tip node *6*,

$$R_6 = \frac{q_6}{k\pi D^2/(4\,\Delta x)} = \theta_5 - \theta_6 + \left[\frac{h(\Delta x)}{k} + \frac{2\bar{h}(\Delta x)^2}{kD}\right](-\theta_6)$$

$$= \theta_5 - 1.3\theta_6$$

The complete set of residual equations is

$$R_1 = 2.54q_b + \theta_2 - 240$$
$$R_2 = 200 + \theta_3 - 2.4\theta_2$$
$$R_3 = \theta_2 + \theta_4 - 2.4\theta_3$$
$$R_4 = \theta_3 + \theta_5 - 2.4\theta_4$$
$$R_5 = \theta_6 + \theta_4 - 2.4\theta_5$$
$$R_6 = \theta_5 - 1.3\theta_6$$

2.88 Consider a $\frac{1}{4}$-in.-thick (y-direction) rectangular stainless-steel fin ($k = 8$ Btu/h · ft · °F) which is 1.0 in. long in the x-direction and very wide in the z-direction (Fig. 2-36). The external convective heat transfer coefficient is $\bar{h} = 96$ Btu/h · ft^2 · °F; the surrounding fluid temperature is $T_\infty = 80$ °F; the fin base temperature is $T_b = 200$ °F; and the end of the fin is insulated. Using the $\frac{1}{8}$-in. square grid shown, write the nodal equations for T_1 through T_{16}.

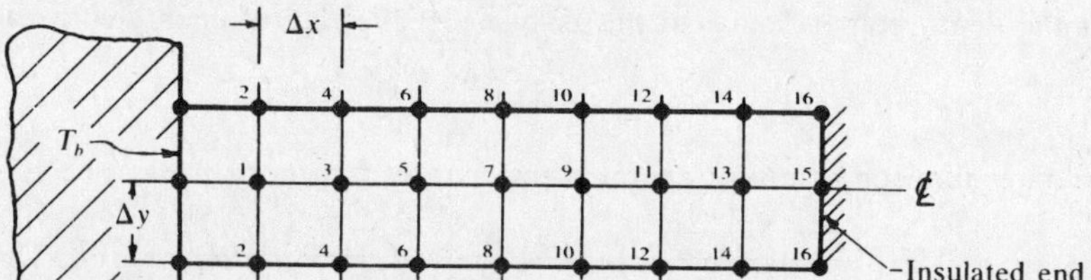

Fig. 2-36

▌ Due to symmetry about the horizontal centerline, there are only 16 different nodal conditions. We treat the nodes by type.

Interior nodes

node *1*	$200 + 2T_2 + T_3 - 4T_1 = 0$
node *3*	$T_1 + 2T_4 + T_5 - 4T_3 = 0$
node *5*	$T_3 + 2T_6 + T_7 - 4T_5 = 0$
node *7*	$T_5 + 2T_8 + T_9 - 4T_7 = 0$
node *9*	$T_7 + 2T_{10} + T_{11} - 4T_9 = 0$
node *11*	$T_9 + 2T_{12} + T_{13} - 4T_{11} = 0$
node *13*	$T_{11} + 2T_{14} + T_{15} - 4T_{13} = 0$

Nodes on a convective boundary

At nodes 2,

$$\frac{1}{2}(200 + 2T_1 + T_4) + \frac{\bar{h}\,\Delta x}{k}(80) - \left(\frac{\bar{h}\,\Delta x}{k} + 2\right)T_2 = 0$$

where

$$\frac{\bar{h}\,\Delta x}{k} = \frac{(96\ \text{Btu/h}\cdot\text{ft}^2\cdot{}^\circ\text{F})[(0.125/12)\ \text{ft}]}{8\ \text{Btu/h}\cdot\text{ft}\cdot{}^\circ\text{F}} = 0.125$$

Thus:

nodes *2*	$220 + 2T_1 + T_4 - 4.25T_2 = 0$
nodes *4*	$T_2 + 2T_3 + T_6 + 20 - 4.25T_4 = 0$
nodes *6*	$T_4 + 2T_5 + T_8 + 20 - 4.25T_6 = 0$
nodes *8*	$T_6 + 2T_7 + T_{10} + 20 - 4.25T_8 = 8$
nodes *10*	$T_8 + 2T_9 + T_{12} + 20 - 4.25T_{10} = 0$
nodes *12*	$T_{10} + 2T_{11} + T_{14} + 20 - 4.25T_{12} = 0$
nodes *14*	$T_{12} + 2T_{13} + T_{16} + 20 - 4.25T_{14} = 0$

Node on an insulated boundary

At node 15, $\frac{1}{2}(T_{16} + T_{16}) + T_{13} - 2T_{15} = 0$, or

node *15* $\quad T_{16} + T_{13} - 2T_{15} = 0$

Corner nodes

By (*2*) of Problem 2.85,

nodes *16* $\quad T_{14} + T_{15} + 10 - 2.125T_{16} = 0$

2.89 For the two-dimensional conduction problem of Fig. 2-37, determine the steady-state nodal temperatures T_1 through T_6 using the relaxation technique. Consider the answers to be satisfactorily accurate when all residuals are equal to or less than 1.0 in absolute value.

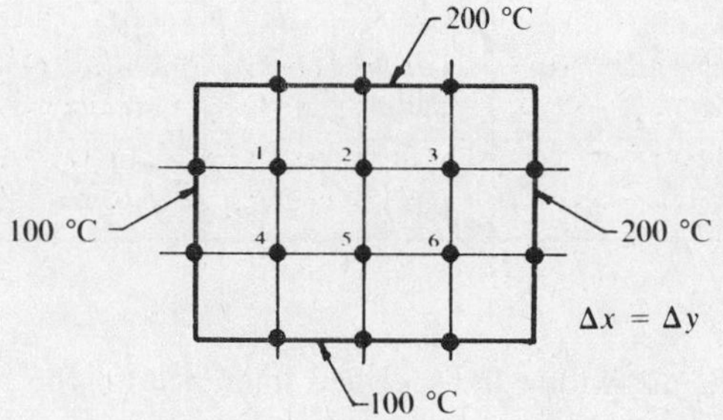

Fig. 2-37

▌ The nodal equations are, in residual form:

node *1*	$300 + T_2 + T_4 - 4T_1 = R_1$
node *2*	$200 + T_1 + T_3 + T_5 - 4T_2 = R_2$
node *3*	$400 + T_2 + T_6 - 4T_3 = R_3$
node *4*	$200 + T_1 + T_5 - 4T_4 = R_4$
node *5*	$100 + T_2 + T_4 + T_6 - 4T_5 = R_5$
node *6*	$300 + T_3 + T_5 - 4T_6 = R_6$

Table 2-12 shows the effects of unit temperature changes. Setting up the relaxation table, Table 2-13, we make an initial guess for each nodal temperature. Using the six initial temperatures (guesses) and the set of six nodal equations, we calculate the initial residuals and record these in the relaxation table. The work proceeds by relaxing, in turn, the residual of largest absolute value. Since all initial residuals were of the same sign, a block 10° reduction in all temperatures was the first step. (The block change is usually advantageous whenever all residuals are of like sign.) Following this, the residual of largest absolute value, R_4, was relaxed by reducing T_4.

TABLE 2-12

	ΔR_1	ΔR_2	ΔR_3	ΔR_4	ΔR_5	ΔR_6
$\Delta T_1 = +1$	−4	1	0	1	0	0
$\Delta T_2 = +1$	1	−4	1	0	1	0
$\Delta T_3 = +1$	0	1	−4	0	0	1
$\Delta T_4 = +1$	1	0	0	−4	1	0
$\Delta T_5 = +1$	0	1	0	1	−4	1
$\Delta T_6 = +1$	0	0	1	0	1	−4
Block = +1	−2	−1	−2	−2	−1	−2

TABLE 2-13

	T_1	R_1	T_2	R_2	T_3	R_3	T_4	R_4	T_5	R_5	T_6	R_6
Initial guesses	160	−20	180	−20	190	−15	140	−50	150	−15	165	−20
Block = −10	$\underline{150}$	0	$\underline{170}$	−10	$\underline{180}$	+5	$\underline{130}$	−30	$\underline{140}$	−5	$\underline{155}$	0
$\Delta T_4 = -8$	150	−8	170	−10	180	+5	$\underline{122}$	+2	140	−13	155	0
$\Delta T_5 = -3$	150	−8	170	−13	180	+5	122	−1	$\underline{137}$	−1	155	−3
$\Delta T_2 = -3$	150	−11	$\underline{167}$	−1	180	+2	122	−1	137	−4	155	−3
$\Delta T_1 = -3$	$\underline{147}$	+1	167	−4	180	+2	122	−4	137	−4	155	−3
Check by equations	147	√ +1	167	√ −4	180	√ +2	122	√ −4	137	√ −4	155	√ −3
$\Delta T_2 = -1$	147	0	$\underline{166}$	0	180	+1	122	−4	137	−5	155	−3
$\Delta T_5 = -1$	147	0	166	−1	180	+1	122	−5	$\underline{136}$	−1	155	−4
$\Delta T_4 = -1$	147	−1	166	−1	180	+1	$\underline{121}$	−1	136	−2	155	−4
$\Delta T_6 = -1$	147	−1	166	−1	180	0	121	−1	136	−3	$\underline{154}$	0
$\Delta T_5 = -1$	147	−1	166	−2	180	0	121	−2	$\underline{135}$	+1	154	−1
$\Delta T_2 = -\frac{1}{2}$	147	$-1\frac{1}{2}$	$\underline{165\frac{1}{2}}$	0	180	$-\frac{1}{2}$	121	−2	135	$+\frac{1}{2}$	154	−1
$\Delta T_4 = -\frac{1}{2}$	147	−2	$165\frac{1}{2}$	0	180	$-\frac{1}{2}$	$\underline{120\frac{1}{2}}$	0	135	0	154	−1
$\Delta T_1 = -\frac{1}{2}$	$\underline{146\frac{1}{2}}$	0	$165\frac{1}{2}$	$-\frac{1}{2}$	180	$-\frac{1}{2}$	$120\frac{1}{2}$	$-\frac{1}{2}$	135	0	154	−1
Check by equations	$146\frac{1}{2}$	√ 0	$165\frac{1}{2}$	√ $-\frac{1}{2}$	180	√ $-\frac{1}{2}$	$120\frac{1}{2}$	√ $-\frac{1}{2}$	135	√ 0	154	√ −1

Continuing, all residuals are within the specified tolerance for the bottom set of temperatures in the relaxation table. The work could have been reduced somewhat by overrelaxing. Notice that T_5, for example, was reduced three times. By reducing it more than apparently needed the first time, the work could have been shortened. Excessive overrelaxation, however, could increase the required effort.

2.90 Estimate the heat transfer rate from the horizontal 100 °C surface in Fig. 2-38. Use the nodal temperatures determined in Problem 2.89. The material is magnesite and the grid size is $\Delta x = \Delta y = 15$ cm.

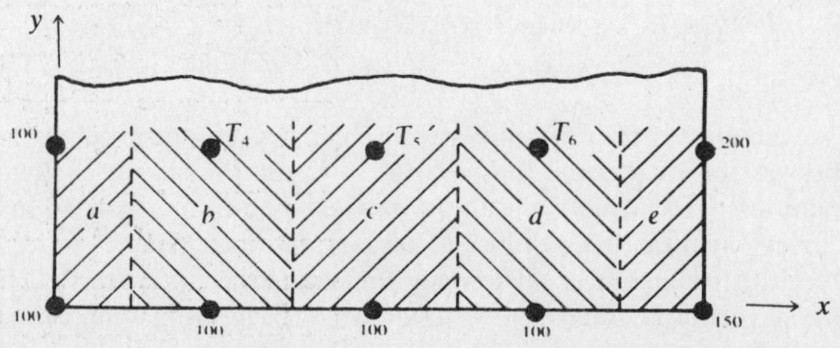

Fig. 2-38

From Appendix B, the thermal conductivity of magnesite at 205 °C (nearest listed temperature to conditions of this problem) is $k = (2.2)(1.7296) = 3.81\ \text{W/m}\cdot\text{K}$. Dividing the body into heat-flow lanes (Fig. 2-38) and assuming the lower right-hand corner to be at the average of the two adjacent surface temperatures, we proceed to approximate the heat transfer rate through each lane by one-dimensional methods. Thus, for depth L perpendicular to the plane of the figure:

lane a $$\left(\frac{q}{L}\right)_a \approx k\left(\frac{A_a/L}{\Delta y}\right)(\Delta T) = (3.81\ \text{W/m}\cdot\text{K})(\tfrac{1}{2})[(100-100)\ \text{K}] = 0\ \text{W/m}$$

lane b $$\left(\frac{q}{L}\right)_b \approx (3.81)(1)(120.5-100) = 78.11\ \text{W/m}$$

lane c $$\left(\frac{q}{L}\right)_c \approx (3.81)(1)(135-100) = 133.35\ \text{W/m}$$

lane d $$\left(\frac{q}{L}\right)_d \approx (3.81)(1)(154-100) = 205.74\ \text{W/m}$$

lane e $$\left(\frac{q}{L}\right)_e \approx (3.81)(\tfrac{1}{2})(200-100) = 190.5\ \text{W/m}$$

for a total $q/L \approx 608\ \text{W/m}$.

2.91 Let T_0 be the temperature at the center of Fig. 2-25. Show that the relation

$$T_1 + T_2 + T_3 + T_4 - 4T_0 = 0$$

is exact.

On the one hand, by Problem 2.82,

$$T_1 + T_2 + T_3 + T_4 = 500 + 200 + 300 + 400 \qquad \text{(exactly)}$$

On the other hand, by Problem 2.7,

$$T_0 = \frac{500 + 200 + 300 + 400}{4} \qquad \text{(exactly)}$$

2.92 A 12-in.-OD steam pipe at 272 °F is buried 5 ft below the 32 °F earth's surface $(k = 0.2\ \text{Btu/h}\cdot\text{ft}\cdot{}^\circ\text{F})$. Determine the heat transfer rate by flux plotting. Compare with the value obtained by use of the conductive shape factor.

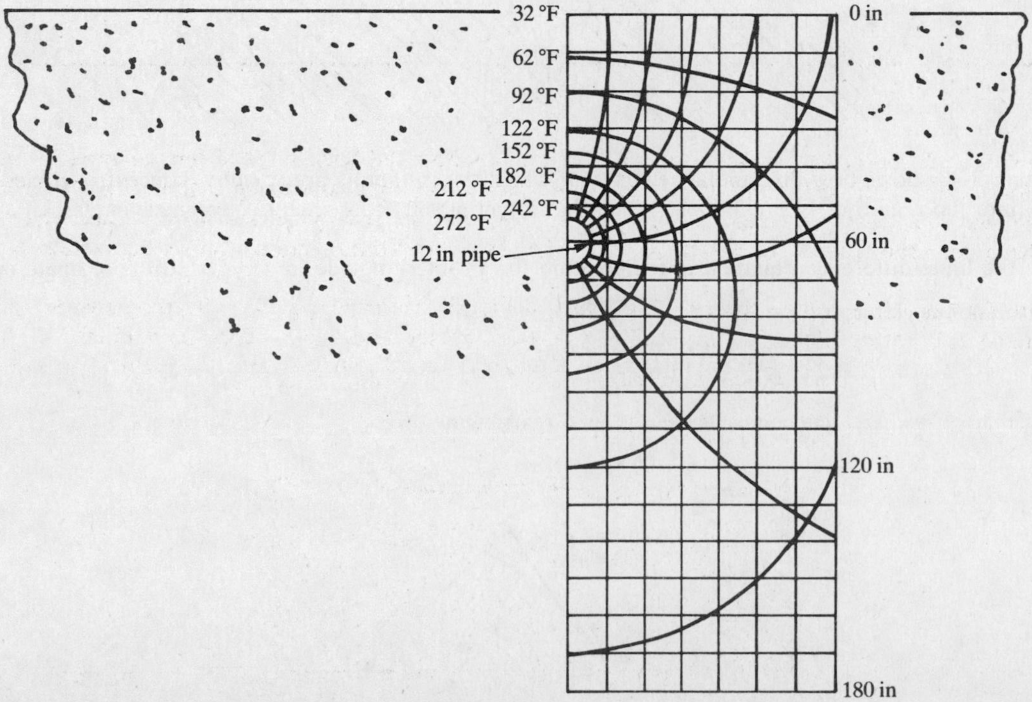

Fig. 2-39

▌ Dividing the overall temperature difference, $272 - 32 = 240$ °F, into eight equal increments of 30 °F each, the isotherms are easy to sketch *since they are circles* (see Problem 2.93). The plot for one half of the symmetrical configuration is shown in Fig. 2-39.

$$\frac{S}{L} = \frac{N}{M} = \frac{18}{8} = 2.25$$

Hence,

$$\frac{q}{L} = \frac{S}{L}k(\Delta T) = (2.25)(0.2 \text{ Btu/h}\cdot\text{ft}\cdot{}^\circ\text{F})[(272-32)\ {}^\circ\text{F}] = 108.0 \text{ Btu/h}\cdot\text{ft}$$

On the other hand, from Appendix C,

$$\frac{S}{L} = \frac{2\pi}{\ln(2D/r)} = \frac{2\pi}{\ln[2(60)/6]} = 2.10$$

giving

$$\frac{q}{L} = \frac{S}{L}k(\Delta T) = (2.10)(0.2)(240) = 100.8 \text{ Btu/h}\cdot\text{ft}$$

The flux plotting answer is 6.7 percent higher than that from the shape factor.

2.93 Use circular inversion (Problem 2.68) to confirm that the isotherms in Fig. 2-39 are a family of eccentric circles.

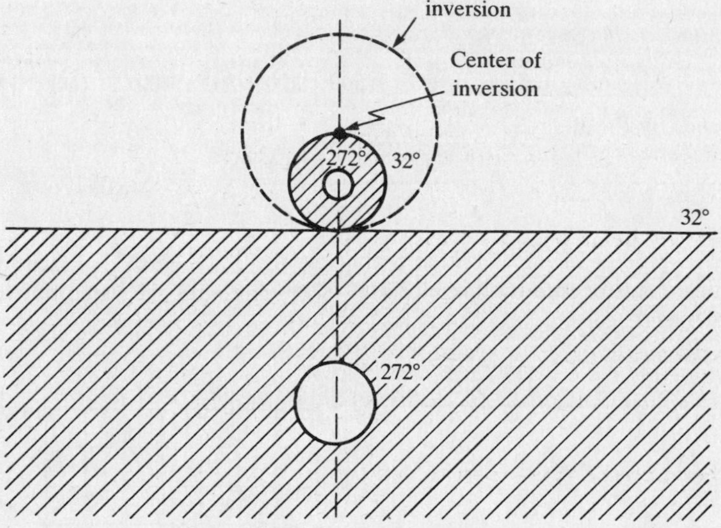

Fig. 2-40

▌ Figure 2-40 shows how an annular region—for which the isotherms are, plainly, concentric circles—inverts into the region of Fig. 2-39—for which the isotherms are, consequently, eccentric circles.

2.94 Derive the finite-difference equation in temperature for an interior node in a cylindrical coordinate system.

▌ In the nomenclature of Fig. 2-41, $q_a + q_b = 0$, or

$$k2\pi\left(r - \frac{\Delta r}{2}\right)L\left(\frac{T_{n-1} - T_n}{\Delta r}\right) + k2\pi\left(r + \frac{\Delta r}{2}\right)L\left(\frac{T_{n+1} - T_n}{\Delta r}\right) = 0$$

Noting that $r = n\,\Delta r$, canceling like terms, and rearranging gives

$$(n - \tfrac{1}{2})T_{n-1} + (n + \tfrac{1}{2})T_{n+1} - 2nT_n = 0$$

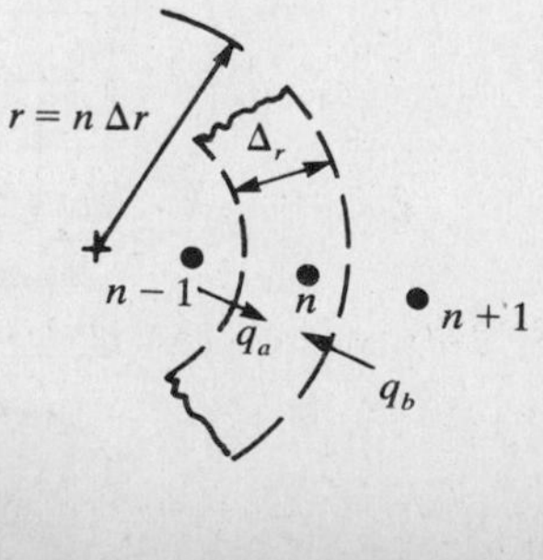

Fig. 2-41

CHAPTER 3
Nonsteady Conduction

3.1 BIOT AND FOURIER MODULI; LUMPED SYSTEMS

3.1 The cylinder of Fig. 3-1 is being cooled by a fluid stream: $T_\infty < T_s < T_c$. Evaluate its *Biot number* (or *modulus*),

$$\mathbf{Bi} \equiv \frac{\text{internal thermal resistance}}{\text{external thermal resistance}} = \frac{\text{internal temperature drop}}{\text{external temperature drop}}$$

in terms of the physical parameters.

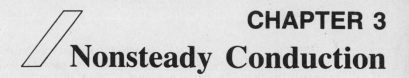

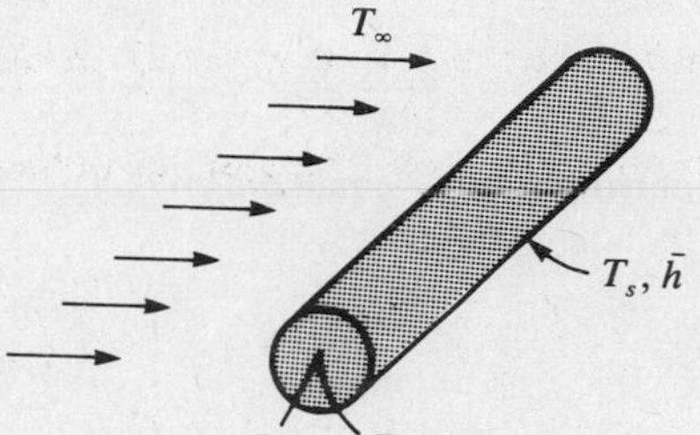

Fig. 3-1

Newton's law of cooling gives the external heat transfer as

$$q_e = \bar{h}A_s(T_s - T_\infty)$$

The internal heat transfer is given by Fourier's law as

$$q_i = -kA_s \left.\frac{dT}{dr}\right|_{r=R} \approx -kA_s \frac{T_s - T_c}{R}$$

The first-law requirement $q_e = q_i$ gives

$$\mathbf{Bi} = \frac{T_c - T_s}{T_s - T_\infty} = \frac{\bar{h}R}{k} \qquad (1)$$

The generalization of (*1*) to an arbitrary body is

$$\mathbf{Bi} = \frac{\bar{h}L}{k} \qquad (2)$$

where L is a characteristic length, usually obtained by dividing the volume of the body, V, by its surface area, A_s. As a rule of thumb, a value $\mathbf{Bi} < 0.1$ indicates negligible internal resistance and therefore the suitability of a lumped-thermal-capacity analysis.

3.2 Find the Biot modulus for a rectangular parallepiped steel with x, y, and z dimensions of 1, 2, and 3 in., respectively, in an annealing operation where $\bar{h} = 10\ \text{Btu/h}\cdot\text{ft}^2\cdot\text{F}$ and $k = 25\ \text{Btu/h}\cdot\text{ft}\cdot{}^\circ\text{F}$.

First determine L:

$$\text{Volume} = 1(2)(3) = 6\ \text{in}^3 \qquad \text{Surface area} = 2[1(2)] + 2[1(3)] + 2[2(3)] = 4 + 6 + 12 = 22\ \text{in}^2$$

whence

$$L = \frac{V}{A_s} = \frac{6}{22}\ \text{in.} = 0.0227\ \text{ft}$$

and

$$\mathbf{Bi} = \frac{\bar{h}L}{k} = \frac{(10)(0.0227)}{25} = 0.009\,08$$

Obviously this problem can be treated by a lumped analysis.

3.3 For a lumped system (small Biot number) determine the relationship between time and body temperature, using the nomenclature of Fig. 3-2. The density and specific heat capacity of the body are ρ and c, and its volume is V.

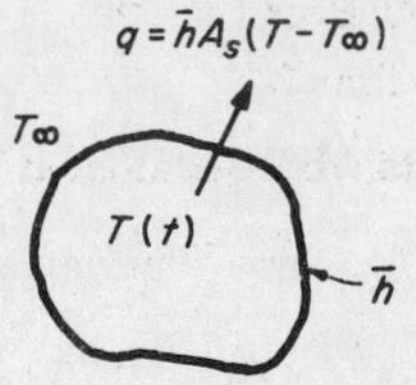

Fig. 3-2

▮ (Heat flow out of body during dt) = −(change of internal energy of body during dt)

$$\bar{h}A_s(T - T_\infty)\,dt = -\rho c V\,dT$$

$$\frac{\bar{h}A_s}{\rho c V}\,dt = -\frac{dT}{T - T_\infty}$$

which is an ordinary first-order differential equation in $T(t)$. To obtain the solution we impose the initial condition $T = T_i$ at $t = 0$; the solution is

$$\frac{T - T_\infty}{T_i - T_\infty} = \exp\left[-\left(\frac{\bar{h}A_s}{\rho c V}\right)t\right] \tag{1}$$

Relation (*1*) is valid both for cooling $(T_i > T_\infty)$ and for heating $(T_i < T_\infty)$; in the latter case, it is more revealing to write the left-hand side as $(T_\infty - T)/(T_\infty - T_i)$.

3.4 In electric capacitor-discharge the voltage decays according to

$$\frac{E}{E_i} = \exp\left[\frac{-t}{(RC)_{\text{elec}}}\right]$$

where the product $(RC)_{\text{elec}}$ is the *time constant* and has units of time. Put (*1*) of Problem 3.3 in analogous form.

▮ For the thermal problem, the thermal capacitance is $C_{\text{therm}} \equiv \rho c V$ and the thermal resistance is $R_{\text{therm}} = 1/\bar{h}A_s$. Thus, the reduced temperature $\theta \equiv T - T_\infty$ varies with time according to

$$\frac{\theta}{\theta_i} = \exp\left[\frac{-t}{(RC)_{\text{therm}}}\right] \tag{1}$$

3.5 Rewrite (*1*) of Problem 3.4 in completely dimensionless form.

▮ Introduce the dimensionless time, or *Fourier modulus*,

$$\mathbf{Fo} \equiv \frac{\alpha t}{L^2} \tag{1}$$

where L is the volume divided by the surface area (as defined for the Biot modulus) and $\alpha \equiv k/\rho c$ is the thermal diffusivity of the body [see (*9*) of Problem 1.54]; also introduce the dimensionless temperature $\Theta \equiv \theta/\theta_i$. Then (*1*) of Problem 3.4 becomes

$$\Theta = \exp(-\mathbf{Bi}\cdot\mathbf{Fo}) \tag{2}$$

3.6 Determine the time constant for a spherical copper-constantan thermocouple (Fig. 3-3) at an average temperature of 32 °F, exposed to a convective environment where $\bar{h} = 8$ Btu/h · ft² · °F, if the bead diameter is (*a*) 0.005 in. and (*b*) 0.010 in.

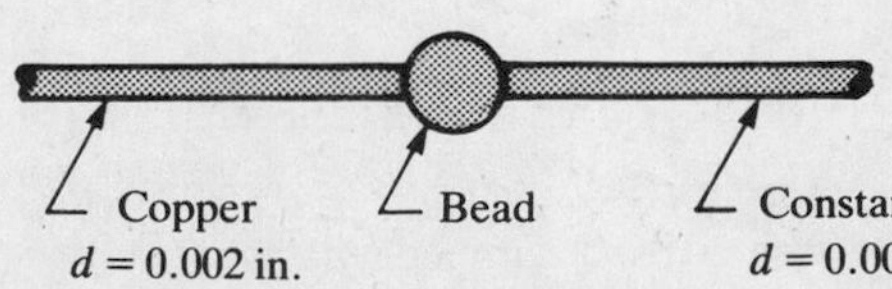

Fig. 3-3

▌ For a spherical object of radius a, $L = V/A_s = a/3$. From Appendix B the properties of the two metals are

$$k_{cu} = 224\ \text{Btu/h}\cdot\text{ft}\cdot{}^\circ\text{F} \qquad k_{con} = 12.4\ \text{Btu/h}\cdot\text{ft}\cdot{}^\circ\text{F}$$
$$c_{cu} = 0.091\ \text{Btu/lbm}\cdot{}^\circ\text{F} \qquad c_{con} = 0.10\ \text{Btu/lbm}\cdot{}^\circ\text{F}$$
$$\rho_{cu} = 558\ \text{lbm/ft}^3 \qquad \rho_{con} = 557\ \text{lbm/ft}^3$$

Assuming linear averaging to be valid, the thermocouple bead properties are

$$k = \frac{224 + 12.4}{2} = 118.2\ \text{Btu/h}\cdot\text{ft}\cdot{}^\circ\text{F}$$
$$c = \frac{0.091 + 0.10}{2} = 0.096\ \text{Btu/lbm}\cdot{}^\circ\text{F}$$
$$\rho = \frac{558 + 557}{2} = 557.5\ \text{lbm/ft}^3$$

(*a*) From Problem 3.4,

$$(RC)_{therm} = \frac{\rho c L}{\bar{h}} = \frac{\rho c a}{3\bar{h}} = \frac{(557.5\ \text{lbm/ft}^3)(0.096\ \text{Btu/lbm}\cdot{}^\circ\text{F})[(0.0025/12)\ \text{ft}]}{3(8\ \text{Btu/h}\cdot\text{ft}^2\cdot{}^\circ\text{F})}$$
$$= 4.646 \times 10^{-4}\ \text{h} = 1.6725\ \text{s}$$

(*b*) The only difference from part (*a*) is that the radius is twice the previous value. Since $(RC)_{therm}$ is linear in a, $(RC)_{therm} = 2(1.6725\ \text{s}) = 3.345\ \text{s}$.

Before leaving this problem, we should verify that the Biot modulus is sufficiently small to insure the validity of a lumped analysis. Checking for the larger thermocouple,

$$\textbf{Bi} = \frac{\bar{h}L}{k} = \frac{(8\ \text{Btu/h}\cdot\text{ft}^2\cdot{}^\circ\text{F})[0.005/(3)(12)]\ \text{ft}}{118.2\ \text{Btu/h}\cdot\text{ft}\cdot{}^\circ\text{F}} = 9.4 \times 10^{-6}$$

and the lumped analysis is certainly suitable.

3.7 If the initial temperature is 32 °F and the surrounding air temperature is 45 °F, how long will it take the 0.005-in.-diam. thermocouple of Problem 3.6 to reach (*a*) 44 °F? (*b*) 44.5 °F? (*c*) 44.9 °F? (*d*) 44.99 °F?

▌ Equation (*1*) of Problem 3.4 gives (for heating)

$$t = -(1.6725\ \text{s}) \ln \frac{-T_\infty - T}{-T_\infty - T_i}$$

where the numerical value of $(RC)_{therm}$ was taken from Problem 3.6(*a*).

(*a*) $$t = -1.6725 \ln \frac{45 - 44}{45 - 32} = 4.29\ \text{s}$$

(*b*) $$t = -1.6725 \ln \frac{45 - 44.5}{45 - 32} = 5.45\ \text{s}$$

(*c*) $$t = -1.6725 \ln \frac{45 - 44.9}{45 - 32} = 8.14\ \text{s}$$

(*d*) $$t = -1.6725 \ln \frac{45 - 44.99}{45 - 32} = 11.99\ \text{s}$$

3.8 A magnesium cube 20 cm along each edge is exposed to a convective flow resulting in $\bar{h} = 300\ \text{W/m}^2\cdot\text{K}$. The initial metal temperature is 100 °C. Determine the Biot number, using the conductivity evaluated at the initial temperature.

▌ The characteristic length L is given by the volume-surface area ratio, i.e.,

$$L = \frac{V}{A_s} = \frac{l^3}{6l^2} = \frac{l}{6} = \frac{0.20\ \text{m}}{6} = 0.033\ \text{m}$$

From Appendix B, $k = (92)(1.7296) = 159.12\ \text{W/m}\cdot\text{K}$; hence

$$\textbf{Bi} = \frac{\bar{h}L}{k} = \frac{(300\ \text{W/m}^2\cdot\text{K})(0.033\ \text{m})}{159.12\ \text{W/m}\cdot\text{K}} = 0.062$$

3.9[D] An iron $(k = 64\ \text{W/m}\cdot\text{K})$ billet measuring $20 \times 16 \times 80$ cm is subjected to free convective heat transfer with $\bar{h} = 2\ \text{Btu/h}\cdot\text{ft}^2\cdot{}^\circ\text{F}$. Determine the Biot number and the suitability of a lumped analysis to represent the cooling rate if the billet is initially hotter than the environment.

▌ The surface area is $A_s = 2(20 \times 16) + 2(16 \times 80) + 2(20 \times 80) = 6400\ \text{cm}^2$, so that

$$L = \frac{V}{A_s} = \frac{(20 \times 16 \times 80)\ \text{cm}^3}{6400\ \text{cm}^2} = 4.0\ \text{cm}$$

Using conversion factors of Appendix A,

$$\bar{h} = (2\ \text{Btu/h}\cdot\text{ft}^2\cdot{}^\circ\text{F})\left(\frac{3.1525\ \text{W/m}^2}{1\ \text{Btu/h}\cdot\text{ft}^2}\right)\left(\frac{9\ {}^\circ\text{F}}{5\ \text{K}}\right) = 11.35\ \text{W/m}^2\cdot\text{K}$$

Thus

$$\mathbf{Bi} = \frac{\bar{h}L}{k} = \frac{(11.35\ \text{W/m}^2\cdot\text{K})(0.04\ \text{m})}{64\ \text{W/m}\cdot\text{K}} = 0.0071$$

and a lumped analysis will represent the transient temperature quite well.

3.10 Consider a container full of frozen mercury that is suddenly exposed to an ambient temperature of 75 °F. The container is cylindrical with a diameter of 2 in. and a length of 6 in. Assume an average heat transfer coefficient of $5\ \text{Btu/h}\cdot\text{ft}^2\cdot{}^\circ\text{F}$. Could a lumped-system analysis be applied to this problem?

▌ The characteristic dimension is

$$L = \frac{V}{A_s} = \frac{\pi r^2 l}{2\pi r l + 2(\pi r^2)} = \frac{lr}{2(l + r)} = \frac{(6)(1)}{2(6 + 1)}$$

$$= \tfrac{3}{7}\ \text{in.} = \tfrac{1}{28}\ \text{ft}$$

Hence
$$\mathbf{Bi} = \frac{\bar{h}L}{k} = \frac{(5\ \text{Btu/h}\cdot\text{ft}^2\cdot{}^\circ\text{F})(\frac{1}{28}\ \text{ft})}{4.8\ \text{Btu/h}\cdot\text{ft}\cdot{}^\circ\text{F}} \approx \frac{1}{28} < \frac{1}{10}$$

Yes, a lumped type of analysis can be applied to this problem.

3.11[D] Determine the time required for a 0.5-in.-diam. mild steel sphere to cool from 1000 °F to 200 °F, if exposed to a cooling air flow at 80 °F resulting in $\bar{h} = 20\ \text{Btu/h}\cdot\text{ft}^2\cdot{}^\circ\text{F}$.

▌ The characteristic linear dimension is $L = r/3$ (Problem 3.6). From Appendix B, the thermal conductivity of mild steel at the average temperature, $(1000 + 200)/2 = 600\ {}^\circ\text{F}$, is approximately $25\ \text{Btu/h}\cdot\text{ft}\cdot{}^\circ\text{F}$. Thus,

$$\mathbf{Bi} = \frac{\bar{h}L}{k} = \frac{(20\ \text{Btu/h}\cdot\text{ft}^2\cdot{}^\circ\text{F})[0.25/(3)(12)]\ \text{ft}}{25\ \text{Btu/h}\cdot\text{ft}\cdot{}^\circ\text{F}} = 0.0056$$

and a lumped analysis is suitable. By (*2*) of Problem 3.4 the time (**Fo**) for $T(t)$ to reach 200 °F satisfies

$$\frac{200 - 80}{1000 - 80} = 0.1304 = e^{-(\mathbf{Bi})(\mathbf{Fo})}$$

Solving,

$$(\mathbf{Bi})(\mathbf{Fo}) = 2.0369$$

$$\mathbf{Fo} = \frac{2.0369}{0.0056} = 363.73 = \frac{\alpha t}{L^2}$$

$$t = \frac{(363.73)(r/3)^2}{\alpha}$$

From Appendix B, at 600 °F,

$$\alpha = \frac{k}{\rho c} \approx \frac{25\ \text{Btu/h}\cdot\text{ft}\cdot{}^\circ\text{F}}{(490\ \text{lbm/ft}^3)(0.11\ \text{Btu/lbm}\cdot{}^\circ\text{F})} = 0.46\ \text{ft}^2/\text{h}$$

Thus
$$t = \frac{(363.73)[0.25/(3)(12)]^2\ \text{ft}^2}{0.46\ \text{ft}^2/\text{h}} = 0.0381\ \text{h} = 2.29\ \text{min}$$

3.12[D] A 10-mm-diam. mild steel sphere is cooled from 550 °C in air flowing at 25 °C. How much time is required for it to reach 90 °C if $\bar{h} = 110\ \text{W/m}^2\cdot\text{K}$?

From Table B-1, $k=(25)(1.7296)=43.24\ \text{W/m}\cdot\text{K}$, and $L=r/3=\frac{5}{3}\ \text{mm}=0.001\,67\ \text{m}$; hence

$$\mathbf{Bi}=\frac{\bar{h}L}{k}=\frac{(110\ \text{W/m}^2\cdot\text{K})(0.001\,67\ \text{m})}{43.24\ \text{W/m}\cdot\text{K}}=0.004\,25<0.1$$

Therefore, a lumped analysis may be made. By (2) of Problem 3.4,

$$\frac{90-25}{550-25}=e^{-(\mathbf{Bi})(\mathbf{Fo})}$$

whence $(\mathbf{Bi})(\mathbf{Fo})=2.089$ and $\mathbf{Fo}=2.089/0.004\,25=491.53$. Then, using Table B-1,

$$t=\frac{(\mathbf{Fo})L^2}{\alpha}=\frac{(491.53)(0.001\,67\ \text{m})^2}{[(0.49)(2.58\times10^{-5})]\ \text{m}^2/\text{s}}=108.8\ \text{s}=1.81\ \text{min}$$

3.13 A 3-in.-diam. orange is subjected to a cold-air environment. Assuming that the orange has properties similar to those of water at 68 °F and that $\bar{h}=2\ \text{Btu/h}\cdot\text{ft}^2\cdot{}^\circ\text{F}$, determine the suitability of a lumped analysis for predicting the temperature of the orange during cooling.

From Appendix B, the thermal conductivity of water at 68 °F is $0.345\ \text{Btu/h}\cdot\text{ft}\cdot{}^\circ\text{F}$. Also, for a sphere, the characteristic dimension is one-third the radius. Hence

$$\mathbf{Bi}=\frac{\bar{h}L}{k}=\frac{(2\ \text{Btu/h}\cdot\text{ft}^2\cdot{}^\circ\text{F})[1.5/(3)(12)]\ \text{ft}}{0.345\ \text{Btu/h}\cdot\text{ft}\cdot{}^\circ\text{F}}=0.24\gg0.1$$

Consequently a lumped analysis would be highly inaccurate.

3.14 What is the maximum edge dimension of a solid aluminum cube at 100 °C subjected to a convective heat transfer with $\bar{h}=25\ \text{W/m}^2\cdot\text{K}$ for a lumped analysis to be accurate to within 5 percent?

For an object of this shape, the lumped approach is accurate to ±5 percent if the Biot modulus is less than 0.1. From Appendix B, $k=205.82\ \text{W/m}\cdot\text{K}$. Thus,

$$\mathbf{Bi}_{max}=0.1=\frac{\bar{h}L_{max}}{k}=\frac{(25\ \text{W/m}^2\cdot\text{K})L_{max}}{205.82\ \text{W/m}\cdot\text{K}}\quad\text{or}\quad L_{max}=0.823\ \text{m}$$

But $L_{max}=l^3_{max}/6l^2_{max}=l_{max}/6=0.823\ \text{m}$, whence

$$l_{max}=6(0.823\ \text{m})=4.94\ \text{m}$$

3.15 A 14-lbm chunk of aluminum is approximately spherical. Between it and the surrounding fluid the convective heat transfer coefficient is $10\ \text{Btu/h}\cdot\text{ft}^2\cdot{}^\circ\text{F}$. Would it be reasonable to neglect its internal resistance at temperatures near 300 °F?

From Table B-1 for aluminum at 300 °F, $k=123\ \text{Btu/h}\cdot\text{ft}\cdot{}^\circ\text{F}$ and $\rho=169\ \text{lbm/ft}^3$. For a sphere,

$$\text{Mass}=\tfrac{4}{3}\pi r^3\rho$$

which may be solved to give $r=0.2704$ ft. Then

$$L=\frac{r}{3}=0.090\,13\ \text{ft}$$

$$\mathbf{Bi}=\frac{\bar{h}L}{k}=\frac{(10)(0.090\,13)}{123}=0.007\,328\ll0.1$$

The internal resistance can be neglected.

3.16 An object in the shape of a right-circular cone has a base of diameter 4 in. and a height of 3 in. It has a thermal conductivity of $10\ \text{Btu/h}\cdot\text{ft}\cdot{}^\circ\text{F}$ and a heat transfer coefficient of $12\ \text{Btu/h}\cdot\text{ft}^2\cdot{}^\circ\text{F}$. Could a lumped-system analysis be applied?

$$L=\frac{V}{A_s}=\frac{(\frac{1}{3})\pi r^2h}{\frac{1}{2}(2\pi r)(r^2+h^2)^{1/2}}=\frac{hr}{3(r^2+h^2)^{1/2}}$$

$$=\frac{(3)(2)}{3(2^2+3^2)^{1/2}}=0.555\ \text{in.}=0.0462\ \text{ft}$$

Hence

$$\mathbf{Bi}=\frac{\bar{h}L}{k}=\frac{(12\ \text{Btu/h}\cdot\text{ft}^2\cdot{}^\circ\text{F})(0.0462\ \text{ft})}{10\ \text{Btu/h}\cdot\text{ft}\cdot{}^\circ\text{F}}=0.0554<0.1$$

A lumped type of analysis could be applied.

3.17 A triangular fin is 1 in. thick and 6 in. long at the base and is 2 in. wide (perpendicular to the base surface). The fin is made of aluminum and has an average heat transfer coefficient along the sides and ends of 15 Btu/h · ft² · °F. There is no convective heat transfer at the base. Calculate the Biot modulus.

I
$$V = \tfrac{1}{2}(1)(2)(6) = 6\ \text{in}^3$$
$$A_s = 2[\tfrac{1}{2}(1)(2) + 6\sqrt{(\tfrac{1}{2})^2 + 2^2}] = 26.74\ \text{in}^2$$
$$L = \frac{V}{A_s} = \frac{6\ \text{in}^3}{26.74\ \text{in}^2} \cdot \frac{1\ \text{ft}}{12\ \text{in.}} = 0.0187\ \text{ft}$$
$$\mathbf{Bi} = \frac{\bar{h}L}{k} = \frac{(15)(0.0187)}{119} = 0.0024$$

3.18 Three resistors of equal resistance are in separate calorimeters, interconnected as in Fig. 3-4; each calorimeter contains the same mass of water. A certain time after closing switch S the water in calorimeter A is 40 °F hotter than before the switch was closed. How much has the temperature of the water in calorimeter C been increased?

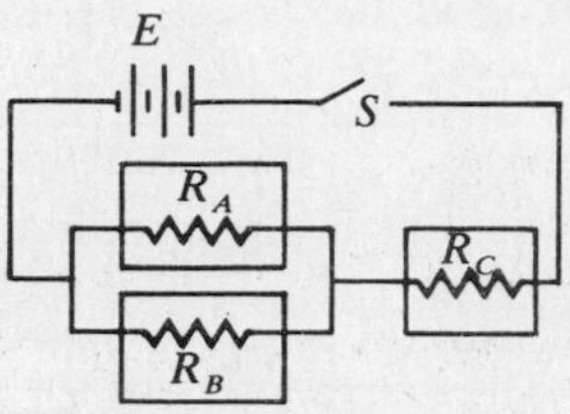

Fig. 3-4

I Because the resistances are equal, $I_A = I_B = \tfrac{1}{2}I_C$ and the power dissipated in any one resistor will be proportional to the square of the current through it. Therefore

$$\frac{\Delta T_C}{40\ °\text{F}} = \frac{I_C^2}{I_A^2} = 4 \qquad \text{or} \qquad \Delta T_C = 160\ °\text{F}$$

3.19 Compare the temperature responses of a 0.016-in.-diam. aluminum wire initially at 400 °F when the wire is suddenly exposed to **(a)** forced air at 100 °F ($\bar{h} = 12$ Btu/h · ft² · °F) and **(b)** still air at 100 °F ($\bar{h} =$ 2 Btu/h · ft² · °F).

I From Table B-1 for aluminum at $T_{avg} = 250$ °F:

$$k = 121\ \text{Btu/h}\cdot\text{ft}\cdot°\text{F} \qquad c = 0.208\ \text{Btu/lbm}\cdot°\text{F} \qquad \rho = 169\ \text{lbm/ft}^3$$

We have

$$\mathbf{Bi}_{\text{still}} = \frac{\bar{h}_{\text{still}}(D/4)}{k} = \frac{2(0.004/12)}{121} = 0.55 \times 10^{-5}$$
$$\mathbf{Bi}_{\text{forced}} = \tfrac{12}{2}(0.55 \times 10^{-5}) = 3.30 \times 10^{-5}$$

Therefore, lumped analysis can be used in either case. The governing equation, (*1*) of Problem 3.3, becomes ($V/A_s = D/4$)

$$\frac{T - T_\infty}{T_i - T_\infty} = \exp\left[\left(\frac{-4\bar{h}}{\rho c D}\right)t\right] \qquad \text{or} \qquad T = T_\infty + (T_i - T_\infty)\exp\left[\left(\frac{-4\bar{h}}{\rho c D}\right)t\right] \tag{1}$$

Substitution of numerical data in (*1*) gives

(*a*) $T = (100\ °\text{F}) + (300\ °\text{F})\exp(-1024.1t/\text{h})$

(*b*) $T = (100\ °\text{F}) + (300\ °\text{F})\exp(-170.69t/\text{h})$

3.20 Sketch the response curves for Problem 3.19.

I See Fig. 3-5.

3.21 A stainless-steel $\tfrac{1}{2}$-in.-diam. ball bearing initially at a temperature of 100 °F is suddenly exposed to a fluid whose temperature is 1000 °F. The average heat transfer coefficient between the bearing and the fluid is 12 Btu/h · ft² · °F. **(a)** Calculate the Biot modulus based on radius (not $R/3$). **(b)** What is the temperature of the bearing after 1 min?

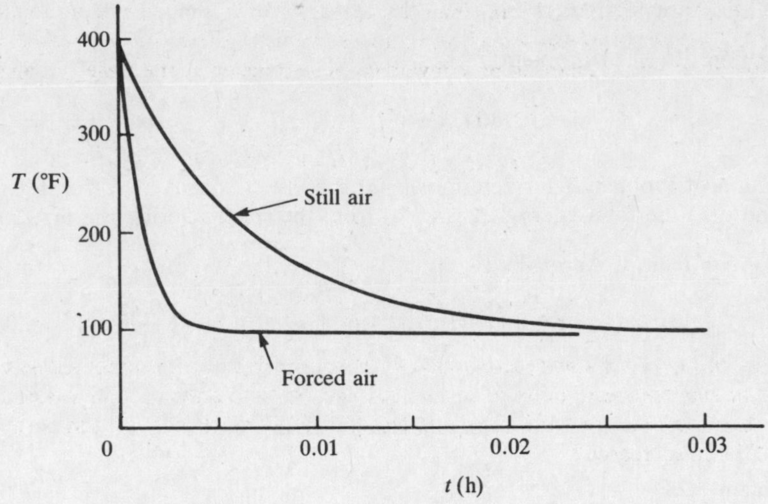

Fig. 3-5

I At $T_{avg} = 550$ °F, Appendix B gives

$$k = 11.0 \text{ Btu/h}\cdot\text{ft}\cdot°\text{F} \qquad c = 0.11 \text{ Btu/lbm}\cdot°\text{F} \qquad \rho = 488 \text{ lbm/ft}^3$$

(*a*)
$$\mathbf{Bi}^* = \frac{\bar{h}R}{k} = \frac{(12 \text{ Btu/h}\cdot\text{ft}^2\cdot°\text{F})[1/(4\times 12)\text{ ft}]}{11.0 \text{ Btu/h}\cdot\text{ft}\cdot°\text{F}} = 0.0227$$

(*b*) Since $\mathbf{Bi}^* < 0.1$ (and $\mathbf{Bi}^*/3 \ll 0.1$) we can use lumped analysis. Therefore,

$$\frac{T_\infty - T}{T_\infty - T_i} = e^{-(\bar{h}A_s/\rho c V)t}$$

Substituting $V/A_s = R/3 = \frac{1}{144}$ ft and the other data, along with $t = \frac{1}{60}$ h,

$$\frac{1000 - T}{1000 - 100} = 0.5851 \qquad \text{or} \qquad T = 473.4 \text{ °F}$$

3.22[D] An iron ingot at a temperature of 2200 °F is rolled into the shape of a rectangular parallelepiped with dimensions 1 ft by 6 in. by 10 ft. The ingot is allowed to cool in a room where the temperature is 120 °F; the average heat transfer coefficient is 10 Btu/h · ft² · °F. Can a lumped analysis be used to find the time required for the center temperature of the ingot to reach 400 °F?

I At $T_{avg} = \frac{1}{2}(2200 + 400) = 1300$ °F, $k = 37$ Btu/h · ft · °F; moreover,

$$V = 1\times\tfrac{1}{2}\times 10 = 5 \text{ ft}^3 \qquad A_s = 2[(1\times 10) + (1\times\tfrac{1}{2}) + (\tfrac{1}{2}\times 10)] = 31 \text{ ft}^2$$

whence
$$L = \frac{V}{A_s} = \frac{5 \text{ ft}^3}{31 \text{ ft}^2} = 0.1613 \text{ ft}$$

and
$$\mathbf{Bi} = \frac{\bar{h}L}{k} = \frac{(10 \text{ Btu/h}\cdot\text{ft}^2\cdot°\text{F})(0.1613 \text{ ft})}{37 \text{ Btu/h}\cdot\text{ft}\cdot°\text{F}} = 0.0436 < 0.1$$

Therefore, the internal resistance can be neglected and the bar considered to maintain a uniform temperature.

3.23 Determine an expression for the instantaneous heat flux from a lumped system.

I From Problems 3.3 and 3.4,

$$\begin{aligned} q(t) &= -\rho c V \frac{dT}{dt} \\ &= -C_{\text{therm}}(T_i - T_\infty)\left(\frac{-1}{R_{\text{therm}}C_{\text{therm}}}\right)\exp\left[\frac{-t}{(RC)_{\text{therm}}}\right] \\ &= \frac{T_i - T_\infty}{R_{\text{therm}}}\exp\left[\frac{-t}{(RC)_{\text{therm}}}\right] \end{aligned} \qquad (1)$$

Note that $q(0)$ has the expected value.

3.24 Find the total heat transferred, over $t=0$ to $t=t_1$, for a lumped system.

By integration of (*1*) of Problem 3.23,

$$Q(t_1) = \int_0^{t_1} q(t)\, dt = C_{\text{therm}}(T_i - T_\infty)\left\{1 - \exp\left[\frac{-t_1}{(RC)_{\text{therm}}}\right]\right\} \tag{1}$$

3.25 For the situation of Problem 3.10, determine (*a*) the instantaneous heat transfer rate 2 min after the start of cooling and (*b*) the total energy transferred from the sphere during the first 2 min.

Using property data of Appendix B,

$$C_{\text{therm}} = \rho c V = (490\ \text{lbm/ft}^3)(0.11\ \text{Btu/lbm}\cdot{}^\circ\text{F})\left(\frac{4\pi}{3}\right)\left(\frac{0.25}{12}\ \text{ft}\right)^3 = 0.002\,04\ \text{Btu/}^\circ\text{F}$$

and, from Problem 3.11,

$$\frac{1}{R_{\text{therm}}C_{\text{therm}}} = \frac{\textbf{Bi}\cdot\textbf{Fo}}{t} = \frac{2.0369}{2.29\ \text{min}}$$

(*a*) By Problem 3.23,

$$q(2\ \text{min}) = [(1000-80)\ {}^\circ\text{F}](0.002\,04\ \text{Btu/}^\circ\text{F})\left(\frac{2.0369}{2.29\ \text{min}}\right)\exp\left[(-2\ \text{min})\left(\frac{2.0369}{2.29\ \text{min}}\right)\right] = 0.282\ \text{Btu/min}$$

(*b*) By Problem 3.24,

$$Q\ (2\ \text{min}) = (0.002\,04)(920)\left\{1 - \exp\left[-2\left(\frac{2.0369}{2.29}\right)\right]\right\} = 1.56\ \text{Btu}$$

3.26 Consider the container of fluid shown in Fig. 3-6. The system is brought to a uniform initial temperature $T_i > T_\infty$ by wrapping it in an insulating blanket (which corresponds to opening switch S_2), running the electric heater (closing switch S_1), and waiting until both T_1 and T_2 reach T_i (at which point open S_1 and close S_2). Give a lumped analysis of the ensuing process, which is one of convective heat transfer from container to environment and from inner fluid to container.

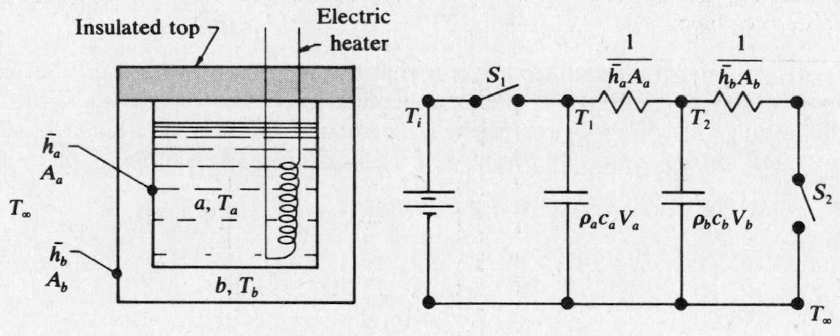

Fig. 3-6

The rate at which thermal energy (heat) leaves each body is equal to the negative of the rate of change of its stored thermal energy. Hence, the energy balances on the two lumps are

$$\textbf{lump } a \qquad \bar{h}_aA_a(T_a - T_b) = -\rho_a c_a V_a \frac{dT_a}{dt} \tag{1}$$

$$\textbf{lump } b \qquad \bar{h}_aA_a(T_b - T_a) + \bar{h}_bA_b(T_b - T_\infty) = -\rho_b c_b V_b \frac{dT_b}{dt} \tag{2}$$

These constitute a pair of simultaneous linear differential equations for $T_a(t)$ and $T_b(t)$. The initial conditions are

$$T_a(0) = T_b(0) = T_i \tag{3}$$

First differentiate (*1*) with respect to t and substitute from (*2*) for dT_b/dt; the result is

$$\frac{d^2T_a}{dt^2} + C_1\frac{dT_a}{dt} + C_2T_a = C_2T_\infty \tag{4}$$

where

$$C_1 \equiv \frac{\bar{h}_a A_a}{\rho_a c_a V_a} + \frac{\bar{h}_a A_a}{\rho_b c_b V_b} + \frac{\bar{h}_b A_b}{\rho_b c_b V_b} \qquad C_2 \equiv \left(\frac{\bar{h}_a A_a}{\rho_a c_a V_a}\right)\left(\frac{\bar{h}_b A_b}{\rho_b c_b V_b}\right)$$

The steady-state solution (particular integral) is clearly $T_a = T_\infty$ while the transient solution (general solution of the homogeneous equation) is

$$T_a = Ae^{m_1 t} + Be^{m_2 t}$$

where A and B are arbitrary constants and

$$m_1 \equiv \frac{-C_1 - (C_1^2 - 4C_2)^{1/2}}{2} \qquad m_2 = \frac{-C_1 + (C_1^2 - 4C_2)^{1/2}}{2}$$

Thus,

$$T_a = Ae^{m_1 t} + Be^{m_2 t} + T_\infty \tag{5}$$

Finally, A and B are determined by applying the initial conditions $T_a(0) = T_i$ and

$$\left.\frac{dT_a}{dt}\right|_{t=0} = 0$$

which is implied by (*1*) and (*3*). The results are

$$A = \frac{m_2}{m_2 - m_1}(T_i - T_\infty) \qquad B = -\frac{m_1}{m_2 - m_1}(T_i - T_\infty)$$

so that, in dimensionless form,

$$\frac{T_a - T_\infty}{T_i - T_\infty} = \frac{m_2}{m_2 - m_1}e^{m_1 t} - \frac{m_1}{m_2 - m_1}e^{m_2 t} \tag{6}$$

The simplest way to obtain the container temperature, $T_b(t)$, is to substitute (*6*) into (*1*).

3.27 Consider a lumped system consisting of two metal blocks in perfect thermal contact (there is negligible contact resistance between the two blocks at their interface). Only block b is exposed to a convective environment at T_∞, as shown in Fig. 3-7. Both blocks are initially at T_i. Give an analytical expression for the temperature of block a.

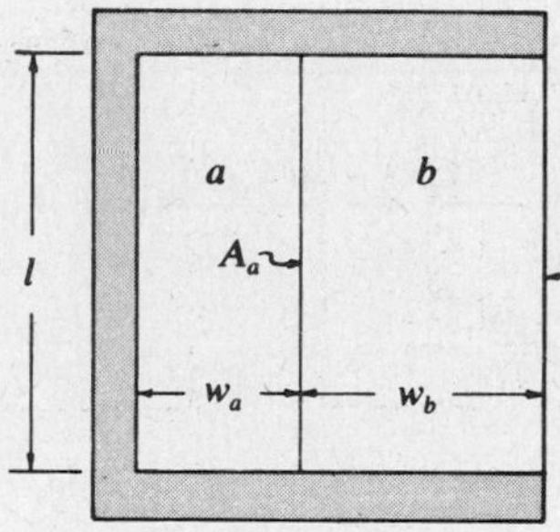

Fig. 3-7

▮ The heat transfer rate from block a and to block b is

$$\frac{q}{A} = -k_a\left(\frac{T_{\text{int}} - T_a}{w_a/2}\right) = -k_b\left(\frac{T_b - T_{\text{int}}}{w_b/2}\right) \tag{1}$$

Eliminating T_{int}, the interfacial temperature, between the two equations (*1*) gives

$$\frac{q}{A} = \frac{q_a}{A_a} = K(T_a - T_b) \tag{2}$$

where

$$K \equiv \frac{2(k_a/w_a)(k_b/w_b)}{(k_a/w_a) + (k_b/w_b)}$$

Using (*2*) to write the energy balances on the two blocks, we find them to be identical to those of Problem 3.26, with h_a replaced by K. Furthermore, the initial conditions are identical to those of Problem 3.26. It follows that all the results of Problem 3.26, in particular (*6*), apply to the present problem if h_a is everywhere replaced by K.

3.28 For the system of Problem 3.27, $T_i = 300$ °F, $T_\infty = 100$ °F, and $\bar{h}_b = 5$ Btu/h · ft² · °F, and the following data apply:

material a (*brass*)	**material b** (*steel*)
$c_a = 0.092$ Btu/lbm · °F	$c_b = 0.11$ Btu/lbm · °F
$k_a = 60.0$ Btu/h · ft · °F	$k_b = 9.4$ Btu/h · ft · °F
$w_a = 3.0$ in.	$w_b = 2.0$ in.
$\rho_a = 532$ lbm/ft³	$\rho_b = 488$ lbm/ft³

(*a*) Is a lumped analysis suitable? (*b*) If the answer to (*a*) is yes, plot the temperature-time history of lump *a* from 300 °F down to 150 °F.

▮ (*a*) An appropriate Biot modulus for material *b* is

$$\mathbf{Bi}_b = \frac{\bar{h}_b w_b}{k_b} = \frac{(5)(\frac{2}{12})}{9.4} = 0.09$$

if we neglect the heat transfer to body *a*. As for $\mathbf{Bi}_a$, the resistance for heat transfer from lump *a* to the environment at T_∞ is clearly greater than that for heat transfer from lump *b*; hence, $\bar{h}_a < \bar{h}_b$. Also,

$$\frac{w_a}{k_a} = \frac{\frac{3}{12}}{60} < \frac{\frac{2}{12}}{9.4} = \frac{w_b}{k_b}$$

so that $\mathbf{Bi}_a < \mathbf{Bi}_b$. As both moduli are smaller than 0.1, a lumped analysis of the system is suitable.

(*b*) Parameter values for use in (*6*) of Problem 3.26 are

$$K = \frac{2(k_a/w_a)(k_b/w_b)}{(k_a/w_a) + (k_b/w_b)} = \frac{2[60(12)/3][9.4(12)/2]}{[60(12)/3] + [9.4(12)/2]} = 91.34 \text{ Btu/h} \cdot \text{ft}^2 \cdot {}^\circ\text{F}$$

$$C_1 = \frac{KA_a}{\rho_a c_a V_a} + \frac{KA_a}{\rho_b c_b V_b} + \frac{\bar{h}_b A_b}{\rho_b c_b V_b} = \frac{K}{\rho_a c_a w_a} + \frac{K}{\rho_b c_b w_b} + \frac{\bar{h}_b}{\rho_b c_b w_b}$$

$$= \frac{(91.34)(12)}{532(0.092)(3)} + \frac{(91.34)(12)}{488(0.11)(2)} + \frac{5(12)}{488(0.11)(2)} = 18.23 \text{ h}^{-1}$$

$$C_2 = (7.46)(0.56) = 4.18 \text{ h}^{-2}$$

$$m_1 = \frac{-18.23 - [(18.23)^2 - 4(4.18)]^{1/2}}{2} = -18.00 \text{ h}^{-1}$$

$$m_2 = \frac{-18.23 + [(18.23)^2 - 4(4.18)]^{1/2}}{2} = -0.23 \text{ h}^{-1}$$

Then,

$$\frac{T_a - 100}{300 - 100} = \frac{-0.23}{18.00 - 0.23} e^{-18.00t} + \frac{18.00}{18.00 - 0.23} e^{-0.23t}$$

or

$$T_a = 100 + 202.59e^{-0.23t} - 2.59e^{-18.0t}$$

where t is in hours and T_a has units of °F. Plotting, we obtain Fig. 3-8.

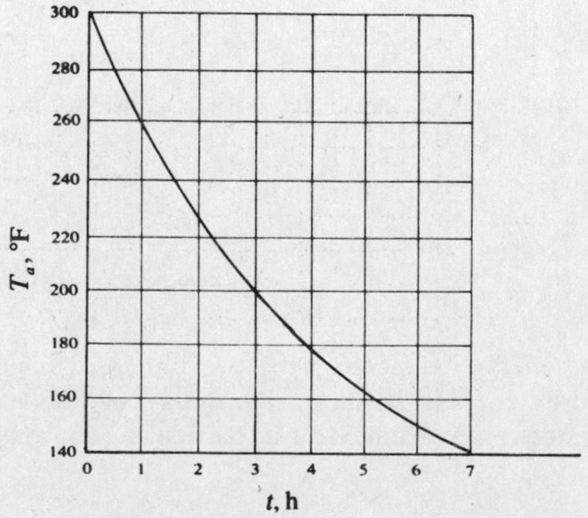

Fig. 3-8

3.2 TRANSIENTS IN SLABS

3.29 The half-space $x > 0$ (with thermal diffusivity α) is initially at uniform temperature T_i (Fig. 3-9). At $t = 0$, the surface $x = 0$ is suddenly brought to the temperature T_s, which is maintained thereafter. Express the resulting temperature distribution. (This is approximated in nature by the earth's surface, large bodies of water, etc.)

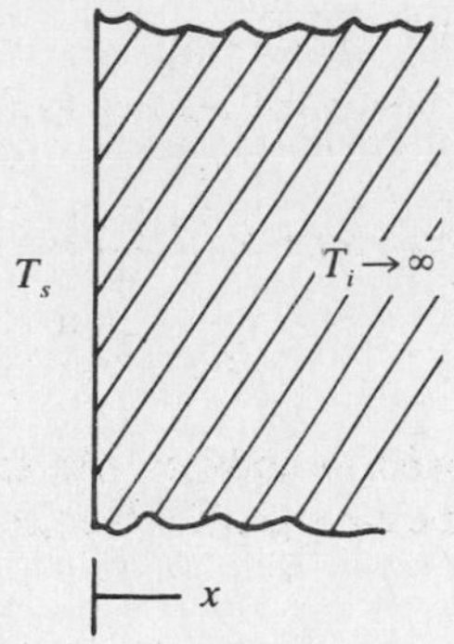

Fig. 3-9

▌ One must solve (*9*) of Problem 1.54,

$$\frac{\partial^2 T}{\partial x^2} = \frac{1}{\alpha}\frac{\partial T}{\partial t}$$

subject to

$$\begin{array}{rll} \textbf{\textit{Boundary condition 1}} & T(0, t) = T_s & (t > 0) \\ \textbf{\textit{Boundary condition 2}} & T(\infty, t) = T_i & \\ \textbf{\textit{Initial condition}} & T(x, 0) = T_i & (x > 0) \end{array}$$

As may be verified by direct substitution, two particular solutions of our equation are

$$T_{(1)} = \text{constant} \qquad T_{(2)} = \int_0^x t^{-1/2} e^{-v^2/4\alpha t}\, dv = \text{constant} \times \text{erf}\, \frac{x}{(4\alpha t)^{1/2}}$$

where

$$\text{erf}\, z \equiv \frac{2}{\sqrt{\pi}} \int_0^z e^{-\xi^2}\, d\xi$$

is the standard notation for the *Gauss error function*, which is extensively tabulated. Thus the linear combination

$$T = C_1 \,\text{erf}\, \frac{x}{(4\alpha t)^{1/2}} + C_2$$

may be taken as the general solution to the present problem, where C_1 and C_2 are arbitrary constants. Now applying the first boundary condition and noting that $\text{erf}\, 0 = 0$, we find that $C_2 = T_s$; the second boundary condition then gives, since $\text{erf}\, \infty = 1$, $C_1 = T_i - T_s$; finally, with these two values, the initial condition is automatically satisfied. Therefore,

$$\frac{T - T_s}{T_i - T_s} = \text{erf}\, \frac{x}{(4\alpha t)^{1/2}} \tag{1}$$

is the complete solution.

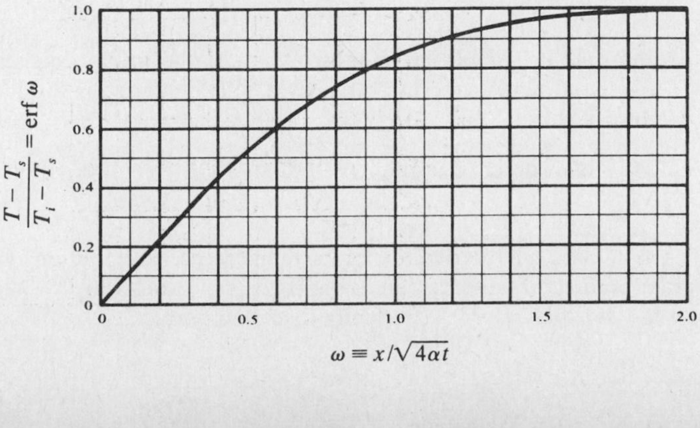

Fig. 3-10

According to (*1*) the temperature impulse $T_s - T_i$ smears out (diffuses) from the boundary into the interior of the semi-infinite region. The closer is location x to the boundary, the sooner does the temperature at x become (essentially) and remain (essentially) T_s. Figure 3-10 is a graph of the error function and, therefore, of the temperature distribution. For a body of finite thickness, $0 < x < L$, (*1*) provides a good approximate solution if $L/\sqrt{4\alpha t} > \frac{1}{2}$; that is,

$$\mathbf{Fo} < 1 \tag{2}$$

3.30 Calculate the instantaneous heat flux at the surface of the semi-infinite body of Problem 3.29.

▮ The heat flow by conduction at any distance x in the body can be calculated by Fourier's equation,

$$\frac{q(x,t)}{A} = -k\,\frac{\partial T}{\partial x}$$

By (*1*) of Problem 3.29,

$$T(x,t) = T_s + (T_i - T_s)\,\frac{2}{\sqrt{\pi}}\int_0^{x/(4\alpha t)^{1/2}} e^{-\xi^2}\,d\xi$$

Performing the partial differentiation by Leibniz' rule yields

$$\frac{\partial T}{\partial x} = \frac{T_i - T_s}{(\pi\alpha t)^{1/2}}\,e^{-x^2/4\alpha t} \tag{1}$$

so that

$$\frac{q(0,t)}{A} = \frac{k(T_s - T_i)}{(\pi\alpha t)^{1/2}} \tag{2}$$

3.31[D] It is essential to locate water pipes below the depth to which freezing can occur in the soil. If the initial soil temperature is 50 °F and the surface temperature drops rapidly to −5 °F, to what depth will the freezing temperature penetrate during a 10-h period? Assume dry soil with $\alpha = 0.01\ \text{ft}^2/\text{h}$.

▮ By (*1*) of Problem 3.29, the dimensionless temperature ratio is

$$\frac{32-(-5)}{50-(-5)} = 0.673 = \operatorname{erf}\frac{x_{\text{crit}}}{(4\alpha t)^{1/2}}$$

Figure 3-10 then gives $x_{\text{crit}}/(4\alpha t)^{1/2} \approx 0.7$, or

$$x_{\text{crit}} \approx (0.7)[4(0.01)(10)]^{1/2} = 0.443\ \text{ft}$$

for the depth to which freezing penetrates.

3.32[D] At what minimal depth should a water pipe be buried in coarse earth initially at 20 °C if the surface temperature drops to −15°C for 30 days?

▮ By (*1*) of Problem 3.29,

$$\frac{0-(-15)}{20-(-15)} = 0.43 = \operatorname{erf}\frac{x_{\text{crit}}}{(4\alpha t)^{1/2}}$$

Then, from Fig. 3-10, $x_{\text{crit}}/(4\alpha t)^{1/2} \approx 0.4$. From Table B-2,

$$\alpha = (0.0054)(2.581\times10^{-5}) = 1.394\times10^{-7}\ \text{m}^2/\text{s}$$

Therefore,

$$x_{\text{crit}} \approx (0.4)[4(1.394\times10^{-7}\ \text{m}^2/\text{s})(3600\ \text{s/h})(24\ \text{h/day})(30\ \text{days})]^{1/2} = 0.48\ \text{m}$$

3.33 In Problems 3.31 and 3.32 it is implicitly supposed that, at the time in question, temperatures are below freezing for $x < x_{\text{crit}}$ and above freezing for $x > x_{\text{crit}}$; here x_{crit} is the location where the temperature is exactly freezing. Justify this.

▮ For $T_s < T_i$, (*1*) of Problem 3.30 implies that $\partial T/\partial x > 0$; i.e., the temperature strictly increases with x.

3.34 From the result of Problem 3.29 derive the exact temperature distribution in the infinite slab $0 < x < L$. The initial condition and the boundary condition at $x = 0$ are as in Problem 3.29, while the boundary condition at $x = L$ is $q(L,t) = 0$ (perfectly insulated surface).

▮ Upon introduction of the reduced temperature $\theta \equiv T - T_i$, the solution (1) of Problem 3.29—which corresponds to unbounded rightward diffusion from the source $T_s - T_i$ at $x = 0$—becomes

$$\theta(x, t) = \theta_s \operatorname{erfc} \frac{x}{\sqrt{4\alpha t}} \tag{1}$$

in which the *complementary error function* is defined as

$$\operatorname{erfc} \omega \equiv 1 - \operatorname{erf} \omega = \frac{2}{\sqrt{\pi}} \int_\omega^\infty e^{-\xi^2}\, d\xi \tag{2}$$

The present initial-value/boundary-value problem is indicated by the shaded region of Fig. 3-11. To solve it, we apply superposition in the form of the *method of images*. Thus (see Fig. 3-11) the function (1) satisfies all conditions (including the *homogeneous* initial condition) save that at $x = L$; we therefore add an image source, $+\theta_s$, propagating to the left, at $x = 2L$. But this destroys the boundary value at $x = 0$; so we must add a second image source, $-\theta_s$, propagating to the right, at $x = -2L$. But this destroys the derivative at $x = L$; so . . . . In this fashion we generate a bi-infinite sequence of sources, regularly spaced $2L$ apart. This sequence is symmetric about $x = L$, causing the boundary condition there to be satisfied, and antisymmetric about $x = 0$, causing the boundary condition *there* to be satisfied. We then have as the solution to our problem:

$$\theta(x, t) = \theta_s\left[\operatorname{erfc} \frac{x}{\sqrt{4\alpha t}} + \left(\operatorname{erfc} \frac{2L - x}{\sqrt{4\alpha t}} - \operatorname{erfc} \frac{2L + x}{\sqrt{4\alpha t}}\right) - \left(\operatorname{erfc} \frac{4L - x}{\sqrt{4\alpha t}} - \operatorname{erfc} \frac{4L + x}{\sqrt{4\alpha t}}\right) + \cdots\right] \tag{3}$$

As was implied above, a superposition like (3) is possible only when all solutions being superposed (in space) obey the same homogeneous initial condition. Note that $0 < x < L$ implies

$$x < 2L - x < 2L + x < 4L - x < 4L + x < \cdots$$

which means that the terms of the series (3) steadily decrease in magnitude.

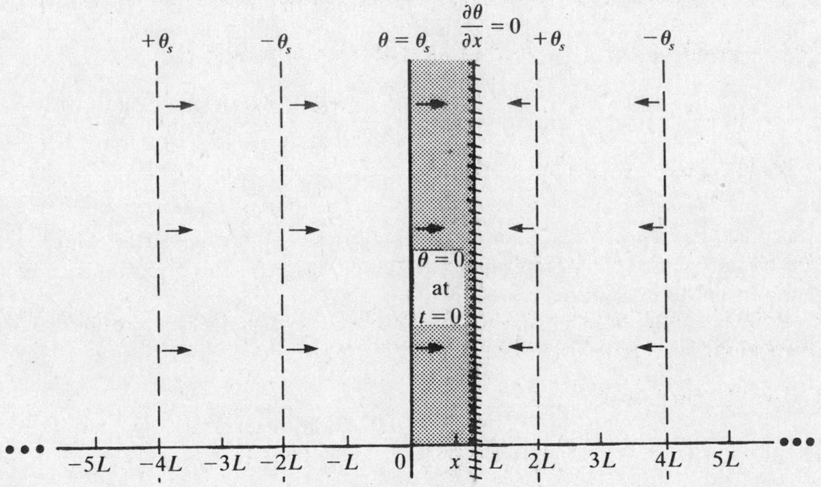

Fig. 3-11

3.35 A mild steel slab 5 cm thick, very wide and very long, is initially at 50 °C. One surface is exposed to a fluid which suddenly causes the surface temperature to increase to and remain at 100 °C. (*a*) What is the maximum time that the slab may be treated as a semi-infinite body? (*b*) Determine the temperature at the center of the slab 1 min after the surface-temperature change.

▮ (*a*) From Appendix B, $\alpha \approx 1.26 \times 10^{-5}\ \text{m}^2/\text{s}$. Then, by (2) of Problem 3.29,

$$t_{max} = \frac{L^2}{\alpha} = \frac{(5\ \text{cm})^2(10^{-2}\ \text{m/cm})^2}{1.26 \times 10^{-5}\ \text{m}^2/\text{s}} = 198.4\ \text{s} = 3.307\ \text{min}$$

(*b*) In view of (*a*), (1) of Problem 3.29 may be applied:

$$\frac{T - (100\ °\text{C})}{(50\ °\text{C}) - (100\ °\text{C})} = \operatorname{erf} \frac{2.5 \times 10^{-2}\ \text{m}}{[4(1.26 \times 10^{-5}\ \text{m}^2/\text{s})(60\ \text{s})]^{1/2}} \approx 0.48$$

Solving, $T \approx 76\ °\text{C}$.

3.36 For a lumped system, show that at the moment at which

$$T = \frac{T_i + T_\infty}{2} \equiv T_{\text{avg}}$$

the heat transfer rate has exactly half its initial value, and exactly half of the total heat transfer has been accomplished.

■ By (*1*) of Problem 3.3, $T = T_{\text{avg}}$ when $\exp[-t/(RC)_{\text{therm}}] = \frac{1}{2}$ [i.e., for $t = (RC)_{\text{therm}} \ln 2$]. But then, by (*1*) of Problem 3.23 and (*1*) of Problem 3.24,

$$q = \frac{T_i - T_\infty}{R_{\text{therm}}}\left(\frac{1}{2}\right) = \frac{1}{2}\,q(0) \qquad Q = C_{\text{therm}}(T_i - T_\infty)\left\{1 - \frac{1}{2}\right\} = \frac{1}{2}\,Q(\infty)$$

3.37 An epoxy which bonds at a temperature of 400 °F is used to join two plates of brass face to face. The dimensions of the plates are $\frac{1}{4}$ by 2 by 4 in., and their initial temperature is 75 °F. For bonding the plates will be placed in an oven, and for mass production it is necessary that the bonding time be less than 5 min. The average heat transfer coefficient to the plates is 12 Btu/h · ft² · °F. What is the minimum temperature of the oven if the bonded faces are in complete thermal contact?

■ Physical data at $T_{\text{avg}} = \frac{1}{2}(75 + 400) = 237.5$ °F are

$$k = 60\ \text{Btu/h}\cdot\text{ft}\cdot{}^\circ\text{F} \qquad c = 0.092\ \text{Btu/lbm}\cdot{}^\circ\text{F} \qquad \rho = 532\ \text{lbm/ft}^3 \qquad \alpha = 1.14\ \text{ft}^2/\text{h}$$

For the bonded assembly, the characteristic dimension is

$$L = \frac{V}{A_s} = \frac{(2)(4)(\frac{1}{2})}{2[(2)(4) + (4)(\frac{1}{2}) + (2)(\frac{1}{2})]} = \frac{2}{11}\ \text{in.} = \frac{1}{66}\ \text{ft}$$

so that

$$\textbf{Bi} = \frac{\bar{h}L}{k} = \frac{12(\frac{1}{66})}{60} = \frac{1}{330} \ll \frac{1}{10}$$

and a lumped analysis applies. Thus, substituting the bonding temperature and the maximum allowable time in (*1*) of Problem 3.3,

$$\frac{(400\ {}^\circ\text{F}) - T_\infty}{(75\ {}^\circ\text{F}) - T_\infty} = \exp\left[-\frac{(12\ \text{Btu/h}\cdot\text{ft}^2\cdot{}^\circ\text{F})(\frac{1}{12}\ \text{h})}{(532\ \text{lbm/ft}^3)(0.092\ \text{Btu/lbm}\cdot{}^\circ\text{F})(\frac{1}{66}\ \text{ft})}\right]$$

Solving, $T_\infty \approx 514$ °F.

3.38 A home fireplace has 4-in. walls made of fireclay brick. A fire is started which causes the surface of the brick to be heated to 500 °F. Determine the temperature on the opposite face of the brick after a period of 4 h, if the initial temperature was 60 °F.

■ Checking applicability of the semi-infinite-body solution (Problem 3.29),

$$\frac{L}{(4\alpha t)^{1/2}} = \frac{\frac{4}{12}\ \text{ft}}{[(4)(0.02\ \text{ft}^2/\text{h})(4\ \text{h})]^{1/2}} = 0.595 > 0.5$$

Therefore, at $x = L$,

$$\frac{T - T_s}{T_i - T_s} = \text{erf}\,\frac{L}{(4\alpha t)^{1/2}} = \text{erf}\,0.595 \approx 0.6$$

Substitution of $T_s = 500$ °F and $T_i = 60$ °F yields $T \approx 236$ °F.

3.39 Water at 60 °F is being used to reduce the temperature of a thick concrete roof that was initially at 180 °F. Assuming the surface to be maintained at 60 °F, how long would the water have to be applied to reduce the temperature 3 in. below the surface to 100 °F?

■ Applying (*1*) of Problem 3.29,

$$\text{erf}\,\frac{x}{\sqrt{4\alpha t}} = \frac{T - T_s}{T_i - T_s} = \frac{100 - 60}{180 - 60} = 0.333$$

so that, from Fig. 3-10 or a table of the error function,

$$\frac{x}{\sqrt{4\alpha t}} = 0.30 \tag{1}$$

Substitute in (*1*) $\alpha = 0.019$ ft²/h (Table B-2) and $x = 0.25$ ft to find $t = 9.14$ h.

3.40 Define the *thermal time constant* τ of a lumped system.

▌ τ is the time required for the temperature difference between system and surroundings—or, equivalently, the heat flux between system and surroundings—to reach $e^{-1} \approx 36.7\%$ of its initial value. It follows from (*1*) of Problem 3.4—or from (*1*) of Problem 3.23—that

$$\tau = (RC)_{\text{therm}} = \frac{\rho c L}{\bar{h}}$$

3.41 Refer to Problem 3.19. How much faster is the forced-air cooling than the still-air cooling?

▌ A natural index of speed is the inverse time constant, $1/\tau$, which is directly proportional to $\bar{h}$ (Problem 3.40). On this understanding, then, the forced-air cooling is $\frac{12}{2} = 6$ times faster.

3.42 Rework Problem 3.34 if the boundary at $x = L$ is maintained at the original temperature T_i; all else remains the same.

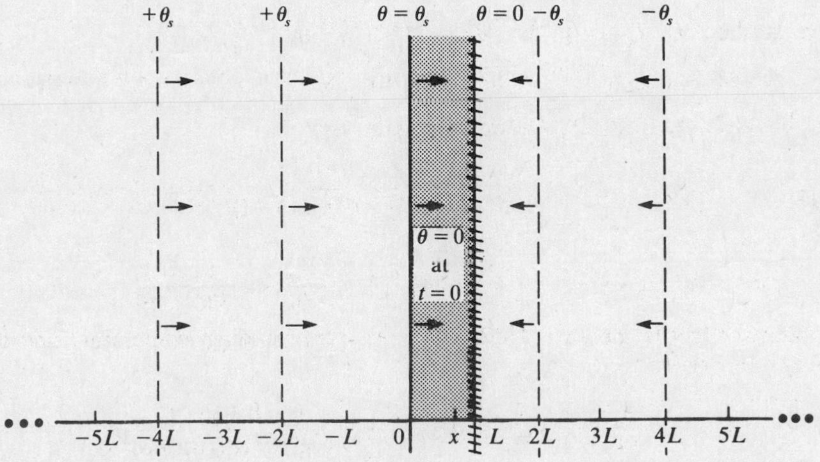

Fig. 3-12

▌ Figure 3-12 shows the solution by the method of images. Now the source distribution is antisymmetric about both $x = 0$ and $x = L$, and so both boundary conditions are satisfied. Explicitly,

$$\theta(x, t) = \theta_s\left[\operatorname{erfc}\frac{x}{\sqrt{4\alpha t}} - \left(\operatorname{erfc}\frac{2L - x}{\sqrt{4\alpha t}} - \operatorname{erfc}\frac{2L + x}{\sqrt{4\alpha t}}\right) - \left(\operatorname{erfc}\frac{4L - x}{\sqrt{4\alpha t}} - \operatorname{erfc}\frac{4L + x}{\sqrt{4\alpha t}}\right) - \cdots\right]$$

3.43 Figure 3-13 shows a semi-infinite body (half-space) that is subjected to a heat flux density q_0/A starting at $t = 0$. Find the temperature distribution.

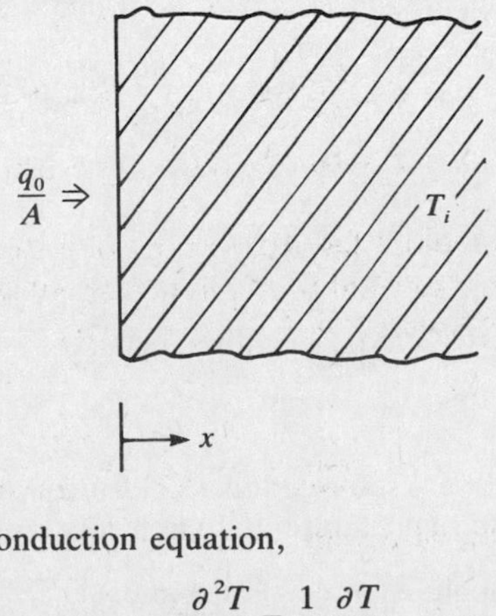

Fig. 3-13

▌ In the case of constant α, the conduction equation,

$$\frac{\partial^2 T}{\partial x^2} = \frac{1}{\alpha}\frac{\partial T}{\partial t}$$

has the important property of admitting as solutions the partial derivatives of any given solution. In particular, the heat flux density $q^* \equiv q/A$, being proportional to $\partial T/\partial x$, must satisfy

$$\frac{\partial^2 q^*}{\partial x^2} = \frac{1}{\alpha}\frac{\partial q^*}{\partial t} \tag{1}$$

subject to the conditions (for the problem at hand)

$$q^*(0, t) = \frac{q_0}{A} \qquad (t>0) \tag{2}$$

$$q^*(\infty, t) = 0 \tag{3}$$

$$q^*(x, 0) = 0 \qquad (x>0) \tag{4}$$

[The initial condition (4) follows from the absence of any interior temperature gradient at $t=0$.] But system (1)-(2)-(3)-(4) has already been solved—in Problem 3.29. Thus, making the replacements

$$T \to q^* \qquad T_s \to \frac{q_0}{A} \qquad T_i \to 0$$

one obtains from (1) of Problem 3.29:

$$q^*(x, t) = \frac{q_0}{A}\,\text{erfc}\,\frac{x}{\sqrt{4\alpha t}} \tag{5}$$

Since $\partial T/\partial x = -q^*/k$, the temperature distribution is obtained by integrating (5) on x:

$$\begin{aligned} T(x, t) &= -\frac{q_0/A}{k}\int_0^x \text{erfc}\,\frac{u}{\sqrt{4\alpha t}}\,du + \Omega(t) \\ &= -\frac{q_0/A}{k}\left(\frac{2}{\sqrt{\pi}}\int_0^x du \int_{u/\sqrt{4\alpha t}}^{\infty} e^{-\xi^2}\,d\xi\right) + \Omega(t) \\ &= -\frac{q_0/A}{k}\left(\frac{2}{\sqrt{\pi}}\int_0^{x/\sqrt{4\alpha t}} d\xi \int_0^{\xi\sqrt{4\alpha t}} e^{-\xi^2}\,du + \frac{2}{\sqrt{\pi}}\int_{x/\sqrt{4\alpha t}}^{\infty} d\xi \int_0^x e^{-\xi^2}\,du\right) + \Omega(t) \\ &= -\frac{q_0/A}{k}\left[\sqrt{\frac{4\alpha t}{\pi}}\int_0^{x^2/4\alpha t} e^{-\xi^2}\,d(\xi^2) + \frac{2x}{\sqrt{\pi}}\int_{x/\sqrt{4\alpha t}}^{\infty} e^{-\xi^2}\,d\xi\right] + \Omega(t) \\ &= -\frac{q_0/A}{k}\left[\sqrt{\frac{4\alpha t}{\pi}}\,(1 - e^{-x^2/4\alpha t}) + x\,\text{erfc}\,\frac{x}{\sqrt{4\alpha t}}\right] + \Omega(t) \end{aligned}$$

Finally, the unknown function of integration is determined from the condition $T(\infty, t) = T_i$; thus,

$$\Omega(t) = T_i + \frac{q_0/A}{k}\sqrt{\frac{4\alpha t}{\pi}}$$

and our solution becomes

$$T(x, t) - T_i = \frac{q_0/A}{k}\left(\sqrt{\frac{4\alpha t}{\pi}}\,e^{-x^2/4\alpha t} - x\,\text{erfc}\,\frac{x}{\sqrt{4\alpha t}}\right) \tag{6}$$

3.44 A large slab of steel initially at a uniform temperature of 20 °C is being heat-treated at its surface by suddenly raising the surface temperature to 250 °C for 5 min. What is the temperature at 2 cm below the surface?

▌ From Table B-1, $k = (26)(1.7296) = 44.97\ \text{W/m}\cdot\text{K}$ and $\alpha = (0.49)(2.581 \times 10^{-5}) = 1.26 \times 10^{-5}\ \text{m}^2/\text{s}$. Using (1) of Problem 3.29 and Fig. 3-10,

$$\frac{T - T_s}{T_i - T_s} = \text{erf}\,\frac{x}{(4\alpha t)^{1/2}}$$

$$\frac{T - 250}{20 - 250} = \text{erf}\,\frac{0.02\ \text{m}}{[4(1.26 \times 10^{-5}\ \text{m}^2/\text{s})(5 \times 60\ \text{s})]^{1/2}} = \text{erf}\,0.16 \approx 0.13$$

$$T \approx 250 - (230)(0.13) = 220.1\ ^\circ\text{C}$$

3.45 If the steel slab of Problem 3.44 were subjected to a sudden constant surface heat flux $q_0/A = 300\ \text{kW/m}^2$, what would be the temperature at 2 cm below the surface after 5 min?

▌ Now (6) of Problem 3.43 applies. From Problem 3.44,

$$\frac{x}{(4\alpha t)^{1/2}} = 0.16 \qquad \text{and} \qquad \text{erfc}\,\frac{x}{(4\alpha t)^{1/2}} \approx 1 - 0.13 = 0.87$$

Therefore,

$$T - 20 = \frac{300 \times 10^3}{44.97}\left[\sqrt{\frac{4(1.26 \times 10^{-5})(5 \times 60)}{\pi}}\, e^{-(0.16)^2} - (0.02)(0.87)\right] = 335.2$$

or $T = 355.2\ °C$.

3.46 How may one solve the following generalization of Problem 3.29?

$$\frac{\partial^2 T}{\partial x^2} = \frac{1}{\alpha}\frac{\partial T}{\partial t}$$

under (β, γ constants)

Boundary condition 1 $\quad T(0, t) + \beta \left.\frac{\partial T}{\partial x}\right|_{x=0} = \gamma \qquad (t > 0)$

Boundary condition 2 $\quad T(\infty, t) = T_i$

Initial condition $\quad T(x, 0) = T_i \qquad (x > 0)$

▮ First note that the solution of the above problem will automatically satisfy

$$\left.\frac{\partial T}{\partial x}\right|_{x=\infty} = 0 \qquad \text{and} \qquad \frac{\partial T(x,0)}{\partial x} = 0 \qquad (x > 0)$$

Therefore, if one defines

$$\Theta(x, t) \equiv T(x, t) + \beta \frac{\partial T(x, t)}{\partial x}$$

there results the problem

$$\frac{\partial^2 \Theta}{\partial x^2} = \frac{1}{\alpha}\frac{\partial \Theta}{\partial t}$$

subject to

$$\Theta(0, t) = \gamma \qquad (t > 0)$$
$$\Theta(\infty, t) = T_i$$
$$\Theta(x, 0) = T_i \qquad (x > 0)$$

But this is precisely Problem 3.29. With Θ a known function, T is recovered by holding t fixed and integrating the ordinary linear differential equation

$$\frac{dT}{dx} + \frac{1}{\beta}T = \frac{\Theta}{\beta}$$

subject to boundary condition 2 above: see any textbook for details.

3.47 Find the temperature distribution in the semi-infinite body $x > 0$, initially at uniform temperature T_i, the surface of which is exposed at $t = 0$ to a convective fluid (T_∞, $\bar{h}$).

▮ The problem corresponds to Problem 3.46, with $\beta = -k/\bar{h}$ and $\gamma = T_\infty$. The solution is found as

$$\frac{T - T_i}{T_\infty - T_i} = \text{erfc}\ \omega - \left[\exp\left(\frac{\bar{h}x}{k} + \frac{\bar{h}^2\alpha t}{k^2}\right)\right]\left[\text{erfc}\left(\omega + \frac{\bar{h}\sqrt{\alpha t}}{k}\right)\right] \qquad (1)$$

where $\omega \equiv x/\sqrt{4\alpha t}$. A plot of (*1*) is given in Fig. 3-14.

3.48 A water pipe is buried 1.2 ft below ground in wet soil ($\alpha = 0.03\ \text{ft}^2/\text{h}$ and $k = 1.5\ \text{Btu/h}\cdot\text{ft}\cdot°\text{F}$). The soil is initially at a uniform temperature of 40 °F. Under sudden application of a convective surface condition due to wind, with $\bar{h} = 10\ \text{Btu/h}\cdot\text{ft}^2\cdot°\text{F}$ and $T_\infty = -5\ °\text{F}$, will the pipe be exposed to freezing temperature in a 10-h period?

▮ Apply Fig. 3-14, with

$$\frac{\bar{h}\sqrt{\alpha t}}{k} = \frac{10\sqrt{(0.03)(10)}}{1.5} = 3.65 \qquad \omega = \frac{x}{\sqrt{4\alpha t}} = \frac{1.2}{\sqrt{4(0.03)(10)}} = 1.10$$

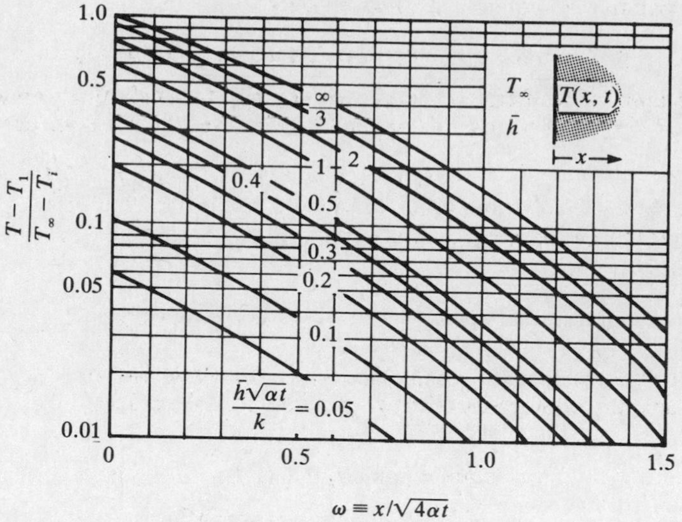

Fig. 3-14

Thus

$$\frac{T - 40}{-5 - 40} \approx 0.1 \qquad \text{or} \qquad T \approx 35.5\ ^\circ\text{F}$$

and freezing will not occur.

3.49 Determine the temperature distribution in a large slab of thickness $2L$ initially at uniform temperature T_i and suddenly exposed to a convective environment at T_∞. The Biot number is too large for a lumped analysis to be appropriate.

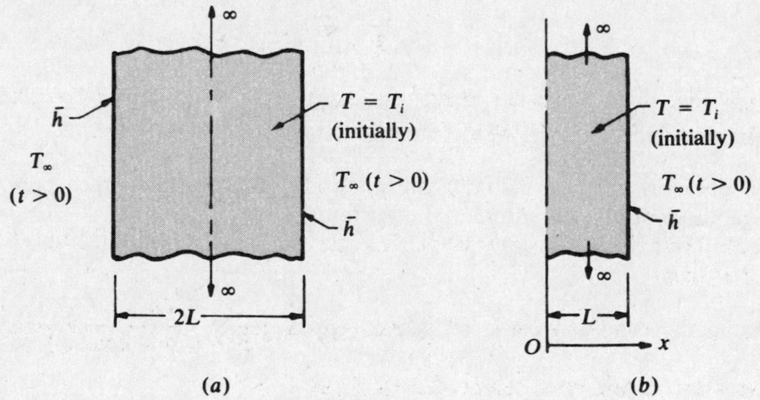

Fig. 3-15

▮ The geometry, boundary conditions, and temperature distribution are symmetrical about the vertical centerline; consequently we may consider one-half of the problem, as shown in Fig. 3-15(*b*). Also, the symmetry assures us that the centerline is adiabatic; thus one boundary condition is that $\partial T/\partial x$ is zero at the centerline. In terms of $\theta \equiv T - T_\infty$, we must solve

$$\frac{\partial^2 \theta}{\partial x^2} = \frac{1}{\alpha}\frac{\partial \theta}{\partial t}$$

subject to

Boundary condition 1 $\quad \frac{\partial \theta}{\partial x} = 0 \quad$ at $\quad x = 0$

Boundary condition 2 $\quad \frac{\partial \theta}{\partial x} = -\frac{\bar{h}}{k}\theta \quad$ at $\quad x = L$

Time condition 1 $\quad \theta = \theta_i \quad$ at $\quad t = 0$

Time condition 2 $\quad \theta \to 0 \quad$ as $\quad t \to \infty$

The classical separation-of-variables technique yields

$$\theta = e^{-\lambda^2 \alpha t}(A \sin \lambda x + B \cos \lambda x) \tag{1}$$

where the separation parameter, $-\lambda^2$, is chosen negative to satisfy time condition 2. Boundary condition 1 requires that $A = 0$, and boundary condition 2 then gives the transcendental equation

$$\cot \lambda L = \frac{\lambda}{\bar{h}/k} \qquad \text{or} \qquad \cot \lambda L = \frac{\lambda L}{\mathbf{Bi}} \tag{2}$$

for λ. Equation (2) has an infinite number of positive roots, $\lambda_1, \lambda_2, \ldots$, and in terms of these eigenvalues the solution for θ is

$$\theta = \sum_{n=1}^{\infty} C_n e^{-\lambda_n^2 \alpha t} \cos \lambda_n x \tag{3}$$

One method of determining the λ_n is by means of an accurate plot of the type shown in Fig. 3-16. Usually the first three or four roots are sufficient to yield an accurate answer.

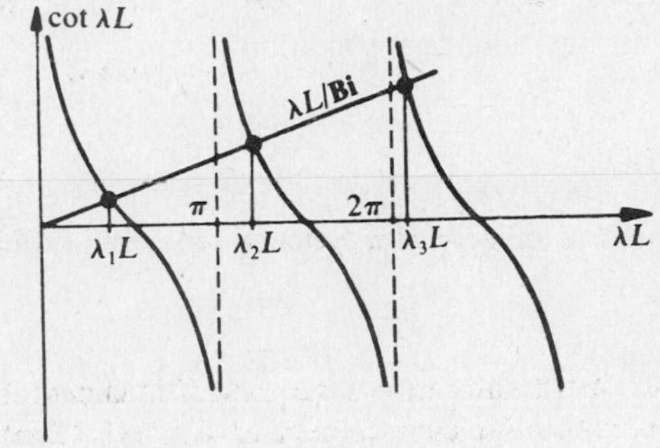

Fig. 3-16

Finally, time condition 1 requires that

$$\theta_i = \sum_{n=1}^{\infty} C_n \cos \lambda_n x$$

which indicates that the C_n must be chosen so that θ_i is represented by an infinite series of cosine terms over the range $0 < x < L$. From the theory of orthogonal functions we find

$$C_n = \frac{2\theta_i \sin \lambda_n L}{\lambda_n L + (\sin \lambda_n L)(\cos \lambda_n L)} \tag{4}$$

Introducing the Fourier modulus (Problem 3.5) and the dimensionless quantities $\delta_n \equiv \lambda_n L$, we obtain our solution in the form

$$\frac{T - T_\infty}{T_i - T_\infty} \equiv \frac{\theta}{\theta_i} = 2 \sum_{n=1}^{\infty} \exp(-\delta_n^2 \, \mathbf{Fo}) \frac{\sin \delta_n}{\delta_n + (\sin \delta_n)(\cos \delta_n)} \cos \frac{\delta_n x}{L} \tag{5}$$

For the dimensionless centerline temperature, (5) gives

$$\frac{T_c - T_\infty}{T_i - T_\infty} = 2 \sum_{n=1}^{\infty} \exp(-\delta_n^2 \, \mathbf{Fo}) \frac{\sin \delta_n}{\delta_n + (\sin \delta_n)(\cos \delta_n)} \tag{6}$$

Equation (6) has been plotted, along with the results for long cylinders and spheres, in the classic Heisler* charts in Appendix D.

The temperature at any position in the plate can be found by applying a position-correction factor to the centerline temperature. Figure D-2 is a plot of θ/θ_c, that is, the ratio of $T - T_\infty$ at any position x/L to $T_c - T_\infty$, at a given dimensionless time (**Fo**).

3.50 Solve Problem 3.49 (finite slab) by a superposition of semi-infinite solutions (Problem 3.47).

▌ To render time condition 1 of Problem 3.49 homogeneous, write

$$\Theta \equiv \frac{T - T_i}{T_\infty - T_i}$$

*From M. D. Heisler, *Trans. ASME*, 69:227 (1947).

producing the problem

$$\frac{\partial^2\Theta}{\partial x^2} = \frac{1}{\alpha}\frac{\partial\Theta}{\partial t}$$

b.c. *1* $\quad \dfrac{\partial\Theta}{\partial x} = 0 \quad$ at $\quad x = 0$

b.c. *2* $\quad \Theta + \beta\dfrac{\partial\Theta}{\partial x} = 1 \quad$ at $\quad x = L$

t.c. *1* $\quad \Theta = 0 \quad$ at $\quad t = 0$

t.c. *2* $\quad \Theta \to 1 \quad$ as $\quad t \to \infty$

with $\beta \equiv k/\bar{h}$. By (*1*) of Problem 3.47, the semi-infinite solution—corresponding to convection at $x = 0^-$—is given by

$$\Theta = \text{erfc}\,\frac{x}{\sqrt{4\alpha t}} - \left[\exp\left(\frac{x}{\beta} + \frac{\alpha t}{\beta^2}\right)\right]\left[\text{erfc}\left(\frac{x}{\sqrt{4\alpha t}} + \frac{\sqrt{\alpha t}}{\beta}\right)\right] \equiv F(x, t) \tag{1}$$

where, at $x = 0$, F satisfies the boundary condition

$$F - \beta\frac{\partial F}{\partial n} = 1 \tag{2}$$

Here, the *inner-normal* derivative $\partial/\partial n$ is the derivative in the $+x$-direction.

The superposition (method of images) of F-functions to solve Problem 3.49 is indicated in Fig. 3-17; clearly, our solution is

$$\frac{T - T_i}{T_\infty - T_i} \equiv \Theta(x, t) = F(L - x, t) + F(L + x, t) - F(3L - x, t)$$
$$- F(3L + x, t) + F(5L - x, t) + F(5L + x, t) - \cdots \tag{3}$$

For $x = 0$, (*3*) implies

$$\frac{T_c - T_\infty}{T_i - T_\infty} = 1 - 2[F(L, t) - F(3L, t) + F(5L, t) - \cdots] \tag{4}$$

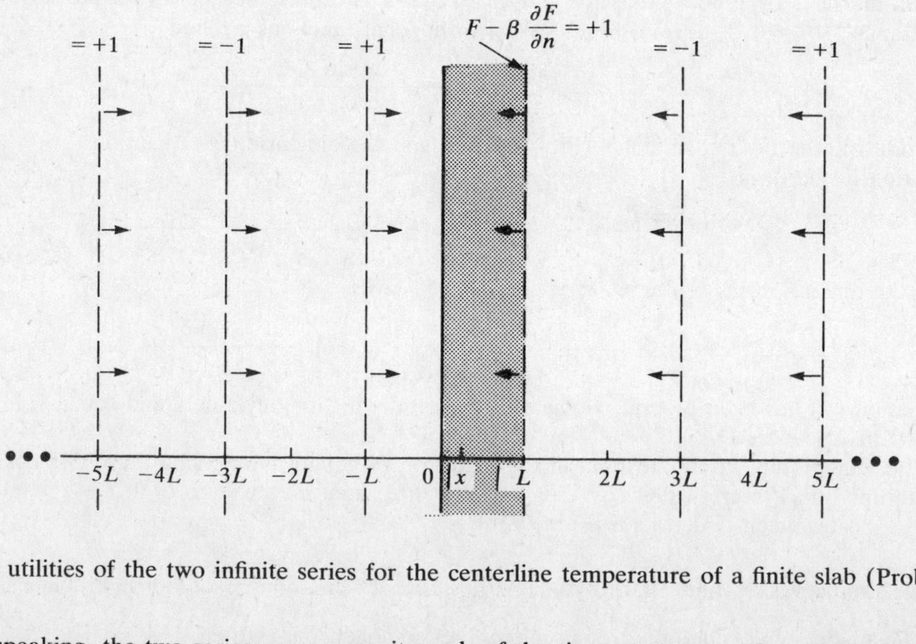

Fig. 3-17

3.51 Compare the utilities of the two infinite series for the centerline temperature of a finite slab (Problems 3.49 and 3.50).

▌ Roughly speaking, the two series cover opposite ends of the time spectrum. The eigenvalue expansion, (*6*) of Problem 3.49, converges fastest for *large* times (large **Fo**); although, because δ_n increases with n, convergence is quite good for small times, too. On the other hand, the image solution, (*4*) of Problem 3.50, converges fastest for *small* times. (It is evident that the smaller the time, the fewer the image sources that make significant contributions at point x.) For large times, the image solution is practically useless; in fact, a special notion of convergence is required to retrieve the limiting value $T_c(\infty) = T_\infty$.

3.52 Calculate the heat transferred to or from the half-slab of Problem 3.49 over the time interval $(0, t)$.

▌ At the convective boundary, $q(L, \tau) = \bar{h}A\theta(L, \tau)$; hence,

$$Q(t) = \bar{h}A \int_0^t \theta(L, \tau)\, d\tau$$

Substituting for the integrand from (*5*) of Problem 3.49, and employing (*2*) of Problem 3.49 in the form

$$\bar{h} \cos \delta_n = \frac{k\delta_n}{L} \sin \delta_n \qquad (n = 1, 2, 3, \ldots)$$

one obtains

$$\frac{Q}{Q_i} = 2 \sum_{n=1}^{\infty} \frac{\sin^2 \delta_n}{\delta_n(\delta_n + \sin \delta_n \cos \delta_n)} [1 - \exp(-\delta_n^2 \,\mathbf{Fo})] \tag{1}$$

where $Q_i \equiv \rho cAL\theta_i$ is the thermal energy above the reference state, T_∞, initially stored in the half-slab per unit depth. The result (*1*) is shown in chart D-3 (also for cylinders and spheres in charts D-5 and D-9) in Appendix D.

3.53 A 4-in.-thick steel slab, very wide and long and initially at 400 °F, is suddenly exposed to a convective-fluid environment at 100 °F. The average convective coefficient is $\bar{h} = 50$ Btu/h · ft² · °F, and the average value of α over the temperature range is 0.452 ft²/h. Determine the centerline temperature and the temperature at a $1\frac{1}{2}$-in. depth after 10 min of exposure.

▌ Figure D-1 of Appendix D will be used to determine the centerline temperature of the slab, and then Figure D-2 will be used to obtain the temperature at a depth of $1\frac{1}{2}$ in. Using $k = 25$ Btu/h · ft · °F (from Table B-1) and $L = 2$ in. $= \frac{1}{6}$ ft, we have

$$\mathbf{Fo} = \frac{\alpha t}{L^2} = \frac{(0.452)(\frac{10}{60})}{(\frac{1}{6})^2} = 2.71 \qquad \frac{1}{\mathbf{Bi}} = \frac{k}{\bar{h}L} = \frac{25}{(50)(\frac{1}{6})} = 3.00 \qquad (<10)$$

From Figure D-1,

$$\frac{T_c - 100}{400 - 100} \approx 0.47 \qquad \text{or} \qquad T_c \approx 241\ \text{°F}$$

From Figure D-2, at $x/L = 0.25$,

$$\frac{T - 100}{241 - 100} \approx 0.98 \qquad \text{or} \qquad T \approx 238\ \text{°F}$$

3.54 How much heat is lost by the slab of Problem 3.53 in the first 10 min?

▌ From Figure D-3, at $\mathbf{Fo} = 2.71$ and $\mathbf{Bi} = 0.333$, $Q/Q_i \approx 0.6$. Using Appendix B, one has for the initial energy capacity per unit area of the *whole slab*

$$Q_i = \rho c(2L)\theta_i = (487\ \text{lbm/ft}^3)(0.113\ \text{Btu/lbm} \cdot \text{°F})(\tfrac{1}{3}\ \text{ft})(300\ \text{°F}) = 5503\ \text{Btu/ft}^2$$

whence $Q = (0.6)(5515) = 3302$ Btu/ft².

3.55 How much time must elapse for the slab of Problem 3.53 to lose 90 percent of its initial heat content?

▌ In Figure D-3 the intersection of the vertical line $\mathbf{Bi} = 1/3.00 = 0.333$ and the horizontal line $Q/Q_i = 0.90$ lies close to the curve $\mathbf{Fo} = 10$. Therefore, since $\mathbf{Fo} = 2.71$ corresponds to 10 min,

$$t \approx \frac{10}{2.71}(10\ \text{min}) = 37\ \text{min}$$

3.56 A concrete wall 1 ft thick completely encloses a paint room in a factory. The wall is initially at 70 °F when a fire erupts in the paint room and causes the inside of the wall to come in contact with hot gases at 1600 °F. The outside of the wall is covered with a material that has a flash-point temperature of 400 °F. The average heat transfer coefficient on the hot side is 5 Btu/h · ft² · °F. Find the time required for the insulating material on the outside of the wall to start burning.

*From H. Gröber, S. Erk, and U. Grigall, *Grundgesetze der Wärmeübertragung*, 3d ed., Springer-Verlag, Berlin, 1955.

▮ Although

$$T_{avg} = \frac{1600 + 70}{2} = 835\ °F$$

we have to use Table B-2 for concrete at 68 °F:

$$k = 0.47\ Btu/h \cdot ft \cdot °F \qquad \alpha = 0.019\ ft^2/h$$

Note that the insulated face corresponds to the centerplane of a slab of thickness $2L$.

$$\frac{1}{\mathbf{Bi}} = \frac{k}{\bar{h}L} = \frac{0.47\ Btu/h \cdot ft \cdot °F}{(5\ Btu/h \cdot ft^2 \cdot °F)(1\ ft)} = 0.094 \qquad (<10)$$

$$\frac{T_c - T_\infty}{T_i - T_\infty} = \frac{(400 - 1600)\ °F}{(70 - 1600)\ °F} = 0.784$$

From Figure D-1, $\mathbf{Fo} = \alpha t/L^2 = 0.22$, so that

$$t = \frac{(0.22)L^2}{\alpha} = \frac{(0.22)(1\ ft)^2}{0.019\ ft^2/h} = 11.58\ h$$

3.3 TRANSIENTS IN LONG CYLINDERS AND SPHERES

3.57 A very long 4-in.-diam. steel cylinder initially at 400 °F is suddenly exposed to a convective-fluid environment at 70 °F. The average convective coefficient is $\bar{h} = 100\ Btu/h \cdot ft^2 \cdot °F$, and $\alpha = 0.452\ ft^2/h$ (average value over the temperature range). Determine the temperature $\frac{1}{2}$ in. from the surface after 5 min.

▮ The Heisler chart, Figure D-4, will be used to find the centerline temperature, and then the position-correction chart, Figure D-6, will be used to determine the temperature $\frac{1}{2}$ in. from the outer surface. Notice that the charts employ the characteristic dimension R, not $R/2$. Using $k = 25\ Btu/h \cdot ft \cdot °F$ (from Table B-1; assumed constant) and $R = 2$ in., we have

$$\mathbf{Fo}^* = \frac{\alpha t}{R^2} = \frac{(0.452)(\frac{1}{12})}{(\frac{1}{6})^2} = 1.35 \qquad \frac{1}{\mathbf{Bi}^*} = \frac{k}{\bar{h}R} = \frac{25}{100(\frac{1}{6})} = 1.50 \qquad (<5)$$

From Figure D-4,

$$\frac{T_c - 70}{400 - 70} \approx 0.25 \qquad \text{or} \qquad T_c \approx 153\ °F$$

From Figure D-6, at $r/R = (2 - \frac{1}{2})/2 = 0.75$,

$$\frac{T - 70}{153 - 70} \approx 0.84 \qquad \text{or} \qquad T \approx 139\ °F$$

3.58 A long 2-in.-diam. aluminum cylinder initially at 500 °F is suddenly exposed to a convective environment at 150 °F. The average heat transfer coefficient is $100\ Btu/h \cdot ft^2 \cdot °F$. Find the energy lost per unit length of the cylinder during the first minute.

▮ For aluminum at the mean temperature, 325 °F,

$$k = 123\ Btu/h \cdot ft \cdot °F \qquad c = 0.29\ Btu/lbm \cdot °F \qquad \rho = 169\ lbm/ft^3 \qquad \alpha = 3.33\ ft^2/h$$

Thus

$$\mathbf{Bi}^* = \frac{\bar{h}R}{k} = \frac{(100)(\frac{1}{12})}{123} = 0.0677$$

$$\mathbf{Fo}^* = \frac{\alpha t}{R^2} = \frac{(3.33\ ft^2/h)(\frac{1}{60}\ h)}{(\frac{1}{12}\ ft)^2} = 8.0$$

and Figure D-5 gives $Q/Q_i \approx 0.6$. But, per unit length,

$$Q_i = \rho c \pi R^2 (T_i - T_\infty) = (169)(0.29)\pi(\tfrac{1}{12})^2(500 - 150) = 375\ Btu/ft$$

so that $Q = (0.6)(375) = 225\ Btu/ft$.

3.59 For Problem 3.58, the Biot number (based on the half-radius) is $\mathbf{Bi} = 0.034 < 0.1$, implying that a lumped analysis should also lead to $Q/Q_i \approx 0.6$. Check this.

▮ For the lumped system, (1) of Problem 3.24 gives

$$\frac{Q}{Q_i} = 1 - \exp(-\mathbf{Bi}\cdot\mathbf{Fo})$$

In the present application, $\mathbf{Bi} = 0.034$ and $\mathbf{Fo} = 4\,\mathbf{Fo}^* = 32.0$; hence

$$\frac{Q}{Q_i} = 1 - \exp(-1.08) \approx 1 - 0.34 = 0.66$$

The agreement is satisfactory, especially in view of (1) the difficulty in inferring the "exact" solution from Figure D-5 and (2) the inconsistent data $(\alpha \neq k/\rho c)$.

3.60 A 3-in.-diam. orange originally at 80 °F is placed in a refrigerator where the air temperature is 35 °F and the average convective heat transfer coefficient over the surface of the orange is $\bar{h} = 10\ \text{Btu/h}\cdot\text{ft}^2\cdot{}^\circ\text{F}$. Estimate the time required for the center of the orange to reach 40 °F.

▮ Since an orange is mainly water, we will take the properties to be those of water at $T_{avg} = (40+80)/2 = 60\ {}^\circ\text{F}$. Thus, from Appendix B, by linear interpolation,

$$\alpha \approx 5.44\times 10^{-3}\ \text{ft}^2/\text{h} \qquad k \approx 0.339\ \text{Btu/h}\cdot\text{ft}\cdot{}^\circ\text{F}$$

The dimensionless parameters for the Heisler chart are

$$\frac{1}{\mathbf{Bi}^*} = \frac{k}{\bar{h}R} \approx \frac{0.339}{(10)(1.5/12)} = 0.27 \qquad \frac{T_c - T_\infty}{T_i - T_\infty} = \frac{40-35}{80-35} = 0.11$$

From Figure D-7, $\mathbf{Fo}^* \approx 0.5$ and thus

$$t = \frac{\mathbf{Fo}^*\,R^2}{\alpha} \approx \frac{(0.5)(1.5/12)^2\ \text{ft}^2}{5.44\times 10^{-3}\ \text{ft}^2/\text{h}} = 1.44\ \text{h}$$

Because $\mathbf{Bi} = \mathbf{Bi}^*/3 = 1/3(0.27) = 1.23 > 0.1$, a lumped-capacity approach is not suitable.

3.61 A 4-in.-diam. iron ball is suddenly exposed to a stream of ice water. It is initially at 70 °F, and $\bar{h} = 100\ \text{Btu/h}\cdot\text{ft}^2\cdot{}^\circ\text{F}$. What is the temperature $\frac{1}{2}$ in. from the center 5 min later?

▮ From Table B-1, $k = 35.8\ \text{Btu/h}\cdot\text{ft}\cdot{}^\circ\text{F}$ and $\alpha = 0.70\ \text{ft}^2/\text{h}$; hence

$$\mathbf{Fo}^* = \frac{\alpha t}{R^2} = \frac{(0.70\ \text{ft}^2/\text{h})(\frac{5}{60}\ \text{h})}{(2/12\ \text{ft})^2} = 2.1 \qquad \frac{1}{\mathbf{Bi}^*} = \frac{k}{\bar{h}R} = \frac{35.8\ \text{Btu/h}\cdot\text{ft}\cdot{}^\circ\text{F}}{(100\ \text{Btu/h}\cdot\text{ft}^2\cdot{}^\circ\text{F})(\frac{2}{12}\ \text{ft})} = 2.15 \qquad (<3.33)$$

From Figure D-7,

$$\frac{T_c - 32}{70-32} \approx 0.076 \qquad \text{or} \qquad T_c \approx 35\ {}^\circ\text{F}$$

From Figure D-8, at $r/R = 0.5/2 = 0.25$,

$$\frac{T-32}{35-32} \approx 0.98 \qquad \text{or} \qquad T \approx 35\ {}^\circ\text{F}$$

Using the charts, one cannot accurately distinguish between the temperatures at the center and $\frac{1}{2}$ in. from the center.

3.62 A solid steel 1.50-in.-diam. sphere is to be heat-treated in the following manner. The temperature of the whole sphere is to be raised uniformly to 1400 °F. Then the sphere is to be plunged into a large lead bath, where its surface is immediately brought to a temperature of 750 °F and kept there until the center temperature of the sphere drops to 900 °F. At this moment, the sphere is to be removed from the lead and quenched in a cold brine bath. The properties of the steel may be taken as follows: $\rho = 485\ \text{lbm/ft}^3$, $c = 0.11\ \text{Btu/lbm}\cdot{}^\circ\text{F}$, $k = 20\ \text{Btu/h}\cdot\text{ft}\cdot{}^\circ\text{F}$. How long should the sphere be kept in the lead bath if the heat transfer coefficient is $200\ \text{Btu/h}\cdot\text{ft}^2\cdot{}^\circ\text{F}$?

▮

$$\alpha = \frac{k}{\rho c} = \frac{20\ \text{Btu/h}\cdot\text{ft}\cdot{}^\circ\text{F}}{(485\ \text{lbm/ft}^3)(0.11\ \text{Btu/lbm}\cdot{}^\circ\text{F})} = 0.375\ \text{ft}^2/\text{h}$$

$$\frac{1}{\mathbf{Bi}^*} = \frac{k}{\bar{h}R} = \frac{20\ \text{Btu/h}\cdot\text{ft}\cdot{}^\circ\text{F}}{(200\ \text{Btu/h}\cdot\text{ft}^2\cdot{}^\circ\text{F})[(0.75/12)\ \text{ft}]} = 1.6 \qquad (<3.33)$$

$$\frac{T_c - T_\infty}{T_i - T_\infty} = \frac{900-750}{1400-750} = 0.231$$

From Figure D-7, $\mathbf{Fo^*} \approx 1.03$; hence

$$t = \frac{\mathbf{Fo^*}\, R^2}{\alpha} \approx \frac{(1.03)(0.75/12)^2}{0.375} = 0.0107 \text{ h} = 0.643 \text{ min}$$

3.63 A 2-in.-diam. copper sphere initially at 300 °F is cooled in a liquid bath at 100 °F; the average heat transfer coefficient is 100 Btu · ft² · °F. How much heat has been removed from the sphere by the time the center temperature reaches 120 °F?

For copper, at $T_{avg} = (300 + 100)/2 = 200\ °F \approx 212\ °F$,

$$k = 218 \text{ Btu/h} \cdot \text{ft} \cdot °\text{F} \qquad c = 0.091 \text{ Btu/lbm} \cdot °\text{F} \qquad \rho = 558 \text{ lbm/ft}^3 \qquad \alpha = 4.42 \text{ ft}^2/\text{h}$$

First find the time $\mathbf{Fo^*}$.

$$\frac{T_c - T_\infty}{T_i - T_\infty} = \frac{120 - 100}{300 - 100} = 0.1 \qquad \frac{1}{\mathbf{Bi^*}} = \frac{k}{\bar{h}R} = \frac{218 \text{ Btu/h} \cdot \text{ft} \cdot °\text{F}}{(100 \text{ Btu/h} \cdot \text{ft}^2 \cdot °\text{F})[(\frac{1}{12}) \text{ ft}]} = 26.16$$

From Figure D-7, $\mathbf{Fo^*} \approx 20$; whence, from Figure D-9, $Q/Q_i \approx 0.85$. But

$$Q_i = \rho c V(T_i - T_\infty) = (558)(0.091)(\tfrac{4}{3}\pi)(\tfrac{1}{12})^3(300 - 100) = 24.6 \text{ Btu}$$

and so $Q = (0.85)(24.6) = 20.9$ Btu.

3.64 Noting that $\mathbf{Bi} = \mathbf{Bi^*}/3 = 1/3(26.16) = 0.0127 < 0.1$ in Problem 3.63, make a lumped analysis.

For the lumped system, (*1*) of Problem 3.24 gives

$$\frac{Q}{Q_i} = 1 - \exp(-\mathbf{Bi} \cdot \mathbf{Fo})$$

Here, $\mathbf{Bi} = 0.0127$ and $\mathbf{Fo} = 9\,\mathbf{Fo^*} = 180$; therefore,

$$\frac{Q}{Q_i} = 1 - \exp(-2.29) \approx 1 - 0.101 = 0.9$$

The agreement with Problem 3.63 is good (cf. Problems 3.58 and 3.59).

3.65 In preparation for a party, the host wishes to chill canned soft drinks from 75 °F to 35 °F. The freezer temperature is 0 °F, and an external heat transfer coefficient of 5 Btu/h · ft² · °F is appropriate. Assume the properties of the soft drink to be similar to those of water; the cans are $2\frac{1}{2}$ in. in diameter and 6 in. long. Neglecting end effects and internal fluid motion due to natural convection, approximate the time required for chilling.

At $T_{avg} = (75 + 35)/2 = 55\ °F$, $k = 0.336$ Btu/h · ft · °F and $\alpha = 5.4 \times 10^{-3}$ ft²/h.

$$\frac{Q}{Q_i} = \frac{\rho c V(T_i - T)}{\rho c V(T_i - T_\infty)} = \frac{75 - 35}{75 - 0} = 0.533$$

$$\mathbf{Bi^*} = \frac{\bar{h}R}{k} = \frac{(5 \text{ Btu/h} \cdot \text{ft}^2 \cdot °\text{F})[(1.25 \text{ ft})/12]}{0.336 \text{ Btu/h} \cdot \text{ft} \cdot °\text{F}} = 1.55$$

Then, from Figure D-5, $\mathbf{Fo^*} = \alpha t/R^2 \approx 0.20$, or

$$t = \frac{(0.20)R^2}{\alpha} = \frac{(0.20)(1.25/12 \text{ ft})^2}{0.0054 \text{ ft}^2/\text{h}} = 0.402 \text{ h} = 24.1 \text{ min}$$

Besides the explicit neglect of end effects in the above solution, there was an implicit neglect of internal temperature gradients involved in writing $Q = \rho c V(T_i - T)$. On the other hand, $\mathbf{Bi} = \mathbf{Bi^*}/2 = 0.78 > 0.1$. Thus the solution must be considered crude. For a better approach, see Problem 3.70.

3.66 In the winter citrus-fruit growers in Florida must concern themselves with frost damage to their crops. On a clear, windless night the average heat transfer coefficient to spherically shaped fruit on trees is approximately 2 Btu/h · ft² · °F. A part of this surface loss is actually due to radiation to the sky, which can be significantly reduced by the use of smudge pots. The ambient temperature is 52 °F, and a cold front suddenly moves into the region, quickly lowering the temperature to 28 °F, at which value it is expected to remain for 8 h. Should smudge pots be used to protect 6-in.-diam. grapefruit? (Examine the surface temperature, assuming fruit properties to be those of water.)

▌ At $T_{avg} = (52 + 28)/2 = 40$ °F, $k = 0.325$ Btu/h · ft · °F and $\alpha = 5.21 \times 10^{-3}$ ft²/h. Hence,

$$\frac{1}{\mathbf{Bi}^*} = \frac{k}{\bar{h}R} = \frac{0.325 \text{ Btu/h}\cdot\text{ft}\cdot{}^\circ\text{F}}{(2 \text{ Btu/h}\cdot\text{ft}^2\cdot{}^\circ\text{F})(\frac{3}{12}\text{ ft})} = 0.65 \qquad (<3.33)$$

$$\mathbf{Fo}^* = \frac{\alpha t}{R^2} = \frac{(0.0052 \text{ ft}^2/\text{h})(8 \text{ h})}{(\frac{3}{12}\text{ ft})^2} = 0.666$$

From Figure D-7,

$$\frac{T_c - 28}{52 - 28} \approx 0.15 \qquad \text{or} \qquad T_c \approx 31.6 \text{ °F}$$

For the surface temperature, Figure D-8 gives $(r/R = 1)$

$$\frac{T - 28}{31.6 - 28} \approx 0.51 \qquad \text{or} \qquad T \approx 29.8 \text{ °F}$$

Thus, smudge pots must be used in order to keep the grapefruit from freezing.

3.4 TWO- AND THREE-DIMENSIONAL TRANSIENTS

3.67 Using the nomenclature of Fig. 3-18, determine an expression for the temperature gradient in a rectangular two-dimensional body. Infer the results for a three-dimensional parallelepiped and cylinder.

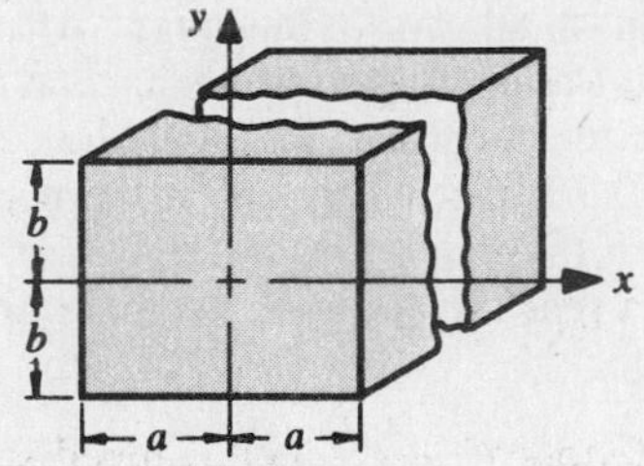

Fig. 3-18

▌ The general two-dimensional conduction equation

$$\frac{\partial^2\theta}{\partial x^2} + \frac{\partial^2\theta}{\partial y^2} = \frac{1}{\alpha}\frac{\partial\theta}{\partial t}$$

may be solved by the separation of variables techniques; i.e., $\theta(x, y, t) = X(x, t)Y(y, t)$. The x- and y-solutions will have the form (5) of Problem 3.49, with $L = a$ in one case and $L = b$ in the other. Thus

$$\frac{\theta}{\theta_i} \equiv \left(\frac{T - T_\infty}{T_i - T_\infty}\right)_{\substack{\text{long}\\\text{bar}}} = \left(\frac{T - T_\infty}{T_i - T_\infty}\right)_{\substack{2a\text{-}\\\text{slab}}}\left(\frac{T - T_\infty}{T_i - T_\infty}\right)_{\substack{2b\text{-}\\\text{slab}}} \tag{1}$$

and the $2a$-slab and the $2b$-slab solutions may be taken from Figures D-1 and D-2. It follows that the temperature distribution in a rectangular parallelepiped (box-shaped object) is given by

$$\left(\frac{T - T_\infty}{T_i - T_\infty}\right)_{\text{box}} = \left(\frac{T - T_\infty}{T_i - T_\infty}\right)_{\substack{2a\text{-}\\\text{slab}}}\left(\frac{T - T_\infty}{T_i - T_\infty}\right)_{\substack{2b\text{-}\\\text{slab}}}\left(\frac{T - T_\infty}{T_i - T_\infty}\right)_{\substack{2c\text{-}\\\text{slab}}} \tag{2}$$

Similarly, the transient temperature in a cylinder with both radial and axial heat transfer is given by

$$\left(\frac{T - T_\infty}{T_i - T_\infty}\right)_{\substack{\text{short}\\\text{cyl.}}} = \left(\frac{T - T_\infty}{T_i - T_\infty}\right)_{\substack{\text{inf.}\\\text{cyl.}}}\left(\frac{T - T_\infty}{T_i - T_\infty}\right)_{\substack{2L\text{-}\\\text{slab}}} \tag{3}$$

where $2L$ is the height of the cylinder.

3.68 A steel roller bearing at 1600 °F is to be immersed in a liquid bath at 100 °F during a heat-treating process. The bearing diameter is $\frac{1}{4}$ in., and its length is $\frac{1}{2}$ in. The heat transfer coefficient $\bar{h}$ may be taken as 100 Btu/h · ft² · °F. Determine the maximum temperature in the bearing 1 min after immersion; $\alpha =$ 0.425 ft²/h.

▌ The maximum temperature will obviously be at the center of the bearing, both radially and axially (at $r=0$ and $\frac{1}{4}$ in. from either end). The temperature at this point can be found from (3) of Problem 3.67, with the two dimensionless temperatures on the right obtained from Figures D-1 and D-4.

Using $k=25$ Btu/h·ft·°F (Table B-1) and $L=\frac{1}{48}$ ft, we have for the slab solution:

$$\mathbf{Fo}^* = \frac{\alpha t}{L^2} = \frac{(0.452)(1/60)}{(\frac{1}{48})^2} = 17.4 \qquad \frac{1}{\mathbf{Bi}^*} = \frac{k}{\bar{h}L} = \frac{25}{(100)(\frac{1}{48})} = 12$$

whence

$$\left(\frac{T_c - T_\infty}{T_i - T_\infty}\right)_{\text{slab}} \approx 0.25$$

For the cylinder solution:

$$\mathbf{Fo}^* = \frac{\alpha t}{R^2} = \frac{(0.452)(\frac{1}{60})}{(\frac{1}{96})^2} = 69.6 \qquad \frac{1}{\mathbf{Bi}^*} = \frac{k}{\bar{h}R} = \frac{25}{(100)(\frac{1}{96})} = 24$$

whence

$$\left(\frac{T_c - T_\infty}{T_i - T_\infty}\right)_{\text{cylinder}} \approx 0.0035$$

Consequently,

$$\left(\frac{T - T_\infty}{T_i - T_\infty}\right)_{\max} \approx (0.25)(0.0035) = 0.000\,875$$

and substitution of $T_i = 1600$ °F and $T_\infty = 100$ °F gives $T_{\max} \approx 101$ °F.

3.69 Tomatoes are processed in 8-in.-diam. cans 8 in. long. After canning, the tomatoes are sterilized in a steam autoclave at 220 °F. The initial temperature of the juice at the start of the sterilizing process is 120 °F. Assume a very large heat transfer coefficient and that the contents of the cans are at rest and have a density of 60 lbm/ft^3, specific heat of 0.9 Btu/lbm·°F, and a thermal conductivity of 0.4 Btu/h·ft·°F. Consider the temperature difference between the steam and the inner surfaces of the cans as negligible. How much time is required for the tomatoes to reach a sterilizing temperature of 200 °F at the center of the can?

▌ By (3) of Problem 3.67,

$$\left(\frac{\theta_c}{\theta_i}\right)_{\substack{\text{short}\\\text{cyl.}}} = \frac{200-220}{120-220} = 0.2 = \left(\frac{\theta_c}{\theta_i}\right)_{\substack{2L\text{-}\\\text{plate}}} \left(\frac{\theta_c}{\theta_i}\right)_{\substack{\text{inf.}\\\text{cyl.}}} \tag{1}$$

A trial-and-error solution is required, but due to the dimensions

$$\mathbf{Fo}^*_{\substack{\text{inf.}\\\text{cyl.}}} = \mathbf{Fo}_{\substack{2L\text{-}\\\text{plate}}}$$

i.e., the abscissas on Figures D-1 and D-4 will be the same. On the line $1/\mathbf{Bi}^* = 0$ (since we are assuming a large $\bar{h}$), at $\mathbf{Fo}^* = 0.26$, we find

$$\left(\frac{\theta_c}{\theta_i}\right)_{\substack{2L\text{-}\\\text{plate}}} = 0.6 \qquad \text{and} \qquad \left(\frac{\theta_c}{\theta_i}\right)_{\substack{\text{inf.}\\\text{cyl.}}} = 0.34$$

the product of which is $0.204 \approx 0.2$. Therefore, (1) is satisfied and $\mathbf{Fo}^* = \alpha t/R^2 = 0.26$. At $T_{\text{avg}} = (200+120)/2 = 160$ °F, $\alpha = 6.32 \times 10^{-3}$ ft^2/h; consequently,

$$t = \frac{(0.26)(\frac{4}{12}\text{ ft})^2}{0.006\,32\text{ ft}^2/\text{h}} = 4.57\text{ h}$$

3.70 At a picnic canned soft drinks which have been left in the sun are initially at a temperature of 100 °F. The cans are then placed in a cooler containing ice water. The average heat transfer coefficient is 50 Btu/h·ft^2·°F. The cans have a $2\frac{1}{2}$-in. diameter and are 6 in. long. Assume the properties of the soft drinks to be similar to those of water and that the fluid in the cans is at rest. Find the time required for the center temperature of the drinks to reach 40 °F, considering both radial and axial conduction.

▌ By (3) of Problem 3.67,

$$\left(\frac{\theta_c}{\theta_i}\right)_{\substack{\text{short}\\\text{cyl.}}} = \frac{40-32}{100-32} = 0.1175 = \left(\frac{\theta_c}{\theta_i}\right)_{\substack{2L\text{-}\\\text{plate}}} \left(\frac{\theta_c}{\theta_i}\right)_{\substack{\text{inf.}\\\text{cyl.}}} \tag{1}$$

A trial-and-error solution is required, but due to the dimensions a relationship can be found between the Fourier moduli of the plate and the infinite cylinder:

$$\frac{\mathbf{Fo}^*}{\mathbf{Fo}} = \frac{L^2}{R^2} = \frac{(3)^2}{(1.25)^2} = 5.76 \qquad (2)$$

Moreover, we can take $1/\mathbf{Bi} = 1/\mathbf{Bi}^* = 0$ (since $\bar{h}$ is very large).

First guess: $\mathbf{Fo} = 0.1$, $\mathbf{Fo}^* = 0.576$. Then,

$$\left(\frac{\theta_c}{\theta_i}\right)_{\substack{2L\text{-}\\ \text{plate}}} \left(\frac{\theta_c}{\theta_i}\right)_{\substack{\text{inf.}\\ \text{cyl.}}} = (0.90)(0.07) = 0.063 \qquad \textit{(too low)}$$

Second guess: $\mathbf{Fo} = 0.075$, $\mathbf{Fo}^* = 0.432$. Then,

$$\left(\frac{\theta_c}{\theta_i}\right)_{\substack{2L\text{-}\\ \text{plate}}} \left(\frac{\theta_c}{\theta_i}\right)_{\substack{\text{inf.}\\ \text{cyl.}}} = (0.92)(0.14) = 0.1288 \qquad \textit{(too high)}$$

Third guess: $\mathbf{Fo} = 0.80$, $\mathbf{Fo}^* = 0.461$. Then,

$$\left(\frac{\theta_c}{\theta_i}\right)_{\substack{2L\text{-}\\ \text{plate}}} \left(\frac{\theta_c}{\theta_i}\right)_{\substack{\text{inf.}\\ \text{cyl.}}} = (0.92)(0.13) = 0.1196 \approx 0.1175$$

Thus we accept $\mathbf{Fo}^* = \alpha t/R^2 = 0.461$. Since, at $T_{\text{avg}} = (100 + 40)/2 = 70\ ^\circ\text{F}$, Table B-3 gives $\alpha = 5.57 \times 10^{-3}\ \text{ft}^2/\text{h}$,

$$t = \frac{(0.461)[(1.25/12)\ \text{ft}]^2}{5.57 \times 10^{-3}\ \text{ft}^2/\text{h}} = 0.897\ \text{h} = 53.8\ \text{min}$$

3.71 A long piece of mild-steel stock, 2 by 3 in., initially at 2200 °F, is heat-treated by quenching in an oil bath. The temperature of the oil bath is 200 °F, and the average heat transfer coefficient is assumed to be 80 Btu/h · ft² · °F. What is the temperature at the geometric center after 2 min of quenching?

▌ Solution (*1*) of Problem 3.67 applies. At $T_{\text{avg}} = (2200 + 200)/2 = 1200\ ^\circ\text{F}$, Table B-1 gives

$$k = 22\ \text{Btu/h}\cdot\text{ft}\cdot{}^\circ\text{F} \qquad \alpha = 0.49\ \text{ft}^2/\text{h}$$

For a 2-in. slab,

$$\frac{1}{\mathbf{Bi}} = \frac{k}{\bar{h}L} = \frac{22\ \text{Btu/h}\cdot\text{ft}\cdot{}^\circ\text{F}}{(80\ \text{Btu/h}\cdot\text{ft}^2\cdot{}^\circ\text{F})(\frac{1}{12}\ \text{ft})} = 3.3 \qquad \mathbf{Fo} = \frac{\alpha t}{L^2} = \frac{(0.49\ \text{ft}^2/\text{h})(\frac{2}{60}\ \text{h})}{(\frac{1}{12})^2\ \text{ft}^2} = 2.35$$

From Figure D-1, $\theta_c/\theta_i = 0.55$.

For a 3-in. slab,

$$\frac{1}{\mathbf{Bi}} = \frac{2}{3}(3.3) = 2.2 \qquad \mathbf{Fo} = \frac{4}{9}(2.35) = 1.045$$

From Figure D-1, $\theta_c/\theta_i = 0.69$.

Therefore,

$$\left(\frac{T_c - T_\infty}{T_i - T_\infty}\right)_{\substack{\text{long}\\ \text{bar}}} = (0.55)(0.69) = 0.3795$$

and substitution of $T_\infty = 200\ ^\circ\text{F}$ and $T_i = 2200\ ^\circ\text{F}$ yields $T_c = 959\ ^\circ\text{F}$.

3.72 A masonry brick 2 by 4 by 8 in. initially at 60 °F is placed in a kiln at a temperature of 1800 °F. Assume the average heat transfer coefficient for the brick to be 10 Btu/h · ft² · °F. Assuming $k = 0.38\ \text{Btu/h}\cdot\text{ft}\cdot{}^\circ\text{F}$ and $\alpha = 0.018\ \text{ft}^2/\text{h}$, what is the temperature at the geometric center after a period of 1 h?

▌ Solution (*2*) of Problem 3.67 pertains. For a 2-in. slab,

$$\frac{1}{\mathbf{Bi}} = \frac{k}{\bar{h}L} = \frac{0.38\ \text{Btu/h}\cdot\text{ft}\cdot{}^\circ\text{F}}{(10\ \text{Btu/h}\cdot\text{ft}^2\cdot{}^\circ\text{F})(\frac{1}{12}\ \text{ft})} = 0.456 \qquad \mathbf{Fo} = \frac{\alpha t}{L^2} = \frac{(0.018\ \text{ft}^2/\text{h})(1\ \text{h})}{(\frac{1}{12})^2\ \text{ft}^2} = 2.59$$

From Figure D-1, $\theta_c/\theta_i = 0.05$.

For a 4-in. slab,

$$\frac{1}{\mathbf{Bi}} = \frac{2}{4}(0.456) = 0.228 \qquad \mathbf{Fo} = \frac{4}{16}(2.59) = 0.65$$

From Figure D-1, $\theta_c/\theta_i = 0.40$.
For an 8-in. slab,

$$\frac{1}{\mathbf{Bi}} = \frac{2}{8}(0.456) = 0.114 \qquad \mathbf{Fo} = \frac{4}{64}(2.59) = 0.16$$

From Figure D-1, $\theta_c/\theta_i = 0.94$.
Therefore,

$$\frac{T_c - T_\infty}{T_i - T_\infty} = (0.05)(0.40)(0.94) = 0.0188$$

and substitution of $T_\infty = 1800$ °F and $T_i = 60$ °F gives $T_c = 1767$ °F.

3.73 A home frozen-food chest has inner dimensions 3 ft by 3 ft by 8 ft. The walls are insulated with $\frac{1}{2}$ in. of fine glass wool. The chest is initially filled, and the stored items may be considered to have average properties two-thirds those tabulated for water. Assume that the initial storage temperature is −5 °F, that the room temperature is 85 °F, and that there is a power failure. The average free-convection coefficient may be taken as 4 Btu/h · ft² · °F. (Note that the resistance due to the insulation is significant.) How long can the power failure persist before the center temperature of the stored product reaches 20 °F?

▌ Solution (*2*) of Problem 3.67 pertains. At $T_{avg} = (85-5)/2 = 40$ °F, Table B-3 gives

$$k_{sp} = (\tfrac{2}{3})(0.325) = 0.216 \text{ Btu/h}\cdot\text{ft}\cdot°\text{F} \qquad \alpha_{sp} = \tfrac{2}{3}(5.21\times10^{-3}) = 3.47\times10^{-3}\ \text{ft}^2/\text{h}$$
$$k_{gw} = 0.024 \text{ Btu/h}\cdot\text{ft}\cdot°\text{F}$$

First we must find an effective heat transfer coefficient:

$$\bar{h}_{eff} = \frac{1}{\dfrac{1}{h_o} + \dfrac{L_{gw}}{k_{gw}}} = \frac{1}{\dfrac{1}{4\text{ Btu/h}\cdot\text{ft}\cdot°\text{F}} + \dfrac{\frac{1}{24}\text{ ft}}{0.024\text{ Btu/h}\cdot\text{ft}\cdot°\text{F}}} = 0.504 \text{ Btu/h}\cdot\text{ft}^2\cdot°\text{F}$$

A trial-and-error procedure will be required to determine the time. It is evident that this is a three-dimensional problem; but as a first approximation we will neglect the 8-ft direction and solve a two-dimensional (square) problem.

$$\left(\frac{T_c - T_\infty}{T_i - T_\infty}\right)_{\text{square}} = \frac{20-85}{-5-85} = 0.722 = \left[\left(\frac{\theta_c}{\theta_i}\right)_{\substack{\text{3-ft}\\\text{slab}}}\right]^2$$

From Figure D-1, we need to choose **Fo** such that, for

$$\left(\frac{1}{\mathbf{Bi}}\right)_{3\text{ ft}} = \frac{0.216}{(0.504)(1.5)} = 0.286$$

$\theta_c/\theta_i = (0.722)^{1/2} = 0.85$; we find $\mathbf{Fo} = 0.275$. Therefore,

$$t = \frac{\mathbf{Fo}\,L^2}{\alpha} = \frac{(0.275)(1.5)^2}{3.47\times10^{-3}} = 178.5 \text{ h}$$

The time for the three-dimensional case will be less than this, because of the additional heat transfer through the 8-ft sides. For the three-dimensional problem,

$$\left(\frac{1}{\mathbf{Bi}}\right)_{3\text{ ft}} = 0.286 \qquad \left(\frac{1}{\mathbf{Bi}}\right)_{8\text{ ft}} = \frac{3}{8}(0.286) = 0.1074$$

$$\frac{(\mathbf{Fo})_{8\text{ ft}}}{(\mathbf{Fo})_{3\text{ ft}}} = \frac{3^2}{8^2} = 0.14$$

and we must find the two **Fo**-values such that

$$\left[\left(\frac{\theta_c}{\theta_i}\right)_{3\text{ ft}}\right]^2 \left(\frac{\theta_c}{\theta_i}\right)_{8\text{ ft}} = 0.722$$

First guess: $(\mathbf{Fo})_{3\text{ ft}} = 0.15$, $(\mathbf{Fo})_{8\text{ ft}} = 0.0211$. Then,

$$(0.94)^2(1.0) = 0.884 \qquad \textit{(too high)}$$

Second guess: $(\mathbf{Fo})_{3\text{ ft}} = 0.20$, $(\mathbf{Fo})_{8\text{ ft}} = 0.028$. Then,

$$(0.90)^2(1.0) = 0.81 \qquad (\textit{too high})$$

Third guess: $(\mathbf{Fo})_{3\text{ ft}} = 0.225$, $(\mathbf{Fo})_{8\text{ ft}} = 0.0316$. Then,

$$(0.875)^2(1.0) = 0.765 \approx 0.722$$

Therefore, we accept $(\mathbf{Fo})_{3\text{ ft}} = \alpha t/L^2 = 0.225$, and obtain

$$t = \frac{(0.225)(1.5\text{ ft})^2}{3.47 \times 10^{-3}\text{ ft}^2/\text{h}} = 146\text{ h} = 6.08\text{ day}$$

3.74 At the time determined in Problem 3.73, will any of the frozen product have spoiled?

▌ In order to check if the stored items have spoiled, we must find the outer surface temperatures (different on the 3-ft sides than on the 8-ft sides). From Figure D-2, for $1/\mathbf{Bi} = 0.286$ and $x/L = 1.0$,

$$\frac{T_s - 85}{20 - 85} = 0.33 \qquad \text{or} \qquad T_s = 63.6\text{ °F} \quad (\textit{spoiled})$$

From Figure D-2, for $1/\mathbf{Bi} = 0.1074$ and $x/L = 1.0$,

$$\frac{T_s - 85}{20 - 85} = 0.15 \qquad \text{or} \qquad T_s = 75.3\text{ °F} \quad (\textit{spoiled})$$

3.75 A solid mild steel 2-in.-diam. by 2.5-in.-long cylinder, initially at 1200 °F, is quenched during heat treatment in a fluid at 200 °F. The surface heat transfer coefficient is 150 Btu/h · ft² · °F. Determine the temperature at the centerpoint 2.7 min after immersion in the fluid.

▌ Checking for suitability of a lumped analysis (length = $2a$),

$$L = \frac{V}{A_s} = \frac{\pi R^2(2a)}{2\pi R(2a) + 2\pi R^2} = \frac{Ra}{2a + R} = \frac{(1)(1.25)}{2.5 + 1} = 0.36\text{ in.}$$

and, using k for mild steel at $T_{\text{avg}} = 572$ °F,

$$\mathbf{Bi} = \frac{\bar{h}L}{k_{572}} = \frac{(150\text{ Btu/h}\cdot\text{ft}^2\cdot\text{°F})[(0.36/12)\text{ ft}]}{25\text{ Btu/h}\cdot\text{ft}\cdot\text{°F}} = 0.18 > 0.1$$

A lumped-thermal-capacity analysis is not suitable; instead one must use (*3*) of Problem 3.67. We shall also need the thermal diffusivity at 572 °F; from Appendix B,

$$\alpha_{572} = \alpha_{32}\frac{k_{572}}{k_{32}} = (0.49)\left(\frac{25}{26.5}\right) = 0.462\text{ ft}^2/\text{h}$$

Cylindrical subproblem

$$\frac{1}{\mathbf{Bi}^*} = \frac{k_{572}}{\bar{h}L} = \frac{25}{(150)(\frac{1}{12})} = 2.0 \qquad \mathbf{Fo}^* = \frac{\alpha_{572}t}{R^2} = \frac{(0.462)(2.7/60)}{(\frac{1}{12})^2} = 2.99$$

From Figure D-4,

$$\left(\frac{T_c - T_\infty}{T_i - T_\infty}\right)_{\substack{\text{inf.}\\ \text{cyl.}}} \approx 0.079$$

Slab subproblem

$$\frac{1}{\mathbf{Bi}} = \frac{k_{572}}{\bar{h}a} = \frac{25}{(150)(1.25/12)} = 1.6 \qquad \mathbf{Fo} = \frac{\alpha_{572}}{a^2} = \frac{(0.462)(2.7/60)}{(1.25/12)^2} = 1.92$$

From Figure D-1,

$$\left(\frac{T_c - T_\infty}{T_i - T_\infty}\right)_{\substack{2a\text{-}\\ \text{slab}}} \approx 0.4$$

Complete solution

$$\left(\frac{T_c - T_\infty}{T_i - T_\infty}\right)_{\substack{\text{short}\\ \text{cyl.}}} \approx (0.079)(0.4) = 0.032$$

which gives $T_c = 200 + (0.032)(1200 - 200) = 232$ °F as the temperature at the axial and radial center of the cylinder, at $t = 2.7$ min.

3.76 For the conditions of Problem 3.75, determine the temperature within the cylinder 0.5 in. from one flat face and at $r = 0.5$ in., 2.7 min after immersion.

▮ ***Cylinder subproblem***
From Problem 3.75,

$$\frac{1}{\mathbf{Bi}^*} = 2.0 \qquad \mathbf{Fo}^* = 2.99 \qquad \frac{T_c - T_\infty}{T_i - T_\infty} = 0.079$$

At $r/R = 0.5/1.0 = 0.5$, from Figure D-4,

$$\frac{T - T_\infty}{T_c - T_\infty} \approx 0.94$$

Thus

$$\frac{T - T_\infty}{T_i - T_\infty} = \left(\frac{T - T_\infty}{T_c - T_\infty}\right)\left(\frac{T_c - T_\infty}{T_i - T_\infty}\right) \approx (0.94)(0.079) = 0.074$$

Slab subproblem
From Problem 3.75,

$$\frac{1}{\mathbf{Bi}} = 1.6 \qquad \mathbf{Fo} = 1.92 \qquad \frac{T_c - T_\infty}{T_i - T_\infty} \approx 0.4$$

Using Figure D-1 with

$$\frac{x}{a} = \frac{1.25 - 0.5}{1.25} = 0.6$$

yields

$$\frac{T - T_\infty}{T_c - T_\infty} \approx 0.905$$

Thus

$$\frac{T - T_\infty}{T_i - T_\infty} \approx (0.905)(0.4) = 0.36$$

Complete solution
By (*3*) of Problem 3.67,

$$\frac{T - T_\infty}{T_i - T_\infty} \approx (0.074)(0.36) = 0.0266$$

or $T \approx 200 + (0.0266)(1200 - 200) = 226.6$ °F; this is the temperature, at $t = 2.7$ min, on the circle of intersection of the cylinder $r = 0.5$ in. and the plane $x = 0.75$ in.

3.77 For the configuration and conditions of Problem 3.75, determine the time required for the temperature at the centerpoint to reach 205 °F.

▮ To find the time required to attain a given temperature at a given location in a transient multidimensional problem, a trial-and-error approach is required. From Problem 3.75, we know that the time is greater than 2.7 min. A logical procedure is to use this as a beginning point and to calculate T_c for several larger values of time; a graph of the results will then give the required answer.

Try $t = 3.25$ min
For the radial subproblem,

$$\frac{1}{\mathbf{Bi}^*} = 2.0 \qquad \mathbf{Fo}^* = \frac{(0.462)(3.25/60)}{(\frac{1}{12})^2} = 3.6$$

and Figure D-4 gives

$$\left(\frac{T_c - T_\infty}{T_i - T_\infty}\right)_{\substack{\text{inf.}\\\text{cyl.}}} \approx 0.047$$

For the axial subproblem,

$$\frac{1}{\mathbf{Bi}} = 1.6 \qquad \mathbf{Fo} = \frac{(0.462)(3.25/60)}{(1.25/12)^2} = 2.3$$

and from Figure D-1,

$$\left(\frac{T_c - T_\infty}{T_i - T_\infty}\right)_{\substack{2a\text{-}\\\text{slab}}} \approx 0.33$$

Hence, the complete solution is

$$\frac{T_c - T_\infty}{T_i - T_\infty} \approx (0.047)(0.33) = 0.016$$

or $T_c \approx 200 + (0.016)(1200 - 200) = 215.5\ ^\circ\text{F}$.

Try $t = 3.9$ ***min***

For the radial subproblem,

$$\frac{1}{\mathbf{Bi}^*} = 2.0 \qquad \mathbf{Fo}^* = \mathbf{Fo}^*_{3.25}\left(\frac{3.9}{3.25}\right) = 4.32$$

and from Figure D-4,

$$\left(\frac{T_c - T_\infty}{T_i - T_\infty}\right)_{\substack{\text{inf.}\\ \text{cyl.}}} \approx 0.025$$

For the axial subproblem,

$$\frac{1}{\mathbf{Bi}} = 1.6 \qquad \mathbf{Fo} = \mathbf{Fo}_{3.25}\left(\frac{3.9}{3.25}\right) = 2.76$$

and from Figure D-1,

$$\left(\frac{T_c - T_\infty}{T_i - T_\infty}\right)_{\substack{2a\text{-}\\ \text{slab}}} \approx 0.26$$

Hence, the complete solution is

$$\frac{T_c - T_\infty}{T_i - T_\infty} \approx (0.025)(0.26) = 0.0065$$

or $T_c \approx 200 + (0.0065)(1200 - 200) = 206.5\ ^\circ\text{F}$.

Try $t = 4.3$ ***min***

For the radial subproblem,

$$\frac{1}{\mathbf{Bi}^*} = 2.0 \qquad \mathbf{Fo}^* = \mathbf{Fo}^*_{3.25}\left(\frac{4.3}{3.25}\right) = 4.76$$

and from Figure D-4,

$$\left(\frac{T_c - T_\infty}{T_i - T_\infty}\right)_{\substack{\text{inf.}\\ \text{cyl.}}} \approx 0.017$$

For the axial subproblem,

$$\frac{1}{\mathbf{Bi}} = 1.6 \qquad \mathbf{Fo} = \mathbf{Fo}_{3.25}\left(\frac{4.3}{3.25}\right) = 3.04$$

and from Figure D-1,

$$\left(\frac{T_c - T_\infty}{T_i - T_\infty}\right)_{\substack{2a\text{-}\\ \text{slab}}} \approx 0.22$$

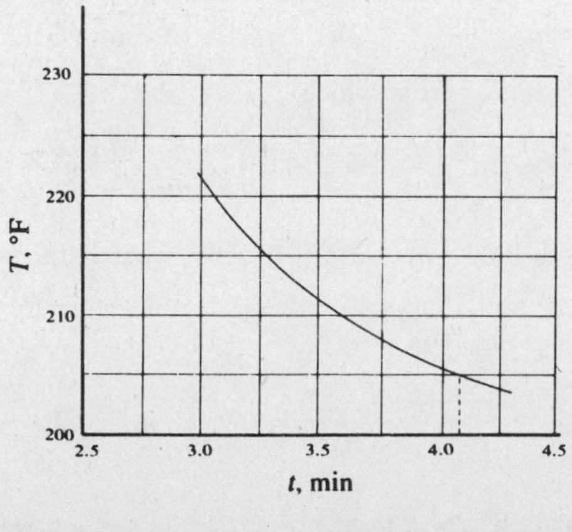

Fig. 3-19

Hence, the complete solution is

$$\frac{T_c - T_\infty}{T_i - T_\infty} \approx (0.017)(0.22) = 0.0037$$

or $T_c \approx 200 + (0.0037)(1200 - 200) = 203.7\ °\text{F}$.

Plotting these results (Fig. 3-19), we find $t \approx 4.07$ min.

3.5 ADDITIONAL ONE-DIMENSIONAL PROBLEMS

3.78 A chemical line is submerged in an outdoor pool of water initially at 40 °F. If the chemical behaves erratically above 50 °F, how deep must the line be situated below the water surface to maintain chemical stability on a day when the surface temperature of the water reaches 90 °F for a 6-h period?

▮ From Table B-3, at $T_{avg} = (50 + 90)/2 = 70\ °\text{F}$, $\alpha = 5.54 \times 10^{-3}\ \text{ft}^2/\text{h}$. Treating the pool as a semi-infinite body, we apply (*1*) of Problem 3.29:

$$\frac{T - T_s}{T_i - T_s} = \text{erf}\,\frac{x}{(4\alpha t)^{1/2}} \qquad \text{or} \qquad 0.80 = \text{erf}\,\frac{x}{0.365}$$

From Fig. 3-10, $x/0.365 = 0.90$, whence $x = 0.33$ ft.

3.79 Change Problem 3.78 so that the pool, initially at 40 °F, is exposed for 6 h to a convective (windy) atmosphere with $\bar{h} = 10\ \text{Btu/h} \cdot \text{ft}^2 \cdot °\text{F}$ and $T_\infty = 90\ °\text{F}$ (note that this is not now the *surface* temperature). How deep must the chemical line be situated?

▮ From Table B-3, $k = 0.345\ \text{Btu/h} \cdot \text{ft} \cdot °\text{F}$. Equation (*1*) of Problem 3.47 holds for this case, and the solution may be obtained readily by using Fig. 3-14. For

$$\frac{T - T_i}{T_\infty - T_i} = \frac{50 - 40}{90 - 40} = 0.2$$

and

$$\frac{\bar{h}(\alpha t)^{1/2}}{k} = \frac{(10)[(5.54 \times 10^{-3})(6)]^{1/2}}{0.345} = 5.28$$

Figure 3.14 gives

$$\frac{x}{(4\alpha t)^{1/2}} \approx 0.82 \qquad \text{or} \qquad x \approx (0.365)(0.82) = 0.30\ \text{ft}$$

Over the 6-h period the mean surface temperature is less than $T_\infty = 90°\ \text{F}$; hence the line need not be so deep as in Problem 3.78.

3.80 A water pipe is buried in a 20-cm-thick (very long and deep) concrete ($k = 0.90\ \text{W/m} \cdot \text{K}$; $\alpha = 5 \times 10^{-7}\ \text{m}^2/\text{s}$) bridge girder, initially at 5 °C, which is exposed to a sudden windstorm ($\bar{h} = 6\ \text{W/m}^2 \cdot \text{K}$). How long can the water withstand freezing when the ambient temperature is −20 °C, if the pipe runs along the centerline of the girder?

▮ We may use the Heisler chart, Figure D-1, upon determining the dimensionless moduli.

$$\frac{T_c - T_\infty}{T_i - T_\infty} = \frac{0 - (-20)}{5 - (-20)} = 0.80$$

$$\frac{1}{\textbf{Bi}} = \frac{k}{\bar{h}L} = \frac{0.90\ \text{W/m} \cdot \text{K}}{6\ \text{W/m}^2 \cdot \text{K}(0.10\ \text{m})} = 1.50$$

We read off $\textbf{Fo} \equiv \alpha t/L^2 \approx 0.7$, from which

$$t \approx (0.7)\left(\frac{L^2}{\alpha}\right) = (0.7)\left[\frac{(0.10\ \text{m})^2}{(5 \times 10^{-7}\ \text{m}^2/\text{s})(3600\ \text{s/h})}\right] = 3.9\ \text{h}$$

3.81 When the water in Problem 3.80 just starts to freeze, what is the temperature 4 cm below the surface of the girder?

▮ Using Figure D-2, at $x/L = 0.4$,

$$\frac{T - (-20)}{0 - (-20)} \approx 0.9 \qquad \text{or} \qquad T \approx -2.0\ °\text{C}$$

3.82 If the pipe of Problem 3.80 were located 2 cm below the surface, how long would be required for it to freeze?

▌ From Figure D-2, at $1/\mathbf{Bi} = 1.50$ and $x/L = 0.2$,

$$\frac{0-(-20)}{T_c-(-20)} \approx 0.955 \qquad \text{or} \qquad T_c \approx 0.94\ ^\circ\text{C}$$

Therefore

$$\frac{T_c - T_\infty}{T_i - T_\infty} = \frac{0.94-(-20)}{5-(-20)} \approx 0.84$$

and, from Figure D-1, $\mathbf{Fo} \equiv \alpha t/L^2 \approx 0.5$, or

$$t = \frac{(0.10\ \text{m})^2(0.5)}{(5\times 10^{-7}\ \text{m}^2/\text{s})(3600\ \text{s/h})} = 2.8\ \text{h}$$

3.83 How much heat is liberated per square meter in Problem 3.80 by the time the water freezes?

▌ From Problem 3.80, $\mathbf{Fo} = 0.7$ and $\mathbf{Bi} = 1/1.50 = 0.67$; Figure D-3 then gives $Q/Q_i \approx 0.3$. But,

$$\frac{Q_i}{A} = \rho c(2L)(T_i - T_\infty)$$

where, from Table B-2,

$$c = (0.21)(4184) = 878.6\ \text{J/kg}\cdot\text{K} \qquad \rho \approx (125)(16.02) = 2002\ \text{kg/m}^3$$

Therefore,

$$\frac{Q_i}{A} = (2002\ \text{kg/m}^3)(878.6\ \text{J/kg}\cdot\text{K})(0.20\ \text{m})[5-(-20)]\ \text{K} = 4795\ \text{kJ/m}^2$$

and $Q/A \approx (0.3)(4795) = 1438\ \text{kJ/m}^2$.

3.84 A 2-in.-diam. steel ($k = 25$ Btu/h · ft · °F; $\alpha = 0.425$ ft/h; $c = 0.11$ Btu/lbm · °F; $\rho = 490$ lbm/ft^3) ball bearing at 2000 °F is heat-treated by immersing in a liquid bath at 100 °F. If $\bar{h} = 100$ Btu/h · ft^2 · °F, what is the maximum temperature in the bearing 1 min after immersion?

▌

$$\frac{1}{\mathbf{Bi}^*} = \frac{k}{\bar{h}R} = \frac{25}{(100)(\frac{1}{12})} = 3$$

$$\mathbf{Fo}^* = \frac{\alpha t}{R^2} = \frac{(0.425)(\frac{1}{60})}{(\frac{1}{12})^2} = 1.02$$

From Figure D-7,

$$\frac{T_c - 100}{2000-100} \approx 0.45 \qquad \text{or} \qquad T_c \approx 955\ ^\circ\text{F}$$

3.85 How much heat is gained by the bath in Problem 3.84?

▌ From Figure D-9 ($\mathbf{Bi}^* = 0.33$), $Q/Q_i \approx 0.55$. But

$$Q_i = \rho c\left(\frac{4}{3}\pi R^3\right)(T_i - T_\infty) = (490)(0.11)\left(\frac{4\pi}{3}\right)\left(\frac{1}{12}\right)^3(2000-100) = 248.25\ \text{Btu}$$

and so $Q \approx (0.55)(248.25) = 137$ Btu.

3.86 If the heat-treatment process on the ball bearing of Problem 3.84 requires that the temperature 0.10 in. below the surface be reduced to 200 °F, how long should the bearing be left in the bath?

▌ From Figure D-8, with $r/R = (1.0-0.1)/1.0 = 0.9$,

$$\frac{200-100}{T_c-100} \approx 0.87 \qquad \text{or} \qquad T_c \approx 215\ ^\circ\text{F}$$

from which

$$\frac{T_c - T_\infty}{T_i - T_\infty} = \frac{215-100}{2000-100} = 0.061$$

Then, from Figure D-7 $(1/\mathbf{Bi^*} = 3)$, $\mathbf{Fo} \equiv \alpha t/R^2 \approx 3.2$; consequently,

$$t \approx \frac{(3.2)(\frac{1}{12}\ \text{ft})^2}{(0.425\ \text{ft}^2/\text{h})(\frac{1}{60}\ \text{h/min})} = 3.1\ \text{min}$$

3.87 In Problem 3.1, it was noted that internal thermal resistance is negligible when the Biot number is less than 0.1; such a body may be assumed to have uniform temperature. What is the error in this approximation, based on the Heisler charts (Figures D-2, D-6, and D-8)?

▌ The condition $\mathbf{Bi} < 0.1$ translates into

$$\textbf{slab} \qquad \frac{1}{\mathbf{Bi}} > 10 \Rightarrow \frac{\theta}{\theta_c} > 0.95 \quad \text{(5\% maximum error)}$$

$$\textbf{cylinder} \qquad \frac{1}{\mathbf{Bi^*}} > 5 \Rightarrow \frac{\theta}{\theta_c} > 0.90 \quad \text{(10\% maximum error)}$$

$$\textbf{sphere} \qquad \frac{1}{\mathbf{Bi^*}} > 3.33 \Rightarrow \frac{\theta}{\theta_c} > 0.85 \quad \text{(15\% maximum error)}$$

If for the cylinder and sphere the condition is strengthened to $\mathbf{Bi^*} < 0.1$, then for all three geometries the nonuniformity in the reduced temperature will be at most 5%. (Don't put too much faith in this 5%: The Heisler charts, too, are only approximate—see Problem 3.97.)

3.88 What error would have resulted in Problem 3.84 if a lumped-thermal-capacity analysis had been made?

▌ As was already seen in Problems 3.59 and 3.64, a lumped analysis via (*2*) of Problem 3.5 requires the use of **Bi** and **Fo** [note that $\mathbf{Bi^*} \cdot \mathbf{Fo^*} \neq t/(RC)_{\text{therm}}$]. For Problem 3.84,

$$\mathbf{Bi} = \frac{\mathbf{Bi^*}}{3} = \frac{1}{9} \qquad \mathbf{Fo} = 9\,\mathbf{Fo^*} = 9(1.02)$$

and the lumped solution is $(T \equiv T_c)$

$$\frac{T_c - 100}{2000 - 100} = \exp(-1.02) \approx 0.36$$

The deviation from the more accurate result of Problem 3.84 is

$$\frac{0.36 - 0.45}{0.45}(100\%) = -20\%$$

This result is compatible with Problem 3.87, which predicts a maximum error of 15%.

3.89 For a lumped-capacity system, express the total heat transferred up to the attainment of temperature T_1.

▌ In (*1*) of Problem 3.24 substitute

$$\exp\left[\frac{-t_1}{(RC)_{\text{therm}}}\right] = \frac{T_1 - T_\infty}{T_i - T_\infty}$$

from the basic solution (*1*) of Problem 3.4, to obtain

$$Q(T_1) = C_{\text{therm}}(T_i - T_1) \tag{1}$$

(cf. Problem 3.65).

3.90 A $\frac{1}{2}$-in.-diam. hybrid-material ball bearing $(\rho = 180\ \text{lbm/ft}^3;\ c = 0.24\ \text{Btu/lbm}\cdot{}^\circ\text{F};\ k = 12\ \text{Btu/h}\cdot\text{ft}\cdot{}^\circ\text{F})$, initially at 750 °F, is cooled in air $(\bar{h} = 6\ \text{Btu/h}\cdot\text{ft}^2\cdot{}^\circ\text{F})$ at 70 °F until the center reaches 600 °F. How much thermal energy is given up by the ball bearing?

▌ For the sphere,

$$\mathbf{Bi} = \frac{\bar{h}R}{3k} = \frac{(6)(\frac{1}{48})}{3(12)} = \frac{1}{288} \ll \frac{1}{10}$$

Therefore, a lumped analysis may be used. By (*1*) of Problem 3.89,

$$Q(600\ {}^\circ\text{F}) = \rho c(\tfrac{4}{3}\pi R^3)[(750 - 600)\ {}^\circ\text{F}] = (180)(0.24)(\tfrac{4}{3}\pi)(\tfrac{1}{48})^3(150) = 0.245\ \text{Btu}$$

Had the Biot modulus exceeded 0.1, we should have had to proceed as in Problem 3.63.

3.91 Evaluate the thermal time constant for the lumped system of Problem 3.90.

▮ From Problem 3.40,

$$\tau = \frac{\rho c(R/3)}{\bar{h}} = \frac{(180\ \text{lbm/ft}^3)(0.24\ \text{Btu/lbm}\cdot{}^\circ\text{F})(\frac{1}{144}\ \text{ft})}{6\ \text{Btu/h}\cdot\text{ft}^2\cdot{}^\circ\text{F}} = \tfrac{1}{20}\ \text{h} = 3.0\ \text{min}$$

3.92 What temperature would the bearing of Problem 3.90 reach after 6.0 min of air-cooling.

▮ At $t = 2\tau$ (see Problem 3.91),

$$T - T_\infty = \left(\frac{1}{e}\right)^2 (T_i - T_\infty)$$

or $T \approx 70 + (0.367)^2(750 - 70) = 161\ {}^\circ\text{F}$.

3.93 Check the result of Problem 3.92 against the Heisler chart.

▮ From Problem 3.90, $1/\overset{*}{\mathbf{Bi}} = 1/3\,\mathbf{Bi} = 288/3 = 96$. Also, since

$$\frac{t}{\tau} = 2 = \mathbf{Bi}\cdot\mathbf{Fo} = \mathbf{Bi}\cdot 9\,\mathbf{Fo}^*$$

we have

$$\mathbf{Fo}^* = \frac{2}{9\,\mathbf{Bi}} = \frac{2(288)}{9} = 64$$

Entering Figure D-7 at these parameter values, we find

$$\frac{T_c - T_\infty}{T_i - T_\infty} \approx 0.13$$

This agrees quite well with the result

$$\frac{T - T_\infty}{T_i - T_\infty} = \left(\frac{1}{e}\right)^2 \approx 0.13$$

of Problem 3.92 (where, of course, $T \equiv T_c$).

3.94 For the semi-infinite body of Problem 3.29, determine the total heat transfer per unit of surface area over the time interval $(0, t_1)$.

▮ Integrate (*2*) of Problem 3.30:

$$\frac{Q(0, t_1)}{A} = \int_0^{t_1} \frac{q(0, t)}{A}\, dt = 2k(T_s - T_i)\left(\frac{t_1}{\pi\alpha}\right)^{1/2}$$

3.95 A semi-infinite slab of aluminum at a uniform temperature of 500 °C has its surface temperature suddenly changed to 100 °C. How much heat is removed per unit area if the internal temperature 10 cm from the surface drops to 200 °C?

▮ From Table B-1,

$$k = (133)(1.7296) = 230\ \text{W/m}\cdot\text{K} \qquad \alpha = (3.33)(2.581\times 10^{-5}) = 8.6\times 10^{-5}\ \text{m}^2/\text{s}$$

From Fig. 3-10 we find that the given temperature change at $x = 0.10$ m,

$$\frac{T - T_s}{T_i - T_s} = \frac{200 - 100}{500 - 100} = 0.25$$

corresponds to $\omega \approx 0.23$, or to

$$t \equiv t_1 = \frac{x^2}{4\alpha\omega^2} \approx \frac{(0.10)^2}{4(8.6\times 10^{-5})(0.23)^2} = 550\ \text{s}$$

The formula of Problem 3.94 now gives

$$\frac{Q(0, 550\ \text{s})}{A} = 2(230)(100 - 500)\left[\frac{550}{\pi(8.6\times 10^{-5})}\right]^{1/2} = -2.6\times 10^8\ \text{J/m}^2$$

The minus sign tells us that the heat is lost from the slab.

3.96 If the semi-infinite slab of Problem 3.95 is suddenly exposed to convective conditions ($\bar{h} = 500\ \text{W/m}^2 \cdot \text{C}$; $T_\infty = 100\ ^\circ\text{C}$), how much time is required for the temperature at a depth of 10 cm to reach 400 °C?

As in similar problems (see Problems 3.69, 3.70, 3.73) a trial-and-error procedure is employed—based this time on (*1*) of Problem 3.47. We are seeking t such that

$$\frac{T(0.10\ \text{m}, t) - T_i}{T_\infty - T_i} = 0.25$$

Table 3-1 presents the solution.

TABLE 3-1

try t	calculate $\bar{h}\sqrt{\alpha t}/k$	calculate $x/\sqrt{4\alpha t}$	$(T - T_i)/(T_\infty - T_i)$ from Fig. 3-14
500 s	0.45	0.24	0.23
1000 s	0.64	0.17	0.33
600 s	0.49	0.22	0.25

Therefore, $t = 600$ s.

3.97 The Heisler and Gröber charts were obtained by truncating the infinite series solutions to a few terms. As must be apparent from Problem 3.51, this limits the applicability of the charts to sufficiently large times; in fact, we require that

$$\mathbf{Fo^*} \equiv \frac{\alpha t}{(L^*)^2} > 0.2$$

For metal bodies and a time 1 h, what is the limiting dimension of (*a*) a $2L$-slab ($L^* = L$)? (*b*) a solid cylinder ($L^* = R$)? (*c*) a solid sphere ($L^* = R$)?

The condition is equivalent to $L^* < [\alpha(1\ \text{h})/0.2]^{1/2}$. The smallest value of α for any metal in Table B-1 is that for stainless steel, $\alpha = 0.15\ \text{ft}^2/\text{h}$. Using this value, we get $L^* < 0.87$ ft. Hence: (*a*) $2L <$ 1.73 ft; (*b*) $R < 0.87$ ft; (*c*) $R < 0.87$ ft.

3.98 A cube of side $2L$ and a sphere of radius R (where $R = L$) are both made of the same material. Both have the same initial temperature and are subjected to the same sudden convective environment ($\bar{h}$, T_∞), where $T_\infty \neq T_i$, for the same time period. Under these conditions,

$$\mathbf{Bi}_{2L\text{-slab}} = \mathbf{Bi}^*_{\text{sphere}} \equiv \mathbf{Bi^*} \qquad \mathbf{Fo}_{2L\text{-slab}} = \mathbf{Fo}^*_{\text{sphere}} \equiv \mathbf{Fo^*}$$

If the time is chosen so as to make $\mathbf{Fo^*} = 20$, for what value of $\mathbf{Bi^*}$ will the two center temperatures be equal?

Recalling (*2*) of Problem 3.67, we are trying to find a $\mathbf{Bi^*}$ such that

$$\left(\frac{T_c - T_\infty}{T_i - T_\infty}\right)_{L\text{-sphere}} = \left(\frac{T_c - T_\infty}{T_i - T_\infty}\right)_{2L\text{-cube}} = \left(\frac{T_c - T_\infty}{T_i - T_\infty}\right)^3_{2L\text{-slab}}$$

Thus, entering Figures D-7 (for the sphere) and D-1 (for the slab) at $\mathbf{Fo^*} = 20$ and a guessed $\mathbf{Bi^*}$, we check whether the ordinate for the sphere is the cube of that for the slab. For instance, guessing $\mathbf{Bi^*} = \frac{1}{10}$,

$$0.0035 \stackrel{?}{=} (0.15)^3$$
$$0.0035 \stackrel{?}{=} 0.003\,375 \qquad \text{(close, but no cigar)}$$

The good guess turns out to be $\mathbf{Bi^*} = \frac{1}{16}$.

3.99 Rework Problem 3.3 if, in addition to the convective transfer, there is internal-energy generation at the constant rate q' (W or Btu/h). Assume an initial state of equilibrium; i.e. $T_i = T_\infty$.

(Heat flow out of body during dt) = −(change of internal energy of body during dt)

$$\bar{h}A_s(T - T_\infty)\,dt = -\rho c V\,dT + q'\,dt$$

$$\frac{dT}{dt} + \frac{\bar{h}A_s}{\rho c V}T = \frac{q' + \bar{h}A_s T_\infty}{\rho c V} \qquad (1)$$

The general solution to (*1*)—the general solution of the homogeneous equation, plus the particular integral $T = T_\infty + (q'/\bar{h}A_s)$—is

$$T(t) = A \exp\left(-\frac{\bar{h}A_s}{\rho c V} t\right) + \left(T_\infty + \frac{q'}{\bar{h}A_s}\right)$$

Evaluating the constant A through the initial condition, one obtains

$$T(t) = T_\infty + \frac{q'}{\bar{h}A_s}\left[1 - \exp\left(-\frac{\bar{h}A_s}{\rho c V} t\right)\right] \tag{2}$$

3.100 A 1-kW stove-top unit consists of a 5-spiral heating element of rectangular cross section ($\frac{3}{16}$ in. by $\frac{3}{8}$ in.) having a total length of 75 in. The unit is initially at 70 °F in an air-conditioned kitchen, where $\bar{h} = 10$ Btu/h · ft² · °F and $T_\infty = 70$ °F. Given $k = 75$ Btu/h · ft · °F, $c = 0.092$ Btu/lbm · °F, $\rho = 53.2$ lbm/ft³, and $\alpha = 1.26$ ft²/h, find the surface temperature 20 s after it is turned on?

▮ Assume that only the top surface and one-half of the side surfaces are involved in the convective process.

$$A_s = \left(\frac{3}{8}\text{ in.} + \frac{3}{16}\text{ in.}\right)(75\text{ in.})\left(\frac{1\text{ ft}^2}{144\text{ in}^2}\right) = 0.293\text{ ft}^2$$

$$V = \left(\frac{3}{8}\text{ in.}\right)\left(\frac{3}{16}\text{ in.}\right)(75\text{ in.})\left(\frac{1\text{ ft}^3}{1728\text{ in}^3}\right) = 0.0031\text{ ft}^3$$

$$L = \frac{V}{A_s} = \frac{0.0031\text{ ft}^3}{0.293\text{ ft}^2} = 0.011\text{ ft}$$

$$\mathbf{Bi} = \frac{\bar{h}L}{k} = \frac{(10\text{ Btu/h}\cdot\text{ft}^2\cdot{}^\circ\text{F})(0.011\text{ ft})}{75\text{ Btu/h}\cdot\text{ft}\cdot{}^\circ\text{F}} = 0.001 < 0.1$$

Therefore, a lumped analysis will give accurate results. Using (*2*) of Problem 3.99, and the conversion 1 Btu/h = 0.293 W,

$$T(20\text{ s}) = (70^\circ\text{ F}) + \frac{1000\text{ W}}{(10\text{ Btu/h}\cdot\text{ft}^2\cdot{}^\circ\text{F})(0.293\text{ W}\cdot\text{h/Btu})(0.293\text{ ft}^2)}$$
$$\times\left\{1 - \exp\left[-\frac{(10\text{ Btu/h}\cdot\text{ft}^2\cdot{}^\circ\text{F})(\frac{20}{3600}\text{ h})}{(53.2\text{ lbm/ft}^3)(0.092\text{ Btu/lbm}\cdot{}^\circ\text{F})(0.011\text{ ft})}\right]\right\}$$
$$= 820\text{ °F}$$

3.101 Rework Problem 3.100 for a solid 8-in.-diam. by $\frac{3}{16}$-in.-thick heating element.

▮ Assume that the top surface and one-half of the rim convect.

$$A_s = \left(\frac{3}{32}\text{ in.}\right)\pi(8\text{ in.}) + \frac{\pi(8\text{ in.})^2}{4} = 52.63\text{ in}^2 = 0.365\text{ ft}^2$$

$$V = \frac{\pi}{4}(8)^2\left(\frac{3}{16}\right) = 9.42\text{ in}^3 = 0.0055\text{ ft}^3$$

$$L = \frac{V}{A_s} = \frac{0.0055\text{ ft}^3}{0.365\text{ ft}^2} = 0.015\text{ ft}$$

$$\mathbf{Bi} = \frac{\bar{h}L}{k} = \frac{(10\text{ Btu/h}\cdot\text{ft}^2\cdot{}^\circ\text{F})(0.015\text{ ft})}{75\text{ Btu/h}\cdot\text{ft}\cdot{}^\circ\text{F}} = 0.002 < 0.1$$

Therefore, use a lumped analysis as in Problem 3.100:

$$T(20\text{ s}) = (70\text{ °F}) + \frac{1000\text{ W}}{(10\text{ Btu/h}\cdot\text{ft}^2\cdot{}^\circ\text{F})(0.293\text{ W}\cdot\text{h/Btu})(0.365\text{ ft}^2)}$$
$$\times\left\{1 - \exp\left[-\frac{(10\text{ Btu/h}\cdot\text{ft}^2\cdot{}^\circ\text{F})(\frac{20}{3600}\text{ h})}{(53.2\text{ lbm/ft}^3)(0.092\text{ Btu/lbm}\cdot{}^\circ\text{F})(0.015\text{ ft})}\right]\right\}$$
$$= 566\text{ °F}$$

Comparing this result with that of Problem 3.100, we see why people complain that solid-top heating elements have a slower response than conventional spiral configurations.

3.6 FINITE-DIFFERENCE METHODS

3.102 Using the nomenclature of Fig. 3-20, develop a finite-difference equation for the temperature at any point (x, y) and any time t in a two-dimensional body. Express the temperature at node n in terms of *earlier* temperatures at node n and its adjacent nodes (*explicit formulation*).

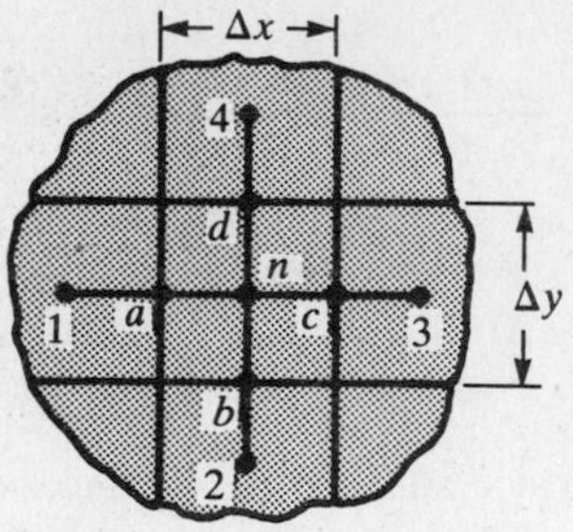

Fig. 3-20

▌ The differential equation for the temperature is

$$\frac{\partial^2 T}{\partial x^2} + \frac{\partial^2 T}{\partial y^2} = \frac{1}{\alpha}\frac{\partial T}{\partial t} \tag{1}$$

Approximate the first derivative of T with respect to x at points c and a as

$$\left(\frac{\partial T}{\partial x}\right)_c \approx \frac{T_3 - T_n}{\Delta x} \qquad \left(\frac{\partial T}{\partial x}\right)_a \approx \frac{T_n - T_1}{\Delta x}$$

The second derivative at n (an interior node) is then by definition

$$\left(\frac{\partial^2 T}{\partial x^2}\right)_n \approx \frac{(\partial T/\partial x)_c - (\partial T/\partial x)_a}{\Delta x} \approx \frac{T_3 + T_1 - 2T_n}{(\Delta x)^2}$$

and in a similar manner,

$$\left(\frac{\partial^2 T}{\partial y^2}\right)_n \approx \frac{(\partial T/\partial y)_d - (\partial T/\partial y)_b}{\Delta y} \approx \frac{T_4 + T_2 - 2T_n}{(\Delta y)^2}$$

The time derivative can be approximated as

$$\left(\frac{\partial T}{\partial t}\right)_n \approx \frac{T_n^{t+1} - T_n^t}{\Delta t}$$

[Superscripts t and $t+1$ stand for the times $r\,\Delta t$ and $(r+1)\,\Delta t$, where r is an integer.] The required approximation of (1) is then

$$\frac{T_3^t + T_1^t - 2T_n^t}{(\Delta x)^2} + \frac{T_4^t + T_2^t - 2T_n^t}{(\Delta y)^2} = \frac{1}{\alpha}\frac{T_n^{t+1} - T_n^t}{\Delta t} \tag{2}$$

In the important special case $\Delta x = \Delta y \equiv \delta$, (2) simplifies to

$$T_n^{t+1} = \frac{\alpha\,\Delta t}{\delta^2}(T_1^t + T_2^t + T_3^t + T_4^t) + \left(1 - \frac{4\alpha\,\Delta t}{\delta^2}\right)T_n^t \tag{3}$$

It is convenient to introduce the dimensionless parameter $M \equiv \delta^2/(\alpha\,\Delta t)$, in terms of which the one-, two-, and three-dimensional explicit finite-difference equations are

$$T_n^{t+1} = \frac{1}{M}(T_1^t + T_3^t) + \left(1 - \frac{2}{M}\right)T_n^t \tag{4}$$

$$T_n^{t+1} = \frac{1}{M}(T_1^t + T_2^t + T_3^t + T_4^t) + \left(1 - \frac{4}{M}\right)T_n^t \tag{5}$$

$$T_n^{t+1} = \frac{1}{M}(T_1^t + T_2^t + T_3^t + T_4^t + T_5^t + T_6^t) + \left(1 - \frac{6}{M}\right)T_n^t \tag{6}$$

with $\Delta x = \Delta y = \Delta z \equiv \delta$ presumed in (6). In general, the smaller are δ and Δt, the more accurate will be the solution, but the slower will be the convergence to that solution. On physical grounds, the coefficient of T_n^t in (4), (5), or (6) cannot be negative; i.e.

$$M \geq \text{twice the dimension number} \tag{7}$$

This means that a selection of δ places an upper limit on that of Δt. When (7) is obeyed, the numerical solution will be stable.

3.103 Using the nomenclature of Fig. 3-21, develop a finite-difference expression for the temperature at a node on a convective boundary in one-dimensional flow.

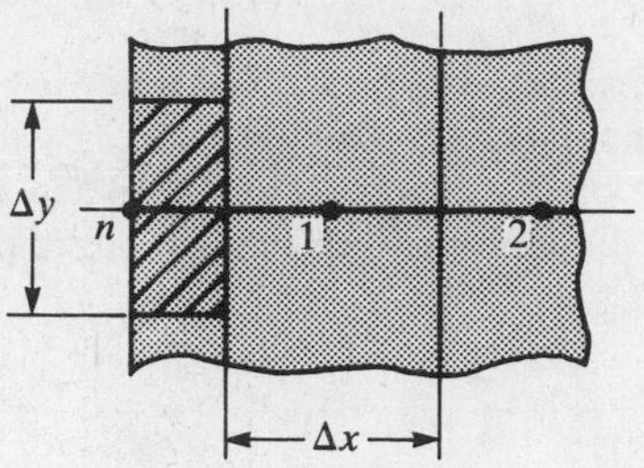

Fig. 3-21

▮ An energy balance on the part of the one-dimensional system represented by the crosshatched area and unit depth is

$$\rho c\left(\frac{\Delta x}{2}\right)(\Delta y)\left(\frac{T^{t+1}-T_n^t}{\Delta t}\right)=k(\Delta y)\left(\frac{T_1^t-T_n^t}{\Delta x}\right)+(\Delta y)\bar{h}(T_\infty-T_n^t) \qquad (1)$$

which for $\Delta x=\Delta y\equiv\delta$ becomes

$$T_n^{t+1}=\left(\frac{2}{M}\right)\left(\frac{\bar{h}\delta}{k}T_\infty+T_1^t\right)+\left[1-\left(\frac{2}{M}\right)\left(\frac{\bar{h}\delta}{k}+1\right)\right]T_n^t \qquad (2)$$

3.104 Solve the two-dimensional analog (Fig. 3-22) of Problem 3.103.

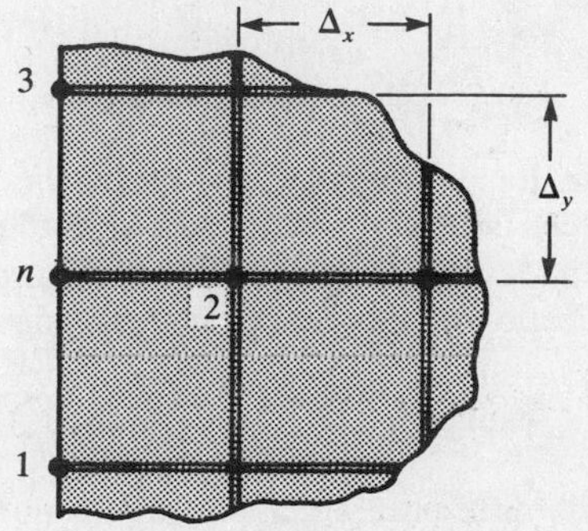

Fig. 3-22

▮ With $T_n=T_1=T_3$, we find, as in Problem 3.103 $(\Delta x=\Delta y\equiv\delta)$:

$$T_n^{t+1}=\left(\frac{1}{M}\right)\left(T_1^t+2T_2^t+T_3^t+2\frac{\bar{h}\delta}{k}T_\infty\right)+\left[1-\left(\frac{2}{M}\right)\left(\frac{\bar{h}\delta}{k}+2\right)\right]T_n^t \qquad (1)$$

To assure stability of the numerical solutions, we require

$$M\ge 2\left(\frac{\bar{h}\delta}{k}+1\right) \qquad \textit{(one-dimensional problem)}$$

$$M\ge 2\left(\frac{\bar{h}\delta}{k}+2\right) \qquad \textit{(two-dimensional problem)}$$

Clearly, (1) would require modification if node n was an interior or exterior corner (see Problem 3.108).

3.105 Reconsider Problem 3.102 in the *implicit formulation*; that is, express the temperature at node n in terms of *contemporary* temperatures at the adjacent nodes (as well as the earlier temperature at node n).

▮ Expressing the two-dimensional conduction equation in terms of central and backward differences,

$$\left(\frac{\partial^2 T}{\partial x^2}\right)_{n,t+\Delta t,\text{cent}}+\left(\frac{\partial^2 T}{\partial y^2}\right)_{n,t+\Delta t,\text{cent}}=\frac{1}{\alpha}\left(\frac{\partial T}{\partial t}\right)_{n,t+\Delta t,\text{bkwd}}$$

we get the implicit form of the nodal equations:

$$\frac{1}{(\Delta x)^2}(T_3^{t+1} - 2T_n^{t+1} + T_1^{t+1}) + \frac{1}{(\Delta y)^2}(T_4^{t+1} - 2T_n^{t+1} + T_2^{t+1}) = \frac{1}{\alpha\,\Delta t}(T_n^{t+1} - T_n^t) \qquad (1)$$

or, when $\Delta x = \Delta y \equiv \delta$,

$$\left(1 + \frac{4}{M}\right)T_n^{t+1} = \frac{1}{M}(T_1^{t+1} + T_2^{t+1} + T_3^{t+1} + T_4^{t+1}) + T_n^t \qquad (2)$$

For each time step, the whole set of equations (2)—one for each node—must be solved simultaneously to obtain the current crop of temperatures. However, because all coefficients are positive in (2), there is no stability restriction on the size of M. Thus, Δt can be made large, so that fewer time steps are required.

In order to treat boundary conditions in the implicit approach, we multiply (1) by $k(\Delta x)(\Delta y)$ and rearrange, to obtain

$$\frac{T_1^{t+1} - T_n^{t+1}}{R_{1n}} + \frac{T_2^{t+1} - T_n^{t+1}}{R_{2n}} + \frac{T_3^{t+1} - T_n^{t+1}}{R_{3n}} + \frac{T_4^{t+1} - T_n^{t+1}}{R_{4n}} = C_n\left(\frac{T_n^{t+1} - T_n^t}{\Delta t}\right) \qquad (3)$$

where the thermal resistances per unit depth are

$$R_{1n} = R_{2n} = R_{3n} = R_{4n} \equiv \frac{\Delta x}{k\,\Delta y} \qquad (4)$$

and the thermal capacitances per unit depth are

$$C_n = \rho c(\Delta x)(\Delta y) \qquad (5)$$

Generalizing (3) through (5) to any number of spatial dimensions and to a variable grid spacing δ—making the C_n position dependent and the R_{jn} position and direction dependent—we have

$$\textbf{implicit} \qquad T_n^{t+1} = T_n^t + (\Delta t)\left(\sum_m \frac{T_m^{t+1} - T_n^{t+1}}{R_{mn}C_n}\right) \qquad (6)$$

where m runs over the nodes adjacent to node n; in (6),

$$R_{mn} \equiv \begin{cases} \delta/kA_{mn} & \text{for conduction} \\ 1/\bar{h}_{mn}\bar{A}_{mn} & \text{for convection} \end{cases} \qquad \text{and} \qquad C_n \equiv \rho c V_n \qquad (7)$$

Here, A_{mn} represents the area for conductive heat transfer between nodes m and n; $\bar{A}_{mn}$ is the area for convective heat transfer between nodes m and n; and V_n is the volume element determined by the value(s) of Δs at node n. Boundary nodes are simply included by proper formulation of the R_{mn} and C_n.

The corresponding generalization of the explicit formulation is

$$\textbf{explicit} \qquad T_n^{t+1} = T_n^t + (\Delta t)\left(\sum_m \frac{T_m^{t+1} - T_n^{t+1}}{R_{mn}C_n}\right) \qquad (8)$$

Both explicit and implicit finite-difference equations are summarized in Table D-1 of Appendix D.

3.106[D] A very long 4-in.-thick plate ($\alpha = 0.174\ \text{ft}^2/\text{h}$) initially at 100 °F is immersed in a fluid ($\bar{h} = \infty$) of temperature 800 °F. By a numerical method calculate the temperature distribution across the plate at 0.5-in. increments after 3 min.

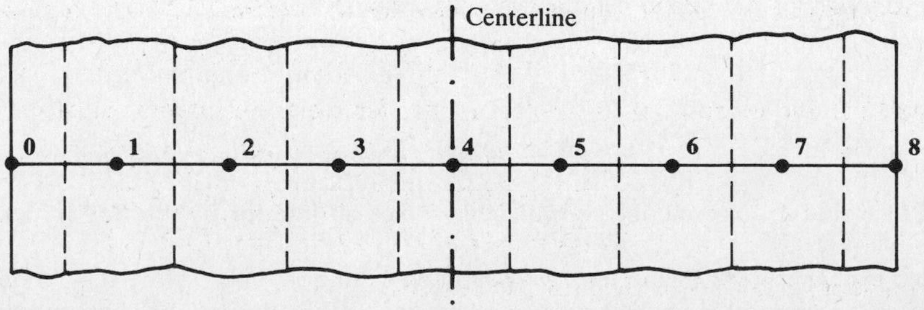

Fig. 3-23

I The appropriate difference equation is (4) of Problem 3.102. We wish to use the smallest allowable parameter value, $M = 2$; this dictates

$$\Delta t = \frac{\delta^2}{\alpha M} = \frac{(0.5/12)^2\ \text{ft}^2}{(0.174\ \text{ft}^2/\text{h})(2)} = 0.005\ \text{h} = 0.3\ \text{min}$$

Our equation is thus $T_n^{t+1} = \frac{1}{2}(T_1^t + T_3^t)$, or in the notation of Fig. 3-23,

$$T_n^{t+1} = \tfrac{1}{2}(T_{n-1}^t + T_{n+1}^t) \qquad (n = 1, 2, \ldots, 7) \tag{1}$$

Table 3-2 presents the numerical results, which exhibit the required symmetry about node **4**. Compare Problem 3.10.

TABLE 3.2

time, min.	temperature, °F				
	nodes 0 & 8	nodes 1 & 7	nodes 2 & 6	nodes 3 & 5	node 4
0.0	800	100	100	100	100
0.3	800	450	100	100	100
0.6	800	450	225	100	100
0.9	800	512.5	225	162.5	100
1.2	800	512.5	337.5	162.5	162.5
1.5	800	568.75	337.5	250	162.5
1.8	800	568.75	409.375	250	250
2.1	800	604.687	409.375	329.687	250
2.4	800	604.687	467.187	329.687	329.687
2.7	800	633.593	467.187	398.437	329.687
3.0	800	633.593	516.015	398.437	398.437

3.107 A $\frac{1}{2}$- by $\frac{1}{2}$- by 1-in. bar ($\alpha = 0.868$ ft²/h) is initially at a temperature of 80 °F. The four rectangular surfaces are suddenly brought to temperatures of 100, 200, 300, and 400 °F, and the two square surfaces to 500 and 600 °F. Using a numerical method, find the temperature at the geometric center of the bar after 15 s.

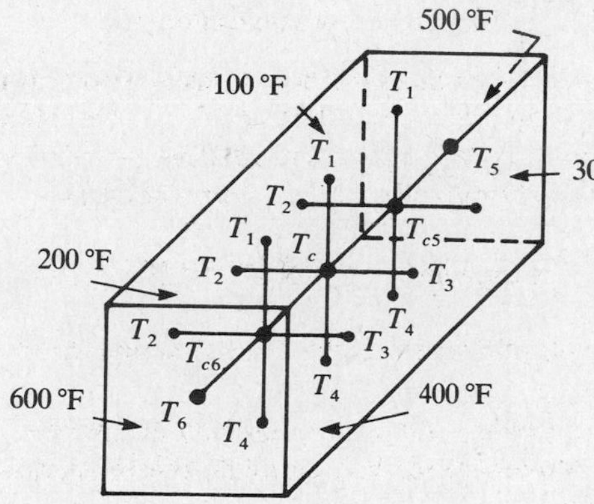

Fig. 3-24

Here one applies (6) of Problem 3.102, which for the minimal $M = 6$ (or $\Delta t = 3$ s) becomes

$$T_n^{t+1} = \tfrac{1}{6}(T_1^t + T_2^t + T_3^t + T_4^t + T_5^t + T_6^t) \tag{1}$$

Adapting (1) to the notation of Fig. 3-24, one has for the three interior nodal temperatures:

$$T_{c6}^{t+1} = \tfrac{1}{6}(T_1^t + T_2^t + T_3^t + T_4^t + T_6^t + T_c^t)$$
$$T_c^{t+1} = \tfrac{1}{6}(T_1^t + T_2^t + T_3^t + T_4^t + T_{c6}^t + T_{c5}^t)$$
$$T_{c5}^{t+1} = \tfrac{1}{6}(T_1^t + T_2^t + T_3^t + T_4^t + T_5^t + T_c^t)$$

The computations (Table 3-3) yield $T_c(15\text{ s}) = 267.42$ °F.

3.108 A 2-in.-square steel bar, initially at 100 °F, is partially immersed in a fluid, as shown in Fig. 3-25; the fluid temperature is 500 °F. The two upper faces of the bar are exposed to ambient air at 80 °F with an average heat transfer coefficient of 10 Btu/h · ft² · °F. (Assume $k = 25$ Btu/h · ft · °F and $\alpha = 0.625$ ft²/h.) Set up a system of explicit difference equations for the temperatures at the indicated nodal points. The bar is very long; there is no heat flow along its length.

TABLE 3-3

time, s	T_{c6}, °F	T_c, °F	T_{c5}, °F
0	80	80	80
3	280	193.33	263.33
6	298.89	257.22	282.22
9	309.54	263.52	292.87
12	310.59	267.07	293.92
15	311.18	267.42	294.51

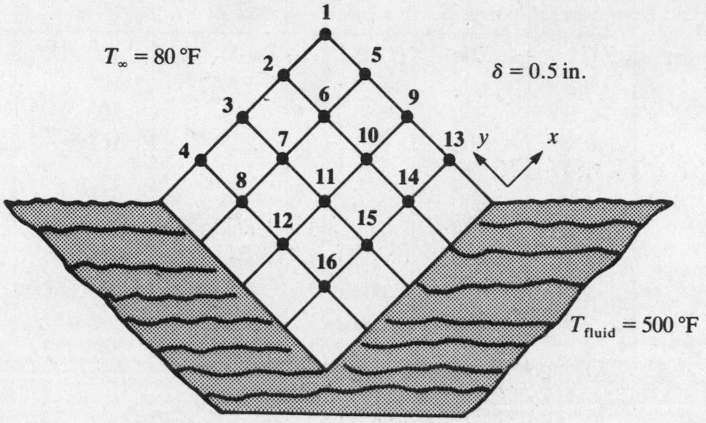

Fig. 3-25

▮ At each interior node, (*5*) of Problem 3.102 holds; at each boundary node except **1**, (*1*) of Problem 3.104 holds; at the corner **1**, an energy balance gives

$$\rho c \frac{\Delta x}{2}\frac{\Delta y}{2}\left(\frac{T_1^{t+1}-T_1^t}{\Delta t}\right)=k\frac{\Delta y}{2}\left(\frac{T_2^t-T_1^t}{\Delta x}\right)+k\frac{\Delta x}{2}\left(\frac{T_5^t-T_1^t}{\Delta y}\right)+\bar{h}\frac{\Delta x}{2}(T_\infty-T_1^t)+\bar{h}\frac{\Delta y}{2}(T_\infty-T_1^t)$$

Letting $\Delta x=\Delta y=\delta$, introducing $M\equiv \delta^2/(\alpha\,\Delta t)$, and solving for T_1^{t+1}, one finds

$$T_1^{t+1}=\frac{2}{M}(T_2^t+T_5^t-2T_1^t)+\frac{4}{M}\frac{\bar{h}\delta}{k}(T_\infty-T_1^t)+T_1^t$$

The complete system of sixteen equations is therefore

$$T_1^{t+1}=\frac{2}{M}\left[T_2^t+T_5^t+2\left(\frac{\bar{h}\delta}{k}\right)T_\infty\right]+\left[1-\frac{4}{M}\left(\frac{\bar{h}\delta}{k}+1\right)T_1^t\right]$$

$$T_2^{t+1}=\frac{1}{M}\left[T_1^t+T_3^t+2T_6^t+2\left(\frac{\bar{h}\delta}{k}\right)T_\infty\right]+\left[1-\frac{2}{M}\left(\frac{\bar{h}\delta}{k}+2\right)\right]T_2^t$$

$$T_3^{t+1}=\frac{1}{M}\left[T_2^t+T_4^t+2T_7^t+2\left(\frac{\bar{h}\delta}{k}\right)T_\infty\right]+\left[1-\frac{2}{M}\left(\frac{\bar{h}\delta}{k}+2\right)\right]T_3^t$$

$$T_4^{t+1}=\frac{1}{M}\left[T_3^t+T_{\text{fl.}}+2T_8^t+2\left(\frac{\bar{h}\delta}{k}\right)T_\infty\right]+\left[1-\frac{2}{M}\left(\frac{\bar{h}\delta}{k}+2\right)\right]T_4^t$$

$$T_5^{t+1}=\frac{1}{M}\left[T_1^t+T_9^t+2T_6^t+2\left(\frac{\bar{h}\delta}{k}\right)T_\infty\right]+\left[1-\frac{2}{M}\left(\frac{\bar{h}\delta}{k}+2\right)\right]T_5^t$$

$$T_6^{t+1}=\frac{1}{M}(T_2^t+T_{10}^t+T_5^t+T_7^t)+\left(1-\frac{4}{M}\right)T_6^t$$

$$T_7^{t+1}=\frac{1}{M}(T_3^t+T_{11}^t+T_6^t+T_8^t)+\left(1-\frac{4}{M}\right)T_7^t$$

$$T_8^{t+1}=\frac{1}{M}(T_4^t+T_{12}^t+T_7^t+T_{\text{fl.}})+\left(1-\frac{4}{M}\right)T_8^t$$

$$T_9^{t+1}=\frac{1}{M}\left[T_5^t+T_{13}^t+2T_{10}^t+2\left(\frac{\bar{h}\delta}{k}\right)T_\infty\right]+\left[1-\frac{2}{M}\left(\frac{\bar{h}\delta}{k}+2\right)\right]T_9^t$$

$$T_{10}^{t+1}=\frac{1}{M}(T_6^t+T_{14}^t+T_9^t++T_{11}^t)+\left(1-\frac{4}{M}\right)T_{10}^t$$

$$T_{11}^{t+1} = \frac{1}{M}\left(T_7^t + T_{15}^t + T_{10}^t + + T_{12}^t\right) + \left(1 - \frac{4}{M}\right)T_{11}^t$$

$$T_{12}^{t+1} = \frac{1}{M}\left(T_8^t + T_{16}^t + T_{11}^t + + T_{\text{fl.}}\right) + \left(1 - \frac{4}{M}\right)T_{12}^t$$

$$T_{13}^{t+1} = \frac{1}{M}\left[T_9^t + T_{\text{fl.}} + 2T_{14}^t + 2\left(\frac{\bar{h}\delta}{k}\right)T_\infty\right] + \left[1 - \frac{2}{M}\left(\frac{\bar{h}\delta}{k} + 2\right)\right]T_{13}^t$$

$$T_{14}^{t+1} = \frac{1}{M}\left(T_{10}^t + T_{\text{fl.}} + T_{13}^t + T_{15}^t\right) + \left(1 - \frac{4}{M}\right)T_{14}^t$$

$$T_{15}^{t+1} = \frac{1}{M}\left(T_{11}^t + T_{\text{fl.}} + T_{14}^t + T_{16}^t\right) + \left(1 - \frac{4}{M}\right)T_{15}^t$$

$$T_{16}^{t+1} = \frac{1}{M}\left(T_{12}^t + T_{15}^t + 2T_{\text{fl.}}\right) + \left(1 - \frac{4}{M}\right)T_{16}^t$$

FORTRAN IV G LEVEL 18

```
          C        NUMERICAL SOLUTION (UNSTEADY-STATE)
          C        TWO-DIMENSIONAL LONG BAR
0001               DIMENSION T(100),TT(100)
0002               AA=5.0
0003               A=1.0/AA
0004               B=(10.0*0.5/12.0)/25.0
0005               C=2.0*B*80.0
0006               D=1.0-2.0/AA*(B+2.0)
0007               E=1.0-4.0/AA
0008               TF=500.0
0009               TIME=0.0
0010               DO 10 J=1,16
0011               T(J)=100.0
0012            10 CONTINUE
0013               PRINT 20
0014            20 FORMAT (2X,'TIME',5X,'T1',5X,'T2',5X,'T3',5X,'T4',5X,'T5',5X,
                  1'T6',5X,'T7',5X,'T8',5X,'T9',5X,'T10',4X,'T11',4X,'T12',4X,
                  2'T13',4X,'T14',4X,'T15',4X,'T16',/)
0015               PRINT 30,TIME,(T(J),J=1,16)
0016            30 FORMAT('0',F7.2,1X,16(F6.2,1X))
0017               DO 40 K=1,60
0018               TT(1)=2.0/AA*(T(2)+T(5)+C)+(1.C-4.0/AA*(B+1.0))*T(1)
0019               TT(2)=A*(T(1)+T(3)+2.0*T(6)+C)+D*T(2)
0020               TT(3)=A*(T(2)+T(4)+2.0*T(7)+C)+D*T(3)
0021               TT(4)=A*(T(3)+TF+2.0*T(8)+C)+D*T(4)
0022               TT(5)=A*(T(1)+T(9)+2.0*T(6)+C)+D*T(5)
0023               TT(6)=A*(T(2)+T(10)+T(5)+T(7))+E*T(6)
0024               TT(7)=A*(T(3)+T(11)+T(6)+T(8))+E*T(7)
0025               TT(8)=A*(T(4)+T(12)+T(7)+TF)+E*T(8)
0026               TT(9)=A*(T(5)+T(13)+2.0*T(10)+C)+D*T(9)
0027               TT(10)=A*(T(6)+T(14)+T(9)+T(11))+E*T(10)
0028               TT(11)=A*(T(7)+T(15)+T(10)+T(12))+E*T(11)
0029               TT(12)=A*(T(8)+T(16)+T(11)+TF)+E*T(12)
0030               TT(13)=A*(T(9)+TF+2.0*T(14)+C)+D*T(13)
0031               TT(14)=A*(T(10)+TF+T(13)+T(15))+E*T(14)
0032               TT(15)=A*(T(11)+TF+T(14)+T(16))+E*T(15)
0033               TT(16)=A*(T(12)+T(15)+2.0*TF)+E*T(16)
0034               DO 50 L=1,16
0035               T(L)=TT(L)
0036            50 CONTINUE
0037               TIME=TIME+2.0
0038               PRINT 30,TIME,(T(L),L=1,16)
0039            40 CONTINUE
0040               CALL EXIT
0041               END
```

Fig. 3-26

3.109 Write a computer program for the calculation of the nodal temperatures of Problem 3.108 after 1 min and after 2 min.

▌ First, the time increment must be chosen. For stability of the given system of difference equations,

$$M \geq \max\left\{2\left(\frac{\bar{h}\delta}{k}+2\right), 4\right\} = \max\{4.033, 4\} = 4.033$$

Therefore, choose $M = 5$, which implies $\Delta t = 2.0$ s.

For one acceptable program, see Fig. 3-26, which employs the symbolism

$$\mathrm{AA} \equiv M \qquad \mathrm{B} \equiv \frac{\bar{h}\delta}{k} \qquad \mathrm{D} \equiv 1 - \frac{2}{M}\left(\frac{\bar{h}\delta}{k}+2\right) \qquad \mathrm{TF} \equiv T_{\mathrm{fl.}}$$

$$\mathrm{A} \equiv \frac{1}{M} \qquad \mathrm{C} \equiv 2\left(\frac{\bar{h}\delta}{k}\right)T_\infty \qquad \mathrm{E} \equiv 1 - \frac{4}{M}$$

Output is displayed in Table 3-4.

TABLE 3-4

	temperatures at nodes, °F															
time, s	**1**	**2**	**3**	**4**	**5**	**6**	**7**	**8**	**9**	**10**	**11**	**12**	**13**	**14**	**15**	**16**
0.0	100	100	100	100	100	100	100	100	100	100	100	100	100	100	100	100
2.00	99.73	99.87	99.87	179.87	99.87	100	100	180.00	99.87	100	100	180.00	179.87	180.00	180.00	260.00
↓	↓	↓	↓	↓	↓	↓	↓	↓	30 Repetitions ↓	↓	↓	↓	↓	↓	↓	↓
60.00	384.51	396.06	420.16	455.24	396.06	407.20	429.70	461.75	420.16	429.70	447.27	471.61	455.24	461.75	471.61	484.80
↓	↓	↓	↓	↓	↓	↓	↓	↓	30 Repetitions ↓	↓	↓	↓	↓	↓	↓	↓
120.00	454.27	461.39	470.58	482.65	461.39	468.38	476.92	487.42	470.58	476.92	483.71	491.43	482.65	487.42	491.43	495.58

3.110^D An 8.0-cm-thick concrete slab $(\alpha = 6.94 \times 10^{-7}\ \mathrm{m^2/s})$ having very large dimensions in the plane normal to the thickness is initially at a uniform temperature of 20 °C. Both surfaces of the slab are suddenly raised to and held at 100 °C. Using a nodal spacing of 1 cm, numerically determine by the explicit method the temperature history in the slab during a $\frac{1}{4}$-h period.

TABLE 3-5

time \ node	0	1	2	3	4
0.0	60	20	20	20	20
0.02	100	40	20	20	20
0.04	100	60	30	20	20
0.06	100	65	40	25	20
0.08	100	70	45	30	25
0.10	100	72.5	50	35	30
0.12	100	75	53.8	40	35
0.14	100	76.9	57.5	44.4	40
0.16	100	78.8	60.6	48.8	44.4
0.18	100	80.3	63.8	52.5	48.8
0.20	100	81.9	66.4	56.3	52.5
0.22	100	83.2	69.1	59.4	56.3
0.24	100	84.6	71.3	62.7	59.4
0.26	100	85.6	73.6	65.4	62.7

▌ The nodal designations are as in Fig. 3-23; since the problem is symmetrical about the centerline, only one-half of the slab is treated. With a minimal $M = 2$—and hence a maximal $\Delta t = 0.020$ h—the system to solve is

$$T_1^{t+1} = \tfrac{1}{2}(T_0^t + T_2^t)$$
$$T_2^{t+1} = \tfrac{1}{2}(T_1^t + T_3^t)$$
$$T_3^{t+1} = \tfrac{1}{2}(T_2^t + T_4^t)$$
$$T_4^{t+1} = \tfrac{1}{2}(T_3^t + T_3^t) = T_3^t$$

Since T_0 is initially 20 °C and is suddenly changed to 100 °C, an appropriate value for T_0 during the first time increment is the average of these two extremes. The solution is carried out in Table 3-5, where the units are h and °C.

CHAPTER 4

Introduction to Convective Processes

4.1 LAMINAR FLOW OVER A FLAT PLATE: "EXACT" TREATMENT

4.1 Specialize the x-direction Navier–Stokes (momentum) equation to steady, incompressible, constant-viscosity, laminar, two-dimensional flow.

▌ The x-direction Navier–Stokes equation for laminar, incompressible, constant-viscosity flow may be written as

$$\rho\left(\frac{\partial u}{\partial t} + u\frac{\partial u}{\partial x} + v\frac{\partial u}{\partial y} + w\frac{\partial u}{\partial z}\right) = \rho g_x - \frac{\partial p}{\partial x} + \mu\left(\frac{\partial^2 u}{\partial x^2} + \frac{\partial^2 u}{\partial y^2} + \frac{\partial^2 u}{\partial z^2}\right) \tag{1}$$

with $\partial u/\partial t = 0$ (steady flow) and $w = 0 = \partial^2 u/\partial z^2$ (two-dimensional flow), (*1*) becomes

$$\rho\left(u\frac{\partial u}{\partial x} + v\frac{\partial u}{\partial y}\right) = \rho g_x - \frac{\partial p}{\partial x} + \mu\left(\frac{\partial^2 u}{\partial x^2} + \frac{\partial^2 u}{\partial y^2}\right) \tag{2}$$

which is the desired result.

4.2 Further simplify the momentum equation (*2*) of Problem 4.1 within the viscous boundary layer resulting from steady, incompressible, laminar, two-dimensional, constant-viscosity fluid flow over a flat plate at zero angle of attack (Fig. 4-1).

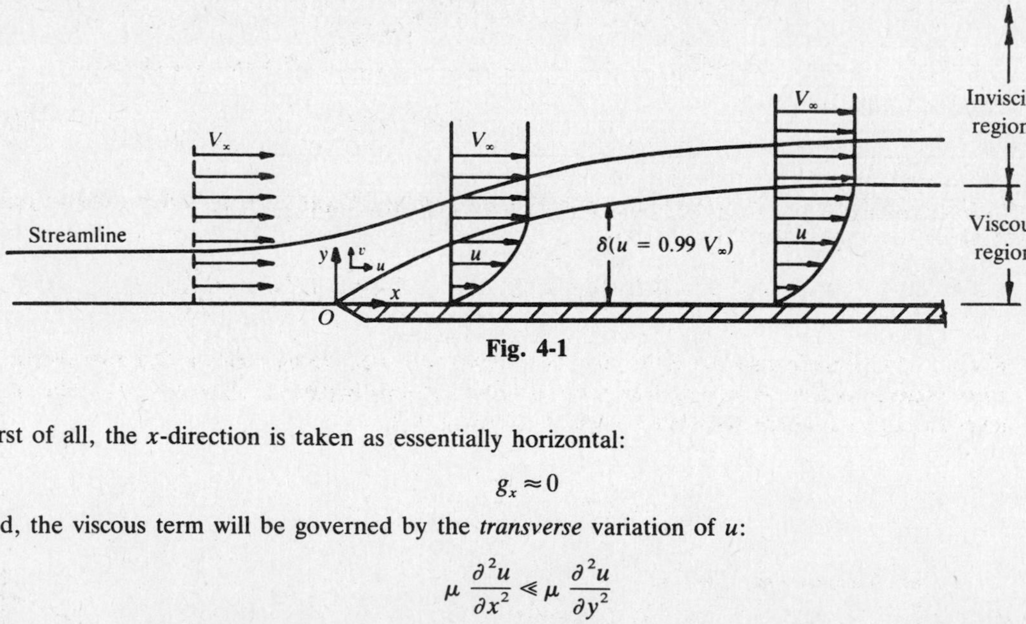

Fig. 4-1

▌ First of all, the x-direction is taken as essentially horizontal:

$$g_x \approx 0 \tag{1}$$

Second, the viscous term will be governed by the *transverse* variation of u:

$$\mu\frac{\partial^2 u}{\partial x^2} \ll \mu\frac{\partial^2 u}{\partial y^2} \tag{2}$$

Third, there is negligible pressure variation across the boundary layer: $\partial p/\partial y \approx 0$. Now, *outside* the boundary layer, Bernoulli's equation gives along a streamline

$$p + \rho g Z + \tfrac{1}{2}\rho V^2 = \text{constant}$$

Because, on the streamline, Z is very nearly constant and V is constant at V_∞, p must be essentially constant. But if p is constant along the streamline, then $(\partial p/\partial y = 0)$ it is constant along the x-axis; so that

$$\frac{\partial p}{\partial x} = 0 \tag{3}$$

Under (*1*), (*2*), and (*3*), the momentum equation becomes

$$u\frac{\partial u}{\partial x} + v\frac{\partial u}{\partial y} = \nu\frac{\partial^2 u}{\partial y^2} \qquad (4)$$

in which the *dynamic viscosity* μ has been replaced by the *kinematic viscosity* $\nu = \mu/\rho$ (or, in British Engineering units, $\nu = \mu g_c/\rho$).

4.3 Figure 4-2 presents some graphical results from the solution by H. Blasius of the simplified momentum equation (*4*) of Problem 4.2. Blasius succeeded in transforming the equation into an ordinary differential equation in the independent variable η. The *local Reynolds number* of the flow is defined as $\mathbf{Re}_x \equiv xV_\infty/\nu$. Using Fig. 4-2, determine an expression for the dimensionless boundary layer thickness, δ/x, where δ is the thickness at the location where $u/V_\infty = 0.99$.

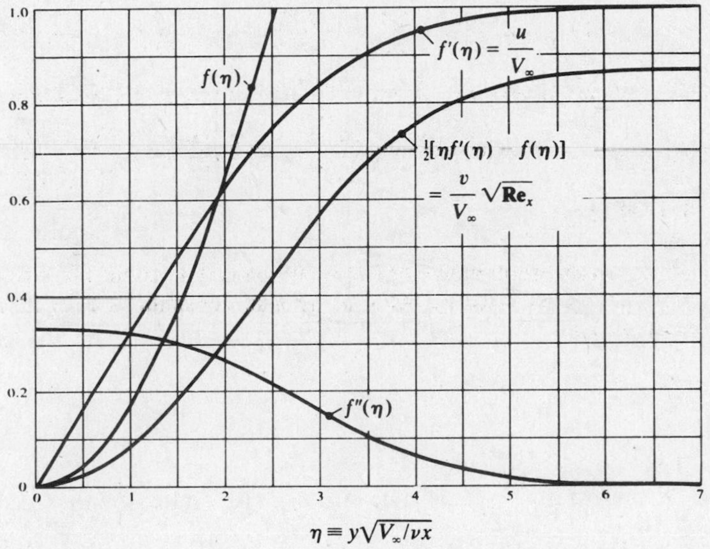

Fig. 4-2

From Fig. 4-2, $u/V_\infty \approx 0.99$ at a value of η between 5.0 and 5.5; the lower value is commonly used in the fluid mechanics literature. Hence

$$5.0 = \delta\sqrt{\frac{V_\infty}{\nu x}} \qquad \text{or} \qquad \frac{\delta}{x} = 5.0\sqrt{\frac{\nu}{V_\infty x}} = \frac{5.0}{\sqrt{\mathbf{Re}_x}}$$

4.4 A fluid of kinematic viscosity $2.0 \times 10^{-5}\ \text{m}^2/\text{s}$ flows with free-stream velocity 2.5 m/s over a flat plate at zero angle of attack. Assuming transition to turbulent flow begins at $\mathbf{Re} = 5 \times 10^5$, what is the boundary layer thickness δ where turbulence begins? See Fig. 4-3.

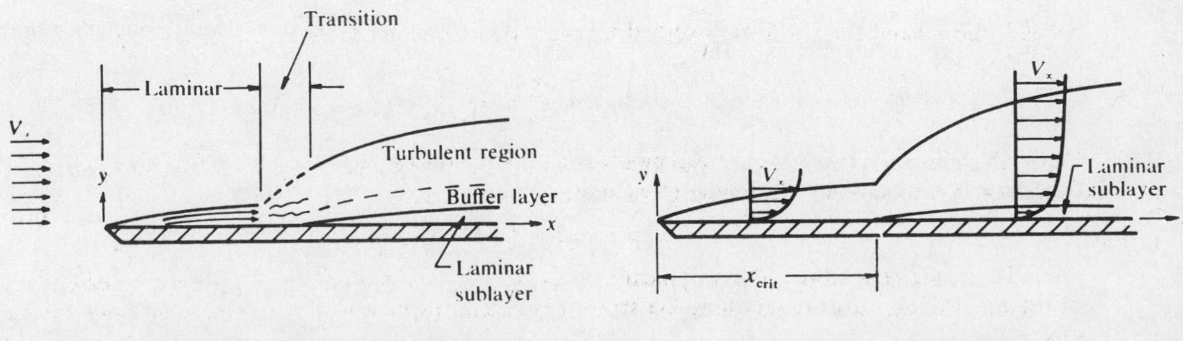

(*a*) Laminar-to-Turbulent Transition

(*b*) Simplified Model

Fig. 4-3

■ $$\mathbf{Re}_{\text{crit}} = 5 \times 10^5 = \frac{V_\infty x_{\text{crit}}}{\nu} \qquad \text{or} \qquad x_{\text{crit}} = \frac{(5 \times 10^5)(2.0 \times 10^{-5}\ \text{m}^2/\text{s})}{2.5\ \text{m/s}} = 4\ \text{m}$$

Problem 4.3 then gives

$$\delta_{\text{crit}} = \frac{(5.0)x_{\text{crit}}}{\sqrt{\mathbf{Re}_{\text{crit}}}} = \frac{(5.0)(4\ \text{m})}{\sqrt{5 \times 10^5}} = 28.28\ \text{mm}$$

4.5 In laminar boundary layer flow, both velocity and temperature boundary layers develop, as indicated in Fig. 4-4 for the case $\delta_t > \delta$. According to the solution by K. Pohlhausen of the energy equation for simplified laminar flow over a heated flat plate, the relationship between δ and δ_t is

$$\delta_t = \frac{\delta}{\mathbf{Pr}^{1/3}} \qquad \text{where} \qquad \mathbf{Pr} = \textit{Prandtl number} \equiv \frac{\nu}{\alpha} \tag{1}$$

For the flow situation of Problem 4.4, find the thermal boundary layer thickness where turbulence begins, if (*a*) $\mathbf{Pr} = 5.0$; (*b*) $\mathbf{Pr} = 0.5$.

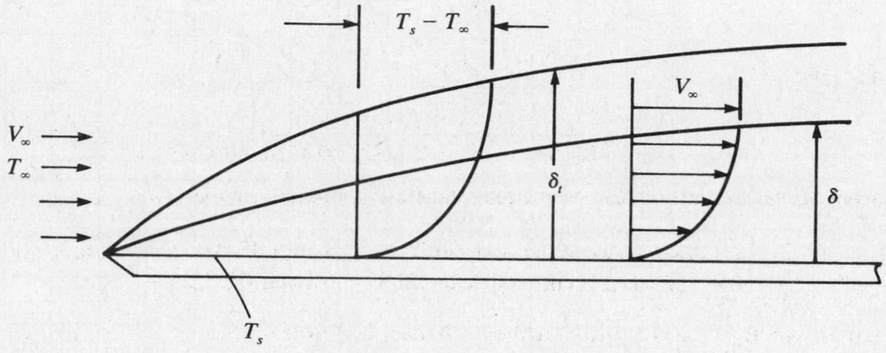

Fig. 4-4

■ (*a*) $\delta_t = (28.28\ \text{mm})/(5.0)^{1/3} = 16.54\ \text{mm}$; (*b*) $\delta_t = (10^{1/3})(16.54\ \text{mm}) = 35.63\ \text{mm}$.

4.6 For a parallel flow over a flat plate ($V_\infty = 10$ ft/s, $T_\infty = 80$ °F), determine the distance from the leading edge to the onset of turbulence, assuming a critical Reynolds number of 5×10^5 and that the fluid is (*a*) H_2 gas at 1 atm; (*b*) H_2O liquid at 1 atm.

■ (*a*) From Table B-4, $\nu = 117.9 \times 10^{-5}\ \text{ft}^2/\text{s}$, so that

$$x_{\text{crit}} = \frac{\mathbf{Re}_{\text{crit}}\,\nu}{V_\infty} = \frac{(5 \times 10^5)(117.9 \times 10^{-5}\ \text{ft}^2/\text{s})}{10\ \text{ft/s}} = 58.95\ \text{ft}$$

(*b*) From Table B-3 (interpolation) $\nu = 0.958 \times 10^{-5}\ \text{ft}^2/\text{s}$, so that

$$x_{\text{crit}} = \frac{0.958}{117.9}(58.95\ \text{ft}) = 0.479\ \text{ft}$$

4.7[D] Engine oil at 140 °F flows over a plate at 1.5 ft/sec; the plate is at 212 °F. At what point does the thermal boundary layer thickness reach 1 in.?

■ Fluid properties are evaluated at the *film temperature*, $T_f = (140 + 212)/2 = 176$ °F. From Table B-3:

$$\nu_f = 0.404 \times 10^{-3}\ \text{ft}^2/\text{sec} \qquad \mathbf{Pr}_f = 490$$

If the flow is laminar, (*1*) of Problem 4.5 gives

$$\delta_t = \frac{\delta}{\mathbf{Pr}_f^{1/3}} \tag{1}$$

On the other hand, Problem 4.3 gives

$$5.0 = \delta\sqrt{\frac{V_\infty}{\nu_f x}} \tag{2}$$

From (*1*) and (*2*),

$$x = \frac{\delta_t^2 \, \mathbf{Pr}_f^{2/3} \, V_\infty}{(25.0)\nu_f} = \frac{(\frac{1}{12}\text{ ft})^2(490)^{2/3}(1.5\text{ ft/sec})}{(25.0)(0.404\times 10^{-3}\text{ ft}^2/\text{sec})} = 64.1\text{ ft}$$

As a check, computation of $\mathbf{Re}_x$ shows that the flow is indeed laminar at x.

4.8$^{\mathbf{D}}$ Engine oil at 60 °C flows over a flat plate at 0.457 m/s; the plate is at 100 °C. At what point does the thermal boundary layer thickness reach 25.4 mm?

▌ For fluid properties use the *film temperature*, $T_f = (60+100)/2 = 80$ °C. From Table B-3:

$$\nu_f = 3.753\times 10^{-5}\text{ m}^2/\text{s} \qquad \mathbf{Pr}_f = 490$$

If the flow is laminar,

$$\delta_t = \frac{\delta}{\mathbf{Pr}_f^{1/3}} \tag{1}$$

by (*1*) of Problem 4.5. On the other hand, Problem 4.3 gives

$$5.0 = \delta\sqrt{\frac{V_\infty}{\nu_f x}} \tag{2}$$

From (*1*) and (*2*),

$$x = \frac{\delta_t^2 \, \mathbf{Pr}_f^{2/3} \, V_\infty}{(25.0)\nu_f} = \frac{(0.0254\text{ m})^2(490)^{2/3}(0.457\text{ m/s})}{(25.0)(3.753\times 10^{-5}\text{ m}^2/\text{s})} = 19.53\text{ m}$$

As a check, computation of $\mathbf{Re}_x$ shows that the flow is indeed laminar at x.

4.9$^{\mathbf{D}}$ Air at 150 °F and 14.7 psia flows along a smooth, flat plate at 40 ft/sec. For laminar flow, at what distance from the leading edge does the boundary layer thickness reach 0.147 in.?

▌ From Figure B-2, $\nu \approx 2\times 10^{-4}\text{ ft}^2/\text{sec}$; hence, by Problem 4.3,

$$x = \frac{\delta^2 V_\infty}{25\nu} \approx \frac{[(0.147/12)\text{ ft}]^2(40\text{ ft/sec})}{25(2\times 10^{-4}\text{ ft}^2/\text{sec})} = 1.20\text{ ft}$$

Check: $$\mathbf{Re}_x = \frac{(1.20\text{ ft})(40\text{ ft/sec})}{2\times 10^{-4}\text{ ft}^2/\text{sec}} = 2.4\times 10^5 < 5\times 10^5 \qquad (\textit{laminar})$$

4.10$^{\mathbf{D}}$ Air at 66 °C and 101.3 kPa (1 atm) pressure flows along a smooth, flat plate at 12.192 m/s. For laminar flow, at what distance from the leading edge does the boundary layer thickness reach 3.73 mm?

▌ From Figure B-2, $\nu \approx 1.86\times 10^{-5}\text{ m}^2/\text{s}$; hence, by Problem 4.3,

$$x = \frac{\delta^2 V_\infty}{25\nu} \approx \frac{(0.003\,73\text{ m})^2(12.192\text{ m/s})}{25(1.86\times 10^{-5}\text{ m}^2/\text{s})} = 0.365\text{ m}$$

Check: $$\mathbf{Re}_x = \frac{(0.365\text{ m})(12.192\text{ m/s})}{1.86\times 10^{-5}\text{ m}^2/\text{s}} = 2.4\times 10^5 < 5\times 10^5 \qquad (\textit{laminar})$$

4.11 Make an order-of-magnitude estimate of the boundary layer thickness δ, assuming constant fluid properties, zero pressure gradient, and two-dimensional flow.

▌ The governing equations are

Continuity $$\frac{\partial u}{\partial x} + \frac{\partial v}{\partial y} = 0$$

x-Momentum $$u\frac{\partial u}{\partial x} + v\frac{\partial u}{\partial y} = \nu\frac{\partial^2 u}{\partial y^2}$$

Except very close to the surface, the velocity u within the boundary layer is of the order of the free-stream velocity; i.e., $u \sim V_\infty$. And the y-dimension within the boundary layer is of the order of the boundary layer thickness, $y \approx \delta$. We can then approximate the continuity equation as

$$\frac{V_\infty}{x} + \frac{v}{\delta} \approx 0 \qquad \text{or} \qquad v \approx \frac{V_\infty \delta}{x}$$

Using the estimates of u, y, and v in the x-momentum equation gives

$$V_\infty \frac{V_\infty}{x} + \frac{V_\infty \delta}{x} \frac{V_\infty}{\delta} \approx \nu \frac{V_\infty}{\delta^2} \qquad \text{or} \qquad \delta^2 \approx \frac{\nu x}{V_\infty}$$

Dividing by x^2 to make dimensionless yields

$$\frac{\delta}{x} \approx \left(\frac{\nu}{V_\infty x}\right)^{1/2} = \frac{1}{\sqrt{\mathbf{Re}_x}}$$

in harmony with Problem 4.3.

4.12 A Pitot tube, located on the undercarriage of an airship 4 in. aft of its leading edge, is to be used to monitor airspeed which varies from 20 to 80 mph. The undercarriage is approximately flat, making the pressure gradient negligible. Air temperature is 40 °F and the pressure is 24.8 in. Hg. To be outside of the boundary layer, at what distance should the Pitot tube be located from the undercarriage?

▌ From Figure. B-1 the dynamic viscosity is $\mu = 3.7 \times 10^{-7}$ lbf · sec/ft^2. The density may be determined from the ideal gas relation,

$$\rho = \frac{p}{RT}$$

where, for air, $R = 53.3$ ft · lbf/lbm · °R and

$$p = (24.8 \text{ in. Hg})\left(\frac{2116 \text{ lbf/ft}^2}{29.92 \text{ in. Hg}}\right) = 1753.9 \text{ lbf/ft}^2$$

Thus

$$\rho = \frac{1753.9 \text{ lbf/ft}^2}{(53.3 \text{ ft} \cdot \text{lbf/lbm} \cdot \text{°R})(500 \text{ °R})} = 0.0658 \text{ lbm/ft}^3$$

and the kinematic viscosity is

$$\nu = \frac{\mu g_c}{\rho} = \frac{(3.7 \times 10^{-7} \text{ lbf} \cdot \text{sec/ft}^2)(32.2 \text{ lbm} \cdot \text{ft/lbf} \cdot \text{sec}^2)}{0.0658 \text{ lbm/ft}^3} = 1.81 \times 10^{-4} \text{ ft}^2\text{/sec}$$

The calculation

$$\mathbf{Re}_{\max} = \frac{(V_\infty)_{\max} x}{\nu} = \frac{(80 \text{ mph})\left(\dfrac{88 \text{ ft/sec}}{60 \text{ mph}}\right)\left(\dfrac{4}{12} \text{ ft}\right)}{1.81 \times 10^{-4} \text{ ft}^2\text{/sec}} = 2.16 \times 10^5 < 5 \times 10^5$$

shows that the flow at the location of the Pitot tube will be laminar over the entire range of airspeed. Hence, Problem 4.3 yields at $x = 4$ in.

$$\delta_{\max} = \frac{(5.0)x}{\sqrt{\mathbf{Re}_{\min}}} = \frac{(5.0)(4 \text{ in.})}{\sqrt{\left(\dfrac{20 \text{ mph}}{80 \text{ mph}}\right)(2.16 \times 10^5)}} = 0.09 \text{ in.}$$

A separation of 0.1 in. would therefore be safe.

4.13 Evaluate the *local skin-friction coefficient* $c_f(x)$ for the Blasius solution (Problems 4.2 and 4.3).

▌ Using the fundamental equation for shear stress, $\tau = \mu(\partial u/\partial y)$, we can get the drag on the plate by evaluating the local shear stress at the surface, $\tau_s(x)$:

$$\tau_s(x) = \mu \left.\frac{\partial u}{\partial y}\right|_{y=0} = \mu V_\infty \left(\frac{V_\infty}{\nu x}\right)^{1/2} f''(0) = \mu V_\infty \left(\frac{V_\infty}{\nu x}\right)^{1/2} (0.332)$$

in which Fig. 4-2 has been used to evaluate $f''(0)$. The local skin-friction coefficient is defined by

$$c_f(x) \equiv \frac{\tau_s(x)}{\rho_\infty V_\infty^2/2} \tag{1}$$

that is, $c_f(x)$ is the ratio of the kinetic energy per unit volume lost to drag at surface location x, to the kinetic energy per unit volume in the free stream. Substituting for $\tau_s(x)$ and using $\nu = \mu/\rho_\infty$ (incompressible flow),

$$c_f = \frac{0.664}{(V_\infty x/\nu)^{1/2}} = \frac{0.644}{(\mathbf{Re}_x)^{1/2}} \tag{2}$$

The average skin-friction coefficient for $0<x<L$ is then (see Problem 4.21)

$$C_f \equiv \frac{1}{L}\int_0^L c_f\,dx = \frac{1.328}{(\mathbf{Re}_L)^{1/2}} \tag{3}$$

4.14[D] Approximate the skin-friction drag on the lateral surface of a 3-ft-long by 2-ft-diameter cylinder, located axially in a wind tunnel, when the airspeed is 15 fps. The pressure is atmospheric and the temperature is 140 °F. (The cylinder is fitted with a streamlined nose cone.)

| From Figure B-2, $\nu \approx 2\times10^{-4}\ \text{ft}^2/\text{sec}$. Since the drag is required, the characteristic length for the calculation of the Reynolds number is the length of the cylinder rather than its diameter; therefore,

$$\mathbf{Re}_L = \frac{V_\infty L}{\nu} \approx \frac{(15\ \text{ft/sec})(3\ \text{ft})}{2\times10^{-4}\ \text{ft}^2/\text{sec}} = 2.25\times10^5$$

and the laminar flow equations are valid. The average skin-friction coefficient is given by (3) of Problem 4.13:

$$C_f = \frac{1.328}{\sqrt{\mathbf{Re}_L}} = \frac{1.328}{\sqrt{225\,000}} = 0.0028$$

The skin-friction drag force, F_f, is given by the product of average shear stress and total area, i.e.,

$$F_f = C_f\left(\frac{\rho_\infty V_\infty^2}{2g_c}\right)A$$

where $\rho_\infty = 0.066\ \text{lbm/ft}^3$ from Table B-4. (Recall that in British engineering units kinetic energy is given by $\frac{1}{2}mv^2/g_c$.)

Hence $$F_f = (0.0028)\frac{(0.066\ \text{lbm/ft}^3)(15\ \text{ft/sec})^2}{2(32.2\ \text{lbm}\cdot\text{ft/lbf}\cdot\text{sec}^2)}[\pi(2\ \text{ft})(3\ \text{ft})] = 0.0122\ \text{lbf}$$

4.15[D] Approximate the skin-friction drag on the lateral surface of a 0.91-m-long by 0.61-m-diameter cylinder located axially in a wind tunnel, when the airspeed is 4.6 m/s. The air pressure is atmospheric and the air and cylinder temperature are both at 60 °C. (The cylinder is fitted with a streamlined nose cone.)

| From Figure B-2, $\nu \approx 1.9\times10^{-5}\ \text{m}^2/\text{s}$, and so

$$\mathbf{Re}_L = \frac{V_\infty L}{\nu} \approx \frac{(4.6\ \text{m/s})(0.91\ \text{m})}{1.9\times10^{-5}\ \text{m}^2/\text{s}} = 2.203\times10^5$$

The laminar flow equations are valid. The average skin-friction coefficient is given by (3) of Problem 4.13:

$$C_f = \frac{1.328}{\sqrt{\mathbf{Re}_L}} = \frac{1.328}{\sqrt{2.203\times10^5}} = 0.0028$$

The skin-friction drag force F_f is

$$F_f = C_f\left(\frac{\rho_\infty V_\infty^2}{2}\right)A$$

From Table B-4, $\rho_\infty = (0.0661)(16.018) = 1.059\ \text{kg/m}^3$; thus,

$$F_f = (0.0028)\frac{(1.059\ \text{kg/m}^3)(4.6\ \text{m/s})^2}{2}[\pi(0.61\ \text{m})(0.91\ \text{m})] = 0.0547\ \text{kg}\cdot\text{m/s}^2 = 0.0547\ \text{N}$$

where the definition of the newton has been used.

4.16 A thermal sensor is to be located 2 m from the leading edge, and 45 mm from the surface, of a flat plate along which a liquid flows at 19 m/s. The pressure is atmospheric and $J = 2.79\times10^{-3}\ \text{m}^2/\text{s}$. Calculate the velocity components, u and v, at the sensor.

| The Reynolds number is

$$\mathbf{Re}_x = \frac{V_\infty x}{\nu} = \frac{(19\ \text{m/s})(2\ \text{m})}{2.79\times10^{-3}\ \text{m}^2/\text{s}} = 13\,620$$

and the flow is laminar. Moreover, at $x = 2$ m,

$$\delta = \frac{5.0x}{\sqrt{\mathbf{Re}_x}} = \frac{5(2\ \text{m})}{\sqrt{13\,620}} = 85.7\ \text{mm} > 45\ \text{mm}$$

so that the sensor is within the boundary layer. The velocity components are obtained from the Blasius solution (Fig. 4-2). At

$$\eta = y\left(\frac{V_\infty}{\nu x}\right)^{1/2} = (0.045\ \text{m})\left[\frac{19\ \text{m/s}}{(2.79\times 10^{-3}\ \text{m}^2/\text{s})(2\ \text{m})}\right]^{1/2} = 2.63$$

we have

$$\frac{u}{V_\infty} \approx 0.775 \qquad \frac{v}{V_\infty}\sqrt{\mathbf{Re}_x} \approx 0.48$$

giving $u \approx 14.73$ m/s, $v \approx 0.078$ m/s.

4.2 LAMINAR FLOW OVER A FLAT PLATE: SIMILARITY SOLUTIONS; INTEGRAL APPROXIMATIONS

4.17 Derive the integral momentum equation (due to von Kármán) for steady, incompressible, laminar flow over a flat plate.

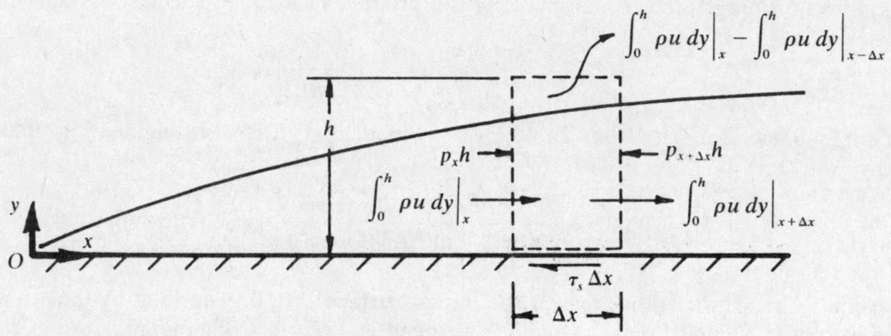

Fig. 4-5

▌ Consider the dashed control volume in Fig. 4-5. The forces acting on the volume are shown (there is no shear at the upper face, which is outside the boundary layer), as well as the mass fluxes through the faces. Notice that the mass efflux through the upper face exactly cancels the net mass influx through the other two faces; this is required by the equation of continuity. The corresponding x-momentum fluxes are

Influx through left face $\quad \displaystyle\int_0^h \rho u^2\, dy\Big|_x$

Efflux through right face $\quad \displaystyle\int_0^h \rho u^2\, dy\Big|_{x+\Delta x}$

Efflux through upper face $\quad \displaystyle V_\infty\left[\int_0^h \rho u\, dy\Big|_x - \int_0^h \rho u\, dy\Big|_{x+\Delta x}\right]$

where, in the last expression, V_∞ is evaluated at some point between x and $x+\Delta x$. Newton's law, $\Sigma F_x = (x\text{-momentum efflux}) - (x\text{-momentum influx})$ gives

$$p_x h - p_{x+\Delta x} h - \tau_s\,\Delta x = \int_0^h \rho u^2\, dy\Big|_{x+\Delta x} + V_\infty\left[\int_0^h \rho u\, dy\Big|_x - \int_0^h \rho u\, dy\Big|_{x+\Delta x}\right] - \int_0^h \rho u^2\, dy\Big|_x$$

Rearranging, dividing by Δx, and taking the limit as Δx approaches zero, we get (ρ = constant):

$$h\frac{dp}{dx} + \tau_s = \rho\left[-\frac{d}{dx}\left(\int_0^h u^2\, dy\right) + V_\infty\frac{d}{dx}\left(\int_0^h u\, dy\right)\right] \tag{1}$$

But, outside the boundary layer, Bernoulli's equation gives

$$\frac{dp}{dx} = -\rho V_\infty \frac{dV_\infty}{dx} \tag{2}$$

Also,

$$V_\infty\frac{d}{dx}\left(\int_0^h u\, dy\right) = \frac{d}{dx}\left(\int_0^h V_\infty u\, dy\right) - \left(\frac{dV_\infty}{dx}\right)\left(\int_0^h u\, dy\right) \tag{3}$$

GENERAL BOOK

FREQUENT BUYER BONUS VOUCHER

GOOD FOR **$2.00**

0165
receipt #

clerk

1/24/94
date

towards any purchase in the General Book Dept. during the month of **February, 1994.**
Good only in the General Book Dept.
Good only between **February 1st and February 28th, 1994.**
Good only when validated by cashier at time of issue.
NOT VALID ON SPECIAL ORDERS
Not valid on U.C. catalogs/schedules.

82

ASUC STORE • BANCROFT AT TELEGRAPH • GENERAL BOOK • 642-7294

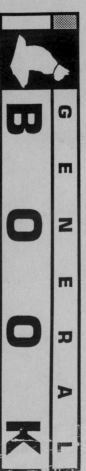

712 CASH-1 0165 0001 143

ACCOUNT NUMBER 4673673744314

VISA/MASTERCARD 21.60

1/31/94 3:15 PM

Substituting (2) and (3) into (1) and recombining terms, we get

$$\tau_s = \rho \frac{d}{dx}\left[\int_0^h (V_\infty - u)u\,dy\right] + \rho\left(\frac{dV_\infty}{dx}\right)\left[\int_0^h (V_\infty - u)u\,dy\right] \qquad (4)$$

The upper limit of integration in (4) may be replaced by δ, yielding, since $(V_\infty - u)$ is zero outside the boundary layer,

$$\tau_s = \rho \frac{d}{dx}\left[\int_0^\delta (V_\infty - u)u\,dy\right] \qquad (5)$$

In the British Engineering system replace ρ in (5) by ρ/g_c.

4.18 Assuming a velocity profile of the form

$$\frac{u}{V_\infty} = C + C_1\frac{y}{\delta} + C_2\left(\frac{y}{\delta}\right)^2 + C_3\left(\frac{y}{\delta}\right)^3$$

within the boundary layer, evaluate the constants subject to the boundary conditions

$$\textbf{At}\quad y = 0 \qquad u = \frac{\partial^2 u}{\partial y^2} = 0 \qquad u = 0$$

$$\textbf{At}\quad y = \delta \qquad u = V_\infty \qquad \frac{\partial u}{\partial y} = 0$$

▌ Since $u = 0$ for $y = 0$, $C = 0$. Applying $u = V_\infty$ when $y = \delta$ gives

$$1 = C_1 + C_2 + C_3 \qquad (1)$$

Differentiating the assumed profile (V_∞ and δ are functions of x alone),

$$\frac{1}{V_\infty}\frac{\partial u}{\partial y} = \frac{C_1}{\delta} + \frac{2C_2}{\delta}\left(\frac{y}{\delta}\right) + \frac{3C_3}{\delta}\left(\frac{y}{\delta}\right)^2 \qquad (2)$$

But $\partial u/\partial y = 0$ when $y = \delta$; therefore,

$$0 = C_1 + 2C_2 + 3C_3 \qquad (3)$$

A second differentiation of the velocity profile gives

$$\frac{1}{V_\infty}\frac{\partial^2 u}{\partial y^2} = \frac{2C_2}{\delta^2} + \frac{6C_3}{\delta^2}\left(\frac{y}{\delta}\right) \qquad (4)$$

from which (at $y = 0$, $\partial^2 u/\partial y^2 = 0$) we get

$$0 = C_2 \qquad (5)$$

Solving (1), (3), and (5) simultaneously,

$$C_1 = \tfrac{3}{2} \qquad C_3 = -\tfrac{1}{2}$$

giving the cubic velocity profile

$$\frac{u}{V_\infty} = \frac{3}{2}\left(\frac{y}{\delta}\right) - \frac{1}{2}\left(\frac{y}{\delta}\right)^3 \qquad (6)$$

4.19 Using the cubic velocity profile from Problem 4.18 in the von Kármán integral equation, (5) of Problem 4.17, determine the boundary layer thickness and the average skin-friction coefficient for laminar flow over a flat plate.

▌ From Newton's law of viscosity and (6) of Problem 4.18

$$\tau_s = \mu \left.\frac{\partial u}{\partial y}\right|_{y=0} = \frac{3\mu V_\infty}{2\delta}$$

and the von Kármán equation becomes ($\phi \equiv y/\delta$)

$$\frac{3\mu V_\infty}{2\delta} = \rho V_\infty^2 \frac{d}{dx}\left[\delta \int_0^1 \left(1 - \frac{3}{2}\phi + \frac{1}{2}\phi^3\right)\left(\frac{3}{2}\phi - \frac{1}{2}\phi^3\right)d\phi\right] \qquad (1)$$

The definite integral in (1) has the value $\frac{39}{280}$, so that ($\nu = \mu/\rho$)

$$\frac{140\nu}{13\delta V_\infty} = \frac{d\delta}{dx} \qquad (2)$$

Separating the variables in (2) and integrating ($\delta = 0$ at $x = 0$; $\mathbf{Re}_x \equiv V_\infty x/\nu$),

$$\frac{\delta}{x} = \frac{4.64}{\sqrt{\mathbf{Re}_x}} \tag{3}$$

The local skin-friction coefficient is given by

$$c_f = \frac{\tau_s}{\frac{1}{2}\rho V_\infty^2} = \frac{3\nu}{V_\infty \delta} = \frac{3}{(\mathbf{Re}_x)(\delta/x)} = \frac{0.646}{\sqrt{\mathbf{Re}_x}} \tag{4}$$

Then

$$C_f = \frac{1}{L}\int_0^L c_f\,dx = \frac{1.292}{\sqrt{\mathbf{Re}_L}} \tag{5}$$

4.20 Air at 20 °C and 101.3 kPa (1 atm) flows parallel to a flat plate at a velocity of 3.5 m/s. Compare the boundary layer thickness and the local skin-friction coefficient at $x = 1$ m as given by the exact Blasius solution with the values from the approximate von Kármán integral technique (Problem 4.19).

▌ From Figure B-2, $\nu = 1.5 \times 10^{-5}$ m²/s, and the Reynolds number is

$$\mathbf{Re}_x = \frac{V_\infty x}{\nu} = \frac{(3.5\ \text{m/s})(1\ \text{m})}{1.5 \times 10^{-5}\ \text{m}^2/\text{s}} = 233\,300$$

The Blasius solution (Problems 4.3 and 4.13) gives

$$\delta = \frac{(5.0)x}{\sqrt{\mathbf{Re}_x}} = \frac{(5.0)(1\ \text{m})}{\sqrt{233\,300}} = 10.35\ \text{mm}$$

$$c_f = \frac{0.664}{\sqrt{\mathbf{Re}_x}} = \frac{0.664}{\sqrt{233\,300}} = 1.37 \times 10^{-3}$$

The von Kármán solution yields

$$\delta = \frac{(4.64)x}{\sqrt{\mathbf{Re}_x}} = 9.61\ \text{mm} \qquad c_f = \frac{0.646}{\sqrt{\mathbf{Re}_x}} = 1.34 \times 10^{-3}$$

We note that the approximate solution deviates from the exact solution by 7.2 percent for the boundary layer thickness and 2.9 percent for the local skin-friction coefficient, deviations which are quite acceptable for the usual engineering accuracy.

4.21 Predict the relation $C_f = 2c_f(L)$ for laminar boundary layer flow over a flat plate.

▌ Problem 4.11 gives

$$c_f \approx \tau_s \approx \left.\frac{\partial u}{\partial y}\right|_{y=0} \approx \frac{1}{\delta} \approx x^{-1/2}$$

or $c_f(x) = Kx^{-1/2}$. Then

$$C_f = \frac{K}{L}\int_0^L x^{-1/2}\,dx = 2KL^{-1/2} = 2c_f(L) \tag{1}$$

A relation of the form (1) clearly holds for any physical parameter that varies as $x^{-1/2}$; one such is the local convective heat transfer coefficient $h(x)$.

4.22 What is the drag per unit width on the leading 1 m of one side of the plate of Problem 4.20?

▌ The drag force per unit width on a section of the length L is

$$\frac{F_f}{W} = C_f\left(\frac{\rho_\infty V_\infty^2}{2}\right)L$$

From Table B-4, by interpolation, $\rho_\infty = 1.210$ kg/m³, and from Problems 4.20 and 4.21 ($L = 1$ m), $C_f = 2(1.37 \times 10^{-3}) = 2.74 \times 10^{-3}$ Therefore,

$$\frac{F_f}{W} = (2.74 \times 10^{-3})\,\frac{(1.210\ \text{kg/m}^3)(3.5\ \text{m/s})^2}{2}\,(1\ \text{m}) = 0.020\ \text{N/m}$$

($1\ \text{N} \equiv 1\ \text{kg}\cdot\text{m/s}^2$).

4.23 In (2) of Problem 4.1, neglect the terms ρg_x and $\partial^2 u/\partial x^2$ to obtain the commonly used momentum equation

$$u\frac{\partial u}{\partial x} + v\frac{\partial u}{\partial y} = -\frac{1}{\rho}\frac{\partial p}{\partial x} + \nu\frac{\partial^2 u}{\partial y^2} \tag{1}$$

Introducing the dimensionless variables

$$p^* = \frac{p}{\rho V_\infty^2} \qquad x^* = \frac{x}{L} \qquad y^* = \frac{y}{L} \qquad u^* = \frac{u}{V_\infty} \qquad v^* = \frac{v}{V_\infty}$$

put (1) in nondimensional form.

I By the chain rule,

$$\frac{\partial u}{\partial x} = \left[\frac{\partial}{\partial x^*}(V_\infty u^*)\right]\frac{\partial x^*}{\partial x} = \left[V_\infty \frac{\partial u^*}{\partial x^*}\right]\frac{1}{L} = \frac{V_\infty}{L}\frac{\partial u^*}{\partial x^*}$$

and similarly for the other derivatives in (1). The final result is

$$u^*\frac{\partial u^*}{\partial x^*} + v^*\frac{\partial u^*}{\partial y^*} = -\frac{\partial p^*}{\partial x^*} + \frac{1}{\mathbf{Re}_L}\frac{\partial^2 u^*}{\partial y^{*2}} \tag{2}$$

4.24 From (2) of Problem 4.23 infer the functional dependence of the local skin-friction coefficient, c_f.

I According to the partial differential equation,

$$u^* = f_1\left(x^*, y^*, \frac{\partial p^*}{\partial x^*}, \mathbf{Re}_L\right) = f_2(x^*, y^*, \mathbf{Re}_L) \tag{1}$$

where the removal of $\partial p^*/\partial x^*$ follows Problem 4.2. Now

$$\tau_s = \mu \left.\frac{\partial u}{\partial y}\right|_{y=0} = \frac{\mu V_\infty}{L}\left.\frac{\partial u^*}{\partial y^*}\right|_{y^*=0}$$

so that

$$c_f = \frac{\tau_s}{\rho V_\infty^2/2} = \frac{2}{\mathbf{Re}_L}\left.\frac{\partial u^*}{\partial y^*}\right|_{y^*=0} = f_3(x^*, \mathbf{Re}_L) \tag{2}$$

where the last step follows upon differentiation of (1). The form of the function f_3 is prescribed by the physical geometry.

4.25^D In a modeling experiment, water at 60 °C and atmospheric pressure flows over a streamlined airfoil without separation. The undisturbed approach velocity is 0.573 m/s, and the measured shear stress 0.1 m from the leading edge is 0.3075 Pa. The airfoil is to be used in an airstream at atmospheric pressure, 77 °C, and approach velocity 25.0 m/s. What will be the shear stress at the same location?

I For water flow, from Table B-3, $\nu_1 = 4.775 \times 10^{-7}\ \text{m}^2/\text{s}$ and $\rho_1 = 985.4\ \text{kg/m}^3$; thus,

$$\mathbf{Re}_{L,1} = \frac{(0.1\ \text{m})(0.573\ \text{m/s})}{4.775 \times 10^{-7}\ \text{m}^2/\text{s}} = 1.2 \times 10^5$$

For air flow, from Table B-4, $\nu_2 = 2.0792 \times 10^{-5}\ \text{m}^2/\text{s}$, and $\rho_2 = 0.9979\ \text{kg/m}^3$; thus,

$$\mathbf{Re}_{L,2} = \frac{(0.1\ \text{m})(25.0\ \text{m/s})}{2.0792 \times 10^{-5}\ \text{m}^2/\text{s}} = 1.2 \times 10^5$$

Hence, for the two cases, $x_1^* = x_2^*$; $\mathbf{Re}_{L,1} = \mathbf{Re}_{L,2}$. (The water velocity was chosen to ensure the latter equality.) Problem 4.24 then gives

$$c_{f,1} = \frac{\tau_{s,1}}{\rho_1 V_{\infty,1}^2/2} = \frac{\tau_{s,2}}{\rho_2 V_{\infty,2}^2/2} = c_{f,2}$$

or

$$\tau_{s,2} = \left(\frac{\rho_2}{\rho_1}\right)\left(\frac{V_{\infty,2}}{V_{\infty,1}}\right)^2 \tau_{s,1} = \left(\frac{0.9979}{985.4}\right)\left(\frac{25.0}{0.573}\right)^2 (0.3075\ \text{Pa}) = 0.593\ \text{Pa}$$

4.26^D In a modeling experiment, water at 140 °F and atmospheric pressure flows over a streamlined airfoil without separation. The undisturbed approach velocity is 1.88 ft/s and the measured shear stress 0.328 ft from the leading edge is 0.0064 lbf/ft². For air flow at 170 °F, atmospheric pressure, and approach velocity 82.0 ft/s, what will be the shear stress at the same location?

For water flow, from Table B-3, $\nu_1 = 0.514 \times 10^{-5}\ \text{ft}^2/\text{s}$ and $\rho_1 = 61.52\ \text{lbm/ft}^3$; thus

$$\mathbf{Re}_{L,1} = \frac{(0.328\ \text{ft})(1.88\ \text{ft/s})}{0.514 \times 10^{-5}\ \text{ft}^2/\text{s}} = 1.2 \times 10^5$$

For air flow, from Table B-4, $\nu_2 = 22.38 \times 10^{-5}\ \text{ft}^2/\text{s}$ and $\rho_2 = 0.0623\ \text{lbm/ft}^3$; thus,

$$\mathbf{Re}_{L,2} = \frac{(0.328\ \text{ft})(82.0\ \text{ft/s})}{22.38 \times 10^{-5}\ \text{ft}^2/\text{s}} = 1.2 \times 10^5$$

Hence, $\mathbf{Re}_{L,1} = \mathbf{Re}_{L,2}$ and $\mathbf{x}_1^* = \mathbf{x}_2^*$. (The water velocity was chosen to ensure the former equality.) Problem 4.24 then gives

$$c_{f,1} = \frac{\tau_{s,1}}{\rho_1 V_{\infty,1}^2/2g_c} = \frac{\tau_{s,2}}{\rho_2 V_{\infty,2}^2/2g_c} = c_{f,2}$$

or

$$\tau_{s,2} = \left(\frac{\rho_2}{\rho_1}\right)\left(\frac{V_{\infty,2}}{V_{\infty,1}}\right)^2 \tau_{s,1} = \left(\frac{0.0623}{61.52}\right)\left(\frac{82.0}{1.88}\right)^2 (0.0064\ \text{lbf/ft}^2) = 0.0123\ \text{lbf/ft}^2$$

4.27^D Water at 68 °F and 1 atm flows parallel to a flat plate at 1.0 ft/s. Using the von Kármán integral approximation, find the drag per unit width on the leading 3.28 ft of one side of the plate.

From Table B-3, $\nu = 1.083 \times 10^{-5}\ \text{ft}^2/\text{s}$ and $\rho = 62.46\ \text{lbm/ft}^3$;

$$\mathbf{Re}_L = \frac{V_\infty L}{\nu} = \frac{(1.0\ \text{ft/s})(3.28\ \text{ft})}{1.083 \times 10^{-5}\ \text{ft}^2/\text{s}} = 3.03 \times 10^5$$

The flow is laminar, and the average skin-friction coefficient over the 3.28 ft is given by (*5*) of Problem 4.19 as

$$C_f = \frac{1.292}{10^2\sqrt{30.3}} = 2.35 \times 10^{-3}$$

Thus, the drag force due to skin friction is

$$\frac{F_f}{W} = C_f\left(\frac{\rho_\infty V_\infty^2}{2g_c}\right)L = (2.35 \times 10^{-3})\frac{(62.46\ \text{lbm/ft}^3)(1.0\ \text{ft/s})^2}{2(32.174\ \text{lbm}\cdot\text{ft/lbf}\cdot\text{s}^2)}(3.28\ \text{ft}) = 7.48 \times 10^{-3}\ \text{lbf/ft}$$

4.28^D Water at 20 °C and 101.3 kPa (1 atm) flows parallel to a flat plate at a velocity of 0.3048 m/s. Using the von Kármán integral approximation, find the drag per unit width on the leading 1 m of one side of the plate.

From Table B-3, $\nu = 1.006 \times 10^{-6}\ \text{m}^2/\text{s}$ and $\rho = 1000.5\ \text{kg/m}^3$;

$$\mathbf{Re}_L = \frac{V_\infty L}{\nu} = \frac{(0.3048\ \text{m/s})(1.0\ \text{m})}{1.006 \times 10^{-6}\ \text{m}^2/\text{s}} = 3.03 \times 10^5$$

The flow is laminar, and the average skin-friction coefficient over the 1 m is given by (*5*) of Problem 4.19 as

$$C_f = \frac{1.292}{10^2\sqrt{30.3}} = 2.35 \times 10^{-3}$$

Thus, the drag force due to skin friction is

$$\frac{F_f}{W} = C_f\left(\frac{\rho_\infty V_\infty^2}{2}\right)L = (2.35 \times 10^{-3})\frac{(1000.5\ \text{kg/m}^3)(0.3048\ \text{m/s})^2}{2}(1\ \text{m}) = 0.109\ \text{N/m}$$

4.29 As a converse to Problem 4.21 show that if

$$F \equiv \frac{1}{L}\int_0^L f(x)\,dx = 2f(L) \tag{1}$$

for arbitrary L, then $f(x) \approx x^{-1/2}$.

Differentiate (*1*) on L:

$$-\frac{1}{L^2}\int_0^L f(x)\,dx + \frac{1}{L}f(L) = 2f'(L)$$

$$-\frac{2}{L}f(L) + \frac{1}{L}f(L) = 2f'(L)$$

$$\frac{f'}{f} = -\frac{1}{2L} \tag{2}$$

Integration of (*2*) gives $f(L) = KL^{-1/2}$, or $f(x) = Kx^{-1/2}$.

4.30^D Water at $T_\infty = 40$ °C flows over an aluminum plate, 10 mm thick. The plate is electrically heated on the side opposite to the water; the surface next to the water is at $T_s = 59.8$ °C and the other surface is at 60 °C. For these conditions at steady state, determine the convective heat transfer coefficient between the water and the plate.

▌ At the unheated surface,

$$\bar{h}(T_s - T_\infty) = -k\frac{dT}{dy} \tag{1}$$

From Table B-1, $k \approx (118)(1.7296) = 204.1$ W/m·K, and

$$\frac{dT}{dy} \approx \frac{\Delta T}{\Delta y}\bigg|_{\text{in alum.}} = \frac{(59.8 - 60)\ °\text{C}}{0.010\ \text{m}} = -20\ °\text{C/m}$$

Hence,

$$\bar{h} = -\left(\frac{1}{T_s - T_\infty}\right)\left(k\frac{dT}{dy}\right) \approx -\frac{1}{(59.8-40)\ °\text{C}}(204.1\ \text{W/m}\cdot\text{K})(-20\ °\text{C/m}) = 206.1\ \text{W/m}^2\cdot\text{K}$$

4.31 Repeat Problem 4.30 with the plate thickness increased to 25 mm; all other parameters are unchanged.

▌ The temperature gradient, and hence $\bar{h}$, changes by the factor $\frac{10}{25} = 0.4$.

4.32^D Water at 104 °F flows over the top surface of an aluminum plate 0.394 in. thick. The plate is heated electrically on the bottom surface; the surface next to the water is at $T_s = 139.64$ °F while the bottom surface is at 140 °F. For these steady-state conditions, determine the convective heat transfer coefficient between the water and the plate.

▌ At the top surface,

$$\bar{h}(T_s - T_\infty) = -k\frac{dT}{dy}$$

From Table B-1, $k \approx 118$ Btu/hr·ft·°F, and

$$\frac{dT}{dy} \approx \frac{\Delta T}{\Delta y}\bigg|_{\text{in alum.}} = \frac{(139.64 - 140)\ °\text{F}}{(0.394/12)\ \text{ft}} = -10.96\ °\text{F/ft}$$

Hence,

$$\bar{h} = -\left(\frac{1}{T_s - T_\infty}\right)\left(k\frac{dT}{dy}\right) \approx -\frac{1}{(139.64-104)\ °\text{F}}(118\ \text{Btu/hr}\cdot\text{ft}\cdot°\text{F})(-10.96\ °\text{F/ft}) = 36.3\ \text{Btu/hr}\cdot\text{ft}^2\cdot°\text{F}$$

4.33 A sheet of $8\frac{1}{2}$- by 11-in. paper lies flat on a smooth table in an office. As the door opens, a draft blows across the table at 3.1 ft/sec. The temperature in the room is 65 °F, and the draft blows along the 11-in. direction of the paper. Air pressure holds the paper down so that the shear bond between the paper and table is 0.0001 lbf/ft². Will the paper slide?

▌ From Table B-4 (by interpolation), $\rho = 0.076$ lbm/ft³ and $\nu = 1.577 \times 10^{-4}$ ft²/sec;

$$\mathbf{Re}_L = \frac{(\frac{11}{12}\ \text{ft})(3.1\ \text{ft/sec})}{0.1577 \times 10^{-3}\ \text{ft}^2/\text{sec}} = 18\,020 \quad (\textit{laminar flow})$$

From Problems 4.20 and 4.21, $C_f = 1.328/\sqrt{\mathbf{Re}_L}$; therefore,

$$\tau_s = \frac{\rho V_\infty^2}{2g_c}C_f = \frac{(0.076\ \text{lbm/ft}^3)(3.1\ \text{ft/sec})^2}{64.4\ \text{lbm}\cdot\text{ft/lbf}\cdot\text{sec}^2}\left(\frac{1.328}{\sqrt{18\,020}}\right) = 0.000\,112\ \text{lbf/ft}^2$$

Yes, the paper will slide.

4.34^D A very small probe is positioned in a boundary layer near a flat plate. At the location of the probe the boundary layer is 0.6 in. thick. Water at 90 °F flows over the plate; $V_\infty = 0.5$ ft/sec. The flow is laminar, and the velocity profile is determined to be

$$\frac{u}{V_\infty} = 2\left(\frac{y}{\delta}\right) - \left(\frac{y}{\delta}\right)^2 \qquad (0 < y < \delta)$$

Find the shear stress and the skin-friction coefficient at the probe location.

▌ From Table B-3, by interpolation, $\nu = 0.853 \times 10^{-5}$ ft²/sec and $\rho = 62.2$ lbm/ft². Then, where $\delta =$ 0.6 in. = 0.05 ft,

$$\tau_s = \mu \left.\frac{\partial u}{\partial y}\right|_{y=0} = \frac{\nu\rho}{g_c}\left.\frac{\partial u}{\partial y}\right|_{y=0} = \frac{2\nu\rho V_\infty}{g_c\delta}$$

$$= \frac{2(0.853\times 10^{-5})(62.2)(0.5)}{(32.2)(0.05)} = 3.3\times 10^{-4}\ \text{lbf/ft}^2$$

and
$$c_f = \frac{\tau_s}{\rho V_\infty^2/2g_c} = \frac{4\nu}{V_\infty\delta} = \frac{4(0.853\times 10^{-5})}{(0.5)(0.05)} = 1.37\times 10^{-3}$$

4.35[D] Repeat Problem 4.34 for water at 32.55 °C flowing at a velocity $V_\infty = 0.1524$ m/s and $\delta = 15.2$ mm.

▮ From Table B-3, by interpolation, $\nu = 7.87\times 10^{-7}$ m²/s and $\rho = 996.8$ kg/m³. Then, where $\delta = 0.0152$ m,

$$\tau_s = \mu \left.\frac{\partial u}{\partial y}\right|_{y=0} = \nu\rho\left.\frac{\partial u}{\partial y}\right|_{y=0} = \frac{2\nu\rho V_\infty}{\delta}$$

$$= \frac{2(7.87\times 10^{-7})(996.8)(0.1524)}{0.0152} = 0.0157\ \text{Pa}$$

and
$$c_f = \frac{\tau_s}{\rho V_\infty^2/2} = \frac{4\nu}{V_\infty\delta} = \frac{4(7.87\times 10^{-7})}{(0.1524)(0.0152)} = 1.36\times 10^{-3}$$

4.36 Check the units on τ_s in Problems 4.34 and 4.35.

▮ In Problem 4.34,

$$[\tau_s] = \frac{[\text{ft}^2/\text{sec}][\text{lbm/ft}^3][\text{ft/sec}]}{[\text{lbm}\cdot\text{ft/lbf}\cdot\text{sec}^2][\text{ft}]} = \frac{[\text{lbm/sec}^2]}{[\text{lbm}\cdot\text{ft}^2/\text{lbf}\cdot\text{sec}^2]} = [\text{lbf/ft}^2]$$

In Problem 4.35,

$$[\tau_s] = \frac{[\text{m}^2/\text{s}][\text{kg/m}^3][\text{m/s}]}{[\text{m}]} = [\text{kg/m}\cdot\text{s}^2] = [\text{N/m}^2] = [\text{Pa}]$$

4.37[D] Again consider a laminar boundary layer, water flow at $T_\infty = 90$ °F and $V_\infty = 0.5$ ft/s over a flat plate, as in Problem 4.34. In this case, the plate is heated to a surface temperature $T_s = 118$ °F, and the probe is a small thermocouple. The probe is used to establish the temperature profile in the water boundary layer, and this is found to be well represented by

$$\Theta \equiv \frac{T - T_s}{T_\infty - T_s} = 2\left(\frac{y}{\delta_t}\right) - \left(\frac{y}{\delta_t}\right)^2 \qquad (1)$$

At a location where $\delta_t = 0.36$ in., calculate (*a*) the heat flux from plate to water; (*b*) the heat transfer coefficient h.

▮ For boundary layer, external-flow heat transfer, the fluid properties are evaluated at the *film temperature*

$$T_f = \frac{T_s + T_\infty}{2} = \frac{118 + 90}{2} = 104\ ^\circ\text{F}$$

From Table B-3, $k_f = 0.363$ Btu/hr · ft · °F.

(*a*)
$$\frac{q}{A} = -k_f\left.\frac{\partial T}{\partial y}\right|_{y=0} = -k_f(T_\infty - T_s)\left.\frac{\partial \Theta}{\partial y}\right|_{y=0} = \frac{2k_f(T_s - T_\infty)}{\delta_t}$$

$$= \frac{2(0.363)(118-90)}{0.36/12} = 678\ \text{Btu/hr}\cdot\text{ft}^2$$

(*b*) Since $q/A = h(T_s - T_\infty)$,

$$h = \frac{2k_f}{\delta_t} = \frac{2(0.363)}{0.03} = 24.2\ \text{Btu/hr}\cdot\text{ft}^2\cdot{}^\circ\text{F}$$

4.38[D] Rework Problem 4.37 for the following parameter values: $T_\infty = 32.2$ °C, $V_\infty = 0.1524$ m/s, $T_s = 47.8$ °C, $\delta_t = 16$ mm. This time use the temperature profile

$$\Theta \equiv \frac{T - T_s}{T_\infty - T_s} = 1.5\left(\frac{y}{\delta_t}\right) - 0.5\left(\frac{y}{\delta_t}\right)^3 \qquad (1)$$

▮ At the film temperature, $T_f = (47.8 + 32.2)/2 = 40.0$ °C, Table B-3 gives $k_f \approx 0.628$ W/m·K.

(a)
$$\frac{q}{A} = -k_f(T_\infty - T_s)\left.\frac{\partial\Theta}{\partial y}\right|_{y=0} = \frac{(1.5)k_f(T_s - T_\infty)}{\delta_t}$$
$$\approx \frac{(1.5)(0.628)(47.8 - 32.2)}{0.016} = 918 \text{ W/m}^2$$

(b)
$$h = \frac{q/A}{T_s - T_\infty} = \frac{(1.5)k_f}{\delta_t} \approx \frac{(1.5)(0.628)}{0.016} = 58.9 \text{ W/m}^2\cdot\text{K}$$

4.39 For the Blasius solution (Problem 4.3), evaluate the local skin-friction coefficient in terms of the local boundary layer thickness.

▮ Substitute $x = \delta^2 V_\infty/25\nu$, from Problem 4.3, in (*2*) of Problem 4.13:

$$c_f = \frac{(3.320)\nu}{V_\infty\delta} \qquad (1)$$

Together, (*1*) and Problem 4.21 imply the useful relation

$$C_f = \frac{(6.640)\nu}{V_\infty\,\delta(L)} \qquad (2)$$

4.40[D] Air at 350 °F is in crossflow over the outside of a large cylinder; the velocity is 0.3 ft/sec. Determine the local skin-friction coefficient where the boundary layer is 1 in. thick.

▮ Assuming a very large radius and laminar flow (to be checked), the Blasius flat-plate solution should apply approximately. From Table B-4, $\nu = 0.3106 \times 10^{-3}$ ft²/sec; hence (*1*) of Problem 4.39 gives

$$c_f = \frac{(3.320)(0.3106 \times 10^{-3}\text{ ft}^2/\text{sec})}{(0.3\text{ ft/sec})(\frac{1}{12}\text{ ft})} = 0.0412$$

[A computation of $\mathbf{Re} = (0.664/c_f)^2$ shows that the flow is in fact laminar.]

4.41[D] Air at 177 °C is in crossflow over the outside of a large cylinder; the velocity is 91.4 mm/s. Approximate the average skin-friction coefficient between the point (line) where the boundary layer begins and the point where it is 25.4 mm thick.

▮ Assuming a very large radius and laminar flow (to be checked), the Blasius flat-plate solution should apply approximately. From Table B-4, $\nu = 2.885 \times 10^{-5}$ m²/s; hence, (*2*) of Problem 4.39 gives

$$C_f = \frac{(6.640)(2.885 \times 10^{-5}\text{ m}^2/\text{s})}{(0.0914\text{ m/s})(0.0254\text{ m})} = 0.0825$$

[A computation of $\mathbf{Re} = (1.328/C_f)^2$ shows that the flow is in fact laminar.]

4.42 Consider the thermal boundary layer for laminar flow over a flat plate. If the temperature profile is to be represented as

$$\Theta \equiv \frac{T - T_s}{T_\infty - T_s} = F\left(\frac{y}{\delta_t}\right) \equiv F(\omega)$$

give three boundary conditions that must be satisfied by the function F.

▮ Because δ_t is independent of y, we have

$$\text{(i)}\quad F(0) = 0 \qquad (\text{at } y = 0,\ T = T_s)$$
$$\text{(ii)}\quad F(1) = 1 \qquad (\text{at } y = \delta_t,\ T = T_\infty)$$
$$\text{(iii)}\quad F'(1) = 0 \qquad (\text{at } y = \delta_t,\ \partial T/\partial y = 0)$$

(A fourth boundary condition follows from energy considerations—see Problem 4.56.)

4.43 Do the temperature profiles (*1*) of Problem 4.37 and (*1*) of Problem 4.38 satisfy the conditions of Problem 4.42?

▮ For $F(\omega) = 2\omega - \omega^2$:

$$\text{(i)}\quad F(0) = 2(0) - (0)^2 = 0$$
$$\text{(ii)}\quad F(1) = 2(1) - (1)^2 = 1$$
$$\text{(iii)}\quad F'(1) = 2 - 2(1) = 0$$

For $F(\omega) = 1.5\omega - 0.5\omega^3$:

$$\text{(i)}\quad F(0) = 1.5(0) - 0.5(0)^3 = 0$$
$$\text{(ii)}\quad F(1) = 1.5(1) - 0.5(1)^3 = 1$$
$$\text{(iii)}\quad F'(1) = 1.5 - 0.5(3)(1)^2 = 0$$

Both profiles are satisfactory.

4.44 For the flow considered in Problems 4.37 and 4.38 (same flow, different locations), is the velocity boundary layer thicker or thinner than the thermal boundary layer?

▮ At $T_f = 104\ °\text{F}$, Table B-3 gives $\mathbf{Pr} = 4.34$. Then the Pohlhausen solution of the laminar boundary layer energy equation gives (for all x)

$$\frac{\delta}{\delta_t} = (\mathbf{Pr})^{1/3} = (4.34)^{1/3} > 1$$

4.45 Solve the von Kármán integral equation, (*5*) of Problem 4.17, under the assumption of the velocity profile

$$\frac{u}{V_\infty} = G\left(\frac{y}{\delta}\right) \equiv G(\phi)$$

▮ Here one is asked to generalize Problem 4.19, which dealt with the particular case $G(\phi) = \frac{3}{2}\phi - \frac{1}{2}\phi^3$. Proceeding exactly as in that problem, one obtains as the generalization of (*3*):

$$\frac{\delta}{x} = \frac{b}{\sqrt{\mathbf{Re}_x}} \qquad \text{where} \qquad b^2 \equiv \frac{2G'(0)}{\int_0^1 [1 - G(\phi)]G(\phi)\, d\phi} \tag{1}$$

4.46 Solve the von Kármán integral equation under the assumption of a linear velocity profile.

▮ Problem 4.45 applies, with $G(\phi) = \phi$. Thus,

$$b^2 = \frac{2}{\int_0^1 (1 - \phi)\phi\, d\phi} = \frac{2}{\frac{1}{6}} = 12$$

$$\frac{\delta}{x} = \frac{\sqrt{12}}{\sqrt{\mathbf{Re}_x}} = \frac{3.464}{\sqrt{\mathbf{Re}_x}}$$

4.47 Solve the von Kármán integral equation if the velocity profile is taken to be

$$\frac{u}{V_\infty} = \sin\frac{\pi y}{2\delta} \qquad (0 \le y \le \delta)$$

▮ For $G(\phi) = \sin(\pi\phi/2)$, Problem 4.45 gives

$$b^2 = \frac{\pi}{\int_0^1 \left(1 - \sin\frac{\pi\phi}{2}\right)\sin\frac{\pi\phi}{2}\, d\phi} = \frac{\pi}{\frac{2}{\pi} - \frac{1}{2}} = 23.00$$

and

$$\frac{\delta}{x} = \frac{\sqrt{23.00}}{\sqrt{\mathbf{Re}_x}} = \frac{4.795}{\sqrt{\mathbf{Re}_x}}$$

4.48 Of the three velocity profiles used in Problems 4.19, 4.46, and 4.47, which is the "best"? the "worst"?—and why?

▮ The exact (Blasius) solution predicts $b \approx 5.0$. Judged against this, the sinusoidal profile is the best of the lot, and the linear profile is the worst. Noting that the sinusoidal and cubic profiles obey

$$\text{(i)}\quad G(0) = 0 \qquad \text{(ii)}\quad G(1) - 1 \qquad \text{(iii)}\quad G'(1) = 0$$

(cf. Problem 4.42) but that the linear profile obeys only (i) and (ii), we should expect poor results from the last.

4.49 Assuming constant fluid properties (k, c_p, μ), determine the energy equation for a steady, incompressible, two-dimensional, laminar boundary layer flow. Neglect conduction in the flow direction.

▌ Make an energy balance on a small control volume, Δx by Δy by unit depth, within the boundary layer (Fig. 4-6). The convective terms q_h may be written in terms of the specific heat, assumed constant; i.e.,

$$\textit{x-direction} \qquad q_{h,x} = (\rho u\, \Delta y) c_p (T - T_{\text{ref}})$$

$$\textit{y-direction} \qquad q_{h,y} = (\rho v\, \Delta x) c_p (T - T_{\text{ref}})$$

The conductive terms (in the y-direction) are, from Fourier's law,

$$q_k = -(k\, \Delta x) \frac{\partial T}{\partial y}$$

The remaining terms, designated $(u\tau)\, \Delta x$, account for the heat generated by fluid friction—a work rate (power) in which $\tau\, \Delta x$ is the force, and u is the velocity at which the shear occurs. This term arises because fluid on the top face of the control volume moves faster than fluid on the bottom face.

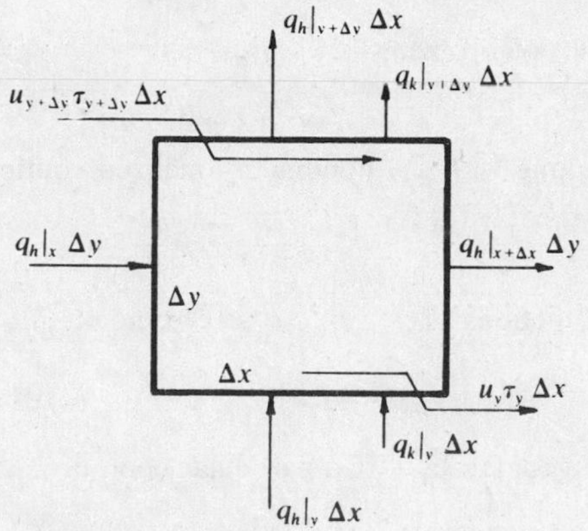

Fig. 4-6

Equating the rate of energy entering the control volume to the rate of energy leaving, we get

$$q_{h,x}(x) + q_{h,y}(y) + q_k(y) + u(y + \Delta y)\tau(y + \Delta y)\, \Delta x$$
$$= q_{h,x}(x + \Delta x) + q_{h,y}(y + \Delta y) + q_k(y + \Delta y) + u(y)\tau(y)\, \Delta x \qquad (1)$$

Substituting in (*1*) the expressions for q_k, $q_{h,x}$ and $q_{h,y}$, grouping terms in the obvious way, dividing through by $\Delta x\, \Delta y$, and taking the limit as Δx and Δy approach zero, the equation simplifies to

$$\rho c_p \frac{\partial}{\partial x} [u(T - T_{\text{ref}})] + \rho c_p \frac{\partial}{\partial y} [v(T - T_{\text{ref}})] = k \frac{\partial^2 T}{\partial y^2} + \frac{\partial}{\partial y} (u\tau) \qquad (2)$$

Now, when the two derivatives on the left of (*2*) are expanded by the product rule, the quantity $T - T_{\text{ref}}$ will emerge with coefficient

$$\rho c_p \left(\frac{\partial u}{\partial x} + \frac{\partial v}{\partial y} \right) = 0 \qquad \textit{(by the continuity equation)}$$

Moreover, the last term on the right,

$$\frac{\partial}{\partial y} (u\tau) = \frac{\partial}{\partial y} \left[u \left(\mu \frac{\partial u}{\partial y} \right) \right] = \mu \frac{\partial^2}{\partial y^2} \left(\frac{u^2}{2} \right) \approx 0$$

except in high-speed flows. Therefore, with $k/\rho c_p \equiv \alpha$, (*2*) becomes

$$u \frac{\partial T}{\partial x} + v \frac{\partial T}{\partial y} = \alpha \frac{\partial^2 T}{\partial y^2} \qquad (3)$$

4.50 Express (*3*) of Problem 4.49 in terms of the dimensionless variables

$$u^* \equiv \frac{u}{V_\infty} \qquad v^* \equiv \frac{v}{V_\infty} \qquad x^* \equiv \frac{x}{L} \qquad y^* \equiv \frac{y}{L} \qquad \Theta = \frac{T - T_s}{T_\infty - T_s}$$

Compare your result with (*2*) of Problem 4.23.

I The substitutions $u = V_\infty u^*$, etc., take (*3*) of Problem 4.49 into

$$u^* \frac{\partial \Theta}{\partial x^*} + v^* \frac{\partial \Theta}{\partial y^*} = \frac{\alpha}{LV_\infty} \frac{\partial^2 \Theta}{\partial y^{*2}}$$

or, since

$$\frac{\alpha}{LV_\infty} = \frac{k}{\rho c_p LV_\infty} = \left(\frac{k}{\mu c_p}\right)\left(\frac{\nu}{LV_\infty}\right) = \frac{1}{\mathbf{Pr} \cdot \mathbf{Re}_L}$$

into

$$u^* \frac{\partial \Theta}{\partial x^*} + v^* \frac{\partial \Theta}{\partial y^*} = \frac{1}{\mathbf{Pr} \cdot \mathbf{Re}_L} \frac{\partial^2 \Theta}{\partial y^{*2}} \qquad (1)$$

The only significant difference between (*1*) and (*2*) of Problem 4.23 resides in the fact that the latter equation is nonhomogeneous. (Even this difference vanishes when the pressure gradient is neglected—see Problem 4.2.)

4.51 Express Θ (Problem 4.50) as a function of dimensionless variables.

I The energy equation, (*1*) of Problem 4.50, gives

$$\Theta = f_1(x^*, y^*, u^*, v^*, \mathbf{Pr} \cdot \mathbf{Re}_L)$$

But (*2*) of Problem 4.23, along with the nondimensionalized continuity equation

$$\frac{\partial u^*}{\partial x^*} + \frac{\partial v^*}{\partial y^*} = 0$$

shows that u^* and v^* are functions of x^*, y^*, $\partial p^*/\partial x^*$, and $\mathbf{Re}_L$. Hence,

$$\Theta = f_2\left(x^*, y^*, \frac{\partial p^*}{\partial x^*}, \mathbf{Re}_L, \mathbf{Pr}\right) \qquad (1)$$

Relation (*1*) holds in all geometries. If a particular contour is specified, $\partial p^*/\partial x^*$ is thereby fixed, and (*1*) becomes

$$\Theta = f_3(x^*, y^*, \mathbf{Re}_L, \mathbf{Pr}) \qquad (2)$$

4.52 For steady, laminar boundary layer flow over a given contour, the *local Nusselt number* is defined by

$$\mathbf{Nu}_x \equiv \frac{h_x x}{k_f}$$

where k_f is the (assumed constant) thermal conductivity of the fluid. Evaluate $\mathbf{Nu}_x$ in terms of other dimensionless parameters.

I A heat balance at location x gives

$$h_x(T_s - T_\infty) = -k_f \left.\frac{\partial T}{\partial y}\right|_{y=0} \qquad (1)$$

In the dimensionless variables of Problem 4.50, (*1*) has the form

$$h_x = \frac{k_f}{L} \left.\frac{\partial \Theta}{\partial y^*}\right|_{y^*=0}$$

or, multiplying both sides of x/k_f,

$$\mathbf{Nu}_x = x^* \left.\frac{\partial \Theta}{\partial y^*}\right|_{y^*=0} \qquad (2)$$

Evaluating the derivative in (*2*) by use of (*2*) of Problem 4.51, one finds

$$\mathbf{Nu}_x = f_4(x^*, \mathbf{Re}_L, \mathbf{Pr}) \qquad (3)$$

In particular, for $x = L$, $\mathbf{Nu}_L = f_5(\mathbf{Re}_L, \mathbf{Pr})$.

4.53 Specify the form of the function f_4 of Problem 4.52 corresponding to flow over a flat plate.

▌ On the one hand, the "exact" Blasius/Pohlhausen treatment yields (see Problem 4.5)

$$\delta_t = \delta\, \mathbf{Pr}^{-1/3} \tag{1}$$

where (see Problem 4.3)

$$\delta = 5.0x\, \mathbf{Re}_x^{-1/2} \tag{2}$$

On the other hand, the approximate cubic temperature profile (*1*) of Problem 4.38 yields [see (*b*)] $h_x = 1.5k_f/\delta_t$, or

$$\mathbf{Nu}_x = \frac{1.5x}{\delta_t} \tag{3}$$

Eliminating δ and δ_t among (*1*), (*2*), and (*3*), one finds

$$\mathbf{Nu}_x = (0.3)\, \mathbf{Re}_x^{1/2}\, \mathbf{Pr}^{1/3} = (0.3)(x^*\, \mathbf{Re}_L)^{1/2}\, \mathbf{Pr}^{1/3} \tag{4}$$

One concludes that $f_4(a, b, c) = C(a, b)^{1/2} c^{1/3}$; the value found for C is uncertain because of the "mixing" of exact and approximate solutions.

4.54 With reference to Problem 4.53, find the values of the constant C corresponding to the assumption of (***a***) the linear temperature profile $\Theta = y/\delta_t$; (***b***) the sinusoidal temperature profile

$$\Theta = \sin\left(\frac{\pi y}{2\delta_t}\right) \qquad (0 \le y \le \delta_t)$$

(cf. Problem 4.47).

▌ (***a***) With $\Theta = Ly^*/\delta_t$, (*2*) of Problem 4.52 gives

$$\mathbf{Nu}_x = \frac{1.0x}{\delta_t}$$

This relation replaces (*3*) of Problem 4.53, leading to $C = 0.2$.

(***b***) With $\Theta = \sin(\pi L y^*/2\delta_t)$, (*2*) of Problem 4.52 gives

$$\mathbf{Nu}_x = \frac{(\pi/2)x}{\delta_t}$$

This relation replaces (*3*) of Problem 4.53, leading to $C = \pi/10 \approx 0.314$.

4.55 How would you rank the results of Problem 4.54?

▌ As in Problem 4.48, the boundary conditions imposed on the velocity or temperature profile suggest that $C = 0.314$ should be a better value than $C = 0.2$. (It turns out that 0.314 is also superior to the value 0.3, found in Problem 4.53; see Problem 5.1, where it is shown that $C = 0.332$.)

4.56 Show that the function $F(\omega)$ of Problem 4.42 must also satisfy

$$\text{(iv)} \quad F''(0) = 0$$

▌ At $y = 0$, both $u = 0$ (no-slip viscous flow) and $v = 0$ (rigid wall); hence, by (*3*) of Problem 4.49,

$$\left.\frac{\partial^2 T}{\partial y^2}\right|_{y=0} = 0$$

which implies (iv).

4.57 Check whether the temperature profiles used in preceding problems satisfy (iv) of Problem 4.56.

▌ Problem 4.37, *no*; Problem 4.38, *yes*; Problem 4.54(*a*), *yes*; Problem 4.54(*b*), *yes*.

4.58 Air at $T_\infty = 227\ ^\circ\text{C}$ flows over a flat plate having surface temperature $T_s = 127\ ^\circ\text{C}$; the air velocity is $V_\infty = 2.886$ m/s. Use the cubic temperature profile of Problem 4.38 to obtain h_x at intervals of approximately one-tenth of the length of the laminar boundary layer.

▌ At $T_f = (227 + 127)/2 = 177\ ^\circ\text{C}$, Table B-4 gives

$$k_f = 0.037\,05\ \text{W/m}\cdot\text{K} \qquad \nu_f = 2.8856 \times 10^{-5}\ \text{m}^2/\text{s} \qquad \mathbf{Pr}_f = 0.683$$

Maximum length for laminar flow is at x where $\mathbf{Re}_x = 500\,000$; thus,

$$\left(\frac{xV_\infty}{\nu_f}\right)_{\max} = 500\,000 \qquad \text{or} \qquad x_{\max} = \frac{(500\,000)(2.8856 \times 10^{-5}\ \text{m}^2/\text{s})}{2.886\ \text{m/s}} = 5.00\ \text{m}$$

From Problem 4.53, $\mathbf{Nu}_x \equiv h_x x/k_f = (0.3)\,\mathbf{Re}_x^{1/2}\,\mathbf{Pr}^{1/3}$, or

$$h_x = (0.3)k_f\,\mathbf{Pr}^{1/3}\,\frac{\mathbf{Re}_x^{1/2}}{x} = Ax^{-1/2} \qquad (1)$$

where, from the data, $A \equiv (0.3)k_f\,\mathbf{Pr}^{1/3}\,\nu_f^{-1/2}V_\infty^{1/2} = 3.10\ \text{W}\cdot\text{m}^{-3/2}\cdot\text{K}^{-1}$.
The evaluation of (*1*) for $x = 0(0.5)5$ m is displayed in Table 4.1.

TABLE 4-1

x, m	h_x, W/m$^2\cdot$K
0	∞
0.5	4.38
1.0	3.10
1.5	2.53
2.0	2.19
2.5	1.96
3.0	1.79
3.5	1.66
4.0	1.55
4.5	1.46
5.0	1.38

4.3 PIPE FLOW

4.59^D Water of density $\rho = 62.4$ lbm/ft^3 and dynamic viscosity $\mu = 2 \times 10^{-5}$ lbf·s/ft^2 flows through a long 2-in.-ID tube at a volumetric rate of 25 gpm. Is the flow laminar or turbulent?

▌ Reynolds number based on inside diameter D and area-averaged velocity V is a measure of whether the flow is laminar or turbulent. If laminar, $\mathbf{Re}_D < 2000$. Here,

$$V = \frac{Q}{A} = \frac{\left(25\ \frac{\text{gal}}{\text{min}}\right)\left(\frac{1\ \text{ft}^3}{7.48\ \text{gal}}\right)\left(\frac{1\ \text{min}}{60\ \text{s}}\right)}{(\pi/4)(\frac{2}{12}\ \text{ft})^2} = 2.55\ \text{ft/s}$$

$$\mathbf{Re}_D = \frac{VD}{\nu} = \frac{VD\rho}{\mu g_c} = \frac{(2.55\ \text{ft/s})(\frac{2}{12}\ \text{ft})(62.4\ \text{lbm/ft}^3)}{(2 \times 10^{-5}\ \text{lbf}\cdot\text{s/ft}^2)(32.2\ \text{lbm}\cdot\text{ft/lbf}\cdot\text{s}^2)} = 41\,180$$

The flow is turbulent.

4.60^D Water of density $\rho = 999.6$ kg/m^3 and dynamic viscosity $\mu = 9.584 \times 10^{-4}$ Pa·s flows through a long 50.8-mm-ID tube at a volumetric rate of 94.64 L/min. Is this flow laminar or turbulent?

▌ For laminar inside flow, $\mathbf{Re}_D < 2000$. In the present case,

$$V = \frac{Q}{A} = \frac{(94.64\ \text{L/min})(10^{-3}\ \text{m}^3/\text{L})(\frac{1}{60}\ \text{min/s})}{(\pi/4)(0.0508\ \text{m})^2} = 0.7782\ \text{m/s}$$

$$\mathbf{Re}_D = \frac{DV}{\nu} = \frac{DV\rho}{\mu} = \frac{(0.0508\ \text{m})(0.7782\ \text{m/s})(999.6\ \text{kg/m}^3)}{9.584 \times 10^{-4}\ \text{Pa}\cdot\text{s}} = 41\,230$$

The flow is turbulent.

4.61 Verify that $1\ \text{Pa}\cdot\text{s} = 1\ \text{kg/m}\cdot\text{s}$ (used implicitly in Problem 4.60).

▌ By definition, $1\ \text{N} = 1\ \text{kg}\cdot\text{m/s}^2$ and $1\ \text{Pa} = 1\ \text{N/m}^2$. Therefore,

$$1\ \text{Pa}\cdot\text{s} = 1\ \frac{\text{kg}\cdot\text{m/s}^2}{\text{m}^2}\cdot\text{s} = 1\ \text{kg/m}\cdot\text{s}$$

4.62[D] What is the maximum discharge rate, in gallons per minute, of fuel oil at atmospheric pressure and 50 °F from a $\frac{3}{4}$-in.-ID tube, if laminar flow is maintained?

▌ The maximum Reynolds number for laminar flow is $(\mathbf{Re}_D)_{max} = 2000$, whence

$$Q_{max} = AV_{max} = \left(\frac{\pi D^2}{4}\right)\left(\frac{2000\nu}{D}\right) = 500\pi D\nu$$

From Figure B-2 at 50 °F, $\nu \approx 1.6 \times 10^{-2}$ ft²/sec; so

$$Q_{max} \approx 500\pi(\tfrac{3}{48}\text{ ft})(1.6 \times 10^{-2}\text{ ft}^2/\text{sec})(7.48\text{ gal/ft}^3)(60\text{ sec/min}) = 705\text{ gpm}$$

4.63 Repeat Problem 4.62 for a fuel oil temperature of 100 °F.

▌ At the higher temperature the viscosity is reduced to $\approx 2.5 \times 10^{-3}$ ft²/sec, so that

$$Q_{max} \approx \frac{2.5 \times 10^{-3}}{1.6 \times 10^{-2}}(705) = 110\text{ gpm}$$

4.64[D] What is the maximum discharge rate, in liters per minute, of fuel oil at atmospheric pressure and 10 °C from a 19-mm-ID tube, if laminar flow is maintained.

▌ The maximum Reynolds number for laminar flow is $(\mathbf{Re}_D)_{max} = 2000$, whence

$$Q_{max} = AV_{max} = \left(\frac{\pi D^2}{4}\right)\left(\frac{2000\nu}{D}\right) = 500\pi D\nu$$

From Figure B-2 at 10° C, $\nu \approx 1.5 \times 10^{-3}$ m²/s; so

$$Q_{max} \approx 500\pi(0.019\text{ m})(1.5 \times 10^{-3}\text{ m}^2/\text{s})(1000\text{ L/m}^3)(60\text{ s/min}) = 2690\text{ L/min}$$

4.65 Find the velocity distribution for fully developed, steady, laminar flow inside a tube by considering the force equilibrium of a cylindrical element of fluid (Fig. 4-7).

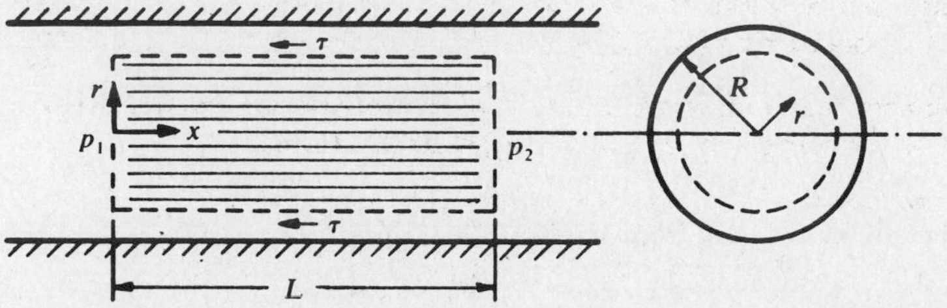

Fig. 4-7

▌ The forces acting on the element are (i) shear on the cylindrical surface and (ii) normal forces due to pressure on the ends. There is no change in momentum, since the velocity profile is the same at stations 1 and 2; therefore

$$(p_1 - p_2)\pi r^2 = \tau(2\pi rL)$$

But

$$\tau = -\mu\frac{du}{dr}$$

where the minus sign is required because u decreases with increasing r. Then,

$$\frac{du}{dr} = -\frac{p_1 - p_2}{2\mu L}r$$

Separating variables and integrating,

$$\int_u^0 du = -\frac{p_1 - p_2}{2\mu L}\int_r^R r\,dr$$

$$-u = -\left(\frac{1}{4\mu}\right)\left(\frac{p_1 - p_2}{L}\right)(R^2 - r^2)$$

or, writing $-(p_1 - p_2)/L = dp/dx$ (steady flow),

$$u = \frac{1}{4\mu}\left(-\frac{dp}{dx}\right)(R^2 - r^2) \qquad (1)$$

Formula (*1*) shows that, as in any flow process, the transfer (of mass, here) is in the direction of *decrease* of the driving force (pressure). A convenient dimensionless form of (*1*) is

$$\frac{u}{u_0} = 1 - \left(\frac{r}{R}\right)^2 \qquad \text{where} \qquad u_0 \equiv \frac{R^2}{4\mu}\left(-\frac{dp}{dx}\right) \tag{2}$$

The centerline velocity u_0 is also the velocity of greatest magnitude.

4.66 For the velocity distribution of Problem 4.65, show that the mean velocity V over the cross section is equal to one-half the centerline velocity u_0.

▮ Using (*2*) of Problem 4.65,

$$V \equiv \frac{1}{A}\int u\,dA = \frac{1}{\pi R^2}\int_0^R u_0\left[1 - \left(\frac{r}{R}\right)^2\right](2\pi r\,dr)$$

$$= \frac{2u_0}{R^2}\left[\frac{r^2}{2} - \frac{r^4}{4R^2}\right]_0^R = \frac{2u_0}{R^2}\cdot\frac{R^2}{4} = \frac{u_0}{2}$$

4.67 The *local pipe-flow friction factor* (a pure number) is defined by

$$-\frac{dp}{dx} \equiv \frac{f}{D}\,\frac{\rho V^2}{2} \tag{1}$$

Find the constant value of f in fully developed, steady, laminar flow.

▮ By Problems 4.65 and 4.66,

$$V = \frac{u_0}{2} = \frac{D^2}{32\mu}\left(-\frac{dp}{dx}\right) = \frac{D^2}{32\rho\nu}\left(\frac{f}{D}\,\frac{\rho V^2}{2}\right)$$

or $f = 64\nu/DV = 64/\mathbf{Re}_D$.

4.68^D What is the pressure drop in 60 ft of smooth $\frac{3}{8}$-in.-ID tubing when 85 °F benzene flows at an average velocity of 0.4 fps?

▮ Assume fully developed flow. The Reynolds number is

$$\mathbf{Re}_D = \frac{VD}{\nu} = \frac{(0.4\ \text{ft/sec})[(0.375/12)\ \text{ft}]}{6.6\times10^{-6}\ \text{ft}^2/\text{sec}} = 1894$$

where the kinematic viscosity is from Figure B-2. Since $\mathbf{Re}_D < 2000$, the friction factor is (Problem 4.67)

$$f = \frac{64}{\mathbf{Re}_D} = \frac{64}{1894} = 0.0338$$

and the pressure drop is given by the *Darcy–Weisbach equation*,

$$-\Delta p = f\,\frac{L}{D}\,\frac{\rho V^2}{2g_c} = f\,\frac{L}{D}\,\frac{\mu V^2}{2\nu}$$

From Figure B-1, $\mu = 1.1\times10^{-5}\ \text{lbf}\cdot\text{sec/ft}^2$; therefore,

$$-\Delta p = (0.0338)\left(\frac{60\times12}{\frac{3}{8}}\right)\frac{(1.1\times10^{-5}\ \text{lbf}\cdot\text{sec/ft}^2)(0.4\ \text{ft/sec})^2}{2(6.6\times10^{-6}\ \text{ft}^2/\text{sec})} = 8.65\ \text{lbf/ft}^2$$

Check the fully developed flow assumption:

$$x_v = (0.05)D\,\mathbf{Re}_D = (0.05)\left(\tfrac{3}{96}\right)(1894) = 2.96\ \text{ft}$$

This entry length is negligible compared with the 60-ft total tube length.

4.69^D A Pitot tube is used to measure the centerline velocity u_o in a long circular tube. With water flow at 40 °C in a 25.4-mm-ID smooth tube, $u_0 = 59.2$ mm/s. Find the pressure loss in 10 m of tube length.

▮ From Table B-3, $\nu = 6.578\times10^{-7}\ \text{m}^2/\text{s}$, $\rho = 994.6\ \text{kg/m}^3$. Assuming fully developed, laminar flow, $V = u_o/2 = 29.6$ mm/s and

$$\mathbf{Re}_D = \frac{DV}{\nu} = \frac{(0.0254\ \text{m})(0.0296\ \text{m/s})}{6.578\times10^{-7}\ \text{m}^2/\text{s}} = 1143 \quad (<2000;\ \text{OK})$$

Then the friction factor is $f = 64/\mathbf{Re}_D = \frac{64}{1143}$ and

$$-\Delta p = f \frac{L}{D} \frac{\rho V^2}{2} = \frac{64}{1143} \left(\frac{10 \text{ m}}{0.0254 \text{ m}} \right) \frac{(994.6 \text{ kg/m}^3)(0.0296 \text{ m/s})^2}{2} = 9.60 \text{ kg/m} \cdot \text{s}^2 = 9.60 \text{ Pa}$$

Check the fully developed flow assumption:

$$x_v = (0.05) D\, \mathbf{Re}_D = (0.05)(0.0254)(1143) = 1.45 \text{ m}$$

which is indeed small compared to 10 m. However, the developing flow region is 14.5% of the total length, and the predicted Δp is slightly too small.

4.70^D What volumetric flow (in L/h) of 50 °C water can be developed in 20 m of smooth 20-mm-ID tubing by a pump furnishing a pressure head of 12.0 Pa?

I From Problems 4.65 and 4.66, for steady, fully developed, laminar flow,

$$Q = AV = (\pi R^2) \left[\frac{R^2}{8\mu} \left(-\frac{dp}{dx} \right) \right] = \frac{\pi R^4}{8\mu} \left(\frac{p_1 - p_2}{L} \right) \tag{1}$$

which is known as the *Hagen–Poiseuille equation.*

From Table B-3,

$$\mu = \rho \nu = [(61.80)(16.018)][(6.11 \times 10^{-6})(0.0929)] = 0.000\,56 \text{ Pa} \cdot \text{s}$$

and so

$$Q = \frac{\pi (0.010 \text{ m})^4}{8(0.000\,56 \text{ Pa} \cdot \text{s})} \left(\frac{12.0 \text{ Pa}}{20 \text{ m}} \right) = 4.2 \times 10^{-6} \text{ m}^3/\text{s}$$

which converts to 15.1 L/h.

To check the assumption of laminar flow, the mean velocity is required:

$$V = \frac{Q}{A} = \frac{4.2 \times 10^{-6} \text{ m}^3/\text{s}}{(\pi/4)(0.020 \text{ m})^2} = 0.0134 \text{ m/s}$$

Since the kinematic viscosity is $\nu = 5.7 \times 10^{-7} \text{ m}^2/\text{s}$, as above,

$$\mathbf{Re}_D = \frac{VD}{\nu} = \frac{(0.0134 \text{ m/s})(0.020 \text{ m})}{5.7 \times 10^{-7} \text{ m}^2/\text{s}} = 470 < 2000$$

Moreover, the flow is fully developed over most of the tube length, since

$$x_v \approx (0.05)\, \mathbf{Re}_D\, D = (0.05)(470)(0.020) = 0.47 \text{ m} \ll 20 \text{ m}$$

4.71^D What volumetric flow (in ft³/hr) of 121 °F water can be developed in 65.5 ft of smooth 0.79-in.-ID tubing by a pump furnishing a pressure head of 0.25 lbf/ft²?

I Assume that the Hagen–Poiseuille equation (Problem 4.70) applies. From Table B-3, $\rho = 61.82 \text{ lbm/ft}^3$ and $\nu = 0.616 \times 10^{-5} \text{ ft}^2/\text{s}$; hence,

$$\mu = \frac{\rho}{g_c} \nu = \frac{61.82}{32.2} (0.616 \times 10^{-5}) = 1.183 \times 10^{-5} \text{ lbf} \cdot \text{sec/ft}^2$$

and

$$Q = \frac{\pi R^4}{8\mu} \left(\frac{p_1 - p_2}{L} \right) = \frac{\pi (0.395/12)^4}{8(1.183 \times 10^{-5})} \left(\frac{0.25}{65.5} \right) = 1.487 \times 10^{-4} \text{ ft}^3/\text{sec} = 0.535 \text{ ft}^3/\text{hr}$$

Checking the assumption of laminar and fully developed flow:

$$V = \frac{Q}{A} = \frac{1.487 \times 10^{-4} \text{ ft}^3/\text{sec}}{(\pi/4)[(0.79/12) \text{ ft}]^2} = 4.37 \times 10^{-2} \text{ ft/sec}$$

$$\mathbf{Re}_D = \frac{DV}{\nu} = \frac{(0.79/12)(4.37 \times 10^{-2})}{0.616 \times 10^{-5}} = 467 < 2000$$

and

$$x_v \approx (0.05)\, \mathbf{Re}_D\, D = (0.05)(467)\left(\frac{0.79}{12} \right) = 1.54 \text{ ft} \ll 65.5 \text{ ft}$$

4.72^D Calculate the friction factor for Problem 4.70 and use it to check the given pressure drop.

▮
$$f = \frac{64}{\mathbf{Re}_D} = \frac{64}{470}$$

Then the Darcy–Weisbach equation (Problem 4.68) gives

$$-\Delta p = f\left(\frac{L}{D}\right)\left(\frac{\rho V^2}{2}\right) = \frac{64}{470}(10^3)\left[\frac{(990.0)(0.0134)^2}{2}\right] = 12.1 \text{ Pa}$$

4.73[D] Calculate the friction factor for Problem 4.71 and use it to check the given pressure drop.

▮
$$f = \frac{64}{\mathbf{Re}_D} = \frac{64}{467}$$

Then the Darcy–Weisbach equation (Problem 4.68) gives

$$-\Delta p = f\left(\frac{L}{D}\right)\left(\frac{\rho V^2}{2g_c}\right) = \frac{64}{467}\left(\frac{65.5 \times 12}{0.79}\right)\left[\frac{(61.82)(4.37 \times 10^{-2})^2}{2(32.2)}\right] = 0.250 \text{ lbf/ft}^2$$

4.74 Refer to Problem 4.70. What pressure head would be required to maintain the same volume flow in a tube of half the diameter?

▮ By (*1*) of Problem 4.70, a sixteen-fold pressure increase is needed to offset a halving of the radius. We must, of course, check that the flow is still laminar and fully developed:

$$\mathbf{Re} \approx \frac{D}{A} \approx \frac{1}{D} \qquad \text{whence} \qquad \mathbf{Re}_{D/2} = 2\,\mathbf{Re}_D = 2(470) < 2000$$

$$x_v \approx \frac{1}{D} \times D \quad (\textit{invariant})$$

4.75 Confirm the result of Problem 4.74 using the friction factor.

▮ Because $f \approx 1/\mathbf{Re} \approx D$ and $V \approx 1/A \approx 1/D^2$, the Darcy–Weisbach equation yields

$$-\Delta p \approx D\left(\frac{1}{D}\right)\left(\frac{1}{D^4}\right) = \frac{1}{D^4}$$

which implies a sixteen-fold increase in pressure to maintain Q when D is halved.

4.76[D] Water flows in a 10-mm-ID tube at a mean temperature $T_{\text{bulk,avg}} = 50$ °C and a volumetric flow rate $Q = 4.2$ mL/s. (The mean temperature is the average of $T_{\text{bulk,in}}$ and $T_{\text{bulk,out}}$.) The wall is electrically heated to make $T_w - T_b$ constant at 10 °C, resulting in heating of the fluid. The average heat transfer coefficient between the wall and the fluid may be taken as 300 W/m$^2 \cdot$K. **(*a*)** What is the bulk fluid temperature rise in a 1.0-m heated tube length? **(*b*)** What is the Reynolds number? **(*c*)** What is the thermal entry length?

▮ **(*a*)** The first law of thermodynamics requires that the rate of energy (heat) transfer from the tube wall to the fluid be equal to the rate of energy gain by the fluid:

$$\bar{h}A_w(T_w - T_b) = \dot{m}c_p(T_{b,\text{out}} - T_{b,\text{in}})$$

where $\dot{m} = \rho Q$; hence,

$$T_{b,\text{out}} - T_{b,\text{in}} = \frac{\bar{h}A_w(T_w - T_b)}{\rho Q c_p} \tag{1}$$

From Table B-3, $\rho = 990$ kg/m^3 and $c_p = 4142$ J/kg·K, giving

$$T_{b,\text{out}} - T_{b,\text{in}} = \frac{(300 \text{ W/m}^2\cdot\text{K})[\pi(0.010 \text{ m})(1.0 \text{ m})](10\text{ °C})}{(990 \text{ kg/m}^3)(4.2 \times 10^{-6} \text{ m}^3/\text{s})(4142 \text{ J/kg}\cdot\text{K})} = 5.5 \text{ °C}$$

(*b*) Table B-3 gives $\nu = 5.67 \times 10^{-7}$ m^2/s; hence,

$$\mathbf{Re}_D = \frac{4Q}{\pi D \nu} = \frac{4(4.2 \times 10^{-6} \text{ m}^3/\text{s})}{\pi(0.010 \text{ m})(5.67 \times 10^{-7} \text{ m}^2/\text{s})} = = 943 < 2000$$

(*c*) From Table B-3, $\mathbf{Pr} = 3.68$.

$$x_t \approx (0.05)\,\mathbf{Re}_D\,\mathbf{Pr}\,D = (0.05)(943)(3.68)(0.010 \text{ m}) = 1.735 \text{ m}$$

The 1-m-long heated tube is all in the developing thermal entry region, with h varying strongly with x. The average value given is only a reasonable approximation.

4.77[D] Repeat Problem 4.76(a) for a tube of 0.394-in. ID, water mean temperature 121 °F, and $T_w - T_b =$ 18 °F. The heated tube length is 3.28 ft. The volumetric water flow rate is 1.487×10^{-4} ft³/s, and $\bar{h} =$ 53 Btu/hr · ft² · °F.

▮ From Table B-3, $\rho = 61.82$ lbm/ft³ and $c_p = 0.999$ Btu/lbm · °F; (1) of Problem 4.76 yields

$$T_{b,\text{out}} - T_{b,\text{in}} = \frac{\bar{h}A_w(T_w - T_b)}{\rho Q c_p}$$

$$= \frac{\left(53 \dfrac{\text{Btu}}{\text{hr}\cdot\text{ft}^2\cdot{}^\circ\text{F}}\right)\left[\pi\left(\dfrac{0.394}{12}\text{ ft}\right)(3.28\text{ ft})\right](18\ {}^\circ\text{F})}{(61.82\text{ lbm/ft}^3)(1.487\times10^{-4}\text{ ft}^3/\text{s})(3600\text{ s/hr})(0.999\text{ Btu/lbm}\cdot{}^\circ\text{F})} = 9.76\ {}^\circ\text{F}$$

4.78[D] Air flows at 80 °F and 100 psia in a 0.25-in.-ID circular tube in a heat exchanger. The mass flow rate is 1.3 lbm/hr. Determine (a) the Reynolds number of the flow and (b) the friction factor for a fully developed velocity profile.

▮ (a) From $\mathbf{Re}_D = DV\rho/\mu g_c$ and $\dot{m} = \rho AV = \rho\pi D^2 V/4$,

$$\mathbf{Re}_D = \frac{4\dot{m}}{\pi(\mu g_c)D} \qquad (1)$$

From Table B-4, $\mu g_c = 1.241 \times 10^{-5}$ lbm/ft · sec, so that

$$\mathbf{Re}_D = \frac{4(1.3\text{ lbm/hr})(1\text{ hr/3600 sec})}{\pi(1.241\times10^{-5}\text{ lbm/ft}\cdot\text{sec})[(0.25/12)\text{ ft}]} = 1778$$

—laminar flow.

(b) For laminar flow,

$$f = \frac{64}{\mathbf{Re}_D} = \frac{64}{1778} = 0.036$$

4.79 Calculate the area-averaged velocity V in Problem 4.78.

▮ From Problem 4.78,

$$V = \frac{(\mu g_c)\,\mathbf{Re}_D}{D\rho_{100}}$$

where, by means of a subscript, we emphasize that the density at the given pressure, 100 psia, is to be used. Table B-4 gives the density *at 14.7 psia* (and at the given temperature) as $\rho_{14.7} = 0.0735$ lbm/ft³. Assuming an ideal gas, density is proportional to pressure (at constant temperature). Thus

$$\rho_{100} = \frac{100}{14.7}(0.0735) = 0.50\text{ lbm/ft}^3$$

and
$$V = \frac{(1.241\times10^{-5}\text{ lbm/ft}\cdot\text{sec})(1778)}{(\frac{1}{48}\text{ ft})(0.50\text{ lbm/ft}^3)} = 2.12\text{ ft/sec}$$

4.80[D] Air at 27 °C and 689.5-kPa pressure flows inside a 6.35-mm-ID tube with average velocity 0.646 m/s. Find (a) the Reynolds number, (b) the friction factor for fully developed flow, and (c) the mass flow rate.

▮ From Table B-4, $\mu = 1.847 \times 10^{-5}$ Pa · s and $\rho_{101.3} = 1.177$ kg/m³ (at 1 atm pressure). Correcting the density through the ideal gas law,

$$\rho_{689.5} = \frac{689.5}{101.3}(1.177) = 8.011\text{ kg/m}^3$$

(a)
$$\mathbf{Re}_D = \frac{DV\rho_{689.5}}{\mu} = \frac{(0.006\,35\text{ m})(0.646\text{ m/s})(8.011\text{ kg/m}^3)}{1.847\times10^{-5}\text{ Pa}\cdot\text{s}} = 1779$$

(b)
$$f = \frac{64}{\mathbf{Re}_D} = \frac{64}{1779} = 0.036$$

(*c*) $$\dot{m} = \rho_{689.5}AV = (8.011\ \text{kg/m}^3)\left[\left(\frac{\pi}{4}\right)(0.006\,35\ \text{m})^2\right](0.646\ \text{m/s}) = 1.639 \times 10^{-4}\ \text{kg/s} = 0.59\ \text{kg/h}$$

4.81 Determine the entry lengths for fully developed velocity and temperature profiles (***a***) for Problem 4.78; (***b***) for Problem 4.80. *Note that* $x_t < x_v$, *resulting for* $\mathbf{Pr} < 1.0$, *is controversial.*

▮ (*a*) $$x_v \approx (0.05)\,\mathbf{Re}_D\,D = (0.05)(1778)(\tfrac{1}{48}\ \text{ft}) = 1.85\ \text{ft}$$

Noting that $\mathbf{Pr} \equiv \mu c_p/k$ is pressure-independent, one uses Table B-4 to find

$$x_t = \mathbf{Pr}\,x_v \approx (0.708)(1.85\ \text{ft}) = 1.31\ \text{ft}$$

(*b*) $$x_v \approx (0.05)\,\mathbf{Re}_D\,D = (0.05)(1779)(0.006\,35\ \text{m}) = 0.565\ \text{m}$$
$$x_t = \mathbf{Pr}\,x_v \approx (0.708)(0.565\ \text{m}) = 0.40\ \text{m}$$

4.82 Nitrogen gas at an average bulk temperature of 77 °C flows inside a 1.0-cm-ID smooth tube at a pressure of 5 bars. Determine, for Reynolds number 2000 (edge of turbulence), (***a***) the average velocity, (***b***) the velocity entry length, and (***c***) the thermal entry length.

▮ From Table B-4 at $T_{b,\text{avg}} = 77$ °C, $\mu = 1.991 \times 10^{-5}$ Pa · s; and the density value, corrected from 1 atm = 1.013 bars to 5 bars, is

$$\rho = \frac{5}{1.013}(0.998\ \text{kg/m}^3) = 4.926\ \text{kg/m}^3$$

(*a*) $$V = \frac{\mu\,\mathbf{Re}_D}{D\rho} = \frac{(1.991 \times 10^{-5}\ \text{Pa}\cdot\text{s})(2000)}{(0.010\ \text{m})(4.926\ \text{kg/m}^3)} = 0.808\ \text{m/s}$$

(*b*) $$x_v \approx (0.05)\,\mathbf{Re}_D\,D = (0.05)(2000)(0.010\ \text{m}) = 1.0\ \text{m}$$

(*c*) $$x_t = \mathbf{Pr}\,x_v \approx (0.702)(1.0\ \text{m}) = 0.702\ \text{m}\ (< x_v\text{—controversial!})$$

where the (pressure-independent) Prandtl number comes from Table B-4.

4.83[D] Rework Problem 4.82 for pure ethylene glycol (an incompressible, viscous liquid) at an average bulk temperature of 80 °C.

▮ From Table B-3 at $T_{b,\text{avg}} = 80$ °C, $\nu = 2.982 \times 10^{-6}\ \text{m}^2/\text{s}$ (independent of pressure) and $\mathbf{Pr} = 32.4$.

(*a*) $$V = \frac{\nu\,\mathbf{Re}_D}{D} = \frac{(2.982 \times 10^{-6}\ \text{m}^2/\text{s})(2000)}{0.010\ \text{m}} = = 0.596\ \text{m/s}$$

(*b*) $x_v \approx 1.0$ m, as in Problem 4.82.

(*c*) $$x_t = \mathbf{Pr}\,x_v \approx (32.4)(1.0\ \text{m}) = 32.4\ \text{m}$$

Observe the great distance required for the establishment of steady temperature conditions in the viscous liquid.

4.84[D] Pure ethylene glycol (an incompressible, viscous liquid) at an average bulk temperature of 176 °F flows inside a 0.394-in.-ID tube at a pressure of 72.5 psia. Determine, for a Reynolds number of 2000 (edge of turbulence), (***a***) the average velocity; (***b***) the velocity entry length; and (***c***) the thermal entry length.

▮ From Table B-3 at 176 °F, $\nu = 3.21 \times 10^{-5}\ \text{ft}^2/\text{sec}$ (independent of pressure) and $\mathbf{Pr} = 32.4$.

(*a*) $$V = \frac{\nu\,\mathbf{Re}_D}{D} = \frac{(3.21 \times 10^{-5}\ \text{ft}^2/\text{sec})(2000)}{(0.394/12)\ \text{ft}} = 1.96\ \text{ft/sec}$$

(*b*) $$x_v \approx (0.05)\,\mathbf{Re}_D\,D = (0.05)(2000)[(0.394/12)\ \text{ft}] = 3.28\ \text{ft}$$

(*c*) $$x_t = \mathbf{Pr}\,x_v \approx (32.4)(3.28\ \text{ft}) = 106.4\ \text{ft}$$

Observe the great distance required for the establishment of steady temperature conditions in the viscous liquid.

4.85[D] Saturated liquid Freon-12 at an average bulk temperature of 14 °F flows inside a circular $\frac{3}{8}$-in.-ID heat-exchanger tube at the rate of 9.38×10^{-3} lbm/sec. Determine (***a***) the Reynolds number and (***b***) the thermal entry length.

▮ From Table B-3, $\rho = 89.24\ \text{lbm/ft}^3$; $\nu = 0.238 \times 10^{-5}\ \text{ft}^2/\text{sec}$; $\mathbf{Pr} = 4.0$.

(*a*) By (*1*) of Problem 4.78,

$$\mathbf{Re}_D = \frac{4\dot{m}}{\pi\rho\nu D} = \frac{4(9.38\times10^{-3})}{\pi(89.24)(0.238\times10^{-5})(\frac{3}{96})} = 1799 \quad (laminar)$$

(*b*) Since laminar flow exists,

$$x_t = (0.05)\,\mathbf{Re}_D\,\mathbf{Pr}\,D = (0.05)(1799)(4.0)(\tfrac{3}{96}\text{ ft}) = 11.24\text{ ft}$$

4.86 For the flow of Problem 4.85, find (*a*) the average velocity; (*b*) the velocity at a radius of 0.1 in. from the centerline in the fully developed velocity region.

▌ (*a*)

$$V = \frac{\nu\,\mathbf{Re}_D}{D} = \frac{(0.238\times10^{-5}\text{ ft}^2/\text{sec})(1799)}{\frac{3}{96}\text{ ft}} = 0.137\text{ ft/sec}$$

(*b*) By Problem 4.66 and (*2*) of Problem 4.65,

$$u = 2V\left[1-\left(\frac{r}{R}\right)^2\right] = 2(0.137\text{ ft/sec})\left[1-\left(\frac{8}{15}\right)^2\right] = 0.196\text{ ft/sec}$$

4.87 Determine the gradient of the Freon bulk temperature in Problem 4.85, if the tube wall temperature is held constant at 32 °F in the developed thermal profile region. The average heat transfer coefficient is 9.4 Btu/hr · ft² · °F; all other conditions are the same as in Problem 4.85.

▌ From Table B-3 at $T_{b,\text{avg}} = 14$ °F, $c_p = 0.2198$ Btu/lbm · °F. Applying the first law to a length Δx in the developed region of the tube, we have (cf. Problem 4.76)

$$\bar{h}(\pi D\,\Delta x)(T_w - T_{b,\text{avg}}) = \dot{m}c_p\,\Delta T_b \tag{1}$$

or

$$\frac{\Delta T_b}{\Delta x} = \frac{\bar{h}\pi D(T_w - T_{b,\text{avg}})}{\dot{m}c_p}$$

$$= \frac{(9.4\text{ Btu/hr}\cdot\text{ft}^2\cdot{}^\circ\text{F})\pi(\frac{3}{96}\text{ ft})[(32-14)\ {}^\circ\text{F}]}{[(9.38\times10^{-3}\text{ lbm/sec})(3600\text{ sec/hr})](0.2198\text{ Btu/lbm}\cdot{}^\circ\text{F})} = 2.24\ {}^\circ\text{F/ft}$$

The above result does not apply to the first 11.24 ft of the tube [Problem 4.85(*b*)]; thus, a long tube is presupposed.

4.88[D] Saturated liquid Freon-12 at an average bulk temperature of −10 °C flows inside a 10-mm-ID smooth tube in a heat exchanger, at a mass flow rate of 4.25×10^{-3} kg/s. For fully developed flow determine (*a*) the friction factor, (*b*) the thermal entry length, and (*c*) the pressure loss per meter of tube length.

▌ From Table B-3 at −10 °C: $\nu = 2.21\times10^{-7}$ m²/s, $\mathbf{Pr} = 4.0$, $\rho = 1429.5$ kg/m³.

(*a*) By (*1*) of Problem 4.78, with $\nu = \mu g_c/\rho$,

$$\mathbf{Re}_D = \frac{4\dot{m}}{\pi D\nu\rho} = \frac{4(4.25\times10^{-3}\text{ kg/s})}{\pi(0.010\text{ m})(2.21\times10^{-7}\text{ m}^2/\text{s})(1429.5\text{ kg/m}^3)} = 1713 \quad (laminar)$$

whence

$$f = \frac{64}{\mathbf{Re}_D} = 0.0374$$

(*b*)

$$x_t \approx (0.05)\,\mathbf{Re}_D\,\mathbf{Pr}\,D = (0.05)(1713)(4.0)(0.010\text{ m}) = 3.43\text{ m}$$

(*c*) Combination of

$$-\frac{\Delta p}{L} = \frac{f}{D}\frac{\rho V^2}{2} \quad \text{and} \quad V = \frac{\dot{m}}{\rho(\pi D^2/4)}$$

yields

$$-\frac{\Delta p}{L} = \frac{8\dot{m}^2 f}{\rho\pi^2 D^5} = \frac{8(4.25\times10^{-3}\text{ kg/s})^2(0.0374)}{(1429.5\text{ kg/m}^3)(\pi)^2(0.010\text{ m})^5}$$

$$= 3.83\text{ kg/s}^2\cdot\text{m}^2 = 3.83\text{ Pa/m}$$

4.89 Repeat Problem 4.88 for a tenfold increase in mass flow.

I (*a*) Now $\mathbf{Re}_D = 10(1713) = 17\,130 > 2000$, and the flow is turbulent. For fully developed turbulent flow inside of smooth tubes,

$$f \approx \begin{cases} (0.316)\,\mathbf{Re}_D^{-1/4} & \mathbf{Re}_D < 2\times 10^4 \\ (0.184)\,\mathbf{Re}_D^{-1/5} & \mathbf{Re}_D > 2\times 10^4 \end{cases} \qquad (1)$$

Thus, $f \approx (0.316)(17\,130)^{-1/4} = 0.0276$.

(*b*) Thermal entry length in turbulent flow inside of tubes is approximately $x_t = 10D = 100$ mm.

(*c*) Pressure loss is directly proportional to $\dot{m}^2 f$; hence,

$$\frac{\Delta p}{L} = \left(\frac{10}{1}\right)^2 \left(\frac{0.0276}{0.0374}\right)(3.83\ \text{Pa/m}) = 283\ \text{Pa/m}$$

This is roughly 75 times the loss in the laminar flow.

4.90 The upper-velocity limit for room-temperature air flow to be treated as an incompressible fluid flow is approximately 200 ft/sec. Consider air at 80 °F and 50 psia flowing in a 2-in.-ID pipe at 200 ft/sec. (*a*) Is this flow laminar or turbulent? (*b*) What is the Darcy friction factor for the fully developed flow?

I (*a*) From Table B-4, $\rho_{14.7} = 0.0735\ \text{lbm/ft}^3$ (at 1 atm) and $\mu g_c = 1.241\times 10^{-5}\ \text{lbm/ft}\cdot\text{s}$. The ideal gas pressure correction gives

$$\rho_{50} = (0.0735\ \text{lbm/ft}^3)\left(\frac{50}{14.7}\right) = 0.25\ \text{lbm/ft}^3$$

and so

$$\mathbf{Re}_D = \frac{DV\rho_{50}}{\mu g_c} = \frac{(\frac{2}{12}\ \text{ft})(200\ \text{ft/s})(0.25\ \text{lbm/ft}^3)}{1.241\times 10^{-5}\ \text{lbm/ft}\cdot\text{s}} = 671\,500$$

The flow is turbulent.

(*b*) From (*1*) of Problem 4.89, $f \approx (0.184)(671\,500)^{-1/5} = 0.0126$.

4.91[D] Saturated steam vapor at $T_{\text{bulk,avg}} = 350$ °F is heated in a 2-in.-ID tube, through which it flows at 0.161 lbm/sec. For fully developed flow, what is (*a*) the Reynolds number? (*b*) the Darcy friction factor f? (*c*) the pressure drop per foot of tubing?

I From Table B-4, for steam vapor at 350 °F (and 1 atm), $\mu g_c = 10.25\times 10^{-6}\ \text{lbm/ft}\cdot\text{sec}$. From steam tables at 350 °F, $p_{\text{sat}} = 134.5\ \text{psia} \approx 9$ atm; however, we need not correct μ. Moreover, the specific volume is given as $v \approx 3.346\ \text{ft}^3/\text{lbm}$, whence

$$\rho = \frac{1}{v} \approx 0.299\ \text{lbm/ft}^3$$

(*a*) As in Problem 4.78,

$$\mathbf{Re}_D = \frac{4\dot{m}}{\pi D(\mu g_c)} = \frac{4(0.161\ \text{lbm/sec})}{\pi(\frac{2}{12}\ \text{ft})(10.25\times 10^{-6}\ \text{lbm/ft}\cdot\text{sec})} = 120\,000 \quad (\textit{turbulent})$$

(*b*) By (*1*) of Problem 4.89,

$$f \approx (0.184)\,\mathbf{Re}_D^{-1/5} = (0.184)(120\,000)^{-1/5} = 0.0177$$

(*c*) By the formula (in gravitational units) of Problem 4.88(*c*),

$$-\frac{\Delta p}{L} = \frac{8\dot{m}^2 f}{\rho\pi^2 D^5 g_c} = \frac{8(0.161\ \text{lbm/sec})^2(0.0177)}{(0.299\ \text{lbm/ft}^3)(\pi)^2(\frac{2}{12}\ \text{ft})^5(32.2\ \text{lbm}\cdot\text{ft/lbf}\cdot\text{sec}^2)}$$
$$= 0.300\ (\text{lbf/ft}^2)/\text{ft}$$

4.92[D] Saturated steam vapor at an average bulk temperature of 177 °C is heated in a 50.8-mm-ID tube, through which it flows at 0.073 kg/s. For fully developed flow, determine (*a*) the Reynolds number, (*b*) the Darcy friction factor f, and (*c*) the pressure drop per meter of tube.

I From Table B-4, for steam vapor at 177 °C (and 1 atm), $\mu = 1.525\times 10^{-5}\ \text{Pa}\cdot\text{s}$. From steam tables at 177 °C, $p_{\text{sat}} = 0.95\ \text{MPa} \approx 9$ atm; however, we need not correct μ. Moreover, the specific volume is given as $v \approx 0.204\ \text{m}^3/\text{kg}$, whence

$$\rho = \frac{1}{v} \approx 4.90\ \text{kg/m}^3$$

(*a*) As in Problem 4.85,

$$\mathbf{Re}_D = \frac{4\dot{m}}{\pi D\mu} = \frac{4(0.073\ \text{kg/s})}{\pi(0.0508\ \text{m})(1.525\times10^{-5}\ \text{Pa}\cdot\text{s})} = 120\,000 \quad (\textit{turbulent})$$

(*b*) Using (*1*) of Problem 4.89,

$$f \approx (0.184)\,\mathbf{Re}_D^{-1/5} = (0.184)(120\,000)^{-1/5} = 0.0177$$

(*c*) From Problem 4.88(*c*),

$$-\frac{\Delta p}{L} = \frac{8\dot{m}^2 f}{\rho\pi^2 D^5} = \frac{8(0.073\ \text{kg/s})^2(0.0177)}{(4.90\ \text{kg/m}^3)\pi^2(0.0508\ \text{m})^5} = 46.12\ \text{Pa/m}$$

4.93 What is the entry length required for the friction factor f to become constant in Problem 4.92?

▌The required distance is given by the *Latzko equation* as

$$x_{v,\text{turb}} = (0.623)\,\mathbf{Re}_D^{1/4}\,D = (0.623)(120\,000)^{1/4}(0.0508\ \text{m}) = 0.589\ \text{m}$$

Beyond this distance, $f \approx 0.0177$, as calculated in Problem 4.92.

4.94 Water at 68 °F flows at the rate of 1.0 ft³/min through a smooth 1-in.-ID drawn copper tube 200 ft long. (*a*) Determine the friction factor and the length required for it to reach a constant value. (*b*) What is the net pressure drop?

▌The average velocity is

$$V = \frac{Q}{A} = \frac{(1.0\ \text{ft}^3/\text{min})(\frac{1}{60}\ \text{min/sec})}{(\pi/4)(\frac{1}{12}\ \text{ft})^2} = 3.056\ \text{ft/sec}$$

which, with ν from Table B-3 at 68 °F, gives the Reynolds number

$$\mathbf{Re}_D = \frac{VD}{\nu} = \frac{(3.056\ \text{ft/sec})(\frac{1}{12}\ \text{ft})}{1.083\times10^{-5}\ \text{ft}^2/\text{sec}} = 23\,515 \quad (\textit{turbulent})$$

(*a*) Equation (*1*) of Problem 4.89 gives the friction factor as

$$f \approx (0.184)\,\mathbf{Re}_D^{-0.2} = (0.184)(23\,515)^{-0.2} = 0.0246$$

which compares very favorably with the value given by Moody's diagram in fluid mechanics books.
The distance required for the friction factor to reach a constant value is given by the Latzko equation as

$$x_{v,\text{turb}} = (0.623)\,\mathbf{Re}_D^{0.25}\,D = (0.623)(23\,515)^{0.25}(\tfrac{1}{12}\ \text{ft}) = 0.64\ \text{ft}$$

a negligible amount in a total of 200 ft; therefore, we shall assume that $f = 0.0246$ throughout.

(*b*) The pressure drop is given by the Darcy–Weisbach equation (Problem 4.68), with ρ from Table B-3 at 68 °F:

$$-\Delta p = f\,\frac{L}{D}\,\frac{\rho V^2}{2g_c} = (0.0246)\left(\frac{200}{\frac{1}{12}}\right)\frac{(62.46\ \text{lbm/ft}^3)(3.056\ \text{ft/sec})^2}{2(32.2\ \text{lbm}\cdot\text{ft/lbf}\cdot\text{sec}^2)}$$

$$= 534.8\ \text{lbf/ft}^2 = 3.71\ \text{psi}$$

4.95 For the water flow in Problem 4.94, the last 12 ft of the 200-ft-long copper tube is heated in such a manner that the tube wall temperature is at each point 10 °F higher than the bulk fluid temperature. What will be the increase in bulk fluid temperature assuming that the fluid bulk temperature entering the heated section is 68 °F? The turbulent flow heat transfer coefficient may be taken as 650 Btu/hr · ft² · °F.

▌Using $T_{b,\text{in}} = 68$ °F, the properties are given by Table B-3 as

$$\rho = 62.46\ \text{lbm/ft}^3 \qquad c_p = 0.9988\ \text{Btu/lbm}\cdot{}^\circ\text{F}$$

The energy balance for the 12-ft section, over which $T_w - T_b = \text{constant} = 10$ °F, is

$$\dot{m}c_p(T_{b,\text{out}} - T_{b,\text{in}}) = \bar{h}A_{\text{conv}}(T_w - T_b) \tag{1}$$

from which ($\dot{m} = \rho Q$)

$$T_{b,\text{out}} - T_{b,\text{in}} = \frac{\bar{h}A_{\text{conv}}(T_w - T_b)}{\rho Q c_p} = \frac{(650\ \text{Btu/hr}\cdot\text{ft}^2\cdot{}^\circ\text{F})\pi(\frac{1}{12}\ \text{ft})(12\ \text{ft})(10\ {}^\circ\text{F})}{(62.46\ \text{lbm/ft}^3)[(1.0\ \text{ft}^3/\text{min})(60\ \text{min/hr})](0.9988\ \text{Btu/lbm}\cdot{}^\circ\text{F})} = 5.46\ {}^\circ\text{F}$$

4.96 Problem 4.95 exemplifies *constant heat flux* because $T_w - T_b$ is maintained constant. The second major possibility is that of *constant wall temperature*. Sketch both cases.

▌ In Fig. 4-8(*a*), the case of constant heat flux, the bulk fluid temperature increases with distance; so must T_w. In Fig. 4-8(*b*), the case of constant T_w, the bulk fluid temperature increases with distance; thus $T_w - T_b$ diminishes for $T_w > T_b$. Note that (*1*) of Problem 4.95 is invalid in this latter case unless T_b on the right is replaced by an average value, as in Problem 4.76.

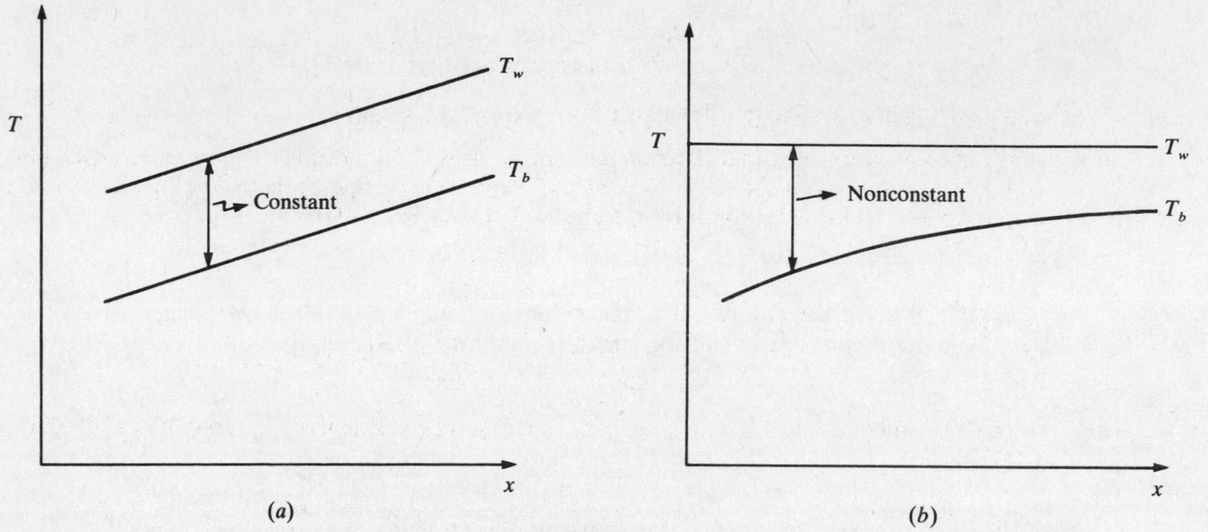

Fig. 4-8

4.97 For turbulent flow inside smooth pipes, the velocity distribution is frequently approximated by the *Nikuradse power law*

$$\frac{\bar{u}}{\bar{u}_0} = \left(\frac{y}{R}\right)^{1/n}$$

where $\bar{u}$ is the local time-averaged velocity, $\bar{u}_0$ is the time-averaged velocity at the centerline, R is the pipe radius, and $y \equiv R - r$ is the distance from the pipe wall. (*a*) For the power-law velocity profile, calculate the ratio of the area-averaged velocity to the maximum velocity. (*b*) Compare the ratio for turbulent flow, with $n = 7$, to that for laminar flow.

▌ (*a*) The area-averaged velocity is, by definition,

$$\bar{V} = \frac{1}{\pi R^2}\int_0^{R/\bar{u}_0}\left(1 - \frac{r}{R}\right)^{1/n} 2\pi r\,dr$$

$$= \frac{2\bar{u}_0}{R^2}\left[\frac{R^2}{\frac{1}{n}+2}\left(1-\frac{r}{R}\right)^{(1/n)+2} - \frac{R^2}{\frac{1}{n}+1}\left(1-\frac{r}{R}\right)^{(1/n)+1}\right]_0^R$$

$$= -2\bar{u}_0\left[\frac{n}{2n+1} - \frac{n}{n+1}\right] = \frac{2n^2\bar{u}_0}{(2n+1)(n+1)}$$

Since the maximum velocity is clearly $\bar{u}_0$, the desired ratio is

$$\frac{\bar{V}}{\bar{u}_0} = \frac{2n^2}{(2n+1)(n+1)}$$

(*b*) For turbulent flow, with $n = 7$,

$$\frac{\bar{V}}{\bar{u}_0} = \frac{2(7)^2}{[2(7)+1](7+1)} = \frac{49}{60}$$

which is considerably greater than the ratio for laminar flow, obtained in Problem 4.66 as

$$\frac{V}{u_0} = \frac{1}{2}$$

4.4 TURBULENT FLOW OVER A FLAT PLATE

4.98[D] Hydrogen gas at 77 °C and 1 atm pressure flows parallel to and along a flat plate at an approach velocity of $V_\infty = 122$ m/s. If the plate is also at 77 °C and is 1.22 m long, determine for a critical (transition to turbulent flow) Reynolds number of 500 000: ***(a)*** the thickness of the hydrodynamic boundary layer at the end of the plate; ***(b)*** the local skin-friction coefficient at the end of the place; ***(c)*** the average skin-friction coefficient over the plate length; and ***(d)*** the drag force per meter of plate width for one side of the plate.

From Table B-4 at $T = 77$ °C, $\rho = 0.0702$ kg/m^3; $\nu = 1.418 \times 10^{-4}$ m^2/s. The Reynolds number at the end of the plate is

$$\mathbf{Re}_L = \frac{V_\infty L}{\nu} = \frac{(122 \text{ m/s})(1.22 \text{ m})}{1.418 \times 10^{-4} \text{ m}^2/\text{s}} = 1\,049\,600$$

Turbulent, but laminar for about $500/1049.6 = 47.6\%$ of the length.

(a) A common assumption is that the boundary layer "grows" as though turbulent from the leading edge. The results are valid only in the actually turbulent portion of the boundary layer—not in the laminar part. Thus, using the turbulence formula (Problem 4.100)

$$\frac{\delta}{x} = \frac{0.376}{\mathbf{Re}_x^{1/5}} \tag{1}$$

we find

$$\delta_L = \frac{(0.376)L}{\mathbf{Re}_L^{1/5}} = \frac{(0.376)(1.22 \text{ m})}{(1\,049\,600)^{1/5}} = 28.7 \text{ mm}$$

(b) In turbulent flow over a flat plate, the local skin-friction coefficient is $c_f = 0.0576/\mathbf{Re}_x^{1/5}$ (Problem 4.101); hence

$$(c_f)_L = \frac{0.0576}{\mathbf{Re}_L^{1/5}} = \frac{0.0576}{(1\,049\,600)^{1/5}} = 0.0036$$

(c) The average skin-friction coefficient must reflect laminar flow over the leading 47.6% of the plate and turbulent flow over the remainder. For a critical **Re** of 500 000, a formula that achieves this is

$$C_f = (1.25)(c_f)_L - (0.003\,34)\left(\frac{x_c}{L}\right) = (1.25)(0.0036) - (0.003\,34)(0.476) = 0.002\,91$$

(d) As in Problem 4.28,

$$\frac{F_f}{W} = C_f \frac{\rho V_\infty^2 L}{2} = \frac{(0.002\,91)(0.0702 \text{ kg/m}^3)(122 \text{ m/s})^2(1.22 \text{ m})}{2} = 1.853 \text{ N/m}$$

4.99[D] Repeat Problem 4.98 in British Engineering units: the gas and plate temperatures are both 170 °F, the pressure is 1 atm, the approach velocity is $V_\infty = 400$ fps, and the plate is 4 ft long.

The appropriate parameters from Table B-4 are $\rho = 0.004\,38$ lbm/ft^3 and $\nu = 152.7 \times 10^{-5}$ ft^2/sec. At the end of the plate the Reynolds number is

$$\mathbf{Re}_L = \frac{V_\infty L}{\nu} = \frac{(400 \text{ ft/sec})(4 \text{ ft})}{1.527 \times 10^{-3} \text{ ft}^2/\text{sec}} = 1.048 \times 10^6$$

and the transition from laminar to turbulent flow occurs at

$$x_c = \frac{\mathbf{Re}_c \nu}{V_\infty} = \frac{(500\,000)(1.527 \times 10^{-3})}{400} = 1.909 \text{ ft}$$

(a) Assuming the hydrodynamic boundary layer grows turbulently throughout the length of the plate (which gives good results in the turbulent regime), the thickness at $x = L$ is given by

$$\delta_L = \frac{(0.376)L}{\mathbf{Re}_L^{1/5}} = \frac{(0.376)(4 \text{ ft})}{(1.048 \times 10^6)^{1/5}} = 0.094 \text{ ft} \approx 1\tfrac{1}{8} \text{ in.}$$

(b) The local skin-friction coefficient at $x = L$ is given by

$$(c_f)_L = \frac{0.0576}{\mathbf{Re}_L^{1/5}} = \frac{0.0576}{(1.048 \times 10^6)^{1/5}} = 0.0036$$

(c)
$$C_f = (1.25)(c_f)_L - (0.003\,34)\left(\frac{x_c}{L}\right) = (1.25)(0.0036) - (0.003\,34)\left(\frac{1.909}{4}\right) = 0.002\,91$$

(d)
$$\frac{F_f}{W} = C_f \frac{\rho V_\infty^2 L}{2g_c} = (0.002\,91)\frac{(0.004\,38 \text{ lbm/ft}^3)(400 \text{ ft/sec})^2(4 \text{ ft})}{2(32.2 \text{ lbm} \cdot \text{ft/lbf} \cdot \text{sec}^2)} = 0.1262 \text{ lbf/ft}$$

4.100 For external turbulent boundary layer flow over a flat plate at zero angle of attack, the velocity profile in the boundary layer may be approximated by the $\frac{1}{7}$-power law

$$\frac{\bar{u}}{V_\infty} = \left(\frac{y}{\delta}\right)^{1/7} \tag{1}$$

where $\bar{u}$ is the time-averaged velocity in the x-direction. Use this together with the integral momentum equation (*5*) of Problem 4.17, and the empirical turbulent shear-stress equation

$$\tau_s = (0.0225)\rho V_\infty^2 \left(\frac{\nu}{V_\infty \delta}\right)^{1/4} \tag{2}$$

to derive (*1*) of Problem 4.98.

▌ With (*1*) and (*2*) substituted, the integral momentum equation becomes

$$(0.0225)\left(\frac{\nu}{V_\infty \delta}\right)^{1/4} = \frac{d}{dx}\left\{\int_0^\delta \left[\left(\frac{y}{\delta}\right)^{1/7} - \left(\frac{y}{\delta}\right)^{2/7}\right] dy\right\}$$

Integrating and simplifying, we get

$$(0.0225)\left(\frac{\nu}{V_\infty \delta}\right)^{1/4} = \frac{7}{72}\frac{d\delta}{dx}$$

Separating variables and integrating:

$$\left(\frac{\nu}{V_\infty}\right)^{1/4} \int_0^x dx = (4.321) \int_0^\delta \delta^{1/4}\, d\delta$$

or

$$\frac{\delta}{x} = \frac{0.376}{(V_\infty x/\nu)^{1/5}} \equiv \frac{0.376}{\mathbf{Re}_x^{1/5}} \tag{3}$$

which is the desired result. It should be noted that in the limits of integration we have tacitly assumed that the turbulent boundary layer begins at the leading edge.

4.101 From the formulas of Problem 4.100, infer an expression for the local skin-friction coefficient for turbulent boundary layer flow over a flat plate at zero angle of attack.

▌ From the definition of c_f and (*2*) of Problem 4.100,

$$c_f \equiv \frac{\tau_s}{\rho V_\infty^2/2} = (0.045)\left(\frac{\nu}{V_\infty \delta}\right)^{1/4}$$

Substituting for δ from (*3*) of Problem 4.100,

$$c_f = (0.045)\left(\frac{\nu\, \mathbf{Re}_x^{1/5}}{0.376 V_\infty x}\right)^{1/4} = (0.045)\left(\frac{1}{0.376\, \mathbf{Re}_x^{4/5}}\right)^{1/4} = \frac{0.0576}{\mathbf{Re}_x^{1/5}}$$

CHAPTER 5

Forced Convection: External Flow

5.1 FLAT PLATE IN LAMINAR FLOW

5.1 Introducing the dimensionless temperature Θ and employing the Blasius coordinate η, where

$$\Theta \equiv \frac{T - T_s}{T_\infty - T_s} \qquad \eta \equiv y\left(\frac{V_\infty}{\nu x}\right)^{1/2}$$

Pohlhausen transformed the energy equation (*3*) of Problem 4.49 into

$$\frac{d^2\Theta}{d\eta^2} + \frac{\mathbf{Pr}}{2}\frac{d\Theta}{d\eta} = 0 \tag{1}$$

He then solved this equation subject to the temperature boundary conditions $\Theta(0) = 0$ and $\Theta(\delta_t) = 1$; the resulting temperature gradient at the wall is given very nearly by

$$\left.\frac{d\Theta}{d\eta}\right|_{\eta=0} = (0.332)\,\mathbf{Pr}^{1/3} \tag{2}$$

for $0.5 < \mathbf{Pr} < 10$. Using these results, determine an expression for $\mathbf{Nu}_x\,(\mathbf{Re}_x, \mathbf{Pr})$ (see Problem 4.55).

▌ By the chain rule,

$$\frac{d\Theta}{d\eta} = \frac{\dfrac{1}{T_\infty - T_s}}{(V_\infty/\nu x)^{1/2}}\frac{dT}{dy} = \frac{1}{T_\infty - T_s}\frac{x}{\mathbf{Re}_x^{1/2}}\frac{dT}{dy}$$

so that

$$(0.332)\,\mathbf{Pr}^{1/3} = \frac{1}{T_\infty - T_s}\frac{x}{\mathbf{Re}_x^{1/2}}\left.\frac{dT}{dy}\right|_{y=0} \tag{3}$$

But an energy balance at $(x, 0)$ gives

$$h_x(T_s - T_\infty) = -k\left.\frac{dT}{dy}\right|_{y=0} \tag{4}$$

Together (*3*)and (*4*) imply

$$\frac{h_x x}{k} \equiv \mathbf{Nu}_x = (0.332)\,\mathbf{Re}_x^{1/2}\,\mathbf{Pr}^{1/3} \tag{5}$$

5.2 In flow over a flat plate, the *conventional mean* Nusselt number over $0 < x < L$ is defined by

$$\overline{\mathbf{Nu}} \equiv \frac{\bar{h}L}{k} \tag{1}$$

where $\bar{h}$ denotes the average heat transfer coefficient over $0 < x < L$. Evaluate $\overline{\mathbf{Nu}}\,(\mathbf{Re}_L, \mathbf{Pr})$.

▌ Since $h_x \approx x^{-1/2}$, Problem 4.21 shows that $\bar{h} = 2h_L$. Hence, by (*5*) of Problem 5.1,

$$\overline{\mathbf{Nu}} = \frac{(2h_L)L}{k} = 2\,\mathbf{Nu}_L = (0.664)\,\mathbf{Re}_L^{1/2}\,\mathbf{Pr}^{1/3} \tag{2}$$

5.3 Is $\overline{\mathbf{Nu}}$ (Problem 5.2) the average of $\mathbf{Nu}_x$ over $0 < x < L$?

▌ Clearly not: $\mathbf{Nu}_x$ varies as $x^{1/2}$ (not as $x^{-1/2}$), and its average value is calculated as $\frac{2}{3}\,\mathbf{Nu}_L$. The ambiguous use of the overbar in (*1*) of Problem 5.2 is useful and sanctioned in elementary textbooks.

5.4[D] Ethylene glycol at 80 °C flows over a flat plate with $V_\infty = 0.46$ m/s. The plate surface temperature is 100 °C. **(*a*)** At what point does the thermal boundary layer thickness reach 5 mm? **(*b*)** What is the local heat transfer coefficient at that x-location?

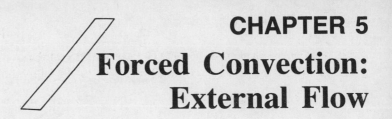

▌ From Table B-3, at $T_{film} = (80 + 100)/2 = 90\ °C$,

$$\nu = 2.503 \times 10^{-6}\ m^2/s \qquad \mathbf{Pr} = 27.4 \qquad k = 0.262\ W/m \cdot K$$

(*a*) Assuming laminar flow, $\delta_t = \delta\ \mathbf{Pr}^{-1/3}$ and $\delta/x = 5.0/(V_\infty x/\nu)^{1/2}$ give

$$x = \frac{\delta_t^2\ \mathbf{Pr}^{2/3}\ V_\infty}{(25.0)\nu} = \frac{(5 \times 10^{-3}\ m)^2(27.4)^{2/3}(0.46\ m/s)}{(25.0)(2.503 \times 10^{-6}\ m^2/s)} = 1.68\ m$$

The flow at x is indeed laminar, because

$$\mathbf{Re}_x = \frac{V_\infty x}{\nu} = \frac{(0.46\ m/s)(1.68\ m)}{2.503 \times 10^{-6}\ m^2/s} = 308\ 750 < 500\ 000$$

(*b*) By (5) of Problem 5.1, for laminar flow,

$$h_x = (0.332)\left(\frac{k}{x}\right)\mathbf{Re}_x^{1/2}\ \mathbf{Pr}^{1/3} = (0.332)\left(\frac{0.262\ W/m \cdot K}{1.68\ m}\right)(308\ 750)^{1/2}(27.4)^{1/3} = 86.7\ W/m^2 \cdot K$$

5.5[D] Ethylene glycol at 176 °F flows over a flat plate with $V_\infty = 1.5$ ft/s. The plate surface temperature is 212 °F. (*a*) At what point does the thermal boundary layer thickness reach 0.2 in.? (*b*) What is the local heat transfer coefficient at that x-location?

▌ From Table B-3 at $T_{film} = (176 + 212)/2 = 194\ °F$,

$$\nu = 2.70 \times 10^{-5}\ ft^2/s \qquad \mathbf{Pr} = 27.3 \qquad k = 0.1515\ Btu/hr \cdot ft \cdot °F$$

(*a*) Assuming laminar flow, $\delta_t = \delta\ \mathbf{Pr}^{-1/3}$ and $\delta/x = 5.0/(V_\infty x/\nu)^{1/2}$ give

$$x = \frac{\delta_t^2\ \mathbf{Pr}^{2/3}\ V_\infty}{(25.0)\nu} = \frac{[(0.2/12)\ ft]^2(27.3)^{2/3}(1.5\ ft/s)}{(25.0)(2.70 \times 10^{-5}\ ft^2/s)} = 5.60\ ft$$

The flow is indeed laminar at x, since

$$\mathbf{Re}_x = \frac{V_\infty x}{\nu} = \frac{(1.5\ ft/s)(5.60\ ft)}{2.70 \times 10^{-5}\ ft^2/s} = 311\ 100 < 500\ 000$$

(*b*) By (5) of Problem 5.1, for laminar flow,

$$h_x = (0.332)\left(\frac{k}{x}\right)\mathbf{Re}_x^{1/2}\ \mathbf{Pr}^{1/3} = (0.332)\left(\frac{0.1515\ Btu/hr \cdot ft \cdot °F}{5.60\ ft}\right)(311\ 100)^{1/2}(27.3)^{1/3} = 15.1\ Btu/hr \cdot ft^2 \cdot °F$$

5.6[D] Air at 80 °F approaches a 3-ft-long flat plate of 2-ft width at $V_\infty = 15$ ft/s. Calculate the local heat transfer coefficient at a distance 1.5 ft from the leading edge, for a plate surface temperature of 260 °F.

▌ From Table B-4, at $T_f = (80 + 260/2) = 170\ °F$,

$$\nu = 22.38 \times 10^{-5}\ ft^2/s \qquad \mathbf{Pr} = 0.697 \qquad k = 0.017\ 35\ Btu/hr \cdot ft \cdot °F$$

Since
$$\mathbf{Re}_x = \frac{V_\infty x}{\nu} = \frac{(15\ ft/s)(1.5\ ft)}{22.38 \times 10^{-5}\ ft^2/s} = 1 \times 10^5 < 5 \times 10^5$$

the flow is laminar; then, as in Problem 5.4(*b*),

$$h_x = (0.332)\left(\frac{k}{x}\right)\mathbf{Re}_x^{1/2}\ \mathbf{Pr}^{1/3} = (0.332)\left(\frac{0.017\ 35\ Btu/hr \cdot ft \cdot °F}{1.5\ ft}\right)(1 \times 10^5)^{1/2}(0.697)^{1/3} = 1.08\ Btu/hr \cdot ft^2 \cdot °F$$

5.7[D] Air at 27 °C approaches a 0.91-m-long by 0.61-m-wide flat plate with an approach velocity $V_\infty = 4.57$ m/s. Determine the local heat transfer coefficient at a distance 0.457 m from the leading edge, for a plate surface temperature of 127 °C.

▌ From Table B-4, at $T_f = (27 + 127)/2 = 77\ °C$,

$$\nu = 2.079 \times 10^{-5}\ m^2/s \qquad \mathbf{Pr} = 0.697 \qquad k = 3.001 \times 10^{-2}\ W/m \cdot K$$

Since
$$\mathbf{Re}_x = \frac{V_\infty x}{\nu} = \frac{(4.57\ m/s)(0.457\ m)}{2.079 \times 10^{-5}\ m^2/s} = 1 \times 10^5 < 5 \times 10^5$$

the flow is laminar at this x-value, and, by Problem 5.4(*b*),

$$h_x = (0.332)\left(\frac{k}{x}\right)\mathbf{Re}_x^{1/2}\ \mathbf{Pr}^{1/3} = (0.332)\left(\frac{3.001 \times 10^{-2}\ W/m \cdot K}{0.457\ m}\right)(1 \times 10^5)^{1/2}(0.697)^{1/3} = 6.11\ W/m^2 \cdot K$$

5.8[D] For the air flow over the flat plate of Problem 5.6, determine the total rate of heat transfer from the plate to the air.

The Reynolds number for overall heat transfer calculations (i.e., the maximal Reynolds number) is

$$\mathbf{Re}_L = \frac{V_\infty L}{\nu} = \frac{(15\ \text{ft/s})(3\ \text{ft})}{22.38 \times 10^{-5}\ \text{ft}^2/\text{s}} = 201\,000$$

As the flow is laminar, Problem 5.2 gives

$$\bar{h} = \frac{k}{L}\,\overline{\mathbf{Nu}} = \frac{k}{L}(0.664)\,\mathbf{Re}_L^{1/2}\,\mathbf{Pr}^{1/3}$$

$$= \frac{0.017\,35\ \text{Btu/hr}\cdot\text{ft}\cdot{}^\circ\text{F}}{3\ \text{ft}}(0.664)(201\,000)^{1/2}(0.697)^{1/3} = 1.526\ \text{Btu/hr}\cdot\text{ft}^2\cdot{}^\circ\text{F}$$

Thus $$q = \bar{h}A\,\Delta T = (1.526\ \text{Btu/hr}\cdot\text{ft}^2\cdot{}^\circ\text{F})(3\ \text{ft})(2\ \text{ft})[(260-80)\ {}^\circ\text{F}] = 1648\ \text{Btu/hr}$$

In this problem and those that follow, $\overline{\mathbf{Nu}}$ functions merely as a nondimensional alias for $\bar{h}$, as is common in the heat transfer literature.

5.9[D] For the air flow over the flat plate of Problem 5.7, determine the total rate of heat transfer from the plate to the air.

The Reynolds number for overall heat transfer calculations (i.e., the maximal Reynolds number) is

$$\mathbf{Re}_L = \frac{V_\infty L}{\nu} = \frac{(4.57\ \text{m/s})(0.91\ \text{m})}{2.079 \times 10^{-5}\ \text{m}^2/\text{s}} = 200\,000$$

As the flow is laminar, Problem 5.2 gives

$$\bar{h} = \frac{k}{L}\,\overline{\mathbf{Nu}} = \frac{k}{L}(0.664)\,\mathbf{Re}_L^{1/2}\,\mathbf{Pr}^{1/3}$$

$$= \frac{3.001 \times 10^{-2}\ \text{W/m}\cdot\text{K}}{0.91\ \text{m}}(0.664)(200\,000)^{1/2}(0.697)^{1/3} = 8.68\ \text{W/m}^2\cdot\text{K}$$

Then $$q = \bar{h}A\,\Delta T = (8.68\ \text{W/m}^2\cdot\text{K})(0.91\ \text{m})(0.61\ \text{m})[(127-27)\ \text{K}] = 482\ \text{W}$$

5.10[D] Water flows over a smooth, flat plate with a free-stream velocity of 0.6 ft/s. The plate is 4 ft long, and the water properties based on the film temperature are

$$\nu = 0.825 \times 10^{-5}\ \text{ft}^2/\text{s} \qquad k = 0.359\ \text{Btu/hr}\cdot\text{ft}\cdot{}^\circ\text{F} \qquad \mathbf{Pr} = 5.13$$

Determine the average rate of heat transfer to the water per square foot of plate surface if the plate temperature is 50 °F higher than the free-stream fluid temperature.

The maximum Reynolds number is

$$\mathbf{Re}_L = \frac{(0.6\ \text{ft/s})(4\ \text{ft})}{0.825 \times 10^{-5}\ \text{ft}^2/\text{s}} = 291\,000 \quad (\textit{laminar flow})$$

Thus, from Problem 5.2,

$$\bar{h} = \frac{k}{L}\,\overline{\mathbf{Nu}} = \frac{k}{L}(0.664)\,\mathbf{Re}_L^{1/2}\,\mathbf{Pr}^{1/3}$$

$$= \frac{0.359\ \text{Btu/hr}\cdot\text{ft}\cdot{}^\circ\text{F}}{4\ \text{ft}}(0.664)(291\,000)^{1/2}(5.13)^{1/3} = 55.4\ \text{Btu/hr}\cdot\text{ft}^2\cdot{}^\circ\text{F}$$

The average heat flux from the plate to the water is therefore

$$\frac{q}{A} = \bar{h}\,\Delta T = (55.4\ \text{Btu/hr}\cdot\text{ft}^2\cdot{}^\circ\text{F})(50\ {}^\circ\text{F}) = 2770\ \text{Btu/hr}\cdot\text{ft}^2$$

5.11[D] Water flows over a smooth, flat plate with a free-stream velocity of 183 mm/s. The plate is 1.22 m long, and the water properties at the fluid film temperature are

$$\nu = 7.66 \times 10^{-7}\ \text{m}^2/\text{s} \qquad k = 0.621\ \text{W/m}\cdot\text{K} \qquad \mathbf{Pr} = 5.13$$

Determine the average rate of heat transfer to the water per square meter of plate surface if the plate temperature is 27.8 °C above the free-stream fluid temperature.

The maximum Reynolds number is

$$\mathbf{Re}_L = \frac{(183 \times 10^{-3}\ \text{m/s})(1.22\ \text{m})}{7.66 \times 10^{-7}\ \text{m}^2/\text{s}} = 291\ 500 \quad (\textit{laminar flow})$$

Thus, from Problem 5.2,

$$\bar{h} = \frac{k}{L}\,\overline{\mathbf{Nu}} = \frac{k}{L}\,(0.664)\,\mathbf{Re}_L^{1/2}\,\mathbf{Pr}^{1/3}$$

$$= \frac{0.621\ \text{W/m}\cdot\text{K}}{1.22\ \text{m}}(0.664)(291\ 500)^{1/2}(5.13)^{1/3} = 314.7\ \text{W/m}^2\cdot\text{K}$$

The average heat flux to the water is

$$\frac{q}{A} = \bar{h}\,\Delta T = (314.7\ \text{W/m}^2\cdot\text{K})(27.8\ \text{K}) = 8.750\ \text{kW/m}^2$$

5.12[D] Water at 20 °C flows through a large rectangular duct, the surfaces of which are maintained at 40 °C. The average water velocity in the duct is 1.83 m/min. Assuming the flow over each one of the duct surfaces is unaffected by the other surfaces, so that it is similar to external flow over a flat plate, what is the local heat transfer coefficient at a length $x = 1.50$ m?

The film temperature is $T_f = (20 + 40)/2 = 30$ °C; from Table B-3, by interpolation,

$$\nu \approx 8.319 \times 10^{-7}\ \text{m}^2/\text{s} \qquad \mathbf{Pr} \approx 5.68 \qquad k \approx 0.612\ \text{W/m}\cdot\text{K}$$

Treating the flow as an external flow over a flat plate

$$\mathbf{Re}_x = \frac{V_{\text{avg}}x}{\nu} = \frac{\left(\dfrac{1.86\ \text{m}}{60\ \text{s}}\right)(1.50\ \text{m})}{8.319 \times 10^{-7}\ \text{m}^2/\text{s}} = 55\ 890 \quad (\textit{laminar})$$

Thus, by (*5*) of Problem 5.1,

$$h_x = \frac{k}{x}\,\mathbf{Nu}_x = \frac{k}{x}\,(0.332)\,\mathbf{Re}_x^{1/2}\,\mathbf{Pr}^{1/3} = \frac{0.612\ \text{W/m}\cdot\text{K}}{1.5\ \text{m}}(0.332)(55\ 890)^{1/2}(5.68)^{1/3} = 57.1\ \text{W/m}^2\cdot\text{K}$$

5.13[D] Water at 68 °F flows through a large rectangular duct, the surfaces of which are held at 104 °F. The average water velocity in the duct is 6 ft/min. Assuming the flow over each duct surface to be unaffected by the other surfaces, so that it is similar to external flow over a flat plate, what is the local heat transfer coefficient at $x = 4.92$ ft?

The film temperature is $T_f = (68 + 104)/2 = 86$ °F; from Table B-3, by interpolation,

$$\nu \approx 0.8955 \times 10^{-5}\ \text{ft}^2/\text{s} \qquad \mathbf{Pr} \approx 5.68 \qquad k \approx 0.354\ \text{Btu/hr}\cdot\text{ft}\cdot{}^\circ\text{F}$$

Thus, treating the flow as if over a flat plate,

$$\mathbf{Re}_x = \frac{V_{\text{avg}}x}{\nu} = \frac{(\frac{6}{60}\ \text{ft/s})(4.92\ \text{ft})}{8.955 \times 10^{-6}\ \text{ft}^2/\text{s}} = 54\ 940 \quad (\textit{laminar})$$

and (*5*) of Problem 5.1 yields

$$h_x = \frac{k}{x}\,\mathbf{Nu}_x = \frac{k}{x}\,(0.332)\,\mathbf{Re}_x^{1/2}\,\mathbf{Pr}^{1/3} = \frac{0.354\ \text{Btu/hr}\cdot\text{ft}\cdot{}^\circ\text{F}}{4.92\ \text{ft}}(0.332)(54\ 910)^{1/2}(5.68)^{1/3} = 9.99\ \text{Btu/hr}\cdot\text{ft}^2\cdot{}^\circ\text{F}$$

5.14 For laminar flow over a flat plate, compare the average heat transfer coefficient $\bar{h}$ with the local value halfway along the plate.

Since $h_x \propto x^{-1/2}$, one has $\bar{h}/h_L = 2$ (Problem 5.2) and $h_{L/2}/h_L = \sqrt{L}/\sqrt{L/2} = \sqrt{2}$. Hence,

$$\frac{\bar{h}}{h_{L/2}} = \frac{2}{\sqrt{2}} = \sqrt{2}$$

5.15 Refer to Problem 5.14. At what fraction of the length L is the ratio $\bar{h}/h_x$ equal to 1?

$$\frac{\bar{h}}{h_{L/4}} = \frac{2h_L}{(\sqrt{L}/\sqrt{L/4})h_L} = 1$$

5.16 Glycerin at 32 °C flows along a 1.83-m-long flat plate with $V_\infty = 3.66$ m/s; the plate is at 48 °C. ***(a)*** Verify that the flow is laminar. ***(b)*** Tabulate h_x versus x, bearing in mind the results of Problems 5.14 and 5.15.

(a) At the film temperature $T_f = (32 + 48)/2 = 40\ °C$, Table B-3 gives

$$\nu = 2.23 \times 10^{-4}\ m^2/s \qquad k = 2.854\ W/m \cdot K \qquad \mathbf{Pr} = 2.45$$

The maximum Reynolds number is

$$\mathbf{Re}_L = \frac{V_\infty L}{\nu} = \frac{(3.66)(1.83)}{2.23 \times 10^{-4}} = 3.00 \times 10^4$$

indicating laminar flow.

(b) From Problem 5.2,

$$\bar{h} = \frac{k}{L}(0.664)\,\mathbf{Re}_L^{1/2}\,\mathbf{Pr}^{1/3} = \frac{2.854}{1.83}(0.664)(3.00 \times 10^4)^{1/2}(2.45)^{1/3} = 242\ W/m^2 \cdot K$$

which value leads to Table 5-1.

TABLE 5-1

x, m	0	0.46	0.92	1.83
h_x, W/m$^2 \cdot$K	∞	242	171	121

5.17^D Air moving at 1.0 ft/sec blows over the top of a chest-type freezer. The top of the freezer is 3 ft by 5 ft and is poorly insulated, so that the surface remains at 50 °F. If the bulk air temperature is 80 °F, what is the maximum heat transfer by forced convection from the top of the freezer?

By interpolation in Table B-4 at $T_f = 65\ °F$:

$$\nu = 15.77 \times 10^{-5}\ ft^2/sec \qquad k = 0.0148\ Btu/hr \cdot ft \cdot °F \qquad \mathbf{Pr} = 0.71$$

The maximum heat transfer occurs if the freezer orientation results in air flow in the 3-ft direction. Thus,

$$\mathbf{Re}_L = \frac{V_\infty L}{\nu} = \frac{(1.0)(3)}{15.77 \times 10^{-5}} = 19\,020 \quad (laminar)$$

$$\overline{\mathbf{Nu}} = (0.664)\,\mathbf{Re}_L^{1/2}\,\mathbf{Pr}^{1/3} = (0.664)(137.9)(0.892) = 81.7$$

$$\bar{h} = \frac{k}{L}\overline{\mathbf{Nu}} = \frac{0.0148}{3}(81.7) = 0.403\ Btu/hr \cdot ft^2 \cdot °F$$

$$q = \bar{h}A\,\Delta T = (0.403)(15)(-30) = -181.3\ Btu/hr$$

The minus sign indicates heat transfer into the freezer.

5.18 For the flow over the top of the freezer in Problem 5.17, what would be the minimum value of the heat transfer rate to the freezer?

Orient the freezer so that the air flows over the top in the 5-ft direction. Then, with the air properties unchanged from Problem 5.17,

$$\mathbf{Re}_L = \frac{V_\infty L}{\nu} = \frac{(1.0)(5)}{15.77 \times 10^{-5}} = 31\,710 \quad (laminar)$$

$$\overline{\mathbf{Nu}} = (0.664)(\mathbf{Re}_L)^{1/2}\,\mathbf{Pr}^{1/3} = (0.664)(31\,710)^{1/2}(0.71)^{1/3} = 105.5$$

$$\bar{h} = \frac{k}{L}\overline{\mathbf{Nu}} = \frac{0.0148}{5}(105.5) = 0.312\ Btu/hr \cdot ft^2 \cdot °F$$

$$q = \bar{h}A\,\Delta T = (0.312)(15)(-30) = -140.5\ Btu/hr$$

5.19^D Air at $T_\infty = 27\ °C$ flows at 0.3 m/s over the top surface of a chest-type freezer. The top of the freezer measures 0.914 by 1.524 m and is poorly insulated so that the surface remains at 10.3 °C. What is the maximum rate of heat transfer by forced convection from the top of the freezer?

From Table B-4, at $T_f = (27 + 10.3)/2 = 18.7\ °C$,

$$\nu = 1.465 \times 10^{-5}\ m^2/s \qquad k = 0.0256\ W/m \cdot K \qquad \mathbf{Pr} = 0.71$$

Maximum average heat transfer coefficient will result if the flow is in the direction of the shorter dimension.

$$\mathbf{Re}_L = \frac{V_\infty L}{\nu} = \frac{(0.3)(0.914)}{1.465 \times 10^{-5}} = 18\,720 \quad (laminar)$$

$$\overline{\mathbf{Nu}} = (0.664)\,\mathbf{Re}_L^{1/2}\,\mathbf{Pr}^{1/3} = (0.664)(18\,720)^{1/2}(0.71)^{1/3} = 81.04$$

$$\bar{h} = \frac{k}{L}\,\overline{\mathbf{Nu}} = \frac{0.0256}{0.914}(81.04) = 2.27\ \mathrm{W/m^2 \cdot K}$$

$$q = \bar{h}A\,\Delta T = (2.27)[(0.914)(1.524)](10.3 - 27) = -52.8\ \mathrm{W}$$

5.20[D] The oil pan of an internal combustion engine protrudes below the framework of an automobile. It approximates a flat plate 0.3 m wide by 0.45 m long. The ambient air temperature is 38 °C, and the engine oil is at 94 °C. Assume that the resistance to conduction through the oil pan is negligible. The car travels at 32 km/h. What is the rate of heat transfer from the oil-pan surface? (Assume two-dimensional flow over a flat plate.)

I From Table B-4, by interpolation at $T_f = (38 + 94)/2 = 66\ °\mathrm{C}$,

$$\nu = 1.967 \times 10^{-5}\ \mathrm{m^2/s} \qquad k = 2.917 \times 10^{-2}\ \mathrm{W/m \cdot K} \qquad \mathbf{Pr} = 0.699$$

Thus

$$\mathbf{Re}_L = \frac{(0.45\ \mathrm{m})(32 \times 10^3\ \mathrm{m/h})}{1.967 \times 10^{-5}\ \mathrm{m^2/s}}\left(\frac{1\ \mathrm{h}}{3600\ \mathrm{s}}\right) = 2.034 \times 10^5$$

and the flow should be laminar everywhere.

$$\overline{\mathbf{Nu}} = (0.664)\,\mathbf{Re}_L^{1/2}\,\mathbf{Pr}^{1/3} = (0.664)(2.034 \times 10^5)^{1/2}(0.699)^{1/3} = 265.7$$

$$\bar{h} = \frac{k}{L}\,\overline{\mathbf{Nu}} = \frac{2.917 \times 10^{-2}}{0.45}(265.7) = 17.22\ \mathrm{W/m^2 \cdot K}$$

$$q = \bar{h}A\,\Delta T = (17.22)[(0.3)(0.45)](94 - 38) = 130.2\ \mathrm{W}$$

5.21[D] The oil pan of an internal combustion engine protrudes below the framework of an automobile. It approximates a flat plate 1 ft wide by 1.5 ft long. The ambient temperature of the air is 100 °F; the engine oil is at 200 °F. Assume that the pan's resistance to conduction is negligible. The car travels at 20 mph. How much energy per hour is convected from the oil-pan surface? (Assume two-dimensional flow and flat-plate theory.)

I From Table B-4, at $T_f = (100 + 200)/2 = 150\ °\mathrm{F}$,

$$\nu = 2.116 \times 10^{-4}\ \mathrm{ft^2/sec} \qquad k = 0.0169\ \mathrm{Btu/hr \cdot ft \cdot °F} \qquad \mathbf{Pr} = 0.699$$

Then

$$\mathbf{Re}_L = \frac{LV_\infty}{\nu} = \frac{(1.5\ \mathrm{ft})(20\ \mathrm{mi/hr})(1.467\ \mathrm{ft \cdot hr/mi \cdot sec})}{2.116 \times 10^{-4}\ \mathrm{ft^2/sec}} = 208\,000 \quad (laminar)$$

$$\overline{\mathbf{Nu}} = (0.664)\,\mathbf{Re}_L^{1/2}\,\mathbf{Pr}^{1/3} = (0.664)(208\,000)^{1/2}(0.699)^{1/3} = 268.8$$

$$\bar{h} = \frac{k}{L}\,\overline{\mathbf{Nu}} = \frac{0.0169}{1.5}(268.8) = 3.03\ \mathrm{Btu/hr \cdot ft^2 \cdot °F}$$

$$q = \bar{h}A\,\Delta T = (3.03)(1.5)(100) = 454\ \mathrm{Btu/hr}$$

5.22[D] A heat-treated steel plate is air-cooled following a hot oil immersion. Air at 10.3 °C blows along both sides of the plate at 9.66 km/h. The plate is initially at 27 °C. The plate measures 3.05 m by 1.0 m and the air flows parallel to the 1-m edge; the boundary layer flow has a maximum length of 1 m. Calculate the initial heat transfer rate from the plate (both sides).

I From Table B-4, at $T_f = (27 + 10.3)/2 = 18.7\ °\mathrm{C}$,

$$\nu = 1.465 \times 10^{-5}\ \mathrm{m^2/s} \qquad k = 0.0256\ \mathrm{W/m \cdot K} \qquad \mathbf{Pr} = 0.71$$

Thus,

$$\mathbf{Re}_L = \frac{V_\infty L}{\nu} = \frac{(9.66 \times 10^3\ \mathrm{m/h})\left(\dfrac{1\ \mathrm{h}}{3600\ \mathrm{s}}\right)(1\ \mathrm{m})}{1.465 \times 10^{-5}\ \mathrm{m^2/s}} = 183\,160 \quad (laminar)$$

$$\overline{\mathbf{Nu}} = (0.664)(\mathbf{Re}_L)^{1/2}(\mathbf{Pr})^{1/3} = (0.664)(183\,160)^{1/2}(0.71)^{1/3} = 253.5$$

$$\bar{h} = \frac{k}{L}\,\overline{\mathbf{Nu}} = \frac{0.0256\ \mathrm{W/m\cdot K}}{1\ \mathrm{m}}(253.5) = 6.49\ \mathrm{W/m^2\cdot K}$$

$$q_i = \bar{h}A\,(\Delta T)_i = (6.49\ \mathrm{W/m^2\cdot K})[2(1.0\ \mathrm{m})(3.05\ \mathrm{m})][(27-10.3)\ ^\circ\mathrm{C}] = 661\ \mathrm{W}$$

5.23[D] A heat-treated steel plate is air-cooled after oil immersion. Air at 50 °F blows along both sides of the plate at 6 mph. The plate is initially at 80 °F. The plate is 10 by 3.3 ft with the air flowing parallel to the 3.3-ft side. Calculate the initial heat transfer rate (two sides).

I From Table B-4, at $T_f = (80+50)/2 = 65\ ^\circ\mathrm{F}$,

$$\nu = 15.77\times10^{-5}\ \mathrm{ft^2/s} \qquad \mathbf{Pr} = 0.71 \qquad k = 0.0148\ \mathrm{Btu/hr\cdot ft\cdot ^\circ F}$$

Recall that 60 mph = 88 ft/s.

$$\mathbf{Re}_L = \frac{V_\infty L}{\nu} = \frac{(8.8)(3.3)}{15.77\times10^{-5}} = 1.84\times10^5 \quad (laminar)$$

$$\overline{\mathbf{Nu}} = (0.664)\,\mathbf{Re}_L^{1/2}\,\mathbf{Pr}^{1/3} = (0.664)(1.84\times10^5)^{1/2}(0.71)^{1/3} = 254$$

$$\bar{h} = \frac{k}{L}\,\overline{\mathbf{Nu}} = \frac{0.0148}{3.3}(254) = 1.139\ \mathrm{Btu/hr\cdot ft^2\cdot ^\circ F}$$

$$q_i = \bar{h}A(\Delta T)_i = (1.139)[2(10)(3.3)](80-50) = 2255\ \mathrm{Btu/hr}$$

5.24[D] A gentle breeze blow along the flat roof of a building at 3 ft/sec. The air is at 40 °F, and the roof is at 80 °F. Calculate the average rate of heat transfer from the roof per square foot if the air moves 15 ft along the roof.

I From Table B-4, at $T_f = (40+80)/2 = 60\ ^\circ\mathrm{F}$,

$$\nu = 15.40\times10^{-5}\ \mathrm{ft^2/sec} \qquad k = 0.014\,65\ \mathrm{Btu/hr\cdot ft\cdot ^\circ F} \qquad \mathbf{Pr} = 0.71$$

Then

$$\mathbf{Re}_L = \frac{V_\infty L}{\nu} = \frac{(3)(15)}{15.40\times10^{-5}} = 2.922\times10^5 \quad (laminar)$$

$$\overline{\mathbf{Nu}} = (0.664)\,\mathbf{Re}_L^{1/2}\,\mathbf{Pr}^{1/3} = (0.664)(2.922\times10^5)^{1/2}(0.71)^{1/3} = 320$$

$$\bar{h} = \frac{k}{L}\,\overline{\mathbf{Nu}} = \frac{0.014\,65}{15}(320) = 0.313\ \mathrm{Btu/hr\cdot ft^2\cdot ^\circ F}$$

$$\frac{q}{A} = \bar{h}\,\Delta T = (0.313\ \mathrm{Btu/hr\cdot ft^2\cdot ^\circ F})(40\ ^\circ\mathrm{F}) = 12.52\ \mathrm{Btu/hr\cdot ft^2}$$

5.25[D] A gentle breeze blows along a flat roof of a building at $T_\infty = 5\ ^\circ\mathrm{C}$ and $V_\infty = 0.91$ m/s. Calculate the average rate of heat transfer per square meter if the roof presents a length of 4.57 m. The roof surface is at 27 °C.

I From Table B-4, by interpolation at $T_f = (27+5)/2 = 16\ ^\circ\mathrm{C}$,

$$\nu = 1.432\times10^{-5}\ \mathrm{m^2/s} \qquad \mathbf{Pr} = 0.71 \qquad k = 2.622\times10^{-2}\ \mathrm{W/m\cdot K}$$

Hence

$$\mathbf{Re}_L = \frac{V_\infty L}{\nu} = \frac{(0.91)(4.57)}{1.432\times10^{-5}} = 2.904\times10^5 \quad (laminar)$$

$$\overline{\mathbf{Nu}} = (0.664)(\mathbf{Re}_L)^{1/2}(\mathbf{Pr})^{1/3} = (0.664)(2.904\times10^5)^{1/2}(0.71)^{1/3} = 319.2$$

$$\bar{h} = \frac{k}{L}\,\overline{\mathbf{Nu}} = \frac{2.622\times10^{-2}}{4.57}(319.2) = 1.83\ \mathrm{W/m^2\cdot K}$$

$$\frac{q}{A} = \bar{h}\,\Delta T = (1.83)(27-5) = 40.3\ \mathrm{W/m^2}$$

5.26 Consider steady laminar flow over a flat plate which has an unheated leading section of length x_i (i.e., the thermal boundary layer starts at $x = x_i$). Can one adapt (*5*) of Problem 5.1 to this situation by the change of coordinate $x \rightarrow x - x_i$?

I The kinematical and thermal problems are coupled—see (*3*) of Problem 4.49—and implicit in the Blasius/Pohlhausen solution is the assumption that *both* boundary layers begin at the same point $(x = 0)$.

Thus, with an unheated leading section, it is a question of new boundary conditions—not just new coordinates for old boundary conditions. When these new conditions are applied (to the von Kármán energy equation, assuming cubic profiles), the result is (*G-3*) of Appendix G.

For a numerical illustration of the difference between (*G-3*) and the naive change of variable, see Problem 5.28.

5.27^D A kitchen in a restaurant has a large, flat burner plate for frying. Since a great deal of heat rises from the plate, the cook decides to let a small fan blow over the burner, which is 4 ft long and positioned some 5 ft down a level smooth counter from the fan. (The total length is 9 ft.) If 90 °F air blows at 7 ft/sec and the plate is at 250 °F, what is the heat transfer rate per square foot at $x = 9$ ft?

▌ From Table B-4, at $T_f = (250 + 90)/2 = 170$ °F,

$$\nu = 22.38 \times 10^{-5}\ \text{ft}^2/\text{sec} \qquad k = 0.017\,35\ \text{Btu/hr}\cdot\text{ft}\cdot{}^\circ\text{F} \qquad \mathbf{Pr} = 0.697$$

At $x = 9\ \text{ft} = L$,

$$\mathbf{Re}_L = \frac{(7)(9)}{22.38 \times 10^{-5}} = 2.815 \times 10^5 \quad (\textit{laminar})$$

Equation (*G-3*) then gives

$$h_L = \frac{k}{L}\,\mathbf{Nu}_L = \frac{0.017\,35}{9}(0.332)(2.815 \times 10^5)^{1/2}(0.697)^{1/3}\left[1 - \left(\frac{5}{9}\right)^{3/4}\right]^{-1/3} = 0.425\ \text{Btu/hr}\cdot\text{ft}^2\cdot{}^\circ\text{F}$$

and from this

$$\left(\frac{q}{A}\right)_L = h_L\,\Delta T = (0.425)(250 - 90) = 68.02\ \text{Btu/hr}\cdot\text{ft}^2$$

5.28 For the situation of Problem 5.27, evaluate the local heat transfer coefficient one-quarter of the way into the burner plate.

▌ At $x = 6$ ft, $\mathbf{Re}_x = \frac{6}{9}\,\mathbf{Re}_L = 1.877 \times 10^5$ and (*G-3*) gives

$$h_x = \frac{k}{x}\,\mathbf{Nu}_x = \frac{k}{x}(0.332)\,\mathbf{Pr}^{1/3}\,\mathbf{Re}_x^{1/2}\left[1 - \left(\frac{x_i}{x}\right)^{3/4}\right]^{-1/3}$$

$$= \frac{0.017\,35}{6}(0.332)(0.697)^{1/3}(1.877 \times 10^5)^{1/2}\left[1 - \left(\frac{5}{6}\right)^{3/4}\right]^{-1/3} = 0.733\ \text{Btu/hr}\cdot\text{ft}^2\cdot{}^\circ\text{F}$$

From Problem 5.27, $h_L = 0.425\ \text{Btu/hr}\cdot\text{ft}^2\cdot{}^\circ\text{F}$; it is seen that h at the one-quarter point is *less than* twice h_L. Compare Problems 5.14, 5.15; see also Problem 5.30.

5.29^D Repeat Problem 5.27 in SI units. The burner length is 1.22 m and it is positioned 1.52 m down a smooth level counter from the fan. (The total length is 2.74 m.) Air at $T_\infty = 32.6$ °C blows at $V_\infty = 2.13$ m/s over the burner plate, which is at 121.4 °C. What is the heat transfer rate per square meter at $x = 2.74$ m?

▌ From Table B-4, at $T_f = (32.6 + 121.4)/2 = 77$ °C,

$$\nu = 2.079 \times 10^{-5}\ \text{m}^2/\text{s} \qquad k = 0.030\,01\ \text{W/m}\cdot\text{K} \qquad \mathbf{Pr} = 0.697$$

At $x = 2.74\ \text{m} = L$,

$$\mathbf{Re}_L = \frac{(2.13)(2.74)}{2.079 \times 10^{-5}} = 2.807 \times 10^5 \quad (\textit{laminar})$$

Equation (*G-3*) then gives

$$h_L = \frac{k}{L}\,\mathbf{Nu}_L = \frac{0.03001}{2.74}(0.332)(2.807 \times 10^5)^{1/2}(0.697)^{1/3}\left[1 - \left(\frac{1.52}{2.74}\right)^{3/4}\right]^{-1/3} = 2.407\ \text{W/m}^2\cdot\text{K}$$

and from this

$$\left(\frac{q}{A}\right)_L = h_L\,\Delta T = (2.407)(121.4 - 32.6) = 213.7\ \text{W/m}^2$$

5.30 Refer to Problem 5.26 and equation (*G-3*). Show that the average heat transfer coefficient over the heated section $x_i < x < L$ is given by

$$\bar{h} = 2\,\frac{1 - (x_i/L)^{3/4}}{1 - (x_i/L)}\,h_L \qquad (1)$$

By (G-3), $h_x = Af(x)$, with

$$A \equiv (0.332)k\,\mathbf{Pr}^{1/3}\left(\frac{V_\infty}{\nu}\right)^{1/2} \qquad f(x) \equiv x^{-1/4}(x^{3/4} - x_i^{3/4})^{-1/3}$$

Therefore,

$$\bar{h} = A\bar{f} = \frac{A}{L - x_i}\int_{x_i}^{L} x^{-1/4}(x^{3/4} - x_i^{3/4})^{-1/3}\,dx \tag{2}$$

In (2), change the integration variable to $\alpha = x^{3/4} - x_i^{3/4}$, obtaining

$$\bar{h} = \frac{\frac{4}{3}A}{L - x_i}\int_0^{L^{3/4}-x_i^{3/4}} \alpha^{-1/3}\,d\alpha = \frac{2A}{L - x_i}(L^{3/4} - x_i^{3/4})^{2/3} \tag{3}$$

But, $h_L = Af(L) = AL^{-1/4}(L^{3/4} - x_i^{3/4})^{-1/3}$; and this together with (3) gives (1). Observe that as $x_i \to 0$,

$$\bar{h} \to 2AL^{-1/2} = 2h_L$$

as expected.

5.31[D] How much heat energy is lost per hour per foot of width by the burner plate of Problem 5.27?

The plate is 4 ft by w (ft); $x_i = 5$ ft, $L = 9$ ft; and, from Problem 5.27, $h_L\,\Delta T = 68.02\ \text{Btu/hr}\cdot\text{ft}^2$. Using (1) of Problem 5.30, we find

$$\frac{q}{w} = (4\ \text{ft})\bar{h}\,\Delta T = (4\ \text{ft})\left[2\,\frac{1 - (\frac{5}{9})^{3/4}}{1 - (\frac{5}{9})}\right](68.02\ \text{Btu/hr}\cdot\text{ft}^2) = 436.5\ \text{Btu/hr}\cdot\text{ft}$$

5.32[D] How much heat energy is lost per second per meter of width by the burner plate of Problem 5.29?

The plate is 1.22 m by w (m); $x_i = 1.52$ m, $L = 2.74$ m; and, from Problem 5.29, $h_L\,\Delta T = 213.7\ \text{J/s}\cdot\text{m}^2$. Using (1) of Problem 5.30, we find

$$\frac{q}{w} = (1.22\ \text{m})\bar{h}\,\Delta T = (1.22\ \text{m})\left[2\,\frac{1 - (1.52/2.74)^{3/4}}{1 - (1.52/2.74)}\right](213.7\ \text{J/s}\cdot\text{m}^2) = 418.3\ \text{J/s}\cdot\text{m}$$

5.33 Heating of benzene is accomplished with a heated section 1.75 m long in a 3-m-wide shallow trough. When the fluid reaches the heating section, it has been flowing along the trough for 1 m; the flow speed is 60 mm/s. (The total length, unheated plus heated, is 2.75 m.) The average temperature of the benzene is 38 °C, and the heated trough is at 50 °C. The trough surface upstream of the heated section is at 38 °C. Approximate the heat flux into the benzene stream, if fluid properties at $T_f = 44$ °C are

$$\nu = 6.04 \times 10^{-7}\ \text{m}^2/\text{s} \qquad k = 0.150\ \text{W/m}\cdot\text{K} \qquad \mathbf{Pr} = 5.1$$

Since

$$\mathbf{Re}_{max} = \mathbf{Re}_L = \frac{V_\infty L}{\nu} = \frac{(0.060\ \text{m/s})(2.75\ \text{m})}{6.04 \times 10^{-7}\ \text{m}^2\text{s}} = 2.732 \times 10^5 < 5 \times 10^5$$

the flow is laminar. Equation (G-3) gives, for $x = L$,

$$h_L = (0.332)\left(\frac{k}{L}\right)\mathbf{Pr}^{1/3}\,\mathbf{Re}_L^{1/2}\left[1 - \left(\frac{x_i}{L}\right)^{3/4}\right]^{-1/3}$$

$$= (0.332)\left(\frac{0.150\ \text{W/m}\cdot\text{K}}{2.75\ \text{m}}\right)(5.1)^{1/3}(2.732 \times 10^5)^{1/2}\left[1 - \left(\frac{1}{2.75}\right)^{3/4}\right]^{-1/3} = 3.89\ \text{W/m}^2\cdot\text{K}$$

Then, by (1) of Problem 5.30,

$$\bar{h} = 2\,\frac{1 - (1/2.75)^{3/4}}{1 - (1/2.75)}(3.89\ \text{W/m}^2\cdot\text{K}) = 6.50\ \text{W/m}^2\cdot\text{K}$$

and the required heat flux is

$$q = \bar{h}A\,\Delta T = (6.50\ \text{W/m}^2\cdot\text{K})[(1.75\ \text{m})(3\ \text{m})][(50 - 38)\ \text{K}] = 409.5\ \text{W}$$

5.34[D] For parallel, two-dimensional flow of air at $T_\infty = 16$ °C and $V_\infty = 3.05$ m/s over a flat plate with a surface temperature of 38 °C, determine (*a*) the maximum length of plate for laminar flow; (*b*) the average heat transfer coefficient for the maximal plate; (*c*) the heat transfer rate per meter of width of the maximal plate.

M

From Table B-4, at $T_f = (16 + 38)/2 = 27\ °\text{C}$,

$$\nu = 1.568 \times 10^{-5}\ \text{m}^2/\text{s} \qquad k = 2.622 \times 10^{-2}\ \text{W/m}\cdot\text{K} \qquad \mathbf{Pr} = 0.708$$

(*a*) Corresponding to $\mathbf{Re}_L = 5 \times 10^5$ (upper limit for laminar flow),

$$L = \frac{(5 \times 10^5)\nu}{V_\infty} = \frac{(5 \times 10^5)(1.568 \times 10^{-5}\ \text{m}^2/\text{s})}{3.05\ \text{m/s}} = 2.57\ \text{m}$$

(*b*)
$$\bar{h} = \frac{k}{L}(0.664)\,\mathbf{Re}_L^{1/2}\,\mathbf{Pr}^{1/3} = \frac{0.026\,22}{2.57}(0.664)(5 \times 10^5)^{1/2}(0.708)^{1/3} = 4.269\ \text{W/m}^2\cdot\text{K}$$

(*c*)
$$\frac{q}{w} = \bar{h}L\,\Delta T = (4.269)(2.57)(38 - 16) = 241.4\ \text{W/m}$$

5.35^{D} For parallel, two-dimensional flow of air at $T_\infty = 60\ °\text{F}$ and $V_\infty = 10$ ft/sec over a flat plate with a surface temperature of 100 °F, determine (*a*) the maximum length of plate for laminar flow; (*b*) the average heat transfer coefficient for the maximal plate; (*c*) the heat transfer rate per meter of width of the maximal plate.

From Table B-4, at $T_f = (60 + 100)/2 = 80\ °\text{F}$,

$$\nu = 16.88 \times 10^{-5}\ \text{ft}^2/\text{sec} \qquad k = 0.015\,16\ \text{Btu/hr}\cdot\text{ft}\cdot°\text{F} \qquad \mathbf{Pr} = 0.708$$

(*a*) Corresponding to $\mathbf{Re}_L = 500\,000$ (upper limit for laminar flow),

$$L = \frac{(500\,000)\nu}{V_\infty} = \frac{(500\,000)(16.88 \times 10^{-5}\ \text{ft}^2/\text{sec})}{10\ \text{ft/sec}} = 8.44\ \text{ft}$$

(*b*)
$$\bar{h} = \frac{k}{L}(0.664)\,\mathbf{Re}_L^{1/2}\,\mathbf{Pr}^{1/3} = \frac{0.015\,16}{8.44}(0.664)(500\,000)^{1/2}(0.708)^{1/3} = 0.752\ \text{Btu/hr}\cdot\text{ft}^2\cdot°\text{F}$$

(*c*)
$$\frac{q}{w} = \bar{h}L\,\Delta T = (0.752)(8.44)(100 - 60) = 254\ \text{Btu/hr}\cdot\text{ft}$$

5.36^{D} Work Problem 5.34 for H_2 gas instead of air.

The physical parameters are now

$$\nu = 1.095 \times 10^{-4}\ \text{m}^2/\text{s} \qquad k = 0.1816\ \text{W/m}\cdot\text{K} \qquad \mathbf{Pr} = 0.706 \quad \text{(essentially unchanged)}$$

(*a*)
$$L = \frac{(5 \times 10^5)\nu}{V_\infty} = \frac{(5 \times 10^5)(1.095 \times 10^{-4}\ \text{m}^2/\text{s})}{3.05\ \text{m/s}} = 17.95\ \text{m}$$

(*b*)
$$\bar{h} = \frac{k}{L}(0.664)\,\mathbf{Re}_L^{1/2}\,\mathbf{Pr}^{1/3} = \frac{0.1816}{17.95}(0.664)(5 \times 10^5)^{1/2}(0.706)^{1/3} = 4.230\ \text{W/m}^2\cdot\text{K}$$

(*c*)
$$\frac{q}{w} = \bar{h}L\,\Delta T = (4.230)(17.95)(38 - 16) = 1670\ \text{W/m}$$

5.37^{D} Work Problem 5.35 for H_2 gas instead of air.

The physical parameters are now

$$\nu = 117.9 \times 10^{-5}\ \text{ft}^2/\text{sec} \qquad k = 0.105\ \text{Btu/hr}\cdot\text{ft}\cdot°\text{F} \qquad \mathbf{Pr} = 0.706 \quad \text{(essentially unchanged)}$$

(*a*)
$$L = \frac{(500\,000)\nu}{V_\infty} = \frac{(500\,000)(117.9 \times 10^{-5}\ \text{ft}^2/\text{sec})}{10\ \text{ft/sec}} = 58.95\ \text{ft}$$

(*b*)
$$\bar{h} = \frac{k}{L}(0.664)\,\mathbf{Re}_L^{1/2}\,\mathbf{Pr}^{1/3} = \frac{0.105}{58.95}(0.664)(500\,000)^{1/2}(0.706)^{1/3} = 0.745\ \text{Btu/hr}\cdot\text{ft}^2\cdot°\text{F}$$

(*c*)
$$\frac{q}{w} = \bar{h}L\,\Delta T = (0.745)(58.95)(100 - 60) = 1757\ \text{Btu/hr}\cdot\text{ft}$$

5.38^{D} Based on Problems 5.34 and 5.36, which is the better coolant, air or hydrogen?

For a meaningful comparison, we must specify a plate length such that both flows are laminar; the natural choice is $L = 2.57$ m. Then, the better-cooling stream will be that of greater $\bar{h}$. From Problem

5.34(b), $(\bar{h})_{\text{air}} = 4.269\ \text{W/m}^2\cdot\text{K}$. We may calculate $(\bar{h})_{\text{hydrogen}}$ by scaling the result of Problem 5.36(b) $(\bar{h} \approx L^{-1/2})$:

$$(\bar{h})_{\text{hydrogen}} = \left(\frac{17.95}{2.57}\right)^{1/2} (4.230\ \text{W/m}^2\cdot\text{K}) = 11.18\ \text{W/m}^2\cdot\text{K}$$

The hydrogen is more than twice as effective as the air.

5.39^D Based on Problems 5.35 and 5.37, which is the better coolant, air or hydrogen?

▮ For a meaningful comparison, we must specify a plate length such that both flows are laminar; the natural choice is $L = 8.44$ ft. Then, the better-cooling stream will be that of greater $\bar{h}$. From Problem 5.35(b), $(\bar{h})_{\text{air}} = 0.752\ \text{Btu/hr}\cdot\text{ft}^2\cdot{}^\circ\text{F}$. We may calculate $(\bar{h})_{\text{hydrogen}}$ by scaling the result of Problem 5.37(b) $(\bar{h} \approx L^{-1/2})$:

$$(\bar{h})_{\text{hydrogen}} = \left(\frac{58.95}{8.44}\right)^{1/2} (0.745\ \text{Btu/hr}\cdot\text{ft}^2\cdot{}^\circ\text{F}) = 1.97\ \text{Btu/hr}\cdot\text{ft}^2\cdot{}^\circ\text{F}$$

The hydrogen is more than twice as effective as the air.

5.40^D Engine oil at 140 °F flows over a flat plate at $V_\infty = 1.5$ ft/sec; the plate surface is maintained at 212 °F. For a plate of length 10 ft and width 2 ft, determine the total rate of heat transfer to the oil.

▮ From Table B-3, at $T_f = (140 + 212)/2 = 176\ {}^\circ\text{F}$,

$$\nu = 0.404 \times 10^{-3}\ \text{ft}^2/\text{sec} \qquad \mathbf{Pr} = 490 \qquad k = 0.080\ \text{Btu/hr}\cdot\text{ft}\cdot{}^\circ\text{F}$$

Because $\mathbf{Re}_L = V_\infty L/\nu = (1.5)(10)/(0.404 \times 10^{-3}) = 37\,130$, the flow is entirely laminar. By (2) of Problem 5.2,

$$\bar{h} = (0.664)\left(\frac{k}{L}\right)\mathbf{Re}_L^{1/2}\,\mathbf{Pr}^{1/3} = (0.664)\left(\frac{0.080}{10}\right)(37\,130)^{1/2}(490)^{1/3} = 8.07\ \text{Btu/hr}\cdot\text{ft}^2\cdot{}^\circ\text{F}$$

$$q = \bar{h}A(T_s - T_\infty) = (8.07)[(10)(2)](212 - 140) = 11\,620\ \text{Btu/hr}$$

5.41^D Engine oil at 60 °C flows over a flat plate at $V_\infty = 0.457$ m/s. The plate surface is held at 100 °C. For a plate of length 3.05 m and width 0.61 m, determine the total rate of heat transfer to the oil.

▮ From Table B-3, at $T_f = (60 + 100)/2 = 80\ {}^\circ\text{C}$,

$$\nu = 3.753 \times 10^{-5}\ \text{m}^2/\text{s} \qquad \mathbf{Pr} = 490 \qquad k = 0.138\ \text{W/m}\cdot\text{K}$$

Because $\mathbf{Re}_L = V_\infty L/\nu = (0.457)(3.05)/(3.753 \times 10^{-5}) = 0.3714 \times 10^5$, the flow is entirely laminar. By (2) of Problem 5.2,

$$\bar{h} = (0.664)\left(\frac{k}{L}\right)\mathbf{Re}_L^{1/2}\,\mathbf{Pr}^{1/3} = (0.664)\left(\frac{0.138}{3.05}\right)(0.3714 \times 10^5)^{1/2}(490)^{1/3} = 45.64\ \text{W/m}^2\cdot\text{K}$$

$$q = \bar{h}A(T_s - T_\infty) = (45.64)[(3.05)(0.61)](100 - 60) = 3397\ \text{W}$$

5.42^D Air at $T_\infty = 150$ °F and 25 psia pressure flows along one side of a smooth, flat plate at $V_\infty = 40$ ft/s. The plate is 1.20 ft long and 1.20 ft wide. What is the rate of heat transfer from the plate if it is maintained at 190 °F?

▮ From Table B-4, at $T_f = (150 + 190)/2 = 170\ {}^\circ\text{F}$ and with the density pressure-corrected as in Problem 4.79,

$$\mu g_c = 1.394 \times 10^{-5}\ \text{lbm/ft}\cdot\text{sec}$$

$$\rho = \frac{25}{14.7}(0.0623) = 0.1059\ \text{lbm/ft}^3$$

$$k = 0.017\,35\ \text{Btu/hr}\cdot\text{ft}\cdot{}^\circ\text{F}$$

$$\mathbf{Pr} = 0.697$$

Then $$\mathbf{Re}_L = \frac{V_\infty \rho L}{\mu g_c} = \frac{(40)(0.1059)(1.20)}{1.394 \times 10^{-5}} = 364\,650 \quad (\textit{laminar})$$

Thus, by (2) of Problem 5.2,

$$\bar{h} = (0.664)\left(\frac{k}{L}\right) \mathbf{Re}_L^{1/2} \mathbf{Pr}^{1/3} = (0.664)\left(\frac{0.017\,35}{1.2}\right)(364\,650)^{1/2}(0.697)^{1/3} = 5.14 \text{ Btu/hr}\cdot\text{ft}^2\cdot{}^\circ\text{F}$$

$$q = \bar{h}A(T_s - T_\infty) = (5.14)[(1.2)(1.2)](190 - 150) = 296 \text{ Btu/hr}$$

5.43[D] Air at $T_\infty = 66$ °C and 172 kPa pressure flows along one side of a smooth, flat plate at $V_\infty = 12.2$ m/s. The plate is 0.366 m long and 0.366 m wide. The plate is heated electrically to maintain a surface temperature of 88 °C. What is the required electric power input?

▮ From Table B-4, at $T_f = (66 + 88)/2 = 77$ °C and with the density pressure-corrected as in Problem 4.79,

$$\mu = 2.074 \times 10^{-5} \text{ Pa}\cdot\text{s}$$

$$\rho = \frac{172}{101.3}(0.9979) = 1.694 \text{ kg/m}^3$$

$$k = 0.030 \text{ W/m}\cdot\text{K}$$

$$\mathbf{Pr} = 0.697$$

Then
$$\mathbf{Re}_L = \frac{\rho V_\infty L}{\mu} = \frac{(1.694)(12.2)(0.366)}{2.074 \times 10^{-5}} = 3.647 \times 10^5$$

The flow being entirely laminar, (2) of Problem 5.2 gives

$$\bar{h} = (0.664)\left(\frac{k}{L}\right) \mathbf{Re}_L^{1/2} \mathbf{Pr}^{1/3} = (0.664)\left(\frac{0.030}{0.366}\right)(3.647 \times 10^5)^{1/2}(0.697)^{1/3} = 29.14 \text{ W/m}^2\cdot\text{K}$$

$$q = \bar{h}A(T_s - T_\infty) = (29.14)[(0.366)(0.366)](88 - 66) = 85.88 \text{ W}$$

5.44 Given a fluid in steady laminar flow over a flat plate of unit width; a given temperature difference is maintained between the plate surface and the fluid. Show that there exists an upper bound on the heat transfer, over all flow speeds and all plate lengths.

▮ We have for the magnitude of the heat flux:

$$|q| = \bar{h}A\,|\Delta T| = [(0.664)\,\frac{k}{L}\,\mathbf{Re}_L^{1/2}\,\mathbf{Pr}^{1/3}][(L)(1)]\,|\Delta T|$$

$$= [(0.664)k\,\mathbf{Pr}^{1/3}\,|\Delta T|]\,\mathbf{Re}_L^{1/2} < (0.664)k\,\mathbf{Pr}^{1/3}\,|\Delta T|\,(5 \times 10^5)^{1/2}$$

$$= (469.5)k\,\mathbf{Pr}^{1/3}\,|\Delta T| \equiv q_{max}$$

It is evident that the factor 469.5 should carry the width unit, whence q_{max} has the unit of power. The maximum is assumed when *the product* $V_\infty L$ is such as to produce the onset of turbulence at the downstream end of the plate.

5.45[D] Evaluate the upper bound found in Problem 5.44 if the fluid is air at 24 °C and 101.3 kPa, and the plate surface is at 108 °C.

▮ From Table B-4, by interpolation at $T_f = \frac{1}{2}(24 + 108) = 66$ °C,

$$k = 0.029\,17 \text{ W/m}\cdot\text{K} \qquad \mathbf{Pr} = 0.699$$

and so

$$q_{max} = (469.5 \text{ m})(0.029\,17 \text{ W/m}\cdot\text{K})(0.699)^{1/3}[(108 - 24) \text{ K}] = 1021 \text{ W}$$

5.46[D] Evaluate the upper bound found in Problem 5.44 if the fluid is air at 75 °F and 14.7 psia, and the plate surface is at 225 °F.

▮ From Table B-4, by interpolation at $T_f = \frac{1}{2}(75 + 225) = 150$ °F,

$$k = 0.016\,86 \text{ Btu/hr}\cdot\text{ft}\cdot{}^\circ\text{F} \qquad \mathbf{Pr} = 0.699$$

and so

$$q_{max} = (469.5 \text{ ft})(0.016\,86 \text{ Btu/hr}\cdot\text{ft}\cdot{}^\circ\text{F})(0.699)^{1/3}[(225 - 75)\ {}^\circ\text{F}] = 1054 \text{ Btu/hr}$$

5.47[D] Approximate the rate of heat transfer from the lateral surface of a 3-ft-long by 2-ft-diameter cylinder located axially in a wind tunnel. The tunnel air speed is $V_\infty = 15$ ft/s, the pressure is 1 atm, and the free-stream air temperature is 60 °F. The cylinder surface is held at 100 °F. The forward end of the cylinder is fitted with an unheated nose cone (Fig. 5-1) and laminar boundary layer flow may be assumed to start at $x = 0$.

▌ Provisionally treat the cylinder as a flat plate. From Table B-4, at $T_f = (100 + 60)/2 = 80\ °\text{F}$,

$$\nu = 16.88 \times 10^{-5}\ \text{ft}^2/\text{s} \qquad k = 0.015\,16\ \text{Btu/hr}\cdot\text{ft}\cdot°\text{F} \qquad \mathbf{Pr} = 0.708$$

The maximum Reynolds number is

$$\mathbf{Re}_L = \frac{V_\infty L}{\nu} = \frac{(15\ \text{ft/s})(3\ \text{ft})}{16.88 \times 10^{-5}\ \text{ft}^2/\text{s}} = 2.666 \times 10^5 \quad (\textit{laminar})$$

Thus

$$\bar{h} = (0.664)\left(\frac{k}{L}\right)\mathbf{Re}_L^{1/2}\,\mathbf{Pr}^{1/3} = (0.664)\left(\frac{0.015\,16}{3}\right)(2.666 \times 10^5)^{1/2}(0.708)^{1/3} = 1.544\ \text{Btu/hr}\cdot\text{ft}^2\cdot°\text{F}$$

$$q = \bar{h}A(T_s - T_\infty) = (1.544)[\pi(2)(3)](100 - 60) = 1164\ \text{Btu/hr}$$

The assumption of a flat plate will be valid if the maximum boundary layer thickness (on that assumption) is small compared to the diameter of the cylinder. Checking:

$$\frac{\delta_{\max}}{D} = \frac{(5.0)L}{D\,\mathbf{Re}_L^{1/2}} = \frac{(5.0)(3)}{(2)(2.666 \times 10^5)^{1/2}} = 1.45\%$$

The assumption is justified.

5.48[D] Approximate the rate of heat transfer from the lateral surface of a 0.91-m-long by 0.61-m-diameter cylinder located axially in a wind tunnel. The tunnel air speed is 4.57 m/s, the pressure is 101.3 kPa, and the free-stream air temperature is 16 °C. The cylindrical surface is held at 38 °C. The forward end of the cylinder is fitted with an unheated nose cone (Fig. 5-1), and laminar boundary layer flow may be assumed to start at $x = 0$ on the cylinder.

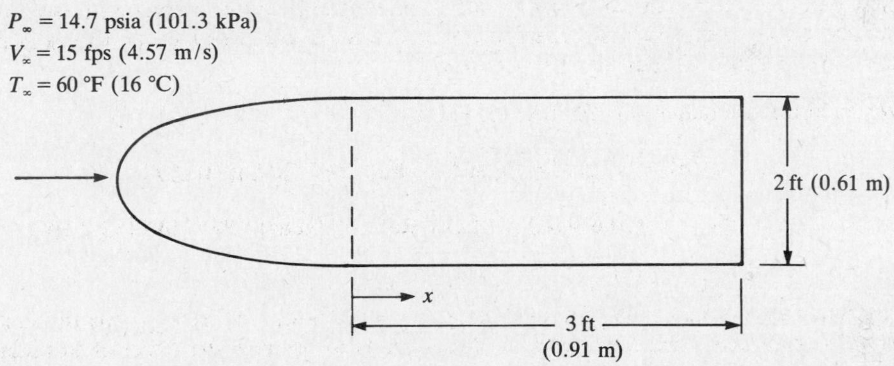

Fig. 5-1

▌ Provisionally treat the cylinder as a flat plate. At a film temperature $T_f = (16 + 38)/2 = 27\ °\text{C}$, from Table B-4,

$$\nu = 1.568 \times 10^{-5}\ \text{m}^2/\text{s} \qquad k = 0.026\,22\ \text{W/m}\cdot\text{K} \qquad \mathbf{Pr} = 0.708$$

The maximum Reynolds number is

$$\mathbf{Re}_L = \frac{(4.57\ \text{m/s})(0.91\ \text{m})}{1.568 \times 10^{-5}\ \text{m}^2/\text{s}} = 2.65 \times 10^5 \quad (\textit{laminar})$$

Then
$$\bar{h} = (0.664)\left(\frac{k}{L}\right)\mathbf{Re}_L^{1/2}\,\mathbf{Pr}^{1/3} = (0.664)\left(\frac{0.026\,22}{0.91}\right)(2.65 \times 10^5)^{1/2}(0.708)^{1/3} = 8.78\ \text{W/m}^2\cdot\text{K}$$

$$q = \bar{h}A(T_s - T_\infty) = (8.78)(\pi)(0.61)(0.91)(38 - 16) = 337\ \text{W}$$

The assumption of a flat plate will be valid if the maximum boundary layer thickness (on that assumption) is small compared to the diameter of the cylinder. Checking:

$$\frac{\delta_{\max}}{D} = \frac{(5.0)L}{D\,\mathbf{Re}_L^{1/2}} = \frac{(5.0)(0.91\ \text{m})}{(0.61\ \text{m})(2.65 \times 10^5)^{1/2}} = 1.45\%$$

The assumption is justified.

5.49 Repeat Problem 5.47 for a wind tunnel pressure of 1.7 atm.

I The ideal gas law predicts that density, and with it the Reynolds number, varies directly with pressure; other parameters remain constant. Then, since $q \approx \bar{h} \approx \mathbf{Re}_L^{1/2}$,

$$q = \left(\frac{1.7}{1}\right)^{1/2} (1164 \text{ Btu/hr}) = 1518 \text{ Btu/hr}$$

The above assumes that the new $\mathbf{Re}_L$ is smaller than 500 000—a fact readily verified.

5.50 The wind tunnel test of Problem 5.48 is repeated with hydrogen gas; all other parameters are unchanged. What will be the heat transfer rate from the cylinder to the gas?

I From Table B-4, for H_2 gas at $T_f = (16 + 38)/2 = 27$ °C,

$$\nu = 1.095 \times 10^{-4} \text{ m}^2/\text{s} \qquad k = 0.1816 \text{ W/m} \cdot \text{K} \qquad \mathbf{Pr} = 0.706$$

Then $$\mathbf{Re}_L = \frac{V_\infty L}{\nu} = \frac{(4.57 \text{ m/s})(0.91 \text{ m})}{1.095 \times 10^{-4} \text{ m}^2/\text{s}} = 3.798 \times 10^4 \quad \textit{(a seven-fold decrease)}$$

$$\bar{h} = (0.664)\left(\frac{k}{L}\right) \mathbf{Re}_L^{1/2} \mathbf{Pr}^{1/3} = (0.664)\left(\frac{0.1816}{0.91}\right)(3.798 \times 10^4)^{1/2}(0.706)^{1/3} = 22.99 \text{ W/m}^2 \cdot \text{K}$$

$$q = \bar{h}A(T_s - T_\infty) = (22.99)[\pi(0.61)(0.91)](38 - 16) = 882 \text{ W}$$

This is nearly three times the heat flux with air under the same conditions. (Compare the result of Problem 5.38.) Because the Reynolds number for hydrogen is much smaller than for air, the increase in q must be due to the much greater thermal conductivity of the lighter gas.

5.51 Carbon dioxide gas is separated from combustion gases and then cooled by flow over a series of flat plates. The flow over one plate 2 m long is at $V_\infty = 3.5$ m/s and $T_\infty = 177$ °C. The pressure is 1 atm (101.3 kPa), and the plate surface is maintained at $T_s = 77$ °C by cooling water on the side opposite to the gas flow. What is the local heat transfer coefficient at $x = L/4 = 0.5$ m?

I From Table B-4, at $T_f = (177 + 77)/2 = 127$ °C,

$$\nu = 1.439 \times 10^{-5} \text{ m}^2/\text{s} \qquad \mathbf{Pr} = 0.738 \qquad k = 2.459 \times 10^{-2} \text{ W/m} \cdot \text{K}$$

The Reynolds number at $x = L/4$ is

$$\mathbf{Re}_{L/4} = \frac{(3.5 \text{ m/s})(0.5 \text{ m})}{1.439 \times 10^{-5} \text{ m}^2/\text{s}} = 1.216 \times 10^5 \quad \textit{(laminar)}$$

Then, by (5) of Problem 5.1,

$$h_{L/4} = (0.332)\left(\frac{2.459 \times 10^{-2} \text{ W/m} \cdot \text{K}}{0.5 \text{ m}}\right)(1.216 \times 10^5)^{1/2}(0.738)^{1/3} = 5.146 \text{ W/m}^2 \cdot \text{K}$$

5.52 For the gaseous carbon dioxide flow of Problem 5.51, what is the heat transfer rate from the entire plate, if it is 1 m wide?

I According to Problems 5.51 and 5.15, $\bar{h} = 5.146$ W/m² · K. [The flow is entirely laminar, because $\mathbf{Re}_L = 4(1.216 \times 10^5) < 5 \times 10^5$.] Hence,

$$q = \bar{h}A(T_s - T_\infty) = (5.146 \text{ W/m}^2 \cdot {}^\circ\text{C})[(2 \text{ m})(1 \text{ m})][(77 - 177) \, {}^\circ\text{C}] = -1029 \text{ W}$$

The minus sign indicates heat transfer from the CO_2.

5.53^D *(a)* Rework Problem 5.51 if the gas is pressurized to 125 kPa. *(b)* For this pressurized flow, is it still true that $\bar{h} = h_{L/4}$?

I *(a)* Correcting the Reynolds number for the pressure change (see Problem 5.49), we have

$$\mathbf{Re}_x = \frac{125}{101.3}(1.216 \times 10^5) = 1.50 \times 10^5 \quad \textit{(laminar)}$$

$$h_x = \left(\frac{125}{101.3}\right)^{1/2} (5.146 \text{ W/m}^2 \cdot \text{K}) = 5.715 \text{ W/m}^2 \cdot \text{K}$$

(b) The Reynolds number at the end of the plate is $\mathbf{Re}_L = 4(1.5 \times 10^5) = 6.0 \times 10^5$; the flow is partly turbulent. Consequently, h_x is not proportional to $x^{-1/2}$ over the whole length of the plate, and so $\bar{h} \neq h_{L/4}$.

5.54[D] Carbon dioxide gas at 18.14 psia, $V_\infty = 11.5$ ft/sec, and $T_\infty = 350$ °F flows over and parallel to a smooth, flat plate which is maintained at $T_s = 170$ °F by cooling water on the side opposite the gas flow. The plate is 6.56 ft long and 3.28 ft wide. (*a*) What is the local heat transfer coefficient at $x = L/4 = 1.64$ ft? (*b*) Is $\bar{h}$ for the entire plate equal to $h_{L/4}$?

From Table B-4, at $T_f = (350 + 170)/2 = 260$ °F and 14.7 psia,

$$\rho = 0.0838 \text{ lbm/ft}^3 \qquad \mu g_c = 12.98 \times 10^{-6} \text{ lbm/ft}\cdot\text{sec} \qquad k = 0.01422 \text{ Btu/hr}\cdot\text{ft}\cdot\text{°F} \qquad \mathbf{Pr} = 0.738$$

By the ideal gas equation of state, the working density is

$$\rho' = 0.0838\left(\frac{18.14}{14.7}\right) = 0.1034 \text{ lbm/ft}^3$$

(*a*)

$$\mathbf{Re}_{L/4} = \frac{\rho' V_\infty (L/4)}{\mu g_c} = \frac{(0.1034 \text{ lbm/ft}^3)(11.5 \text{ ft/sec})(1.64 \text{ ft})}{12.98 \times 10^{-6} \text{ lbm/ft}\cdot\text{sec}} = 150\,200 \quad (\textit{laminar})$$

$$h_{L/4} = (0.332)\left(\frac{k}{L/4}\right)\mathbf{Re}_{L/4}^{1/2}\,\mathbf{Pr}^{1/3}$$

$$= (0.332)\left(\frac{0.01422 \text{ Btu/hr}\cdot\text{ft}\cdot\text{°F}}{1.64 \text{ ft}}\right)(150\,200)^{1/2}(0.738)^{1/3}$$

$$= 1.008 \text{ Btu/hr}\cdot\text{ft}^2\cdot\text{°F}$$

(*b*) The Reynolds number at the end of the plate is $\mathbf{Re}_L = 4(150\,200) = 600\,800$; hence, as in Problem 5.53(*b*), $\bar{h} \neq h_{L/4}$.

5.55 Water at 60 °C and 1 atm (101.3 kPa) pressure flows over a streamlined airfoil without separation. The undisturbed approach velocity is 0.573 m/s. Evaluate h 0.1 m downstream of the leading edge, if the airfoil is maintained at $T_s = 20$ °C.

Flat-plate theory applies. From Table B-3, at $T_f = (60 + 20)/2 = 40$ °C,

$$\nu = 6.578 \times 10^{-7} \text{ m}^2/\text{s} \qquad \mathbf{Pr} = 4.34 \qquad k = 0.628 \text{ W/m}\cdot\text{K}$$

At $x = 0.1$ m,

$$\mathbf{Re}_x = \frac{V_\infty x}{\nu} = \frac{(0.573 \text{ m/s})(0.1 \text{ m})}{6.578 \times 10^{-7} \text{ m}^2/\text{s}} = 8.71 \times 10^4 \quad (\textit{laminar})$$

$$h_x = (0.332)\left(\frac{k}{x}\right)\mathbf{Re}_x^{1/2}\,\mathbf{Pr}^{1/3} = (0.332)\left(\frac{0.628 \text{ W/m}\cdot\text{K}}{0.1 \text{ m}}\right)(8.71 \times 10^4)^{1/2}(4.34)^{1/3} = 1004 \text{ W/m}^2\cdot\text{K}$$

5.56[D] What length of airfoil in the laminar flow of Problem 5.55 produces the greatest absorption of heat per meter of width?

According to Problem 5.44, the condition is $V_\infty L = (5 \times 10^5)\nu$. In this case, V_∞ is given; so that

$$L = \frac{(5 \times 10^5)\nu}{V_\infty} = \frac{(5 \times 10^5)(6.578 \times 10^{-7} \text{ m}^2/\text{s})}{0.573 \text{ m/s}} = 0.574 \text{ m}$$

5.57[D] Water at 140 °F and 1 atm (14.7 psia) pressure is in attached, laminar flow over a 1-ft-wide streamlined airfoil; the surface of the airfoil is kept at 68 °F. What is the maximum possible rate of heat into the airfoil?

The airfoil may be treated as a flat plate. From Table B-3, at $T_f = (140 + 68)/2 = 104$ °F,

$$k = 0.363 \text{ Btu/hr}\cdot\text{ft}\cdot\text{°F} \qquad \mathbf{Pr} = 4.34$$

By Problem 5.44,

$$q_{max} = (469.5 \text{ ft})(0.363 \text{ Btu/hr}\cdot\text{ft}\cdot\text{°F})(4.34)^{1/3}[(140 - 68) \text{ °F}] = 20\,000 \text{ Btu/hr}$$

5.58 To realize the maximum heat flux in Problem 5.57, what should be the flow speed over a 3.76-ft-long airfoil?

By Problem 5.44, with $\nu = 0.708 \times 10^{-5}$ ft²/sec (Table B-3),

$$V_\infty = \frac{(5 \times 10^5)\nu}{L} = \frac{(5 \times 10^5)(0.708 \times 10^{-5} \text{ ft}^2/\text{sec})}{3.76 \text{ ft}} = 0.94 \text{ ft/sec}$$

5.59 You need to convert an h-value from British Engineering units to SI, but conversion tables (such as are found in Appendix A) are not at hand; you can only recall the basic conversions

$$1 \text{ cal} = 4.184 \text{ J} \qquad 1 \text{ in.} = 2.54 \text{ cm} \qquad 1 \text{ lbm} \approx 454 \text{ g}$$

How do you proceed?

■ You recollect the parallel definitions of the calorie and the Btu:

$$1 \text{ cal} \equiv \text{heat to raise 1 g of water 1 °C} \quad \text{(above some base temperature)}$$
$$1 \text{ Btu} \equiv \text{heat to raise 1 lbm of water 1 °F} \quad \text{(above some base temperature)}$$

It follows that

$$1 \text{ Btu} \approx (454)\left(\frac{1\text{ °F}}{1\text{ °C}}\right) \text{ cal}$$

from which

$$1 \frac{\text{Btu}}{\text{hr}\cdot\text{ft}^2\cdot\text{°F}} \approx \frac{(454)\left(\frac{1\text{ °F}}{1\text{ °C}}\right)(4.184\text{ J})}{(3600\text{ s})(12\times 0.0254\text{ m})^2(1\text{ °F})} = 5.68\text{ W/m}^2\cdot\text{°C}$$

(A more precise calculation gives the numerical factor as 5.6745.)

5.60 It is useful to know ranges of average heat transfer coefficients for certain classes of convective flows. Tabulate $\bar{h}$ for the previous problems in *gaseous* laminar flow over a flat plate.

■ See Table 5-2.

TABLE 5-2

problem no.	gas	$\bar{h}$ (W/m$^2\cdot$K)	$\bar{h}$ (Btu/hr$\cdot$ft$^2\cdot$°F)
5.8, 9	Air	8.681	1.526
5.17, 19	Air	2.27	0.403
5.18	Air	—	0.312
5.20, 21	Air	17.22	3.03
5.22, 23	Air	6.49	1.139
5.24, 25	Air	1.83	0.313
5.34, 35	Air	4.269	0.752
5.36, 37	H_2	4.230	0.745
5.38, 39	H_2	11.18	1.97
5.42, 43	Air	29.14	5.14
5.47, 48	Air	8.78	1.544
5.50	H_2	22.99	
5.51	CO_2	5.146	

In general: laminar, external, attached, boundary layer, gaseous flows result in $\bar{h}$'s of less than 1.0 up to 10 in British Engineering units, and $\bar{h}$'s of the order of 1.0 to 50 in SI units.

5.61 Repeat Problem 5.60 for *liquid* laminar flow.

■ See Table 5-3.

TABLE 5-3

problem no.	fluid	$\bar{h}$ (W/m$^2\cdot$K)	$\bar{h}$ (Btu/hr$\cdot$ft$^2\cdot$°F)
5.10, 11	Water	314.7	55.4
5.16	Glycerin	242	
5.40, 41	Engine oil	45.64	8.07

In general: laminar, external, attached, boundary layer, liquid flows result in $\bar{h}$'s of the order of 1.0 to several hundred in British Engineering units, and $\bar{h}$'s of the order of 1.0 to several thousand in SI units.

5.2 FLAT PLATE IN MIXED FLOW

5.62[D] Hydrogen gas at $T_\infty = 140\ °\text{F}$ and a pressure of 1 atm flows along a flat plate at $V_\infty = 400$ ft/sec. The plate is at 200 °F and is 4 ft long. Assuming a critical Reynolds number of 500 000, determine the local convective heat transfer coefficient at the end of the plate.

From Table B-4, at $T_f = (200 + 140)/2 = 170\ °\text{F}$,

$$\nu = 152.7 \times 10^{-5}\ \text{ft}^2/\text{sec} \qquad \mathbf{Pr} = 0.697 \qquad k = 0.119\ \text{Btu/hr}\cdot\text{ft}\cdot°\text{F}$$

Since
$$\mathbf{Re}_L = \frac{V_\infty L}{\nu} = \frac{(400\ \text{ft/sec})(4\ \text{ft})}{152.7 \times 10^{-5}\ \text{ft}^2/\text{sec}} = 1.048 \times 10^6$$

the end of the plate is in the turbulent regime. The simple *Reynolds analogy* between fluid friction and heat transfer,

$$\mathbf{St}_x \equiv \frac{\mathbf{Nu}_x}{\mathbf{Re}_x\,\mathbf{Pr}} = \frac{c_f}{2} \qquad (\mathbf{Pr} = 1) \tag{1}$$

does not hold at $x = L$, because $\mathbf{Pr} \neq 1$. The simplest applicable equation is the *Colburn modification* of the Reynolds analogy:

$$\mathbf{j}_H \equiv \mathbf{St}_x\,\mathbf{Pr}^{2/3} = \frac{c_f}{2} \qquad (0.6 < \mathbf{Pr} < 50) \tag{2}$$

Upon substitution of $\mathbf{Nu}_x \equiv h_x x/k$ and $c_f \approx 0.0592/\mathbf{Re}_x^{1/5}$ (a refinement of the result of Problem 4.101), (2) yields

$$h_x = (0.0296)\left(\frac{k}{x}\right)\mathbf{Re}_x^{4/5}\,\mathbf{Pr}^{1/3} \qquad (0.6 < \mathbf{Pr} < 50) \tag{3}$$

Thus, at $x = L$,

$$h_L = (0.0296)\left(\frac{0.119\ \text{Btu/hr}\cdot\text{ft}\cdot°\text{F}}{4\ \text{ft}}\right)(1.048 \times 10^6)^{4/5}(0.697)^{1/3} = 51.14\ \text{Btu/hr}\cdot\text{ft}^2\cdot°\text{F}$$

5.63[D] Hydrogen gas at $T_\infty = 60.3\ °\text{C}$ and a pressure of 1 atm flows along a flat plate at $V_\infty = 122$ m/s. The plate is maintained at 93.7 °C, and it is 1.22 m long. Assuming a critical Reynolds number of 5×10^5, determine the local convective heat transfer coefficient at the end of the plate.

From Table B-4, at $T_f = (60.3 + 93.7)/2 = 77\ °\text{C}$,

$$\nu = 1.419 \times 10^{-4}\ \text{m}^2/\text{s} \qquad \mathbf{Pr} = 0.697 \qquad k = 0.2058\ \text{W/m}\cdot\text{K}$$

Since
$$\mathbf{Re}_L = \frac{V_\infty L}{\nu} - \frac{(122\ \text{m/s})(1/22\ \text{m})}{1.419 \times 10^{-4}\ \text{m}^2/\text{s}} = 1.048 \times 10^6 \quad (\textit{turbulent})$$

the simplest equation that applies is the Colburn equation, (3) of Problem 5.62. Hence,

$$h_L = (0.0296)\left(\frac{0.2058\ \text{W/m}\cdot\text{K}}{1.22\ \text{m}}\right)(1.048 \times 10^6)^{4/5}(0.697)^{1/3} = 290\ \text{W/m}^2\cdot\text{K}$$

5.64 Refer to Problem 5.62. If the Reynolds analogy (1) were (improperly) used to determine h_x at the actual Prandtl number (different from 1), what would be the percent error relative to the Colburn expression (3)?

(1) gives $h_x = \gamma\,\mathbf{Pr}$, and (3) gives $h_x = \gamma\,\mathbf{Pr}^{1/3}$; here, γ is the selfsame function of x. The percent error is then

$$\frac{\gamma\,\mathbf{Pr} - \gamma\,\mathbf{Pr}^{1/3}}{\gamma\,\mathbf{Pr}^{1/3}}(100\%) = (\mathbf{Pr}^{2/3} - 1)(100\%)$$

independent of x.

5.65 Evaluate the percent error committed in using the unmodified Reynolds analogy for h_x when $\mathbf{Pr} = 0.697$ (the value in Problem 5.62).

By Problem 5.64,

$$\text{Percent error} = [(0.697)^{2/3} - 1](100\%) = -21.4\%$$

When $\mathbf{Pr} < 1$, the unmodified Reynolds analogy uniformly *underestimates* h_x.

5.66 An improved analogy between fluid friction and heat transfer for turbulent flow over a flat plate, due to von Kármán, is

$$\mathbf{St}_x \equiv \frac{\mathbf{Nu}_x}{\mathbf{Re}_x\,\mathbf{Pr}} = \frac{c_f/2}{1+5(c_f/2)^{1/2}\left[(\mathbf{Pr}-1) + \ln\left(\frac{5\,\mathbf{Pr}+1}{6}\right)\right]} \qquad (1)$$

Compare this with the Colburn equation, (*2*) of Problem 5.62, for Prandtl numbers slightly below 1; i.e., $\mathbf{Pr} = 1-\epsilon$ $(0<\epsilon \ll 1)$. What is the common limit as $\epsilon \to 0$?

▌ Substituting $1-\epsilon$ for **Pr** in (*1*) and making use of the familiar series

$$\ln(1+\alpha) = \alpha - \frac{\alpha^2}{2} + \cdots \qquad \text{and} \qquad \frac{1}{1+\beta} = 1-\beta+\cdots$$

one obtains

$$\textit{von Kármán} \qquad \frac{\mathbf{St}_x}{c_f/2} = 1 + \frac{55}{6}\left(\frac{c_f}{2}\right)^{1/2}\epsilon - \cdots$$

Now, in the turbulent regime,

$$c_f = \frac{0.0592}{\mathbf{Re}_x^{1/5}} \approx \frac{5\times 10^{-2}}{(5\times 10^5)^{1/5}} = \frac{1}{200\times 5^{1/5}}$$

so that

$$\text{Coefficient of } \epsilon \approx \left(\frac{55}{6}\right)\left(\frac{1}{20\times 5^{1/10}}\right) \approx \frac{55}{120} \approx \frac{1}{2}$$

On the other hand, the generalized binomial expansion gives

$$\textit{Colburn} \qquad \frac{\mathbf{St}_x}{c_f/2} = \mathbf{Pr}^{-2/3} = (1-\epsilon)^{-2/3} = 1 + \tfrac{2}{3}\epsilon - \cdots$$

It is seen that the two analogies are not too far apart, with the von Kármán probably yielding a smaller $\mathbf{St}_x$, and hence a smaller h_x (see Problem 5.67). In the limit of vanishing ϵ, both reduce to

$$\frac{\mathbf{St}_x}{c_f/2} = 1$$

which is just the Reynolds analogy.

5.67 Rework Problem 5.62 using the von Kármán analogy.

▌ Substitution of $c_f = 0.0592/\mathbf{Re}_x^{0.2}$ in (*1*) of Problem 5.66 produces the working formula

$$\mathbf{Nu}_x = \frac{(0.0296)\,\mathbf{Re}_x^{4/5}\,\mathbf{Pr}}{1+(0.860)\,\mathbf{Re}_x^{-1/10}\left[(\mathbf{Pr}-1) + \ln\left(\frac{5\,\mathbf{Pr}+1}{6}\right)\right]} \qquad (1)$$

which, for the data of Problem 5.62, gives

$$\mathbf{Nu}_L = \frac{(0.0296)(1.048\times 10^6)^{4/5}(0.697)}{1+(0.860)(1.048\times 10^6)^{-1/10}\left[(0.697-1) + \ln\left(\frac{5\times 0.697\;+\;1}{6}\right)\right]} = 1.549\times 10^3$$

Then $$h_L = \frac{k}{L}\,\mathbf{Nu}_L = \left(\frac{0.119\ \text{Btu/hr}\cdot\text{ft}\cdot{}^\circ\text{F}}{4\ \text{ft}}\right)(1.549\times 10^3) = 46.09\ \text{Btu/hr}\cdot\text{ft}^2\cdot{}^\circ\text{F}$$

This answer, which is probably more accurate, is about 10% smaller than that given by the far easier to use—therefore, far more popular—Colburn equation.

5.68 For the hydrogen gas flow of Problem 5.62, determine the average heat transfer coefficient over the entire plate.

▌ The computation

$$x_{\text{crit}} = \frac{\mathbf{Re}_{\text{crit}}}{\mathbf{Re}_L}(L) = \frac{5\times 10^5}{1.048\times 10^6}(4\ \text{ft}) = 1.908\ \text{ft}$$

shows that slightly less than half the plate experiences laminar flow, the remaining 2.09 ft being in transition to turbulent flow. To such mixed flows, the so-called *averaged Colburn modification of the Reynolds analogy applies*:

$$\bar{\mathbf{J}}_H \equiv \overline{\mathbf{St}}\,\mathbf{Pr}^{2/3} = \frac{C_f}{2} \qquad (1)$$

In (*1*) the term C_f is obtained by (I) calculating the turbulent drag for the entire length, (II) subtracting the turbulent drag for the laminar leading section, and (III) adding the laminar drag for the laminar leading section:

$$C_f \equiv \overset{\text{(I)}}{\frac{0.074}{\mathbf{Re}_L^{1/5}}} - \overset{\text{(II)}}{\frac{0.074}{\mathbf{Re}_{\text{crit}}^{1/5}}\left(\frac{x_{\text{crit}}}{L}\right)} + \overset{\text{(III)}}{\frac{1.328}{\mathbf{Re}_{\text{crit}}^{1/2}}\left(\frac{x_{\text{crit}}}{L}\right)} \qquad (2)$$

Because the turbulent boundary layer does not in fact start at the leading edge, it is clear that—contrary to what the notation might suggest—C_f *is not* the average of c_f over the length of the plate. Instead, it is a pseudoaverage or "conventional mean," similar to $\overline{\mathbf{Nu}}$. On the other hand, the quantity $\overline{\mathbf{St}}$ in (*1*) *really is* an average: since $\mathbf{St}_x = (\nu/kV_\infty \mathbf{Pr})h_x$, we have

$$\overline{\mathbf{St}} \equiv \frac{1}{L}\int_0^L \mathbf{St}_x\,dx = \frac{\nu}{kV_\infty \mathbf{Pr}}\,\bar{h} = \frac{\bar{h}L/k}{(V_\infty L/\nu)\,\mathbf{Pr}} \equiv \frac{\overline{\mathbf{Nu}}}{\mathbf{Re}_L\,\mathbf{Pr}} \qquad (3)$$

Application of (*1*), (*2*), and (*3*) to the data of Problem 5.62 yields

$$\bar{h} = \frac{k}{L}\overline{\mathbf{Nu}} = \frac{k}{L}\mathbf{Re}_L\,\mathbf{Pr}\,\overline{\mathbf{St}} = \frac{k}{2L}\mathbf{Re}_L\,\mathbf{Pr}^{1/3}\,C_f$$

$$= \frac{0.119}{2(4)}(1.048\times10^6)(0.697)^{1/3}\left[\frac{0.074}{(1.048\times10^6)^{1/5}} - \frac{0.074}{(5\times10^5)^{1/5}}\left(\frac{1.908}{4}\right) + \frac{1.328}{(5\times10^5)^{1/2}}\left(\frac{1.908}{4}\right)\right]$$

$$= 41\ \text{Btu/hr}\cdot\text{ft}^2\cdot{}^\circ\text{F}$$

5.69 In view of the discussion in Problem 5.68, what can be said about the relation between the Colburn correlation [(*2*) of Problem 5.62] and the averaged Colburn correlation [(*1*) of Problem 5.68]?

▌ Very little: The two expressions are of similar form, but the averaged Colburn $\bar{\mathbf{J}}_H$ factor is not obtained by mathematically averaging the local factor $\mathbf{j}_H$.

5.70 From Problem 5.68 infer that, for mixed flow over a flat plate,

$$\overline{\mathbf{Nu}} \equiv \frac{\bar{h}L}{k} = \mathbf{Pr}^{1/3}(0.037\,\mathbf{Re}_L^{4/5} - \mathcal{A}) \qquad (1)$$

where $\mathcal{A}$ is a pure number that depends only on the critical Reynolds number for transition to turbulent flow.

▌

$$\overline{\mathbf{Nu}} = \frac{1}{2}\mathbf{Re}_L\,\mathbf{Pr}^{1/3}\,C_f = \mathbf{Pr}^{1/3}\left(\frac{\mathbf{Re}_L}{2}\,C_f\right) = \mathbf{Pr}^{1/3}(0.037\,\mathbf{Re}_L^{4/5} - \mathcal{A})$$

where

$$\mathcal{A} \equiv 0.037\frac{\mathbf{Re}_L}{\mathbf{Re}_{\text{crit}}^{1/5}}\frac{x_{\text{crit}}}{L} - 0.664\frac{\mathbf{Re}_L}{\mathbf{Re}_{\text{crit}}^{1/2}}\frac{x_{\text{crit}}}{L} = 0.037\,\mathbf{Re}_{\text{crit}}^{4/5} - 0.664\,\mathbf{Re}_{\text{crit}}^{1/2} \qquad (2)$$

The last step in (*2*) depends on the identity

$$\frac{\mathbf{Re}_L}{\mathbf{Re}_{\text{crit}}} = \frac{L}{x_{\text{crit}}}$$

5.71 Calculate the parameter $\mathcal{A}$ of Problem 5.70, for $\mathbf{Re}_{\text{crit}}\times10^{-5} = 2, 3, 4, 5, 6, 10$; display the values as a table and as a graph.

▌ See Table 5-4 and Fig. 5-2. The latter shows that linear interpolation in the former is valid.

TABLE 5-4

$\mathbf{Re}_{\text{crit}}\times10^{-5}$	2	3	4	5	6	10
$\mathcal{A}$	347	527	702	871	1037	1671

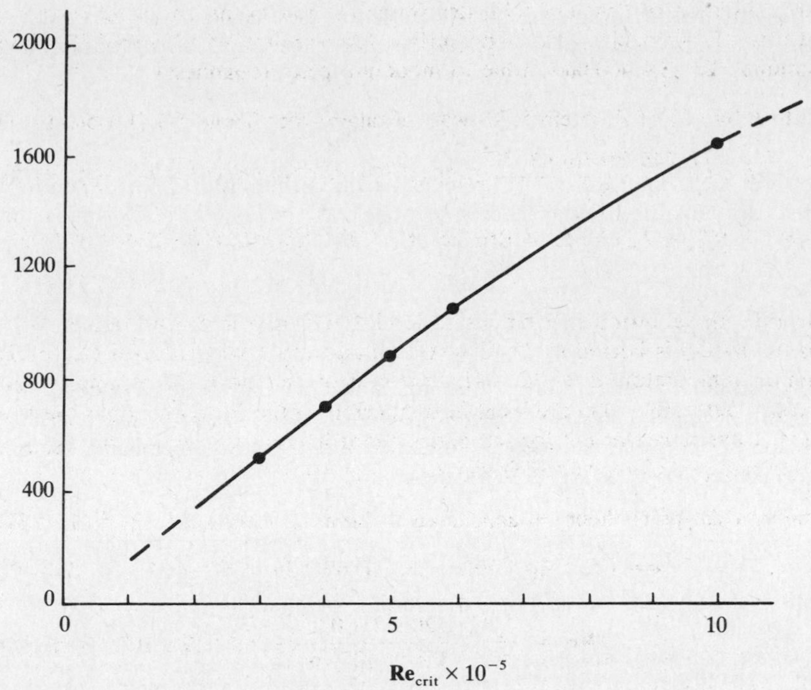

Fig. 5-2

5.72[D] Water at $T_f = 20\ °\text{C}$ and 1 atm (101.3 kPa) pressure flows over and parallel to a flat plate at $V_\infty = 0.305\ \text{m/s}$; the plate is 2.44 m long. Determine the local heat transfer coefficient at the downstream end of the plate.

▮ From Table B-3, at $T_f = 20\ °\text{C}$,

$$\nu = 1.006 \times 10^{-6}\ \text{m}^2/\text{s} \qquad \mathbf{Pr} = 7.02 \qquad k = 0.597\ \text{W/m} \cdot \text{K}$$

giving

$$\mathbf{Re}_L = \frac{V_\infty L}{\nu} = \frac{(0.305\ \text{m/s})(2.44\ \text{m})}{1.006 \times 10^{-6}\ \text{m}^2/\text{s}} = 7.4 \times 10^5 \quad (\textit{turbulent})$$

Proceeding as in Problem 5.62 $(\mathbf{Pr} \neq 1)$,

$$h_L = (0.0296)\left(\frac{0.597\ \text{W/m} \cdot \text{K}}{2.44\ \text{m}}\right)(7.4 \times 10^5)^{4/5}(7.02)^{1/3} = 688\ \text{W/m}^2 \cdot \text{K}$$

5.73[D] For the flow of Problem 5.72, determine the rate of heat transfer to the water, if the fluid temperature is $T_\infty = 10\ °\text{C}$, the plate surface temperature is $T_s = 30\ °\text{C}$, and the plate is 1.0 m wide. Assume $\mathbf{Re}_{\text{crit}} = 4 \times 10^5$ (due to moderate plate roughness).

▮ Compute $\bar{h}$ by (*1*) of Problem 5.70, with $\mathscr{A}$ taken from Table 5-4 (Problem 5.71):

$$\bar{h} = \left(\frac{0.597\ \text{W/m} \cdot \text{K}}{2.44\ \text{m}}\right)(7.02)^{1/3}[0.037(7.4 \times 10^5)^{4/5} - 702] = 531\ \text{W/m}^2 \cdot \text{K}$$

Then

$$q = \bar{h}A(T_s - T_\infty) = (531\ \text{W/m}^2 \cdot \text{K})[(2.44\ \text{m})(1.0\ \text{m})][(30 - 10)\ \text{K}] = 25.9\ \text{kW}$$

5.74[D] Water at $T_f = 68\ °\text{F}$ and 1 atm (14.7 psia) pressure flows over and parallel to a flat plate with a velocity $V_\infty = 1.0\ \text{ft/sec}$; the plate is 8 ft long. Determine the heat transfer coefficient at the downstream end of the plate.

▮ From Table B-3, at $T_f = 68\ °\text{F}$,

$$\nu = 1.083 \times 10^{-5}\ \text{ft}^2/\text{sec} \qquad \mathbf{Pr} = 7.02 \qquad k = 0.345\ \text{Btu/hr} \cdot \text{ft}^2 \cdot °\text{F}$$

giving

$$\mathbf{Re}_L = \frac{V_\infty L}{\nu} = \frac{(1.0\ \text{ft/sec})(8\ \text{ft})}{1.083 \times 10^{-5}\ \text{ft}^2/\text{sec}} = 740\,000 \quad (\textit{turbulent})$$

Proceeding as in Problem 5.62 $(\mathbf{Pr} \neq 1)$,

$$h_L = (0.0296)\left(\frac{0.345\ \text{Btu/hr} \cdot \text{ft} \cdot °\text{F}}{8\ \text{ft}}\right)(740\,000)^{4/5}(7.02)^{1/3} = 121.2\ \text{Btu/hr} \cdot \text{ft}^2 \cdot °\text{F}$$

5.75[D] For the flow situation of Problem 5.74, determine the rate of heat transfer to the water. The fluid temperature is $T_\infty = 49.4\ °F$, the plate surface temperature is $T_s = 85.4\ °F$, and the plate is 3.28 ft wide. Assume $\mathbf{Re}_{crit} = 400\ 000$ (due to moderate plate roughness).

▌ Compute $\bar{h}$ by (*1*) of Problem 5.70, with $\mathscr{A}$ taken from Table 5-4 (Problem 5.71):

$$\bar{h} = \left(\frac{0.345\ \text{Btu/hr}\cdot\text{ft}\cdot°\text{F}}{8\ \text{ft}}\right)(7.02)^{1/3}[0.037(740\ 000)^{4/5} - 702] = 93.54\ \text{Btu/hr}\cdot\text{ft}^2\cdot°\text{F}$$

Then $$q = \bar{h}A(T_s - T_\infty) = (93.54\ \text{Btu/hr}\cdot\text{ft}^2\cdot°\text{F})[(8\ \text{ft})(3.28\ \text{ft})][(85.4 - 49.4)\ °\text{F}] = 88\ 360\ \text{Btu/hr}$$

5.76[D] A smooth, thin model airfoil is to be tested for lift and drag in a wind tunnel. To obtain the desired conditions the model is heated to 37 °C surface temperature with internal electric heaters while the free-stream air temperature is 17 °C. The model chord length is 1.5 m, and the airstream velocity is $V_\infty = 20$ m/s. Determine (*a*) the local heat transfer coefficient 0.8 m from the leading edge; (*b*) the average heat transfer coefficient for the entire chord length, assuming that $\mathbf{Re}_{crit} = 5 \times 10^5$; and (*c*) the heat transfer rate per meter of airfoil width.

▌ Assume that flat-plate theory is applicable. From Table B-4, at $T_f = (17 + 37)/2 = 27\ °C$,

$$\nu = 1.57 \times 10^{-5}\ \text{m}^2/\text{s} \qquad k = 0.026\ 22\ \text{W/m}\cdot\text{K} \qquad \mathbf{Pr} = 0.708$$

(*a*) $$\mathbf{Re}_x = \frac{V_\infty x}{\nu} = \frac{(20\ \text{m/s})(0.8\ \text{m})}{1.57 \times 10^{-5}\ \text{m}^2/\text{s}} = 1.019 \times 10^6 \quad (\textit{turbulent})$$

and by (*3*) of Problem 5.62,

$$h_x = (0.0296)\left(\frac{0.026\ 22\ \text{W/m}\cdot\text{K}}{0.8\ \text{m}}\right)(1.019 \times 10^6)^{4/5}(0.708)^{1/3} = 55.4\ \text{W/m}^2\cdot\text{K}$$

(*b*) $$\mathbf{Re}_L = \frac{L}{x}\mathbf{Re}_x = \frac{15}{8}(1.019 \times 10^6) = 1.91 \times 10^6 \quad (\textit{turbulent})$$

From (*1*) of Problem 5.70, with $\mathscr{A}$ taken from Table 5-4 (Problem 5.71),

$$\bar{h} = \left(\frac{0.026\ 22\ \text{W/m}\cdot\text{K}}{1.5\ \text{m}}\right)(0.708)^{1/3}[0.037(1.91 \times 10^6)^{4/5} - 871] = 47.5\ \text{W/m}^2\cdot\text{K}$$

(*c*) Taking into account both sides,

$$\frac{q}{w} = \bar{h}(2L)(T_s - T_\infty) = (47.5\ \text{W/m}^2\cdot\text{K})(3.0\ \text{m})[(37 - 17)\ \text{K}] = 2850\ \text{W/m}$$

5.77 Rework Problem 5.76(*b*) and (*c*) for $\mathbf{Re}_{crit} = 3 \times 10^5$ (a surface roughening).

▌ (*b*) $$\bar{h} = \left(\frac{0.026\ 22\ \text{W/m}\cdot\text{K}}{1.5\ \text{m}}\right)(0.708)^{1/3}[0.037(1.91 \times 10^6)^{4/5} - 527] = 52.9\ \text{W/m}^2\cdot\text{K}$$

(*c*) $$\frac{q}{w} = \bar{h}(2L)(T_s - T_\infty) = (52.9\ \text{W/m}^2\cdot\text{K})(3.0\ \text{m})[(37 - 17)\ \text{K}] = 3174\ \text{W/m}$$

5.78 Rework Problem 5.76(*b*) and (*c*) assuming that a "tripping wire" to cause immediate transition to turbulence is placed 1 mm from the leading edge on each surface (side).

▌ (*b*) Because the laminar region is of negligible length, (*1*) of Problem 5.70 may be applied with $\mathscr{A} = 0$:

$$\bar{h} = \frac{k}{L}\mathbf{Pr}^{1/3}(0.037)\mathbf{Re}_L^{4/5} = \left(\frac{0.026\ 22\ \text{W/m}\cdot\text{K}}{1.5\ \text{m}}\right)(0.708)^{1/3}(0.037)(1.91 \times 10^6)^{4/5} = 61.1\ \text{W/m}^2\cdot\text{K}$$

(*c*) $$\frac{q}{w} = \bar{h}(2L)(T_s - T_\infty) = (61.1\ \text{W/m}^2\cdot\text{K})(3.0\ \text{m})[(37 - 17)\ \text{K}] = 3.67\ \text{kW/m}$$

5.79[D] A smooth, thin model airfoil is to be tested for lift and drag in a wind tunnel. To obtain the desired conditions, the model is heated to 98 °F surface temperature with internal electric heaters while the free-stream air temperature is 62 °F. The model chord length is 59.05 in., and the airstream velocity is

$V_\infty = 65.6$ ft/sec. Determine (*a*) the local heat transfer coefficient 31.5 in. from the leading edge; (*b*) the average heat transfer coefficient for the entire chord length; and (*c*) the rate of heat transfer per foot of airfoil width. Assume that $\mathbf{Re}_{\text{crit}} = 5 \times 10^5$.

▍ Assume that flat-plate theory applies. From Table B-4, at $T_f = (98 + 62)/2 = 80$ °F,

$$k = 0.015\,16 \text{ Btu/hr}\cdot\text{ft}\cdot\text{°F} \qquad \nu = 16.88 \times 10^{-5} \text{ ft}^2/\text{sec} \qquad \mathbf{Pr} = 0.708$$

(*a*)
$$\mathbf{Re}_x = \frac{V_\infty x}{\nu} = \frac{(65.6 \text{ ft/sec})[(31.5/12) \text{ ft}]}{16.88 \times 10^{-5} \text{ ft}^2/\text{sec}} = 1.020 \times 10^6 \quad (\textit{turbulent})$$

and by (*3*) of Problem 5.62,

$$h_x = (0.0296)\left[\frac{0.015\,16 \text{ Btu/hr}\cdot\text{ft}\cdot\text{°F}}{(31.5/12) \text{ ft}}\right](1.02 \times 10^6)^{4/5}(0.708)^{1/3} = 9.768 \text{ Btu/hr}\cdot\text{ft}^2\cdot\text{°F}$$

(*b*)
$$\mathbf{Re}_L = \frac{V_\infty L}{\nu} = \frac{(65.6 \text{ ft/sec})[(59.05/12) \text{ ft}]}{16.88 \times 10^{-5} \text{ ft}^2/\text{sec}} = 1.91 \times 10^6 \quad (\textit{turbulent})$$

From (*1*) of Problem 5.70, with $\mathscr{A}$ taken from Table 5-4 (Problem 5.71),

$$\bar{h} = \frac{0.015\,16 \text{ Btu/hr}\cdot\text{ft}\cdot\text{°F}}{(59.05/12) \text{ ft}}(0.708)^{1/3}[0.037(1.91 \times 10^6)^{4/5} - 871] = 8.37 \text{ Btu/hr}\cdot\text{ft}^2\cdot\text{°F}$$

(*c*) Accounting for both sides,

$$\frac{q}{w} = \bar{h}(2L)(T_s - T_\infty) = (8.37)\left(\frac{118.10}{12}\right)(98 - 62) = 2970 \text{ Btu/hr}\cdot\text{ft}$$

5.80^{D} In an industrial process, liquid ammonia flows over one surface of a 1-m-long heated flat plate at free-stream conditions $V_\infty = 3$ m/s and $T_\infty = 10$ °C. If the plate is held at $T = 30$ °C, determine the heat transfer rate per meter width, assuming that $\mathbf{Re}_{\text{crit}} = 5 \times 10^5$.

▍ From Table B-3, at $T_f = (30 + 10)/2 = 20$ °C,

$$\nu = (0.0929)(0.386 \times 10^{-5}) = 3.59 \times 10^{-7} \text{ m}^2/\text{s}$$
$$k = (1.729\,577)(0.301) = 0.521 \text{ W/m}\cdot\text{K}$$
$$\mathbf{Pr} = 2.02$$

The Reynolds number at the end of the plate is

$$\mathbf{Re}_L = \frac{V_\infty L}{\nu} = \frac{(3 \text{ m/s})(1 \text{ m})}{3.59 \times 10^{-7} \text{ m}^2/\text{s}} = 8.36 \times 10^6 \quad (\textit{highly turbulent})$$

By (*1*) of Problem 5.70 and Table 5-4 of Problem 5.71,

$$\bar{h} = \left(\frac{0.521 \text{ W/m}\cdot\text{K}}{1 \text{ m}}\right)(2.02)^{1/3}[0.037(8.36 \times 10^6)^{4/5} - 871] = 7.833 \text{ kW/m}^2\cdot\text{K}$$

whence
$$\frac{q}{w} = \bar{h}L(T_s - T_\infty) = (7.833 \text{ kW/m}^2\cdot\text{K})(1 \text{ m})[(30 - 10) \text{ K}] = 156.7 \text{ kW/m}$$

5.81 Because the Reynolds number at the end of the plate in Problem 5.80 is quite high when compared with $\mathbf{Re}_{\text{crit}}$, the leading laminar portion ought to have little effect on $\bar{h}$ and q/w. Examine this supposition.

▍ The critical length, which is the length experiencing laminar flow, is, for the assumed $\mathbf{Re}_{\text{crit}} = 5 \times 10^5$,

$$x_{\text{crit}} = \frac{\nu}{V_\infty}\mathbf{Re}_{\text{crit}} = \frac{3.59 \times 10^{-7} \text{ m}^2/\text{s}}{3 \text{ m/s}}(5 \times 10^5) = 60.0 \text{ mm}$$

or 6 percent of the total length. Assuming turbulence over the entire length ($\mathscr{A} = 0$ in Problem 5.70),

$$\bar{h} = \left(\frac{0.521 \text{ W/m}\cdot\text{K}}{1 \text{ m}}\right)(2.02)^{1/3}[0.037(8.36 \times 10^6)^{4/5}] = 8.406 \text{ kW/m}^2\cdot\text{K}$$

which is in error by only

$$\frac{8.406 - 7.833}{7.833}(100\%) = 7.3\%$$

The percent error in q/w is the same.

5.82^{D} In an industrial process, liquid ammonia flows over one surface of a 3.28-ft-long heated flat plate at free-stream conditions $V_\infty = 9.8$ ft/sec and $T_\infty = 50$ °F. If the plate is held at $T_s = 86$ °F, determine the rate of heat loss of the plate per foot width, assuming that $\mathbf{Re}_{\text{crit}} = 500\,000$.

From Table B-3, at $T_f = (86 + 50)/2 = 68\ °F$,

$$\nu = 0.386 \times 10^{-5}\ \text{ft}^2/\text{sec} \qquad k = 0.301\ \text{Btu/hr}\cdot\text{ft}\cdot°\text{F} \qquad \mathbf{Pr} = 2.02$$

The Reynolds number at the end of the plate is

$$\mathbf{Re}_L = \frac{V_\infty L}{\nu} = \frac{(9.8\ \text{ft/sec})(3.28\ \text{ft})}{0.386 \times 10^{-5}\ \text{ft}^2/\text{sec}} = 8.33 \times 10^6 \quad (\textit{highly turbulent})$$

By (*1*) of Problem 5.70 and Table 5-4 of Problem 5.71,

$$\bar{h} = \left(\frac{0.301\ \text{Btu/hr}\cdot\text{ft}\cdot°\text{F}}{3.28\ \text{ft}}\right)(2.02)^{1/3}[0.037(8.33 \times 10^6)^{4/5} - 871] = 1375\ \text{Btu/hr}\cdot\text{ft}^2\cdot°\text{F}$$

and $$\frac{q}{w} = \bar{h}L(T_s - T_\infty) = (1375\ \text{Btu/hr}\cdot\text{ft}^2\cdot°\text{F})(3.28\ \text{ft})[(86 - 50)\ °\text{F}] = 162\,400\ \text{Btu/hr}\cdot\text{ft}$$

5.83[D] A corn chip cooker has the form of a trough with a flat bottom 15 ft long by 3 ft wide. The bottom is heated to a uniform surface temperature of 410 °F, and the cooking oil flows along the heated surface with $V_\infty = 0.8$ ft/sec and $T_\infty = 390$ °F. Calculate the convective heat transfer rate to the oil ($\nu = 2 \times 10^{-5}$ ft²/sec, $k = 0.070$ Btu/hr · ft · °F, and $\mathbf{Pr} = 31$). Assume moderate plate roughness causing $\mathbf{Re}_{crit} = 400\,000$.

The maximum Reynolds number,

$$\mathbf{Re}_L = \frac{V_\infty L}{\nu} = \frac{(0.8\ \text{ft/sec})(15\ \text{ft})}{2 \times 10^{-5}\ \text{ft}^2/\text{sec}} = 600\,000$$

indicates a mixed flow, to which (*1*) of Problem 5.70 and Table 5-4 (Problem 5.71) apply.

$$\bar{h} = \left(\frac{0.07\ \text{Btu/hr}\cdot\text{ft}\cdot°\text{F}}{15\ \text{ft}}\right)(31)^{1/3}[0.037(6 \times 10^5)^{4/5} - 702] = 12.45\ \text{Btu/hr}\cdot\text{ft}^2\cdot°\text{F}$$

and $$q = \bar{h}A(T_s - T_\infty) = (12.45\ \text{Btu/hr}\cdot\text{ft}^2\cdot°\text{F})[(15\ \text{ft})(3\ \text{ft})][(410 - 390)\ °\text{F}] = 11\,205\ \text{Btu/hr}$$

Note the very small value of $\bar{h}$ for this problem. The Reynolds number is close to those of typical laminar, external-flow, flat-plate problems. See the remarks at the end of Problem 5.85.

5.84[D] A corn chip cooker has the form of a trough with a flat bottom 4.6 m long by 0.9 m wide. The bottom is heated to a uniform surface temperature of 210 °C, and the cooking oil flows along the heated surface with $V_\infty = 0.24$ m/s and $T_\infty = 199$ °C. Calculate the heat transfer rate to the oil by convection, if $\mathbf{Re}_{crit} = 4 \times 10^5$, $\nu = 1.86 \times 10^{-6}$ m²/s, $k = 0.123$ W/m · K, and $\mathbf{Pr} = 31$.

The maximum Reynolds number,

$$\mathbf{Re}_L = \frac{V_\infty L}{\nu} = \frac{(0.24\ \text{m/s})(4.6\ \text{m})}{1.86 \times 10^{-6}\ \text{m}^2/\text{s}} = 5.94 \times 10^5$$

indicates a mixed flow. Applying (*1*) of Problem 5.70 and Table 5-4 (Problem 5.71):

$$\bar{h} = \left(\frac{0.123\ \text{W/m}\cdot\text{K}}{4.6\ \text{m}}\right)(31)^{1/3}[0.037(5.94 \times 10^5)^{4/5} - 702] = 70.30\ \text{W/m}^2\cdot\text{K}$$

whence $$q = \bar{h}A(T_s - T_\infty) = (70.3\ \text{W/m}^2\cdot\text{K})[(4.6\ \text{m})(0.9\ \text{m})][(210 - 199)\ \text{K}] = 3.2\ \text{kW}$$

Refer to the last paragraph of Problem 5.83.

5.85 Rework Problem 5.83 if a turbulence promoter is installed so that the flow is turbulent over the entire length. All other conditions remain unchanged.

Now set $\mathscr{A} = 0$ in (*1*) of Problem 5.70:

$$\bar{h} = \frac{0.07}{15}(31)^{1/3}[0.037(6 \times 10^5)^{4/5}] = 22.74\ \text{Btu/hr}\cdot\text{ft}^2\cdot°\text{F}$$

$$q = (22.74)[(15)(3)](410 - 390) = 20\,470\ \text{Btu/hr}$$

This is an increase of 83% over the answer to Problem 5.83. Clearly, further increases in the convective heat transfer to the oil require increased velocity, a longer trough, etc. Note that the boiling resulting from driving off the water in the chips will *greatly* increase the heat transfer rate.

5.86[D] How rapidly is "heat" convected from the hood of an automobile traveling at an average speed of 65 mph on a day when the ambient air temperature is 60 °F? Approximate the hood as a 4-ft by 4-ft flat plate at 100 °F.

▮ From Table B-4, at $T_f = (60 + 100)/2 = 80\ °F$,

$$\nu = 16.88 \times 10^{-5}\ \text{ft}^2/\text{sec} \qquad k = 0.015\,16\ \text{Btu/hr}\cdot\text{ft}\cdot°\text{F} \qquad \textbf{Pr} = 0.708$$

The maximum Reynolds number is

$$\textbf{Re}_L = \frac{V_\infty L}{\nu} = \frac{[(\frac{65}{60})(88)\ \text{ft/sec}](4\ \text{ft})}{16.88 \times 10^{-5}\ \text{ft}^2/\text{sec}} = 2.259 \times 10^6 \quad (\textit{turbulent})$$

Assuming "average" conditions, $\textbf{Re}_{\text{crit}} = 5 \times 10^5$ and, by (*1*) of Problem 5.70 and Table 5-4,

$$\bar{h} = \left(\frac{0.015\,16\ \text{Btu/hr}\cdot\text{ft}\cdot°\text{F}}{4\ \text{ft}}\right)(0.708)^{1/3}[0.037(2.259 \times 10^6)^{4/5} - 871] = 12.19\ \text{Btu/hr}\cdot\text{ft}^2\cdot°\text{F}$$

$$q = \bar{h}A(T_s - T_\infty) = (12.19\ \text{Btu/hr}\cdot\text{ft}^2\cdot°\text{F})[(4\ \text{ft})(4\ \text{ft})][(100 - 60)\ °\text{F}] = 7800\ \text{Btu/hr}$$

5.87[D] A BMW is traveling at a steady 105 km/h on the Autobahn; the ambient air temperature is 16 °C and the hood of the vehicle is at 38 °C. If the hood approximates a 1.2-m-square flate plate, how much "heat" does it lose per kilometer?

▮ From Table B-4, at $T_f = (16 + 38)/2 = 27\ °C$,

$$\nu = 1.568 \times 10^{-5}\ \text{m}^2/\text{s} \qquad k = 0.026\,22\ \text{W/m}\cdot\text{K} \qquad \textbf{Pr} = 0.708$$

The maximum Reynolds number is

$$\textbf{Re}_L = \frac{V_\infty L}{\nu} = \frac{\left(\dfrac{105 \times 10^3\ \text{m}}{3600\ \text{s}}\right)(1.2\ \text{m})}{1.568 \times 10^{-5}\ \text{m}^2/\text{s}} = 2.232 \times 10^6 \quad (\textit{turbulent})$$

Assuming "average" flow and surface conditions, $\textbf{Re}_{\text{crit}} = 5 \times 10^5$ and, by (*1*) of Problem 5.70 and Table 5-4,

$$\bar{h} = \left(\frac{0.026\,22\ \text{W/m}\cdot\text{K}}{1.2\ \text{m}}\right)(0.708)^{1/3}[0.037(2.232 \times 10^6)^{4/5} - 871] = 69.46\ \text{W/m}^2\cdot\text{K}$$

Then
$$\frac{q}{V_\infty} = \frac{\bar{h}A\,\Delta T}{V_\infty} = \frac{(69.46\ \text{J/s}\cdot\text{m}^2\cdot\text{K})(1.2\ \text{m})^2(22\ \text{K})}{(105\ \text{km})/(3600\ \text{s})} = 75.4\ \text{kJ/km}$$

5.88 An alternative equation to (*1*) of Problem 5.70, for use with *liquids* and based on $\textbf{Re}_{\text{crit}} = 5 \times 10^5$, is the *Whitaker equation*,

$$\overline{\textbf{Nu}} = (0.036)\,\textbf{Pr}^{0.43}\,(\textbf{Re}_L^{4/5} - 9200)\left(\frac{\mu_\infty}{\mu_s}\right)^{1/4} \qquad (1)$$

Here, μ_∞ is evaluated at T_∞; μ_s at T_s; all other properties at T_f. Apply the Whitaker equation to the process of Problem 5.80.

▮ From Table B-3, at $T_\infty = 10\ °C$,

$$\mu_\infty g_c = \rho\nu = (39.09\ \text{lbm/ft}^3)(0.396 \times 10^{-5}\ \text{ft}^2/\text{sec}) = 1.548 \times 10^{-4}\ \text{lbm/ft}\cdot\text{sec}$$

and at $T_s = 30\ °C$,

$$\mu_s g_c = \rho\nu = (37.23\ \text{lbm/ft}^3)(0.376 \times 10^{-5}\ \text{ft}^2/\text{sec}) = 1.400 \times 10^{-4}\ \text{lbm/ft}\cdot\text{sec}$$

Then, by (*1*) (in which the viscosity units will cancel),

$$\bar{h} = \frac{k}{L}(0.036)\,\textbf{Pr}^{0.43}\,(\textbf{Re}_L^{4/5} - 9200)\left(\frac{\mu_\infty}{\mu_s}\right)^{1/4}$$

$$= \left(\frac{0.521\ \text{W/m}\cdot\text{K}}{1\ \text{m}}\right)(0.036)(2.02)^{0.43}[(8.36 \times 10^6)^{4/5} - 9200]\left(\frac{1.548}{1.400}\right)^{1/4} = 8737\ \text{W/m}^2\cdot\text{K}$$

$$\frac{q}{w} = \bar{h}L(T_s - T_\infty) = (8737\ \text{W/m}^2\cdot\text{K})(1\ \text{m})[(30 - 10)\ \text{K}] = 174.7\ \text{kW/m}$$

5.89 Alcohol is heated by passing it over a 25-ft-long flat plate at 100 °F; the alcohol is initially at 40 °F and flows at 3 ft/sec. Assume that the free-stream temperature of the alcohol remains at 40 °F and that the film temperature remains constant. Determine the average heat transfer coefficient, given

$$\nu = 4.01 \times 10^{-5}\ \text{ft}^2/\text{sec} \qquad k = 0.097\ \text{Btu/hr}\cdot\text{ft}\cdot{}^\circ\text{F} \qquad \mathbf{Pr} = 42.2$$

at $T_f = 70\ ^\circ\text{F}$.

▮
$$\mathbf{Re}_L = \frac{V_\infty L}{\nu} = \frac{(3\ \text{ft/sec})(25\ \text{ft})}{4.01 \times 10^{-5}\ \text{ft}^2/\text{sec}} = 1.87 \times 10^6 > 5 \times 10^5 = \mathbf{Re}_{\text{crit}}$$

so that Problems 5.70 and 5.71 apply.

$$\bar{h} = \left(\frac{0.097\ \text{Btu/hr}\cdot\text{ft}\cdot{}^\circ\text{F}}{25\ \text{ft}}\right)(42.2)^{1/3}[0.037(1.87 \times 10^6)^{4/5} - 871] = 40.3\ \text{Btu/hr}\cdot\text{ft}^2\cdot{}^\circ\text{F}$$

5.90[D] For liquid benzene at 70 °F flowing with a velocity of 1 ft/sec over a 20-ft-long flat plate which is maintained at 90 °F, determine the average heat transfer coefficient. The fluid properties at $T_f = 80\ ^\circ\text{F}$ are

$$\nu = 0.725 \times 10^{-5}\ \text{ft}^2/\text{sec} \qquad k = 0.092\ \text{Btu/hr}\cdot\text{ft}\cdot{}^\circ\text{F} \qquad \mathbf{Pr} = 6.5$$

▮
$$\mathbf{Re}_L = \frac{V_\infty L}{\nu} = \frac{(1\ \text{ft/sec})(20\ \text{ft})}{0.725 \times 10^{-5}\ \text{ft}^2/\text{sec}} = 2.759 \times 10^6 > 5 \times 10^5 = \mathbf{Re}_{\text{crit}}$$

so that Problems 5.70 and 5.71 apply.

$$\bar{h} = \left(\frac{0.092\ \text{Btu/hr}\cdot\text{ft}\cdot{}^\circ\text{F}}{20\ \text{ft}}\right)(6.5)^{1/3}[0.037(2.759 \times 10^6)^{4/5} - 871] = 37.66\ \text{Btu/hr}\cdot\text{ft}^2\cdot{}^\circ\text{F}$$

5.91[D] For liquid benzene at $21\ ^\circ\text{C} = T_\infty$, flowing at 0.6 m/s over a 6-m-long flat plate maintained at 33 °C, determine the average heat transfer coefficient. The fluid properties at $T_f = 27^\circ\ \text{C}$ are

$$\nu = 6.735 \times 10^{-7}\ \text{m}^2/\text{s} \qquad k = 0.159\ \text{W/m}\cdot\text{K} \qquad \mathbf{Pr} = 6.5$$

▮
$$\mathbf{Re}_L = \frac{V_\infty L}{\nu} = \frac{(0.6\ \text{m/s})(6\ \text{m})}{6.735 \times 10^{-7}\ \text{m}^2/\text{s}} = 5.345 \times 10^6 > 5 \times 10^5 = \mathbf{Re}_{\text{crit}}$$

so that Problems 5.70 and 5.71 apply.

$$\bar{h} = \left(\frac{0.159\ \text{W/m}^2\cdot\text{K}}{6\ \text{m}}\right)(6.5)^{1/3}[0.037(5.345 \times 10^6)^{4/5} - 871] = 398.3\ \text{W/m}^2\cdot\text{K}$$

5.92 For Problem 5.91, calculate the critical length. Then approximate the rate of heat transfer by treating the problem as though the entire plate length experiences turbulent flow.

▮ The critical length is

$$x_{\text{crit}} = L\,\frac{\mathbf{Re}_{\text{crit}}}{\mathbf{Re}_L} = (6\ \text{m})\left(\frac{5 \times 10^5}{5.345 \times 10^6}\right) = 0.561\ \text{m}$$

which is less than 10% of plate length. By (*1*) of Problem 5.70, with $\mathscr{A} = 0$,

$$\bar{h} = \left(\frac{0.159\ \text{W/m}\cdot\text{K}}{6\ \text{m}}\right)(6.5)^{1/3}[0.037(5.345 \times 10^6)^{4/5}] = 441\ \text{W/m}^2\cdot\text{K}$$

This value, while acceptable for many engineering applications, is about 10.7 percent above the better value obtained in Problem 5.91.

5.93[D] Saturated liquid Freon-12 at $T_\infty = 92\ ^\circ\text{F}$ flows over a smooth plate at 12 ft/sec. The plate is 5 ft long and held at 116 °F. What is the average rate of heat transfer per square foot to the Freon?

▮ From Table B-3, at $T_f = (116 + 92)/2 = 104\ ^\circ\text{F}$,

$$\nu = 0.206 \times 10^{-5}\ \text{ft}^2/\text{sec} \qquad k = 0.040\ \text{Btu/hr}\cdot\text{ft}\cdot{}^\circ\text{F} \qquad \mathbf{Pr} = 3.5$$

Assuming $\mathbf{Re}_{\text{crit}} = 5 \times 10^5$,

$$\frac{x_{\text{crit}}}{L} = \frac{\nu\,\mathbf{Re}_{\text{crit}}}{V_\infty L} = \frac{(0.206 \times 10^{-5})(5 \times 10^5)}{(12)(5)} = 0.0172$$

With only 1.7 percent of the plate experiencing laminar flow, the flow can be considered to be turbulent over the entire length. Then, (*1*) of Problem 5.70, with $\mathscr{A} = 0$ and

$$\mathbf{Re}_L = \frac{\mathbf{Re}_{\text{crit}}}{x_{\text{crit}}/L} = \frac{5 \times 10^5}{0.0172} = 2.91 \times 10^7$$

yields

$$\bar{h} = (0.037)\left(\frac{0.040\ \text{Btu/hr}\cdot\text{ft}\cdot{}^\circ\text{F}}{5\ \text{ft}}\right)(2.91\times10^7)^{4/5}(3.5)^{1/3} = 420\ \text{Btu/hr}\cdot\text{ft}^2\cdot{}^\circ\text{F}$$

whence

$$\frac{q}{A} = \bar{h}(T_s - T_\infty) = (420)(116-92) = 10\,080\ \text{Btu/hr}\cdot\text{ft}^2$$

5.94[D] Saturated liquid Freon-12 at $T_\infty = 33\ ^\circ\text{C}$ flows over a smooth plate at $V_\infty = 3.66\ \text{m/s}$. The plate is 1.52 m long and held at 47 °C. What is the average rate of heat transfer per square meter of plate to the Freon?

▌ From Table B-3, at $T_f = (33+47)/2 = 40\ ^\circ\text{C}$,

$$\nu = 1.914\times10^{-7}\ \text{m}^2/\text{s} \qquad k = 0.0692\ \text{W/m}\cdot\text{K} \qquad \mathbf{Pr} = 3.5$$

Assuming $\mathbf{Re}_{\text{crit}} = 5\times10^5$,

$$\frac{x_{\text{crit}}}{L} = \frac{\nu\,\mathbf{Re}_{\text{crit}}}{V_\infty L} = \frac{(1.914\times10^{-7})(5\times10^5)}{(3.66)(1.52)} = 0.0172$$

With only 1.7 percent of the plate exposed to laminar flow, we may treat the flow as turbulent over the entire length. Then (*1*) of Problem 5.70, with $\mathscr{A} = 0$ and

$$\mathbf{Re}_L = \frac{\mathbf{Re}_{\text{crit}}}{x_{\text{crit}}/L} = \frac{5\times10^5}{0.0172} = 2.91\times10^7$$

gives

$$\bar{h} = \left(\frac{0.0692\ \text{W/m}\cdot\text{K}}{1.52\ \text{m}}\right)(3.5)^{1/3}[0.037(2.91\times10^7)^{4/5}] = 2.39\ \text{kW/m}^2\cdot\text{K}$$

$$\bar{h} = 2391\ \text{W/m}^2\cdot\text{K} \approx 421.4\ \text{Btu/hr}\cdot\text{ft}^2\cdot{}^\circ\text{F}$$

whence

$$\frac{q}{A} = \bar{h}(T_s - T_\infty) = (2.39\ \text{kW/m}^2\cdot\text{K})[(47-33)\ \text{K}] = 33.46\ \text{kW/m}^2$$

5.95 For parallel flow of air at 1.2 atm pressure, $T_\infty = 16\ ^\circ\text{C}$, and $V_\infty = 3.05\ \text{m/s}$ over a flat plate with a surface temperature of 38 °C, determine the average heat transfer coefficient for a plate length of 3.5 m.

▌ From Table B-4, at $T_f = (16+38)/2 = 27\ ^\circ\text{C}$,

$$\rho_1 = 1.177\ \text{kg/m}^3\ \ \text{(at 1 atm)} \qquad \mu = 1.847\times10^{-5}\ \text{Pa}\cdot\text{s}$$
$$k = 2.622\times10^{-2}\ \text{W/m}\cdot\text{K} \qquad \mathbf{Pr} = 0.708$$

With the air density at T_f given by

$$\rho = \rho_1\left(\frac{p}{p_1}\right) = (1.177)(1.2) = 1.412\ \text{kg/m}^3$$

the maximum Reynolds number has the value

$$\mathbf{Re}_L = \frac{\rho V_\infty L}{\mu} = \frac{(1.412\ \text{kg/m}^3)(3.05\ \text{m/s})(3.5\ \text{m})}{1.847\times10^{-5}\ \text{Pa}\cdot\text{s}} = 8.161\times10^5 > 5\times10^5 = \mathbf{Re}_{\text{crit}}$$

Thus, by Problems 5.70 and 5.71,

$$\bar{h} = \left(\frac{2.622\times10^{-2}\ \text{W/m}\cdot\text{K}}{3.5\ \text{m}}\right)(0.708)^{1/3}[0.037(8.161\times10^5)^{4/5} - 871] = 7.43\ \text{W/m}^2\cdot\text{K}$$

5.96 Determine the local heat transfer coefficient at the end of the plate of Problem 5.95.

▌ By (*3*) of Problem 5.62,

$$h_L = (0.0296)\left(\frac{0.026\,22\ \text{W/m}\cdot\text{K}}{3.5\ \text{m}}\right)(8.161\times10^5)^{4/5}(0.708)^{1/3} = 10.60\ \text{W/m}^2\cdot\text{K}$$

5.97[D] CO_2 gas at 16.2 psia, $V_\infty = 12\ \text{ft/sec}$, and $T_\infty = 300\ ^\circ\text{F}$ flows over a flat plate which is maintained at $T_s = 220\ ^\circ\text{F}$ by a cooling liquid flow on the side opposite the gas flow; the plate is 7 ft long. Transition to turbulence is observed (by use of a hot-wire anemometer measuring the boundary layer velocity profile) 3.2 ft from the leading edge. What is the average heat transfer coefficient over the length of the plate?

From Table B-4, at $T_f = (300 + 220)/2 = 260\ °F$,

$$\rho_1 = 0.0838\ \text{lbm/ft}^3 \quad \text{(at 14.7 psia)} \qquad k = 0.014\,22\ \text{Btu/hr}\cdot\text{ft}\cdot°\text{F}$$
$$\mu g_c = 12.98 \times 10^{-6}\ \text{lbm/ft}\cdot\text{sec} \qquad \mathbf{Pr} = 0.738$$

Then, by the ideal gas law,

$$\rho = \rho_1\left(\frac{p}{p_1}\right) = (0.0838)\left(\frac{16.2}{14.7}\right) = 0.0924\ \text{lbm/ft}^3$$

whence
$$\mathbf{Re}_{\text{crit}} = \frac{\rho V_\infty x_{\text{crit}}}{\mu g_c} = \frac{(0.0924\ \text{lbm/ft}^3)(12\ \text{ft/sec})(3.2\ \text{ft})}{12.98 \times 10^{-6}\ \text{lbm/ft}\cdot\text{sec}} = 273\,200$$

To the mixed flow we apply (*1*) of Problem 5.70, wherein

$$\mathbf{Re}_L = \mathbf{Re}_{\text{crit}}\left(\frac{L}{x_{\text{crit}}}\right) = (273\,200)\left(\frac{7}{3.2}\right) = 597\,650$$

and $\mathscr{A} = (0.037)(273\,200)^{4/5} - (0.664)(273\,200)^{1/2} = 480$, from (*2*) of Problem 5.70. Then

$$\bar{h} = \left(\frac{0.014\,22\ \text{Btu/hr}\cdot\text{ft}\cdot°\text{F}}{7.0\ \text{ft}}\right)(0.738)^{1/3}[0.037(597\,650)^{4/5} - 480] = 1.96\ \text{Btu/hr}\cdot\text{ft}^2\cdot°\text{F}$$

5.98[D] CO_2 gas at 111.7 kPa, $V_\infty = 3.66$ m/s, and $T_\infty = 149\ °C$ flows over a flat plate which is maintained at $T_s = 105\ °C$ by a cooling liquid flow on the side opposite to the gas flow; the plate is 2.13 m long. Transition to turbulence occurs 0.975 m from the leading edge. What is the average heat transfer coefficient over the entire plate length?

From Table B-4, at $T_f = (149 + 105)/2 = 127\ °C$,

$$\rho_1 = (0.0838)(1.6018 \times 10) = 1.342\ \text{kg/m}^3 \quad \text{(at 101.3 kPa)} \qquad k = (0.014\,22)(1.729\,577) = 0.024\,59\ \text{W/m}\cdot\text{K}$$
$$\mu = (12.98 \times 10^{-6})(1.488\,164) = 1.931 \times 10^{-5}\ \text{Pa}\cdot\text{s} \qquad \mathbf{Pr} = 0.738$$

Then the pressure-corrected density is

$$\rho = (1.342)\left(\frac{111.7}{101.3}\right) = 1.479\ \text{kg/m}^3$$

and
$$\mathbf{Re}_{\text{crit}} = \frac{\rho V_\infty x_{\text{crit}}}{\mu} = \frac{(1.479\ \text{kg/m}^3)(3.66\ \text{m/s})(0.975\ \text{m})}{1.931 \times 10^{-5}\ \text{Pa}\cdot\text{s}} = 2.733 \times 10^5$$

To the mixed flow we apply (*1*) of Problem 5.70, wherein

$$\mathbf{Re}_L = \mathbf{Re}_{\text{crit}}\left(\frac{L}{x_{\text{crit}}}\right) = (2.733 \times 10^5)\left(\frac{2.13}{0.975}\right) = 5.971 \times 10^5$$

and $\mathscr{A} = (0.037)(2.733 \times 10^5)^{4/5} - (0.664)(2.733 \times 10^5)^{1/2} = 480$, from (*2*) of Problem 5.70. Then

$$\bar{h} = \left(\frac{0.024\,59\ \text{W/m}\cdot\text{K}}{2.13\ \text{m}}\right)(0.738)^{1/3}[0.037(5.971 \times 10^5)^{4/5} - 480] = 11.11\ \text{W/m}^2\cdot\text{K}$$

5.99 What fractional error in $\bar{h}$ results if, in Problem 5.97, $\mathbf{Re}_{\text{crit}}$ is rounded to 300 000?

By (*1*) of Problem 5.70,

$$\frac{\bar{h}(300\,000) - \bar{h}(273\,200)}{\bar{h}(273\,200)} = \frac{\mathscr{A}(273\,200) - \mathscr{A}(300\,000)}{0.037\ \mathbf{Re}_L^{4/5}} = \frac{480 - 527}{(0.037)(597\,650)^{4/5}} = -0.0304$$

5.100[D] For a parallel flow of N_2 gas at $T_\infty = 66\ °C$ and $V_\infty = 4.0$ m/s over one side of a 6-m-long flat plate having $T_s = 88\ °C$, determine $\bar{h}$ and the rate of heat transfer from the plate per meter of width. The transition to turbulent flow is at $\mathbf{Re}_{\text{crit}} = 4.5 \times 10^5$.

From Table B-4, by interpolation at $T_f = (66 + 88)/2 = 77\ °C$,

$$\nu = (22.26 \times 10^{-5})(9.2903 \times 10^{-2}) = 2.099 \times 10^{-6}\ \text{m}^2/\text{s}$$
$$k = (0.017\,21)(1.729\,577) = 0.029\,77\ \text{W/m}\cdot\text{K}$$
$$\mathbf{Pr} = 0.702$$

As the maximum Reynolds number is

$$\mathbf{Re}_L = \frac{V_\infty L}{\nu} = \frac{(4.0\ \text{m/s})(6\ \text{m})}{2.099 \times 10^{-6}\ \text{m}^2/\text{s}} = 1.143 \times 10^7$$

the flow is mixed, and (*1*) of Problem 5.70 applies. By linear interpolation in Table 5-4,

$$\mathscr{A} \approx \frac{702 + 871}{2} \approx 786$$

and so

$$\bar{h} \approx \left(\frac{0.029\,77\ \text{W/m}\cdot\text{K}}{6\ \text{m}}\right)(0.702)^{1/3}[(1.143 \times 10^7)^{4/5} - 786] = 8.0\ \text{W/m}^2\cdot\text{K}$$

$$\frac{q}{w} = \bar{h}L(T_s - T_\infty) \approx (8.0)(6)(88 - 66) = 1056\ \text{W/m}$$

5.101[D] For a parallel flow of N_2 gas at $T_\infty = 150\ °\text{F}$ and $V_\infty = 13.12\ \text{ft/sec}$ over one side of a 19.7-ft-long plate having $T_s = 190\ °\text{F}$, determine $\bar{h}$ and the rate of heat transfer per foot of width. Assume transition to turbulent flow at $\mathbf{Re}_{\text{crit}} = 450\,000$.

▌ From Table B-4, by interpolation at $T_f = (150 + 190)/2 = 170\ °\text{F}$,

$$\nu = 22.26 \times 10^{-5}\ \text{ft}^2/\text{sec} \qquad k = 0.017\,21\ \text{Btu/hr}\cdot\text{ft}\cdot°\text{F} \qquad \mathbf{Pr} = 0.702$$

The maximum Reynolds number is

$$\mathbf{Re}_L = \frac{V_\infty L}{\nu} = \frac{(13.12\ \text{ft/sec})(19.7\ \text{ft})}{22.26 \times 10^{-5}\ \text{ft}^2/\text{sec}} = 1\,161\,000$$

so the flow is mixed, and (*1*) of Problem 5.70 applies. By linear interpolation in Table 5-4,

$$\mathscr{A} \approx \frac{702 + 871}{2} \approx 786$$

Thus,

$$\bar{h} \approx \left(\frac{0.017\,21\ \text{Btu/hr}\cdot\text{ft}\cdot°\text{F}}{19.7\ \text{ft}}\right)(0.702)^{1/3}[0.037(1\,161\,000)^{4/5} - 786] = 1.43\ \text{Btu/hr}\cdot\text{ft}^2\cdot°\text{F}$$

$$\frac{q}{w} = \bar{h}L(T_s - T_\infty) = (1.43)(19.7)(190 - 150) = 1024\ \text{Btu/hr}\cdot\text{ft}$$

5.102 Parallel to Problem 5.60, tabulate $\bar{h}$ for the previous problems in *gaseous* laminar/turbulent flow over a flat plate.

▌ See Table 5-5.

TABLE 5-5

problem no.	gas	$\bar{h}$ (W/m² · K)	$\bar{h}$ (Btu/hr · ft² · °F)
5.68	H_2	—	41
5.76, 79	Air	47.5	8.37
5.77	Air	52.9	
5.78	Air	61.1	
5.86, 87	Air	69.46	12.19
5.95	Air	7.43	
5.97, 98	CO_2	11.11	1.96
5.100, 101	N_2	8.0	1.43

The entries in Table 5-5 are on the order of 10 times as large as those in Table 5-2.

5.103 Parallel to Problem 5.61, tabulate $\bar{h}$ for the previous problems in *liquid* laminar/turbulent flow over a flat plate.

▌ See Table 5-6. The values exceed those in Table 5-3, but only by a factor of 2 or 3.

TABLE 5-6

problem no.	liquid	$\bar{h}$ (W/m² · K)	$\bar{h}$ (Btu/hr · ft² · °F)
5.73, 75	Water	531	93.5
5.80, 82	Ammonia	7.833	1375
5.83, 84	Cooking oil	70.3	12.4
5.85	Cooking oil	—	22.7
5.88	Ammonia	8737	
5.89	Alcohol	—	40.3
5.90	Benzene	—	37.7
5.91	Benzene	398	
5.92	Benzene	441	
5.93, 94	Freon-12	2390	420

5.104[D] A parallel flow of liquid water at $T_f = 27\ ^\circ\text{C}$ and $V_\infty = 1.2$ m/s over a flat plate 2 m long is such that the heat flux from the plate is constant over the laminar region. Compute h_x for $x = x_{\text{crit}}$, if $\mathbf{Re}_{\text{crit}} = 5 \times 10^5$.

▌ From Table B-3, by interpolation at 27 °C,

$$\nu = 8.842 \times 10^{-7}\ \text{m}^2/\text{s} \qquad k = 0.6076\ \text{W/m} \cdot \text{K} \qquad \mathbf{Pr} = 6.08$$

and so

$$x_{\text{crit}} = \frac{\mathbf{Re}_{\text{crit}}\ \nu}{V_\infty} = \frac{(5 \times 10^5)(8.842 \times 10^{-7}\ \text{m}^2/\text{s})}{1.2\ \text{m/s}} = 0.37\ \text{m}$$

At this location, for a constant q_s over the laminar portion, $(G\text{-}4)$ of Appendix G yields

$$h_{x,\text{crit}} = (0.453)\left(\frac{k}{x_{\text{crit}}}\right)\mathbf{Re}_{\text{crit}}^{1/2}\,\mathbf{Pr}^{1/3} = (0.453)\left(\frac{0.6076\ \text{W/m} \cdot \text{K}}{0.37\ \text{m}}\right)(5 \times 10^5)^{1/2}(6.08)^{1/3} = 960\ \text{W/m}^2 \cdot \text{K}$$

5.105[D] For the flow situation of Problem 5.104, determine h_x for $x = 1.5$ m, again assuming $\mathbf{Re}_{\text{crit}} = 5 \times 10^5$.

▌ The specified location is in the turbulent regime, and the local Reynolds number is

$$\mathbf{Re}_x = \frac{V_\infty x}{\nu} = \frac{(1.2\ \text{m/s})(1.5\ \text{m})}{8.842 \times 10^{-7}\ \text{m}^2/\text{s}} = 2.03 \times 10^6$$

The procedure is first to calculate h_x as if the wall *temperature* were constant, then to correct to the actual condition of constant *heat flux*. Thus, $(G\text{-}5)$ gives

$$h_x = (0.0296)\left(\frac{0.6076\ \text{W/m} \cdot \text{K}}{1.5\ \text{m}}\right)(2.03 \times 10^6)^{4/5}(6.08)^{1/3} = 2435\ \text{W/m}^2 \cdot \text{K}$$

and then $(G\text{-}9)$ yields $h_x = (1.04)(2435) = 2533\ \text{W/m}^2 \cdot \text{K}$.

5.106[D] Consider parallel flow of N_2 gas, at $T_f = 170\ ^\circ\text{F}$, 1 atm pressure, and $V_\infty = 20$ ft/sec, over one side of a 12-ft-long plate. Assume transition to turbulent flow at $\mathbf{Re}_{\text{crit}} = 4 \times 10^5$, and determine the local heat transfer coefficient at $x = 3$ ft for the case of constant heat flux from the plate.

▌ From Table B-4, by interpolation at $T_f = 170\ ^\circ\text{F}$,

$$\nu = 22.26 \times 10^{-5}\ \text{ft}^2/\text{sec} \qquad \mathbf{Pr} = 0.702 \qquad k = 0.017\,21\ \text{Btu/hr} \cdot \text{ft} \cdot {}^\circ\text{F}$$

At $x = 3$ ft:

$$\mathbf{Re}_x = \frac{V_\infty x}{\nu} = \frac{(20)(3)}{22.26 \times 10^{-5}} = 2.695 \times 10^5 \quad (laminar)$$

and equation $(G\text{-}4)$ gives

$$h_x = (0.453)\left(\frac{0.017\,21\ \text{Btu/hr} \cdot \text{ft} \cdot {}^\circ\text{F}}{3\ \text{ft}}\right)(2.695 \times 10^5)^{1/2}(0.702)^{1/3} = 1.199\ \text{Btu/hr} \cdot \text{ft}^2 \cdot {}^\circ\text{F}$$

5.107[D] Repeat Problem 5.106 for $x = 12$ ft.

At $x = 12$ ft, $\mathbf{Re}_x = 1.078 \times 10^6$ (*turbulent*); thus, equations (*G-5*) and (*G-9*) may be combined to give

$$h_x = (1.04)(0.0296)\left(\frac{0.017\,21 \text{ Btu/hr}\cdot\text{ft}\cdot{}^\circ\text{F}}{12 \text{ ft}}\right)(1.078\times10^6)^{4/5}(0.702)^{1/3} = 2.629 \text{ Btu/hr}\cdot\text{ft}^2\cdot{}^\circ\text{F}$$

5.3 CROSS FLOW OVER A CYLINDER

5.108^D Air at 27 °C flows normal to a 30-mm-OD hot-water pipe at 1.0 m/s; the water and pipe are at 77 °C. Estimate the heat transfer rate per lineal meter.

Assuming atmospheric pressure, the required properties (from Table B-4), evaluated at $T_f = (27 + 77)/2 = 52$ °C, are

$$\nu = (19.63\times10^{-5})(0.0929) = 1.824\times10^{-5} \text{ m}^2/\text{s}$$
$$k = (0.016\,255)(1.7296) = 0.0281 \text{ W/m}\cdot\text{K} \qquad \mathbf{Pr} = 0.702$$

The Reynolds number is

$$\mathbf{Re}_D = \frac{V_\infty D}{\nu} = \frac{(1 \text{ m/s})(0.030 \text{ m})}{1.824\times10^{-5} \text{ m}^2/\text{s}} = 1645$$

which defines the approximate constants from Table G-1 to be used in equation (*G-10*); therefore,

$$\bar{h} = \left(\frac{k}{D}\right)(0.683)\,\mathbf{Pr}^{1/3}\,\mathbf{Re}_D^{0.466} = \left(\frac{0.0281 \text{ W/m}\cdot\text{K}}{0.030 \text{ m}}\right)(0.683)(0.702)^{1/3}(1645)^{0.466} = 17.93 \text{ W/m}^2\cdot\text{K}$$

$$\frac{q}{L} = \bar{h}(\pi D)(T_s - T_\infty) = (17.93 \text{ W/m}^2\cdot\text{K})[\pi(0.030 \text{ m})](50 \text{ K}) = 84.49 \text{ W/m}$$

5.109 Repeat Problem 5.108, using the *Zhukauskas equation* (*G-11*).

Assuming atmospheric pressure, properties at $T_\infty = 27$ °C are, from Table B-4:

$$\nu_\infty = (16.88\times10^{-5})(0.0929) = 1.568\times10^{-5} \text{ m}^2/\text{s}$$
$$k_\infty = (0.015\,16)(1.7296) = 0.026\,22 \text{ W/m}\cdot\text{K} \qquad \mathbf{Pr}_\infty = 0.708$$

and, at $T_s = 77$ °C, $\mathbf{Pr}_s = 0.697$. The Reynolds number is now

$$\mathbf{Re}_{D,\infty} = \frac{V_\infty D}{\nu_\infty} = \frac{(1 \text{ m/s})(0.030 \text{ m})}{1.568\times10^{-5} \text{ m}^2/\text{s}} = 1913$$

From Table G-2, $C = 0.26$, $n = 0.6$; and, since $\mathbf{Pr}_\infty < 10$, $m = 0.37$.

$$\bar{h} = \left(\frac{k_\infty}{D}\right)(0.26)\,\mathbf{Re}_{D,\infty}^{0.6}\,\mathbf{Pr}_\infty^{0.37}\left(\frac{\mathbf{Pr}_\infty}{\mathbf{Pr}_s}\right)^{1/4} = \left(\frac{0.026\,22 \text{ W/m}\cdot\text{K}}{0.030 \text{ m}}\right)(0.26)(1913)^{0.6}(0.708)^{0.37}\left(\frac{0.708}{0.697}\right)^{1/4}$$
$$= 18.69 \text{ W/m}^2\cdot\text{K}$$
$$\frac{q}{L} = \frac{18.69}{17.93}(84.49) = 88.07 \text{ W/m}$$

5.110^D Air at 1 atm pressure, $T_\infty = 80$ °F, and $V_\infty = 3.28$ ft/sec flows normal to a hot-water pipe with an OD of 1.2 in. and a surface temperature of 170 °C. Estimate the rate of heat transfer from the pipe to the air per lineal foot.

From Table B-4, by interpolation at $T_f = (80 + 170)/2 = 125$ °F,

$$\nu = 19.63\times10^{-5} \text{ ft}^2/\text{sec} \qquad k = 0.016\,26 \text{ Btu/hr}\cdot\text{ft}^2\cdot{}^\circ\text{F} \qquad \mathbf{Pr} = 0.702$$

The Reynolds number is

$$\mathbf{Re}_D = \frac{V_\infty D}{\nu} = \frac{(3.28 \text{ ft/sec})\left(\frac{1.2}{12} \text{ ft}\right)}{19.63\times10^{-5} \text{ ft}^2/\text{sec}} = 1671$$

For this value, C and n from Table G-1 are $C = 0.683$ and $n = 0.466$; the Hilpert equation, (*G-10*), then gives

$$\bar{h} = \left(\frac{k}{D}\right)(0.683)\,\mathbf{Pr}^{1/3}\,\mathbf{Re}_D^{0.466} = \left(\frac{0.016\,26 \text{ Btu/hr}\cdot\text{ft}\cdot{}^\circ\text{F}}{0.1 \text{ ft}}\right)(0.683)(0.702)^{1/3}(1671)^{0.466} = 3.135 \text{ Btu/hr}\cdot\text{ft}^2\cdot{}^\circ\text{F}$$

$$\frac{q}{L} = \bar{h}(\pi D)(T_s - T_\infty) = (3.135 \text{ Btu/hr}\cdot\text{ft}^2\cdot{}^\circ\text{F})[(\pi)(0.1 \text{ ft})](90 \text{ °F}) = 88.64 \text{ Btu/hr}\cdot\text{ft}$$

5.111^D Air at 27 °C flows normal to a side of a 50-mm by 50-mm square conduit having surface temperature 77 °C; the air moves at $V_\infty = 2$ m/s. Estimate the heat transfer rate per unit length, using the Hilpert equation.

▌ Assuming atmospheric pressure, properties at $T_f = (27 + 77)/2 = 52$ °C are

$$\nu = (19.63 \times 10^{-5})(0.0929) = 1.824 \times 10^{-5}\ \text{m}^2/\text{s}$$
$$k = (0.016\,255)(1.7296) = 0.0281\ \text{W/m}\cdot\text{K}$$
$$\mathbf{Pr} = 0.702$$

The Reynolds number is

$$\mathbf{Re}_D = \frac{V_\infty D}{\nu} = \frac{(2\ \text{m/s})(0.050\ \text{m})}{1.824 \times 10^{-5}\ \text{m}^2/\text{s}} = 5482$$

For this value, the constants from Table G-1 are $C = 0.102$ and $n = 0.675$; equation (*G-10*) gives

$$\bar{h} = \left(\frac{k}{D}\right)(0.102)\,\mathbf{Pr}^{1/3}\,\mathbf{Re}_D^{0.675} = \left(\frac{0.0281\ \text{W/m}\cdot\text{K}}{0.050\ \text{m}}\right)(0.102)(0.702)^{1/3}(5482)^{0.675} = 17.02\ \text{W/m}^2\cdot\text{K}$$

$$\frac{q}{L} = \bar{h}(4D)(T_s - T_\infty) = (17.02\ \text{W/m}^2\cdot\text{K})[4(0.050\ \text{m})][(77 - 27)\ \text{K}] = 170.2\ \text{W/m}$$

5.112 Rework Problem 5.111 for a 50-mm-OD circular conduit.

▌ Properties and the Reynolds number are unchanged from Problem 5.111. From Table G-1, $C = 0.193$ and $n = 0.618$; then, by equation (*G-10*),

$$\bar{h} = \left(\frac{k}{D}\right)(0.193)\,\mathbf{Pr}^{1/3}\,\mathbf{Re}_D^{0.618} = \left(\frac{0.0281\ \text{W/m}\cdot\text{K}}{0.050\ \text{m}}\right)(0.193)(0.702)^{1/3}(5482)^{0.618} = 19.71\ \text{W/m}^2\cdot\text{K}$$

$$\frac{q}{L} = \bar{h}(\pi D)(T_s - T_\infty) = (19.71\ \text{W/m}^2\cdot\text{K})[\pi(0.050\ \text{m})][(77 - 27)\ \text{K}] = 154.8\ \text{W/m}$$

The higher $\bar{h}$ of the circular geometry is overbalanced by its smaller area.

5.113 Rework Problem 5.111 if the square conduit is positioned as in the second line of Table G-1.

▌ Now, by the Pythagorean theorem, $D = (0.050)\sqrt{2} = 0.0707$ m, whence $\mathbf{Re}_D = (5482)\sqrt{2} = 7752$. From Table G-1 we obtain $C = 0.246$, $n = 0.588$; equation (*G-10*) then gives

$$\bar{h} = \left(\frac{k}{D}\right)(0.246)\,\mathbf{Pr}^{1/3}\,\mathbf{Re}_D^{0.588} = \left(\frac{0.0281\ \text{W/m}\cdot\text{K}}{0.0707\ \text{m}}\right)(0.246)(0.702)^{1/3}(7752)^{0.588} = 16.83\ \text{W/m}^2\cdot\text{K}$$

$$\frac{q}{L} = \bar{h}(\text{perimeter})(T_s - T_\infty) = (16.83\ \text{W/m}^2\cdot\text{K})(0.200\ \text{m})(50\ \text{K}) = 168.3\ \text{W/m}$$

5.114^D Air at 80 °F flows normal to a side of a 2-in. by 2-in. square conduit with the surface temperature 170 °F; the air moves at $V_\infty = 6.5$ ft/sec. Estimate the rate of heat transfer per unit length, using the Hilpert equation.

▌ Assuming atmospheric pressure, properties from Table B-4, at $T_f = 125$ °F, are

$$\nu = 19.63 \times 10^{-5}\ \text{ft}^2/\text{sec} \qquad k = 0.016\,255\ \text{Btu/hr}\cdot\text{ft}\cdot{}^\circ\text{F} \qquad \mathbf{Pr} = 0.702$$

The Reynolds number is

$$\mathbf{Re}_D = \frac{V_\infty D}{\nu} = \frac{(6.5\ \text{ft/sec})(\frac{2}{12}\ \text{ft})}{19.63 \times 10^{-5}\ \text{ft}^2/\text{sec}} = 5519$$

At this value of $\mathbf{Re}_D$, the constants for (*G-10*) are given by Table G-1 as $C = 0.102$ and $n = 0.675$; thus,

$$\bar{h} = \left(\frac{k}{D}\right)(0.102)\,\mathbf{Pr}^{1/3}\,\mathbf{Re}_D^{0.675} = \left(\frac{0.016\,255\ \text{Btu/hr}\cdot\text{ft}\cdot{}^\circ\text{F}}{\frac{2}{12}\ \text{ft}}\right)(0.102)(0.702)^{1/3}(5519)^{0.675} = 2.967\ \text{Btu/hr}\cdot\text{ft}^2\cdot{}^\circ\text{F}$$

$$\frac{q}{L} = \bar{h}(4D)(T_s - T_\infty) = (2.967\ \text{Btu/hr}\cdot\text{ft}^2\cdot{}^\circ\text{F})[4(\tfrac{2}{12}\ \text{ft})][(170 - 80)\ {}^\circ\text{F}] = 178.02\ \text{Btu/hr}\cdot\text{ft}$$

5.115^D Atmospheric air at 80 °F flows normal to a steampipe of 4.0 in. outside diameter and with $T_s = 260$ °F; the air velocity is $V_\infty = 5$ ft/sec. Estimate the average heat transfer coefficient, using the Hilpert equation, (*G-10*).

▮ From Table B-4, at 1 atm pressure and $T_f = 170\ °F$,

$$\nu = 22.38 \times 10^{-5}\ \text{ft}^2/\text{sec} \qquad k = 0.017\,35\ \text{Btu/hr}\cdot\text{ft}\cdot°\text{F} \qquad \mathbf{Pr} = 0.697$$

The Reynolds number is

$$\mathbf{Re}_D = \frac{DV_\infty}{\nu} = \frac{(\frac{4}{12}\ \text{ft})(5\ \text{ft/sec})}{22.38 \times 10^{-5}\ \text{ft}^2/\text{sec}} = 7447$$

and, from Table G-1, $C = 0.193$ and $n = 0.618$. Thus,

$$\bar{h} = \left(\frac{k}{D}\right)(0.193)\,\mathbf{Pr}^{1/3}\,\mathbf{Re}_D^{0.618} = \left(\frac{0.017\,35\ \text{Btu/hr}\cdot\text{ft}\cdot°\text{F}}{\frac{4}{12}\ \text{ft}}\right)(0.193)(0.697)^{1/3}(7447)^{0.618} = 2.201\ \text{Btu/hr}\cdot\text{ft}^2\cdot°\text{F}$$

5.116 Compare the average heat transfer coefficient obtained in Problem 5.115 with that from the Zhukauskas equation, (*G-11*).

▮ From Table B-4, at $T_\infty = 80\ °F$ and pressure 1 atm,

$$\nu_\infty = 16.88 \times 10^{-5}\ \text{ft}^2/\text{sec} \qquad k_\infty = 0.015\,16\ \text{Btu/hr}\cdot\text{ft}\cdot°\text{F} \qquad \mathbf{Pr}_\infty = 0.708$$

Also, at $T_s = 260\ °F$, $\mathbf{Pr}_s = 0.689$. The Reynolds number is

$$\mathbf{Re}_{D,\infty} = \frac{DV_\infty}{\nu_\infty} = \frac{(\frac{4}{12}\ \text{ft})(5\ \text{ft/sec})}{16.88 \times 10^{-5}\ \text{ft}^2/\text{sec}} = 9874$$

Then, from Table G-2, $C = 0.26$, $n = 0.6$, and equation (*G-11*) yields ($\mathbf{Pr}_\infty < 10$):

$$\bar{h} = (0.26)\left(\frac{k_\infty}{D}\right)\mathbf{Re}_{D,\infty}^{0.6}\,\mathbf{Pr}_\infty^{0.37}\left(\frac{\mathbf{Pr}_\infty}{\mathbf{Pr}_s}\right)^{1/4} = (0.26)\left(\frac{0.015\,16\ \text{Btu/hr}\cdot\text{ft}\cdot°\text{F}}{\frac{4}{12}\ \text{ft}}\right)(9874)^{0.6}(0.708)^{0.37}\left(\frac{0.708}{0.689}\right)^{1/4}$$

$$= 2.612\ \text{Btu/hr}\cdot\text{ft}^2\cdot°\text{F}$$

This value is

$$\frac{2.612 - 2.201}{2.201} \times 100\% = 18.7\%$$

higher than that given by the Hilpert equation. Although the expected error in either correlation exceeds 25%, (*G-11*) is probably better than (*G-10*).

5.117^D Compare the average heat transfer coefficients obtained in Problems 5.115 and 5.116 with that from the *Churchill–Bernstein equation*, (*G-12*).

▮ By Problem 5.115, $\mathbf{Re}_D\,\mathbf{Pr} = (7447)(0.697) \gg 0.2$, so that (*G-12*) indeed applies.

$$\overline{\mathbf{Nu}} = 0.3 + \frac{(0.62)(7447)^{1/2}(0.697)^{1/3}}{[1 + (0.4/0.697)^{2/3}]^{1/4}}\left[1 + \left(\frac{7447}{28\,200}\right)^{5/8}\right]^{4/5} = 54.05$$

$$\bar{h} = \overline{\mathbf{Nu}}\left(\frac{k}{D}\right) = (54.05)\left(\frac{0.017\,35\ \text{Btu/hr}\cdot\text{ft}\cdot°\text{F}}{\frac{4}{12}\ \text{ft}}\right) = 2.813\ \text{Btu/hr}\cdot\text{ft}^2\cdot°\text{F}$$

This value is 8% larger than the value obtained from the Zhukauskas equation, and it is 28% larger than the value obtained from the Hilpert equation. Such differences are not uncommon for flow over a blunt body.

5.118^D Atmospheric air at 27 °C flows normal to a steampipe of 102 mm outside diameter and with $T_s = 127\ °C$; the air velocity is $V_\infty = 1.52$ m/s. Estimate the average heat transfer coefficient, using the Hilpert equation, (*G-10*).

▮ From Table B-4, at 1 atm pressure and $T_f = 77\ °C$,

$$\nu = (22.38 \times 10^{-5})(9.290\,304 \times 10^{-2}) = 2.079 \times 10^{-5}\ \text{m}^2/\text{s}$$
$$k = (0.017\,35)(1.729\,577) = 0.030\ \text{W/m}\cdot\text{K}$$
$$\mathbf{Pr} = 0.697$$

The Reynolds number is

$$\mathbf{Re}_D = \frac{V_\infty D}{\nu} = \frac{(1.52\ \text{m/s})(0.102\ \text{m})}{2.079 \times 10^{-5}\ \text{m}^2/\text{s}} = 7457$$

Then, from Table G-1, $C = 0.193$, $n = 0.618$, and (G-10) yields

$$\bar{h} = \left(\frac{k}{D}\right)(0.193)\,\mathbf{Pr}^{1/3}\,\mathbf{Re}_D^{0.618} = \left(\frac{0.030\ \text{W/m}\cdot\text{K}}{0.102\ \text{m}}\right)(0.193)(0.697)^{1/3}(7457)^{0.618} = 12.45\ \text{W/m}^2\cdot\text{K}$$

5.119[D] Compare the average heat transfer coefficient obtained in Problem 5.118 with that from the Zhukauskas equation, (G-11).

▌ Assume standard atmospheric pressure. From Table B-4, at $T_\infty = 27\ °\text{C}$,

$$\nu_\infty = (16.88 \times 10^{-5})(9.290\,304 \times 10^{-2}) = 1.568 \times 10^{-5}\ \text{m}^2/\text{s}$$
$$k_\infty = (0.015\,16)(1.729\,577) = 0.026\,22\ \text{W/m}\cdot\text{K} \qquad \mathbf{Pr}_\infty = 0.708$$

Also, at $T_s = 127\ °\text{C}$, $\mathbf{Pr}_s = 0.689$. The Reynolds number is

$$\mathbf{Re}_{D,\infty} = \frac{DV_\infty}{\nu_\infty} = \frac{(0.102\ \text{m})(1.52\ \text{m/s})}{1.568 \times 10^{-5}\ \text{m}^2/\text{s}} = 9887$$

The Zhukauskas equation yields, for $\mathbf{Pr}_\infty < 10$ and with $C = 0.26$ and $n = 0.6$ from Table G-2:

$$\bar{h} = \left(\frac{0.026\,22\ \text{W/m}\cdot\text{K}}{0.102\ \text{m}}\right)(0.26)(9887)^{0.6}(0.708)^{0.37}\left(\frac{0.708}{0.689}\right)^{1/4} = 14.78\ \text{W/m}^2\cdot\text{K}$$

The answer is about 18.7% higher than that obtained with the simpler Hilpert equation in Problem 5.118.

5.120[D] Compare the average heat transfer coefficients obtained in Problems 5.118 and 5.119 with that obtainable with the Churchill–Bernstein equation, (G-12).

▌ From Problem 5.118, $\mathbf{Re}_D\,\mathbf{Pr} = (7457)(0.697) \gg 0.2$, so that ($G$-$12$) indeed applies.

$$\overline{\mathbf{Nu}} = 0.3 + \frac{(0.62)(7457)^{1/2}(0.697)^{1/3}}{[1 + (0.4/0.697)^{2/3}]^{1/4}}\left[1 + \left(\frac{7457}{28\,200}\right)^{5/8}\right]^{4/5} = 55.89$$

$$\bar{h} = \frac{k}{D}\,\overline{\mathbf{Nu}} = \frac{0.030\ \text{W/m}\cdot\text{K}}{0.102\ \text{m}}(55.89) = 16.44\ \text{W/m}^2\cdot\text{K}$$

This value is 32% larger than that given by the Hilpert equation and 11% larger than that given by the Zhukauskas equation.

5.121 Saturated water at $T_\infty = 10\ °\text{C}$ flows at $V_\infty = 3.33\ \text{m/s}$ across a 15-mm-OD cylinder which has a temperature of 60 °C. Determine the heat transfer rate per unit length of the cylinder, using the Churchill–Bernstein equation.

▌ From Table B-3, for saturated water at $T_f = 35\ °\text{C}$,

$$\nu = (0.8018 \times 10^{-5})(9.290\,304 \times 10^{-2}) = 0.074\,49 \times 10^{-5}\ \text{m}^2/\text{s}$$
$$k = (0.3585)(1.729\,577) = 0.6201\ \text{W/m}\cdot\text{K} \qquad \mathbf{Pr} = 5.01$$

The Reynolds number is

$$\mathbf{Re}_D = \frac{V_\infty D}{\nu} = \frac{(3.33\ \text{m/s})(0.015\ \text{m})}{0.074\,49 \times 10^{-5}\ \text{m}^2/\text{s}} = 67\,055$$

Then, by (G-12),

$$\overline{\mathbf{Nu}} = 0.3 + \frac{(0.62)(67\,055)^{1/2}(5.01)^{1/3}}{[1 + (0.4/5.01)^{2/3}]^{1/4}}\left[1 + \left(\frac{67\,055}{28\,200}\right)^{5/8}\right]^{4/5} = 586$$

$$\bar{h} = \left(\frac{k}{D}\right)\overline{\mathbf{Nu}} = \left(\frac{0.6201\ \text{W/m}\cdot\text{K}}{0.015\ \text{m}}\right)(586) = 24.24\ \text{kW/m}^2\cdot\text{K}$$

$$\frac{q}{L} = \bar{h}(\pi D)(T_s - T_\infty) = (24.24\ \text{kW/m}^2\cdot\text{K})[\pi(0.015\ \text{m})][(60 - 10)\ \text{K}] = 57.1\ \text{kW/m}$$

5.122 Repeat Problem 5.121 for engine oil instead of water. (Note that at 35 °C, engine oil has a very large **Pr**, and the Zhukauskas equation is not recommended for $\mathbf{Pr} > 500$.)

▌ From Table B-3, for saturated engine oil at $T_f = 35\ °\text{C}$,

$$\nu = (0.004\,375)(9.290\,304 \times 10^{-2}) = 4.064 \times 10^{-4}\ \text{m}^2/\text{s}$$
$$k = (0.083\,25)(1.729\,577) = 0.1445\ \text{W/m}\cdot\text{K} \qquad \mathbf{Pr} = 4752$$

The Reynolds number is

$$\mathbf{Re}_D = \frac{V_\infty D}{\nu} = \frac{(3.33\ \text{m/s})(0.015\ \text{m})}{4.064\times 10^{-4}\ \text{m}^2/\text{s}} = 123$$

Clearly, $\mathbf{Re}_D\,\mathbf{Pr} > 0.2$; so the Churchill–Bernstein equation applies.

$$\overline{\mathbf{Nu}} = 0.3 + \frac{(0.62)(123)^{1/2}(4752)^{1/3}}{[1+(0.4/4752)^{2/3}]^{1/4}}\left[1+\left(\frac{123}{28\,200}\right)^{5/8}\right]^{4/5} = 119$$

$$\bar{h} = \frac{k}{D}\,\overline{\mathbf{Nu}} = \frac{0.1445\ \text{W/m}\cdot\text{K}}{0.015\ \text{m}}(119) = 1146\ \text{W/m}^2\cdot\text{K}$$

$$\frac{q}{L} = \bar{h}(\pi D)(T_s - T_\infty) = (1146\ \text{W/m}^2\cdot\text{K})[\pi(0.015\ \text{m})][(60-10)\ \text{K}] = 2700\ \text{W/m}$$

5.123[D] A circular metal cylinder of 30 mm outside diameter and initially of temperature 160 °C is quenched in a bath of engine oil at 60 °C. The oil moves in cross flow over the cylinder at 1.67 m/s. Determine the initial rate of heat transfer from the cylinder per unit length.

▌ On the assumption that a steady state is immediately established, (*G-10*), (*G-11*), and (*G-12*) apply. Choosing the last, we have, by interpolation in Table B-3 at $T_f = 110$ °C,

$$\nu = (0.176\times 10^{-3})(9.290\,304\times 10^{-2}) = 1.635\times 10^{-5}\ \text{m}^2/\text{s}$$

$$k = (0.0785)(1.729\,577) = 0.1358\ \text{W/m}\cdot\text{K} \qquad \mathbf{Pr} = 226$$

Then, the Reynolds number is

$$\mathbf{Re}_D = \frac{DV_\infty}{\nu} = \frac{(0.030\ \text{m})(1.67\ \text{m/s})}{1.635\times 10^{-5}\ \text{m}^2/\text{s}} = 3064$$

and (*G-12*) gives

$$\overline{\mathbf{Nu}} = 0.3 + \frac{(0.62)(3064)^{1/2}(226)^{1/3}}{[1+(0.4/226)^{2/3}]^{1/4}}\left[1+\left(\frac{3064}{28\,200}\right)^{5/8}\right]^{4/5} = 249.3$$

$$\bar{h} = \left(\frac{k}{D}\right)\overline{\mathbf{Nu}} = \left(\frac{0.1358\ \text{W/m}\cdot\text{K}}{0.030\ \text{m}}\right)(249.3) = 1128\ \text{W/m}^2\cdot\text{K}$$

$$\frac{q}{L} = \bar{h}(\pi D)(T_s - T_\infty) = (1128\ \text{W/m}^2\cdot\text{K})[\pi(0.030\ \text{m})][(160-60)\ \text{K}] = 10.6\ \text{kW/m}$$

5.124[D] Repeat Problem 5.123 using **(*a*)** (*G-10*), and **(*b*)** (*G-11*).

▌ **(*a*)** For $\mathbf{Re}_D = 3064$, Table G-1 gives $C = 0.683$, $n = 0.466$; therefore,

$$\bar{h} = \left(\frac{0.1358\ \text{W/m}\cdot\text{K}}{0.030\ \text{m}}\right)(0.683)(226)^{1/3}(3064)^{0.466} = 793.5\ \text{W/m}^2\cdot\text{K}$$

$$\frac{q}{L} = \bar{h}(\pi D)(T_s - T_\infty) = (793.5\ \text{W/m}^2\cdot\text{K})[\pi(0.030\ \text{m})](100\ \text{K}) = 7.48\ \text{kW/m}$$

(*b*) For the Zhukauskas equation, all properties except $\mathbf{Pr}_s$ are evaluated at $T_\infty = 60$ °C:

$$\nu_\infty = (0.903\times 10^{-3})(9.290\,304\times 10^{-2}) = 8.389\times 10^{-5}\ \text{m}^2/\text{s}$$

$$k_\infty = (0.081)(1.729\,577) = 0.1401\ \text{W/m}\cdot\text{K} \qquad \mathbf{Pr}_\infty = 1050$$

At $T_s = 160$ °C, $\mathbf{Pr}_s = 84$. For

$$\mathbf{Re}_{D,\infty} = \frac{V_\infty D}{\nu_\infty} = \frac{(1.67\ \text{m/s})(0.030\ \text{m})}{8.389\times 10^{-5}\ \text{m}^2/\text{s}} = 597.2$$

Table G-2 gives $C = 0.51$, $n = 0.5$; therefore

$$\bar{h} = \left(\frac{k_\infty}{D}\right)(0.51)\,\mathbf{Re}_{D,\infty}^{0.5}\,\mathbf{Pr}_\infty^{0.36}\left(\frac{\mathbf{Pr}_\infty}{\mathbf{Pr}_s}\right)^{1/4} = \left(\frac{0.1401\ \text{W/m}\cdot\text{K}}{0.030\ \text{m}}\right)(597)^{0.5}(1050)^{0.36}\left(\frac{1050}{84}\right)^{1/4} = 1339\ \text{W/m}\cdot\text{K}$$

$$\frac{q}{L} = \bar{h}(\pi D)(T_s - T_\infty) = (1339\ \text{W/m}\cdot\text{K})[\pi(0.030\ \text{m})](100\ \text{K}) = 12.6\ \text{kW/m}$$

The heat flux is suspiciously large. Checking back, one finds that $\mathbf{Pr}_\infty$ exceeds the maximum recommended for use of the Zhukauskas equation. The Churchill–Bernstein equation probably yields the best result for this problem.

5.125ᴰ A circular cylinder 0.10 ft in outside diameter and initially of temperature 320 °F is quenched in a bath of engine oil at 140 °F. The oil moves in cross flow over the cylinder at 5.5 ft/sec. Determine the initial rate of heat transfer from the cylinder per unit length.

▌ From Table B-3, at $T_f = 230$ °F,

$$\nu = 0.176 \times 10^{-3}\ \text{ft}^2/\text{sec} \qquad k = 0.0785\ \text{Btu/hr}\cdot\text{ft}\cdot°\text{F} \qquad \textbf{Pr} = 226$$

Also, from Table B-3 at $T_\infty = 140$ °F, we have $\textbf{Pr}_\infty = 1050$, which invalidates the use of equation (*G-11*). Using the Churchill–Bernstein equation, with

$$\textbf{Re}_D = \frac{V_\infty D}{\nu} = \frac{(5.5\ \text{ft/sec})(0.10\ \text{ft})}{0.176 \times 10^{-3}\ \text{ft}^2/\text{sec}} = 3125$$

we find

$$\overline{\textbf{Nu}} = 0.3 + \frac{(0.62)(3125)^{1/2}(226)^{1/3}}{[1 + (0.4/226)^{2/3}]^{1/4}}\left[1 + \left(\frac{3125}{28\,200}\right)^{5/8}\right]^{4/5} = 252.2$$

$$\bar{h} = \frac{k}{D}\overline{\textbf{Nu}} = \frac{0.0785\ \text{Btu/hr}\cdot\text{ft}\cdot°\text{F}}{0.10\ \text{ft}}(252.2) = 198\ \text{Btu/hr}\cdot\text{ft}^2\cdot°\text{F}$$

$$\frac{q}{L} = \bar{h}(\pi D)(T_s - T_\infty) = (198\ \text{Btu/hr}\cdot\text{ft}^2\cdot°\text{F})[\pi(0.10\ \text{ft})][(320 - 140)\ °\text{F}] = 11\,196\ \text{Btu/hr}\cdot\text{ft}$$

5.126ᴰ Repeat Problem 5.125 using the Hilpert correlation equation.

▌ At $\textbf{Re}_D = 3125$, $C = 0.683$ and $n = 0.466$, from Table G-1. Hence,

$$\bar{h} = \left(\frac{k}{D}\right)(0.683)\,\textbf{Pr}^{1/3}\,(\textbf{Re}_D)^{0.466} = \left(\frac{0.0785}{0.10}\right)(0.683)(226)^{1/3}(3125)^{0.466} = 139\ \text{Btu/hr}\cdot\text{ft}^2\cdot°\text{F}$$

$$\frac{q}{L} = \bar{h}(\pi D)(T_s - T_\infty) = (139)[\pi(0.10)](320 - 140) = 7860\ \text{Btu/hr}\cdot\text{ft}$$

5.127ᴰ Hot-wire anemometry depends on the convective heat transfer from forced flow across a small-diameter wire. What is the estimated velocity of an airstream at 27 °C and atmospheric pressure across a 0.1-mm wire, if an electric energy input of 10 W/m is required to maintain a surface temperature of 77 °C? Use the Hilpert equation.

▌ By an energy balance on the wire,

$$\bar{h} = \frac{k}{D}\,C\,\textbf{Pr}^{1/3}\,V_\infty^n\left(\frac{D}{\nu}\right)^n = \frac{q/L}{\pi D(T_s - T_\infty)}$$

Solving for V_∞,

$$V_\infty = \left[\frac{q/L}{\pi(T_s - T_\infty)kC\,\textbf{Pr}^{1/3}}\right]^{1/n}\frac{\nu}{D} \tag{1}$$

Because C and n in (*1*) depend on V_∞ via the Reynolds number, the equation must be solved by iteration. From Table B-4, at atmospheric pressure and $T_f = 52$ °C,

$$\nu = 1.824 \times 10^{-5}\ \text{m}^2/\text{s} \qquad k = 0.028\,11\ \text{W/m}\cdot\text{K} \qquad \textbf{Pr} = 0.703$$

For the first iteration guess a low Reynolds number for the fine wire—say, $4 < \textbf{Re}_D^{(0)} < 40$—to obtain $C = 0.911$ and $n = 0.385$ from Table G-1.

$$V_\infty^{(1)} = \left[\frac{10}{\pi(50)(0.028\,11)(0.911)(0.703)^{1/3}}\right]^{1/0.385}\frac{1.824 \times 10^{-5}}{0.0001} = 2.63\ \text{m/s}$$

Because $\textbf{Re}_D^{(1)} = V_\infty^{(1)}D/\nu = (2.63)(0.0001)/(1.824 \times 10^{-5}) = 14.4$ does in fact lie between 4 and 40, $V_\infty^{(1)}$ may be accepted as the approximate solution.

5.128ᴰ Use the Zhukauskas equation to confirm the approximate velocity calculated in Problem 5.127.

▌ From the Zhukauskas equation one readily obtains the following analog to (*1*) of Problem 5.127:

$$V_\infty = \left[\frac{q/L}{\pi(T_s - T_\infty)Ck_\infty\,\textbf{Pr}_\infty^m\,(\textbf{Pr}_\infty/\textbf{Pr}_s)^{1/4}}\right]^{1/n}\frac{\nu_\infty}{D} \tag{1}$$

On the basis of Problem 5.127 assume that $1 < \mathbf{Re}_{D,\infty}^{(0)} < 40$. From Table G-1, $C = 0.75$, $n = 0.4$. From Table B-4, at atmospheric pressure and $T_\infty = 27\ °C$,

$$\nu_\infty = 1.568 \times 10^{-5}\ m^2/s \qquad k_\infty = 0.026\,22\ W/m \cdot K \qquad \mathbf{Pr}_\infty = 0.708 \quad (\text{so} \quad m = 0.37)$$

and, at $T_s = 77\ °C$, $\mathbf{Pr}_s = 0.697$. Substituting in (*1*),

$$V_\infty^{(1)} = \left[\frac{10}{\pi(50)(0.75)(0.026\,22)(0.708)^{0.37}(0.708/0.697)^{1/4}}\right]^{1/0.4} \frac{1.568 \times 10^{-5}}{0.0001} = 4.03\ m/s$$

Because $\mathbf{Re}_{D,\infty}^{(1)} = V_\infty^{(1)} D/\nu_\infty = (4.03)(0.0001)/(1.568 \times 10^{-5}) = 25.7$ does in fact lie between 1 and 40, $V_\infty^{(1)}$ may be accepted as the approximate solution. The larger V_∞ given by the Zhukauskas equation is probably more accurate than the value calculated from the Hilpert equation.

5.129^{D} Hot-wire anemometry is based on the convective heat transfer from a thin wire to a gas in cross flow. Estimate the velocity of an 80 °F airstream at 1 atm pressure flowing across a 0.004-in.-diameter wire which is maintained at 170 °F by an electrical input of 3.05 W/ft of length. Use the Zhukauskas equation.

I Use (*1*) of Problem 5.128, with the initial guess $1 < \mathbf{Re}_{D,\infty}^{(0)} < 40$; from Table G-2, $C = 0.75$ and $n = 0.4$. Also, from Table B-4 at $T_\infty = 80\ °F$ and $T_s = 170\ °F$,

$$\nu_\infty = 16.88 \times 10^{-5}\ ft^2/sec \qquad k_\infty = 0.015\,16\ Btu/hr \cdot ft \cdot °F \qquad \mathbf{Pr}_\infty = 0.708 \qquad \mathbf{Pr}_s = 0.697$$

Then,

$$V_\infty^{(1)} = \left[\frac{(3.05\ W/ft)(3.414\ Btu/hr \cdot W)}{\pi(90\ °F)(0.75)(0.015\,16\ Btu/hr \cdot ft \cdot °F)(0.708)^{0.37}(0.708/0.697)^{1/4}}\right]^{1/0.4} \frac{16.88 \times 10^{-5}\ ft^2/sec}{(0.004/12)\ ft}$$
$$= 13.0\ ft/sec$$

Because $\mathbf{Re}_{D,\infty}^{(1)} = V_\infty^{(1)} D/\nu_\infty = (13.0)(0.004/12)/16.88 \times 10^{-5} = 25.7$ does in fact lie between 1 and 40, $V_\infty^{(1)}$ may be accepted as the approximate solution.

5.130^{D} For the cross flow of air at 80 °F and $V_\infty = 8.63\ ft/sec$ over a 0.004-in.-diameter wire, use the Hilpert equation to estimate the heat transfer rate per foot of length, if the wire is maintained at 170 °F. (As 8.63 ft/sec = 2.63 m/s, this problem is a converse to Problem 5.127.)

I From Table B-4, assuming 1 atm pressure and with $T_f = 125\ °F$,

$$\nu = 19.63 \times 10^{-5}\ ft^2/sec \qquad k = 0.016\,26\ Btu/hr \cdot ft \cdot °F \qquad \mathbf{Pr} = 0.703$$

The Reynolds number is

$$\mathbf{Re}_D = \frac{DV_\infty}{\nu} = \frac{\left(\frac{0.004}{12\ ft}\right)(8.63\ ft/sec)}{19.63 \times 10^{-5}\ ft^2/sec} = 14.65$$

From Table G-1, $C = 0.911$ and $n = 0.385$; hence

$$\bar{h} = \frac{k}{D}(0.911)(\mathbf{Pr})^{1/3}(\mathbf{Re}_D)^{0.385} = \frac{0.016\,26\ Btu/hr \cdot ft \cdot °F}{(0.004/12)\ ft}(0.911)(0.703)^{1/3}(14.65)^{0.385} = 111.1\ Btu/hr \cdot ft^2 \cdot °F$$

$$\frac{q}{L} = \bar{h}(\pi D)(T_s - T_\infty) = (111.1)\left[\pi\left(\frac{0.004}{12}\right)\right](90) = 10.47\ Btu/hr \cdot ft \quad (= 10.0\ W/m)$$

5.131^{D} A lightly clad jogger can be modeled as a cylinder 0.35 m in diameter by 1.9 m tall, with a surface temperature of 37 °C. For jogging at 8.05 km/h on a day with no wind and with ambient air temperature 19 °C, estimate the rate of heat loss by convection.

I From Table B-4, at 1 atm and $T_f = 28\ °C$,

$$\nu = 1.578 \times 10^{-5}\ m^2/s \qquad k = 0.026\,30\ W/m \cdot K \qquad \mathbf{Pr} = 0.708$$

Treating the jogger as a stationary vertical cylinder with air cross flow at 8.05 km/h, the Reynolds number is

$$\mathbf{Re}_D = \frac{DV_\infty}{\nu} = \frac{(0.35\ m)[(8.05 \times 1000/3600)\ m/s]}{1.578 \times 10^{-5}\ m^2/s} = 49\,600$$

By the Churchill–Bernstein equation (*G-12*),

$$\overline{\mathbf{Nu}} = 0.3 + \frac{(0.62)(49\,600)^{1/2}(0.708)^{1/3}}{[1 + (0.4/0.708)^{2/3}]^{1/4}}\left[1 + \left(\frac{49\,600}{28\,200}\right)^{5/8}\right]^{4/5} = 219.6$$

$$\bar{h} = \frac{k}{D}\,\overline{\mathbf{Nu}} = \frac{0.026\,30}{0.35}(219.6) = 16.5\ \text{W/m}^2\cdot\text{K}$$

$$q = \bar{h}(\pi DL)(T_s - T_\infty) = (16.5)[\pi(0.35)(1.9)](37 - 19) = 620\ \text{W}$$

5.132 A lightly clad jogger can be modeled as a cylinder 1.15 ft in diameter by $6\frac{1}{4}$ ft tall, with a surface temperature of 99 °F. For jogging at 5.0 mph on a day with no wind and with ambient air temperature 66 °F, estimate the rate of heat loss by convection.

I From Table B-4 at 1 atm and $T_f = 82.5$ °F,

$$\nu = 17.03 \times 10^{-5}\ \text{ft}^2/\text{sec} \qquad k = 0.015\,63\ \text{Btu/hr}\cdot\text{ft}\cdot{}^\circ\text{F} \qquad \mathbf{Pr} = 0.708$$

The Reynolds number is

$$\mathbf{Re}_D = \frac{DV_\infty}{\nu} = \frac{(1.15\ \text{ft})[(\frac{88}{12})\ \text{ft/sec}]}{17.03 \times 10^{-5}\ \text{ft}^2/\text{sec}} = 49\,520$$

By the Churchill–Bernstein equation (*G-12*),

$$\overline{\mathbf{Nu}} = 0.3 + \frac{(0.62)(49\,520)^{1/2}(0.708)^{1/3}}{[1 + (0.4/0.708)^{2/3}]^{1/4}}\left[1 + \left(\frac{49\,520}{28\,200}\right)^{5/8}\right]^{4/5} = 219.4$$

$$\bar{h} = \frac{k}{D}\,\overline{\mathbf{Nu}} = \frac{0.015\,63}{1.15}(219.4) = 2.982\ \text{Btu/hr}\cdot\text{ft}^2\cdot{}^\circ\text{F}$$

$$q = \bar{h}(\pi DL)(T_s - T_\infty) = (2.982)[\pi(1.15)(6.25)](99 - 66) = 2222\ \text{Btu/hr}$$

5.4 FLOW OVER A SPHERE

5.133 A 40-W incandescent bulb at 127 °C is immersed in a 27 °C airstream moving at 0.3 m/s. Approximating the light bulb as a 50-mm-diameter sphere, determine the percentage of electric power lost via forced convection.

I From Table B-4, with $T_\infty = 27$ °C and $T_s = 77$ °C, at an assumed pressure of 1 atm,

$$\nu_\infty = 1.568 \times 10^{-5}\ \text{m}^2/\text{s} \qquad k_\infty = 0.026\,22\ \text{W/m}\cdot\text{K} \qquad \mathbf{Pr}_\infty = 0.708$$

$$\mu_\infty = 1.8468 \times 10^{-5}\ \text{Pa}\cdot\text{s} \qquad \mu_s = 2.2858 \times 10^{-5}\ \text{Pa}\cdot\text{s}$$

The Reynolds number is

$$\mathbf{Re}_{D,\infty} = \frac{V_\infty D}{\nu_\infty} = \frac{(0.3\ \text{m/s})(50 \times 10^{-3}\ \text{m})}{1.568 \times 10^{-5}\ \text{m}^2/\text{s}} = 957$$

Since $\mathbf{Pr}_\infty \approx 0.71$ and $\mathbf{Re}_{D,\infty}$ is in the proper range, the Whitaker equation (*G-13*) applies:

$$\overline{\mathbf{Nu}}_\infty = 2 + [0.4(957)^{1/2} + 0.06(957)^{2/3}](0.708)^{0.4}\left(\frac{1.8468}{2.2858}\right)^{1/4} = 17.03$$

$$\bar{h} = \left(\frac{k_\infty}{D}\right)\overline{\mathbf{Nu}}_\infty = \left(\frac{0.026\,22\ \text{W/m}\cdot\text{K}}{50 \times 10^{-3}\ \text{m}}\right)(17.03) = 8.93\ \text{W/m}^2\cdot\text{K}$$

$$q = \bar{h}(\pi D^2)(T_s - T_\infty) = (8.93\ \text{W/m}^2\cdot\text{K})[\pi(50 \times 10^{-3}\ \text{m})^2][(127 - 27)\ \text{K}] = 7.0\ \text{W}$$

Thus the percentage lost is $(7.0/40)(100\%) \approx 18\%$.

5.134[D] A 40-W incandescent light bulb at 260 °F is immersed in a 80 °F airstream moving at 1 ft/sec. Approximating the bulb as a 2-in.-diameter sphere, determine the percentage of electric power lost via forced convection.

I From Table B-4, with $T_\infty = 80$ °F and $T_s = 260$ °F, at an assumed pressure of 1 atm,

$$\nu_\infty = 16.88 \times 10^{-5}\ \text{ft}^2/\text{sec} \qquad k_\infty = 0.015\,16\ \text{Btu/hr}\cdot\text{ft}\cdot{}^\circ\text{F} \qquad \mathbf{Pr}_\infty = 0.708$$

$$\mu_\infty g_c = 1.241 \times 10^{-5}\ \text{lbm/ft}\cdot\text{sec} \qquad \mu_s g_c = 1.536 \times 10^{-5}\ \text{lbm/ft}\cdot\text{sec}$$

The Reynolds number is

$$\mathbf{Re}_{D,\infty} = \frac{DV_\infty}{\nu_\infty} = \frac{(\frac{2}{12}\ \text{ft})(1\ \text{ft/sec})}{16.88 \times 10^{-5}\ \text{ft}^2/\text{sec}} = 992$$

With $\mathbf{Pr} \approx 0.71$ and $\mathbf{Re}_{D,\infty}$ in the right range, equation (*G-13*) applies:

$$\overline{\mathbf{Nu}}_\infty = 2 + [0.4(992)^{1/2} + 0.06(992)^{2/3}](0.708)^{0.4}\left(\frac{1.241}{1.536}\right)^{1/4} = 17.33$$

Thus,

$$\bar{h} = \left(\frac{k_\infty}{D}\right)\overline{\mathbf{Nu}}_\infty = \left(\frac{0.015\,16\ \text{Btu/hr}\cdot\text{ft}\cdot°\text{F}}{\frac{2}{12}\ \text{ft}}\right)(17.33) = 1.576\ \text{Btu/hr}\cdot\text{ft}^2\cdot°\text{F}$$

$$q = \bar{h}(\pi D^2)(T_s - T_\infty) = (1.576\ \text{Btu/hr}\cdot\text{ft}^2\cdot°\text{F})[\pi(\tfrac{2}{12}\ \text{ft})^2][(260-80)\ °\text{F}] = 24.76\ \text{Btu/hr} \approx 7.25\ \text{W}$$

The percentage lost is then $(7.25/40)(100\%) \approx 18\%$.

5.135 Water droplets falling in a cooling tower have an average temperature of 180 °F and an average diameter of 0.06 in. The air is at atmospheric pressure and 60 °F. At the moment when the droplet velocity relative to the air is 3 ft/sec, determine the surface-averaged heat transfer coefficient between the droplets and the air.

▌ At atmospheric pressure and $T_\infty = 60\ °\text{F}$, the properties from Table B-4 are

$$\nu_\infty = 15.40 \times 10^{-5}\ \text{ft}^2/\text{sec} \qquad k_\infty = 0.014\,65\ \text{Btu/hr}\cdot\text{ft}\cdot°\text{F} \qquad \mathbf{Pr}_\infty = 0.711$$

The Reynolds number is

$$\mathbf{Re}_{D,\infty} = \frac{V_\infty D}{\nu_\infty} = \frac{(3\ \text{ft/sec})\left(\frac{0.06}{12}\ \text{ft}\right)}{15.40 \times 10^{-5}\ \text{ft}^2/\text{sec}} = 97.4$$

Then, from (*G-14*) with oscillations/distortions neglected,

$$\overline{\mathbf{Nu}}_\infty = 2 + (0.6)(97.4)^{1/2}(0.711)^{1/3} = 2 + 5.29 = 7.29$$

$$\bar{h} = \frac{k_\infty}{D}\overline{\mathbf{Nu}}_\infty = \frac{0.014\,65\ \text{Btu/hr}\cdot\text{ft}\cdot°\text{F}}{(0.06/12)\ \text{ft}}(7.29) = 21.36\ \text{Btu/hr}\cdot\text{ft}^2\cdot°\text{F}$$

5.136 For the situation of Problem 5.135, what percentage change in $\bar{h}$ results from inclusion of the oscillations factor?

▌ Because, for free fall, $x = V_\infty^2/2g$, the factor has the value

$$25\left(\frac{D}{x}\right)^{0.7} = 25\left(\frac{2gD}{V_\infty^2}\right)^{0.7} = 25\left[\frac{2(32.2\ \text{ft/sec}^2)(0.005\ \text{ft})}{(3\ \text{ft/sec})^2}\right]^{0.7} \approx 2.42$$

so that the percent change in $\overline{\mathbf{Nu}}_\infty$ or $\bar{h}$ is

$$\frac{5.29(2.42-1)}{7.29}(100\%) = 103\%$$

Thus Problem 5.135 greatly underestimates $\bar{h}$.

5.137 Let $Q(\tau)$ represent the total thermal energy transferred to or from a free-falling droplet in the time interval $(0, \tau)$. Assuming a constant temperature difference between the droplet and the surrounding air, show that (*G-14*) implies a law of the form

$$Q(\tau) = \alpha\tau + \beta\tau^{0.1} \qquad (\alpha, \beta = \text{const.})$$

▌ For free-fall from rest, $V_\infty = gt$ and $x = \frac{1}{2}gt^2$; so that (*G-14*) assumes the form

$$\overline{\mathbf{Nu}}_\infty = 2 + \gamma t^{-0.9}$$

For a constant temperature difference, the heat flux q is proportional to $\bar{h}$ and hence to $\overline{\mathbf{Nu}}_\infty$; thus one can write

$$q = \alpha + \frac{\beta}{10}t^{-0.9}$$

and integration gives

$$Q(\tau) = \int_0^\tau q\,dt = \alpha\tau + \beta\tau^{0.1}$$

When τ is small—in which case the constant-temperature assumption is most justified—$\tau \ll \tau^{0.1}$, giving a $\frac{1}{10}$-power law.

5.138^D Saturated liquid water at 40 °C flows over a 25-mm-diameter sphere with a velocity of 2 m/s. The surface temperature of the sphere is 80 °C. What is the rate of heat transfer from the sphere to the water?

As will be seen, the Whitaker equation (*G-13*) applies (barely). From Table B-3, at $T_\infty = 40$ °C and $T_s = 80$ °C,

$$\nu_\infty = (0.708 \times 10^{-5})(9.290\,304 \times 10^{-2}) = 6.577 \times 10^{-7}\ \text{m}^2/\text{s}$$
$$k_\infty = (0.363)(1.729\,577) = 0.628\ \text{W/m} \cdot \text{K}$$
$$\mathbf{Pr}_\infty = 4.34 \quad (OK)$$
$$\rho_\infty = (62.09)(1.601\,846 \times 10^1) = 994.6\ \text{kg/m}^3$$
$$\rho_s = (60.81)(1.601\,846 \times 10^1) = 974.1\ \text{kg/m}^3$$
$$\nu_s = (0.392 \times 10^{-5})(9.290\,304 \times 10^{-2}) = 3.642 \times 10^{-7}\ \text{m}^2/\text{s}$$

The Reynolds number is

$$\mathbf{Re}_{D,\infty} = \frac{DV_\infty}{\nu_\infty} = \frac{(25 \times 10^{-3}\ \text{m})(2\ \text{m/s})}{6.577 \times 10^{-7}\ \text{m}^2/\text{s}} \approx 7.6 \times 10^4 \quad (barely\ OK)$$

Thus, the conditions on $\mathbf{Pr}_\infty$ and $\mathbf{Re}_{D,\infty}$ are met and equation (*G-13*) may be used. Calculating the viscosity ratio

$$\frac{\mu_\infty}{\mu_s} = \frac{\nu_\infty \rho_\infty}{\nu_s \rho_s} = \frac{(6.577 \times 10^{-7})(994.6)}{(3.642 \times 10^{-7})(974.1)} = 1.844$$

and so

$$\overline{\mathbf{Nu}}_\infty = 2 + [0.4(7.6 \times 10^4)^{1/2} + 0.06(7.6 \times 10^4)^{2/3}](4.34)^{0.4}(1.844)^{1/4} = 459$$
$$\bar{h} = \frac{k_\infty}{D}\,\overline{\mathbf{Nu}} = \frac{0.628\ \text{W/m} \cdot \text{K}}{0.025\ \text{m}}(459) = 11.5\ \text{kW/m}^2 \cdot \text{K}$$

so

$$q = \bar{h}(\pi D^2)(T_s - T_\infty) = (11.5\ \text{kW/m}^2 \cdot \text{K})[\pi(0.025\ \text{m})^2][(80 - 40)\ \text{K}] = 0.9\ \text{kW}$$

5.139^D Saturated liquid water at 104 °F flows over a 1.0-in.-diameter sphere with a velocity of 6.5 ft/sec. The surface temperature of the sphere is 176 °F. What is the rate of heat transfer from the sphere to the water?

As will be seen, the Whitaker equation (*G-13*) applies (barely). From Table B-3, at $T_\infty = 104$ °F and $T_s = 176$ °F,

$$\nu_\infty = 0.708 \times 10^{-5}\ \text{ft}^2/\text{sec} \qquad \nu_s = 0.392 \times 10^{-5}\ \text{ft}^2/\text{sec}$$
$$k_\infty = 0.363\ \text{Btu/hr} \cdot \text{ft} \cdot °\text{F} \qquad \rho_s = 60.81\ \text{lbm/ft}^3$$
$$\mathbf{Pr}_\infty = 4.34 \quad (OK) \qquad \rho_\infty = 62.09\ \text{lbm/ft}^3$$

The Reynolds number is

$$\mathbf{Re}_{D,\infty} = \frac{DV_\infty}{\nu_\infty} = \frac{(\frac{1}{12}\ \text{ft})(6.5\ \text{ft/sec})}{0.708 \times 10^{-5}\ \text{ft}^2/\text{sec}} \approx 76\,000 \quad (barely\ OK)$$

The viscosity ratio is

$$\frac{\mu_\infty}{\mu_s} = \frac{\rho_\infty \nu_\infty}{\rho_s \nu_s} = \frac{(62.09)(0.708 \times 10^{-5})}{(60.81)(0.392 \times 10^{-5})} = 1.844$$

and

$$\overline{\mathbf{Nu}}_\infty = 2 + [0.4(76\,000)^{1/2} + 0.06(76\,000)^{2/3}](4.34)^{0.4}(1.844)^{1/4} = 459$$
$$\bar{h} = \frac{k_\infty}{D}\,\overline{\mathbf{Nu}}_\infty = \frac{0.363\ \text{Btu/hr} \cdot \text{ft} \cdot °\text{F}}{(\frac{1}{12})\ \text{ft}}(459) = 1999\ \text{Btu/hr} \cdot \text{ft}^2 \cdot °\text{F}$$
$$q = \bar{h}(\pi D^2)(T_s - T_\infty) = (1999\ \text{Btu/hr} \cdot \text{ft}^2 \cdot °\text{F})[\pi(\tfrac{1}{12}\ \text{ft})^2][(176 - 104)\ °\text{F}] = 3140\ \text{Btu/hr}$$

5.5 FLOW THROUGH TUBE BUNDLES

5.140^D An in-line tube bundle consists of 19 rows of 1-in.-OD tubes with 12 tubes in each row (in the direction of flow). The tube spacing is 1.5 in. in the direction normal to the flow, and 2.0 in. parallel to it. The tube surfaces are maintained at 260 °F. Air at 80 °F and 14.7 psia flows through the bundle with a maximum velocity of 30 ft/sec. Calculate the total heat transfer from the bundle per foot of length.

▮ From Table B-4, at $T_f = 170\ °F$ and 1 atm,

$$\nu = 22.38 \times 10^{-5}\ \text{ft}^2/\text{sec} \qquad k = 0.017\,35\ \text{Btu/hr}\cdot\text{ft}\cdot°\text{F} \qquad \mathbf{Pr} = 0.697 \approx 0.7$$

whence

$$\mathbf{Re}_{D,\max} = \frac{V_{\max} D}{\nu} = \frac{(30\ \text{ft/sec})(\frac{1}{12}\ \text{ft})}{22.38 \times 10^{-5}\ \text{ft}^2/\text{sec}} = 11\,170$$

In terms of Figure G-1(a), the geometric configuration is

$$\frac{a}{D} = \frac{1.5}{1} = 1.5 \qquad \frac{b}{D} = \frac{2.0}{1} = 2$$

Table G-3 gives $C_1 = 0.299$ and $n = 0.602$, and (G-15) gives

$$\bar{h} = \frac{k}{D} C_1 \mathbf{Re}^n_{D,\max} = \frac{0.017\,35\ \text{Btu/hr}\cdot\text{ft}\cdot°\text{F}}{(\frac{1}{12})\ \text{ft}} (0.299)(11\,170)^{0.602} = 17.0\ \text{Btu/hr}\cdot\text{ft}^2\cdot°\text{F}$$

The total heat transfer per unit length is $q/L = \bar{h}\pi DN(T_s - T_\infty)$, where N is the total number of tubes.

$$\frac{q}{L} = (17.0\ \text{Btu/hr}\cdot\text{ft}^2\cdot°\text{F})\pi(\tfrac{1}{12}\ \text{ft})[(19)(12)][(260 - 80)\ °\text{F}] = 1.83 \times 10^5\ \text{Btu/hr}\cdot\text{ft}$$

5.141^D An in-line tube bundle consists of 19 rows of 25-mm-OD tubes with 12 tubes in each row (in the direction of flow). The tube spacing is 37.5 mm in the direction normal to the flow and 50 mm parallel to it; the tube surfaces are maintained at 127 °C. Air at 27 °C and 101.3 kPa (1 atm) flows through the bundle with a maximum velocity of 9.15 m/s. Calculate the total heat transfer from the bundle per meter of length.

▮ At $T_f = 77\ °C$, the fluid properties are (from Table B-4)

$$\nu = (22.38 \times 10^{-5})(9.2903 \times 10^{-2}) = 2.079 \times 10^{-5}\ \text{m}^2/\text{s}$$
$$k = (0.017\,35)(1.729\,577) = 0.030\,01\ \text{W/m}\cdot\text{K}$$
$$\mathbf{Pr} = 0.697$$

whence

$$\mathbf{Re}_{D,\max} = \frac{V_{\max} D}{\nu} = \frac{(9.15\ \text{m/s})(0.025\ \text{m})}{2.079 \times 10^{-5}\ \text{m}^2/\text{s}} = 1.1 \times 10^4$$

In terms of Figure G-1(a), the geometric configuration is

$$\frac{a}{D} = \frac{37.5}{25} = 1.5 \qquad \frac{b}{D} = \frac{50}{25} = 2$$

Table G-3 gives $C_1 = 0.299$ and $n = 0.602$, and (G-15) gives

$$\bar{h} = \frac{k}{D} C_1 \mathbf{Re}^n_{D,\max} = \frac{0.030\,01\ \text{W/m}\cdot\text{K}}{0.025\ \text{m}} (0.299)(1.1 \times 10^4)^{0.602} = 97.0\ \text{W/m}^2\cdot\text{K}$$

The total heat transfer per unit length is $q/L = \bar{h}\pi DN(T_s - T_\infty)$, where N is the total number of tubes.

$$\frac{q}{L} = (97.0\ \text{W/m}^2\cdot\text{K})\pi(0.025\ \text{m})[(19)(12)][(127 - 27)\ \text{K}] = 174\ \text{kW/m}$$

5.142^D Pressurized liquid water at 104 °F flows (without phase change) across an 18-in.-wide (tube-length direction), staggered tube bank (Fig. 5-3), carrying combustion gases which keep the tube surfaces at 248 °F. For each 1 ft in height of tube bank, water is supplied in a 6-in.-ID pipe, flowing at a velocity of 3 ft/sec. Estimate the temperature of the water after passing through the tube bank.

▮ At $T_f = 176\ °F$, the fluid parameters are, from Table B-3,

$$\nu = 0.392 \times 10^{-5}\ \text{ft}^2/\text{sec} \qquad k = 0.386\ \text{Btu/hr}\cdot\text{ft}\cdot°\text{F} \qquad \mathbf{Pr} = 2.22$$

The minimum flow area per foot of height in the tube bank is equal to the free space between the tubes; therefore,

$$A_{\min} = [(12\ \text{in.}) - 6(1\ \text{in.})](18\ \text{in.}) = 108\ \text{in}^2$$

The maximum velocity in the tube bank is given by $V_{\max} A_{\min} = V_{\text{pipe}} A_{\text{pipe}}$, or

$$V_{\max} = (3\ \text{ft/sec})\frac{(\pi/4)(6\ \text{in.})^2}{108\ \text{in}^2} = 0.785\ \text{ft/sec}$$

whence

$$\mathbf{Re}_{D,\max} = \frac{V_{\max} D}{\nu} = \frac{(0.785\ \text{ft/sec})(\frac{1}{12}\ \text{ft})}{0.392 \times 10^{-5}\ \text{ft}^2/\text{sec}} = 16\,690$$

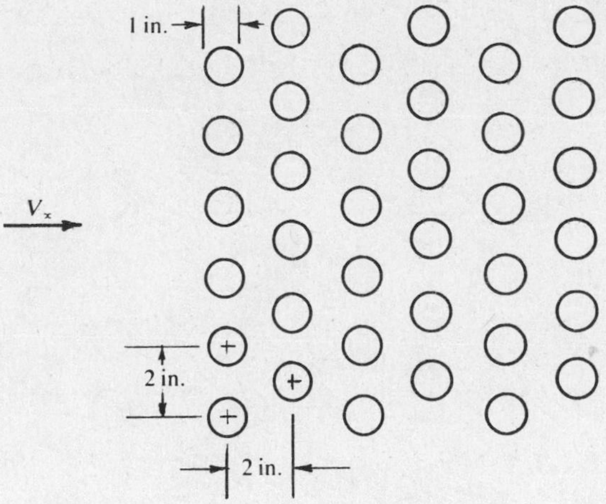

Fig. 5-3

From the tube bank's geometry,

$$\frac{a}{D} = \frac{b}{D} = 2$$

which gives $C_1 = 0.482$ and $n = 0.556$, from Table G-3. Equation (*G-16*) gives

$$\bar{h}_{10} = (1.13)\frac{k}{D} C_1 \mathbf{Re}^n_{D,\max} \mathbf{Pr}^{1/3} = (1.13)\left[\frac{0.386 \text{ Btu/hr}\cdot\text{ft}\cdot{}^\circ\text{F}}{(\frac{1}{12})\text{ ft}}\right](0.482)(16\,690)^{0.556}(2.22)^{1/3}$$
$$= 733 \text{ Btu/hr}\cdot\text{ft}^2\cdot{}^\circ\text{F}$$

This value must be modified by the appropriate factor from Table G-4 to account for the fact that the heat transfer coefficient has not reached its constant value produced by passing over a minimum of 10 tubes; therefore, for six tubes in the flow direction,

$$\bar{h} = (0.95)(733) = 696 \text{ Btu/hr}\cdot\text{ft}^2\cdot{}^\circ\text{F}$$

For $N = 36$ tubes, the heat transferred per foot of height is given by

$$q = \bar{h}AN(T_\infty - T_s) = (696 \text{ Btu/hr}\cdot\text{ft}^2\cdot{}^\circ\text{F})[\pi(\tfrac{1}{12}\text{ ft})(1.5\text{ ft})](36)[(248-104)\,{}^\circ\text{F}] = 1.42\times10^6 \text{ Btu/hr}$$
$$= 394 \text{ Btu/sec}$$

Since the water gains energy at this rate, a heat balance on it gives its exit temperature.

$$q = \dot{m}c_{p,\text{avg}}(T_{\text{out}} - T_{\text{in}}) = (\rho AV)_{\text{in}} c_{p,\text{avg}}(T_{\text{out}} - T_{\text{in}})$$

At the inlet temperature, $\rho_{\text{in}} = 62.09 \text{ lbm/ft}^3$ and $c_{p,\text{in}} = 0.9980 \text{ Btu/lbm}\cdot{}^\circ\text{F}$; therefore, with $c_{p,\text{avg}} \approx c_{p,\text{in}}$,

$$T_{\text{out}} \approx \frac{q}{(\rho AV)_{\text{in}} c_{p,\text{in}}} + T_{\text{in}} = \frac{394 \text{ Btu/sec}}{(62.09 \text{ lbm/ft}^3)(\pi/4)(0.5\text{ ft})^2(3\text{ ft/sec})(0.9980 \text{ Btu/lbm}\cdot{}^\circ\text{F})} + (104\,{}^\circ\text{F}) = 115\,{}^\circ\text{F}$$

This answer could be improved by iteration to evaluate c_p at the mean water temperature, but the improvement would be slight.

5.143[D] Pressurized liquid water at 40 °C flows (without phase change) across a 46-cm-wide (tube-length direction) staggered tube bank (Fig. 5-4), carrying combustion gases which keep the tube surfaces at 120 °C. For each 30 cm in height of tube bank, water is supplied in a 15-cm-ID pipe, flowing at a velocity of 1 m/s. Estimate the temperature of the water after passing through the tube bank.

▌ At $T_f = 80$ °C, the fluid parameters are, from Table B-3,

$$\nu = 0.364\times10^{-6}\text{ m}^2/\text{s} \qquad k = 0.668 \text{ W/m}\cdot\text{K} \qquad \mathbf{Pr} = 2.22$$

The minimum flow area per 0.30 m of height in the tube bank is equal to the free space between the tubes; therefore,

$$A_{\min} = [(30\text{ cm}) - 6(2.5\text{ cm})](46\text{ cm}) = 690\text{ cm}^2$$

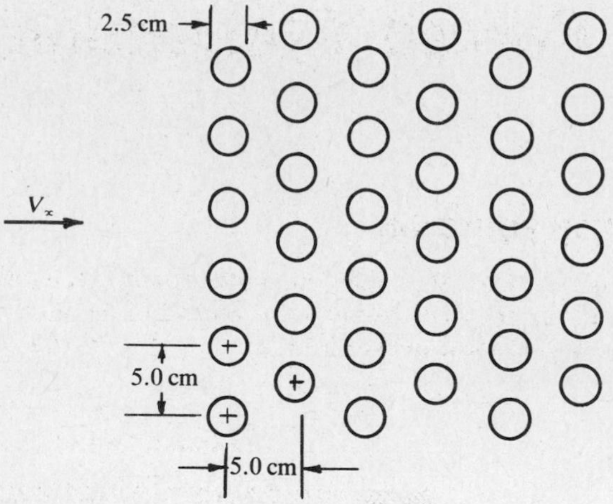

Fig. 5-4

The maximum velocity in the tube bank is given by $V_{max}A_{min} = V_{pipe}A_{pipe}$, or

$$V_{max} = (1 \text{ m/s}) \frac{(\pi/4)(15 \text{ cm})^2}{690 \text{ cm}^2} = 0.256 \text{ m/s}$$

whence
$$\mathbf{Re}_{D,max} = \frac{V_{max}D}{\nu} = \frac{(0.256 \text{ m/s})(0.025 \text{ m})}{0.364 \times 10^{-6} \text{ m}^2/\text{s}} = 1.76 \times 10^4$$

From the tube bank's geometry,

$$\frac{a}{D} = \frac{b}{D} = 2$$

which gives $C_1 = 0.482$ and $n = 0.556$, from Table G-3. Equation (*G-16*) gives

$$\bar{h}_{10} = (1.13)\frac{k}{D} C_1 \mathbf{Re}^n_{D,max} \mathbf{Pr}^{1/3} = (1.13)\left(\frac{0.668 \text{ W/m}\cdot\text{K}}{0.025 \text{ m}}\right)(0.482)(1.76 \times 10^4)^{0.556}(2.22)^{1/3} = 4350 \text{ W/m}^2\cdot\text{K}$$

This value must be modified by the appropriate factor from Table G-4 to account for the fact that the heat transfer coefficient has not reached its constant value produced by passing over a minimum of 10 tubes; therefore, for six tubes in the flow direction,

$$\bar{h} = (0.95)(4350) = 4130 \text{ W/m}^2\cdot\text{K}$$

For $N = 36$ tubes, the heat transferred per meter of height is given by

$$q = \bar{h}AN(T_\infty - T_s) = (4130 \text{ W/m}^2\cdot\text{K})[\pi(0.025 \text{ m})(0.46 \text{ m})](36)[(120 - 40) \text{ K}] = 4.3 \times 10^5 \text{ W}$$

Since the water gains at this rate, a heat balance on it gives its exit temperature.

$$q = \dot{m}c_{p,avg}(T_{out} - T_{in}) = (\rho AV)_{in}c_{p,avg}(T_{out} - T_{in})$$

At the inlet temperature, $\rho_{in} = 994.6 \text{ kg/m}^3$ and $c_{p,in} = 4175.6 \text{ J/kg}\cdot\text{K}$; therefore, with $c_{p,avg} \approx c_{p,in}$,

$$T_{out} \approx \frac{q}{(\rho AV)_{in}c_{p,in}} + T_{in} = \frac{4.30 \times 10^5 \text{ W}}{(994.6 \text{ kg/m}^3)(\pi/4)(0.15 \text{ m})^2(1 \text{ m/s})(4175.6 \text{ J/kg}\cdot{}^\circ\text{C})} + (40\ {}^\circ\text{C}) = 46\ {}^\circ\text{C}$$

An iteration to evaluate c_p at the mean water temperature would yield little improvement.

CHAPTER 6

Forced Convection: Internal Flow

6.1 BASIC TEMPERATURE EQUATION

6.1 Consider steady mass flow $\dot{m}$ inside a tube of constant perimeter P, accompanied by convective heat transfer. Express dT_b/dx, the local axial rate of change of the fluid bulk temperature, in terms of $\dot{m}$, P, the specific heat c_p, and the local surface heat flux $q_s'' \equiv h_x(T_s - T_b)$.

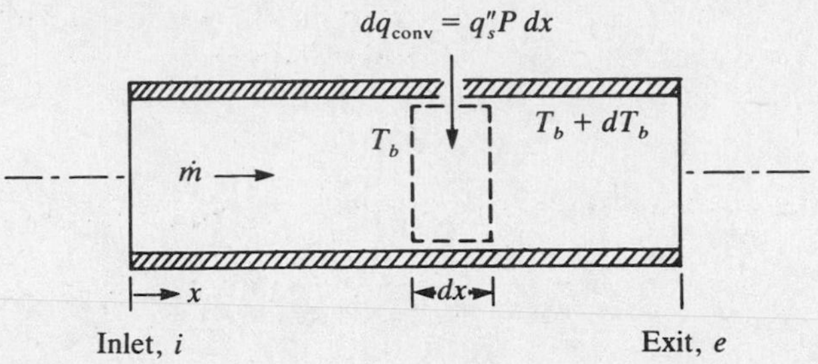

Fig. 6-1

For steady flow through the infinitesimal control volume of Fig. 6-1, with no change in the mechanical energy of the fluid, the first law of thermodynamics gives

$$dq_{\text{conv}} = \dot{m}\,dH \tag{1}$$

where H denotes the specific enthalpy of the fluid. If the fluid is an ideal gas—which is hereby assumed—then

$$dH = c_p\,dT_b \qquad (c_p = \text{const.}) \tag{2}$$

Together, (1), (2), and

$$dq_{\text{conv}} = q_s''P\,dx \tag{3}$$

imply the desired differential equation

$$\frac{dT_b}{dx} = \frac{q_s''P}{\dot{m}c_p} = \frac{P}{\dot{m}c_p}h_x(T_s - T_b) \tag{4}$$

If the constant c_p in (4) is replaced by $c_{p,b}$, the specific heat *at the mean bulk temperature* of the fluid, the resulting differential equation [(*H-1*) of Appendix H] will remain linear in T_b for the much wider class of fluids for which $c_p = c_p(T_b)$. It should be noted that, in general, both h_x and T_s vary with x.

6.2 For the case of constant surface heat flux, q_s'', obtain an expression for $T_b(x)$. Plot this qualitatively.

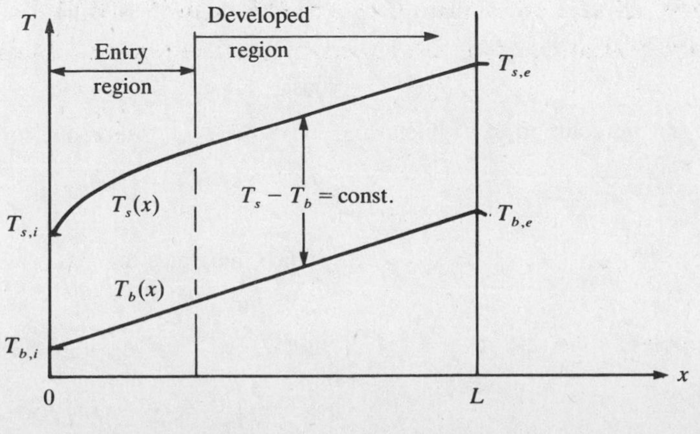

Fig. 6-2

For $q_s'' = \text{const.}$, (*H-1*) gives

$$\int_{T_{b,i}}^{T_b} dT_b = \frac{q_s''P}{\dot{m}c_{p,b}} \int_0^x dx \qquad \text{or} \qquad T_b - T_{b,i} = \frac{q_s''P}{\dot{m}c_{p,b}} x$$

which is (*H-2*) of Appendix H. The linear function is sketched, along with T_s, in Fig. 6-2, on the assumption $q_s'' > 0$. Note that thermally developed flow is characterized by the constant difference $T_s - T_b$.

6.3 For the case of constant tube surface temperature, T_s, obtain an expression for $T_b(x)$. Plot this qualitatively.

Integrate (*H-1*) for fixed T_s, using the notation

$$\bar{h}(0, x) \equiv \frac{1}{Px} \int_0^x h_x \, d(Px) = \frac{1}{x} \int_0^x h_x \, dx$$

for the running average heat transfer coefficient; thus,

$$\int_{T_{b,i}}^{T_b} \frac{dT_b}{T_b - T_s} = -\frac{P}{\dot{m}c_{p,b}} \int_0^x h_x \, dx$$

$$\ln \frac{T_b - T_s}{T_{b,i} - T_s} = -\frac{P}{\dot{m}c_{p,b}} x\bar{h}(0, x)$$

$$T_b = T_s - (T_s - T_{b,i}) \exp\left[\frac{-Px\bar{h}(0, x)}{\dot{m}c_{p,b}}\right]$$

The last equation is (*H-3*). The graph in Fig. 6-3 is for the case $T_{b,i} < T_s$.

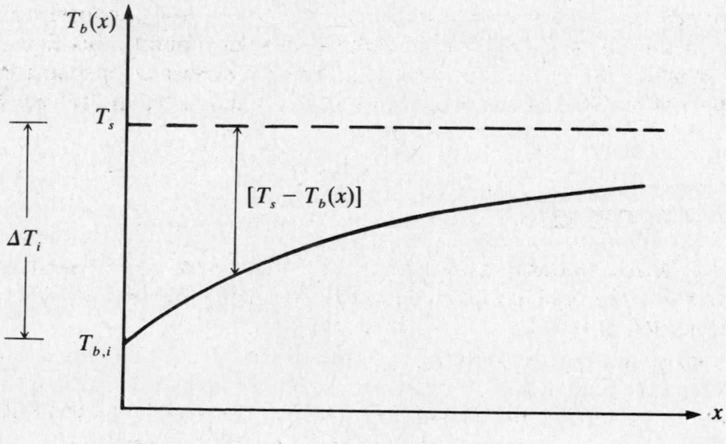

Fig. 6-3

6.4 For a tube of length L, obtain the total heat transfer **(*a*)** in Problem 6.2, **(*b*)** in Problem 6.3.

(*a*) $q_{\text{conv}} = \text{flux} \times \text{area} = q_s''PL$. While this relation is exact, the constant value of q_s'' may be unknown. To the extent that the flow may be taken to be fully developed over the entire length of the tube, we can write

$$T_s - T_b = \text{const.} \qquad \text{and} \qquad q_s'' = \bar{h}(T_s - T_b)$$

where, in the notation of Problem 6.3, $\bar{h} = \bar{h}(0, L)$. The above formula becomes

$$q_{\text{conv}} = \bar{h}PL(T_s - T_b)$$

which is (*H-4*).

(*b*) By Problem 6.1, $dq_{\text{conv}} = \dot{m}c_{p,b}\, dT_b$, which integrates to

$$q_{\text{conv}} = \dot{m}c_{p,b}(T_{b,e} - T_{b,i}) \tag{1}$$

But Problem 6.3 gives, at $x = L$,

$$\ln \frac{T_{b,e} - T_s}{T_{b,i} - T_s} = -\frac{P}{\dot{m}c_{p,b}} L\bar{h} \tag{2}$$

Elimination of the quantity $\dot{m}c_{p,b}$ between (1) and (2) gives

$$q_{\text{conv}} = \bar{h}PL \frac{T_{b,e} - T_{b,i}}{\ln \dfrac{T_s - T_{b,i}}{T_s - T_{b,e}}} \equiv \bar{h}PL(\Delta T)_{\text{l.m.}} \qquad (3)$$

which is reproduced as (*H-5*) in Appendix H. Here $(\Delta T)_{\text{l.m.}}$ stands for the *logarithmic mean temperature difference* across the tube.

6.5 A tube experiences a constant heat influx of 4 kW/m^2 due to electric resistance heating. The tube is 50-mm-ID, has a length of 8 m, and is well-insulated outside. Water flows through the tube at $\dot{m} = 0.40$ kg/s, and its inlet temperature is 20 °C. ***(a)*** What is the outlet temperature? ***(b)*** What is the local convective heat transfer coefficient at $x = L$, if $T_s(L) = 40$ °C?

▮ Assume (1) steady-state conditions; (2) uniform electrical heating along the tube, so that $q_s'' =$ constant; (3) constant fluid properties; (4) adiabatic outer wall, so that all electrical heating is to the fluid.

(a) From Table B-3, at 20 °C, $c_{p,b} = (0.9988)(4.184 \times 10^3) = 4.179$ kJ/kg · K. Then, with $x = L = 8$ m, (*H-2*) yields

$$T_b(L) = (20\text{ °C}) + \frac{(4\text{ kW/m}^2)\pi(0.050\text{ m})(8\text{ m})}{(0.40\text{ kg/s})(4.179\text{ kJ/kg}\cdot\text{K})} = 23.0\text{ °C}$$

(b) At $x = L$, the relation $q_s'' = h_x(T_s - T_b)$ yields

$$h_L = \frac{q_s''}{T_s(L) - T_b(L)} = \frac{4000\text{ W/m}^2}{(40 - 23.0)\text{ K}} = 235\text{ W/m}^2\cdot\text{K}$$

6.6 Criticize ***(a)*** the statement, ***(b)*** the solution, of Problem 6.5.

▮ ***(a)*** The problem as stated is overdetermined: The given flow parameters and the given heat flux may be incompatible. See Problem 6.75 before accepting the values found. ***(b)*** Equation (*H-2*) calls for $c_{p,b}$, but $c_{p,i}$ was actually used. Since the specific heat of water is almost independent of temperature, this is a small matter as compared to (*a*). In any event, an iterative solution (as in Problem 6.17) would converge very rapidly.

6.2 LAMINAR-FLOW PROBLEMS

6.7^D Water flows in a long, electrically heated tube with $q_s'' =$ constant. The velocity and temperature profiles are fully developed. The tube has inside diameter 30 mm and, at $x = \alpha$, $T_b(\alpha) = 20$ °C and $T_s(\alpha) =$ 30 °C. The mass flow is 0.04 kg/s. Determine ***(a)*** h_α, ***(b)*** q_s''.

▮ ***(a)*** From Table B-3, at $T_b = 20$ °C,

$$\nu_b = (1.083 \times 10^{-5})(9.290\,304 \times 10^{-2}) = 1.006 \times 10^{-6}\text{ m}^2\text{/s}$$
$$\rho_b = (62.46)(16.018\,46) = 1000.5\text{ kg/m}^3$$
$$k_b = (0.345)(1.729\,577) = 0.597\text{ W/m}\cdot\text{K}$$

From the continuity equation, $\dot{m} = \rho_b V_b(\pi D^2/4)$, the Reynolds number is

$$\mathbf{Re}_{D,b} = \frac{V_b D}{\nu_b} = \frac{4\dot{m}}{\pi\rho_b D\nu_b} = \frac{4(0.04\text{ kg/s})}{\pi(1000.5\text{ kg/m}^3)(0.030\text{ m})(1.006 \times 10^{-6}\text{ m}^2\text{/s})} = 1688$$

indicating laminar flow. By (*H-6*),

$$h_\alpha = h_\infty = \frac{k_b}{D}(4.36) = \frac{(0.597)(4.36)}{(0.030)} = 86.8\text{ W/m}^2\cdot\text{K}$$

(b)
$$q_s'' = q_s''(\alpha) = h_\alpha[T_s(\alpha) - T_b(\alpha)] = (86.8)(30 - 20) = 868\text{ W/m}^2$$

6.8^D Water flows through a long, electrically heated tube with $q_s'' =$ constant. The velocity and temperature profiles are fully developed. The tube has inside diameter 1.18 in. and, at $x = \alpha$, $T_b(\alpha) = 68$ °F and $T_s(\alpha) = 86$ °F. The mass flow is 0.088 lbm/sec. Determine ***(a)*** h_α, ***(b)*** q_s''.

▮ ***(a)*** From Table B-3, for saturated water at 68 °F,

$$\nu_b = 1.083 \times 10^{-5}\text{ ft}^2\text{/sec} \qquad \rho_b = 62.46\text{ lbm/ft}^3 \qquad k_b = 0.345\text{ Btu/hr}\cdot\text{ft}\cdot\text{°F}$$

The continuity equation, $\dot{m} = \rho_b V_b(\pi D^2/4)$, gives the Reynolds number as

$$\mathbf{Re}_{D,b} = \frac{V_b D}{\nu_b} = \frac{4\dot{m}}{\pi \rho_b D \nu_b} = \frac{4(0.088\ \text{lbm/sec})}{\pi(62.46\ \text{lbm/ft}^3)[(1.18/12)\ \text{ft}](1.083 \times 10^{-5}\ \text{ft}^2/\text{sec})} = 1686$$

indicating laminar flow. By (*H-6*),

$$h_\alpha = h_\infty = \frac{k_b}{D}(4.36) = \left[\frac{0.345\ \text{Btu/hr}\cdot\text{ft}\cdot{}^\circ\text{F}}{(1.18/12)\ \text{ft}}\right](4.36) = 15.30\ \text{Btu/hr}\cdot\text{ft}^2\cdot{}^\circ\text{F}$$

(*b*)
$$q_s'' = q_s''(\alpha) = h_\alpha[T_s(\alpha) - T_b(\alpha)] = (15.30)(86 - 68) = 275\ \text{Btu/hr}\cdot\text{ft}^2$$

6.9[D] Air at $\bar{T}_b \equiv \frac{1}{2}(T_{b,i} + T_{b,e}) = 80\ {}^\circ\text{F}$ and 30 psia flows in a 1.0-in.-ID tube with a bulk average velocity $V_b = 0.5$ ft/sec. The tube wall temperature is constant at 170 °F, (*a*) Assuming fully developed velocity and temperature profiles, determine the heat transfer coefficient. (*b*) Estimate the velocity and temperature profile entry lengths.

▌ From Table B-4, at 80 °F, and with a pressure correction of the density,

$$\mu_b g_c = 1.241 \times 10^{-5}\ \text{lbm/ft}\cdot\text{sec} \qquad \mathbf{Pr}_b = 0.708$$
$$k_b = 0.015\,16\ \text{Btu/hr}\cdot\text{ft}\cdot{}^\circ\text{F} \qquad \rho_b = (0.0735)(30/14.7) = 0.150\ \text{lbm/ft}^3$$

(*a*) The Reynolds number is

$$\mathbf{Re}_{D,b} = \frac{\rho_b D V_b}{\mu_b g_c} = \frac{(0.150\ \text{lbm/ft}^3)(\frac{1}{12}\ \text{ft})(0.5\ \text{ft/sec})}{1.241 \times 10^{-5}\ \text{lbm/ft}\cdot\text{sec}} = 504$$

For this laminar flow with T_s = constant, (*H-7*) gives

$$h_\infty = \frac{k_b}{D}(3.66) = \left(\frac{0.015\,16}{\frac{1}{12}}\right)(3.66) = 0.666\ \text{Btu/hr}\cdot\text{ft}^2\cdot{}^\circ\text{F}$$

(*b*) From expressions in Problem 4.81,

$$x_v \approx D(0.05\,\mathbf{Re}_{D,b}) = (\tfrac{1}{12})(0.05)(504) = 2.1\ \text{ft}$$
$$x_t \approx x_v\,\mathbf{Pr}_b = (2.1)(0.708) = 1.49\ \text{ft}$$

6.10[D] Air at $\bar{T}_b \equiv \frac{1}{2}(T_{b,i} + T_{b,e}) = 27\ {}^\circ\text{C}$ and 206.7 kPa flows inside a 25.4-mm-ID tube with a bulk average velocity 0.152 m/s. The tube wall temperature is constant at 77 °C. (*a*) Assuming fully developed velocity and temperature profiles in the fluid, determine the heat transfer coefficient. (*b*) Estimate the velocity and temperature entry lengths.

▌ From Table B-4, at 27 °C and with a pressure correction of the density,

$$\mu_b = (1.241 \times 10^{-5})(1.488\,164) = 1.847 \times 10^{-5}\ \text{Pa}\cdot\text{s}$$
$$k_b = (0.015\,16)(1.729\,577) = 0.026\,22\ \text{W/m}\cdot\text{K}$$
$$\mathbf{Pr}_b = 0.708$$
$$\rho_b = (0.0735)(16.018\,46)(206.7/101.3) = 2.402\ \text{kg/m}^3$$

(*a*) The Reynolds number is

$$\mathbf{Re}_{D,b} = \frac{\rho_b D V_b}{\mu_b} = \frac{(2.402\ \text{kg/m}^3)(0.0254\ \text{m})(0.152\ \text{m/s})}{1.847 \times 10^{-5}\ \text{Pa}\cdot\text{s}} = 502$$

For this laminar flow with T_s = constant, (*H-7*) gives

$$h_\infty = \frac{k_b}{D}(3.66) = \left(\frac{0.026\,22\ \text{W/m}\cdot\text{K}}{0.0254\ \text{m}}\right)(3.66) = 3.78\ \text{W/m}^2\cdot\text{K}$$

(*b*) From Problem 4.81,

$$x_v \approx D(0.05\,\mathbf{Re}_{D,b}) = (0.0254)(0.05)(502) = 0.64\ \text{m}$$
$$x_t \approx x_v\,\mathbf{Pr}_b = (0.64)(0.708) = 0.451\ \text{m}$$

6.11[D] Water at $\bar{T}_b = 68\ {}^\circ\text{F}$ flows at 0.2 ft/sec in a long, 1.0-in.-ID tube. The tube wall is at a constant 80 °F. Determine (*a*) the (constant) heat transfer coefficient in the fully developed velocity and temperature profile region; (*b*) the velocity and temperature entry lengths.

From Table B-3, at 68 °F,

$$\nu_b = 1.083 \times 10^{-5}\ \text{ft}^2/\text{sec} \qquad k_b = 0.345\ \text{Btu/hr}\cdot\text{ft}\cdot°\text{F} \qquad \mathbf{Pr}_b = 7.02$$

(*a*) The Reynolds number is

$$\mathbf{Re}_{D,b} = \frac{DV_b}{\nu_b} = \frac{(\frac{1}{12}\ \text{ft})(0.2\ \text{ft/sec})}{1.083 \times 10^{-5}\ \text{ft}^2/\text{sec}} = 1539$$

For this laminar flow at constant T_s, (*H-7*) gives

$$h_\infty = \frac{(3.66)k_b}{D} = \frac{(3.66)(0.345\ \text{Btu/hr}\cdot\text{ft}\cdot°\text{F})}{\frac{1}{12}\ \text{ft}} = 15.15\ \text{Btu/hr}\cdot\text{ft}^2\cdot°\text{F}$$

(*b*) From Problem 4.81,

$$x_v \approx D(0.05\,\mathbf{Re}_{D,b}) = (\tfrac{1}{12})(0.05)(1539) = 6.4\ \text{ft}$$
$$x_t \approx x_v\,\mathbf{Pr}_b = (6.4)(7.02) = 45\ \text{ft}$$

The very large thermal entry length indicates that it may be impractical to have fully developed conditions; this is frequently true with *liquid* flows. See Problem 6.14 for the results with a fluid of greater viscosity (higher Prandtl number).

6.12 If the tube of Problem 6.11 is only 4 ft long, calculate the average heat transfer coefficient.

By Problem 6.10(*b*), the entire tube lies in the developing region. From Table B-3, at $\bar{T}_b = 68$ °F and $T_s = 80$ °F,

$$\mu_b g_c = \nu_b \rho_b = (1.083 \times 10^{-5})(62.46) = 67.6 \times 10^{-5}\ \text{lbm/ft}\cdot\text{sec}$$
$$\mu_s g_c = \nu_s \rho_s = (0.958 \times 10^{-5})(62.33) = 59.7 \times 10^{-5}\ \text{lbm/ft}\cdot\text{sec}$$

and (*H-8*) gives

$$\overline{\mathbf{Nu}} = (1.86)\left[\frac{(1539)(7.02)}{\frac{48}{1}}\right]^{1/3}\left(\frac{67.6 \times 10^{-5}}{59.7 \times 10^{-5}}\right)^{0.14} = 11.51$$

Thus

$$\bar{h} = \left(\frac{0.345\ \text{Btu/hr}\cdot\text{ft}\cdot°\text{F}}{\frac{1}{12}\ \text{ft}}\right)(11.51) = 47.7\ \text{Btu/hr}\cdot\text{ft}^2\cdot°\text{F}$$

6.13^D Water at $\bar{T}_b = 20$ °C flows at 61 mm/s in a long, 25.4-mm-ID tube. The tube wall is at a constant 27 °C. Determine (*a*) The (constant) heat transfer coefficient in the fully developed velocity and temperature profile region; (*b*) The velocity and temperature entry lengths.

From Table B-3, at 20 °C,

$$\nu_b = 1.006 \times 10^{-6}\ \text{m}^2/\text{s} \qquad k_b = 0.597\ \text{W/m}\cdot\text{K} \qquad \mathbf{Pr}_b = 7.02$$

(*a*) The Reynolds number is

$$\mathbf{Re}_{D,b} = \frac{DV_b}{\nu_b} = \frac{(0.0254\ \text{m})(0.061\ \text{m/s})}{1.006 \times 10^{-6}\ \text{m}^2/\text{s}} = 1540$$

For this laminar flow at constant T_s, (*H-7*) gives

$$h_\infty = \frac{k_b}{D}(3.66) = \left(\frac{0.597\ \text{W/m}\cdot\text{K}}{0.0254\ \text{m}}\right)(3.66) = 86.02\ \text{W/m}^2\cdot\text{K}$$

(*b*) From Problem 4.81,

$$x_v \approx (0.05)D\,\mathbf{Re}_{D,b} = (0.05)(0.0254\ \text{m})(1540) = 1.96\ \text{m}$$
$$x_t \approx x_v\,\mathbf{Pr}_b = (1.96\ \text{m})(7.02) = 13.73\ \text{m}$$

6.14^D Glycerin flows through a 1.0-in.-ID horizontal tube which is 1 mile long; the average bulk velocity is 12 ft/sec. The average bulk temperature of the glycerin is 68 °F and the tube wall temperature is constant at 130 °F. Calculate (*a*) the (constant) value of the heat transfer coefficient in the fully developed thermal and velocity region; (*b*) the approximate lengths of the thermal and velocity inlet regions.

From Table B-3, at $\bar{T}_b = 68$ °F,

$$\nu_b = 0.0127\ \text{ft}^2/\text{sec} \qquad k_b = 0.165\ \text{Btu/hr}\cdot\text{ft}\cdot°\text{F} \qquad \mathbf{Pr}_b = 12.5 \times 10^3$$

The Reynolds number is

$$\mathbf{Re}_{D,b} = \frac{DV_b}{\nu_b} = \frac{(\frac{1}{12}\text{ ft})(12\text{ ft/sec})}{0.0127\text{ ft}^2/\text{sec}} = 78.7 \quad (\textit{laminar})$$

(*a*) By (*H-7*),

$$h_\infty = \frac{k_b}{D}(3.66) = \left(\frac{0.165\text{ Btu/hr}\cdot\text{ft}\cdot{}^\circ\text{F}}{\frac{1}{12}\text{ ft}}\right)(3.66) = 7.25\text{ Btu/hr}\cdot\text{ft}^2\cdot{}^\circ\text{F}$$

(*b*) From Problem 4.81,

$$x_v \approx (0.05)\,\mathbf{Re}_{D,b}\,D = (0.05)(78.7)(\tfrac{1}{12}) = 0.33\text{ ft}$$
$$x_t \approx x_v\,\mathbf{Pr}_b = (0.33)(12.5\times 10^3) = 4100\text{ ft}$$

Note that the highly viscous glycerin, with its large value of **Pr**, requires an exorbitant length for development of the temperature profile. Thus, part (*a*) applies only to the last quarter-mile of tube.

6.15[D] Glycerin at $\bar{T}_b = 20\ ^\circ\text{C}$ flows in a 1.6-km-long, 25.4-mm-ID horizontal tube; the average bulk velocity is 3.66 m/s. The tube wall temperature is constant at 55 °C. Calculate (*a*) the (constant) value of the heat transfer coefficient in the fully developed (thermal and velocity) region; (*b*) the approximate lengths of the thermal and velocity inlet regions.

From Table B-3 at $\bar{T}_b = 20\ ^\circ\text{C}$,

$$\nu_b = (0.0127)(9.2903\times 10^{-2}) = 1.18\times 10^{-3}\text{ m}^2/\text{s}$$
$$k_b = (0.165)(1.729\,577) = 0.285\text{ W/m}\cdot\text{K}$$
$$\mathbf{Pr}_b = 12.5\times 10^3$$

The Reynolds number is

$$\mathbf{Re}_{D,b} = \frac{DV_b}{\nu_b} = \frac{(0.0254\text{ m})(3.66\text{ m/s})}{1.18\times 10^{-3}\text{ m}^2/\text{s}} = 78.8 \quad (\textit{laminar})$$

(*a*) By (*H-7*),

$$h_\infty = \frac{k_b}{D}(3.66) = \left(\frac{0.285\text{ W/m}\cdot\text{K}}{0.0254\text{ m}}\right)(3.66) = 41.07\text{ W/m}^2\cdot\text{K}$$

(*b*) From Problem 4.81,

$$x_v \approx (0.05)\,\mathbf{Re}_{D,b}\,D = (0.05)(78.8)(25.4\text{ mm}) = 100\text{ mm}$$
$$x_t \approx x_v\,\mathbf{Pr}_b = (0.100\text{ m})(12.5\times 10^3) = 1.25\text{ km}$$

As in Problem 6.13 we have an enormous thermal entry length.

6.16[D] For a fully developed velocity profile prior to a heated tube section (2.5 ft long, $T_s = \text{const.} = 150\ ^\circ\text{F}$), determine the average heat transfer coefficient for a flow of saturated benzene at $\bar{T}_b = 80\ ^\circ\text{F}$ and $V_b = 1.6\text{ ft/sec}$ in a 0.10-in.-ID tube.

Fluid properties at 80 °F are

$$\rho_b = 54.6\text{ lbm/ft}^3 \qquad k_b = 0.092\text{ Btu/hr}\cdot\text{ft}\cdot{}^\circ\text{F}$$
$$c_{p,b} = 0.42\text{ Btu/lbm}\cdot{}^\circ\text{F} \qquad \mu_b g_c = 3.96\times 10^{-4}\text{ lbm/ft}\cdot\text{sec}$$
$$\mathbf{Pr}_b = 6.5$$

The Reynolds number is

$$\mathbf{Re}_{D,b} = \frac{V_b D\rho_b}{\mu_b g_c} = \frac{(1.6\text{ ft/sec})[(0.10/12)\text{ ft}](54.6\text{ lbm/ft}^3)}{3.96\times 10^{-4}\text{ lbm/ft}\cdot\text{sec}} = 1838 \quad (\textit{laminar})$$

The thermal entry length is (Problem 4.81)

$$x_t \approx (0.05)D\,\mathbf{Re}_{D,b}\,\mathbf{Pr}_b = (0.05)\left(\frac{0.10}{12}\text{ ft}\right)(1838)(6.5) = 4.98\text{ ft}$$

Since the heated length is smaller than x_t, the temperature profile is developing; whereas the velocity profile is fully developed (parabolic). In this situation, (*H-9*) and Table H-1 apply:

$$\overline{\mathbf{Nu}} = 3.66 + \frac{(0.0668)[(\frac{1}{300})(1838)(6.5)]}{1 + (0.04)[(\frac{1}{300})(1838)(6.5)]^{2/3}} = 5.47$$

so that

$$\bar{h} \equiv \bar{h}(0, L) = \frac{k_b}{D}\,\mathbf{Nu} = \left(\frac{0.092}{0.10/12}\right)(5.47) = 60.4\ \text{Btu/hr}\cdot\text{ft}^2\cdot{}^\circ\text{F}$$

6.17 If the benzene inlet temperature in Problem 6.15 is $T_{b,i} = 60\ ^\circ\text{F}$, what is the exit temperature, $T_{b,e}$?

An iterative solution is necessary. To exploit the results of Problem 6.15, let the initial guess be $T_{b,e}^{(0)} = 100\ ^\circ\text{F}$; i.e., $\bar{T}_b^{(0)} = 80\ ^\circ\text{F}$. Then, with

$$\frac{-PL\bar{h}(0, L)}{\dot{m}c_{p,b}} = \frac{-\pi DL\bar{h}}{\rho_b(\pi D^2/4)V_b c_{p,b}} = \frac{-4L\bar{h}}{\rho_b DV_b c_{p,b}}$$

$$= \frac{-4(2.5\ \text{ft})(60.4\ \text{Btu/hr}\cdot\text{ft}^2\cdot{}^\circ\text{F})}{(54.6\ \text{lbm/ft}^3)(\frac{1}{120}\ \text{ft})(1.6\times 3600\ \text{ft/hr})(0.42\ \text{Btu/lbm}\cdot{}^\circ\text{F})} = -0.548$$

evaluated from Problem 6.15, equation (*H-3*) yields the next approximation

$$T_{b,e}^{(1)} = T_b(L) = (150\ ^\circ\text{F}) - [(150 - 60)\ ^\circ\text{F}]\exp(-0.548) = 98.0\ ^\circ\text{F}$$

To continue the iteration, one obtains fluid properties and $\bar{h}$ for the new mean bulk temperature, $\bar{T}_b^{(1)} = \frac{1}{2}(T_{b,i} + T_{b,e}^{(1)})$; calculates $T_{b,e}^{(2)}$ from (*H-3*); and so on. Luckily, in the present problem, $\bar{T}_b^{(1)} = 79\ ^\circ\text{F} \approx 80\ ^\circ\text{F} = \bar{T}_b^{(0)}$; so that $T_{b,e}^{(1)} = 98.0\ ^\circ\text{F}$ can rest as the final answer.

6.18^D For a fully developed velocity profile prior to a heated tube section (0.76 m long, $T_s = 66\ ^\circ\text{C}$), determine the average heat transfer coefficient for a flow of saturated benzene at $\bar{T}_b = 27\ ^\circ\text{C}$ and $V_b = 0.49\ \text{m/s}$ inside a 2.54-mm-ID tube.

Fluid properties at 27 °C are

$$\rho_b = 874.6\ \text{kg/m}^3 \qquad k_b = 0.159\ \text{W/m}\cdot\text{K}$$
$$c_{p,b} = 1757\ \text{J/kg}\cdot\text{K} \qquad \mu_b = 5.89\times 10^{-4}\ \text{Pa}\cdot\text{s}$$
$$\mathbf{Pr}_b = 6.5$$

The Reynolds number is

$$\mathbf{Re}_{D,b} = \frac{DV_b\rho_b}{\mu_b} = \frac{(2.54\times 10^{-3}\ \text{m})(0.49\ \text{m/s})(874.6\ \text{kg/m}^3)}{5.89\times 10^{-4}\ \text{Pa}\cdot\text{s}} = 1848 \quad (\textit{laminar})$$

The entry length is (Problem 4.81)

$$x_t \approx (0.05)D\,\mathbf{Re}_{D,b}\,\mathbf{Pr}_b = (0.05)(2.54\times 10^{-3}\ \text{m})(1848)(6.5) = 1.52\ \text{m}$$

The heated length, 0.76 m, is therefore within the region of thermal development; the velocity profile, on the other hand, will be taken as fully developed (parabolic). In this situation, (*H-9*) and Table H-1 apply:

$$\mathbf{Nu} = 3.66 + \frac{(0.0668)[(\frac{1}{300})(1848)(6.5)]}{1 + (0.04)[(\frac{1}{300})(1848)(6.5)]^{2/3}} = 5.48$$

or

$$\bar{h} \equiv \bar{h}(0, L) = \frac{k_b}{D}\,\mathbf{Nu} = \left(\frac{0.159}{2.54\times 10^{-3}}\right)(5.48) = 343\ \text{W/m}^2\cdot\text{K}$$

6.19^D In the preceding problems requiring evaluation of $\bar{h}$, the average bulk temperature was given. This is usually unknown; the common application/design problem instead provides $T_{b,i}$ and requires determination of $T_{b,e}$ (or $\bar{T}_b$ or length to achieve a required $T_{b,e}$). For heating water from 20 °C to 60 °C, an electrically heated tube is *proposed*. The tube has inside diameter 25 mm and has constant heat flux 5 kW/m². The mass flow is to be such that $\mathbf{Re}_{D,b} = 2000$ (i.e., the flow is to remain laminar). Determine **(*a*)** the length of tube required, and **(*b*)** if this proposed heating system is feasible.

From Table B-3, at $\bar{T}_b = (20 + 60)/2 = 40\ ^\circ\text{C}$,

$$\nu_b = 6.577\times 10^{-7}\ \text{m}^2/\text{s} \quad \mathbf{Pr}_b = 4.34 \quad \rho_b = 994.5\ \text{kg/m}^3 \quad k_b = 0.628\ \text{W/m}\cdot\text{K} \quad c_{p,b} = 4.176\ \text{kJ/kg}\cdot\text{K}$$

(*a*) At $\mathbf{Re}_{D,b} = 2000$,

$$V_b = \frac{2000\nu_b}{D} = \frac{2000(6.577\times 10^{-7}\ \text{m/s})}{0.025\ \text{m}} = 0.0526\ \text{m/s}$$

Then, with $\dot{m} = \rho_b V_b(\pi D^2/4)$, (*H-2*) gives, for $x = L$,

$$L = \frac{\rho_b V_b D c_{p,b}(T_{b,e} - T_{b,i})}{4q_s''} = \frac{(994.5\ \text{kg/m}^3)(0.0526\ \text{m/s})(0.025\ \text{m})(4.176\ \text{kJ/kg}\cdot\text{K})[(60-20)\ \text{K}]}{4(5\ \text{kW/m}^2)}$$

$$= 10.92\ \text{m}$$

(*b*) The question is: For a 10.92-m flow to absorb 5 kW/m^2 in heat, must the wall temperature be unrealistically high? Let us, then, test T_s at the midpoint, $x_{\text{mid}} = 5.46$ m, of the tube. Since (Problem 4.81)

$$x_v \approx (0.05) D\,\mathbf{Re}_{D,b} = (0.05)(0.025\ \text{m})(2000) = 2.5\ \text{m}$$

$$x_t \approx x_v\,\mathbf{Pr}_b = (2.5\ \text{m})(4.34) = 10.85\ \text{m}$$

the flow at x_{mid} is undeveloped in temperature and developed in velocity. Hence, by (*H-9*) and Table H-1,

$$\mathbf{Nu}\,(x_{\text{mid}}) = 4.36 + \frac{(0.023)[(0.025/5.46)(2000)(4.34)]}{1 + (0.0012)[(0.025/5.46)(2000)(4.34)]^{1.0}} = 5.23$$

and
$$h(x_{\text{mid}}) = \frac{k_b}{D}\,\mathbf{Nu}\,(x_{\text{mid}}) = \frac{0.628}{0.025}(5.23) = 131.4\ \text{W/m}^2\cdot\text{K}$$

Now we can solve the relation

$$q_s'' = h(x_{\text{mid}})[T_s(x_{\text{mid}}) - T_b(x_{\text{mid}})] \approx h(x_{\text{mid}})[T_s(x_{\text{mid}}) - \bar{T}_b]$$

to obtain $T_s(x_{\text{mid}}) \approx 78$ °C, a reasonable wall temperature for a *pressurized* water flow. The proposed design is feasible.

6.20^{D} A 1.0-in.-ID tube is electrically heated to provide a constant heat flux $q_s'' = 465$ W/ft^2. Water flows through the tube and has a fully developed (parabolic) velocity profile prior to the heated length. The water is to be heated from $T_{b,i} = 68$ °F to $T_{b,e} = 140$ °F. The water mass flow is 0.0575 lbm/sec. Determine (*a*) the length of tube required, and (*b*) if the proposed heating scheme is feasible.

▌ From Table B-3, at $\bar{T}_b = (68 + 140)/2 = 104$ °F,

$$\nu_b = 0.708 \times 10^{-5}\ \text{ft}^2/\text{sec} \qquad \mathbf{Pr}_b = 4.34 \qquad \rho_b = 62.09\ \text{lbm/ft}^3 \qquad k_b = 0.363\ \text{Btu/hr}\cdot\text{ft}\cdot{}^\circ\text{F}$$
$$c_{p,b} = 0.998\ \text{Btu/lbm}\cdot{}^\circ\text{F}$$

(*a*) The Reynolds number is

$$\mathbf{Re}_{D,b} = \frac{V_b D}{\nu_b} = \frac{4\dot{m}}{\pi D \rho_b \nu_b} = \frac{4(0.0575\ \text{lbm/sec})}{\pi(\frac{1}{12}\ \text{ft})(62.09\ \text{lbm/ft}^3)(0.708 \times 10^{-5}\ \text{ft}^2/\text{sec})} = 1999 \quad (\textit{still laminar})$$

By (*H-2*), for $x = L$,

$$L = \frac{\dot{m} c_{p,b}(T_{b,e} - T_{b,i})}{\pi D q_s''} = \frac{(0.0575\ \text{lbm/sec})(0.998\ \text{Btu/lbm}\cdot{}^\circ\text{F})[(140-68)\ {}^\circ\text{F}]}{\pi(\frac{1}{12}\ \text{ft})(465\ \text{W/ft}^2)\left(\dfrac{3.414\ \text{Btu}}{1\ \text{W}\cdot\text{hr}}\right)\left(\dfrac{1\ \text{hr}}{3600\ \text{sec}}\right)} = 35.8\ \text{ft}$$

(*b*) The question is: For a 35.8-ft flow to absorb 465 W/ft^2 in heat, is an excessive wall temperature demanded? Let us check the wall temperature at the midpoint, $x_{\text{mid}} = 17.9$ ft, of the heated section. From the entry length (Problem 4.81)

$$x_t \approx (0.05) D\,\mathbf{Re}_{D,b}\,\mathbf{Pr}_b = (0.05)(\tfrac{1}{12}\ \text{ft})(1999)(4.34) \approx 36\ \text{ft}$$

The entire length is in the developing region. For developing temperature and fully developed velocity profiles in laminar flow, (*H-9*) and Table H-1 give

$$\mathbf{Nu}\,(x_{\text{mid}}) = 4.36 + \frac{(0.023)[(\frac{1}{12}/17.9)(1999)(4.34)]}{1 + (0.0012)[(\frac{1}{12}/17.9)(1999)(4.34)]^{1.0}} = 5.25$$

and
$$h(x_{\text{mid}}) = \frac{k_b}{D}\,\mathbf{Nu}\,(x_{\text{mid}}) = \frac{0.363\ \text{Btu/hr}\cdot\text{ft}\cdot{}^\circ\text{F}}{\frac{1}{12}\ \text{ft}}(5.25) = 22.87\ \text{Btu/hr}\cdot\text{ft}^2\cdot{}^\circ\text{F}$$

Now we can solve the relation

$$q_s'' = h(x_{\text{mid}})[T_s(x_{\text{mid}}) - T_b(x_{\text{mid}})] \approx h(x_{\text{mid}})[T_s(x_{\text{mid}}) - \bar{T}_b]$$

to obtain $T_s(x_{\text{mid}}) \approx 173$ °F, which is not excessive. The proposed scheme is feasible.

6.21^D For a flow of 705 gpm of fuel oil at $\bar{T}_b = 50$ °F inside a 5.0-in.-ID tube, determine (*a*) the thermal and velocity entry lengths and (*b*) the value of $\bar{h}$ for a 5-ft-long tube. The following approximate values of properties may be assumed:

$$\nu_s = 0.76 \times 10^{-2}\ \text{ft}^2/\text{sec} \qquad \rho_s = 54.5\ \text{lbm/ft}^3$$
$$\nu_b = 1.6 \times 10^{-2}\ \text{ft}^2/\text{sec} \qquad \rho_b = 55.5\ \text{lbm/ft}^3$$
$$k_b = 0.08\ \text{Btu/hr}\cdot\text{ft}\cdot°\text{F} \qquad c_{p,b} = 0.499\ \text{Btu/lbm}\cdot°\text{F}$$

The tube wall temperature is constant, $T_s = 80$ °F, over the entire 5-ft length.

▮ (*a*) The Prandtl number is

$$\mathbf{Pr}_b = \frac{\rho_b \nu_b c_{p,b}}{k_b} = \frac{(55.5\ \text{lbm/ft}^3)(1.6\times10^{-2}\ \text{ft}^2/\text{sec})(3600\ \text{sec/hr})(0.499\ \text{Btu/lbm}\cdot°\text{F})}{0.08\ \text{Btu/hr}\cdot\text{ft}\cdot°\text{F}} = 19\,940$$

The Reynolds number is (with $Q \equiv$ volumetric flow)

$$\mathbf{Re}_{D,b} = \frac{4Q}{\pi D \nu_b} = \frac{4(705\ \text{gal/min})\left(\dfrac{1\ \text{min}}{60\ \text{sec}}\right)\left(\dfrac{1\ \text{ft}^3}{7.48\ \text{gal}}\right)}{\pi[(5.0/12)\ \text{ft}](1.6\times10^{-2}\ \text{ft}^2/\text{sec})} = 300 \quad (\textit{laminar})$$

Thus, from Problem 4.81,

$$x_v \approx (0.05)D\,\mathbf{Re}_{D,b} = (0.05)\left(\frac{5.0}{12}\ \text{ft}\right)(300) = 6.25\ \text{ft}$$

$$x_t \approx x_v\,\mathbf{Pr}_b = (6.25\ \text{ft})(19\,940) \approx 24\ \text{miles}$$

(*b*) Under the conditions established in (*a*), (*H-8*) holds. Thus, using $\mu \equiv \nu\rho/g_c$,

$$\bar{h} = (1.86)\left(\frac{k_b}{D}\right)\left(\frac{\mathbf{Re}_{D,b}\,\mathbf{Pr}_b}{L/D}\right)^{1/3}\left(\frac{\nu_b\rho_b}{\nu_s\rho_s}\right)^{0.14}$$

$$= (1.86)\left(\frac{0.08}{5.0/12}\right)\left[\frac{(300)(19\,940)}{12}\right]^{1/3}\left[\frac{(1.6)(55.5)}{(0.76)(54.5)}\right]^{0.14} = 31.8\ \text{Btu/hr}\cdot\text{ft}^2\cdot°\text{F}$$

6.22^D For a flow of 2.668 m³/min of fuel oil at $\bar{T}_b = 10$ °C in a 127-mm-ID tube, determine (*a*) the thermal and velocity entry lengths and (*b*) the average heat transfer coefficient for a 1.5-m-long tube kept at $T_s = 27$ °C. The following approximate property values may be used:

$$\nu_s = 7.06\times10^{-4}\ \text{m}^2/\text{s} \qquad \rho_s = 873\ \text{kg/m}^3$$
$$\nu_b = 14.9\times10^{-4}\ \text{m}^2/\text{s} \qquad \rho_b = 889\ \text{kg/m}^3$$
$$k_b = 0.138\ \text{W/m}\cdot\text{K} \qquad c_{p,b} = 2088\ \text{J/kg}\cdot\text{K}$$

▮ (*a*) The Reynolds number is ($Q \equiv$ volumetric flow)

$$\mathbf{Re}_{D,b} = \frac{4Q}{\pi D\nu_b} = \frac{4(2.668\ \text{m}^3/\text{min})}{\pi(0.127\ \text{m})(14.9\times10^{-4}\ \text{m}^2/\text{s})(60\ \text{s/min})} = 299 \quad (\textit{laminar})$$

The Prandtl number is

$$\mathbf{Pr}_b = \frac{\nu_b\rho_b c_{p,b}}{k_b} = \frac{(14.9\times10^{-4}\ \text{m}^2/\text{s})(889\ \text{kg/m}^3)(2088\ \text{J/kg}\cdot\text{K})}{0.138\ \text{J/s}\cdot\text{m}\cdot\text{K}} = 20\,040$$

Thus, from Problem 4.81,

$$x_v \approx (0.05)D\,\mathbf{Re}_{D,b} = (0.05)(0.019\ \text{m})(2000) = 1.9\ \text{m}$$
$$x_t \approx x_v\,\mathbf{Pr}_b = (1.9\ \text{m})(20\,040) \approx 38\ \text{km}$$

(*b*) Under the conditions established in (*a*), (*H-8*) holds. Thus, using $\mu = \nu\rho$,

$$\bar{h} = (1.86)\left(\frac{k_b}{D}\right)\left(\frac{\mathbf{Re}_{D,b}\,\mathbf{Pr}_b}{L/D}\right)^{1/3}\left(\frac{\nu_b\rho_b}{\nu_s\rho_s}\right)^{0.14}$$

$$= (1.86)\left(\frac{0.138}{0.127}\right)\left[\frac{(299)(20\,040)}{1.5/0.127}\right]^{1/3}\left[\frac{(14.9)(889)}{(7.06)(873)}\right]^{0.14} = 179.3\ \text{W/m}^2\cdot\text{K}$$

6.23^D For the fuel-oil flow of Problem 6.22, determine the rise in T_b that occurs in the 1.5-m length.

▌ (*H-3*) gives, for $x = L$,

$$T_{b,e} = T_s - (T_s - T_{b,i}) \exp\left(\frac{-\pi D L \bar{h}}{\rho_b Q c_{p,b}}\right) \tag{1}$$

Also, by definition,

$$T_{b,e} + T_{b,i} = 2\bar{T}_b \tag{2}$$

Equations (*1*) and (*2*) are two linear equations in $T_{b,i}$ and $T_{b,e}$, with coefficients and constant terms calculable from Problem 6.22. Solving, one finds $T_{b,i} = 9.99$ °C and $T_{b,e} = 10.01$ °C, for a temperature rise of only 0.02 °C.

This problem illustrates the fact that a "reasonably large" (≈180 W/m^2 · K) heat transfer coefficient may result in an insignificant change in fluid temperature if the fluid is highly viscous and the Reynolds number not too small (≈300). [To see this, write the argument of the exponential function in (*1*) as $-4L\bar{h}/\mu_b \, \mathbf{Re}_{D,b} c_{p,b}$.]

6.24^D Gasoline at $\bar{T}_b = 80$ °F flows in a $\frac{3}{4}$-in.-ID tube with an average velocity of 0.2 fps. The tube is 5 ft long and has a constant wall temperature $T_s = 100$ °F. The fluid properties may be taken as

$$\mu_s g_c = 35.1 \times 10^{-5} \text{ lbm/ft} \cdot \text{sec} \qquad \mathbf{Pr}_b = 6.5$$
$$\mu_b g_c = 39.6 \times 10^{-5} \text{ lbm/ft} \cdot \text{sec} \qquad k_b = 0.092 \text{ Btu/hr} \cdot \text{ft} \cdot \text{°F}$$
$$\nu_b = 0.725 \times 10^{-5} \text{ ft}^2\text{/sec} \qquad c_{p,b} = 0.42 \text{ Btu/lbm} \cdot \text{°F}$$

Determine (*a*) the velocity and thermal entry lengths and (*b*) the average heat transfer coefficient over the 5-ft length.

▌ (*a*) The Reynolds number is

$$\mathbf{Re}_{D,b} = \frac{DV_b}{\nu_b} = \frac{[(0.75/12)\text{ ft}](0.2\text{ ft/sec})}{0.725 \times 10^{-5}\text{ ft}^2\text{/sec}} = 1724 \quad (\textit{laminar})$$

By Problem 4.81,

$$x_v \approx (0.05) D\, \mathbf{Re}_{D,b} = (0.05)\left(\frac{0.75}{12}\text{ ft}\right)(1724) = 5.4 \text{ ft}$$

$$x_t \approx x_v \, \mathbf{Pr}_b = (5.4 \text{ ft})(6.5) = 35 \text{ ft}$$

(*b*) For laminar flow and developing velocity and thermal profiles, equation (*H-8*) applies:

$$\overline{\mathbf{Nu}} = (1.86)\left[\frac{(1724)(6.5)}{80/1}\right]^{1/3}\left(\frac{39.6}{35.1}\right)^{0.14} = 9.82$$

$$\bar{h} = \frac{k_b}{D}\overline{\mathbf{Nu}} = \left[\frac{0.092 \text{ Btu/hr} \cdot \text{ft} \cdot \text{°F}}{(0.75/12)\text{ ft}}\right](9.82) = 14.46 \text{ Btu/hr} \cdot \text{ft}^2 \cdot \text{°F}$$

6.25^D For the gasoline flow of Problem 6.24, determine the rise in T_b that occurs in the 5-ft length.

▌ (*H-3*) gives, for $x = L$,

$$T_{b,e} = T_s - (T_s - T_{b,i}) \exp\left[\frac{-4L\bar{h}}{(\mu_b g_c)\, \mathbf{Re}_{D,b}\, c_{p,b}}\right] \tag{1}$$

Also, by definition,

$$T_{b,e} + T_{b,i} = 2\bar{T}_b \tag{2}$$

Equations (*1*) and (*2*) are two linear equations in $T_{b,i}$ and $T_{b,e}$, with coefficients and constant terms calculable from Problem 6.24. Solving, one finds $T_{b,i} = 77.2$ °F and $T_{b,e} = 82.8$ °F, for a significant (cf. Problem 6.23) rise of 5.6 °F.

6.26^D Liquid Freon-12 at $T_{b,i} = -30$ °C enters a 0.61-m-long heated tube having a constant $T_s = -10$ °C; the mass flow is 0.012 kg/s and the tube ID is 25 mm. What is the discharge temperature $T_{b,e}$?

▌ The solution is iterative, as in Problem 6.17.

First iteration. As initial guess, take $T_{b,e}^{(0)} = -24$ °C, or $\bar{T}_b^{(0)} = -27$ °C. Then, from Table B-3,

$$\nu_b \approx (0.267 \times 10^{-5})(9.2903 \times 10^{-2}) = 2.48 \times 10^{-7}\ \text{m}^2/\text{s}$$
$$\rho_b \approx (92.43)(16.018) = 1481\ \text{kg/m}^3$$
$$k_b \approx (0.0402)(1.729\,577) = 0.0695\ \text{W/m}\cdot\text{K}$$
$$\nu_s \approx (0.238 \times 10^{-5})(9.2903 \times 10^{-2}) = 2.21 \times 10^{-7}\ \text{m}^2/\text{s}$$
$$\rho_s \approx (89.24)(16.018) = 1429\ \text{kg/m}^3$$
$$\mathbf{Pr}_b \approx 4.7$$
$$c_{p,b} \approx (0.2148)(4184) = 898.7\ \text{J/kg}\cdot\text{K}$$

The Reynolds number is

$$\mathbf{Re}_{D,b}^{(0)} = \frac{4\dot{m}}{\pi D \nu_b \rho_b} = \frac{4(0.012\ \text{kg/s})}{\pi(0.025\ \text{m})(2.48 \times 10^{-7}\ \text{m}^2/\text{s})(1481\ \text{kg/m}^3)} = 1664 \quad (\textit{laminar})$$

and the entry lengths are

$$x_v^{(0)} \approx (0.05) D\, \mathbf{Re}_{D,b}^{(0)} = (0.05)(0.025\ \text{m})(1664) \approx 2.08\ \text{m}$$
$$x_t^{(0)} \approx x_v^{(0)}\, \mathbf{Pr}_b \approx (2.08\ \text{m})(4.7) \approx 9.78\ \text{m}$$

Thus, the entire flow is in the developing velocity, developing temperature profile, region; (*H-8*) gives, with $\mu = \nu\rho$,

$$\overline{\mathbf{Nu}}^{(0)} = (1.86)\left[\frac{(1664)(4.7)}{0.61/0.025}\right]^{1/3}\left[\frac{(2.48)(1481)}{(2.21)(1429)}\right]^{0.14} = 13.0$$

$$\bar{h}^{(0)} = \left(\frac{0.0695}{0.025}\right)(13.0) = 36.14\ \text{W/m}^2\cdot\text{K}$$

With $\bar{h}^{(0)} = \bar{h}(0, L)$ determined, all parameters in (*H-3*) are known; that equation then yields

$$T_{b,e}^{(1)} = T_b(L) = -27.04\ ^\circ\text{C} \qquad \text{and} \qquad \bar{T}_b^{(1)} \approx -28.5\ ^\circ\text{C}$$

Second iteration. At $\bar{T}_b^{(1)}$, Table B-3 gives

$$\nu_b \approx (0.269 \times 10^{-5})(9.2903 \times 10^{-2}) = 2.49 \times 10^{-7}\ \text{m}^2/\text{s}$$
$$\rho_b \approx (92.71)(16.018) = 1485\ \text{kg/m}^3$$
$$k_b \approx (0.0402)(1.729\,577) = 0.0695\ \text{W/m}\cdot\text{K}$$
$$c_{p,b} \approx (0.2143)(4184) = 896.6\ \text{J/kg}\cdot\text{K}$$
$$\mathbf{Pr}_b \approx 4.74$$

The Reynolds number is

$$\mathbf{Re}_{D,b}^{(1)} = \frac{4\dot{m}}{\pi D \nu_b \rho_b} = \frac{4(0.012\ \text{kg/s})}{\pi(0.025\ \text{m})(2.49 \times 10^{-7}\ \text{m}^2/\text{s})(1485\ \text{kg/m}^3)} = 1653 \quad (\textit{laminar})$$

Clearly there is little need of proceeding because the answers will be very close to those of the first iteration. If we proceed, we obtain

$$x_v^{(1)} \approx 2.07\ \text{m} \qquad x_t^{(1)} \approx 9.79\ \text{m} \qquad \bar{h}^{(1)} = 36.22\ \text{W/m}^2\cdot\text{K}$$

and $T_{b,e}^{(2)} = -27.02\ ^\circ\text{C}$, which clearly is an insignificant change. In fact, the correlation equation (*H-8*) is probably no more accurate than ±20 percent.

6.27 Repeat the calculation of $\bar{h}^{(1)}$ in Problem 6.26, using the Hausen equation for the situation where the velocity profile is fully developed (parabolic profile) prior to the start of the 0.61-m heated tube length.

▌ For the parameter values belonging to the second iteration of Problem 6.26, (*H-9*) and Table H-1 give

$$\overline{\mathbf{Nu}} = 3.66 + \frac{(0.0668)[(0.025/0.61)(1653)(4.74)]}{1 + (0.04)[(0.025/0.61)(1653)(4.74)]^{2/3}} = 11.12$$

$$\bar{h}^{(1)} = \left(\frac{0.0695}{0.025}\right)(11.12) = 30.9\ \text{W/m}^2\cdot\text{K}$$

This decreased $\bar{h}$ may be due to the fully developed velocity profile in the Hausen equation; a developing profile would enhance $\bar{h}$.

6.28ᴰ Liquid Freon-12 at $T_{b,i} = -22$ °F enters a 2-ft-long, 1.0 in.-ID heated tube having a constant wall temperature $T_s = +14$ °F. The Freon-12 mass flowrate is 0.026 lbm/sec. What is the discharge temperature $T_{b,e}$?

I The solution is iterative, as in Problem 6.17.

First iteration. Guessing $T_{b,e}^{(0)} = -18$ °F, $\bar{T}_b^{(0)} = -20$ °F; from Table B-3:

$$\nu_b = 0.2699 \times 10^{-5}\ \text{ft}^2/\text{sec}$$
$$\rho_b = 92.78\ \text{lbm/ft}^3$$
$$k_b = 0.0401\ \text{Btu/hr}\cdot\text{ft}\cdot{}^\circ\text{F}$$
$$c_{p,b} = 0.2142\ \text{Btu/lbm}\cdot{}^\circ\text{F}$$
$$\mathbf{Pr}_b = 4.75$$
$$\nu_s = 0.238 \times 10^{-5}\ \text{ft}^2/\text{sec}$$
$$\rho_s = 89.24\ \text{lbm/ft}^3$$

The Reynolds number is

$$\mathbf{Re}_{D,b}^{(0)} = \frac{4\dot{m}}{\pi D \nu_b \rho_b} = \frac{4(0.026\ \text{lbm/sec})}{\pi(\frac{1}{12}\ \text{ft})(0.270 \times 10^{-5}\ \text{ft}^2/\text{sec})(92.78\ \text{lbm/ft}^3)} = 1586 \quad (\textit{laminar})$$

The velocity and thermal entry lengths are

$$x_v^{(0)} \approx (0.05) D\, \mathbf{Re}_{D,b}^{(0)} = (0.05)(\tfrac{1}{12}\ \text{ft})(1586) = 6.61\ \text{ft}$$
$$x_t^{(0)} \approx x_v^{(0)}\, \mathbf{Pr}_b = (6.61\ \text{ft})(4.75) = 31.4\ \text{ft}$$

Clearly, the 2-ft-long tube has flow in the region of developing velocity and thermal profile regions. By (*H-8*), with $\mu = \nu\rho$,

$$\overline{\mathbf{Nu}}^{(0)} = (1.86)\left[\frac{(1586)(4.75)}{\frac{24}{1}}\right]^{1/3}\left[\frac{(0.270)(92.78)}{(0.238)(89.24)}\right]^{0.14} = 12.94$$

$$\bar{h}^{(0)} = \left(\frac{0.0401}{\frac{1}{12}}\right)(12.94) = 6.23\ \text{Btu/hr}\cdot\text{ft}^2\cdot{}^\circ\text{F}$$

With $\bar{h}^{(0)} = \bar{h}(0, L)$ determined, all parameters in (*H-3*) are known; that equation then yields

$$T_{b,e}^{(1)} = T_b(L) = -16.6\ {}^\circ\text{F} \qquad \text{or} \qquad \bar{T}_b^{(1)} = -19.3\ {}^\circ\text{F}$$

The value of $\bar{T}_b^{(1)}$ is sufficiently close to $\bar{T}_b^{(0)}$ to make a second iteration unnecessary. Thus, $T_{b,e} \approx -17$ °F.

6.29ᴰ Gasoline at $\bar{T}_b = 27$ °C flows inside a circular tube of 19 mm ID. The bulk average velocity is 0.061 m/s and the tube is 1.5 m long. The flow starts at the heated tube inlet (no upstream developing section), and the inside tube surface temperature is constant, $T_s = 38$ °C. The fluid properties may be taken as

$$\mu_s = 5.223 \times 10^{-4}\ \text{Pa}\cdot\text{s} \qquad k_b = 0.1591\ \text{W/m}\cdot\text{K}$$
$$\mu_b = 5.892 \times 10^{-4}\ \text{Pa}\cdot\text{s} \qquad \rho_b = 874.6\ \text{kg/m}^3$$
$$\mathbf{Pr}_b = 6.5 \qquad c_{p,b} = 1757\ \text{J/kg}\cdot\text{K}$$

Determine (*a*) the velocity and thermal entry lengths, and (*b*) the average heat transfer coefficient over the 1.5-m length.

I (*a*) The Reynolds number is

$$\mathbf{Re}_{D,b} = \frac{D V_b \rho_b}{\mu_b} = \frac{(0.019\ \text{m})(0.061\ \text{m/s})(874.6\ \text{kg/m}^3)}{5.892 \times 10^{-4}\ \text{kg/m}\cdot\text{s}} = 1720 \quad (\textit{laminar})$$

By Problem 4.81,

$$x_v \approx (0.05) D\, \mathbf{Re}_{D,b} = (0.05)(0.019\ \text{m})(1720) = 1.63\ \text{m}$$
$$x_t \approx x_v\, \mathbf{Pr}_b = (1.63)(6.5) = 10.6\ \text{m}$$

(*b*) Since both the velocity and temperature profiles are "developing" in the 1.5-m-long tube, (*H-8*) applies:

$$\overline{\mathbf{Nu}} = (1.86)\left[\frac{(1720 \times 6.5)}{1.5/0.019}\right]^{1/3}\left(\frac{5.892}{5.223}\right)^{0.14} = 9.86$$

$$\bar{h} = \frac{k_b}{D}\,\overline{\mathbf{Nu}} = \left(\frac{0.1591\ \text{W/m}\cdot\text{K}}{0.019\ \text{m}}\right)(9.86) = 82.56\ \text{W/m}^2\cdot\text{K}$$

6.30 Suppose that the heated tube section of Problem 6.29 is preceded by an unheated straight section of the same diameter for a distance of 2.0 m. **(a)** Estimate the average heat transfer coefficient and **(b)** the rise in T_b over the 1.5-m heated length.

I (*a*) It is clear from Problem 6.29 that the laminar flow with a 2.0-m unheated starting length has a fully developed, parabolic velocity profile. Thus, by (*H-9*) and Table H-1,

$$\overline{\mathbf{Nu}} = 3.66 + \frac{(0.0668)[(D/L)\,\mathbf{Re}_D\,\mathbf{Pr}]}{1+(0.04)[(D/L)\,\mathbf{Re}_D\,\mathbf{Pr}]^{2/3}} = 3.66 + \frac{(0.0668)[(0.019/1.5)(1720)(6.5)]}{1+0.04[(0.019/1.5)(1720)(6.5)]^{2/3}} = 8.19$$

whence $(T_s = \text{const.})$

$$\bar{h} = \bar{h}(0, L) = \frac{k_b}{D}\,\overline{\mathbf{Nu}} = \left(\frac{0.1591\ \text{W/m}\cdot\text{K}}{0.019\ \text{m}}\right)(8.19) = 68.6\ \text{W/m}^2\cdot\text{K}$$

(*b*) Solve (*H-3*),

$$T_{b,e} = T_s - (T_s - T_{b,i})\exp\left(\frac{-4L\bar{h}}{\mu_b\,\mathbf{Re}_{D,b}\,c_{p,b}}\right)$$

and the definition

$$T_{b,e} + T_{b,i} \equiv 2\bar{T}_b$$

simultaneously for $T_{b,e}$ and $T_{b,i}$, taking all needed data (except $\bar{h}$) from Problem 6.29:

$$T_{b,e} = 28.3\ \text{°C} \qquad T_{b,i} = 25.7\ \text{°C} \qquad T_{b,e} - T_{b,i} = 2.6\ \text{°C}$$

6.31[D] Saturated water, at mean temperature $\bar{T}_b = 68$ °F, flows inside a 1.0-in.-ID circular tube that is electrically heated to provide a constant heat flux over its length. The mass flow is 0.065 lbm/sec. At a location 9.0 ft downstream of the tube inlet, $T_s = 100$ °F. Determine the total heat transfer rate for a 10-ft-long tube.

I From Table B-3, for saturated water at 68 °F,

$$\nu_b = 1.083\times10^{-5}\ \text{ft}^2/\text{sec} \qquad \mathbf{Pr}_b = 7.02$$
$$\rho_b = 62.46\ \text{lbm/ft}^3 \qquad c_{p,b} = 0.9988\ \text{Btu/lbm}\cdot\text{°F}$$
$$k_b = 0.345\ \text{Btu/hr}\cdot\text{ft}\cdot\text{°F}$$

The Reynolds number is

$$\mathbf{Re}_{D,b} = \frac{4\dot{m}}{\pi D\nu_b\rho_b} = \frac{4(0.065\ \text{lbm/sec})}{\pi(\frac{1}{12}\ \text{ft})(1.083\times10^{-5}\ \text{ft}^2/\text{sec})(62.46\ \text{lbm/ft}^3)} = 1468 \quad (\textit{laminar})$$

The velocity and temperature entry lengths are (Problem 4.81)

$$x_v \approx (0.05)D\,\mathbf{Re}_{D,b} = (0.05)(\tfrac{1}{12}\ \text{ft})(1468) \approx 6\ \text{ft}$$
$$x_t \approx x_v\,\mathbf{Pr}_b = (6\ \text{ft})(7.02) \approx 42\ \text{ft}$$

Clearly, $x = 9$ ft lies in the developing profile region for temperature, but the velocity profile is fully developed. For this case, (*H-9*) and Table H-1 yield

$$h_{x=9} = \frac{k_b}{D}\left\{4.36 + \frac{(0.023)[(D/x)\,\mathbf{Re}_D\,\mathbf{Pr}]}{1+(0.0012)[(D/x)\,\mathbf{Re}_D\,\mathbf{Pr}]^{1.0}}\right\}$$
$$= \left(\frac{0.345}{\frac{1}{12}}\right)\left\{4.36 + \frac{(0.023)[(\frac{1}{108})(1468)(7.02)]}{1+(0.0012)[(\frac{1}{108})(1468)(7.02)]}\right\} = 26.2\ \text{Btu/hr}\cdot\text{ft}^2\cdot\text{°F}$$

Another needed parameter value is

$$B \equiv \frac{P}{\dot{m}c_{p,b}} = \frac{\pi(\frac{1}{12}\ \text{ft})}{(0.065\ \text{lbm/sec})(3600\ \text{sec/hr})(0.9988\ \text{Btu/lbm}\cdot\text{°F})} = 1.12\times10^{-3}\ \text{hr}\cdot\text{ft}\cdot\text{°F/Btu}$$

We are now prepared to write a system of four linear equations in the four unknowns q_s'', $T_{b,i}$, $T_b(9\ \text{ft})$, and $T_{b,e}$:

$$1q_s'' + 0T_{b,i} + h_{x=9}T_b(9) + 0T_{b,e} = h_{x=9}T_s(9) \qquad (1)$$
$$(9\ \text{ft})Bq_s'' + 1T_{b,i} - 1T_b(9) + 0T_{b,e} = 0 \qquad (2)$$

$$(10\text{ ft})Bq_s'' + 1T_{b,i} + \quad 0T_b(9) - 1T_{b,e} = \qquad 0 \tag{3}$$

$$0q_s'' + 1T_{b,i} + \quad 0T_b(9) + 1T_{b,e} = \qquad 2\bar{T}_b \tag{4}$$

[Here, (*1*) is essentially the definition of $h_{x=9}$; (*2*) is (*H-2*) at $x = 9$ ft; (*3*) is (*H-2*) at $x = 10$ ft; and (*4*) is the definition of $\bar{T}_b$.] Solving for q_s'',

$$q_s'' = \frac{T_s(9) - \bar{T}_b}{(1/h_{x=9}) + (4\text{ ft})B} = \frac{(100 - 68)\ ^\circ\text{F}}{[(1/26.2) + (4)(1.12)]\ \text{hr}\cdot\text{ft}^2\cdot{}^\circ\text{F/Btu}} = 750.3\ \text{Btu/hr}\cdot\text{ft}^2 \tag{5}$$

whence $$q_{\text{conv}} = q_s''\pi DL = (750.3)\pi(\tfrac{1}{12})(10) = 1964\ \text{Btu/hr}$$

Looking at the "exact" solution (*5*), we realize that we have been overscrupulous: B is negligibly small, and we should have written $q_s'' = h_{x=9}[T_s(9) - \bar{T}_b]$, i.e., have approximated $T_b(9)$ by $\bar{T}_b$ straight off. This would have led to $q_{\text{conv}} = 2190$ Btu/hr.

6.32[D] Saturated water, at mean temperature $\bar{T}_b = 20\ ^\circ\text{C}$, flows inside a 25-mm-ID circular tube which is electrically heated to provide a constant heat flux over its entire length. The mass flowrate is 0.0295 kg/s. At a location 2.74 m downstream of the tube inlet, $T_s = 38\ ^\circ\text{C}$. Determine the total heat transfer rate for a 3.05-m-long tube.

From Table B-3, for saturated water at 20 °C,

$$\nu_b = (1.083\times 10^{-5})(9.2903\times 10^{-2}) = 1.006\times 10^{-6}\ \text{m}^2/\text{s} \qquad k_b = (0.345)(1.7295) = 0.597\ \text{W/m}\cdot\text{K}$$
$$\rho_b = (62.46)(16.018) = 1000.5\ \text{kg/m}^3 \qquad \mathbf{Pr}_b = 7.02$$

The Reynolds number is

$$\mathbf{Re}_{D,b} = \frac{4\dot{m}}{\pi D\nu_b\rho_b} = \frac{4(0.0295\ \text{kg/s})}{\pi(0.025\ \text{m})(1.006\times 10^{-6}\ \text{m}^2/\text{s})(1000.5\ \text{kg/m}^3)} = 1493 \quad (\textit{laminar})$$

The velocity and thermal entry lengths are

$$x_v \approx (0.05)D\,\mathbf{Re}_{D,b} = (0.05)(0.025\ \text{m})(1493) = 1.87\ \text{m}$$
$$x_t \approx x_v\,\mathbf{Pr}_b = (1.87\ \text{m})(7.02) = 13.13\ \text{m}$$

At $x = 2.74$ m, the flow is fully developed in velocity, developing in temperature. Thus (*H-9*) and Table H-1 yield

$$h_{2.74} = \frac{k_b}{D}\left\{4.36 + \frac{(0.023)[(D/x)\,\mathbf{Re}_D\,\mathbf{Pr}]}{1 + (0.0012)[(D/x)\,\mathbf{Re}_D\,\mathbf{Pr}]^{1.0}}\right\}$$
$$= \left(\frac{0.597}{0.025}\right)\left\{4.36 + \frac{(0.023)[(0.025/2.74)(1493)(7.02)]}{1 + (0.0012)[(0.025/2.74)(1493)(7.02)]}\right\} = 151.2\ \text{W/m}^2\cdot\text{K}$$

The discussion of Problem 6.31 justifies the approximation $T_{b,2.74} \approx \bar{T}_b$; hence

$$q_s'' \approx h_{2.74}[T_s(2.74) - \bar{T}_b] = (151.2)(38 - 20) = 2722\ \text{W/m}^2$$
$$q_{\text{conv}} = q_s''\pi DL \approx (2722)\pi(0.025)(3.05) = 652\ \text{W}$$

6.33[D] Air at $\bar{T}_b = 80\ ^\circ\text{F}$ and a pressure of 90 psia flows inside a 1.0-in.-ID circular tube with average bulk velocity $V_b = 0.5$ ft/sec. The tube wall temperature is $T_s = 170\ ^\circ\text{F}$, a constant. For a tube length of 3.0 ft, determine (*a*) the average heat transfer coefficient and (*b*) the air bulk temperature rise.

From Table B-4, at $\bar{T}_b = 80\ ^\circ\text{F}$ and $T_s = 170\ ^\circ\text{F}$,

$$\mu_b g_c = 1.241\times 10^{-5}\ \text{lbm/ft}\cdot\text{sec} \qquad \rho_{b,1} = 0.0735\ \text{lbm/ft}^3\ \ (\text{at 1 atm pressure})$$
$$k_b = 0.015\,16\ \text{Btu/hr}\cdot\text{ft}\cdot{}^\circ\text{F} \qquad \mu_s g_c = 1.394\times 10^{-5}\ \text{lbm/ft}\cdot\text{sec}$$
$$\mathbf{Pr}_b = 0.708 \qquad c_{p,b} = 0.2402\ \text{Btu/lbm}\cdot{}^\circ\text{F}$$

(*a*) The pressure-corrected air density is

$$\rho_b = \rho_{b,1}\left(\frac{p}{p_1}\right) = (0.0735\ \text{lbm/ft}^3)\left(\frac{90}{14.7}\right) = 0.450\ \text{lbm/ft}^3$$

so that

$$\mathbf{Re}_{D,b} = \frac{\rho_b D V_b}{\mu_b g_c} = \frac{(0.45\ \text{lbm/ft}^3)(\frac{1}{12}\ \text{ft})(0.5\ \text{ft/sec})}{1.241\times 10^{-5}\ \text{lbm/ft}\cdot\text{sec}} = 1511 \quad (\textit{laminar})$$

The entry lengths are (see Problem 4.81)

$$x_v \approx (0.05)(\tfrac{1}{12}\text{ ft})(1511) = 6.3\text{ ft}$$

$$x_t \approx (6.3\text{ ft})(0.708) = 4.46\text{ ft}$$

Thus, the 3-ft-long heated tube is experiencing developing velocity and thermal profiles; (*H-8*) gives

$$\overline{\mathbf{Nu}} = (1.86)\left[\frac{(1511 \times 0.708)}{36/1}\right]^{1/3}\left(\frac{1.241}{1.394}\right)^{0.14} = 5.67$$

$$\bar{h} = \frac{k_b}{D}\overline{\mathbf{Nu}} = \left(\frac{0.015\,16\text{ Btu/hr}\cdot\text{ft}\cdot{}^\circ\text{F}}{\frac{1}{12}\text{ ft}}\right)(5.67) = 1.03\text{ Btu/hr}\cdot\text{ft}^2\cdot{}^\circ\text{F}$$

(*b*) The inlet and exit air temperatures are determined by simultaneous solution of (*H-3*),

$$T_{b,e} = T_b(L) = T_s - (T_s - T_{b,i})\exp\left[\frac{-4L\bar{h}}{(\mu_b g_c)\,\mathbf{Re}_{D,b}\,c_{p,b}}\right]$$

and the definition

$$T_{b,e} + T_{b,i} \equiv 2\bar{T}_b$$

Thus, $T_{b,i} = 47.3\ {}^\circ\text{F}$ and $T_{b,e} = 112.7\ {}^\circ\text{F}$, for a rise of 65.4 °F.

6.34[D] Air at $\bar{T}_b = 27\ {}^\circ\text{C}$ and a pressure of 620.5 kPa flows inside a 25.4-mm-ID circular tube with average bulk velocity 0.15 m/s. The tube wall temperature is $T_s = 77\ {}^\circ\text{C}$, a constant. For a tube length of 0.91 m, determine (*a*) the average heat transfer coefficient and (*b*) the rise in the fluid bulk temperature.

I From Table B-4, at $\bar{T}_b = 27\ {}^\circ\text{C}$ and $T_s = 77\ {}^\circ\text{C}$,

$$\mu_b = (1.241 \times 10^{-5})(1.488\,164) = 1.847 \times 10^{-5}\text{ Pa}\cdot\text{s}$$
$$k_b = (0.015\,16)(1.729\,577) = 0.026\,22\text{ W/m}\cdot\text{K}$$
$$\rho_{b,1} = (0.0735)(1.6018 \times 10^1) = 1.177\text{ kg/m}^3 \quad (\text{at } 101.325\text{ kPa})$$
$$c_{p,b} = (0.2402)(4184) = 1005\text{ J/kg}\cdot\text{K}$$
$$\mathbf{Pr}_b = 0.708$$
$$\mu_s = (1.394 \times 10^{-5})(1.488\,164) = 2.074 \times 10^{-5}\text{ Pa}\cdot\text{s}$$

(*a*) The pressure-corrected air density is

$$\rho_b = \rho_{b,1}\left(\frac{p}{p_1}\right) = \left(1.177\text{ kg/m}^3\right)\left(\frac{620.5}{101.325}\right) = 7.208\text{ kg/m}^3$$

so that

$$\mathbf{Re}_{D,b} = \frac{\rho_b D V_b}{\mu_b} = \frac{(7.208\text{ kg/m}^3)(0.0254\text{ m})(0.15\text{ m/s})}{1.847 \times 10^{-5}\text{ kg/m}\cdot\text{s}} = 1487 \quad (\textit{laminar})$$

The entry lengths are (see Problem 4.81)

$$x_v \approx (0.05)(0.0254\text{ m})(1487) = 1.89\text{ m}$$

$$x_t \approx (1.89\text{ m})(0.708) = 1.34\text{ m}$$

Thus the 0.91-m tube is in the developing velocity, developing thermal profile region; (*H-8*) gives

$$\overline{\mathbf{Nu}} = (1.86)\left[\frac{(1487)(0.708)}{0.91/0.0254}\right]^{1/3}\left(\frac{1.847}{2.074}\right)^{0.14} = 5.65$$

$$\bar{h} = \frac{k_b}{D}\overline{\mathbf{Nu}}\left(\frac{0.026\,22\text{ W/m}\cdot\text{K}}{0.0254\text{ m}}\right)(5.65) = 5.83\text{ W/m}^2\cdot\text{K}$$

(*b*) The inlet and exit air temperatures are determined by simultaneous solution of (*H-3*),

$$T_{b,e} = T_b(L) = T_s - (T_s - T_{b,i})\exp\left(\frac{-4L\bar{h}}{\mu_b\,\mathbf{Re}_{D,b}\,c_{p,b}}\right)$$

and the definition

$$T_{b,e} + T_{b,i} = 2\bar{T}_b$$

Thus, $T_{b,i} = 8.65\ {}^\circ\text{C}$ and $T_{b,e} = 45.35\ {}^\circ\text{C}$, for a rise of 36.7 °C.

6.35^D Rework Problem 6.33(*a*) if the 3-ft heated section is preceded by an 8-ft unheated section.

▮ With $x_v \approx 6.3\text{ ft} < 8\text{ ft}$ and $x_t \approx 4.46\text{ ft} > 3\text{ ft}$, the Hausen equation, (*H-9*), applies to the heated section:

$$\overline{\mathbf{Nu}} = 3.66 + \frac{(0.0668)[(D/L)\,\mathbf{Re}_D\,\mathbf{Pr}]}{1+(0.04)[(D/L)\,\mathbf{Re}_D\,\mathbf{Pr}]^{2/3}} = 3.66 + \frac{(0.0668)[(\frac{1}{36})(1511)(0.708)]}{1+(0.04)[(\frac{1}{36})(1511)(0.708)]^{2/3}} = 5.094$$

$$\bar{h} = \frac{k_b}{D}\,\overline{\mathbf{Nu}} = \left(\frac{0.015\,16\text{ Btu/hr}\cdot\text{ft}\cdot{}^\circ\text{F}}{\frac{1}{12}\text{ ft}}\right)(5.094) = 0.927\text{ Btu/hr}\cdot\text{ft}^2\cdot{}^\circ\text{F}$$

The more "active" developing velocity profile in Problem 6.33 gave a somewhat higher $\bar{h}$.

6.36^D Rework Problem 6.34(*a*) if the 0.91-m heated section is preceded by a 2.5-m unheated section.

▮ With $x_v \approx 1.89\text{ m} < 2.5\text{ m}$ and $x_t \approx 1.34\text{ m} > 0.91\text{ m}$, the Hausen equation, (*H-9*), applies to the heated section:

$$\overline{\mathbf{Nu}} = 3.66 + \frac{(0.0668)[(D/L)\,\mathbf{Re}_D\,\mathbf{Pr}]}{1+(0.04)[(D/L)\,\mathbf{Re}_D\,\mathbf{Pr}]^{2/3}} = 3.66 + \frac{(0.0668)[(0.0254/0.91)(1487)(0.708)]}{1+(0.04)[(0.0254/0.91)(1487)(0.708)]^{2/3}} = 5.08$$

$$\bar{h} = \frac{k_b}{D}\,\overline{\mathbf{Nu}} = \left(\frac{0.026\,22\text{ W/m}\cdot\text{K}}{0.0254\text{ m}}\right)(5.08) = 5.24\text{ W/m}^2\cdot\text{K}$$

6.37^D Rework Problem 6.33(*a*), using the Hausen equation to determine $\bar{h}$. Note that $\mathbf{Pr}_b = 0.708$, so that the Hausen equation for constant T_s, developing velocity profile, $\mathbf{Pr} = 0.7$ may be applied.

▮
$$\overline{\mathbf{Nu}} = 3.66 + \frac{(0.104)[(\frac{1}{36})(1511)(0.708)]}{1+(0.016)[(\frac{1}{36})(1511)(0.708)]^{0.8}} = 6.15$$

$$\bar{h} = \frac{k_b}{D}\,\overline{\mathbf{Nu}} = \left(\frac{0.015\,16\text{ Btu/hr}\cdot\text{ft}\cdot{}^\circ\text{F}}{\frac{1}{12}\text{ ft}}\right)(6.15) = 1.12\text{ Btu/hr}\cdot\text{ft}^2\cdot{}^\circ\text{F}$$

The Hausen equation yields an answer 8.6 percent larger than did the Sieder–Tate equation for the same conditions in Problem 6.33. Bear in mind that no convection correlation equation is more accurate than ±15–20 percent.

6.38^D Rework Problem 6.34(*a*) using the Hausen equation to determine $\bar{h}$. Note that $\mathbf{Pr}_b = 0.708$, so that the Hausen equation for constant T_s, developing velocity profile, $\mathbf{Pr} = 0.7$ may be applied.

▮
$$\overline{\mathbf{Nu}} = 3.66 + \frac{(0.104)[(0.0254/0.91)(1487)(0.708)]}{1+(0.016)[(0.0254/0.91)(1487)(0.708)]^{0.8}} = 6.13$$

$$\bar{h} = \frac{k_b}{D}\,\overline{\mathbf{Nu}} = \left(\frac{0.026\,22\text{ W/m}\cdot\text{K}}{0.0254\text{ m}}\right)(6.13) = 6.33\text{ W/m}^2\cdot\text{K}$$

This answer is 8.6 percent higher than that given by the Sieder–Tate equation for the same conditions in Problem 6.34. The difference is not unusual since neither correlation is expected to be better than within 15–20 percent error.

6.39^D In Problem 6.34 the condition $T_s = 77\ {}^\circ\text{C} = \text{const.}$ is changed to $q_s'' = 327.5\text{ W/m}^2 = \text{const.}$ (uniform electric resistance heating). Determine (*a*) the local heat transfer coefficient at the midsection of the tube, and (*b*) the temperature rise of the air.

▮ (*a*) Using data from Problem 6.34 in (*H-9*) yields

$$\mathbf{Nu}\left(\frac{L}{2}\right) = 4.36 + \frac{(0.036)[(0.0254/0.455)(1487)(0.708)]}{1+(0.0011)[(0.0254/0.455)(1487)(0.0708)]^{1.0}} = 6.35$$

$$h_{L/2} = \frac{k_b}{D}\,\mathbf{Nu}\left(\frac{L}{2}\right) = \left(\frac{0.026\,22\text{ W/m}\cdot\text{K}}{0.0254\text{ m}}\right)(6.35) = 6.55\text{ W/m}^2\cdot\text{K}$$

(*b*) By (*H-2*),

$$T_{b,e} - T_{b,i} = \frac{q_s''PL}{\dot{m}c_{p,b}} = \frac{4q_s''L}{\rho_b D V_b c_{p,b}}$$

$$= \frac{4(327.5\text{ W/m}^2)(0.91\text{ m})}{(7.208\text{ kg/m}^3)(0.0254\text{ m})(0.15\text{ m/s})(1005\text{ J/kg}\cdot{}^\circ\text{C})} = 43.2\ {}^\circ\text{C}$$

6.40[D] In Problem 6.33 the condition $T_s = 170\ °\text{F} = \text{const.}$ is changed to $q_s'' = 10^4\ \text{Btu/hr}\cdot\text{ft}^2 = \text{const.}$ (uniform electric resistance heating). Determine (*a*) the local heat transfer coefficient at the midsection of the tube, and (*b*) the temperature rise of the air.

▌ (*a*) Using data from Problem 6.33 in (*H-9*) yields

$$\mathbf{Nu}\,(L/2) = 4.36 + \frac{(0.036)[(\frac{1}{18})(1511)(0.708)]}{1 + (0.0011)[(\frac{1}{18})(1511)(0.708)]^{1.0}} = 6.37$$

$$h_{L/2} = \frac{k_b}{D}\,\mathbf{Nu}\,(L/2) = \left(\frac{0.015\,16\ \text{Btu/hr}\cdot\text{ft}\cdot°\text{F}}{\frac{1}{12}\ \text{ft}}\right)(6.37) = 1.16\ \text{Btu/hr}\cdot\text{ft}^2\cdot°\text{F}$$

(*b*) By (*H-2*),

$$T_{b,e} - T_{b,i} = \frac{q_s'' PL}{\dot m c_{p,b}} = \frac{4q_s'' L}{\rho_b V_b D c_{p,b}}$$

$$= \frac{4(104\ \text{Btu/hr}\cdot\text{ft}^2)(3.0\ \text{ft})}{(0.45\ \text{lbm/ft}^3)(0.5\ \text{ft/sec})(3600\ \text{sec/hr})(\frac{1}{12}\ \text{ft})(0.2402\ \text{Btu/lbm}\cdot°\text{F})} = 77\ °\text{F}$$

6.41 In terms of the Graetz number,

$$\mathbf{Gz} \equiv \frac{\dot m c_{p,b}}{k_b x} = \frac{\mathbf{Re}_D\ \mathbf{Pr}}{x/D}$$

where the second form is restricted to a circular tube, (*H-9*) takes the form

$$\mathbf{Nu}\,(x) = \mathbf{Nu}_{D,\infty} + \frac{K_1\,\mathbf{Gz}}{1 + K_2\,\mathbf{Gz}^n}$$

Another empirical correlation is graphed in Fig. 6-4. Use the graph to check the result of Problem 6.39(*a*).

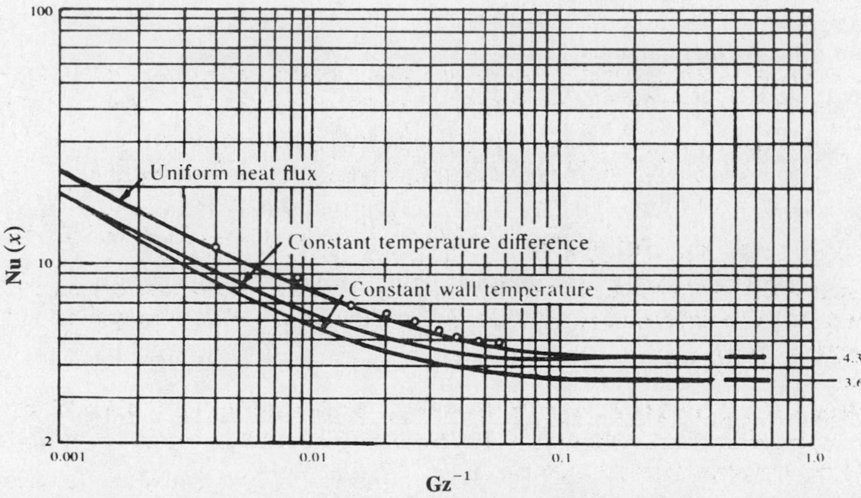

Fig. 6-4. Nusselt number in entry region of tube with developing velocity profile. (Adapted by permission from Knudsen, J. G. and D. L. Katz, *Fluid Dynamics and Heat Transfer*, McGraw-Hill, New York, 1958.)

▌ Using data from Problem 6.34, we have for $x = 455$ mm:

$$\mathbf{Gz}^{-1} = \frac{x/D}{\mathbf{Re}_D\ \mathbf{Pr}} = \frac{455/25.4}{(1487)(0.708)} = 0.017$$

Figure 6-4 gives $\mathbf{Nu}\,(455) \approx 6.7$, whence

$$h_{455} = \frac{k_b}{D}\,\mathbf{Nu}\,(455) \approx \left(\frac{0.026\,22}{0.0254}\right)(6.7) = 6.9\ \text{W/m}^2\cdot\text{K}$$

which is about 5 percent higher than as given by the Hausen equation.

6.42 In Problem 6.33 the condition of *constant wall temperature* T_s is replaced by one of *constant temperature difference* $T_s - T_b$. Determine $h_{L/2}$.

▮ Use the appropriate curve in Fig. 6-4. With data from Problem 6.33,

$$\mathbf{Gz}^{-1} = \frac{x/D}{\mathbf{Re}_D\,\mathbf{Pr}} = \frac{\frac{18}{1}}{(1511)(0.708)} = 0.0168$$

From Fig. 6-4, $\mathbf{Nu}\,(1.5) \approx 5.2$; and so

$$h_{1.5} = \frac{k_b}{D}\,\mathbf{Nu}\,(1.5) \approx \left(\frac{0.015\,16}{\frac{1}{12}}\right)(5.2) = 0.95\ \text{Btu/hr}\cdot\text{ft}^2\cdot{}^\circ\text{F}$$

6.43 Rework Problem 6.39 if a 2.5-m unheated entry section is attached.

▮ **(a)** The heated section now receives a fully developed (parabolic) velocity profile; (*H-9*) gives

$$\mathbf{Nu}\left(\frac{L}{2}\right) = 4.36 + \frac{(0.023)[(0.0254/0.455)(1487)(0.708)]}{1 + (0.012)[(0.0254/0.455)(1487)(0.708)]^{1.0}} = 5.62$$

$$h_{L/2} = \frac{k_b}{D}\,\mathbf{Nu}\left(\frac{L}{2}\right) = \left(\frac{0.026\,22\ \text{W/m}\cdot\text{K}}{0.0254\ \text{m}}\right)(5.62) = 5.80\ \text{W/m}^2\cdot\text{K}$$

(b) Exactly as in Problem 6.39(*b*), $T_{b,e} - T_{b,i} = 43.2\ {}^\circ\text{C}$.

6.44 Compare the results of Problems 6.39 (developing velocity) and 6.43 (parabolic velocity).

▮ The parabolic case shows a lower heat transfer coefficient at $x = L/2$ (and, it is easily seen, at any other location x). Hence, at each x, $T_s - T_b = q''_s/h_x$ is larger in the parabolic case. But (*H-2*) implies the same $T_b(x)$ for the two cases. Therefore, at each point, *the wall temperature is higher for the velocity-developed flow*.

6.45^D Gaseous nitrogen at $\bar{T}_b = 77\ {}^\circ\text{C}$ flows inside a 20-mm-ID circular tube at a pressure of 150 kPa and with a mean velocity of 0.50 m/s. If the tube is heated to $T_s = 127\ {}^\circ\text{C}$ (constant) over its thermal entry length (developing temperature region) only, determine **(a)** the entry lengths and **(b)** the average heat transfer coefficient over the thermal entry length.

▮ From Table B-4, at $\bar{T}_b = 77\ {}^\circ\text{C}$ and $T_s = 127\ {}^\circ\text{C}$,

$$\mu_b = (13.38 \times 10^{-6})(1.488\,164) = 19.91\ \mu\text{Pa}\cdot\text{s}$$
$$\rho_{b,1} = (0.0623)(16.018\,46) = 0.998\ \text{kg/m}^3 \quad (\text{at } 101.3\ \text{kPa})$$
$$k_b = (0.017\,205)(1.729\,577) = 0.029\,76\ \text{W/m}\cdot\text{K}$$
$$\mathbf{Pr}_b = 0.702$$
$$c_{p,b} = (0.2492)(4.184 \times 10^3) = 1043\ \text{J/kg}\cdot\text{K}$$
$$\mu_s = 14.77 \times 10^{-6} \times 1.488\,164 = 21.98\ \mu\text{Pa}\cdot\text{s}$$

The corrected density is $\rho_b = (150/101.3)(0.998) = 1.478\ \text{kg/m}^3$, giving a Reynolds number

$$\mathbf{Re}_{D,b} = \frac{DV_b\rho_b}{\mu_b} = \frac{(0.020\ \text{m})(0.50\ \text{m/s})(1.478\ \text{kg/m}^3)}{19.91 \times 10^{-6}\ \text{Pa}\cdot\text{s}} = 742.3 \quad (\textit{laminar})$$

(a)

$$x_v = (0.05)D\,\mathbf{Re}_{D,b} = (0.05)(0.020\ \text{m})(742.3) = 0.74\ \text{m}$$
$$x_t = x_v\,\mathbf{Pr} = (0.74\ \text{m})(0.702) = 0.52\ \text{m}$$

(b) With both profiles in development and a constant T_s, (*H-8*) gives

$$\overline{\mathbf{Nu}} = (1.86)\left[\frac{(742.3)(0.702)}{0.52/0.02}\right]^{1/3}\left(\frac{19.91}{21.98}\right)^{0.14} = 4.98$$

$$\bar{h} = \frac{k_b}{D}\,\overline{\mathbf{Nu}} = \left(\frac{0.029\,76\ \text{W/m}\cdot\text{K}}{0.02\ \text{m}}\right)(4.98) = 7.41\ \text{W/m}^2\cdot\text{K}$$

6.46^D Repeat Problem 6.45 for the following data: $\bar{T}_b = 170\ {}^\circ\text{F}$, $T_s = 260\ {}^\circ\text{F}$, tube ID = 0.79 in., $V_b =$ 1.64 ft/sec, tube pressure = 21.8 psia.

▮ From Table B-4, at $\bar{T}_b = 170\ {}^\circ\text{F}$ and $T_s = 260\ {}^\circ\text{F}$,

$$\mu_b g_c = 13.38 \times 10^{-6}\ \text{lbm/ft}\cdot\text{sec}$$
$$\rho_{b,1} = 0.0623\ \text{lbm/ft}^3 \quad \text{[at 14.7 psia]}$$
$$k_b = 0.017\,205\ \text{Btu/hr}\cdot\text{ft}\cdot{}^\circ\text{F}$$
$$\mathbf{Pr}_b = 0.702$$
$$\mu_s g_c = 14.77 \times 10^{-6}\ \text{lbm/ft}\cdot\text{sec}$$

The N_2 density is $\rho_b = (21.8/14.7)(0.0623) = 0.0924\ \text{lbm/ft}^3$, and so the Reynolds number is

$$\mathbf{Re}_{D,b} = \frac{DV_b\rho_b}{\mu_b g_c} = \frac{[(0.79/12)\ \text{ft}](1.64\ \text{ft/sec})(0.0924\ \text{lbm/ft}^3)}{13.38 \times 10^{-6}\ \text{lbm/ft}\cdot\text{sec}} = 746 \quad (laminar)$$

(*a*)
$$x_v \approx (0.05)D\,\mathbf{Re}_{D,b} = (0.05)(0.79/12)(746) = 2.45\ \text{ft}$$
$$x_t \approx x_v\,\mathbf{Pr}_b = (2.45)(0.702) = 1.72\ \text{ft}$$

(*b*) With both profiles in development and a constant T_s, (*H-8*) gives

$$\overline{\mathbf{Nu}} = (1.86)\left[\frac{(746)(0.702)}{(1.72 \times 12)/0.79}\right]^{1/3}\left(\frac{13.38}{14.77}\right)^{0.14} = 4.98$$

$$\bar{h} = \frac{k_b}{D}\overline{\mathbf{Nu}} = \left[\frac{0.017\,205\ \text{Btu/hr}\cdot\text{ft}\cdot{}^\circ\text{F}}{(0.79/12)\ \text{ft}}\right](4.98) = 1.3\ \text{Btu/hr}\cdot\text{ft}^2\cdot{}^\circ\text{F}$$

6.47[D] For Problem 6.45, (*a*) find $\bar{h}$ (over the heated length) from the Hausen equation; (*b*) calculate the net temperature rise (over the heated length) of the N_2.

▌ (*a*) By (*H-9*), with parameter values from Problem 6.45,

$$\overline{\mathbf{Nu}} = 3.66 + \frac{(0.104)[(0.020/0.52)(742.3)(0.702)]}{1 + (0.016)[(0.020/0.52)(742.3)(0.702)]^{0.8}} = 5.43$$

$$\bar{h}(0, L) = \frac{k_b}{D}\overline{\mathbf{Nu}} = \frac{0.029\,76\ \text{W/m}\cdot\text{K}}{0.020\ \text{m}}(5.43) = 8.08\ \text{W/m}^2\cdot\text{K}$$

which is about 9 percent higher than the result of Problem 6.45(*b*).

(*b*) Together, (*H-3*) for $x = L$,

$$T_{b,e} = T_s - (T_s - T_{b,i})\exp\left[\frac{-4L\bar{h}(0, L)}{DV_b\rho_b c_{p,b}}\right]$$

and the definition

$$T_{b,e} + T_{b,i} \equiv 2\bar{T}_b$$

determine $T_{b,i} = 52.2\ {}^\circ\text{C}$, $T_{b,e} = 101.8\ {}^\circ\text{C}$, and hence $T_{b,e} - T_{b,i} = 49.6\ {}^\circ\text{C}$.

6.48 Consider (as in Problem 6.45) the heating of N_2 gas inside a 20-mm-ID tube. Again, the gas pressure is 150 kPa; the mass flow is 0.56 kg/h. If the inlet temperature is $T_{b,i} = 60\ {}^\circ\text{C}$ and the wall temperature is a constant $T_s = 127\ {}^\circ\text{C}$, what length L of tube is required to make $T_{b,e} = 90\ {}^\circ\text{C}$?

▌ From Table B-4, at $\bar{T}_b = 75\ {}^\circ\text{C}$ and $T_s = 127\ {}^\circ\text{C}$,

$$\mu_b = (13.32 \times 10^{-6})(1.488\,164) = 19.82\ \mu\text{Pa}\cdot\text{s}$$
$$k_b = (0.017\,12)(1.729\,577) = 0.029\,61\ \text{W/m}\cdot\text{K}$$
$$\rho_{b,1} = (0.0627)(16.018\,46) = 1.004\ \text{kg/m}^3 \quad \text{(at 101.3 kPa)}$$
$$c_{p,b} = (0.2492)(4.184 \times 10^3) = 1043\ \text{J/kg}\cdot\text{K}$$
$$\mathbf{Pr}_b = 0.702$$
$$\mu_s = (14.77 \times 10^{-6})(1.488\,164) = 21.98\ \mu\text{Pa}\cdot\text{s}$$

The pressure-corrected density is $\rho_b = (150/101.3)(1.004) = 1.487\ \text{kg/m}^3$; the Reynolds number is

$$\mathbf{Re}_{D,b} = \frac{4\dot{m}}{\pi D\mu_b} = \frac{4(0.56\ \text{kg/h})}{\pi(0.020\ \text{m})(19.82 \times 10^{-6}\ \text{Pa}\cdot\text{s})(3600\ \text{s/h})} = 500 \quad (laminar)$$

The entry lengths are

$$x_v \approx (0.05)D\,\mathbf{Re}_{D,b} = (0.05)(0.020\ \text{m})(500) = 0.5\ \text{m}$$
$$x_t \approx x_v\,\mathbf{Pr}_b = (0.5\ \text{m})(0.702) = 0.35\ \text{m}$$

We proceed by trial and error: Guessing the value of L, we compute $\bar{h} \equiv \bar{h}(0, L)$ and see whether these two values jointly produce the desired exit temperature.

Try $L = 0.35$ m. Then both profiles are developing, and (*H-8*) gives

$$\bar{h} = \frac{k_b}{D}\,\overline{\mathbf{Nu}} = \left(\frac{0.029\,61}{0.020}\right)(1.86)\left[\frac{(500)(0.702)}{\frac{350}{20}}\right]^{1/3}\left(\frac{19.82}{21.98}\right)^{0.14} = 7.37\ \mathrm{W/m^2\cdot K}$$

Now (*H-3*) gives

$$T_{b,e} = T_s - (T_s - T_{b,i})\exp\left(\frac{-\pi D L \bar{h}}{\dot{m} c_{p,b}}\right)$$

$$= 127 - (127 - 60)\exp\left[\frac{-\pi(0.020)(0.35)(7.37)}{(0.56/3600)(1043)}\right] = 102.3\ ^\circ\mathrm{C} \quad (\textit{too high})$$

Try $L = 0.20$ m. Using (*H-8*) and (*H-3*) as above, we find

$$\bar{h} = 8.88\ \mathrm{W/m^2\cdot K} \qquad T_{b,e} = 93.3\ ^\circ\mathrm{C} \quad (\textit{still too high})$$

Try $L = 0.16$ m. Similarly,

$$\bar{h} = 9.58\ \mathrm{W/m^2\cdot K} \qquad T_{b,e} = 89.99\ ^\circ\mathrm{C} \quad (\textit{final})$$

6.49 Check the result of Problem 6.48 by algebraic solution of (*H-8*) and (*H-3*).

▮ Eliminate $\bar{h}$ between the two equations to obtain

$$L = \left[\frac{\dot{m} c_{p,b}}{\pi(1.86)k_b(D\,\mathbf{Re}_{D,b}\,\mathbf{Pr}_b)^{1/3}(\mu_b/\mu_s)^{0.14}}\ln\frac{T_s - T_{b,i}}{T_s - T_{b,e}}\right]^{3/2}$$

$$= \left\{\frac{(0.56/3600)(1043)}{\pi(1.86)(0.029\,61)[(0.020)(500)(0.702)]^{1/3}(19.82/21.98)^{0.14}}\ln\frac{127-60}{127-90}\right\}^{3/2}$$

$$= 0.160\ \mathrm{m}$$

6.50 As in Problem 6.46, gaseous N_2 at $\bar{T}_b = 170$ °F flows in a 0.79-in.-ID circular tube at a pressure of 21.8 psia and with a mean velocity of 1.64 ft/sec. The tube is heated with $T_s = 260$ °F = const. over the thermal entry length only. Use the appropriate form of the Hausen equation to determine (*a*) the average heat transfer coefficient over the heated tube length, and (*b*) the temperature rise $T_{b,e} - T_{b,i}$.

▮ All parameters have been evaluated in Problem 6.46; the heated tube length is $L = x_t = 1.72$ ft.

(*a*) By (*H-9*),

$$\overline{\mathbf{Nu}} = 3.66 + \frac{(0.104)\left[\left(\frac{0.79/12}{1.72}\right)(746)(0.702)\right]}{1 + (0.016)\left[\left(\frac{0.79/12}{1.72}\right)(746)(0.702)\right]^{0.8}} = 5.43$$

$$\bar{h}(0, L) = \frac{k_b}{D}\,\overline{\mathbf{Nu}} = \left[\frac{0.017\,205\ \mathrm{Btu/hr\cdot ft\cdot °F}}{(0.79/12)\ \mathrm{ft}}\right](5.43) = 1.419\ \mathrm{Btu/hr\cdot ft^2\cdot °F}$$

(*b*) Together, (*H-3*) for $x = L$,

$$T_{b,e} = T_s - (T_s - T_{b,i})\exp\left[\frac{-4L\bar{h}(0, L)}{D V_b \rho_b c_{p,b}}\right]$$

and the definition

$$T_{b,e} + T_{b,i} \equiv 2\bar{T}_b$$

determine $T_{b,i} = 125.3$ °F, $T_{b,e} = 214.7$ °F, and hence $T_{b,e} - T_{b,i} = 89.4$ °F.

6.51 Would the results of Problem 6.50 be changed if the pressure and/or bulk velocity of the N_2 changed?

▮ No, but L or x_t might change. First of all, notice that the choice $L = x_t = (0.05)D\,\mathbf{Re}_D\,\mathbf{Pr}$ makes

$$\left(\frac{D}{L}\right)\mathbf{Re}_D\,\mathbf{Pr} = 20 \qquad \text{and so} \qquad \mathbf{Nu}\,(L) = 5.43$$

absolutely; in particular, for all p and V_b. Consequently, $\bar{h}(0, L) = 1.419\ \text{Btu/hr}\cdot\text{ft}^2\cdot{}^\circ\text{F}$, for all p and V_b. Second, the argument of the exponential function in (*H-3*) can be written—cf. Problem 6.33(*b*)—

$$\frac{-4\bar{h}(0, L)}{(\mu_b g_c)(\mathbf{Re}_D/L)c_{p,b}} = \frac{-4\bar{h}(0, L)}{(\mu_b g_c)(20/D\,\mathbf{Pr})c_{p,b}}$$

which is also independent of p and V_b. The same is then true of the inlet and exit bulk temperatures.

6.52 Reconsider Problem 6.45 with all conditions unchanged, except (1) the gas pressure is 400 kPa and (2) the tube heated length is 0.52 m. Determine (*a*) the entry lengths; (*b*) the average heat transfer coefficient over the tube length, by the Sieder–Tate equation; and (*c*) the average heat transfer coefficient, by the Hausen equation.

I (*a*) The entry lengths $x \propto \mathbf{Re}_{D,b} \propto \rho_b \propto p$; therefore, by Problem 6.45,

$$x_v = \left(\tfrac{400}{150}\right)(0.74\ \text{m}) = 1.99\ \text{m}$$
$$x_t = \left(\tfrac{400}{150}\right)(0.52\ \text{m}) = 1.38\ \text{m}$$

Clearly, the entire tube length of 0.52 m is well within the developing velocity and thermal profile region, and either the Sieder–Tate or the Hausen equation may be used.

(*b*) By (*H-8*), $\bar{h} \propto \mathbf{Re}_D^{1/3} \propto p^{1/3}$; hence, from Problem 6.45(*b*),

$$\bar{h} = \left(\tfrac{400}{150}\right)^{1/3}(7.41\ \text{W/m}^2\cdot\text{K}) = 10.3\ \text{W/m}^2\cdot\text{K}$$

(*c*) By (*H-9*),

$$\overline{\mathbf{Nu}} = 3.66 + \frac{(0.104)[(D/L)\,\mathbf{Re}_D\,\mathbf{Pr}]}{1 + (0.016)[(D/L)\,\mathbf{Re}_D\,\mathbf{Pr}]^{0.8}}$$

$$= 3.66 + \frac{(0.104)[(0.020/0.52)(\tfrac{400}{150})(742.3)(0.702)]}{1 + (0.016)[(0.020/0.52)(\tfrac{400}{150})(742.3)(0.702)]^{0.8}} = 7.67$$

$$\bar{h} = \frac{k_b}{D}\,\overline{\mathbf{Nu}} = \left(\frac{0.029\,76\ \text{W/m}\cdot\text{K}}{0.020\ \text{m}}\right)(7.67) = 11.41\ \text{W/m}^2\cdot\text{K}$$

6.53[D] In Problem 4.76 it was supposed that $\bar{h} = 300\ \text{W/m}^2\cdot\text{K}$. Is this value consistent with the other data, assuming a fully developed velocity profile at the start of the heated section?

I From Table B-3, at $\bar{T}_b = 50\ ^\circ\text{C}$,

$$\nu_b = 5.67 \times 10^{-7}\ \text{m}^2/\text{s} \qquad k_b = 0.639\ \text{W/m}\cdot\text{K} \qquad \mathbf{Pr}_b = 3.68$$

The Reynolds number is

$$\mathbf{Re}_{D,b} = \frac{4Q}{\pi D \nu_b} = \frac{4(4.2 \times 10^{-6}\ \text{m}^3/\text{s})}{\pi(0.010\ \text{m})(5.67 \times 10^{-7}\ \text{m}^2/\text{s})} = 943 \quad (\textit{laminar})$$

The thermal entry length is

$$x_t \approx (0.05)D\,\mathbf{Re}_{D,b}\,\mathbf{Pr} = (0.05)(0.010\ \text{m})(943)(3.68) = 1.74\ \text{m}$$

and clearly the 1-m-long tube is in the developing temperature profile region.

None of our equations covers this situation (laminar flow, developing in temperature, with $T_s - T_b =$ const.). However, under the *additional* assumption that $h_x = \bar{h} =$ const. along the tube—which implies that $q_s''(x) = h_x(T_s - T_b)$ also is constant—(*H-9*) may be used to evaluate $h_{L/2}$, which should closely approximate $\bar{h}$.

$$\mathbf{Nu}\left(\frac{L}{2}\right) = 4.36 + \frac{(0.023)[(\tfrac{1}{50})(943)(3.68)]}{1 + (0.0012)[(\tfrac{1}{50})(943)(3.68)]^{1.0}} = 5.83$$

$$h_{L/2} = \frac{k_b}{D}\,\mathbf{Nu}\left(\frac{L}{2}\right) = \left(\frac{0.639\ \text{W/m}\cdot\text{K}}{0.010\ \text{m}}\right)(5.83) = 373\ \text{W/m}^2\cdot\text{K}$$

So, the value used in Problem 4.76 was a reasonable approximation.

6.54[D] In Problem 4.77 it was supposed that $\bar{h} = 53\ \text{Btu/hr}\cdot\text{ft}^2\cdot{}^\circ\text{F}$. Is this value consistent with the other data, assuming a fully developed velocity profile at the start of the heated tube length?

I From Table B-3, at $\bar{T}_b = 121\ ^\circ\text{F}$,

$$\nu_b = 0.616 \times 10^{-5}\ \text{ft}^2/\text{sec} \qquad k_b = 0.369\ \text{Btu/hr}\cdot\text{ft}\cdot{}^\circ\text{F} \qquad \mathbf{Pr}_b = 3.72$$

The Reynolds number is

$$\mathbf{Re}_{D,b}=\frac{4Q}{\pi D \nu_b}=\frac{4(1.49\times10^{-4}\ \text{ft}^3/\text{sec})}{\pi[(\frac{0.39}{12})\ \text{ft}](0.616\times10^{-5}\ \text{ft}^2/\text{sec})}=948 \quad (laminar)$$

The thermal entry length is

$$x_t \approx (0.05)D\,\mathbf{Re}_{D,b}\,\mathbf{Pr}_b=(0.05)\left(\frac{0.39}{12}\ \text{ft}\right)(948)(3.72)=5.73\ \text{ft}$$

and the entire heated length of 3.28 ft is within the developing thermal region.

None of our equations covers this situation (laminar flow, developing in temperature, with $T_s - T_b =$ const.). However, under the *additional* assumption that $h_x = \bar{h} =$ const. along the tube—which implies that $q_s''(x) = h_x(T_s - T_b)$ also is constant—(*H-9*) may be used to evaluate $h_{L/2}$, which should closely approximate $\bar{h}$.

$$\mathbf{Nu}\left(\frac{L}{2}\right)=4.36+\frac{(0.023)\left[\left(\frac{0.39/12}{1.64}\right)(948)(3.72)\right]}{1+(0.0012)\left[\left(\frac{0.39/12}{1.64}\right)(948)(3.72)\right]^{1.0}}=5.84$$

$$h_{L/2}=\frac{k_b}{D}\,\mathbf{Nu}\left(\frac{L}{2}\right)=\left[\frac{0.369\ \text{Btu/hr}\cdot\text{ft}\cdot{}^\circ\text{F}}{(0.39/12)\ \text{ft}}\right](5.84)=66.3\ \text{Btu/hr}\cdot\text{ft}^2\cdot{}^\circ\text{F}$$

So, the value used in Problem 4.77 was reasonable.

6.55 What would be the answer to Problem 6.53 if the tube diameter were halved?

▌ Since $D\,\mathbf{Re}_{D,b} = 4Q/\pi\nu_b$ is unaffected (for fixed Q), so is $\mathbf{Nu}\,(L/2)$. Therefore, $h_{L/2} \approx 1/D$ is doubled—to 746 W/m$^2\cdot$K.

6.56 Water flows inside a 10-mm-ID circular tube at $\bar{T}_b = 50$ °C with a volumetric flow of 4.2 mL/s (as in Problem 4.76). The tube wall temperature is constant at $T_s = 60$ °C, and the tube has a heated length of 0.4 m. Determine $\bar{h}$ for this flow.

▌ From Table B-3, at $\bar{T}_b = 50$ °C and $T_s = 60$ °C,

$$\nu_b = 5.67\times10^{-7}\ \text{m}^2/\text{s} \qquad \mathbf{Pr}_b = 3.68$$
$$k_b = 0.639\ \text{W/m}\cdot\text{K} \qquad \rho_b = 990\ \text{kg/m}^3$$
$$\nu_s = 4.77\times10^{-7}\ \text{m}^2/\text{s} \qquad \rho_s = 985.4\ \text{kg/m}^3$$

and from Problem 6.53,

$$\mathbf{Re}_{D,b} = 943 \qquad x_t = 1.74\ \text{m} \qquad x_v = \frac{x_t}{\mathbf{Pr}_b} = 0.47\ \text{m}$$

Thus, the entire tube length is experiencing developing velocity, developing temperature profile flow; (*H-8*) applies. Recalling that $\mu = \nu\rho$,

$$\overline{\mathbf{Nu}}=(1.86)\left(\frac{\mathbf{Re}_{D,b}\,\mathbf{Pr}_b}{L/D}\right)^{1/3}\left(\frac{\mu_b}{\mu_s}\right)^{0.14}=(1.86)\left[\frac{(943)(3.68)}{40/1}\right]^{1/3}\left[\frac{(5.67)(990)}{(4.77)(985.4)}\right]^{0.14}=8.44$$

$$\bar{h}=\frac{k_b}{D}\,\overline{\mathbf{Nu}}=\left(\frac{0.639\ \text{W/m}\cdot\text{K}}{0.010\ \text{m}}\right)(8.44)=539\ \text{W/m}^2\cdot\text{K}$$

6.57 Calculate the temperature rise of the fluid in Problem 6.56.

▌ From Table B-3, at $\bar{T}_b = 50$ °C, $c_{p,b} = 4178.6$ J/kg$\cdot$K. Then, by simultaneous solution of (*H-3*),

$$T_{b,e}=T_s-(T_s-T_{b,i})\exp\left(\frac{-4L\bar{h}}{\rho_b V_b D c_{p,b}}\right)$$

and the definition $T_{b,e} + T_{b,i} = 2\bar{T}_b$, one finds $T_{b,e} = 51.93$ °C and $T_{b,i} = 48.07$ °C, for a rise of 3.86 °C.

6.58 Ethylene glycol flows at 0.013 kg/s through a 5-mm-diameter, thin-walled copper tube. The tube is coiled, with the coil diameter being 0.5 m, and submerged in a stirred water bath maintained at 20 °C. If the fluid enters the coil at 80 °C, what tube length L is required for it to exit at 40 °C?

▌ For a thin-walled tube in a well-stirred bath, the tube wall temperature will be approximately constant and equal to the bath temperature. From Table B-3, at $\bar{T}_b = 60\ °C$,

$$\nu_b = 4.747 \times 10^{-6}\ m^2/s \qquad k_b = 0.259\ W/m \cdot K \qquad \rho_b = 1087.6\ kg/m^3 \qquad c_{p,b} = 2561\ J/kg \cdot K \qquad \mathbf{Pr}_b = 51$$

The Reynolds number is

$$\mathbf{Re}_{D,b} = \frac{4\dot{m}}{\pi D \nu_b \rho_b} = \frac{4(0.013\ kg/s)}{\pi(0.005\ m)(4.747 \times 10^{-6}\ m^2/s)(1087.6\ kg/m^3)} = 641 \quad (laminar)$$

and so the entry lengths are

$$x_v \approx (0.05) D\, \mathbf{Re}_{D,b} = (0.05)(0.005\ m)(641) = 0.16\ m$$
$$x_t \approx x_v\, \mathbf{Pr}_b = (0.16\ m)(51) = 8.17\ m$$

Because x_v is quite small and x_t is quite large, let us assume temporarily that the flow over the unknown length L is developed in velocity but undeveloped in temperature. Then an equation for L may be found by eliminating $\bar{h}$ between (2) of Problem 6.4 and (*H-9*) (with parameters from the third line of Table H-1); thus,

$$L = \frac{\dot{m} c_{p,b} \ln[(T_{b,i} - T_s)/(T_{b,e} - T_s)]}{\pi k_b \left\{ 3.66 + \dfrac{(0.0668)[(D/L)\, \mathbf{Re}_{D,b}\, \mathbf{Pr}_b]}{1 + (0.04)[(D/L)\, \mathbf{Re}_{D,b}\, \mathbf{Pr}_b]^{2/3}} \right\}}$$

$$= \frac{(0.013)(2561) \ln 3}{\pi(0.259) \left\{ 3.66 + \dfrac{(0.0668)[(0.005/L)(641)(51)]}{1 + (0.04)[(0.005/L)(641)(51)]^{2/3}} \right\}} \equiv F(L)$$

This equation may be solved iteratively in the usual manner: $L^{(j+1)} = F(L^{(j)})$. In view of the value of x_t, we choose $L^{(0)} = 8$ m; the sequence is found to converge to $L \approx 10$ m. Since the temperature profile is developing over 80% of this length, we accept it as an approximate solution. Note that the coil has been treated as a straight tube: The coil-diameter-to-tube-diameter ratio,

$$\frac{0.5\ m}{5 \times 10^{-3}\ m} = 100$$

renders the straight-tube assumption reasonable.

6.59^D Calculate $\bar{h}$ for the flow of Problems 4.76 and 6.53 if the wall condition is changed to T_s = const.

▌ With properties and x_t unchanged from Problem 6.53, (*H-9*) gives

$$\overline{\mathbf{Nu}} = 3.66 + \frac{(0.0668)[(\frac{1}{100})(943)(3.68)]}{1 + (0.04)[(\frac{1}{100})(943)(3.68)]^{2/3}} = 5.29$$

$$\bar{h}(0, L) \equiv \bar{h} = \frac{k_b}{D}\, \overline{\mathbf{Nu}} = \left(\frac{0.639\ W/m \cdot K}{0.010\ m}\right)(5.29) = 338\ W/m^2 \cdot K$$

which is some 9% smaller than the estimate obtained in Problem 6.53. In general, constant heat flux in laminar tube flow results in a higher $\bar{h}$ than does constant wall temperature.

6.60^D Calculate $\bar{h}$ for the flow of Problems 4.77 and 6.54 if the wall condition is changed to T_s = const.

▌ With properties and x_t unchanged from Problem 6.54, (*H-9*) gives

$$\overline{\mathbf{Nu}} = 3.66 + \frac{(0.0668)\left[\dfrac{0.39}{(3.28)(12)}(948)(3.72)\right]}{1 + (0.04)\left[\dfrac{0.39}{(3.28)(12)}(948)(3.72)\right]^{2/3}} = 5.30$$

$$\bar{h}(0, L) \equiv \bar{h} = \frac{k_b}{D}\, \overline{\mathbf{Nu}} = \left[\frac{0.369\ Btu/hr \cdot ft \cdot °F}{(0.39/12)\ ft}\right](5.30) = 60.2\ Btu/hr \cdot ft^2 \cdot °F$$

which is some 9% smaller than the estimate obtained in Problem 6.54. In general, constant heat flux in laminar tube flow results in a higher $\bar{h}$ than does constant wall temperature.

6.61[D] Use the upper curve of Fig. 6-4 to check the value $h(x_{mid}) = 131.4\ \text{W/m}^2\cdot\text{K}$ found in Problem 6.19(*b*).

▌ Corresponding to $x = 5.46$ m,

$$\mathbf{Gz}^{-1} \equiv \frac{x/D}{\mathbf{Re}_D\ \mathbf{Pr}} = \frac{5.46/0.025}{(2000)(4.34)} = 0.025$$

$$\mathbf{Nu}\,(x) \approx 5.5 \quad \text{[from the graph]}$$

$$h_x \approx \left(\frac{0.628\ \text{W/m}\cdot\text{K}}{0.025\ \text{m}}\right)(5.5) = 138\ \text{W/m}^2\cdot\text{K}$$

6.62[D] Use the upper curve of Fig. 6-4 to check the value $h(9\ \text{ft}) = 26.2\ \text{Btu/hr}\cdot\text{ft}^2\cdot{}^\circ\text{F}$ found in Problem 6.31.

▌ Corresponding to $x = 9$ ft,

$$\mathbf{Gz}^{-1} \equiv \frac{x/D}{\mathbf{Re}_D\ \mathbf{Pr}} = \frac{108/1}{(1468)(7.02)} = 0.0105$$

$$\mathbf{Nu}\,(x) \approx 6.7 \quad \text{[from the graph]}$$

$$h_x \approx \left(\frac{0.345\ \text{Btu/hr}\cdot\text{ft}\cdot{}^\circ\text{F}}{\frac{1}{12}\ \text{ft}}\right)(6.7) = 27.7\ \text{Btu/hr}\cdot\text{ft}^2\cdot{}^\circ\text{F}$$

in good agreement for this class of problem.

6.63[D] Saturated liquid water at $\bar{T}_b = 20\ ^\circ\text{C}$ flows inside the annular region formed by two concentric circular tubes (Fig. 6-5). The outer tube has ID = 25 mm, while the inner tube has OD = 10 mm. The mass flow is 0.02 kg/s. The outer surface is insulated and the inner surface is kept at constant $T_{s,i} = 50\ ^\circ\text{C}$. Determine the average heat transfer coefficient from the inner surface to the water, h_i, for fully developed velocity and temperature flow.

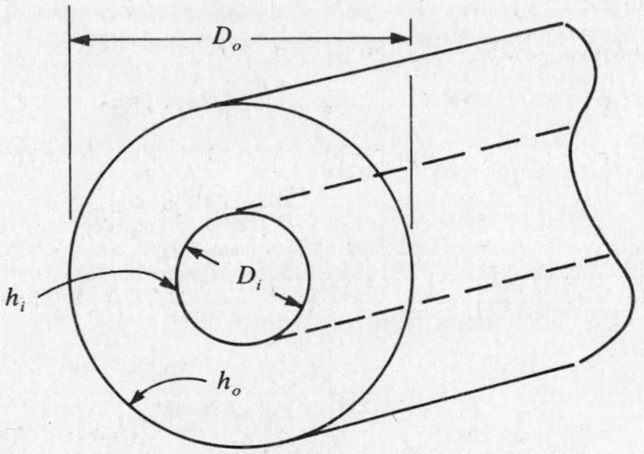

Fig. 6-5

▌ From Table B-3, for saturated water at $\bar{T}_b = 20\ ^\circ\text{C}$,

$$\nu_b = 1.006 \times 10^{-6}\ \text{m}^2/\text{s} \qquad \rho_b = 1000.5\ \text{kg/m}^3 \qquad k_b = 0.597\ \text{W/m}\cdot\text{K}$$

The Reynolds number is based on the *hydraulic diameter*,

$$D_h \equiv \frac{4(\text{cross-sectional flow area})}{\text{wetted perimeter}} \equiv \frac{4A}{P} \tag{1}$$

so that

$$\mathbf{Re}_{D,h} \equiv \frac{D_h V_b}{\nu_b} = \frac{4\dot{m}}{P\rho_b \nu_b} \tag{2}$$

For the present geometry, $D_h = D_o - D_i = 0.015$ m and $P = \pi(D_o + D_i)$. Thus,

$$\mathbf{Re}_{D,h} = \frac{4(0.02\ \text{kg/s})}{\pi(0.035\ \text{m})(1000.5\ \text{kg/m}^3)(1.006\times 10^{-6}\ \text{m}^2/\text{s})} = 723$$

and the flow is laminar. Nusselt numbers

$$\overline{\mathbf{Nu}}_i \equiv \frac{\bar{h}_i D_h}{k_b} \qquad \text{and} \qquad \overline{\mathbf{Nu}}_o \equiv \frac{\bar{h}_o D_h}{k_b}$$

TABLE 6-1

D_i/D_o	$\overline{Nu}_i$	$\overline{Nu}_o$
0	—	3.66
0.05	17.46	4.06
0.10	11.56	4.11
0.25	7.37	4.23
0.50	5.74	4.43
1.00	4.86	4.86

From W. M. Kays and H. C. Perkins, in *Handbook of Heat Transfer*, chap. 7, W. M. Rohsenow and J. P. Hartnett (eds.), McGraw-Hill, New York, 1972. Used with permission.

for laminar, fully developed (velocity and temperature) flow are tabulated in Table 6-1. Thus, with $D_i/D_o = \frac{10}{25} = 0.4$, interpolation in Table 6-1 yields $\overline{\mathbf{Nu}}_i \approx 6.39$; or

$$\bar{h}_i = \frac{k_b \overline{\mathbf{Nu}}_i}{D_h} \approx \frac{(0.597\ \text{W/m}\cdot\text{K})(6.39)}{0.015\ \text{m}} = 254\ \text{W/m}\cdot\text{K}$$

Of course, $\overline{\mathbf{Nu}}_o = \bar{h}_o = 0$ for an insulated outer surface.

6.64[D] Saturated liquid water at $\bar{T}_b = 68\ °\text{F}$ flows inside the annular region formed by two concentric tubes (Fig. 6-5). The outer tube has $\text{ID} = 1.0$ in. and the inner tube has $\text{OD} = 0.39$ in. The mass flow of water is 159 lbm/hr. The outer surface is insulated and the inner surface is kept at constant $T_{s,i} = 121\ °\text{F}$. Determine the heat transfer coefficient from the inner tube surface to the water, h_i, for fully developed velocity and temperature flow.

▌ From Table B-3, for saturated water at $\bar{T}_b = 68\ °\text{F}$,

$$\nu_b = 1.083 \times 10^{-5}\ \text{ft}^2/\text{s} \qquad \rho_b = 62.46\ \text{lbm/ft}^3 \qquad k_b = 0.345\ \text{Btu/hr}\cdot\text{ft}\cdot°\text{F}$$

The hydraulic diameter (see Problem 6.63) is $D_h = D_o - D_i = 0.61\ \text{in.} = 0.051\ \text{ft}$, and

$$\mathbf{Re}_{D,h} = \frac{4\dot{m}}{\pi(D_0 + D_i)\nu_b\rho_b} = \frac{4(\frac{159}{3600}\ \text{lbm/s})}{\pi(0.116\ \text{ft})(1.083\times10^{-5}\ \text{ft}^2/\text{s})(62.46\ \text{lbm/ft}^3)} = 717 \quad (\textit{laminar})$$

and Table 6-1 applies. Interpolation at $D_i/D_o = 0.39/1.0 = 0.39$ yields

$$\overline{\mathbf{Nu}}_i \approx 6.46$$

$$\bar{h}_i = \frac{k_b}{D_h}\overline{\mathbf{Nu}}_i \approx \left(\frac{0.345\ \text{Btu/hr}\cdot\text{ft}\cdot°\text{F}}{0.051\ \text{ft}}\right)(6.46) = 43.7\ \text{Btu/hr}\cdot\text{ft}^2\cdot°\text{F}$$

Of course, $\overline{\mathbf{Nu}}_o = \bar{h}_o = 0$ for an insulated outer surface.

6.65[D] Engine oil at $\bar{T}_b = 40\ °\text{C}$ flows inside a square tube 20 mm on a side, at $V_b = 2\ \text{m/s}$. For fully developed flow, determine $\bar{h}$ for (*a*) constant q_s'', and (*b*) constant T_s.

▌ From Table B-3, at $\bar{T}_b = 40\ °\text{C}$, $\nu_b = 2.42 \times 10^{-4}\ \text{m}^2/\text{s}$ and $k_b = 0.144\ \text{W/m}\cdot\text{K}$. The hydraulic diameter is

$$D_h \equiv \frac{4(\text{cross-sectional flow area})}{\text{wetted perimeter}} = \frac{4(20\ \text{mm})(20\ \text{mm})}{4(20\ \text{mm})} = 20\ \text{mm}$$

giving the Reynolds number as

$$\mathbf{Re}_{D,h} = \frac{D_h V_b}{\nu_b} = \frac{(0.020\ \text{m})(2\ \text{m/s})}{2.42\times10^{-4}\ \text{m}^2/\text{s}} = 165 \quad (\textit{laminar})$$

(*a*) From Table 6-2, for fully developed flow, constant q_s'', and $b/a = 1.0$,

$$\overline{\mathbf{Nu}}_h = 3.61$$

$$\bar{h} = \frac{k_b}{D_h}\overline{\mathbf{Nu}}_h = \left(\frac{0.144\ \text{W/m}\cdot\text{K}}{0.020\ \text{m}}\right)(3.61) = 25.99\ \text{W/m}^2\cdot\text{K}$$

TABLE 6-2

fully developed flow in cross-sectional shape	b/a	$\overline{Nu}_h \equiv \bar{h}D_h/k_b$	
		const. q_s''	const. T_s
○		4.364	3.66
a □ b	1.0	3.61	2.98
a ▭ b	1.43	3.73	3.08
a ▭ b	2.0	4.12	3.39
a ▭ b	3.0	4.79	3.96
a ▭ b	4.0	5.33	4.44
a ▭ b	8.0	6.49	5.60
═══	∞	8.235	7.54
▨		5.385	4.86
△		3.00	2.35

From W. M. Kays and M. Crawford, *Convective Heat and Mass Transfer*, McGraw-Hill, New York, 1980. Used with permission.

(b) From Table 6-2, for constant T_s and $b/a = 1.0$,

$$\overline{Nu}_h = 2.98$$

$$\bar{h} = \frac{k_b}{D_h}\overline{Nu}_h = \left(\frac{0.144\ \text{W/m} \cdot \text{K}}{0.020\ \text{m}}\right)(2.98) = 21.46\ \text{W/m}^2 \cdot \text{K}$$

6.66[D] Engine oil at $\bar{T}_b = 104\ °F$ flows in a 0.79-in.-inside-measurement square tube; the bulk velocity is 6.56 ft/sec. For fully developed velocity and thermal profiles, determine $\bar{h}$ for **(a)** constant q_s'', and **(b)** constant T_s.

❚ From Table B-3, at $\bar{T}_b = 104\ °F$, $\nu_b = 0.0026\ \text{ft}^2/\text{sec}$ and $k_b = 0.083\ \text{Btu/hr} \cdot \text{ft} \cdot °F$. The hydraulic diameter is

$$D_h \equiv \frac{4(\text{cross-sectional flow area})}{\text{wetted perimeter}} = \frac{4(0.79\ \text{in.})(0.79\ \text{in.})}{4(0.79\ \text{in.})} = 0.79\ \text{in.} = 0.066\ \text{ft}$$

and the Reynolds number is

$$\text{Re}_{D,h} = \frac{D_h V_b}{\nu_b} = \frac{(0.066\ \text{ft})(6.56\ \text{ft/sec})}{0.0026\ \text{ft}^2/\text{sec}} = 167\quad (laminar)$$

(a) From Table 6-2, for constant q_s'' and $b/a = 1.0$,

$$\overline{Nu}_h = 3.61$$

$$\bar{h} = \frac{k_b}{D_h}\overline{Nu}_h = \left(\frac{0.083\ \text{Btu/hr} \cdot \text{ft} \cdot °F}{0.066\ \text{ft}}\right)(3.61) = 4.54\ \text{Btu/hr} \cdot \text{ft}^2 \cdot °F$$

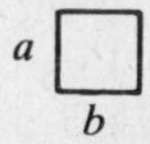

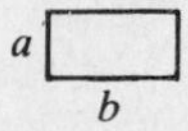

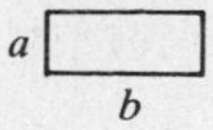

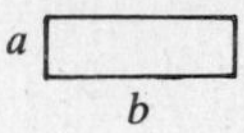

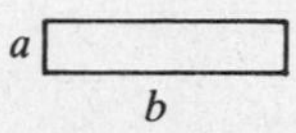

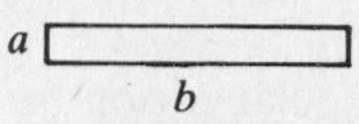

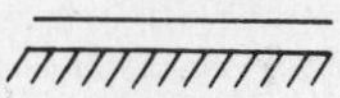

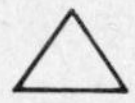

(*b*) From Table 6-2, for constant T_s and $b/a = 1.0$,

$$\overline{\mathbf{Nu}}_h = 2.98$$

$$\bar{h} = \frac{k_b}{D_h}\overline{\mathbf{Nu}}_h = \left(\frac{0.083\ \text{Btu/hr}\cdot\text{ft}\cdot{}^\circ\text{F}}{0.066\ \text{ft}}\right)(2.98) = 3.75\ \text{Btu/hr}\cdot\text{ft}^2\cdot{}^\circ\text{F}$$

6.67[D] Air at a mean temperature $\bar{T}_b = 80\ {}^\circ\text{F}$ and a pressure of 1 atm flows inside a very wide channel with a bulk average velocity of 1.1 ft/sec; the channel depth is $a = 2.0$ in. For fully developed velocity and thermal profiles, find $\bar{h}$ for (*a*) constant q_s'' (same for both surfaces) and (*b*) constant T_s (same for both surfaces).

▌ From Table B-4, at $\bar{T}_b = 80\ {}^\circ\text{F}$ and 1 atm pressure,

$$\mu_b g_c = 1.241 \times 10^{-5}\ \text{lbm/ft}\cdot\text{sec} \qquad k_b = 0.015\,16\ \text{Btu/hr}\cdot\text{ft}\cdot{}^\circ\text{F} \qquad \rho_b = 0.0735\ \text{lbm/ft}^3$$

This being the case $b/a = \infty$ of Table 6-2, the hydraulic diameter is

$$D_h = \lim \frac{4ab}{2(a+b)} = \lim \frac{2a}{(a/b)+1} = 2a = \tfrac{1}{3}\ \text{ft}$$

and
$$\mathbf{Re}_{D,h} = \frac{\rho_b V_b D_h}{\mu_b g_c} = \frac{(0.0735\ \text{lbm/ft}^3)(1.1\ \text{ft/sec})(\frac{1}{3}\ \text{ft})}{1.241 \times 10^{-5}\ \text{lbm/ft}\cdot\text{sec}} = 2172 < 2300 \quad (\textit{laminar})$$

(*a*)
$$\overline{\mathbf{Nu}}_h = 8.23$$

$$\bar{h} = \frac{k_b}{D_h}\overline{\mathbf{Nu}}_h = \left(\frac{0.015\,16\ \text{Btu/hr}\cdot\text{ft}\cdot{}^\circ\text{F}}{\frac{1}{3}\ \text{ft}}\right)(8.23) = 0.374\ \text{Btu/hr}\cdot\text{ft}^2\cdot{}^\circ\text{F}$$

(*b*)
$$\overline{\mathbf{Nu}}_h = 7.54$$

$$\bar{h} = \frac{k_b}{D_h}\overline{\mathbf{Nu}}_h = \left(\frac{0.015\,16\ \text{Btu/hr}\cdot\text{ft}\cdot{}^\circ\text{F}}{\frac{1}{3}\ \text{ft}}\right)(7.54) = 0.343\ \text{Btu/hr}\cdot\text{ft}^2\cdot{}^\circ\text{F}$$

6.68[D] Air at $\bar{T}_b = 27\ {}^\circ\text{C}$ and a pressure of 1 atm (101 kPa) flows inside a very wide channel, with an average velocity of 0.33 m/s; the channel depth is $a = 51$ mm. For fully developed velocity and temperature profile flow, what is $\bar{h}$ for both surfaces held at the same fixed T_s?

▌ From Table B-4, at $\bar{T}_b = 27\ {}^\circ\text{C}$ and 1 atm pressure, $\nu_b = 1.568 \times 10^{-5}\ \text{m}^2/\text{s}$ and $k_b = 0.0262\ \text{W/m}\cdot\text{K}$. This being the case $b/a = \infty$ of Table 6-2, the hydraulic diameter is

$$D_h = \lim \frac{4ab}{2(a+b)} = \lim \frac{2a}{(a/b)+1} = 2a = 0.102\ \text{m}$$

and
$$\mathbf{Re}_{D,h} = \frac{D_h V_b}{\nu_b} = \frac{(0.102\ \text{m})(0.33\ \text{m/s})}{1.568 \times 10^{-5}\ \text{m}^2/\text{s}} = 2147 < 2300 \quad (\textit{laminar})$$

From Table 6-2,

$$\overline{\mathbf{Nu}}_h = 7.54$$

$$\bar{h} = \frac{k_b}{D_h}\overline{\mathbf{Nu}}_h = \left(\frac{0.0262\ \text{W/m}\cdot\text{K}}{0.102\ \text{m}}\right)(7.54) = 1.94\ \text{W/m}^2\cdot\text{K}$$

6.69[D] In a laboratory demonstration ethylene glycol, at $\bar{T}_b = 60\ {}^\circ\text{C}$ and $\dot{m} = 0.045\ \text{kg/s}$, flows through a triangular duct having equal sides $s = 20$ mm. The flow is fully developed with respect to velocity and temperature. What is the average heat transfer coefficient if the duct surfaces are maintained at a constant $T_s = 80\ {}^\circ\text{C}$?

▌ From Table B-3, at $\bar{T}_b = 60\ {}^\circ\text{C}$,

$$\nu_b = 4.747 \times 10^{-6}\ \text{m}^2/\text{s} \qquad \rho_b = 1087.6\ \text{kg/m}^3 \qquad k_b = 0.259\ \text{W/m}\cdot\text{K}$$

By (*2*) of Problem 6.63,

$$\mathbf{Re}_{D,h} = \frac{4\dot{m}}{P\rho_b \nu_b} = \frac{4(0.045\ \text{kg/s})}{(0.060\ \text{m})(1087.6\ \text{kg/m}^3)(4.747 \times 10^{-6}\ \text{m}^2/\text{s})} = 581 \quad (\textit{laminar})$$

Then, from Table 6-2, with $D_h = s/\sqrt{3} = 0.011\,55$ m,

$$\overline{\mathbf{Nu}}_h = 2.35$$

$$\bar{h} = \frac{k_b}{D_h}\overline{\mathbf{Nu}}_h = \left(\frac{0.259\ \text{W/m}\cdot\text{K}}{0.011\ 55\ \text{m}}\right)(2.35) = 52.7\ \text{W/m}^2\cdot\text{K}$$

6.70[D] In a laboratory demonstration ethylene glycol, at $\bar{T}_b = 140\ °\text{F}$ and $\dot{m} = 0.10\ \text{lbm/sec}$, flows through a triangular duct having equal sides $s = 0.79$ in. The flow is fully developed with respect to both velocity and temperature, and the duct surfaces are maintained at constant wall temperature $T_s = 176\ °\text{F}$. What is the average value of the heat transfer coefficient?

▮ From Table B-3, at $\bar{T}_b = 140\ °\text{F}$,

$$\nu_b = 5.11 \times 10^{-5}\ \text{ft}^2/\text{sec} \qquad \rho_b = 67.90\ \text{lbm/ft}^3 \qquad k_b = 0.150\ \text{Btu/hr}\cdot\text{ft}\cdot°\text{F}$$

By (2) of Problem 6.63,

$$\mathbf{Re}_{D,h} = \frac{4\dot{m}}{P\rho_b\nu_b} = \frac{4(0.10\ \text{lbm/sec})}{[(2.37/12)\ \text{ft}](67.90\ \text{lbm/ft}^3)(5.11 \times 10^{-5}\ \text{ft}^2/\text{sec})} = 584 \quad (\textit{laminar})$$

Then, from Table 6-2, with $D_h = s/\sqrt{3} = 0.038$ ft,

$$\overline{\mathbf{Nu}}_h = 2.35$$

$$\bar{h} = \frac{k_b}{D_h}\overline{\mathbf{Nu}}_h = \left(\frac{0.150\ \text{Btu/hr}\cdot\text{ft}\cdot°\text{F}}{0.038\ \text{ft}}\right)(2.35) = 9.28\ \text{Btu/hr}\cdot\text{ft}^2\cdot°\text{F}$$

6.3 TURBULENT-FLOW PROBLEMS

6.71[D] In a "cold climate" chamber, liquid Freon-12, at $T_{b,i} = -30\ °\text{C}$, enters a 4.0-m-long heated circular tube having a constant wall inside temperature $T_s = -10\ °\text{C}$. The Freon mass flow is 0.12 kg/s, and the tube ID is 25 mm. What is the discharge temperature $T_{b,e}$?

▮ A fluid mean temperature $\bar{T}_b$ must be estimated for property evaluation. Clearly, $-30 < T_{b,e} < -10\ °\text{C}$; thus let us start an iterative solution with the guess $T_{b,e}^{(0)} = -16\ °\text{C}$, or $\bar{T}_b^{(0)} = -23\ °\text{C}$. From Table B-3,

$$\nu_b = (0.259 \times 10^{-5})(9.2903 \times 10^{-2}) = 2.41 \times 10^{-7}\ \text{m}^2/\text{s} \qquad c_{p,b} = (0.2159)(4184) = 903.3\ \text{J/kg}\cdot\text{K}$$
$$\rho_b = (91.72)(16.018) = 1469\ \text{kg/m}^3 \qquad \mathbf{Pr}_b = 4.52$$
$$k_b = (0.0407)(1.729\ 577) = 0.0704\ \text{W/m}\cdot\text{K}$$

The Reynolds number is

$$\mathbf{Re}_{D,b} = \frac{4\dot{m}}{P\nu_b\rho_b} = \frac{(4)(0.12\ \text{kg/s})}{\pi(0.025\ \text{m})(2.41 \times 10^{-7}\ \text{m}^2/\text{s})(1469\ \text{kg/m}^3)} = 1.7263 \times 10^4 \quad (\textit{turbulent})$$

and $$\frac{L}{D} = \frac{4\ \text{m}}{0.025\ \text{m}} = 160 > 60 \qquad \bar{h} = \bar{h}_\infty$$

Thus we can find $\bar{h}_\infty$, the $\bar{h}$ for $(L/D) > 60$, from the Dittus–Boelter equation for heating the fluid (even though the value of $T_s - T_b$ is slightly larger than recommended for this correlation), and then use (H-3), at $x = L$, to obtain $T_{b,e}^{(1)}$:

$$\mathbf{Nu}_{D,\infty} = (0.023)(17\ 263)^{0.8}(4.52)^{0.4} = 103.2$$

$$\bar{h}_\infty = \left(\frac{0.0704\ \text{W/m}\cdot\text{K}}{0.025\ \text{m}}\right)(103.2) = 291\ \text{W/m}^2\cdot\text{K}$$

$$T_{b,e}^{(1)} = -10 - (-10 + 30)\exp\left[\frac{-\pi(0.025)(4.0)(291)}{(0.12)(903.3)}\right] = -18.6\ °\text{C}$$

or $\bar{T}_b^{(1)} = -24.3\ °\text{C}$. A very slightly improved answer would result from evaluation of properties at $-24.3\ °\text{C}$, rather than $-23\ °\text{C}$. But neither the accuracy of the property values nor the Dittus–Boelter equation, especially the latter with an error of $\pm 20\%$, would justify such rework.

6.72[D] Rework Problem 6.71 using the Sieder–Tate equation.

▮ Again choosing $\bar{T}_b^{(0)} = -23\ °\text{C}$, we have first-iteration properties from Problem 6.71; in addition, from Table B-3 at $-10\ °\text{C}$,

$$\nu_s = (0.238 \times 10^{-5})(9.2903 \times 10^{-2}) = 2.21 \times 10^{-7}\ \text{m}^2/\text{s} \qquad \rho_s = (89.24)(16.018) = 1429\ \text{kg/m}^3$$

Thus, from (*H-12*) and (*H-3*):

$$\mathbf{Nu}_{D,\infty} = (0.027)(17\,263)^{0.8}(4.52)^{1/3}\left[\frac{(2.41\times10^{-7})(1469)}{(2.21\times10^{-7})(1429)}\right]^{0.14} = 111.3$$

$$\bar{h}_\infty = \left(\frac{0.0704\ \text{W/m}\cdot\text{K}}{0.025\ \text{m}}\right)(111.3) = 313\ \text{W/m}^2\cdot\text{K}$$

$$T_{b,e}^{(1)} = -10 - (-10+30)\exp\left[\frac{-\pi(0.025)(4.0)(313)}{(0.12)(903.3)}\right] = -18.1\ °\text{C}$$

or $\bar{T}_b^{(1)} = -24.0$ °C. Again, further iteration is not justified. The two correlations do not generally agree so well.

6.73^D In a "cold climate" chamber, liquid Freon-12, at $T_{b,i} = -22$ °F, enters a 13-ft-long heated tube having constant wall temperature $T_s = +14$ °F. The Freon mass flow is 0.26 lbm/sec, and the tube ID is 1.0 in. What is the discharge temperature $T_{b,e}$?

▌ A fluid mean temperature $\bar{T}_b$ must be estimated for property evaluation. Since $-22 < T_{b,e} < 14$ °F, we make the starting guess $T_{b,e}^{(0)} = 2$ °F, or $\bar{T}_b^{(0)} = -10$ °F. From Table B-3,

$$\nu_b = 0.259\times10^{-5}\ \text{ft}^2/\text{sec} \qquad c_{p,b} = 0.2158\ \text{Btu/lbm}\cdot°\text{F}$$
$$\rho_b = 91.78\ \text{lbm/ft}^3 \qquad \mathbf{Pr}_b = 4.53$$
$$k_b = 0.0407\ \text{Btu/hr}\cdot\text{ft}\cdot°\text{F}$$

The Reynolds number is

$$\mathbf{Re}_{D,b} = \frac{4\dot{m}}{\pi D\nu_b\rho_b} = \frac{(4)(0.26\ \text{lbm/sec})}{\pi(\frac{1}{12}\ \text{ft})(0.259\times10^{-5}\ \text{ft}^2/\text{sec})(91.78\ \text{lbm/ft}^3)} = 16\,711 \quad (\textit{turbulent})$$

and
$$\frac{L}{D} = \frac{13\ \text{ft}}{\frac{1}{12}\ \text{ft}} = 156 > 60 \qquad \bar{h} = \bar{h}_\infty$$

Thus we can find $\bar{h}_\infty$ from the Dittus–Boelter heating equation (even though $|T_s - T_b| > 10$ °F), and then use (*H-3*), at $x = L$, to obtain $T_{b,e}^{(1)}$:

$$\bar{h}_\infty = (0.023)\left(\frac{k_b}{D}\right)\mathbf{Re}_{D,B}^{0.8}\,\mathbf{Pr}_b^{0.4} = (0.023)\left[\frac{0.0407\ \text{Btu/hr}\cdot\text{ft}\cdot°\text{F}}{\frac{1}{12}\ \text{ft}}\right](16\,711)^{0.8}(4.53)^{0.4} = 49.1\ \text{Btu/hr}\cdot\text{ft}^2\cdot°\text{F}$$

$$T_{b,e}^{(1)} = 14 - (14+22)\exp\left[\frac{-\pi(\frac{1}{12})(13)(49.1)}{(0.26\times3600)(0.2158)}\right] = -1.7\ °\text{F}$$

or $\bar{T}_b^{(1)} = -12$ °F. As in Problem 6.71, it would be senseless to continue the iteration.

6.74^D Rework Problem 6.73 using the Sieder–Tate equation.

▌ Again choosing $\bar{T}_b^{(0)} = -10$ °F, we have first-iteration properties from Problem 6.73; in addition, from Table B-3 at 14 °F,

$$\rho_s = 89.24\ \text{lbm/ft}^3 \qquad \nu_s = 0.238\times10^{-5}\ \text{ft}^2/\text{sec}$$

Thus, from (*H-12*) and (*H-3*),

$$\bar{h}_\infty = \frac{k_b}{D}(0.027)\,\mathbf{Re}_{D,b}^{0.8}\,\mathbf{Pr}_b^{1/3}\left(\frac{\mu_b}{\mu_s}\right)^{0.14}$$
$$= \left(\frac{0.0407}{\frac{1}{12}}\right)(0.027)(16\,711)^{0.8}(4.53)^{1/3}\left[\frac{(0.259)(91.78)}{(0.238)(89.24)}\right]^{0.14} = 52.98\ \text{Btu/hr}\cdot\text{ft}^2\cdot°\text{F}$$

$$T_{b,e}^{(1)} = 14 - (14+22)\exp\left[\frac{-\pi(\frac{1}{12})(13)(52.98)}{(0.26\times3600)(0.2158)}\right] = -0.7\ °\text{F}$$

or $\bar{T}_b^{(1)} = -11.4$ °F.

6.75 For the water flow of Problem 6.5, estimate the average heat transfer coefficient.

▌ From Problem 6.5, $T_{b,i} = 20$ °C and $T_{b,e} = 23$ °C; so $\bar{T}_b = 21.5$ °C. From Table B-3, at 21.5 °C,

$$\nu_b = (1.055\times10^{-5})(9.2903\times10^{-2}) = 9.801\times10^{-7}\ \text{m}^2/\text{s} \qquad \rho_b = (62.43)(1.601\,846\times10^1) = 1000.0\ \text{kg/m}^3$$
$$k_b = (0.346)(1.729\,577) = 0.598\ \text{W/m}\cdot\text{K} \qquad \mathbf{Pr}_b = 6.82$$

The Reynolds number is

$$\mathbf{Re}_{D,b} = \frac{4\dot{m}}{\pi D \nu_b \rho_b} = \frac{4(0.40\ \text{kg/s})}{\pi(0.050\ \text{m})(9.801 \times 10^{-7}\ \text{m}^2/\text{s})(1000.0\ \text{kg/m}^3)} = 1.04 \times 10^4 \quad (\textit{turbulent})$$

and $L/D = 8/0.050 = 160 > 60$ and $\bar{h} = \bar{h}_\infty$. Hence, by (*H-11*),

$$\bar{h}_\infty = \left(\frac{0.598\ \text{W/m}\cdot\text{K}}{0.050\ \text{m}}\right)(0.023)(1.04 \times 10^4)^{0.8}(6.82)^{0.4} = 970\ \text{W/m}^2\cdot\text{K}$$

But Problem 6.5 gave $\bar{h} = h_L = 235\ \text{W/m}^2\cdot\text{K}$ (h_x is constant for developed flow and constant q_s''). We must conclude that *the data for Problem 6.5 are inconsistent.*

6.76^D In a laboratory demonstration, water flows in a long electrically heated tube with $q_s'' = \text{const}$. The velocity and temperature profiles are fully developed. The tube has inside diameter 30 mm and, at a certain point $x = \lambda$, $T_b(\lambda) = 20\ °\text{C}$ and $T_s(\lambda) = 25\ °\text{C}$. The mass flow is 0.40 kg/s. **(*a*)** What is the value of h_λ? **(*b*)** What is the value of q_s''?

I (*a*) From Table B-3, at $T_b(\lambda) = 20\ °\text{C}$,

$\nu_b = (1.083 \times 10^{-5})(9.290\,304 \times 10^{-2}) = 1.006 \times 10^{-6}\ \text{m}^2/\text{s}$ $\quad\mathbf{Pr}_b = 7.02$

$\rho_b = (62.46)(16.018\,46) = 1000.5\ \text{kg/m}^3$ $\quad k_b = (0.345)(1.729\,577) = 0.597\ \text{W/m}\cdot\text{K}$

The Reynolds number is

$$\mathbf{Re}_{D,b} = \frac{4\dot{m}}{\pi D \rho_b \nu_b} = \frac{4(0.40)}{\pi(0.030)(1000.5)(1.006 \times 10^{-6})} = 1.687 \times 10^4$$

For this turbulent developed flow, the Dittus–Boelter equation gives, for $x = \lambda$,

$$h_\lambda = \frac{k_b}{D}(0.023)\,\mathbf{Re}_{D,b}^{0.8}\,\mathbf{Pr}_b^{0.4} = \left(\frac{0.597}{0.030}\right)(0.023)(1.687 \times 10^4)^{0.8}(7.02)^{0.4} = 2404\ \text{W/m}^2\cdot\text{K}$$

(*b*)
$$q_s''(\lambda) = h_\lambda[T_s(\lambda) - T_b(\lambda)] = (2.404\ \text{kW/m}^2\cdot\text{K})[(25-20)\ \text{K}] = 12.02\ \text{kW/m}^2$$

6.77^D In a laboratory demonstration, water flows through a long electrically heated tube with $q_s'' = \text{const}$. The velocity and temperature are fully developed. The tube has inside diameter 1.18 in., and at a certain point $x = \lambda$, $T_b(\lambda) = 68\ °\text{F}$ and $T_s(\lambda) = 77\ °\text{F}$. The mass flow is 0.88 lbm/sec. **(*a*)** What is the value of h_λ? **(*b*)** What is the value of q_s''?

I (*a*) From Table B-3, for saturated water at 68 °F,

$\nu_b = 1.083 \times 10^{-5}\ \text{ft}^2/\text{sec}$ $\quad\rho_b = 62.46\ \text{lbm/ft}^3$ $\quad k_b = 0.345\ \text{Btu/hr}\cdot\text{ft}\cdot°\text{F}$ $\quad\mathbf{Pr}_b = 7.02$

The Reynolds number is

$$\mathbf{Re}_{D,b} = \frac{4\dot{m}}{\pi D \rho_b \nu_b} = \frac{4(0.88)}{\pi(1.18/12)(62.46)(1.083 \times 10^{-5})} = 16\,855$$

For fully developed turbulent flow, the Dittus–Boelter equation gives, for $x = \lambda$,

$$h_\lambda = \frac{k_b}{D}(0.023)\,\mathbf{Re}_{D,b}^{0.8}\,\mathbf{Pr}_b^{0.4} = \left[\frac{0.345\ \text{Btu/hr}\cdot\text{ft}\cdot°\text{F}}{(1.18/12)\ \text{ft}}\right](0.023)(16\,855)^{0.8}(7.02)^{0.4} = 423\ \text{Btu/hr}\cdot\text{ft}^2\cdot°\text{F}$$

(*b*)
$$q_s''(\lambda) = h_\lambda[T_s(\lambda) - T_b(\lambda)] = (423)(77-68) = 3807\ \text{Btu/hr}\cdot\text{ft}^2$$

6.78^D Compressed air at $\bar{T}_b = 80\ °\text{F}$ and 30 psia flows in a long ($L/D > 60$) 1.0-in.-ID tube with a bulk average velocity $V_b = 10.0\ \text{ft/sec}$. The tube wall temperature is $T_s = 170\ °\text{F}$, constant. **(*a*)** Assuming fully developed velocity and temperature profiles, determine the average heat transfer coefficient. **(*b*)** What are the velocity and temperature profile entry lengths?

I From Table B-4, at $\bar{T}_b = 80\ °\text{F}$ and 14.7 psia,

$\mu_b g_c = 1.241 \times 10^{-5}\ \text{lbm/ft}\cdot\text{sec}$ $\quad k_b = 0.015\,16\ \text{Btu/hr}\cdot\text{ft}\cdot°\text{F}$ $\quad\mathbf{Pr}_b = 0.708$ $\quad\rho_{b,1} = 0.0735\ \text{lbm/ft}^3$

(*a*) The corrected density is $\rho_b = (30/14.7)(0.0735) = 0.150\ \text{lbm/ft}^3$, and so the Reynolds number is

$$\mathbf{Re}_{D,b} = \frac{\rho_b D V_b}{\mu_b g_c} = \frac{(0.15\ \text{lbm/ft}^3)(\frac{1}{12}\ \text{ft})(10.0\ \text{ft/sec})}{1.241 \times 10^{-5}\ \text{lbm/ft}\cdot\text{sec}} = 10\,070 \quad (\textit{turbulent})$$

From (*H-10*), with $(L/D) > 60$,

$$\mathbf{Nu}_{D,\infty} = (0.023)(10\,070)^{0.8}(0.708)^{1/3} = 32.67$$

$$\bar{h} = \bar{h}_\infty = \left(\frac{0.015\,16\ \text{Btu/hr}\cdot\text{ft}\cdot{}^\circ\text{F}}{\frac{1}{12}\ \text{ft}}\right)(32.67) = 5.94\ \text{Btu/hr}\cdot\text{ft}^2\cdot{}^\circ\text{F}$$

(*b*) In turbulent flow inside a tube, the velocity and temperature profiles are usually fully developed within ten diameters:

$$x_{\text{ent}} \approx 10(\tfrac{1}{12}\ \text{ft}) = 0.83\ \text{ft}$$

6.79[D] Compressed air at $\bar{T}_b = 27\ ^\circ\text{C}$ and 206.7 kPa flows inside of a long $(L/D > 60)$ 2.54-cm-ID tube with a bulk average velocity 3.04 m/s. The tube wall temperature is held constant at $T_s = 77\ ^\circ\text{C}$. (*a*) Assuming fully developed velocity and temperature profiles in the fluid, determine the average heat transfer coefficient. (*b*) Estimate the velocity and temperature entry lengths.

I From Table B-4, at 27 °C and 101.3 kPa,

$$\mu_b = (1.241\times10^{-5})(1.488\,164) = 1.847\times10^{-5}\ \text{Pa}\cdot\text{s} \qquad \mathbf{Pr}_b = 0.708$$
$$k_b = (0.015\,16)(1.729\,577) = 0.026\,22\ \text{W/m}\cdot\text{K} \qquad \rho_{b,1} = (0.0735)(16.018\,46) = 1.177\ \text{kg/m}^3$$

(*a*) The corrected density is $\rho_b = (206.7/101.3)(1.177) = 2.402\ \text{kg/m}^3$; the Reynolds number is

$$\mathbf{Re}_{D,b} = \frac{\rho_b D V_b}{\mu_b} = \frac{(2.402\ \text{kg/m}^3)(0.0254\ \text{m})(3.04\ \text{m/s})}{1.847\times10^{-5}\ \text{Pa}\cdot\text{s}} = 1.004\times10^4 \quad (\textit{just turbulent})$$

From (*H-10*),

$$\mathbf{Nu}_{D,\infty} = (0.023)(1.004\times10^4)^{0.8}(0.708)^{1/3} = 32.6$$

$$\bar{h}_\infty = \left(\frac{0.026\,22\ \text{W/m}\cdot\text{K}}{0.0254\ \text{m}}\right)(32.6) = 33.65\ \text{W/m}^2\cdot\text{K}$$

(*b*) The velocity or thermal entry length is approximately ten diameters.

$$x_{\text{ent}} \approx 10(0.0254\ \text{m}) \approx 0.25\ \text{m}$$

6.80[D] Saturated liquid water at $\bar{T}_b = 68\ ^\circ\text{F}$ flows at 2.0 ft/sec in a 1.0-in.-ID tube; the tube wall is heated by solar irradiation to $T_s = 78\ ^\circ\text{F} = \text{const}$. Find the average heat transfer coefficient in the fully developed velocity and temperature profile region, assuming that $L/D > 60$.

I From Table B-3, at $\bar{T}_b = 68\ ^\circ\text{F}$,

$$\nu_b = 1.083\times10^{-5}\ \text{ft}^2/\text{sec} \qquad k_b = 0.345\ \text{Btu/hr}\cdot\text{ft}\cdot{}^\circ\text{F} \qquad \mathbf{Pr}_b = 7.02$$

The Reynolds number is

$$\mathbf{Re}_{D,b} = \frac{DV_b}{\nu_b} = \frac{(\frac{1}{12}\ \text{ft})(2.0\ \text{ft/sec})}{1.083\times10^{-5}\ \text{ft}^2/\text{sec}} = 15\,390 \quad (\textit{turbulent})$$

Of the three correlations given in Appendix H, we choose (*H-11*):

$$\mathbf{Nu}_{D,\infty} = (0.023)(15\,390)^{0.8}(7.02)^{0.4} = 112$$

$$\bar{h}_\infty = \left(\frac{0.345\ \text{Btu/hr}\cdot\text{ft}\cdot{}^\circ\text{F}}{\frac{1}{12}\ \text{ft}}\right)(112) = 464\ \text{Btu/hr}\cdot\text{ft}^2\cdot{}^\circ\text{F}$$

6.81 Rework Problem 6.80 if the tube is 4 ft long.

I Now $L/D = 48/1$, and the entry length,

$$x_{\text{ent}} \approx 10D = 10\ \text{in.}$$

amounts to some 20% of L. Thus, we need a correction factor to modify $\bar{h}$ for the first ten diameters (0.83 ft). For $L/D = 48$, the Nusselt equation, (*H-15*), applies:

$$\overline{\mathbf{Nu}}_D = (0.036)(15\,390)^{0.8}(7.02)^{1/3}(\tfrac{1}{48})^{0.055} = 124.7$$

Thus,

$$\bar{h}_{\text{ent}} = \frac{k_b}{D}\overline{\mathbf{Nu}}_D = \left(\frac{0.345\ \text{Btu/hr}\cdot\text{ft}\cdot{}^\circ\text{F}}{\frac{1}{12}\ \text{ft}}\right)(124.7) = 516\ \text{Btu/hr}\cdot\text{ft}^2\cdot{}^\circ\text{F}$$

which is the average value for the entire tube heated length. This heat transfer coefficient is about 11 percent above $\bar{h}$ as found in Problem 6.80.

6.82[D] Saturated liquid water at $\bar{T}_b = 20\ °C$ flows at 0.61 m/s in a 25.4-mm-ID tube; the tube wall is heated by solar irradiation to $T_s = 26\ °C$, constant. Find the average heat transfer coefficient in the fully developed velocity and temperature profile region assuming that $L/D > 60$.

From Table B-3, at $\bar{T}_b = 20\ °C$,

$$\nu_b = 1.006 \times 10^{-6}\ \text{m}^2/\text{s} \qquad k_b = 0.597\ \text{W/m}\cdot\text{K} \qquad \mathbf{Pr}_b = 7.02$$

The Reynolds number is

$$\mathbf{Re}_{D,b} = \frac{DV_b}{\nu_b} = \frac{(0.0254\ \text{m})(0.61\ \text{m/s})}{1.006 \times 10^{-6}\ \text{m}^2/\text{s}} = 1.54 \times 10^4 \quad (\textit{turbulent})$$

Choosing the Dittus–Boelter equation for fluid heating, we have

$$\mathbf{Nu}_{D,\infty} = (0.023)(1.54 \times 10^4)^{0.8}(7.02)^{0.4} = 112$$

$$\bar{h}_\infty = \left(\frac{0.597\ \text{W/m}\cdot\text{K}}{0.0254\ \text{m}}\right)(112) = 2.632\ \text{kW/m}^2\cdot\text{K}$$

6.83[D] In an internal combustion engine laboratory, the effects of fuel preheating on engine starting are evaluated. A supply of benzene is heated, stored in an insulated tank, and used throughout the day in a number of engines. For a fully developed velocity profile prior to the heating section, and a subsequent constant-temperature heated tube length of 4.0 ft, determine the average heat transfer coefficient for a flow of saturated benzene at $\bar{T}_b = 80\ °F$, $V_b = 1.6$ ft/sec, in a 1.0-in.-ID tube with $T_s = 150\ °F$.

Fluid properties at $\bar{T}_b = 80\ °F$ and $T_s = 150\ °F$ are

$$\rho_b = 54.6\ \text{lbm/ft}^3 \qquad k_b = 0.092\ \text{Btu/hr}\cdot\text{ft}\cdot°\text{F}$$
$$c_{p,b} = 0.42\ \text{Btu/lbm}\cdot°\text{F} \qquad \mu_b g_c = 3.96 \times 10^{-4}\ \text{lbm/ft}\cdot\text{sec}$$
$$\mathbf{Pr}_b = 6.5 \qquad \mu_s g_c = 2.6 \times 10^{-4}\ \text{lbm/ft}\cdot\text{sec}$$

The Reynolds number is

$$\mathbf{Re}_{D,b} = \frac{V_b D \rho_b}{\mu_b g_c} = \frac{(1.6\ \text{ft/sec})[(1.0/12)\ \text{ft}](54.6\ \text{lbm/ft}^3)}{3.96 \times 10^{-4}\ \text{lbm/ft}\cdot\text{sec}} = 18\,380 \quad (\textit{turbulent})$$

As in Problem 6.81 an entry correction seems called for. However, the entrance region effects should be minimal because of the fully developed velocity prior to heating. Thus we apply (*H-12*)—$T_s - T_b$ being large—as though $L/D > 60$:

$$\mathbf{Nu}_{D,\infty} = (0.027)(18\,380)^{0.8}(6.5)^{1/3}\left(\frac{3.96}{2.6}\right)^{0.14} = 138$$

$$\bar{h}_\infty = \frac{k_b}{D}\mathbf{Nu}_{D,\infty} = \left(\frac{0.092}{\frac{1}{12}}\right)(138) = 152\ \text{Btu/hr}\cdot\text{ft}^2\cdot°\text{F}$$

6.84 If the benzene inlet temperature in Problem 6.83 is $T_{b,i} = 72\ °F$, what is the exit temperature, $T_{b,e}$?

By (*H-3*),

$$T_{b,e} = T_s - (T_s - T_{b,i})\exp\left[\frac{-4L\bar{h}}{(\mu_b g_c)\mathbf{Re}_{D,b}\,c_{p,b}}\right]$$
$$= 150 - (150 - 72)\exp\left[\frac{-4(4.0)(152)(1/3600)}{(3.96 \times 10^{-4})(18\,380)(0.42)}\right] = 87.4\ °\text{F}$$

By luck, $\frac{1}{2}(T_{b,e} + T_{b,i}) = \frac{1}{2}(87.4 + 72) = 79.7 \approx 80\ °F$, which was the assumed value of $\bar{T}_b$; therefore the value of $T_{b,e}$ does not have to be refined by iteration.

6.85[D] To evaluate the effects of fuel preheating on engine cold-starting, benzene is heated in an I.C. engines test laboratory. For a fully developed velocity profile prior to the heating tube length of 1.22 m, determine the average heat transfer coefficient for a flow of saturated benzene at $\bar{T}_b = 27\ °C$, $V_b = 0.49$ m/s inside a 25.4-mm-ID tube with $T_s = 66\ °C$.

Fluid properties at $\bar{T}_b = 27\ °C$ and $T_s = 66\ °C$ are

$$\rho_b = 874.6\ \text{kg/m}^3 \qquad k_b = 0.159\ \text{W/m}\cdot\text{K}$$
$$c_{p,b} = 1757\ \text{J/kg}\cdot\text{K} \qquad \mu_b = 5.89\times10^{-4}\ \text{Pa}\cdot\text{s}$$
$$\mathbf{Pr}_b = 6.5 \qquad \mu_s = 3.87\times10^{-4}\ \text{Pa}\cdot\text{s}$$

The Reynolds number is

$$\mathbf{Re}_{D,b} = \frac{DV_b\rho_b}{\mu_b} = \frac{(0.0254\ \text{m})(0.49\ \text{m/s})(874.6\ \text{kg/m}^3)}{5.89\times10^{-4}\ \text{Pa}\cdot\text{s}} = 1.848\times10^4 \quad (\textit{turbulent})$$

Because the entry length (ten-diameter rule of thumb) is $x_{\text{ent}} \approx 10(0.0254\ \text{m}) \approx 0.25\ \text{m}$ and because

$$\frac{L}{D} = \left(\frac{1.22}{0.0254}\right) = 48 < 60$$

an entry correction would normally be required. However, the entry region effect should be minimal due to the fully developed velocity profile upstream of the heated section. Thus, one can apply (*H-12*)—given the large $T_s - T_b$—as though $L/D > 60$:

$$\mathbf{Nu}_{D,\infty} = (0.027)(18\,480)^{0.8}(6.5)^{1/3}\left(\frac{5.89}{3.87}\right)^{0.14} = 138$$

$$\bar{h}_\infty = \frac{k_b}{D}\mathbf{Nu}_{D,\infty} = \left(\frac{0.159\ \text{W/m}\cdot\text{K}}{0.0254\ \text{m}}\right)(138) = 864\ \text{W/m}^2\cdot\text{K}$$

6.86[D] In the preceding turbulent-flow problems the average bulk temperature has been given. This is usually unknown; the common application/design problem provides $T_{b,i}$ and requires determination either of $T_{b,e}$ or of length to achieve a required $T_{b,e}$. For heating water from 20 °C to 60 °C an electrically heated tube is available. The tube has inside diameter 25 mm and provides a constant heat flux 50 kW/m². The mass flow is to be such that $\mathbf{Re}_{D,b} = 20\,000$. Determine (*a*) the length of tube required, and (*b*) whether this *proposed* heating system is feasible.

▌ At $\bar{T}_b = 40$ °C, Table B-3 gives

$$\nu_b = 6.577\times10^{-7}\ \text{m}^2/\text{s} \qquad \mathbf{Pr}_b = 4.34 \qquad \rho_b = 994.5\ \text{kg/m}^3 \qquad k_b = 0.628\ \text{W/m}\cdot\text{K} \qquad c_{p,b} = 4.176\ \text{kJ/kg}\cdot\text{K}$$

(*a*) Writing (*H-2*) for $x = L$ and expressing $\dot{m}$ in terms of $\mathbf{Re}_{D,b}$, one arrives at the formula

$$L = \frac{\rho_b\nu_b\mathbf{Re}_{D,b}c_{p,b}}{4q_s''}(T_{b,e} - T_{b,i})$$
$$= \left[\frac{(994.5\ \text{kg/m}^3)(6.577\times10^{-7}\ \text{m}^2/\text{s})(20\,000)(4.176\ \text{kJ/kg}\cdot\text{K})}{4(50\ \text{kW/m}^2)}\right](40\ \text{K}) = 10.92\ \text{m}$$

independent of the tube diameter for specified $\mathbf{Re}_{D,b}$.

(*b*) The question is (cf. Problem 6.19): For a 10.92-m flow to absorb 50 kW/m² in heat, is an excessive wall temperature required? In answer, we observe that, with $x_{\text{ent}} \approx 10D = 0.25$ m and $L/D = 10.92/0.025 = 437$, one of the correlations for fully developed turbulent flow at constant q_s'' may be used to calculate $\bar{h}_\infty$. Since $\mathbf{Pr}_b$ is moderate, select (*H-10*):

$$\bar{h}_\infty = \frac{k_b}{D}(0.023)\,\mathbf{Re}_{D,b}^{0.8}\,\mathbf{Pr}_b^{1/3} = \left(\frac{0.628}{0.025}\right)(0.023)(20\,000)^{0.8}(4.34)^{1/3} = 2.6\ \text{kW/m}^2\cdot\text{K}$$

Then, throughout the tube length, except in the first ten diameters,

$$T_s - T_b = \frac{q_s''}{\bar{h}} = \frac{50\ \text{kW/m}^2}{2.6\ \text{kW/m}^2\cdot\text{K}} \approx 19\ ^\circ\text{C}$$

which is a readily achieved temperature increment. The proposal is thus found to be feasible.

6.87[D] A 1.0-in.-ID tube is electrically heated to provide a constant heat flux $q_s'' = 4645\ \text{W/ft}^2$. Water flowing through the tube at 0.575 lbm/sec is to be heated from $T_{b,i} = 68$ °F to $T_{b,e} = 140$ °F. Determine (*a*) the length of tube required, and (*b*) whether the *proposed* heating scheme is feasible.

▌ At the desired $\bar{T}_b = 104$ °F, Table B-3 gives

$$\nu_b = 0.708\times10^{-5}\ \text{ft}^2/\text{sec} \qquad k_b = 0.363\ \text{Btu/hr}\cdot\text{ft}\cdot{}^\circ\text{F}$$
$$\mathbf{Pr}_b = 4.34 \qquad c_{p,b} = 0.998\ \text{Btu/lbm}\cdot{}^\circ\text{F}$$
$$\rho_b = 62.09\ \text{lbm/ft}^3$$

(*a*) By (*H-2*), at $x = L$,

$$L = \frac{\dot{m}c_{p,b}(T_{b,e} - T_{b,i})}{\pi D q_s''} = \frac{(0.575\ \text{lbm/sec})(3600\ \text{sec/hr})(0.998\ \text{Btu/lbm}\cdot{}^\circ\text{F})[(140-68)\ {}^\circ\text{F}]}{\pi(\frac{1}{12}\ \text{ft})(4645\ \text{W/ft}^2)(3.414\ \text{Btu/hr}\cdot\text{W})} = 35.82\ \text{ft}$$

(*b*) The question is (cf. Problem 6.20): For a 35.82-ft flow to absorb 4645 W/ft² in heat, is an excessive wall temperature required? In answer, we observe that with

$$\mathbf{Re}_{D,b} = \frac{4\dot{m}}{\pi D \rho_b \nu_b} = \frac{4(0.575)}{\pi(\frac{1}{12})(62.09)(0.708\times 10^{-5})} = 19\,985$$

$$x_{\text{ent}} \approx 10D = 10(\tfrac{1}{12}) = 0.833\ \text{ft}$$

$$\frac{L}{D} = 35.82/\tfrac{1}{12} \approx 430$$

one of the correlations for fully developed turbulent flow at constant q_s'' may be used to calculate $\bar{h}_\infty$. Because $\mathbf{Pr}_b$ is not too large, (*H-10*) is appropriate:

$$\bar{h}_\infty = \frac{k_b}{D}(0.023)\,\mathbf{Re}_{D,b}^{0.8}\,\mathbf{Pr}_b^{1/3} = \frac{0.363}{1/12}(0.023)(19\,985)^{0.8}(4.34)^{1/3} = 450\ \text{Btu/hr}\cdot\text{ft}^2\cdot{}^\circ\text{F}$$

Then, throughout the tube length,

$$T_s - T_b = \frac{q_s''}{\bar{h}_\infty} = \frac{(4645\ \text{W/ft}^2)(3.414\ \text{Btu/W}\cdot\text{hr})}{450\ \text{Btu/hr}\cdot\text{ft}^2\cdot{}^\circ\text{F}} \approx 35\ {}^\circ\text{F}$$

which is a readily achieved temperature increment. The proposed scheme is feasible.

6.88 An automotive fuel at $\bar{T}_b = 80\ {}^\circ\text{F}$ flows in a $\frac{3}{4}$-in.-ID tube with an average velocity of 2.0 fps. The tube is 50 ft long and has a constant wall temperature $T_s = 100\ {}^\circ\text{F}$. The fluid properties may be taken as

$\mu_s g_c = 35.1\times 10^{-5}\ \text{lbm/ft}\cdot\text{sec}$ $\qquad k_b = 0.092\ \text{Btu/hr}\cdot\text{ft}\cdot{}^\circ\text{F}$

$\mu_b g_c = 39.6\times 10^{-5}\ \text{lbm/ft}\cdot\text{sec}$ $\qquad \rho_b = 54.6\ \text{lbm/ft}^3$

$\nu_b = 0.725\times 10^{-5}\ \text{ft}^2/\text{sec}$ $\qquad c_{p,b} = 0.42\ \text{Btu/hr}\cdot{}^\circ\text{F}$

$\mathbf{Pr}_b = 6.5$

Determine (*a*) the entry length, and (*b*) the average heat transfer coefficient over the 50-ft length by three appropriate correlations.

▌ (*a*) The Reynolds number is

$$\mathbf{Re}_{D,b} = \frac{DV_b}{\nu_b} = \frac{[(0.75/12)\ \text{ft}](2.0\ \text{ft/sec})}{0.725\times 10^{-5}\ \text{ft}^2/\text{sec}} = 17\,240$$

The flow being turbulent, the entry length is $x_{\text{ent}} \approx 10D = 10(0.75/12) = 0.625$ ft.

(*b*) Since $L/D = (50\times 12)/0.75 = 80 > 60$, $\bar{h} = \bar{h}_\infty$ and $\bar{h}_\infty$ can be found from any of (*H-10*), (*H-11*), and (*H-12*). Using (*H-10*),

$$\mathbf{Nu}_{D,\infty} = (0.023)(17\,240)^{0.8}(6.5)^{1/3} = 105.2$$

$$\bar{h}_\infty = \left[\frac{0.092\ \text{Btu/hr}\cdot\text{ft}\cdot{}^\circ\text{F}}{(0.75/12)\ \text{ft}}\right](105.2) = 155\ \text{Btu/hr}\cdot\text{ft}^2\cdot{}^\circ\text{F}$$

Using (*H-11*)—even though $T_s - T_b \approx 20\ {}^\circ\text{F}$—

$$\mathbf{Nu}_{D,\infty} = (0.023)(17\,240)^{0.8}(6.5)^{0.4} = 119.2$$

$$\bar{h}_\infty = \left[\frac{0.092\ \text{Btu/hr}\cdot\text{ft}\cdot{}^\circ\text{F}}{(0.75/12)\ \text{ft}}\right](119) = 175\ \text{Btu/hr}\cdot\text{ft}^2\cdot{}^\circ\text{F}$$

Using (*H-12*),

$$\mathbf{Nu}_{D,\infty} = (0.027)(17\,240)^{0.8}(6.5)^{1/3}\left(\frac{39.6}{35.1}\right)^{0.14} = 125.6$$

$$\bar{h}_\infty = \left[\frac{0.092\ \text{Btu/hr}\cdot\text{ft}\cdot{}^\circ\text{F}}{(0.75/12)\ \text{ft}}\right](125.6) = 185\ \text{Btu/hr}\cdot\text{ft}^2\cdot{}^\circ\text{F}$$

Observe that the Chilton–Colburn equation resulted in the lowest $\bar{h}$ (about 11 percent below the $\bar{h}$ from the Dittus–Boelter equation and about 16 percent below the $\bar{h}$ from the Sieder–Tate equation). This is a frequently observed trend.

6.89 What rise in T_b occurs in Problem 6.88?

▌ Simultaneous solution of (*H-3*) (at $x = L$ and with the intermediate, Dittus–Boelter value of $\bar{h}_\infty$),

$$T_{b,e} = T_s - (T_s - T_{b,i}) \exp\left[\frac{-4L\bar{h}_\infty}{(\mu_b g_c)\,\mathbf{Re}_{D,b}\, c_{p,b}}\right]$$

and the definition

$$T_{b,e} + T_{b,i} \equiv 2\bar{T}_b$$

yields $T_{b,i} = 77.9$ °F and $T_{b,e} = 82.1$ °F, for an increase of 4.2 °F.

6.90^D Liquid Freon-12 at $T_{b,i} = -30$ °C enters a 0.61-m-long heated tube having a constant $T_s = -10$ °C. The mass flow is 0.12 kg/s and the tube ID is 25 mm. What is the discharge temperature $T_{b,e}$?

▌ The situation is that of Problem 6.71, except for a much shorter tube.

First iteration. Choose $T_{b,e}^{(0)} = -24$ °C, or $\bar{T}_b^{(0)} = -27$ °C; then, from Table B-3,

$$\nu_b \approx (0.267 \times 10^{-5})(9.2903 \times 10^{-2}) = 2.48 \times 10^{-7}\ \text{m/s} \qquad \mathbf{Pr}_b \approx 4.7$$

$$\rho_b \approx (92.43)(16.018) = 1481\ \text{kg/m}^3 \qquad c_{p,b} \approx (0.2148)(4184) = 898.7\ \text{J/kg}\cdot\text{K}$$

$$k_b \approx (0.0402)(1.729\,577) = 0.0695\ \text{W/m}\cdot\text{K}$$

The Reynolds number is

$$\mathbf{Re}_{D,b} = \frac{4\dot{m}}{\pi D \nu_b \rho_b} = \frac{4(0.12\ \text{kg/s})}{\pi(0.025\ \text{m})(2.48 \times 10^{-7}\ \text{m}^2/\text{s})(1481\ \text{kg/m}^3)} = 1.664 \times 10^4 \quad (\textit{turbulent})$$

Since $L/D = (0.61\ \text{m})/(0.025\ \text{m}) = 24.4 < 60$ and since $x_{\text{ent}} \approx 10D = 0.25$ m is significant relative to $L = 0.61$ m, an entry correction for $\bar{h}$ is needed. Using (*H-15*),

$$\overline{\mathbf{Nu}}_D = (0.036)(1.664 \times 10^4)^{0.8}(4.7)^{1/3}\left(\frac{1}{24.4}\right)^{0.055} = 120.5$$

$$\bar{h}_{\text{ent}} = \left(\frac{0.0695\ \text{W/m}\cdot\text{K}}{0.025\ \text{m}}\right)(120.5) = 335\ \text{W/m}^2\cdot\text{K}$$

Now (*H-3*) gives

$$T_{b,e}^{(1)} = -10 - (-10 + 30)\exp\left[\frac{-\pi(0.025)(0.61)(335)}{(0.12)(898.7)}\right] = -27.23\ \text{°C}$$

or $\bar{T}_b^{(1)} = -28.6$ °C.

Second iteration. At $\bar{T}_b^{(1)}$, Table B-3 gives

$$\nu_b \approx (0.269 \times 10^{-5})(9.2903 \times 10^{-2}) = 2.50 \times 10^{-7}\ \text{m}^2/\text{s} \qquad c_{p,b} \approx (0.2143)(4184) = 896.6\ \text{J/kg}\cdot\text{K}$$

$$\rho_b \approx (92.74)(16.018) = 1485\ \text{kg/m}^3 \qquad \mathbf{Pr}_b \approx 4.74$$

$$k_b \approx (0.0401)(1.729\,577) = 0.0694\ \text{W/m}\cdot\text{K}$$

The Reynolds number is

$$\mathbf{Re}_{D,b} = \frac{(4)(0.12\ \text{kg/s})}{\pi(0.025\ \text{m})(2.50 \times 10^{-7}\ \text{m}^2/\text{s})(1485\ \text{kg/m}^3)} = 1.646 \times 10^4$$

Proceeding as in the first iteration: $\bar{h}_{\text{ent}} = 333\ \text{W/m}^2\cdot\text{K}$ and $T_{b,e}^{(2)} = -27.24$ °C. Clearly, it is unnecessary to continue.

6.91 Assuming a fully developed velocity profile at the start of the 0.61-m heated section, repeat the calculation of $\bar{h}$ in Problem 6.90 using (*a*) the Dittus–Boelter and (*b*) the Petukhov equation.

▌ Take the mean bulk temperature and corresponding parameter values from the second iteration in Problem 6.90; also, at $T_s = -10$ °C,

$$\nu_s = 2.21 \times 10^{-7}\ \text{m/s} \qquad \rho_s = 1429\ \text{kg/m}^3$$

(*a*) From (*H-11*),

$$\mathbf{Nu}_{D,\infty} = (0.023)(1.646 \times 10^4)^{0.8}(4.74)^{0.4} = 101.2$$

$$\bar{h}_\infty = \frac{k_b}{D}\mathbf{Nu}_{D,\infty} = \left(\frac{0.0694\ \text{W/m}^2\cdot\text{K}}{0.025\ \text{m}}\right)(101.2) = 281\ \text{W/m}^2\cdot\text{K}$$

Note that if a *developing* velocity profile (as in Problem 6.90) had been assumed, a larger $\bar{h}$ would have been found. In fact, with $L/D = 24.4$, $(H\text{-}16)$ would have given

$$\bar{h}_{\text{ent}} = \bar{h}_\infty\left(1 + \frac{6}{L/D}\right) = (281)\left(1 + \frac{6}{24.4}\right) = 351 \text{ W/m}^2 \cdot \text{K}$$

(**b**) From $(H\text{-}18)$,

$$\phi = [(5.15)(4 + \log 1.646) - 4.64]^{-1} = 0.0586$$

$$\overline{\text{Nu}}_{D,\infty} = \left[\frac{(0.0586)^2(1.646 \times 10^4)(4.72)}{1.07 + (12.7)(0.0586)(4.72^{2/3} - 1)}\right]\left[\left(\frac{2.49}{2.21}\right)\left(\frac{1484}{1429}\right)\right]^{0.11} = 111.8$$

$$\bar{h}_\infty = \left(\frac{0.0694 \text{ W/m} \cdot \text{K}}{0.025 \text{ m}}\right)(111.8) = 310 \text{ W/m}^2 \cdot \text{K}$$

Again, for a *developing* velocity profile, this would be modified to

$$\bar{h}_{\text{ent}} = (310)\left(1 + \frac{6}{24.4}\right) = 388 \text{ W/m}^2 \cdot \text{K}$$

This last value represents, it is believed, the best of our three approximations of $\bar{h}$ for the conditions of Problem 6.90.

6.92^{D} Liquid Freon-12 at $T_{b,i} = -22$ °F enters a 2-ft-long, 1.0-in.-ID heated tube having a constant wall temperature $T_s = +14$ °F. The mass flow is 0.26 lbm/sec. What is the discharge temperature $T_{b,e}$?

$\blacksquare$ The situation is that of Problem 6.73, except for a much shorter tube.

First iteration. Guess $T_{b,e}^{(0)} = -18$ °F, or $\bar{T}_b^{(0)} = -20$ °F. From Table B-3:

$$\nu_b = 0.2699 \times 10^{-5} \text{ ft}^2/\text{sec} \qquad c_{p,b} = 0.2142 \text{ Btu/lbm} \cdot \text{°F}$$
$$\rho_b = 92.78 \text{ lbm/ft}^3 \qquad \text{Pr}_b = 4.75$$
$$k_b = 0.0401 \text{ Btu/hr} \cdot \text{ft} \cdot \text{°F}$$

The Reynolds number is

$$\text{Re}_{D,b} = \frac{4\dot{m}}{\pi D \nu_b \rho_b} = \frac{4(0.260 \text{ lbm/sec})}{\pi(\frac{1}{12} \text{ ft})(0.270 \times 10^{-5} \text{ ft}^2/\text{sec})(92.78 \text{ lbm/ft}^3)} = 15\,860 \quad (turbulent)$$

Since $L/D = 24/1 < 60$ and since $x_{\text{ent}} \approx 10D = 10$ in. is significant relative to $L = 24$ in., an entry correction for $\bar{h}$ is needed. By $(H\text{-}15)$,

$$\overline{\text{Nu}}_D = (0.036)(15\,860)^{0.8}(4.75)^{1/3}(\tfrac{1}{24})^{0.055} = 116$$

$$\bar{h}_{\text{ent}} = \left(\frac{0.0401 \text{ Btu/hr} \cdot \text{ft} \cdot \text{°F}}{\frac{1}{12} \text{ ft}}\right)(116) = 56.0 \text{ Btu/hr} \cdot \text{ft}^2 \cdot \text{°F}$$

Now $(H\text{-}3)$ gives

$$T_{b,e}^{(1)} = 14 - (14 + 22)\exp\left[\frac{-\pi(\frac{1}{12})(2)(56.0)}{(0.26 \times 3600)(0.2142)}\right] = -17.1 \text{ °F}$$

or $\bar{T}_b^{(1)} = -19.5$ °F, which is so close to $\bar{T}_b^{(0)} = -20$ °F that a second iteration may be dispensed with.

6.93^{D} An automotive fuel at $\bar{T}_b = 27$ °C flows inside a smooth circular tube 1.2 m long and having a 19-mm ID; the bulk average velocity is 0.90 m/s. The flow starts at the tube inlet from a reservoir (no upstream developing section), and the inside tube surface temperature is constant, $T_s = 33$ °C. The fluid properties may be taken as

$$\mu_s = 5.22 \times 10^{-4} \text{ Pa} \cdot \text{s} \qquad k_b = 0.1591 \text{ W/m} \cdot \text{K}$$
$$\mu_b = 5.892 \times 10^{-4} \text{ Pa} \cdot \text{s} \qquad \rho_b = 874.6 \text{ kg/m}^3$$
$$\text{Pr}_b = 6.5 \qquad c_{p,b} = 1757 \text{ J/kg} \cdot \text{K}$$

Determine (**a**) the entry length, and (**b**) the average heat transfer coefficient over the 1.2-m length.

$\blacksquare$ The Reynolds number is

$$\text{Re}_{D,b} = \frac{DV_b\rho_b}{\mu_b} = \frac{(0.019 \text{ m})(0.90 \text{ m/s})(874.6 \text{ kg/m}^3)}{5.892 \times 10^{-4} \text{ Pa} \cdot \text{s}} = 2.538 \times 10^4 \quad (turbulent)$$

(**a**)
$$x_{\text{ent}} = 10(0.019) = 0.19 \text{ m}$$

(*b*) As $L/D = 1.2/0.019 = 63.2 > 60$ (and x_{ent}/L is small), (*H-10*), (*H-11*), and (*H-12*) may be applied without any entry-region correction. Thus, (*H-11*) yields (for fluid heating)

$$\bar{h}_\infty = \frac{k_b}{D}(0.023)\,\mathbf{Re}_{D,b}^{0.8}\,\mathbf{Pr}_b^{0.4} = \left(\frac{0.1591\ \text{W/m}\cdot\text{K}}{0.019\ \text{m}}\right)(0.023)(2.538\times 10^4)^{0.8}(6.5)^{0.4} = 1360\ \text{W/m}^2\cdot\text{K}$$

while (*H-12*) gives

$$\bar{h}_\infty = \frac{k_b}{D}(0.027)\,\mathbf{Re}_{D,b}^{0.8}\,\mathbf{Pr}_b^{1/3}\left(\frac{\mu_b}{\mu_s}\right)^{0.14} = \left(\frac{0.1591\ \text{W/m}\cdot\text{K}}{0.019\ \text{m}}\right)(0.027)(2.538\times 10^4)^{0.8}(6.5)^{1/3} = \left(\frac{5.892}{5.22}\right)^{0.14}$$

$$= 1433\ \text{W/m}^2\cdot\text{K}$$

this value being about 5 percent higher than that obtained with the Dittus–Boelter equation.

6.94[D] Repeat Problem 6.93 for a tube ID of 50 mm. All other parameters and fluid properties are unchanged.

▮ The Reynolds number is $\mathbf{Re}_{D,b} = (\frac{50}{19})(2.538\times 10^4) = 6.68\times 10^4$.

(*a*)
$$x_{ent} \approx 10D = 10(0.050) = 0.50\ \text{m}$$

(*b*) Since x_{ent}/L is sizable and $L/D = 24 < 60$, an entry-region correlation is needed. Equation (*H-15*) applies $(10 < L/D < 400)$:

$$\overline{\mathbf{Nu}}_D = (0.036)\,\mathbf{Re}_{D,b}^{0.8}\,\mathbf{Pr}_b^{1/3}\left(\frac{D}{L}\right)^{0.055} = (0.036)(6.68\times 10^4)^{0.8}(6.5)^{1/3}(\tfrac{1}{24})^{0.055} = 408$$

$$\bar{h}_{ent} = \frac{k_b}{D}\overline{\mathbf{Nu}}_D = \left(\frac{0.1591\ \text{W/m}\cdot\text{K}}{0.050\ \text{m}}\right)(408) = 1300\ \text{W/m}^2\cdot\text{K}$$

This result may be compared with that yielded by (*H-11*)—according to which $\bar{h}_\infty \approx D^{-0.2}$—as corrected by (*H-16*):

$$\bar{h}_\infty = (\tfrac{19}{50})^{0.2}(1360) = 1120\ \text{W/m}^2\cdot\text{K} \qquad \bar{h}_{ent} = \bar{h}_\infty\left(1 + \frac{6}{L/D}\right) = (1120)\left(1 + \frac{6}{24}\right) = 1400\ \text{W/m}^2\cdot\text{K}$$

or a 7.7 percent increase over the value from (*H-15*).

6.95[D] Repeat Problem 6.94(*b*) for flow through a 50-mm-ID tube immediately downstream of a 90° bend, rather than flow from a fluid reservoir. All other parameters and fluid properties are unchanged.

▮ The calculation follows the final paragraph of Problem 6.94, with the appropriate entry correction, (*H-14*), replacing (*H-16*):

$$\bar{h}_{ent} = \bar{h}_\infty\left(1 + \frac{C}{L/D}\right) = (1120)\left(1 + \frac{7}{24}\right) = 1447\ \text{W/m}^2\cdot\text{K}$$

which represents an 11 percent increase over the straight-entry result of the Nusselt equation (Problem 6.94).

6.96[D] An automotive fuel at $\bar{T}_b = 80\ °\text{F}$ flows inside a 0.75-in.-ID smooth circular tube. The bulk average velocity is 2.95 ft/sec. The tube is heated from its beginning, and the flow originates with an abrupt contraction from a reservoir. The inside tube surface temperature is 90 °F, constant. The fluid properties may be taken as

$$\nu_b = 7.249\times 10^{-6}\ \text{ft}^2/\text{sec} \qquad \mathbf{Pr}_b = 6.5$$
$$k_b = 0.091\,99\ \text{Btu/hr}\cdot\text{ft}\cdot°\text{F} \qquad \mu_s g_c = 3.508\times 10^{-4}\ \text{lbm/ft}\cdot\text{sec}$$
$$\mu_b g_c = 3.959\times 10^{-4}\ \text{lbm/ft}\cdot\text{sec}$$

Determine (*a*) the minimum entry length and (*b*) the average heat transfer coefficient over a 4.0-ft length.

▮ The Reynolds number is

$$\mathbf{Re}_{D,b} = \frac{DV_b}{\nu_b} = \frac{[(0.75/12)\ \text{ft})](2.95\ \text{ft/sec})}{7.249\times 10^{-6}\ \text{ft}^2/\text{sec}} = 25\,460 \quad (\textit{turbulent})$$

(*a*)
$$x_{ent} \approx 10D = 10(0.75\ \text{in.}) = 7.5\ \text{in.}$$

which is small compared to $L = 48.0$ in.

(*b*) In any case, $L/D = 48/0.75 = 64 > 60$, and no entry-region correction is necessary. Using the Dittus–Boelter equation (heating version),

$$\bar{h}_\infty = \frac{k_b}{D}(0.023)\,\mathbf{Re}_{D,b}^{0.8}\,\mathbf{Pr}_b^{0.4} = \left[\frac{0.091\,99\ \text{Btu/hr}\cdot\text{ft}\cdot{}^\circ\text{F}}{(0.75/12)\ \text{ft}}\right](0.023)(25\,460)^{0.8}(6.5)^{0.4} = 240\ \text{Btu/hr}\cdot\text{ft}^2\cdot{}^\circ\text{F}$$

Likewise, the Sieder–Tate equation yields $\bar{h}_\infty = 252\ \text{Btu/hr}\cdot\text{ft}^2\cdot{}^\circ\text{F}$.

To confirm that entry corrections are unnecessary when L/D exceeds 60, we apply the Nusselt equation (*H-15*) to find $\bar{h}_{\text{ent}} = 263\ \text{Btu/hr}\cdot\text{ft}^2\cdot{}^\circ\text{F}$, which is only 4 percent above the Sieder–Tate value.

6.97[D] Repeat Problem 6.96 for a tube ID of 2.0 in. All other parameters and fluid properties are unchanged.

▌ The Reynolds number is $\mathbf{Re}_{D,b} = (2.0/0.75)(25\,460) = 67\,825$.

(*a*)

$$x_{\text{ent}} \approx 10D = 10(\tfrac{2}{12}) = 1.67\ \text{ft}$$

(*b*) With x_{ent}/L sizable and $L/D = 24 < 60$, an inlet-region correction is required. The Nusselt equation, (*H-15*), may be used $(10 < L/D < 400)$:

$$\overline{\mathbf{Nu}}_D = (0.036)\,\mathbf{Re}_{D,b}^{0.8}\,\mathbf{Pr}_b^{1/3}\left(\frac{D}{L}\right)^{0.055} = (0.036)(67\,825)^{0.8}(6.5)^{1/3}\left(\frac{1}{24}\right)^{0.055} = 413$$

$$\bar{h}_{\text{ent}} = \frac{k_b}{D}\,\overline{\mathbf{Nu}}_D = \left(\frac{0.091\,99\ \text{Btu/hr}\cdot\text{ft}\cdot{}^\circ\text{F}}{\frac{2}{12}\ \text{ft}}\right)(413) = 228\ \text{Btu/hr}\cdot\text{ft}^2\cdot{}^\circ\text{F}$$

This result may be compared with that yielded by (*H-11*)—according to which $\bar{h}_\infty \approx D^{-0.2}$—as corrected by (*H-16*):

$$\bar{h}_\infty = (0.75/2.0)^{0.2}(240) = 197\ \text{Btu/hr}\cdot\text{ft}^2\cdot{}^\circ\text{F}$$

$$\bar{h}_{\text{ent}} = \bar{h}_\infty\left(1 + \frac{6}{L/D}\right) = (197)\left(1 + \frac{6}{24}\right) = 246\ \text{Btu/hr}\cdot\text{ft}^2\cdot{}^\circ\text{F}$$

or an 8 percent increase over the value from (*H-15*).

6.98[D] Repeat Problem 6.97 for flow through a 2.0-in.-ID tube immediately downstream of a 90° bend. All other properties and fluid conditions remain unchanged.

▌ The calculation follows the final paragraph of Problem 6.97, with the appropriate entry correction, (*H-14*), replacing (*H-16*). Note that $(L/D)_c = 10.0$—check this—and $10 < L/D < 60$.

$$\bar{h}_{\text{ent}} = \bar{h}_\infty\left(1 + \frac{C}{L/D}\right) = (197)\left(1 + \frac{7}{24}\right) = 254\ \text{Btu/hr}\cdot\text{ft}^2\cdot{}^\circ\text{F}$$

6.99 For the flow of automotive fuel of Problem 6.93, assume that the heated tube section—now at $T_s = 38\ {}^\circ\text{C}$—is preceded by an unheated straight section of length 2.0 m. Estimate the rise in T_b over the 1.2-m heated length.

▌ As calculated in Problem 6.93 (it is still assumed that $\bar{T}_b = 27\ {}^\circ\text{C}$),

$$\mathbf{Re}_{D,b} = 2.538 \times 10^4 \qquad x_{\text{ent}} \approx 0.19\ \text{m} \qquad \frac{L}{D} = 63.2\ \text{m} \qquad \bar{h}_\infty = 1433\ \text{W/m}^2\cdot\text{K}$$

Here, the Sieder–Tate, rather than the Dittus–Boelter, heat transfer coefficient is quoted, because $|T_s - T_b| > 6\ {}^\circ\text{C}$. Substituting these data, and others as needed from Problem 6.93, in (*H-3*),

$$T_{b,e} = T_s - (T_s - T_{b,i})\exp\left[\frac{-4(L/D)\bar{h}_\infty}{\rho_b V_b c_{p,b}}\right] \tag{1}$$

and solving simultaneously with

$$T_{b,e} + T_{b,i} \equiv 2\bar{T}_b \tag{2}$$

one obtains $T_{b,i} = 25.6\ {}^\circ\text{C}$ and $T_{b,e} = 28.2\ {}^\circ\text{C}$; whence $\Delta T_b = 2.6\ {}^\circ\text{C}$.

6.100 Again consider the automotive fuel flow of Problem 6.93. The only changes are (1) $T_s = 38\ {}^\circ\text{C}$, and (2) the smooth tube is replaced with one that has been roughened. The measured friction factor, from

$$\Delta p \equiv f \frac{L}{D} \frac{\rho_b V_b^2}{2}$$

is $f = 0.035$. **(a)** What is the appropriate value of $\bar{h}$? **(b)** What will be the temperature rise, $T_{b,e} - T_{b,i}$?

▌ **(a)** Use (*H-17*), with data from Problem 6.93. Thus, with $\bar{h}_{\text{smooth}}$ from the Sieder–Tate equation and

$$f_{\text{smooth}} = \frac{0.184}{(2.538 \times 10^4)^{0.2}} = 0.024 \qquad n = (0.68)(6.5)^{0.215} = 1.017$$

we have

$$\bar{h}_{\text{rough}} = (1433\ \text{W/m}^2 \cdot \text{K})\left(\frac{0.035}{0.024}\right)^{1.017} = 2106\ \text{W/m}^2 \cdot \text{K}$$

(b) Simultaneous solution of (*1*) and (*2*) of Problem 6.99, with $\bar{h}_\infty$ replaced by $\bar{h}_{\text{rough}}$, yields $T_{b,i} = 24.9\ °\text{C}$ and $T_{b,e} = 29.1\ °\text{C}$, or $\Delta T_b = 4.2\ °\text{C}$. This is some 62 percent higher than the smooth-tube change determined in Problem 6.99.

6.101 As a final problem in the series beginning with Problem 6.93, consider again the flow of fuel, with all parameters unchanged from Problem 6.93, but in a roughened tube. Based on the Sieder–Tate value of $\bar{h}_{\text{smooth}}$, what is the largest $\bar{h}$ attainable with the roughened tube?

▌ By (*H-17*), in which $n \equiv (0.68)(6.5)^{0.215} = 1.0169$,

$$\bar{h}_{\text{rough}} < (3^n)\bar{h}_{\text{smooth}} = (3^{1.0169})(1433) = 4371\ \text{W/m}^2 \cdot \text{K}$$

6.102[D] Saturated water flows inside a 1.0-in.-ID circular tube which is electrically heated over its entire length to provide a constant heat flux, q_s''. The mass flow is 0.65 lbm/sec. At a location 9.0 ft downstream of the tube inlet, $T_b = 68\ °\text{F}$ and $T_s = 78\ °\text{F}$. Calculate the total heat transfer rate for a 10-ft-long tube. (Compare Problem 6.31.)

▌ From Table B-3, for saturated water at $T_b = 68\ °\text{F}$,

$$\nu_b = 1.083 \times 10^{-5}\ \text{ft}^2/\text{sec} \qquad \rho_b = 62.46\ \text{lbm/ft}^3 \qquad k_b = 0.345\ \text{Btu/hr} \cdot \text{ft} \cdot °\text{F} \qquad \mathbf{Pr}_b = 7.02$$

The Reynolds number is

$$\mathbf{Re}_{D,b} = \frac{4\dot{m}}{\pi D \nu_b \rho_b} = \frac{4(0.65\ \text{lbm/sec})}{\pi(\frac{1}{12}\ \text{ft})(1.083 \times 10^{-5}\ \text{ft}^2/\text{sec})(62.46\ \text{lbm/ft}^3)} = 14\,680 \quad (\textit{turbulent})$$

Because $L/D = 96/1 > 60$, we can treat the flow as thermally developed and can therefore apply (*H-4*), with $h_{x=q} = \bar{h} = \bar{h}_\infty$ calculated from (*H-11*). Thus,

$$h_{x=q} = \bar{h} = \bar{h}_\infty = \frac{k_b}{D}\mathbf{Nu}_{D,\infty} = \left(\frac{0.345\ \text{Btu/hr} \cdot \text{ft} \cdot °\text{F}}{\frac{1}{12}\ \text{ft}}\right)(0.023)(14\,680)^{0.8}(7.02)^{0.4} = 447\ \text{Btu/hr} \cdot \text{ft}^2 \cdot °\text{F}$$

$$q_{\text{conv}} = \bar{h}PL(T_s - T_b) \approx (447)\pi(\tfrac{1}{12})(10)(78 - 68) = 11\,700\ \text{Btu/hr}$$

Note that constant h and constant q_s'' require constant $T_s - T_b$. In this problem $h_{x=q} = \bar{h}_\infty$ (a constant), whereas in Problem 6.31 the expression for $h_{x=q}$ gives a local value.

6.103[D] Saturated water, at mean temperature $\bar{T}_b = 20\ °\text{C}$, flows inside a 25-mm-ID circular tube which is electrically heated over its entire length to provide a constant heat flux. The mass flow is 0.295 kg/s. At a location 2.74 m downstream of the tube inlet, $T_s = 26\ °\text{C}$. Determine the net heat transfer if the total tube length is 3.05 m. (Compare Problem 6.32.)

▌ From Table B-3, for saturated water at 20 °C,

$$\nu_b = (1.083 \times 10^{-5})(9.2903 \times 10^{-2}) = 1.006 \times 10^{-6}\ \text{m}^2/\text{s} \qquad k_b = (0.345)(1.7295) = 0.597\ \text{W/m} \cdot \text{K}$$
$$\rho_b = (62.46)(16.018) = 1000.5\ \text{kg/m}^3 \qquad \mathbf{Pr}_b = 7.02$$

The Reynolds number is

$$\mathbf{Re}_{D,b} = \frac{4\dot{m}}{\pi D \nu_b \rho_b} = \frac{4(0.295\ \text{kg/s})}{\pi(0.025\ \text{m})(1.006 \times 10^{-6}\ \text{m}^2/\text{s})(1000.5\ \text{kg/m}^3)} = 1.493 \times 10^4 \quad (\textit{turbulent})$$

Because $L/D = 2.74/0.025 = 109.6 > 60$, we can treat the flow as thermally developed and can therefore apply (*H-4*), with $\bar{h} = \bar{h}_\infty$ calculated from (*H-11*). Thus,

$$\mathbf{Nu}_{D,\infty} = (0.023)(1.493 \times 10^4)^{0.8}(7.02)^{0.4} = 110$$

$$\bar{h}_\infty = \frac{k_b}{D}\mathbf{Nu}_{D,\infty} = \left(\frac{0.597\ \text{W/m}\cdot\text{K}}{0.025\ \text{m}}\right)(110) = 2627\ \text{W/m}^2\cdot\text{K}$$

$$q_{\text{conv}} = \bar{h}_\infty PL(T_s - T_b) \approx (2627)\pi(0.025)(3.05)(26-20) = 3775\ \text{W}$$

6.104 For the water flow of Problem 6.103, determine the average heat transfer coefficient if the 25-mm-ID tube originates with a contraction from a larger reservoir (tube header) and is heated for only 150 mm of length.

▮ From Problem 6.103, we have

$$\mathbf{Re}_{D,b} = 1.493 \times 10^4 \qquad k_b = 0.597\ \text{W/m}\cdot\text{K} \qquad \mathbf{Pr}_b = 7.02$$

For the present problem, $L/D = \frac{150}{25} = 6$; an entry correction is necessary. Checking the applicability of (*H-13*):

$$\left(\frac{L}{D}\right)_{\text{crit}} = (0.623)\,\mathbf{Re}_{D,b}^{1/4} = (0.623)(1.493 \times 10^4)^{1/4} = 6.89 > 6$$

Hence
$$\frac{\bar{h}_{\text{ent}}}{\bar{h}_\infty} = (1.11)\left[\frac{\mathbf{Re}_{D,b}^{1/5}}{(L/D)^{4/5}}\right]^{0.275} = (1.11)\left[\frac{(1.493 \times 10^4)^{1/5}}{6^{4/5}}\right]^{0.275} = 1.27$$

which indicates the strong effect of the first few diameters. Appropriating $\bar{h}_\infty$ from Problem 6.103, we have

$$\bar{h}_{\text{ent}} = (1.27)(2627\ \text{W/m}^2\cdot\text{K}) = 3336\ \text{W/m}^2\cdot\text{K}$$

6.105[D] Air at $\bar{T}_b = 80\ °\text{F}$ and a pressure of 90 psia flows inside a 1.0-in.-ID circular tube with average bulk velocity $V_b = 7.0$ ft/sec. There is a 2.8-in.-long unheated (calming) section leading into a heated 3.0-ft-long section which has constant $T_s = 170\ °\text{F}$. For this heated section determine ***(a)*** the average heat transfer coefficient and ***(b)*** the air bulk temperature rise.

▮ From Table B-4 at $\bar{T}_b = 80\ °\text{F}$ and $T_s = 170\ °\text{F}$,

$$\mu_b g_c = 1.241 \times 10^{-5}\ \text{lbm/ft}\cdot\text{sec} \qquad \mu_s g_c = 1.394 \times 10^{-5}\ \text{lbm/ft}\cdot\text{sec}$$
$$k_b = 0.015\,16\ \text{Btu/hr}\cdot\text{ft}\cdot°\text{F} \qquad c_{p,b} = 0.2402\ \text{Btu/lbm}\cdot°\text{F}$$
$$\mathbf{Pr}_b = 0.708 \qquad \rho_{b,1} = 0.0735\ \text{lbm/ft}^3\ \text{(at 1 atm)}$$

(a) The pressure-corrected density is

$$\rho_b = (90/14.7)(0.0735) = 0.450\ \text{lbm/ft}^3$$

whence
$$\mathbf{Re}_{D,b} = \frac{\rho_b D V_b}{\mu_b g_c} = \frac{(0.45\ \text{lbm/ft}^3)(\frac{1}{12}\ \text{ft})(7.0\ \text{ft/sec})}{1.241 \times 10^{-5}\ \text{lbm/ft}\cdot\text{sec}} = 21\,150 \quad (\textit{turbulent})$$

Because
$$\left(\frac{L}{D}\right)_{\text{crit}} = (0.623)\,\mathbf{Re}_{D,b}^{1/4} = (0.623)(21\,150)^{1/4} = 7.51 < \frac{36}{1} = \frac{L}{D} < 60$$

we shall use (*H-14*) and Table H-2 (calming $L/D = 2.8$) to correct the developed value $\bar{h}_\infty$. This latter value will be estimated from the Sieder–Tate correlation, (*H-12*), using average properties as though $L/D > 60$. Thus:

$$\mathbf{Nu}_{D,\infty} = (0.027)(21\,150)^{0.8}(0.708)^{1/3}\left(\frac{1.241}{1.394}\right)^{0.14} = 68.32$$

$$\bar{h}_\infty = \frac{k_b}{D}\mathbf{Nu}_{D,\infty} = \left(\frac{0.015\,16\ \text{Btu/hr}\cdot\text{ft}\cdot°\text{F}}{\frac{1}{12}\ \text{ft}}\right)(68.32) = 12.43\ \text{Btu/hr}\cdot\text{ft}^2\cdot°\text{F}$$

$$\bar{h}_{\text{ent}} = \bar{h}_\infty\left(1 + \frac{C}{L/D}\right) = (12.43)\left(1 + \frac{3.0}{36}\right) = 13.47\ \text{Btu/hr}\cdot\text{ft}^2\cdot°\text{F}$$

(b) Equation (*H-3*), for $x = L$,

$$T_{b,e} = T_s - (T_s - T_{b,i})\exp\left(\frac{-4L\bar{h}_{\text{ent}}}{DV_b\rho_b c_{p,b}}\right)$$

and the definition $T_{b,e} + T_{b,i} \equiv 2\bar{T}_b$ may be solved simultaneously to yield $T_{b,i} = 49.3\ °\text{F}$ and $T_{b,e} = 110.7\ °\text{F}$; hence $\Delta T_b = 61.4\ °\text{F}$.

6.106[D] Air at $\bar{T}_b = 27$ °C and a pressure of 620.5 kPa flows inside a 25.4-mm-ID circular tube with average bulk velocity 2.10 m/s. The tube wall temperature is $T_s = 77$ °C, constant. There is a 71.1-mm-long unheated (calming) section upstream of the heated 0.91-m-long tube. Over the heated length, determine **(a)** the average heat transfer coefficient; **(b)** the rise in the fluid bulk temperature.

From Table B-4, at $\bar{T}_b = 27$ °C and $T_s = 77$ °C,

$$\mu_b = (1.241 \times 10^{-5})(1.488\,164) = 1.847 \times 10^{-5} \text{ Pa}\cdot\text{s}$$
$$k_b = (0.015\,16)(1.729\,577) = 0.026\,22 \text{ W/m}\cdot\text{K}$$
$$\rho_{b,1} = (0.0735)(1.6018 \times 10^1) = 1.177 \text{ kg/m}^3 \quad \text{(at 101.3 kPa)}$$
$$c_{p,b} = (0.2402)(4184) = 1005 \text{ J/kg}\cdot\text{K}$$
$$\mathbf{Pr}_b = 0.708$$
$$\mu_s = (1.394 \times 10^{-5})(1.488\,164) = 2.074 \times 10^{-5} \text{ Pa}\cdot\text{s}$$

(a) The pressure-corrected density is

$$\rho_b = \left(\frac{620.5}{101.3}\right)(1.177) = 7.208 \text{ kg/m}^3$$

whence
$$\mathbf{Re}_{D,b} = \frac{\rho_b D V_b}{\mu_b} = \frac{(7.208 \text{ kg/m}^3)(0.0254 \text{ m})(2.10 \text{ m/s})}{1.847 \times 10^{-5} \text{ Pa}\cdot\text{s}} = 2.082 \times 10^4 \quad (turbulent)$$

Because $L/D = 0.91/0.0254 = 35.8$ and

$$\left(\frac{L}{D}\right)_{\text{crit}} = (0.623)\,\mathbf{Re}_{D,b}^{1/4} = (0.623)(2.082 \times 10^4)^{1/4} = 7.48 < 35.8 < 60$$

we shall use (*H-14*) and Table H-2 (calming $L/D = 2.8$) to correct the developed value $\bar{h}_\infty$. This latter value will be estimated from the Sieder–Tate correlation, (*H-12*), using average properties as though $L/D > 60$. Thus:

$$\mathbf{Nu}_{D,\infty} = (0.027)(2.082 \times 10^4)^{0.8}(0.708)^{1/3}\left(\frac{1.847}{2.074}\right)^{0.14} = 67.5$$

$$\bar{h}_\infty = \frac{k_b}{D}\,\mathbf{Nu}_{D,\infty} = \left(\frac{0.026\,22 \text{ W/m}\cdot\text{K}}{0.0254 \text{ m}}\right)(67.5) = 69.7 \text{ W/m}^2\cdot\text{K}$$

$$\bar{h}_{\text{ent}} = \bar{h}_\infty\left(1 + \frac{C}{L/D}\right) = (69.7)\left(1 + \frac{3.0}{35.8}\right) = 75.54 \text{ W/m}^2\cdot\text{K}$$

(b) Equation (*H-3*), for $x = L$,

$$T_{b,e} = T_s - (T_s - T_{b,i})\exp\left(\frac{-4L\bar{h}_{\text{ent}}}{DV_b\rho_b c_{p,b}}\right)$$

and the definition $T_{b,e} + T_{b,i} \equiv 2\bar{T}_b$ may be solved simultaneously to give $T_{b,i} = 9.9$ °C and $T_{b,e} = 44.1$ °C, for a rise $\Delta T_b = 44.1 - 9.9 = 34.2$ °C.

6.107[D] In Problem 6.105 the calming section is lengthened to 1.0 ft. Determine **(a)** the average heat transfer coefficient for a smooth tube, and **(b)** the value of $\bar{h}_{\text{ent}}$ if the measured friction factor is $f = 0.06$.

(a) The solution is unchanged from Problem 6.105(*a*) up to the last step, which must now involve the *C*-value for the new calming length; thus,

$$\bar{h}_{\text{ent}} = (12.43)\left(1 + \frac{1.4}{36}\right) = 12.91 \text{ Btu/hr}\cdot\text{ft}^2\cdot{}^\circ\text{F}$$

(b) By (*H-17*), with

$$f_{\text{smooth}} = (0.184)(21\,150)^{-0.2} = 0.025 \qquad f_{\text{rough}} = 0.06 \qquad n = (0.68)(0.708)^{0.215} = 0.63$$

and $\bar{h}_{\text{smooth}} = 12.91$ Btu/hr · ft² · °F,

$$\bar{h}_{\text{rough}} = (12.91)\left(\frac{0.06}{0.025}\right)^{0.63} = 22.4 \text{ Btu/hr}\cdot\text{ft}^2\cdot{}^\circ\text{F}$$

6.108[D] Repeat Problem 6.106 if a 90° bend replaces the straight calming section.

(a) The solution is unchanged from Problem 6.106(*a*) up to the last step, which must now involve the *C*-value for the new inlet configuration; thus,

$$\bar{h}_{\text{ent}} = (69.7)\left(1 + \frac{7.0}{35.8}\right) = 83.4\ \text{W/m}^2 \cdot \text{K}$$

(b) The solution is the same as that in Problem 6.106(*b*), except that now the value $\bar{h}_{\text{ent}} = 83.4\ \text{W/m}^2 \cdot \text{K}$ is used. The final results are $T_{b,i} = 8.3\ °\text{C}$, $T_{b,e} = 45.7\ °\text{C}$, and $\Delta T_b = 37.4\ °\text{C}$.

6.109[D] Rework Problem 6.105(*a*) using **(*a*)** the Petukhov equation, (*H-18*); **(*b*)** the Dittus–Boelter equation, (*H-11*).

▮ **(*a*)**

$$\phi = (5.15 \log 21\,150 - 4.64)^{-1} = 0.0566$$

$$\overline{\mathbf{Nu}}_{D,\infty} = \left[\frac{(0.0566)^2(21\,150)(0.708)}{1.07 + (12.7)(0.0566)(0.708^{2/3} - 1)}\right](1) = 52.18$$

$$\bar{h}_\infty = \left(\frac{0.015\,16}{\frac{1}{12}}\right)(52.18) = 9.493\ \text{Btu/hr} \cdot \text{ft}^2 \cdot °\text{F}$$

$$\bar{h}_{\text{ent}} = (9.493)\left(1 + \frac{3.0}{36}\right) = 10.28\ \text{Btu/hr} \cdot \text{ft}^2 \cdot °\text{F}$$

This smaller answer is considered to be more accurate than the Sieder–Tate result.

(*b*)

$$\mathbf{Nu}_{D,\infty} = (0.023)(21\,150)^{0.8}(0.708)^{0.4} = 57.8$$

$$\bar{h}_\infty = \left(\frac{0.015\,16}{\frac{1}{12}}\right)(57.8) = 10.82\ \text{Btu/hr} \cdot \text{ft}^2 \cdot °\text{F}$$

$$\bar{h}_{\text{ent}} = (10.82)\left(1 + \frac{3.0}{36}\right) = 11.72\ \text{Btu/hr} \cdot \text{ft}^2 \cdot °\text{F}$$

which is intermediate between the Sieder–Tate and Petukhov results.

6.110[D] Rework Problem 6.106(*a*), using the Petukhov equation, (*H-18*).

▮

$$\phi = [(5.15)(4 + \log 2.082) - 4.64]^{-1} = 0.0568$$

$$\overline{\mathbf{Nu}}_{D,\infty} = \left[\frac{(0.0568)^2(2.082 \times 10^4)(0.708)}{1.07 + (12.7)(0.0568)(0.708^{2/3} - 1)}\right](1) = 51.6$$

$$\bar{h}_\infty = \left(\frac{0.026\,22}{0.0254}\right)(51.6) = 53.3\ \text{W/m}^2 \cdot \text{K}$$

$$\bar{h}_{\text{ent}} = (53.3)\left(1 + \frac{3.0}{35.8}\right) = 57.8\ \text{W/m}^2 \cdot \text{K}$$

This is probably more accurate than the Sieder–Tate result.

6.111[D] Air at $T_{b,i} = 10\ °\text{C}$ and a pressure of 620.5 kPa flows inside a circular 25.4-mm-ID tube with average bulk velocity 1.5 m/s. The tube is electrically heated over its entire length of 1.6 m in such a manner as to provide a uniform heat flux $q_s'' = 1470\ \text{W/m}^2$ to the air flowing inside. Determine **(*a*)** the temperature rise of the air; **(*b*)** the local heat transfer coefficient at $x = 1.0$ m.

▮ **(*a*)** If the initial guess $\bar{T}_b^{(0)} = 27\ °\text{C}$ is made, Problem 6.106 will provide initial mean property values. Then, from (*H-2*),

$$T_{b,e}^{(1)} - T_{b,i} = \frac{4q_s''L}{\rho_b D V_b c_{p,b}} = \frac{4(1470)(1.6)}{(7.208)(0.0254)(1.5)(1005)} = 34.1\ °\text{C}$$

and $\bar{T}_b^{(1)} = 10 + \frac{1}{2}(34.1) = 27.05\ °\text{C}$, which, by great good luck, is so close to $\bar{T}_b^{(0)}$ that further iteration is unnecessary.

(*b*) Taking property values at $T_b = 27\ °\text{C}$ from Problem 6.106, we have

$$\mathbf{Re}_{D,b} = \frac{\rho_b D V_b}{\mu_b} = \frac{(7.208\ \text{kg/m}^3)(0.0254\ \text{m})(1.5\ \text{m/s})}{1.847 \times 10^{-5}\ \text{Pa} \cdot \text{s}} = 1.487 \times 10^4 \quad (turbulent)$$

Since $(L/D)_{\text{ent}} \approx 10$ and since (see Fig. 6-2) h_x is constant along with q_s'' over the fully developed region, h_x at $x = 1.0$ m, or $(x/D) = 39$, should be well approximated by $\bar{h}_\infty$ as given by (*H-10*), (*H-11*), or (*H-12*). Choosing (*H-11*):

$$h_x = \bar{h}_\infty = \frac{k_b}{D}(0.023)\,\mathbf{Re}_{D,b}^{0.8}\,\mathbf{Pr}_b^{0.4} = \left(\frac{0.026\,22\ \mathrm{W/m\cdot K}}{0.0254\ \mathrm{m}}\right)(0.023)(1.487\times 10^4)^{0.8}(0.708)^{0.4} = 45\ \mathrm{W/m^2\cdot K}$$

6.112^D Air at $T_{b,i} = 50$ °F and a pressure of 90 psia flows inside a 1.0-in.-ID circular tube with average bulk velocity 5.0 ft/sec. The tube is electrically heated over its entire length of 5.25 ft in such a manner as to provide a uniform heat flux $q_s'' = 465$ Btu/hr · ft² to the air flowing inside. Determine ***(a)*** the temperature rise of the air; ***(b)*** the local heat transfer coefficient at $x = 3.0$ ft.

I If the initial guess $\bar{T}_b^{(0)} = 80$ °F is made, Problem 6.105 will provide initial mean property values. Then, from (*H-2*),

$$T_{b,e}^{(1)} - T_{b,i} = \frac{4q_s''L}{\rho_b D V_b c_{p,b}} = \frac{4(465)(5.25)}{(0.450)(\frac{1}{12})(5.0\times 3600)(0.2402)} = 60.2\ ^\circ\mathrm{F}$$

and $\bar{T}_b^{(1)} = 50 + \frac{1}{2}(60.2) = 80.1$ °F, which, most fortuitously, is so close to $\bar{T}_b^{(0)}$ that further iteration is unnecessary.

$$\mathbf{Re}_{D,b} = \frac{\rho_b D V_b}{\mu_b g_c} = \frac{(0.45\ \mathrm{lbm/ft^3})(\frac{1}{12}\ \mathrm{ft})(5.0\ \mathrm{ft/sec})}{1.241\times 10^{-5}\ \mathrm{lbm/ft\cdot sec}} = 15\,110\quad (\textit{turbulent})$$

Since $(L/D)_{\mathrm{ent}} \approx 10$ and since (see Fig. 6-2) h_x is constant along with q_s'' over the fully developed region, h_x at $x = 3.0$ ft, or $(x/D) = 36$, should be well approximated by $\bar{h}_\infty$ as given by (*H-10*), (*H-11*), or (*H-12*). Choosing (*H-11*):

$$h_x = \bar{h}_\infty = \frac{k_b}{D}(0.023)\,\mathbf{Re}_{D,b}^{0.8}\,\mathbf{Pr}_b^{0.4} = \left(\frac{0.015\,16\ \mathrm{Btu/hr\cdot ft\cdot ^\circ F}}{\frac{1}{12}\ \mathrm{ft}}\right)(0.023)(15\,110)^{0.8}(0.708)^{0.4} = 8.04\ \mathrm{Btu/hr\cdot ft^2\cdot ^\circ F}$$

6.113^D For the air flow of Problem 6.111, estimate the average heat transfer coefficient if the heated tube is only 0.8 m long and it is immediately downstream of a reservoir resulting in a flow contraction at the inlet.

I Because the major change from Problem 6.111 is a 50 percent reduction in tube length, the value of $T_{b,e} - T_{b,i}$ will be approximately 50 percent of the value calculated in that problem; so, as a first guess, we write

$$T_{b,e}^{(0)} - T_{b,i} = 17\ ^\circ\mathrm{C} \qquad \text{and} \qquad \bar{T}_b^{(0)} = 10 + \frac{17}{2} = 18.5\ ^\circ\mathrm{C}$$

The property values from Table B-4, at $\bar{T}_b = 18.5$ °C and 101.3-kPa pressure, are

$$\rho_{b,1} = (0.0759)(1.6018\times 10^1) = 1.217\ \mathrm{kg/m^3} \qquad c_{p,b} = (0.2402)(4.184\times 10^3) = 1005\ \mathrm{J/kg\cdot K}$$

Then, at 620.5 kPa, $\rho_b = (1.217)(620.5/101.3) = 7.454$ kg/m³; hence, by (*H-2*),

$$T_{b,e}^{(1)} - T_{b,i} = \frac{4q_s''L}{\rho_b D V_b c_{p,b}} = \frac{4(1470)(0.8)}{(7.457)(0.0254)(1.5)(1005)} = 16.5\ ^\circ\mathrm{C}$$

or $\bar{T}_b^{(1)} = 10 + \frac{1}{2}(16.5) = 18.25$ °C.

Neglecting the difference between $\bar{T}_b^{(0)}$ and $\bar{T}_b^{(1)}$, we have the additional mean property values

$$\mu_b = 1.803\times 10^{-5}\ \mathrm{Pa\cdot s} \qquad k_b = 0.025\,53\ \mathrm{W/m\cdot K} \qquad \mathbf{Pr}_b = 0.710$$

The Reynolds number is

$$\mathbf{Re}_{D,b} = \frac{\rho_b D V_b}{\mu_b} = \frac{(7.454\ \mathrm{kg/m^3})(0.0254\ \mathrm{m})(1.5\ \mathrm{m/s})}{1.803\times 10^{-5}\ \mathrm{Pa\cdot s}} = 1.575\times 10^4\quad (\textit{turbulent})$$

Because $L/D = 0.8/0.0254 = 31.5 < 60$, making $x_{\mathrm{ent}}/L \approx \frac{1}{3}$, we shall compute $\bar{h}_\infty$ from (*H-11*) and then correct it by (*H-16*):

$$\bar{h}_\infty = \frac{k_b}{D}(0.023)\,\mathbf{Re}_{D,b}^{0.8}\,\mathbf{Pr}_b^{0.4} = \left(\frac{0.025\,53\ \mathrm{W/m\cdot K}}{0.0254\ \mathrm{m}}\right)(0.023)(1.575\times 10^4)^{0.8}(0.710)^{0.4} = 45.95\ \mathrm{W/m^2\cdot K}$$

$$\bar{h}_{\mathrm{ent}} = \bar{h}_\infty\left(1 + \frac{6}{L/D}\right) = (45.95)\left(1 + \frac{6}{31.5}\right) = 54.7\ \mathrm{W/m^2\cdot K}$$

Checking the applicability of (*H-11*):

$$T_s - T_b = \frac{q_s''}{\bar{h}_{\mathrm{ent}}} = \frac{1470\ \mathrm{W/m^2}}{54.7\ \mathrm{W/m^2\cdot K}} = 26.9\ ^\circ\mathrm{C}\quad (\mathrm{OK})$$

6.114[D] For the airflow of Problem 6.112, estimate the average heat transfer coefficient if the heated tube is only 2.65 ft long, and it is immediately downstream of a reservoir resulting in a flow contraction at the inlet.

▮ Because the major change from Problem 6.112 is a 50 percent reduction in tube length, the value of $T_{b,e} - T_{b,i}$ will be approximately 50 percent of the value calculated in that problem; so, as a first guess, we write

$$T_{b,e}^{(0)} - T_{b,i} = 30\ °\text{F} \qquad \text{and} \qquad \bar{T}_b^{(0)} = 50 + \tfrac{30}{2} = 65\ °\text{F}$$

From Table B-4, at 65 °F and 14.7 psia,

$$\rho_{b,1} = 0.075\,95\ \text{lbm/ft}^3 \qquad c_{p,b} = 0.2402\ \text{Btu/lbm}\cdot°\text{F}$$

Then, at 90 psia, $\rho_b = (0.075\,95)(90/14.7) = 0.465\ \text{lbm/ft}^3$; hence, by (*H-2*),

$$T_{b,e}^{(1)} - T_{b,i} = \frac{4q_s''L}{\rho_b DV_b c_{p,b}} = \frac{4(465)(2.65)}{(0.465)(\frac{1}{12})(5.0\times 3600)(0.2402)} = 29.4\ °\text{F}$$

or $\bar{T}_b^{(1)} = 50 + \frac{1}{2}(29.4) = 64.7\ °\text{F}$.

Neglecting the difference between $\bar{T}_b^{(0)}$ and $\bar{T}_b^{(1)}$, we have the additional mean property values

$$\mu_b g_c = 1.213\times 10^{-5}\ \text{lbm/ft}\cdot\text{sec} \qquad k_b = 0.014\,78\ \text{Btu/hr}\cdot\text{ft}\cdot°\text{F} \qquad \mathbf{Pr}_b = 0.710$$

The Reynolds number is

$$\mathbf{Re}_{D,b} = \frac{\rho_b DV_b}{\mu_b g_c} = \frac{(0.465\ \text{lbm/ft}^3)(\frac{1}{12}\ \text{ft})(5.0\ \text{ft/sec})}{1.213\times 10^{-5}\ \text{lbm/ft}\cdot\text{sec}} = 15\,750 \quad (\textit{turbulent})$$

Because $L/D = 2.65/(\frac{1}{12}) = 31.8 < 60$, making $x_{\text{ent}}/L \approx 1/3$, we shall compute $\bar{h}_\infty$ from (*H-11*) and then correct it by (*H-16*):

$$\bar{h}_\infty = \frac{k_b}{D}(0.023)\,\mathbf{Re}_{D,b}^{0.8}\,\mathbf{Pr}_b^{0.4} = \left(\frac{0.014\,78\ \text{Btu/hr}\cdot\text{ft}\cdot°\text{F}}{\frac{1}{12}\ \text{ft}}\right)(0.023)(15\,750)^{0.8}(0.710)^{0.4} = 8.11\ \text{Btu/hr}\cdot\text{ft}^2\cdot°\text{F}$$

$$\bar{h}_{\text{ent}} = \bar{h}_\infty\left(1 + \frac{6}{L/D}\right) = (8.11)\left(1 + \frac{6}{31.8}\right) = 9.65\ \text{Btu/hr}\cdot\text{ft}^2\cdot°\text{F}$$

Checking the applicability of (*H-11*),

$$T_s - T_b = \frac{q_s''}{\bar{h}_{\text{ent}}} = \frac{465\ \text{Btu/hr}\cdot\text{ft}^2}{9.65\ \text{Btu/hr}\cdot\text{ft}^2\cdot°\text{F}} = 48.2\ °\text{F} \quad (\text{OK})$$

6.115[D] Heated air flows through a 150-mm-ID sheet metal duct that is poorly insulated. At a certain location $x = \alpha$, the air bulk temperature, is $T_b(\alpha) = 44$ °C; at that same location, the pressure is 1 atm, $T_s(\alpha) =$ 41°C, and the volumetric flow is $Q_\alpha = 0.20\ \text{m}^3/\text{s}$. Assuming no entry-region effects, (*a*) use the Dittus–Boelter equation to determine h_α; (*b*) estimate the energy loss per meter of duct length at $x = \alpha$; and (*c*) estimate the air bulk temperature drop per meter of duct length at $x = \alpha$.

▮ From Table B-4, at $T_b(\alpha)$ and atmospheric pressure,

$$\nu_b = (18.75\times 10^{-5})(9.2903\times 10^{-2}) = 1.742\times 10^{-5}\ \text{m}^2/\text{s} \qquad c_{p,b} = (0.2405)(4184) = 1006\ \text{J/kg}\cdot\text{K}$$

$$k_b = (0.015\,90)(1.729\,577) = 0.0275\ \text{W/m}\cdot\text{K} \qquad \mathbf{Pr}_b = 0.704$$

$$\rho_b = (0.069\,69)(16.018\,46) = 1.116\ \text{kg/m}^3$$

(*a*) The Reynolds number at $x = \alpha$ is

$$\mathbf{Re}_{D,b} = \frac{4Q_\alpha}{\pi D\nu_b} = \frac{4(0.20\ \text{m}^3/\text{s})}{\pi(0.150\ \text{m})(1.742\times 10^{-5}\ \text{m}^2/\text{s})} = 9.745\times 10^4 \quad (\textit{turbulent})$$

Applying the Dittus–Boelter equation for cooling at uniform q_s'',

$$h_\alpha = \frac{k_b}{D}(0.023)\,\mathbf{Re}_{D,b}^{0.8}\,\mathbf{Pr}_b^{0.3} = \left(\frac{0.0275\ \text{W/m}\cdot\text{K}}{0.150\ \text{m}}\right)(0.023)(9.745\times 10^4)^{0.8}(0.704)^{0.3} = 37.18\ \text{W/m}^2\cdot\text{K}$$

(*b*) Over a small length, dx, of duct, centered on $x = \alpha$, we have—by definition of h—

$$\frac{dq}{(\pi D)\,dx} = h_\alpha[T_s(\alpha) - T_b(\alpha)]$$

whence
$$\left.\frac{dq}{dx}\right|_{x=\alpha} = \pi Dh_\alpha[T_s(\alpha) - T_b(\alpha)] = \pi(0.150)(37.18)(41 - 44) = -52.6\ \text{W/m}$$

with the minus sign signifying an energy loss.

(*c*) Applying the first law at $x = \alpha$ (see Problem 6.1),

$$\left.\frac{dq}{dx}\right|_{x=\alpha} = \dot{m}c_{p,b}\left.\frac{dT_b}{dx}\right|_{x=\alpha}$$

which gives, since $\rho_b Q_\alpha = \dot{m}$,

$$\left.\frac{dT_b}{dx}\right|_{x=\alpha} = \frac{1}{\rho_b Q_\alpha c_{p,b}}\left.\frac{dq}{dx}\right|_{x=\alpha} = \frac{1}{(1.116)(0.20)(1006)}(-52.6) = -0.23\ ^\circ\text{C/m}$$

with the minus sign signifying a temperature drop.

6.116[D] Heated air flows through a 6.0-in.-ID sheet metal duct that is poorly insulated. At a certain location $x = \alpha$, the air bulk temperature is $T_b(\alpha) = 111\ ^\circ\text{F}$; at that same location, the pressure is 1 atm, $T_s(\alpha) = 105.6\ ^\circ\text{F}$, and the volumetric flow is $Q_\alpha = 7.0\ \text{ft}^3/\text{sec}$. Assuming no entry-region effects, (***a***) use the Dittus–Boelter equation to determine h_α; (***b***) estimate the energy loss per foot of duct length at $x = \alpha$; (***c***) estimate the air bulk temperature drop per foot of duct length at $x = \alpha$.

From Table B-4, at $T_b^{(\alpha)}$ and 1 atm pressure,

$$\nu_b = 18.75 \times 10^{-5}\ \text{ft}^2/\text{sec} \qquad c_{p,b} = 0.2405\ \text{Btu/lbm}\cdot{}^\circ\text{F}$$
$$k_b = 0.015\,90\ \text{Btu/hr}\cdot\text{ft}\cdot{}^\circ\text{F} \qquad \mathbf{Pr}_b = 0.704$$
$$\rho_b = 0.069\,69\ \text{lbm/ft}^3$$

(*a*) The Reynolds number at $x = \alpha$ is

$$\mathbf{Re}_{D,b} = \frac{4Q}{\pi D \nu_b} = \frac{4(7.0\ \text{ft}^3/\text{sec})}{\pi(0.5\ \text{ft})(18.75 \times 10^{-5}\ \text{ft}^2/\text{sec})} = 95\,070 \quad (\textit{turbulent})$$

Applying the Dittus–Boelter equation for cooling at uniform q_s'',

$$h_\alpha = \frac{k_b}{D}(0.023)\,\mathbf{Re}_{D,b}^{0.8}\,\mathbf{Pr}_b^{0.3} = \left(\frac{0.015\,90\ \text{Btu/hr}\cdot\text{ft}\cdot{}^\circ\text{F}}{0.5\ \text{ft}}\right)(0.023)(95\,070)^{0.8}(0.704)^{0.3} = 6.32\ \text{Btu/hr}\cdot\text{ft}^2\cdot{}^\circ\text{F}$$

(*b*) By the formula of Problem 6.115(*b*),

$$\left.\frac{dq}{dx}\right|_{x=\alpha} = \pi(0.5)(6.32)(105.6 - 111) = -53.6\ \text{Btu/hr}\cdot\text{ft}$$

(*c*) By the formula of Problem 6.115(*c*),

$$\left.\frac{dT_b}{dx}\right|_{x=\alpha} = \frac{1}{(0.069\,69)(7.0 \times 3600)(0.2405)}(-53.6) = -0.127\ ^\circ\text{F/ft}$$

6.117[D] Blood passing through a 2.5-mm-ID steel tube is to be warmed from 32 °C to 37.3 °C. The volumetric flow is 15 mL/s, and the tube wall is heated electrically to provide a uniform q_s''. It is important to prevent the wall temperature from exceeding 43 °C, to prevent damage to the blood. Approximating blood properties by those of water, what is the minimum tube length required?

From Table B-3, for saturated water at $\bar{T}_b = \frac{1}{2}(32 + 37.3) = 34.6\ ^\circ\text{C}$,

$$\rho_b = 996.2\ \text{kg/m}^3 \qquad c_{p,b} = 4176\ \text{J/kg}\cdot\text{K} \qquad \nu_b = 7.51 \times 10^{-7}\ \text{m}^2/\text{s} \qquad k_b = 0.619\ \text{W/m}\cdot\text{K} \qquad \mathbf{Pr}_b = 5.06$$

The rate of heat transfer to the blood flow is

$$q_{\text{conv}} = \dot{m}c_{p,b}(T_{b,e} - T_{b,i}) = \rho_b Q c_{p,b}(T_{b,e} - T_{b,i})$$
$$= (996.2\ \text{kg/m}^3)(15 \times 10^{-6}\ \text{m}^3/\text{s})(4176\ \text{J/kg}\cdot\text{K})[(37.3 - 32)\ \text{K}] = 330\ \text{W}$$

The Reynolds number is

$$\mathbf{Re}_{D,b} = \frac{4Q}{\pi D \nu_b} = \frac{4(15 \times 10^{-6})}{\pi(2.5 \times 10^{-3})(7.51 \times 10^{-7})} = 1.017 \times 10^4$$

Assuming that the length of the minimal tube exceeds 60 diameters, its average heat transfer coefficient can be obtained from the Dittus–Boelter equation (the flow is turbulent, at constant q_s'', and $T_s - T_b$ is constant at the maximum value $43 - 37.3 = 5.7\ ^\circ\text{C}$):

$$\bar{h}_\infty = \frac{k_b}{D}(0.023)\,\mathbf{Re}_{D,b}^{0.8}\,\mathbf{Pr}_b^{0.4} = \left(\frac{0.619\ \text{W/m}\cdot\text{K}}{2.5 \times 10^{-3}\ \text{m}}\right)(0.023)(1.017 \times 10^4)^{0.8}(5.06)^{0.4} = 17\,500\ \text{W/m}^2\cdot\text{K}$$

Now (*H-4*) gives

$$L_{\min} = \frac{q_{\text{conv}}}{\bar{h}_\infty \pi D(T_s - T_b)_{\max}} = \frac{330\ \text{W}}{(17\,500\ \text{W/m}^2\cdot\text{K})\pi(2.5\times10^{-3}\ \text{m})(5.7\ \text{K})} = 0.42\ \text{m}$$

which does in fact exceed $60D = 0.15$ m.

6.118[D] Blood passing through a 0.10-in.-ID stainless steel tube is to be warmed from 89 °F to 98.5 °F. The volumetric flow is 1.0 in³/sec, and the tube wall is heated electrically to provide a uniform q_s''. It is important to prevent the wall temperature from exceeding 109 °F, to prevent damage to the blood. Approximate blood properties by those of water. What is the minimum tube length required?

▮ From Table B-3, for saturated water at $\bar{T}_b = \frac{1}{2}(89 + 98.5) = 93.8$ °F,

$$\rho_b = 62.19\ \text{lbm/ft}^3 \qquad c_{p,b} = 0.9982\ \text{Btu/lbm}\cdot°\text{F} \qquad \nu_b = 0.809\times10^{-5}\ \text{ft}^2/\text{sec}$$
$$k_b = 0.358\ \text{Btu/hr}\cdot\text{ft}\cdot°\text{F} \qquad \mathbf{Pr}_b = 5.06$$

The rate of heat transfer to the blood flow is

$$q_{\text{conv}} = \dot{m}c_{p,b}(T_{b,e} - T_{b,i}) = \rho_b Q c_{p,b}(T_{b,e} - T_{b,i})$$
$$= (62.19\ \text{lbm/ft}^3)(1.0\times\tfrac{3600}{1728}\ \text{ft}^3/\text{hr})(0.9982\ \text{Btu/lbm}\cdot°\text{F})[(98.5 - 89)\ °\text{F}] = 1229\ \text{Btu/hr}$$

The Reynolds number is

$$\mathbf{Re}_{D,b} = \frac{4Q}{\pi D\nu_b} = \frac{4(1.0\ \text{in}^3/\text{sec})}{\pi(0.10\ \text{in.})(0.809\times10^{-5}\times144\ \text{in}^2/\text{sec})} = 10\,930$$

Assuming that the length of the minimal tube exceeds 60 diameters, its average heat transfer coefficient can be obtained from the Dittus–Boelter equation (the flow is turbulent, at constant q_s'', and $T_s - T_b$ is constant at the maximum value $109 - 98.5 = 10.5$ °F):

$$\bar{h}_\infty = \frac{k_b}{D}(0.023)\,\mathbf{Re}_{D,b}^{0.8}\,\mathbf{Pr}_b^{0.4} = \left[\frac{0.358\ \text{Btu/hr}\cdot\text{ft}\cdot°\text{F}}{(0.1/12)\ \text{ft}}\right](0.023)(10\,930)^{0.8}(5.06)^{0.4} = 3216\ \text{Btu/hr}\cdot\text{ft}^2\cdot°\text{F}$$

Now (*H-4*) gives

$$L_{\min} = \frac{q_{\text{conv}}}{\bar{h}_\infty \pi D(T_s - T_b)_{\max}} = \frac{1229\ \text{Btu/hr}}{(3216\ \text{Btu/hr}\cdot\text{ft}^2\cdot°\text{F})\pi[0.1/12)\ \text{ft}](10.5\ °\text{F})} = 1.39\ \text{ft}$$

which does in fact exceed $60D = 0.5$ ft.

6.119[D] Benzene at an average bulk temperature of 100 °F flows inside a 2.0-in.-ID tube; at each point, the local tube wall temperature is 8 °F below the local T_b (constant q_s''). The tube cooled length is 6 ft, and it is downstream of a bell-mouth inlet with a screen. The benzene properties may be taken as

$$\nu_b = 0.65\times10^{-5}\ \text{ft}^2/\text{sec} \qquad k_b = 0.087\ \text{Btu/hr}\cdot\text{ft}\cdot°\text{F} \qquad \mu_b g_c = 35.1\times10^{-5}\ \text{lbm/ft}\cdot\text{sec} \qquad \mathbf{Pr}_b = 5.1$$

If the average bulk velocity is 8 ft/sec, determine the rate of heat transfer from the benzene.

▮ According to (*H-4*), the problem reduces to the calculation of $\bar{h}$. The Reynolds number is

$$\mathbf{Re}_{D,b} = \frac{DV_b}{\nu_b} = \frac{(\frac{2}{12}\ \text{ft})(8\ \text{ft/sec})}{0.65\times10^{-5}\ \text{ft}^2/\text{sec}} = 205\,100$$

This is highly turbulent, with the Reynolds number exceeding the limit for any of the listed correlation equations. Nonetheless, the Dittus–Boelter equation will be used to estimate $\bar{h}_\infty$. [Note that $|T_s - T_b| < 10$ °F. Now, by (*H-2*), the bulk temperature 100 °F is assumed at $x = L/2 = 3$ ft; hence (*H-11*) will actually furnish $h_{L/2}$. But h is constant for the conditions of this flow; whence $h_{L/2} = \bar{h}_\infty$.]

$$\bar{h}_\infty = \frac{k_b}{D}(0.023)\,\mathbf{Re}_{D,b}^{0.8}\,\mathbf{Pr}_b^{0.3} = \left(\frac{0.087\ \text{Btu/hr}\cdot\text{ft}\cdot°\text{F}}{\frac{2}{12}\ \text{ft}}\right)(0.023)(205\,100)^{0.8}(5.1)^{0.3} = 348\ \text{Btu/hr}\cdot\text{ft}^2\cdot°\text{F}$$

Since $L/D = 72/2 = 36$ and $(L/D)_{\text{crit}} = (0.623)(205\,100)^{1/4} = 13.26$, (*H-14*) and Table H-2 give the corrected value

$$\bar{h}_{\text{ent}} = \bar{h}_\infty\left(1 + \frac{C}{L/D}\right) = (348)\left(1 + \frac{1.4}{36}\right) = 362\ \text{Btu/hr}\cdot\text{ft}^2\cdot°\text{F}$$

Now (*H-4*) yields

$$q_{\text{conv}} = \bar{h}_{\text{ent}}\pi DL(T_s - T_b) = (362)\pi(\tfrac{2}{12})(6)(-8) = -9100\ \text{Btu/hr}$$

6.120^D Benzene at an average bulk temperature of 38 °C flows inside a 50-mm-ID tube; at each point the tube wall is held 4.5 °C below the fluid bulk temperature, resulting in a constant heat efflux q_s''. The cooled length of the tube is 1.8 m and it is located downstream of a bell-mouth inlet with a screen. The average bulk velocity is 2.44 m/s, and mean property values are

$$\nu_b = 6.04 \times 10^{-7}\ \text{m}^2/\text{s} \qquad k_b = 0.1505\ \text{W/m}\cdot\text{K} \qquad \mathbf{Pr}_b = 5.1$$

Calculate the rate of heat transfer from the benzene.

I According to (*H-4*) the problem reduces to the determination of $\bar{h}$. The Reynolds number is

$$\mathbf{Re}_{D,b} = \frac{DV_b}{\nu_b} = \frac{(0.050\ \text{m})(2.44\ \text{m/s})}{6.04 \times 10^{-7}\ \text{m}^2/\text{s}} = 2.02 \times 10^5$$

This is highly turbulent, with the Reynolds number beyond the limits of the correlation equations. However, for this small value of $|T_s - T_b|$, the Dittus–Boelter equation should be reasonably accurate [refer to Problem 6.119].

$$\bar{h}_\infty = \frac{k_b}{D}(0.023)\,\mathbf{Re}_{D,b}^{0.8}\,\mathbf{Pr}_b^{0.3} = \left(\frac{0.1505\ \text{W/m}\cdot\text{K}}{0.050\ \text{m}}\right)(0.023)(2.02\times 10^5)^{0.8}(5.1)^{0.3} = 1.98\ \text{kW/m}^2\cdot\text{K}$$

Since $L/D = 1.8/0.050 = 36$ and $(L/D)_{\text{crit}} = (0.623)(2.02 \times 10^5)^{1/4} = 13.21$, (*H-14*) and Table H-2 give the corrected value

$$\bar{h}_{\text{ent}} = \bar{h}_\infty\left(1 + \frac{C}{L/D}\right) = (1.98)\left(1 + \frac{1.4}{36}\right) = (2.06)\ \text{kW/m}^2\cdot\text{K}$$

and (*H-4*) gives

$$q_{\text{conv}} = \bar{h}_{\text{ent}}\,\pi DL(T_s - T_b) = (2.06)\pi(0.050)(1.8)(-4.5) = -2.62\ \text{kW}$$

6.121 Water at average bulk temperature 80 °C flows inside a 25-mm-ID circular tube with average bulk velocity 1.5 m/s. The pipe wall temperature is at each point 40 °C below the local value of T_b, and the cooling takes place at constant q_s''. Estimate the rate of energy loss from a 6-m length of tubing.

I The fluid properties at $\bar{T}_b = 80$ °C and $T_s = 40$ °C (these temperatures are assumed at the midsection of the tube) are, from Table B-3,

$$\nu_b = 3.64 \times 10^{-7}\ \text{m}^2/\text{s} \qquad \mathbf{Pr}_b = 2.22$$
$$k_b = 0.6676\ \text{W/m}\cdot\text{K} \qquad \nu_s = 6.58 \times 10^{-7}\ \text{m}^2/\text{s}$$
$$\rho_b = 974\ \text{kg/m}^3 \qquad \rho_s = 995\ \text{kg/m}^3$$

The Reynolds number is

$$\mathbf{Re}_{D,b} = \frac{DV_b}{\nu_b} = \frac{(0.025\ \text{m})(1.5\ \text{m/s})}{3.64 \times 10^{-7}\ \text{m}^2/\text{s}} = 1.03 \times 10^5 \quad (\textit{highly turbulent})$$

Since $|T_s - T_b| = 40$ °C, we need to apply the Sieder–Tate turbulent-flow equation. Noting that $L/D = 6/0.025 = 240 > 60$, we have

$$\bar{h}_\infty = \frac{k_b}{D}(0.027)\,\mathbf{Re}_{D,b}^{0.8}\,\mathbf{Pr}_b^{1/3}\left(\frac{\mu_b}{\mu_s}\right)^{0.14} = \left(\frac{0.6676\ \text{W/m}\cdot\text{K}}{0.025\ \text{m}}\right)(0.027)(1.03\times 10^5)^{0.8}(2.22)^{1/3}\left[\left(\frac{3.64}{6.58}\right)\left(\frac{974}{995}\right)\right]^{0.14}$$

$$= 8.83\ \text{kW/m}^2\cdot\text{K}$$

and (*H-4*) gives

$$q_{\text{conv}} = \bar{h}_\infty\,\pi DL(T_s - T_b) = (8.83)\pi(0.025)(6)(-40) = -166.5\ \text{kW}$$

6.122^D Repeat Problem 6.121 using the more modern Petukhov equation to determine $\bar{h}$.

I

$$\phi = [(5.15)(5 + \log 1.03) - 4.64]^{-1} = 0.0472$$

$$\overline{\mathbf{Nu}}_{D,\infty} = \left\{\frac{(0.0472)^2(1.03\times 10^5)(2.22)}{1.07 + (12.7)(0.0472)[(2.22)^{2/3} - 1]}\right\}\left[\left(\frac{3.64}{6.58}\right)\left(\frac{974}{995}\right)\right]^{0.25} = 293$$

$$\bar{h}_\infty = \frac{k_b}{D}\overline{\mathbf{Nu}}_{D,\infty} = \left(\frac{0.6676\ \text{W/m}\cdot\text{K}}{0.025\ \text{m}}\right)(293) = 7.82\ \text{kW/m}^2\cdot\text{K}$$

Scaling the result of Problem 6.121,

$$q_{\text{conv}} = \left(\frac{7.82}{8.83}\right)(-166.5) = -147.4\ \text{kW}$$

6.123[D] A liquid is to be cooled from $T_{b,i} = 300$ °F to $T_{b,e} = 280$ °F while passing through a 2.0-in.-ID tube having a constant wall temperature of 60 °F. The average bulk fluid velocity is 2.0 ft/sec. Determine the length of tube required, if the fluid properties at $\bar{T}_b = 290$ °F and $T_s = 60$ °F are

$$\nu_b = 1.185 \times 10^{-5}\ \text{ft}^2/\text{s} \qquad \mathbf{Pr}_b = 13.8$$
$$k_b = 0.0972\ \text{Btu/hr}\cdot\text{ft}\cdot°\text{F} \qquad \mu_b g_c = 71.75 \times 10^{-5}\ \text{lbm/ft}\cdot\text{sec}$$
$$c_{p,b} = 0.512\ \text{Btu/lbm}\cdot°\text{F} \qquad \mu_s g_c = 325 \times 10^{-5}\ \text{lbm/ft}\cdot\text{sec}$$
$$\rho_b = 61.05\ \text{lbm/ft}^3$$

▌ The Reynolds number is

$$\mathbf{Re}_{D,b} = \frac{DV_b}{\nu_b} = \frac{(\frac{2}{12}\ \text{ft})(2\ \text{ft/sec})}{1.185 \times 10^{-5}\ \text{ft}^2/\text{sec}} = 28\,130 \quad (\textit{turbulent})$$

Since the $|T_s - T_b|$ is large, the Sieder–Tate equation, (*H-12*), seems appropriate. Thus, on the assumption that $L/D > 60$,

$$\bar{h} = \frac{k_b}{D}(0.027)\,\mathbf{Re}_{D,b}^{0.8}\,\mathbf{Pr}_b^{1/3}\left(\frac{\mu_b}{\mu_s}\right)^{0.14} = \left(\frac{0.0972\ \text{Btu/hr}\cdot\text{ft}\cdot°\text{F}}{\frac{2}{12}\ \text{ft}}\right)(0.027)(28\,130)^{0.8}(13.8)^{1/3}\left(\frac{71.75}{325}\right)^{0.14}$$
$$= 111\ \text{Btu/hr}\cdot\text{ft}^2\cdot°\text{F}$$

Substitute this value, along with $x = L$, in (*H-3*) and solve for L:

$$L = \frac{\rho_b V_b D c_{p,b}}{4\bar{h}} \ln\frac{T_{b,i} - T_s}{T_{b,e} - T_s} = \frac{(61.05)(2.0 \times 3600)(2/12)(0.512)}{4(111)} \ln\frac{300 - 60}{280 - 60} = 7.36\ \text{ft} \qquad (1)$$

But then $L/D = 7.36/\frac{2}{12} = 44.2 < 60$, so that our use of (*H-12*) was improper.

Applying instead (*H-15*), valid for $10 < L/D < 400$, we obtain

$$\bar{h} = \frac{k_b}{D}(0.036)\,\mathbf{Re}_{D,b}^{0.8}\,\mathbf{Pr}_b^{1/3}\left(\frac{D}{L}\right)^{0.055} \qquad (2)$$

Now eliminate $\bar{h}$ between (2) and (1) to obtain

$$L = \left\{\frac{\rho_b V_b D^{1.945} c_{p,b} \ln[(T_{b,i} - T_s)/(T_{b,e} - T_s)]}{4k_b(0.036)\,\mathbf{Re}_{D,b}^{0.8}\,\mathbf{Pr}_b^{1/3}}\right\}^{1/0.945} = \left\{\frac{(61.05)(2.0 \times 3600)(\frac{2}{12})^{1.945}(0.512)\ln(\frac{12}{11})}{4(0.0972)(0.036)(28\,130)^{0.8}(13.8)^{1/3}}\right\}^{1/0.945}$$
$$= 5.41\ \text{ft} \qquad (3)$$

6.124 Problem 6.123 determined L from (*H-3*) and (*H-15*). What answer would have been found if (*H-5*) and (*H-15*) had been used?

▌ Exactly the same answer. By Problem 6.4(*b*), (*H-3*) and (*H-5*) are equivalent.

6.125[D] Repeat Problem 6.123, using (*H-16*) to correct the result of (*H-12*). Assume in this case that the cooling begins with the 2-in.-ID tube attached to a "header," as is common in engineering practice.

▌ In Problem 6.123 the Sieder–Tate equation, (*H-12*), gave $\bar{h}_\infty = 111$ Btu/hr · ft² · °F. Although L is unknown at this point, it is clear from Problem 6.123 that $20 < L/D < 60$; hence (*H-16*) gives

$$\bar{h} = \bar{h}_\infty\left(1 + \frac{6D}{L}\right) \qquad (1)$$

Elimination of $\bar{h}$ between (1) above and (1) of Problem 6.123 yields

$$L = \left(\frac{\rho_b V_b D c_{p,b}}{4\bar{h}_\infty} \ln\frac{T_{b,i} - T_s}{T_{b,e} - T_s}\right) - 6D = (7.36\ \text{ft}) - 6(\tfrac{2}{12}\ \text{ft}) = 6.36\ \text{ft} \qquad (2)$$

Thus, the effect of the entry correction is to reduce the Sieder–Tate length by six diameters (=1 ft).

6.126[D] A liquid is to be cooled from $T_{b,i} = 150$ °C to $T_{b,e} = 138$ °C while passing through a 50-mm-ID tube having a constant wall temperature of 16 °C. The average bulk fluid velocity is 0.6 m/s. Determine the length of tube required if the fluid properties at $\bar{T}_b = 144$ °C and $T_s = 16$ °C are

$$\nu_b = 1.101 \times 10^{-6}\ \text{m}^2/\text{s} \qquad \mathbf{Pr}_b = 13.8$$
$$k_b = 0.168\ \text{W/m}\cdot\text{K} \qquad \mu_b = 106.7 \times 10^{-5}\ \text{Pa}\cdot\text{s}$$
$$c_{p,b} = 2142\ \text{J/kg}\cdot\text{K} \qquad \mu_s = 483.6 \times 10^{-5}\ \text{Pa}\cdot\text{s}$$
$$\rho_b = 977.9\ \text{kg/m}^3$$

The Reynolds number is

$$\mathbf{Re}_{D,b} = \frac{DV_b}{\nu_b} = \frac{(0.050 \text{ m})(0.6 \text{ m/s})}{1.101 \times 10^{-6} \text{ m}^2/\text{s}} = 2.725 \times 10^4 \quad (\textit{turbulent})$$

An analytical formula for the required L, (3) of Problem 6.123, was derived from (H-3) and (H-15). Substituting present data,

$$L = \left[\frac{(977.9)(0.6)(0.050)^{1.945}(2142)\ln\left(\frac{134}{122}\right)}{4(0.168)(0.036)(2.725\times 10^4)^{0.8}(13.8)^{1/3}}\right]^{1/0.945} = 1.75 \text{ m}$$

6.127^D Repeat Problem 6.126, calculating $\bar{h}$ from (H-12) and (H-16). Assume in this case that the cooling begins with the 50-mm-ID tube attached to a "header," as is common in engineering practice.

For the data of Problem 6.126, the Sieder–Tate equation gives $\bar{h}_\infty = 622 \text{ W/m}^2\cdot\text{K}$. Then formula (2) of Problem 6.125 yields

$$L = \left[\frac{(977.9)(0.6)(0.050)(2142)}{4(622)}\right]\left(\ln\frac{134}{122}\right) - 6(0.050) = 2.07 \text{ m}$$

6.128^D Saturated water at $T_{b,i} = 60$ °F enters a 1.5-in.-ID tube with an abrupt contraction from a reservoir. The tube wall inside temperature is fixed at $T_s = 120$ °F. Estimate the discharge temperature if the heated tube is 30 ft long and the average bulk velocity is 5 ft/sec.

First iteration. Guess a discharge temperature: $T_{b,e}^{(0)} = 76$ °F; then $\bar{T}_b^{(0)} = 68$ °F. From Table B-3, at $\bar{T}_b^{(0)}$ and T_s,

$$\nu_b = 1.083\times 10^{-5} \text{ ft}^2/\text{sec} \qquad \mathbf{Pr}_b = 7.02$$
$$k_b = 0.345 \text{ Btu/hr}\cdot\text{ft}\cdot{}^\circ\text{F} \qquad \nu_s = 0.622\times 10^{-5} \text{ ft}^2/\text{sec}$$
$$c_{p,b} = 0.9988 \text{ Btu/lbm}\cdot{}^\circ\text{F} \qquad \rho_s = 61.84 \text{ lbm/ft}^3$$
$$\rho_b = 62.46 \text{ lbm/ft}^3$$

The Reynolds number is

$$\mathbf{Re}_{D,b} = \frac{DV_b}{\nu_b} = \frac{[(1.5/12)\text{ ft}](5 \text{ ft/sec})}{1.083\times 10^{-5} \text{ ft}^2/\text{sec}} = 57\,710 \quad (\textit{turbulent})$$

For this relatively large value of $|T_s - T_b|$, we apply the Sieder–Tate equation for turbulent flow, obtaining $(L/D = 360/1.5 = 240)$

$$\bar{h}_\infty = \frac{k_b}{D}(0.027)\,\mathbf{Re}_{D,b}^{0.8}\,\mathbf{Pr}_b^{1/3}\left(\frac{\mu_b}{\mu_s}\right)^{0.14} = \left[\frac{0.345 \text{ Btu/hr}\cdot\text{ft}\cdot{}^\circ\text{F}}{(1.5/12)\text{ ft}}\right](0.027)(57\,710)^{0.8}(7.02)^{1/3}\left[\left(\frac{1.083}{0.622}\right)\left(\frac{62.46}{61.84}\right)\right]^{0.14}$$
$$= 995 \text{ Btu/hr}\cdot\text{ft}^2\cdot{}^\circ\text{F}$$

Then (H-3) at $x = L$ gives

$$T_{b,e}^{(1)} = T_s - (T_s - T_{b,i})\exp\left(\frac{-4L\bar{h}_\infty}{DV_b\rho_b c_{p,b}}\right)$$
$$= 120 - (120-60)\exp\left[\frac{-4(30)(995)}{(1.5/12)(5\times 3600)(62.46)(0.9988)}\right] = 94.4 \text{ °F}$$

or $\bar{T}_b^{(1)} = 77.2$ °F.

Second iteration. From Table B-3, at $\bar{T}_b^{(1)}$,

$$\nu_b = 0.9872\times 10^{-5} \text{ ft}^2/\text{sec} \qquad \rho_b = 62.36 \text{ lbm/ft}^3$$
$$k_b = 0.350 \text{ Btu/hr}\cdot\text{ft}\cdot{}^\circ\text{F} \qquad \mathbf{Pr}_b = 6.34$$
$$c_{p,b} = 0.9986 \text{ Btu/lbm}\cdot{}^\circ\text{F}$$

and both ν_s and ρ_s are unchanged from the previously listed values at 120 °F.

$$\mathbf{Re}_{D,b} = \frac{[(1.5/12)\text{ ft}](5 \text{ ft/sec})}{0.9872\times 10^{-5} \text{ ft}^2/\text{sec}} = 63\,310$$

$$\bar{h}_\infty = \left[\frac{0.350 \text{ Btu/hr}\cdot\text{ft}\cdot{}^\circ\text{F}}{(1.5/12)\text{ ft}}\right](0.027)(63\,310)^{0.8}(6.34)^{1/3}\left[\left(\frac{0.9872}{0.622}\right)\left(\frac{62.36}{61.84}\right)\right]^{0.14} = 1037 \text{ Btu/hr}\cdot\text{ft}^2\cdot{}^\circ\text{F}$$

$$T_{b,e}^{(2)} = 120 - (120-60)\exp\left[\frac{-4(30)(1037)}{(1.5/12)(5\times 3600)(62.36)(0.9986)}\right] = 95.3 \text{ °F}$$

or $\bar{T}_b^{(2)} = 77.6$ °F. This is so close to $\bar{T}_b^{(1)}$ that further iteration is unnecessary.

6.129^D Saturated water at $T_{b,i} = 16\ °C$ enters a 37.5-mm-ID tube with an abrupt contraction from a reservoir. The tube wall inside temperature is fixed at $T_s = 49\ °C$. Estimate the discharge temperature if the heated tube is 9 m long and the average bulk velocity is 1.5 m/s.

▌ **First iteration.** Guess a discharge temperature: $T_{b,e}^{(0)} = 25°C$; then $\bar{T}_b^{(0)} = 20.5\ °C$. From Table B-3, at $\bar{T}_b^{(0)}$ and T_s,

$$\nu_b = 1.006 \times 10^{-6}\ \text{m}^2/\text{s} \qquad \mathbf{Pr}_b = 7.02$$
$$k_b = 0.597\ \text{W/m}\cdot\text{K} \qquad \nu_s = 0.577 \times 10^{-6}\ \text{m}^2/\text{s}$$
$$c_{p,b} = 4179\ \text{J/kg}\cdot\text{K} \qquad \rho_s = 990.6\ \text{kg/m}^3$$
$$\rho_b = 1000.5\ \text{kg/m}^3$$

The Reynolds number is

$$\mathbf{Re}_{D,b} = \frac{DV_b}{\nu_b} = \frac{(0.0375\ \text{m})(1.5\ \text{m/s})}{1.006 \times 10^{-6}\ \text{m/s}} = 5.591 \times 10^4 \quad (\textit{turbulent})$$

For this relatively large value of $|T_s - T_b|$, we apply the Sieder–Tate equation for turbulent flow, obtaining $(L/D = 360/1.5 = 240)$

$$\bar{h}_\infty = \frac{k_b}{D}(0.027)\,\mathbf{Re}_{D,b}^{0.8}\,\mathbf{Pr}_b^{1/3}\left(\frac{\mu_b}{\mu_s}\right)^{0.14}$$
$$= \left(\frac{0.597\ \text{W/m}\cdot\text{K}}{0.0375\ \text{m}}\right)(0.027)(5.591 \times 10^4)^{0.8}(7.02)^{1/3}\left[\left(\frac{1.006}{0.577}\right)\left(\frac{1000.5}{990.6}\right)\right]^{0.14} = 5595\ \text{W/m}^2\cdot\text{K}$$

Then $(H\text{-}3)$ at $x = L$ gives

$$T_{b,e}^{(1)} = T_s - (T_s - T_{b,i})\exp\left(\frac{-4L\bar{h}_\infty}{DV_b\rho_b c_{p,b}}\right)$$
$$= 49 - (49 - 16)\exp\left[\frac{-4(9)(5595)}{(0.0375)(1.5)(1000.5)(4179)}\right] = 35.0\ °C$$

or $\bar{T}_b^{(1)} = 25.5\ °C$.

Second iteration. From Table B-3, at $\bar{T}_b^{(1)}$,

$$\nu_b = 0.917 \times 10^{-6}\ \text{m}^2/\text{s} \qquad k_b = 0.605\ \text{W/m}\cdot\text{K} \qquad c_{p,b} = 4178\ \text{J/kg}\cdot\text{K} \qquad \rho_b = 998.9\ \text{kg/m}^3 \qquad \mathbf{Pr}_b = 6.34$$

and both ν_s and ρ_s are unchanged from the previously listed values at 49 °C.

$$\mathbf{Re}_{D,b} = \frac{(0.0375\ \text{m})(1.5\ \text{m/s})}{0.917 \times 10^{-6}\ \text{m}^2/\text{s}} = 6.134 \times 10^4$$

$$\bar{h}_\infty = \left(\frac{0.605\ \text{W/m}\cdot\text{K}}{0.0375\ \text{m}}\right)(0.027)(6.134 \times 10^4)^{0.8}(6.34)^{1/3}\left[\left(\frac{0.917}{0.577}\right)\left(\frac{998.9}{990.6}\right)\right]^{0.14} = 5825\ \text{W/m}^2\cdot\text{K}$$

$$T_{b,e}^{(2)} = 49 - (49 - 16)\exp\left[\frac{-4(9)(5825)}{(0.0375)(1.5)(998.9)(4178)}\right] = 35.5\ °C$$

or $\bar{T}_b^{(2)} = 25.75\ °C$. This is so close to $\bar{T}_b^{(1)}$ that further iteration is unnecessary.

6.130^D Engine oil at $T_{b,i} = 248\ °F$ flows through a coiled copper tube of 2.0-in. ID at the rate of 105 gpm. The tube is in a forced-flow, cooling-water bath resulting in a uniform $T_s = 122\ °F$. For a discharge temperature $T_{b,e} = 212\ °F$, what length of coiled tube is required?

▌ From Table B-3, at $\bar{T}_b = 230\ °F$ and $T_s = 122\ °F$,

$$\rho_b = 52.09\ \text{lbm/ft}^3 \qquad \nu_s = 1.75 \times 10^{-3}\ \text{ft}^2/\text{sec}$$
$$c_{p,b} = 0.5405\ \text{Btu/lbm}\cdot°\text{F} \qquad k_b = 0.0785\ \text{Btu/hr}\cdot\text{ft}\cdot°\text{F}$$
$$\nu_b = 0.176 \times 10^{-3}\ \text{ft}^2/\text{sec} \qquad \mathbf{Pr}_b = 225.5$$
$$\rho_s = 54.31\ \text{lbm/ft}^3$$

The Reynolds number is

$$\mathbf{Re}_{D,b} = \frac{4Q}{\pi D\nu_b} = \frac{4\left(105\ \dfrac{\text{gal}}{\text{min}}\right)\left(\dfrac{1\ \text{min}}{60\ \text{sec}}\right)\left(\dfrac{1\ \text{ft}^3}{7.48\ \text{gal}}\right)}{\pi(\frac{2}{12}\ \text{ft})(0.176 \times 10^{-3}\ \text{ft}^3/\text{sec})} = 10\,160 \quad (\textit{turbulent})$$

With $|T_s - T_b| \approx 108$ °F and $\mathbf{Pr}_b$ also large, the turbulent-flow Sieder–Tate equation is a good choice. Thus, (*H-12*) is a good choice for the estimation of $\bar{h}$. Thus, on the assumption that the unknown length exceeds 60 diameters and can be treated as uncoiled,

$$\bar{h}_\infty = \frac{k_b}{D}(0.027)\,\mathbf{Re}_{D,b}^{0.8}\,\mathbf{Pr}_b^{1/3}\left(\frac{\mu_b}{\mu_s}\right)^{0.14}$$

$$= \left(\frac{0.0785\ \text{Btu/hr}\cdot\text{ft}\cdot{}^\circ\text{F}}{\frac{2}{12}\ \text{ft}}\right)(0.027)(10\,160)^{0.8}(225.5)^{1/3}\left[\left(\frac{52.09}{54.31}\right)\left(\frac{0.176}{1.75}\right)\right]^{0.14} = 89.6\ \text{Btu/hr}\cdot\text{ft}^2\cdot{}^\circ\text{F}$$

Then, adapting (*1*) of Problem 6.123,

$$L = \frac{\rho_b \nu_b \mathbf{Re}_{D,b}\, c_{p,b}}{4\bar{h}_\infty}\ln\frac{T_{b,i}-T_s}{T_{b,e}-T_s} = \frac{(52.09)(0.176\times 3.6)(10\,160)(0.5405)}{4(89.6)}\ln\frac{248-122}{212-122} = 170\ \text{ft}$$

which indeed exceeds $60D = 10$ ft.

6.131^D Engine oil at $T_{b,i} = 120$ °C flows through a coiled copper tube of 50-mm ID at 6.66 L/s. The tube is in a forced-flow, cooling-water bath resulting in a uniform $T_s = 50$ °C. For a discharge temperature $T_{b,e} = 100$ °C, what length of coiled tube is required?

▮ From Table B-3, at $\bar{T}_b = 110$ °C and $T_s = 50$ °C,

$\rho_b = 834.5\ \text{kg/m}^3$ $\qquad \nu_s = 1.627\times 10^{-4}\ \text{m}^2/\text{s}$

$c_{p,b} = 2261\ \text{J/kg}\cdot\text{K}$ $\qquad k_b = 0.1358\ \text{W/m}\cdot\text{K}$

$\nu_b = 0.1635\times 10^{-4}\ \text{m}^2/\text{s}$ $\qquad \mathbf{Pr}_b = 225.5$

$\rho_s = 870.0\ \text{kg/m}^3$

The Reynolds number is

$$\mathbf{Re}_{D,b} = \frac{4Q}{\pi D \nu_b} = \frac{4(6.66\times 10^{-3}\ \text{m}^3/\text{s})}{\pi(0.050\ \text{m})(0.1635\times 10^{-4}\ \text{m}^3/\text{s})} = 1.038\times 10^4 \quad (\textit{turbulent})$$

With $|T_s - T_b| \approx 60$ °C and $\mathbf{Pr}_b$ also large, (*H-12*) is a good choice for the estimation of $\bar{h}$. Thus, on the assumption that the unknown length exceeds 60 diameters and can be treated as uncoiled,

$$\bar{h}_\infty = \frac{k_b}{D}(0.027)\,\mathbf{Re}_{D,b}^{0.8}\,\mathbf{Pr}_b^{1/3}\left(\frac{\mu_b}{\mu_s}\right)^{0.14}$$

$$= \left(\frac{0.1358\ \text{W/m}\cdot\text{K}}{0.050\ \text{m}}\right)(0.027)(1.038\times 10^4)^{0.8}(225.5)^{1/3}\left[\left(\frac{834.5}{870.0}\right)\left(\frac{0.1635}{1.627}\right)\right]^{0.14} = 525.3\ \text{W/m}^2\cdot\text{K}$$

Then, adapting (*1*) of Problem 6.123,

$$L = \frac{\rho_b \nu_b \mathbf{Re}_{D,b}\, c_{p,b}}{4\bar{h}_\infty}\ln\frac{T_{b,i}-T_s}{T_{b,e}-T_s} = \frac{(834.5)(0.1635\times 10^{-4})(1.038\times 10^4)(2261)}{4(525.3)}\ln\frac{120-50}{100-50} = 51.3\ \text{m}$$

which does indeed exceed $60D = 3$ m.

6.132^D Gaseous nitrogen at $T_{b,i} = 44$ °C flows inside a 100-mm-ID circular duct at a pressure of 150 kPa, with a mean velocity of 2.0 m/s. If the tube is heated, with $T_s = 127$ °C (constant) over a length of 7 m, determine the average heat transfer coefficient and the exit temperature, $T_{b,e}$.

▮ **First iteration.** From Table B-4, at $T_b^{(0)} = 77$ °C (this temperature was chosen for the first trial since it is the arithmetic mean of listed temperatures 27 °C and 127 °C),

$$\mu_b = (13.38\times 10^{-6})(1.488\,164) = 19.91\ \mu\text{Pa}\cdot\text{s}$$
$$\rho_b = (0.0623)(16.018\,46) = 0.998\ \text{kg/m}^3 \quad (\text{at } 101.3\ \text{kPa})$$
$$= 1.478\ \text{kg/m}^3 \quad (\text{at } 150\ \text{kPa})$$
$$k_b = (0.017\,205)(1.729\,577) = 0.029\,76\ \text{W/m}\cdot\text{K}$$
$$c_{p,b} = 1042.6\ \text{J/kg}\cdot\text{K}$$
$$\mathbf{Pr}_b = 0.702$$

and at $T_s = 127$ °C, $\mu_s = (14.77\times 10^{-6})(1.488\,164) = 21.98\ \mu\text{Pa}\cdot\text{s}$. The Reynolds number is therefore

$$\mathbf{Re}_{D,b} = \frac{DV_b\rho_b}{\mu_b} = \frac{(0.100\ \text{m})(2.0\ \text{m/s})(1.478\ \text{kg/m}^3)}{19.91\times 10^{-6}\ \text{Pa}\cdot\text{s}} = 1.485\times 10^4 \quad (\textit{turbulent})$$

For turbulent flow with $L/D = 7/0.100 = 70 > 60$ and $|T_s - T_b| > 10$ °C, (*H-12*) is suitable for the estimation of $\bar{h}$:

$$\bar{h}_\infty^{(1)} = \frac{k_b}{D}(0.027)\,\mathbf{Re}_{D,b}^{0.8}\,\mathbf{Pr}_b^{1/3}\left(\frac{\mu_b}{\mu_s}\right)^{0.14} = \left(\frac{0.029\,76\text{ W/m}\cdot\text{K}}{0.100\text{ m}}\right)(0.027)(1.485\times 10^4)^{0.8}(0.702)^{1/3}\left(\frac{19.91}{21.98}\right)^{0.14}$$

$$= 15.32\text{ W/m}^2\cdot\text{K}$$

Substituting this value and $x = L$ in (*H-3*),

$$T_{b,e}^{(1)} = T_s - (T_s - T_{b,i})\exp\left(\frac{-4L\bar{h}}{DV_b\rho_b c_{p,b}}\right) = 127 - (127-44)\exp\left[\frac{-4(7)(15.32)}{(0.100)(2.0)(1.478)(1042.6)}\right] = 106.4\text{ °C}$$

or $\bar{T}_b^{(1)} = 75.3$ °C. Further iteration is precluded by the inaccuracy of (*H-12*).

6.133[D] Repeat Problem 6.132 for nitrogen gas flow, with $T_{b,i} = 110$ °F, $T_s = 260$ °F, duct ID = 4.0 in., duct length = 23 ft, $V_b = 6.5$ ft/sec, and pressure 21.8 psia.

▌ **First iteration.** Choose $T_{b,e}^{(0)} = 230$ °F; then $\bar{T}_b^{(0)} = 170$ °F, at which (Table B-4)

$$\mu_b g_c = 13.38\times 10^{-6}\text{ lbm/ft}\cdot\text{sec} \qquad k_b = 0.017\,205\text{ Btu/hr}\cdot\text{ft}\cdot\text{°F}$$
$$\rho_b = 0.0623\text{ lbm/ft}^3\ \text{(at 14.7 psia)} \qquad c_{p,b} = 0.2492\text{ Btu/lbm}\cdot\text{°F}$$
$$= 0.0924\text{ lbm/ft}^3\ \text{(at 21.8 psia)} \qquad \mathbf{Pr}_b = 0.702$$

Also from Table B-4, at $T_s = 260$ °F, $\mu_s g_c = 14.77\times 10^{-6}$ lbm/ft · sec. The Reynolds number is then

$$\mathbf{Re}_{D,b} = \frac{DV_b\rho_b}{\mu_b g_c} = \frac{[(4.0/12)\text{ ft}](6.5\text{ ft/sec})(0.0924\text{ lbm/ft}^3)}{13.38\times 10^{-6}\text{ lbm/ft}\cdot\text{sec}} = 14\,960\quad (\textit{turbulent})$$

For turbulent flow with $L/D = (23\times 12)/4.0 = 69$ and $|T_s - T_b| > 10$ °F, (*H-12*) serves for the estimation of $\bar{h}$:

$$\bar{h}_\infty^{(1)} = \frac{k_b}{D}(0.027)\,\mathbf{Re}_{D,b}^{0.8}\,\mathbf{Pr}_b^{1/3}\left(\frac{\mu_b}{\mu_s}\right)^{0.14} = \left(\frac{0.0172\text{ Btu/hr}\cdot\text{ft}\cdot\text{°F}}{\frac{4}{12}\text{ ft}}\right)(0.027)(14\,960)^{0.8}(0.702)^{1/3}\left(\frac{13.38}{14.77}\right)^{0.14}$$

$$= 2.67\text{ Btu/hr}\cdot\text{ft}^2\cdot\text{°F}$$

Substituting this value and $x = L$ in (*H-3*),

$$T_{b,e}^{(1)} = T_s - (T_s - T_{b,i})\exp\left(\frac{-4L\bar{h}}{DV_b\rho_b c_{p,b}}\right) = 260 - (260-110)\exp\left[\frac{-4(23)(2.67)}{(4.0/12)(6.5\times 3600)(0.0924)(0.2492)}\right]$$

$$= 221.8\text{ °F}$$

or $\bar{T}_b^{(1)} = 165.9$ °F. Further iteration is precluded by the inaccuracy of (*H-12*).

6.134 Repeat Problem 6.132, using (*H-18*) for $\bar{h}$.

▌ **First iteration** (same starting values as in Problem 6.132).

$$\phi = [(5.15)(\log 1.485 + 4) - 4.64]^{-1} = 0.0594$$

$$\mathbf{Nu}_{D,\infty} = \left\{\frac{(0.0594)^2(1.485\times 10^4)(0.702)}{1.07 + (12.7)(0.0594)[(0.702)^{2/3} - 1]}\right\}(1) = 40.5$$

$$\bar{h}_\infty^{(1)} = \left(\frac{0.029\,76\text{ W/m}\cdot\text{K}}{0.100\text{ m}}\right)(40.5) = 12.0\text{ W/m}^2\cdot\text{K}$$

$$T_{b,e}^{(1)} = 127 - (127-44)\exp\left[\frac{-4(7)(12.0)}{(0.100)(2.0)(1.478)(1042.6)}\right] = 99.1\text{ °C}$$

or $\bar{T}_b^{(1)} = 71.5$ °C.

Second iteration (probably unjustified). A repetition of the preceding, using property values at 71.5 °C, leads to $\bar{h}_\infty^{(2)} = 12.08$ W/m² · K and $T_{b,e}^{(2)} = 98.8$ °C.

6.135 Saturated liquid water at $T_{b,i} = 25$ °C flows inside an electrically heated 25-mm-ID tube having constant heat flux $q_s'' = 14$ kW/m² to the water over its entire length. Determine the length of tube for a water discharge temperature of 35 °C, if the local tube surface temperature, $T_s(x)$, is 5 °C above the local water bulk temperature, $T_b(x)$.

Here, $\bar{T}_b = \frac{1}{2}(25 + 35) = 30\ °\mathrm{C}$; from Table B-3,

$$k_b = 0.6123\ \mathrm{W/m \cdot K} \qquad \mathbf{Pr}_b = 5.68$$

We can now write two simultaneous equations in the two unknowns L and $\mathbf{Re}_{D,b}$ (the Reynolds number is unknown because neither $\dot{m}$, Q, nor V_b is given):

$$L = \frac{k_b\, \mathbf{Re}_{D,b}\, \mathbf{Pr}_b\, (T_{b,e} - T_{b,i})}{4q_s''} \tag{1}$$

$$\left(\frac{\bar{h}_\infty D}{k_b} = \right) \frac{q_s'' D}{k_b(T_s - T_b)} = (0.023)\, \mathbf{Re}_{D,b}^{0.8}\, \mathbf{Pr}_b^{0.4} \tag{2}$$

Equation (*1*) is simply a form of (*H-2*) at $x = L$; equation (*2*) is (*H-11*) for heating, which applies because $|T_s - T_b| = 5\ °\mathrm{C}$ and because, presumably, $L/D > 60$. By elimination of $\mathbf{Re}_{D,b}$, one finds:

$$L = \frac{1}{4}\left(\frac{q_s''\, \mathbf{Pr}_b^2}{k_b}\right)^{1/4}\left[\frac{D}{(0.023)(T_s - T_b)}\right]^{5/4}(T_{b,e} - T_{b,i})$$

$$= \frac{1}{4}\left[\frac{(14 \times 10^3)(5.68)^2}{0.6123}\right]^{1/4}\left[\frac{0.025}{(0.023)(5)}\right]^{5/4}(35 - 25) = 10.9\ \mathrm{m}$$

Check: $L/D = 10.9/0.025 = 436 \gg 60$.

CHAPTER 7
Natural Convection

7.1 BOUNDARY-LAYER ANALYSIS: PLANAR WALL

7.1 Using Pohlhausen's similarity parameter,

$$\eta = \frac{y}{x}\left(\frac{\mathbf{Gr}_x}{4}\right)^{1/4} \tag{I-6}$$

and a stream function $\Psi(x, y)$ that involves an unknown function $f(\eta)$,

$$\psi(x, y) = f(\eta)\left[4\nu\left(\frac{\mathbf{Gr}_x}{4}\right)^{1/4}\right]$$

outline the reduction of the three partial differential equations (*I-3*), (*I-4*), and (*I-5*) to two ordinary differential equations. Cite the boundary conditions.

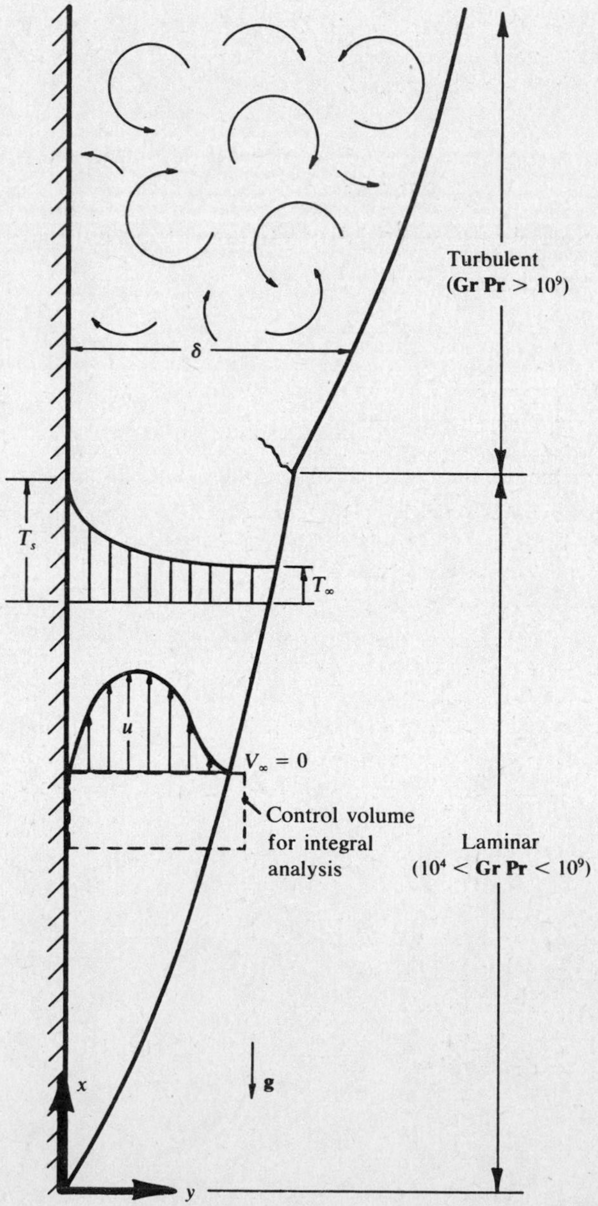

Fig. 7-1

From the definition of a stream function Ψ,

$$u = \frac{\partial \Psi}{\partial y} \quad \text{and} \quad v = -\frac{\partial \Psi}{\partial x}$$

so that (*I-5*) is satisfied. The (x, y)-coordinate system is defined in Fig. 7-1. Expressing the velocity components in terms of x and η, we get

$$u = \frac{\partial \Psi}{\partial y} = \frac{\partial \Psi}{\partial \eta}\frac{\partial \eta}{\partial y} = f'(\eta)\left[4\nu\left(\frac{\mathbf{Gr}_x}{4}\right)^{1/4}\right]\frac{1}{x}\left(\frac{\mathbf{Gr}_x}{4}\right)^{1/4} = \frac{2\nu}{x}\mathbf{Gr}_x^{1/2} f'$$

$$-v = \frac{\partial \Psi}{\partial x} = f'(\eta)\frac{\partial \eta}{\partial x}\left[4\nu\left(\frac{\mathbf{Gr}_x}{4}\right)^{1/4}\right] + f(\eta)\frac{d}{dx}\left[4\nu\left(\frac{\mathbf{Gr}_x}{4}\right)^{1/4}\right] = -\frac{\nu}{x}\left(\frac{\mathbf{Gr}_x}{4}\right)^{1/4}(\eta f' - 3f)$$

Defining a dimensionless temperature,

$$\theta \equiv \frac{T - T_\infty}{T_s - T_\infty} \qquad (I\text{-}7)$$

taking the respective derivatives, and substituting into (*I-3*) and (*I-4*), we get

$$\textit{momentum} \quad f''' + 3ff'' - 2(f')^2 + \theta = 0$$
$$\textit{energy} \quad \theta'' + 3\,\mathbf{Pr}\, f\theta' = 0$$

where the primes indicate differentiation with respect to η. Although θ can be eliminated between these two equations, it is simpler to consider them as simultaneous equations coupled through the function f. Each solution must be for a particular Prandtl number, since **Pr** appears as a parameter. The boundary conditions are

$$\textit{at} \quad \eta = 0 \quad f = 0 \quad f' = 0 \quad \theta = 1$$
$$\textit{as} \quad \eta \to \infty \quad f' \to 0 \quad \theta \to 0$$

Figures 7-2 and 7-3 give the solution (due to Ostrach) for a wide range of Prandtl numbers.

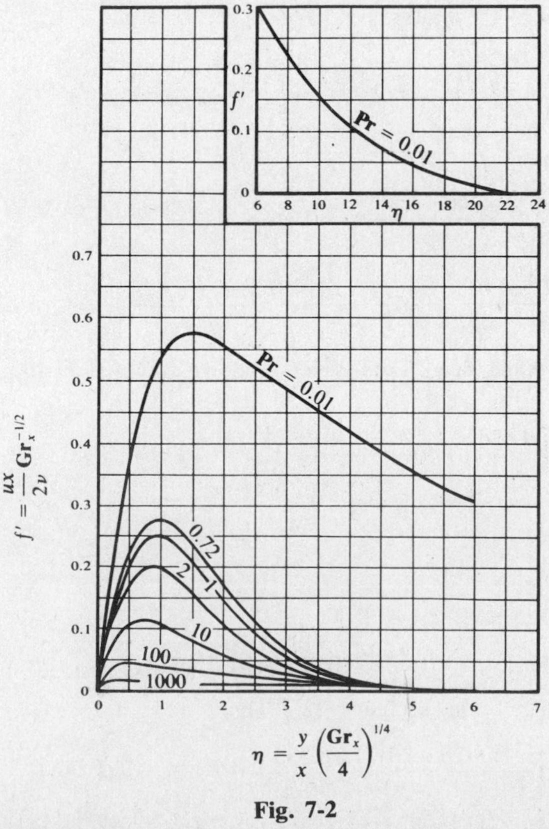

Fig. 7-2

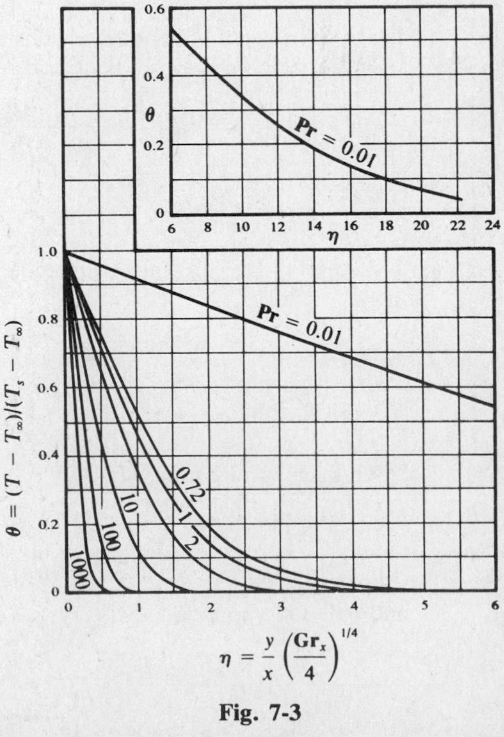

Fig. 7-3

7.2[D] Water is heated by a 4-in. by 4-in. vertical flat plate which is maintained at 126 °F. Using the similarity solution of Problem 7.1, find the heat transfer rate when the water is at 68 °F.

▌ At the reference temperature $T_{\text{ref}} = T_s + (0.38)(T_\infty - T_s) = 104\ °\text{F}$, pertinent parameters from Table B-3 are

$$\nu = 0.708 \times 10^{-5}\ \text{ft}^2/\text{sec} \qquad k = 0.363\ \text{Btu/hr}\cdot\text{ft}\cdot°\text{F} \qquad \mathbf{Pr} = 4.34$$

By inspection of Fig. 7-3, the temperature gradient at the wall is, for $\mathbf{Pr} \approx 4$,

$$\left.\frac{d\theta}{d\eta}\right|_{\eta=0} \approx \frac{-0.2}{0.24} = -0.9 = -F_1(\mathbf{Pr})$$

The coefficient of volumetric expansion β may be approximated by differencing in Table B-3. According to (*I-2*),

$$\beta_{104} \approx -\frac{1}{\rho_{104}}\left(\frac{\rho_{104} - \rho_{68}}{104 - 68}\right) = \frac{1}{62.09}\left(\frac{62.46 - 62.09}{104 - 68}\right) \approx \frac{1}{6200\ °\text{F}}$$

Then $$\mathbf{Gr}_L = \frac{g\beta(T_s - T_\infty)L^3}{\nu^2} = \frac{(32.2\ \text{ft/sec}^2)[(126 - 68)\ °\text{F}](0.333\ \text{ft})^3}{(6200\ °\text{F})(0.708 \times 10^{-5}\ \text{ft}^2/\text{sec})^2} = 2.22 \times 10^8$$

and $\mathbf{Gr}_L\,\mathbf{Pr} = (2.22 \times 10^8)(4.34) = 9.6 \times 10^8$ (laminar flow). Substituting known values in (*I-10*), we obtain

$$\bar{h} = \frac{4}{3}\left(\frac{k}{L}\right)[F_1(\mathbf{Pr})]\left(\frac{\mathbf{Gr}_L}{4}\right)^{1/4} = \frac{4}{3}\left(\frac{0.363}{0.333}\right)(0.9)\left(\frac{2.22 \times 10^8}{4}\right)^{1/4} = 113\ \text{Btu/hr}\cdot\text{ft}^2\cdot°\text{F}$$

$$q = \bar{h}A(T_s - T_\infty) = (113)(0.333)^2(126 - 68) = 727\ \text{Btu/hr}$$

7.3[D] Water is heated by a 10-cm by 10-cm vertical flat plate which is maintained at 52.2 °C. Using the similarity solution of Problem 7.1, find the heat transfer rate when the water is at 20.2 °C.

▌ At the reference temperature $T_{\text{ref}} = T_s + (0.38)(T_\infty - T_s) = 40\ °\text{C}$, pertinent parameters from Table B-3 are

$$\nu = 0.657 \times 10^{-6}\ \text{m}^2/\text{s} \qquad k = 0.628\ \text{W/m}\cdot\text{K} \qquad \mathbf{Pr} = 4.34$$

By inspection of Fig. 7-3, the temperature gradient at the wall is, for $\mathbf{Pr} \approx 4$,

$$\left.\frac{d\theta}{d\eta}\right|_{\eta=0} \approx \frac{-0.2}{0.24} \approx -0.9 = -F_1(\mathbf{Pr})$$

The coefficient of volumetric expansion β may be approximated by differencing in Table B-3. According to (*I-2*), in which ρ may be expressed in arbitrary units,

$$\beta_{40} \approx -\frac{1}{\rho_{40}}\left(\frac{\rho_{40} - \rho_{20}}{40 - 20}\right) = \frac{1}{62.09}\left(\frac{62.46 - 62.09}{40 - 20}\right) = 2.916 \times 10^{-4}\ \text{K}^{-1}$$

Then $$\mathbf{Gr}_L = \frac{g\beta(T_s - T_\infty)L^3}{\nu^2} = \frac{(9.807\ \text{m/s}^2)(2.979 \times 10^{-4}\ \text{K}^{-1})[(52.2 - 20.2)\ \text{K}](0.10\ \text{m})^3}{(0.657 \times 10^{-6}\ \text{m}^2/\text{s})^2} = 2.16 \times 10^8$$

and $\mathbf{Gr}_L\,\mathbf{Pr} = (2.12 \times 10^8)(4.34) = 9.20 \times 10^8$ (laminar flow). Substituting known values in (*I-10*), we obtain

$$\bar{h} = \frac{4}{3}\left(\frac{k}{L}\right)[F_1(\mathbf{Pr})]\left(\frac{\mathbf{Gr}_L}{4}\right)^{1/4} = \frac{4}{3}\left(\frac{0.628}{0.10}\right)(0.9)\left(\frac{2.16 \times 10^8}{4}\right)^{1/4} = 646\ \text{W/m}^2\cdot\text{K}$$

$$q = \bar{h}A(T_s - T_\infty) = (646)(0.10)^2(52.2 - 20.2) = 207\ \text{W}$$

7.4[D] Solve Problem 7.2 using (*I-17*).

▌ At the film temperature $T_f = (T_s + T_\infty)/2 = (126 + 68)/2 = 97.0\ °\text{F}$,

$$k = 0.360\ \text{Btu/hr}\cdot\text{ft}\cdot°\text{F} \qquad \mathbf{Pr} = 4.86 \qquad \nu = 0.781 \times 10^{-5}\ \text{ft}^2/\text{sec}$$

and one can approximate β_{97} by $\beta_{104} \approx 1/(6200\ °\text{F})$, from Problem 7.2. Then

$$\mathbf{Gr}_L = \frac{g\beta(T_s - T_\infty)L^3}{\nu^2} = \frac{(32.2\ \text{ft/sec}^2)[(126 - 68)\ °\text{F}](0.333\ \text{ft})^3}{(6200\ °\text{F})(0.781 \times 10^{-5}\ \text{ft}^2/\text{sec})^2} = 1.82 \times 10^8$$

and $\mathbf{Gr}_L\,\mathbf{Pr} = (1.82 \times 10^8)(4.86) = 8.85 \times 10^8$ (laminar flow). Table I-3 and (*I-17*) give $(\Lambda = L)$:

$$\bar{h} = \frac{k}{L}(0.59)(\mathbf{Gr}_L\,\mathbf{Pr})^{1/4} = \frac{0.360\ \text{Btu/hr}\cdot\text{ft}\cdot{}^\circ\text{F}}{0.333\ \text{ft}}(0.59)(8.85\times 10^8)^{1/4} = 110\ \text{Btu/hr}\cdot\text{ft}^2\cdot{}^\circ\text{F}$$

This value for the heat transfer is smaller than that obtained in Problem 7.2, i.e.,

$$q_{\text{emp}} = \tfrac{110}{113}(727) = 708\ \text{Btu/hr}$$

It is worth noting here that this value is perhaps more reliable than that of Problem 7.2, since the similarity solution required an estimation of the slope in Fig. 7-3. But the two solutions compare well with each other.

7.5^{D} Solve Problem 7.3 using (*I-17*).

▮ At the film temperature $T_f = (T_s + T_\infty)/2 = 36.2\ {}^\circ\text{C}$,

$$k = 0.622\ \text{W/m}\cdot\text{K} \qquad \mathbf{Pr} = 4.85 \qquad \nu = 0.723\times 10^{-6}\ \text{m}^2/\text{s}$$

and one can approximate $\beta_{36.2}$ by $\beta_{40} \approx 2.979\times 10^{-4}\ \text{K}^{-1}$, from Problem 7.3. Then

$$\mathbf{Gr}_L = \frac{g\beta(T_s - T_\infty)L^3}{\nu^2} = \frac{(9.80\ \text{m/s}^2)(2.979\times 10^{-4}\ \text{K}^{-1})[(52.2-20.2)\ \text{K}](0.10\ \text{m})^3}{(0.723\times 10^{-6}\ \text{m}^2/\text{s})^2} = 1.79\times 10^8$$

and $\mathbf{Gr}_L\,\mathbf{Pr} = (1.79\times 10^8)(4.85) = 8.68\times 10^8$ (laminar flow). Table I-3 and (*I-17*) give $(\Lambda = L)$:

$$\bar{h} = \frac{k}{L}(0.59)(\mathbf{Gr}_L\,\mathbf{Pr})^{1/4} = \frac{0.622\ \text{W/m}\cdot\text{K}}{0.10\ \text{m}}(0.59)(8.68\times 10^8)^{1/4} = 629\ \text{W/m}^2\cdot\text{K}$$

This value for the heat transfer is smaller than that obtained in Problem 7.3, i.e.,

$$q_{\text{emp}} = \tfrac{629}{646}(208) = 201.5\ \text{W}$$

It is worth noting here that this value is perhaps more reliable than that of Problem 7.3, since the similarity solution required an estimation of the slope in Fig. 7-3. But the two solutions compare well with each other.

7.6^{D} Repeat Problem 7.4 if the square heating element of Problem 7.2 is replaced by a 2-in. (vertical dimension) by 8-in. rectangle.

▮ It is easy to verify that (*I-17*), with the same C and the same a, is applicable for both heating elements, which, moreover, have the same area. Thus,

$$q_{\text{emp}} \propto \bar{h} \propto (\mathbf{Gr}_L^{1/4})\left(\frac{1}{L}\right) \propto L^{3/4}\left(\frac{1}{L}\right) \propto L^{-1/4}$$

so that, for the rectangle, $q_{\text{emp}} = (2^{1/4})(708) = 842\ \text{Btu/hr}$.

7.7^{D} Repeat Problem 7.5 if the square heating element of Problem 7.3 is replaced by a square of half the dimension.

▮ It is easy to see that (*I-17*), with the same C and the same a, is applicable for both heating elements. Thus,

$$q_{\text{emp}} \propto \bar{h}A \propto (\mathbf{Gr}_L^{1/4})\left(\frac{1}{L}\right)(L^2) \propto L^{3/4}\left(\frac{1}{L}\right)(L^2) \propto L^{7/4}$$

so that, for the smaller square, $q_{\text{emp}} = (\tfrac{1}{2})^{7/4}(201.5) = 59.9\ \text{W}$.

In Problems 7.8 through 7.18, all gas properties, including β, are evaluated at T_f in accordance with widespread practice in engineering texts. However, β should be evaluated at T_∞ for more accurate results. Also, all laminar equations resulting from boundary-layer analyses should be used with properties at T_{ref} from (*I-8*), with $C = 0.38$.

7.8^{D} Estimate the heat loss from a vertical wall exposed to gaseous nitrogen at 1 atm and 40 °F. The wall is 6 ft high and 8 ft wide, and is maintained at 120 °F.

▮ From Table B-4, at $T_f = 80\ {}^\circ\text{F}$,

$$\nu = 16.82\times 10^{-5}\ \text{ft}^2/\text{sec} \qquad k = 0.015\,14\ \text{Btu/hr}\cdot\text{ft}\cdot{}^\circ\text{F} \qquad \mathbf{Pr} = 0.713$$

For the gas, $\beta = 1/T_f = 1(460+80) = 1/(540\ {}^\circ\text{R})$, giving a Grashof number of

$$\mathbf{Gr}_L = \frac{g\beta(T_s - T_\infty)L^3}{\nu^2} = \frac{(32.2\ \text{ft/sec}^2)[(120-40)\ °\text{R}](6\ \text{ft})^3}{(540\ °\text{R})(16.82\times10^{-5}\ \text{ft}^2/\text{sec})^2} = 3.64\times10^{10}$$

and $\mathbf{Gr}_L\,\mathbf{Pr} = (3.64\times10^{10})(0.713) = 2.59\times10^{10}$. The flow is turbulent; (*I-17*) with the appropriate constants from Table I-3 gives

$$\bar{h} = \frac{k}{L}(0.13)(\mathbf{Gr}_L\,\mathbf{Pr})^{1/3} = \frac{0.015\,14\ \text{Btu/hr}\cdot\text{ft}\cdot°\text{F}}{6\ \text{ft}}(0.13)(2.59\times10^{10})^{1/3} = 0.9714\ \text{Btu/hr}\cdot\text{ft}^2\cdot°\text{F}$$

$$q = \bar{h}A(T_s - T_\infty)(0.9714\ \text{Btu/hr}\cdot\text{ft}^2\cdot°\text{F})(48\ \text{ft}^2)[(120-40)\ °\text{F}] = 3730\ \text{Btu/hr}$$

7.9^D Estimate the heat loss from a vertical wall exposed to gaseous nitrogen at 1 atm and 5 °C. The wall is 1.80 m high and 2.40 m wide, and is maintained at 49 °C.

▌ From Table B-4, at $T_f = 27$ °C,

$$\nu = 15.63\times10^{-6}\ \text{m}^2/\text{s} \qquad k = 0.026\,19\ \text{W/m}\cdot\text{K} \qquad \mathbf{Pr} = 0.713$$

For the gas, $\beta = 1/T_f = 1/(273+27) = 1/(300\ \text{K})$, giving a Grashof number of

$$\mathbf{Gr}_L = \frac{g\beta(T_s - T_\infty)L^3}{\nu^2} = \frac{(9.80\ \text{m/s}^2)[(49-5)\ \text{K}](1.8\ \text{m})^3}{(300\ \text{K})(15.63\times10^{-6}\ \text{m}^2/\text{s})^2} = 3.43\times10^{10}$$

and $\mathbf{Gr}_L\,\mathbf{Pr} = (3.43\times10^{10})(0.713) = 2.45\times10^{10}$. The flow is turbulent; (*I-17*) with the appropriate constants from Table I-3 gives

$$\bar{h} = \frac{k}{L}(0.13)(\mathbf{Gr}_L\,\mathbf{Pr})^{1/3} = \frac{0.026\,19\ \text{W/m}\cdot\text{K}}{1.80\ \text{m}}(0.13)(2.45\times10^{10})^{1/3} = 5.49\ \text{W/m}^2\cdot\text{K}$$

$$q = \bar{h}A(T_s - T_\infty)(5.49\ \text{W/m}^2\cdot\text{K})[(1.80\ \text{m})(2.40\ \text{m})][(49-5)\ \text{K}] = 1044\ \text{W}$$

7.10^D Find the maximum vertical velocity in the boundary-layer flow of Problem 7.8 at $x = 1.2$ ft (from the bottom of the wall). At what distance y from the wall does this occur?

▌ From Problem 7.8, $\mathbf{Gr}_6 = 3.64\times10^{10}$; hence,

$$\mathbf{Gr}_{1.2} = (3.64\times10^{10})\left(\frac{1.2}{6}\right)^3 = 2.91\times10^8$$

making $\mathbf{Gr}_{1.2}\,\mathbf{Pr} < 10^9$ (laminar flow). The Ostrach results of Fig. 7-2 apply. For $\mathbf{Pr} = 0.713 \approx 0.72$ and $x = 1.2$ ft, we read

$$(f')_{\max} = \frac{u_{\max}(1.2\ \text{ft})}{2(16.82\times10^{-5}\ \text{ft}^2/\text{sec})}(2.91\times10^8)^{-1/2} \approx 0.275 \qquad \text{whence} \qquad u_{\max} \approx 1.31\ \text{ft/sec}$$

the maximum being at

$$\eta = \frac{y}{1.2\ \text{ft}}\left(\frac{2.91\times10^8}{4}\right)^{1/4} \approx 1.0 \qquad \text{whence} \qquad y \approx 0.013\ \text{ft}$$

This is well within the boundary layer; see Problem 7.12.

7.11^D Find the maximum vertical velocity in the boundary-layer flow of Problem 7.9 at $x = 0.366$ m (from the bottom of the wall). At what distance y from the wall does this occur?

▌ From Problem 7.9, $\mathbf{Gr}_{1.80} = 3.43\times10^{10}$; hence,

$$\mathbf{Gr}_{0.366} = (3.43\times10^{10})\left(\frac{0.366}{1.80}\right)^3 = 2.88\times10^8$$

and $\mathbf{Gr}_{0.366}\,\mathbf{Pr} < 10^9$ (laminar flow). The Ostrach results of Fig. 7-2 apply. For $\mathbf{Pr} = 0.713 \approx 0.72$ and $x = 0.366$ m, we read

$$(f')_{\max} = \frac{u_{\max}(0.366\ \text{m})}{2(15.63\times10^{-6}\ \text{m}^2/\text{s})}(2.88\times10^8)^{-1/2} \approx 0.275 \qquad \text{whence} \qquad u_{\max} \approx 0.398\ \text{m/s}$$

the maximum being at

$$\eta = \frac{y}{366\ \text{mm}}\left(\frac{2.88\times10^8}{4}\right)^{1/4} \approx 1.0 \qquad \text{whence} \qquad y \approx 3.97\ \text{mm}$$

This is well within the boundary layer; see Problem 7.13.

7.12^D For the vertical flow of Problems 7.8 and 7.10 (and in all similar situations) the *kinematic-boundary-layer thickness* δ at a given height x is conventionally taken as the (larger) value of y for which $u(x, y) = \frac{1}{100} u_{max}(x)$. Estimate δ at $x = 1.2$ ft.

▌ Refer to Fig. 7-2. At fixed x, f' and u are proportional; therefore, find where the curve $\mathbf{Pr} = 0.72$ reaches 1 percent of its peak value. By inspection, the (larger) abscissa corresponding to the ordinate $0.275/100 \approx 0.003$ is $\eta \approx 5.5$. Thus, since η and y also are proportional,

$$\delta \approx \frac{5.5}{1.0}(0.013 \text{ ft}) = 0.071 \text{ ft}$$

7.13^D For the vertical flow of Problems 7.9 and 7.11, estimate the kinematic-boundary-layer thickness at $x = 0.366$ m.

▌ By the method of Problem 7.12,

$$\delta \approx \frac{5.5}{1.0}(3.97 \text{ mm}) = 21.8 \text{ mm}$$

7.14 Refer to Problem 7.12. Estimate the *thermal-boundary-layer thickness* δ_t at $x = 1.2$ ft.

▌ For each x, one defines δ_t as the value of y for which $\theta \approx 0$. From Fig. 7-3, at $\mathbf{Pr} = 0.72$, $\theta \approx 0$ for $\eta \approx 5$; hence

$$5 \approx \frac{\delta_t}{1.2}\left(\frac{2.91 \times 10^8}{4}\right)^{1/4} \qquad \text{or} \qquad \delta_t \approx 0.065 \text{ ft}$$

By comparison with Problem 7.12, δ_t and δ have approximately the same value. This is true for **Pr** in the vicinity of 1.0; for large **Pr**, δ_t becomes much smaller than δ.

7.15^D For a plane vertical wall 1 ft high held at 140 °F and exposed to air at 60 °F, determine whether the problem is one of free, forced, or mixed convection if the forced-approach or free-stream velocity in the vertical direction is (*a*) 10 ft/sec, (*b*) 2 ft/sec, (*c*) 100 ft/sec, (*d*) 0.1 ft/sec, (*e*) 0.01 ft/sec, (*f*) 1 ft/sec.

▌ With

$$\mathbf{Gr}_L = \frac{g\beta(T_s - T_\infty)L^3}{\nu^2} \qquad \text{and} \qquad \mathbf{Re}_L = \frac{V_\infty L}{\nu}$$

the rule of thumb for determining the main mechanism of heat transfer is

1. $\mathbf{Gr}_L \gg \mathbf{Re}_L^2$—Free convection dominates
2. $\mathbf{Gr}_L \ll \mathbf{Re}_L^2$—Forced convection dominates
3. $\mathbf{Gr}_L \approx \mathbf{Re}_L^2$—Mixed forced and free convection

Using properties from Table B-4, at $T_f = 100$ °F, along with $\beta = 1/(560\ °\text{R})$, we have

$$\mathbf{Gr}_L = 1.408 \times 10^8 \qquad \text{and} \qquad \mathbf{Re}_L^2 = (3.09 \times 10^7 \text{ sec}^2/\text{ft}^2)V_\infty^2$$

Hence: (*a*) Forced; (*b*) Mixed; (*c*) Forced; (*d*) Free; (*e*) Free; (*f*) Borderline free.

7.16^D For a plane vertical wall 0.3 m high held at 60 °C and exposed to air at 16 °C, determine whether the problem is one of free, forced, or mixed convection if the forced-approach or free-stream velocity in the vertical direction is (*a*) 3 m/s, (*b*) 0.6 m/s, (*c*) 30 m/s, (*d*) 0.03 m/s, (*e*) 0.003 m/s, (*f*) 0.30 m/s.

▌ With properties from Table B-4, at $T_f = 38$ °C, along with $\beta = 1/(311 \text{ K})$, we have

$$\mathbf{Gr}_L = \frac{g\beta(T_s - T_\infty)L^3}{\nu^2} = 1.33 \times 10^8 \qquad \text{and} \qquad \mathbf{Re}_L^2 = \left(\frac{L}{\nu}\right)^2 V_\infty^2 = (3.19 \times 10^8 \text{ s}^2/\text{m}^2)V_\infty^2$$

Then the rule of Problem 7.15 yields: (*a*) Forced; (*b*) Mixed; (*c*) Forced; (*d*) Free; (*e*) Free; (*f*) Borderline free.

7.17^D For a vertical wall at 140 °F exposed to still air at 60 °F and 1 atm pressure, determine the maximum height for laminar free convective flow.

▮ Transition from laminar to turbulent flow occurs at

$$\mathbf{Gr}_x\,\mathbf{Pr} \equiv \frac{g\beta(T_s - T_\infty)}{\nu^2}\,x^3\,\mathbf{Pr} \approx 10^9$$

Substituting data from Table B-4, at $T_f = 100\ °\text{F} = 560\ °\text{R}\ (=1/\beta)$, and solving for x, one obtains $x \approx 2.15$ ft.

7.18^D For a vertical wall at 60 °C exposed to still air at 1 atm pressure and 16 °C, determine the maximum height for laminar free convective flow.

▮ Transition from laminar to turbulent flow occurs at

$$\mathbf{Gr}_x\,\mathbf{Pr} \equiv \frac{g\beta(T_s - T_\infty)}{\nu^2}\,x^3\,\mathbf{Pr} \approx 10^9$$

Substituting data from Table B-4, at $T_f = 38\ °\text{C} = 311\ \text{K}\ (=1/\beta)$, and solving for x, one obtains $x \approx 0.66$ m.

7.19 Derive (*I-3*) in the coordinates of Fig. 7-1.

▮ In (*2*) of Problem 4.1, $g_x = -g$ and

$$\left|\frac{\partial^2 u}{\partial y^2}\right| \gg \left|\frac{\partial^2 u}{\partial x^2}\right|$$

Hence, dividing through by ρ,

$$u\,\frac{\partial u}{\partial x} + v\,\frac{\partial u}{\partial y} = -g - \frac{1}{\rho}\,\frac{\partial p}{\partial x} + \nu\,\frac{\partial^2 u}{\partial y^2} \tag{1}$$

Now, in the still fluid outside the boundary layer the pressure follows the familiar linear law

$$p = p_0 - \rho_\infty g x \qquad \text{or} \qquad \frac{\partial p}{\partial x} = -\rho_\infty g \tag{2}$$

and, by the argument of Problem 4.2, (*2*) remains valid within the boundary layer. Thus, (*1*) becomes

$$u\,\frac{\partial u}{\partial x} + v\,\frac{\partial u}{\partial y} = -\frac{\rho - \rho_\infty}{\rho}\,g + \nu\,\frac{\partial^2 u}{\partial y^2} \tag{3}$$

But, $|T - T_\infty|$ being small (except at the wall), a Taylor expansion at constant p, together with definition (*I-2*), yields

$$\rho \approx \rho_\infty + \left(\frac{\partial \rho}{\partial T}\right)_p (T - T_\infty) \qquad \text{or} \qquad -\frac{\rho - \rho_\infty}{\rho} \approx \beta(T - T_\infty) \tag{4}$$

and (*4*) and (*3*) imply (*I-3*).

7.20^D For the case of a 2-ft-high vertical wall with $T_s = 155\ °\text{F}$ exposed to atmospheric air at 77 °F, determine at $x = 1$ ft (*a*) the distance y from the wall to the point of maximum vertical velocity (u_{max}); (*b*) the value of u_{max}.

▮ At $T_{ref} = T_s + (0.38)(T_\infty - T_s) = 155 + (0.38)(77 - 155) \approx 125\ °\text{F}$,

$$\nu_{ref} = 19.63 \times 10^{-5}\ \text{ft}^2/\text{s} \qquad \mathbf{Pr}_{ref} = 0.703$$

Also, $\beta = 1/T_\infty = 1/(537\ °\text{R})$. Thus,

$$\mathbf{Gr}_x = \frac{g\beta(T_s - T_\infty)x^3}{\nu^2} = \frac{(32.2\ \text{ft/s}^2)[(155 - 77)\ °\text{R}](1\ \text{ft})^3}{(537\ °\text{R})(19.63 \times 10^{-5}\ \text{ft}^2/\text{s})^2} = 1.214 \times 10^8$$

Check: $\mathbf{Gr}_x\,\mathbf{Pr} = (1.214 \times 10^8)(0.703) = 8.532 \times 10^8 < 10^9$, so the problem involves laminar flow and the similarity solutions apply.

(*a*) From Fig. 7-2 at $\mathbf{Pr} \approx 0.7$, $f'_{max} \approx 0.275$ at $\eta \approx 1.0$. Thus,

$$1.0 \approx \frac{y}{1\ \text{ft}}\left(\frac{1.214 \times 10^8}{4}\right)^{1/4} \qquad \text{or} \qquad y \approx 0.0135\ \text{ft}$$

(*b*)

$$0.275 \approx \frac{u_{max}(1\ \text{ft})}{2(19.63 \times 10^{-5}\ \text{ft}^2/\text{s})}\,(1.214 \times 10^8)^{-1/2} \qquad \text{or} \qquad u_{max} \approx 1.19\ \text{ft/s}$$

7.21[D] For the case of a 0.6-m-high vertical wall with $T_s = 69\ °C$ exposed to atmospheric air at 25 °C, determine at $x = 0.3$ m (*a*) the distance y from the wall to the point of maximum vertical velocity; (*b*) the value of u_{max}.

$T_{ref} = T_s + (0.38)(T_\infty - T_s) = 69 + (0.38)(25 - 69) = 52\ °C,$

$$\nu_{ref} = 18.24 \times 10^{-6}\ m^2/s \qquad \mathbf{Pr}_{ref} = 0.703$$

Also, $\beta = 1/T_\infty = 1/(298\ K)$. Thus,

$$\mathbf{Gr}_x = \frac{g\beta(T_s - T_\infty)x^3}{\nu^2} = \frac{(9.8\ m/s^2)[(69 - 25)\ K](0.3\ m)^3}{(298\ K)(18.24 \times 10^{-6}\ m^2/s)^2} = 1.176 \times 10^8$$

Check: $\mathbf{Gr}_x\,\mathbf{Pr} = (1.176 \times 10^8)(0.703) = 8.265 \times 10^8 < 10^9$, so the problem involves laminar flow and the similarity solutions apply.

(*a*) From Fig. 7-2 at $\mathbf{Pr} \approx 0.7$, $f'_{max} \approx 0.275$ at $\eta \approx 1.0$. Thus,

$$1.0 \approx \frac{y}{0.3\ m}\left(\frac{1.176 \times 10^8}{4}\right)^{1/4} \qquad \text{or} \qquad y \approx 4.07\ mm$$

(*b*)
$$0.275 \approx \frac{u_{max}(0.3\ m)}{2(18.24 \times 10^{-6}\ m^2/s)}(1.176 \times 10^8)^{-1/2} \qquad \text{or} \qquad u_{max} \approx 0.363\ m/s$$

7.22[D] For the situation of Problem 7.17, determine the heat transfer per foot of wall width over the laminar-flow height of 2.15 ft by (*a*) (*I-10*); (*b*) (*I-14*).

(*a*) From Table B-4, for air at $T_{ref} = 140 + (0.38)(60 - 140) = 110\ °F$,

$$\nu = 18.7 \times 10^{-5}\ ft^2/sec \qquad \mathbf{Pr} = 0.704 \qquad k = 0.0159\ Btu/hr \cdot ft \cdot °F$$

Also, $\beta = 1/T_\infty = 1/(520\ °R)$; thus

$$\mathbf{Gr}_L = \frac{g\beta(T_s - T_\infty)L^3}{\nu^2} = \frac{(32.2\ ft/sec^2)(80\ °R)(2.15\ ft)^3}{(520\ °R)(0.187 \times 10^{-3}\ ft^2/sec)^2} = 1.41 \times 10^9$$

and $\mathbf{Gr}_L\,\mathbf{Pr} = (1.41 \times 10^9)(0.704) = 1.0 \times 10^9$ (laminar, but bordering on turbulent). From (*I-11*), $F_1(0.704) \approx 0.500$; (*I-10*) gives

$$\bar{h} = \frac{4}{3}\left(\frac{k}{L}\right)[F_1(\mathbf{Pr})]\left(\frac{\mathbf{Gr}_L}{4}\right)^{1/4} = \frac{4}{3}\left(\frac{0.0159}{2.15}\right)(0.500)\left(\frac{1.41 \times 10^9}{4}\right)^{1/4} = 0.676\ Btu/hr \cdot ft^2 \cdot °F$$

$$\frac{q}{w} = \bar{h}L(T_s - T_\infty) = (0.676)(2.15)(80) = 116\ Btu/hr \cdot ft$$

(*b*)
$$\bar{h} = \frac{4}{3}\left(\frac{k}{L}\right)(0.508)\left(\frac{\mathbf{Gr}_L\,\mathbf{Pr}^2}{0.952 + \mathbf{Pr}}\right)^{1/4}$$

$$= \frac{4}{3}\left(\frac{0.0159}{2.15}\right)(0.508)\left[\frac{(1.41 \times 10^9)(0.704)^2}{0.952 + 0.704}\right]^{1/4} = 0.718\ Btu/hr \cdot ft^2 \cdot °F$$

$$\frac{q}{w} = \bar{h}L(T_s - T_\infty) = (0.718)(2.15)(80) = 123.5\ Btu/hr \cdot ft$$

7.23[D] For the situation of Problem 7.18, determine the heat transfer per meter of wall width over the laminar-flow height of 0.66 m by (*a*) (*I-10*); (*b*) (*I-14*).

(*a*) From Table B-4, for air at $T_{ref} = 60 + (0.38)(16 - 60) = 43\ °C$,

$$\nu = 0.171 \times 10^{-4}\ m^2/s \qquad \mathbf{Pr} = 0.704 \qquad k = 0.0274\ W/m \cdot K$$

Also, $\beta = 1/T_\infty = 1/(289\ K)$; thus

$$\mathbf{Gr}_L = \frac{g\beta(T_s - T_\infty)L^3}{\nu^2} = \frac{(9.8\ m^2/s)(44\ K)(0.66\ m)^3}{(289\ K)(0.171 \times 10^{-4}\ m^2/s)^2} = 1.46 \times 10^9$$

whence $\mathbf{Gr}_L\,\mathbf{Pr} = (1.46 \times 10^9)(0.704) = 1.0 \times 10^9$ (laminar, but bordering on turbulent). From (*I-11*), $F_1(0.704) \approx 0.500$; (*I-10*) gives

$$\bar{h} = \frac{4}{3}\left(\frac{k}{L}\right)[F_1(\mathbf{Pr})]\left(\frac{\mathbf{Gr}_L}{4}\right)^{1/4} = \frac{4}{3}\left(\frac{0.0274}{0.66}\right)(0.500)\left(\frac{1.46 \times 10^9}{4}\right)^{1/4} = 3.83\ W/m^2 \cdot K$$

$$\frac{q}{w} = \bar{h}L(T_s - T_\infty) = (3.83)(0.66)(44) = 112.2\ W/m$$

(*b*)
$$\bar{h}L = \frac{4}{3}k(0.508)\left(\frac{\mathbf{Gr}_L\,\mathbf{Pr}^2}{0.952+\mathbf{Pr}}\right)^{1/4}$$
$$= \frac{4}{3}(0.0274)(0.508)\left[\frac{(1.46\times10^9)(0.704)^2}{0.952+0.704}\right]^{1/4} = 2.683\ \text{W/m}\cdot\text{K}$$
$$\frac{q}{w} = (\bar{h}L)(T_s - T_\infty) = (2.683)(44) = 118.1\ \text{W/m}$$

7.24[D] A vertical square plate 1 ft on a side is maintained at a temperature of 275 °F and is exposed to still air at 75 °F and 1 atm pressure. Using Ostrach's results, determine the heat loss from the plate.

▌ From Table B-4, at $T_{\text{ref}} = 275 + (0.38)(75 - 275) \approx 200$ °F,

$$k = 0.018\,05\ \text{Btu/hr}\cdot\text{ft}\cdot\text{°F} \qquad \nu = 2.42\times10^{-4}\ \text{ft}^2/\text{sec} \qquad \mathbf{Pr} = 0.694$$

Also, $\beta = 1/T_\infty = 1/(535\ \text{°R})$. Therefore,

$$\mathbf{Gr}_L = \frac{g\beta(T_s - T_\infty)L^3}{\nu^2} = \frac{(32.2\ \text{ft/sec}^2)(200\ \text{°R})(1\ \text{ft})^3}{(535\ \text{°R})(2.42\times10^{-4}\ \text{ft}^2/\text{sec})^2} = 2.06\times10^8$$

and so $\mathbf{Gr}_L\,\mathbf{Pr} = (2.06\times10^8)(0.694) = 1.43\times10^8$ (*laminar*). From Table I-1, $F_1(\mathbf{Pr}) \approx 0.5046$; (*I-10*) yields

$$\bar{h} = \frac{4}{3}\left(\frac{k}{L}\right)[F_1(\mathbf{Pr})]\left(\frac{\mathbf{Gr}_L}{4}\right)^{1/4} = \frac{4}{3}\left(\frac{0.018\,05}{1}\right)(0.5046)\left(\frac{2.06\times10^8}{4}\right)^{1/4} = 1.029\ \text{Btu/hr}\cdot\text{ft}^2\cdot\text{°F}$$
$$q = \bar{h}A(T_s - T_\infty) = (1.029)(1)^2(200) = 205.8\ \text{Btu/hr}$$

7.25[D] A vertical square plate 0.3 m on a side is maintained at a temperature of 135 °C and is exposed to still air at 1 atm pressure and 24 °C. Using Ostrach's results, determine the heat loss from the plate.

▌ From Table B-4, at $T_{\text{ref}} = 135 + (0.38)(24 - 135) \approx 93$ °C,

$$k = 0.0312\ \text{W/m}\cdot\text{K} \qquad \nu = 0.224\times10^{-4}\ \text{m}^2/\text{s} \qquad \mathbf{Pr} = 0.694$$

Also, $\beta = 1/T_\infty = 1/(297\ \text{K})$. Therefore,

$$\mathbf{Gr}_L = \frac{g\beta(T_s - T_\infty)L^3}{\nu^2} = \frac{(9.80\ \text{m/s}^2)[(135-24)\ \text{K}](0.3\ \text{m})^3}{(297\ \text{K})(0.224\times10^{-4}\ \text{m}^2/\text{s})^2} = 1.97\times10^8$$

and so $\mathbf{Gr}_L\,\mathbf{Pr} = (1.97\times10^8)(0.694) = 1.37\times10^8$ (*laminar*). From Table I-1, $F_1(\mathbf{Pr}) \approx 0.50$; (*I-10*) yields

$$\bar{h} = \frac{4}{3}\left(\frac{k}{L}\right)[F_1(\mathbf{Pr})]\left(\frac{\mathbf{Gr}_L}{4}\right)^{1/4} = \frac{4}{3}\left(\frac{0.0312}{0.3}\right)(0.50)\left(\frac{1.97\times10^8}{4}\right)^{1/4} = 5.82\ \text{W/m}^2\cdot\text{K}$$
$$q = \bar{h}A(T_s - T_\infty) = (5.82)(0.3)^2(135 - 24) = 58.1\ \text{W}$$

7.26 A vertical cylinder which is 0.3 m in diameter and 0.60 m high is maintained at a temperature of 77 °C and is exposed to air at 24 °C. Determine the heat transfer from the cylinder, using flat-plate theory.

▌ From Table B-4, at $T_{\text{ref}} = 77 + (0.38)(24 - 77) = 57$ °C,

$$\nu = 1.87\times10^{-5}\ \text{m}^2/\text{s} \qquad k = 0.028\,49\ \text{W/m}\cdot\text{K} \qquad \mathbf{Pr} = 0.701$$

and $\beta = 1/T_\infty = 1/(297\ \text{K})$. Hence

$$\mathbf{Gr}_L = \frac{g\beta(T_s - T_\infty)L^3}{\nu^2} = \frac{(9.80\ \text{m/s}^2)(53\ \text{K})(0.60\ \text{m})^3}{(297\ \text{K})(1.87\times10^{-5}\ \text{m}^2/\text{s})^2} = 1.08\times10^9$$

and so $\mathbf{Gr}_L\,\mathbf{Pr} = (1.08\times10^9)(0.701) = 7.58\times10^8$ (*laminar*). From Table I-1, $F_1(\mathbf{Pr}) \approx 0.50$; (*I-10*) now gives

$$\bar{h}L = \frac{4}{3}kF_1(\mathbf{Pr})\left(\frac{\mathbf{Gr}_L}{4}\right)^{1/4} = \frac{4}{3}(0.028\,49)(0.50)\left(\frac{1.08\times10^9}{4}\right)^{1/4} = 2.43\ \text{W/m}\cdot\text{K}$$
$$q = (\bar{h}L)(\pi D)(T_s - T_\infty) = (2.43)(\pi)(0.3)(53) = 121\ \text{W}$$

7.27 Justify the application of flat-plate theory in Problem 7.26 by showing that (*a*) the boundary-layer thickness is everywhere small compared to the radius of the cylinder; (*b*) condition (*I-18*) is satisfied.

▌ (*a*) By Problem 7.12, for $\mathbf{Pr} \approx 0.7$,

$$5.5 = \frac{\delta}{x}\left(\frac{\mathbf{Gr}_x}{4}\right)^{1/4} \qquad \text{or} \qquad \delta \propto \frac{x}{\mathbf{Gr}_x^{1/4}} \propto x^{1/4}$$

Thus δ is a maximum for $x = L$:

$$\delta_{max} = (5.5)L\left(\frac{4}{\mathbf{Gr}_L}\right)^{1/4} = (5.5)(0.60\ \text{m})\left(\frac{4}{1.08 \times 10^9}\right)^{1/4} = 0.026\ \text{m}$$

and so $\delta / r < \delta_{max} / r = 0.026/0.15 = 0.17$.

(*b*)
$$\frac{35}{\mathbf{Gr}_L^{1/4}} = \frac{35}{(1.08 \times 10^9)^{1/4}} = 0.193 < \frac{1}{2} = \frac{D}{L}$$

7.28[D] A vertical cylinder 1 ft high and 1.25 ft in diameter is heated electrically to provide a uniform q/A. It has an average surface temperature of 118 °F and is exposed to air at 70 °F and atmospheric pressure. Using the results of Sparrow and Gregg, determine the rate of heat transfer from the cylinder.

▌ Assume that flat-plate theory is applicable. From Table B-4, at $T_{ref} = 118 + (0.38)(70 - 118) = 100$ °F,

$$\nu = 1.81 \times 10^{-4}\ \text{ft}^2/\text{sec} \qquad k = 0.0156\ \text{Btu/hr} \cdot \text{ft} \cdot {}^\circ\text{F} \qquad \mathbf{Pr} = 0.705$$

Also, $\beta = 1/T_\infty = 1/(530\ {}^\circ\text{R})$. Then

$$\mathbf{Gr}_L = \frac{g\beta(T_s - T_\infty)L^3}{\nu^2} = \frac{(32.2\ \text{ft/sec}^2)[(118 - 70)\ {}^\circ\text{R}](1\ \text{ft})^3}{(530\ {}^\circ\text{R})(1.81 \times 10^{-4}\ \text{ft}^2/\text{sec})^2} = 8.92 \times 10^7$$

and $\mathbf{Gr}_L\,\mathbf{Pr} = (8.92 \times 10^7)(0.705) = 6.29 \times 10^7$ (*laminar*). Interpolation in Table I-2 yields

$$F_2(\mathbf{Pr}) = 0.335 + \frac{0.605}{0.9}(0.811 - 0.335) = 0.655$$

By equation (*I-12*),

$$\bar{h}L = kF_2(\mathbf{Pr})\left(\frac{\mathbf{Gr}_L}{4}\right)^{1/4} = (0.0156)(0.655)\left(\frac{8.92 \times 10^7}{4}\right)^{1/4} = 0.702\ \text{Btu/hr} \cdot \text{ft} \cdot {}^\circ\text{F}$$

$$q = (\bar{h}L)\pi D(T_s - T_\infty) = (0.702)\pi(1.25)(118 - 70) = 132.3\ \text{Btu/hr}$$

Checking the flat-plate assumption by (*I-18*):

$$\frac{35}{\mathbf{Gr}_L^{1/4}} = \frac{35}{(8.92 \times 10^7)^{1/4}} = 0.36 < 1.25 = \frac{D}{L} \quad (OK)$$

7.29[D] A vertical cylinder 0.3 m high and 0.38 m in diameter is heated electrically to provide uniform heat flux. It has an average surface temperature of 48 °C and is exposed to air at 21 °C and atmospheric pressure. Using the results of Sparrow and Gregg, determine the rate of heat transfer from the cylinder.

▌ Assume that flat-plate theory is applicable. From Table B-4 at $T_{ref} = 48 + (0.38)(21 - 48) = 38$ °C,

$$\nu = 1.68 \times 10^{-5}\ \text{m}^2/\text{s} \qquad k = 0.0270\ \text{W/m} \cdot \text{K} \qquad \mathbf{Pr} = 0.705$$

Also, $\beta = 1/T_\infty = 1/(294\ \text{K})$. Then

$$\mathbf{Gr}_L = \frac{g\beta(T_s - T_\infty)L^3}{\nu^2} = \frac{(9.8\ \text{m/s}^2)[(48 - 21)\ \text{K}](0.3\ \text{m})^3}{(294\ \text{K})(1.68 \times 10^{-5}\ \text{m}^2/\text{s})^2} = 8.61 \times 10^7$$

and $\mathbf{Gr}_L\,\mathbf{Pr} = (8.61 \times 10^7)(0.705) = 6.07 \times 10^7$ (*laminar*). Interpolation in Table I-2 yields

$$F_2(\mathbf{Pr}) = 0.335 + \frac{0.605}{0.9}(0.811 - 0.335) = 0.655$$

Then (*I-12*) gives

$$\bar{h}L = kF_2(\mathbf{Pr})\left(\frac{\mathbf{Gr}_L}{4}\right)^{1/4} = (0.0270)(0.655)\left(\frac{8.61 \times 10^7}{4}\right)^{1/4} = 1.20\ \text{W/m} \cdot \text{K}$$

$$q = (\bar{h}L)\pi D(T_s - T_\infty) = (1.20)\pi(0.38)(27) = 39\ \text{W}$$

Checking the flat-plate assumption by (*I-18*):

$$\frac{35}{\mathbf{Gr}_L^{1/4}} = \frac{35}{(8.61 \times 10^7)^{1/4}} = 0.36 < \frac{38}{30} = \frac{D}{L} \quad (OK)$$

7.30[D] A vertical flat rectangular plate which is 5 ft wide and 2 ft high is heated with electric ribbon heaters to provide a uniform q/A. The average surface temperature is 193 °F, and the plate is exposed to nitrogen gas at 80 °F and atmospheric pressure. Determine the heat loss from the plate, using (*I-12*).

▮ From Table B-4, at $T_{\text{ref}} = 193 + (0.38)(80 - 193) = 150\ °\text{F}$,

$$\mathbf{Pr} = 0.704 \qquad k = 0.0167\ \text{Btu/hr}\cdot\text{ft}\cdot°\text{F} \qquad \nu = 21.06 \times 10^{-5}\ \text{ft}^2/\text{sec}$$

Also, $\beta = 1/T_\infty = 1/(540\ °\text{R})$. Thus

$$\mathbf{Gr}_L = \frac{g\beta(T_s - T_\infty)L^3}{\nu^2} = \frac{32.2\ \text{ft/sec}^2}{(540\ °\text{R})(21.06 \times 10^{-5}\ \text{ft}^2/\text{sec})^2}[(193 - 80)\ °\text{R}](2\ \text{ft})^3 = 1.21 \times 10^9$$

and $\mathbf{Gr}_L\,\mathbf{Pr} = (1.21 \times 10^9)(0.704) = 8.5 \times 10^8$ (*laminar*). Interpolation in Table I-2 yields

$$F_2(\mathbf{Pr}) = 0.335 + \frac{0.604}{0.9}(0.811 - 0.335) = 0.654$$

and (*I-12*) yields

$$\bar{h}L = kF_2(\mathbf{Pr})\left(\frac{\mathbf{Gr}_L}{4}\right)^{1/4} = (0.0167)(0.654)\left(\frac{12.1 \times 10^8}{4}\right)^{1/4} = 1.44\ \text{Btu/hr}\cdot\text{ft}\cdot°\text{F}$$

$$q = (\bar{h}L)w(T_s - T_\infty) = (1.44)(5)(193 - 80) = 813\ \text{Btu/hr}$$

7.31[D] A vertical flat rectangular plate which is 1.5 m wide and 0.61 m high is heated with electric ribbon heaters to provide a uniform q/A. The average surface temperature is 90 °C and the plate is exposed to nitrogen gas at 27 °C and atmospheric pressure. Determine the heat loss from the plate, using (*I-12*).

▮ From Table B-4, at $T_{\text{ref}} = 90 + (0.38)(27 - 90) = 66\ °\text{C}$,

$$\mathbf{Pr} = 0.704 \qquad k = 0.0289\ \text{W/m}\cdot\text{K} \qquad \nu = 1.96 \times 10^{-5}\ \text{m}^2/\text{s}$$

Also, $\beta = 1/T_\infty = 1/(300\ \text{K})$. Therefore

$$\mathbf{Gr}_L = \frac{g\beta(T_s - T_\infty)L^3}{\nu^2} = \frac{9.8\ \text{m/s}^2}{(300\ \text{K})(1.96 \times 10^{-5}\ \text{m}^2/\text{s})^2}[(90 - 27)\ \text{K}](0.61\ \text{m})^3 = 1.21 \times 10^9$$

and $\mathbf{Gr}_L\,\mathbf{Pr} = (1.21 \times 10^9)(0.704) = 8.5 \times 10^8$ (*laminar*). Interpolation in Table I-2 yields

$$F_2(\mathbf{Pr}) = 0.335 + \frac{0.604}{0.9}(0.811 - 0.335) = 0.654$$

so that (*I-12*) gives

$$\bar{h}L = kF_2(\mathbf{Pr})\left(\frac{\mathbf{Gr}_L}{4}\right)^{1/4} = (0.0289)(0.654)\left(\frac{12.1 \times 10^8}{4}\right)^{1/4} = 2.49\ \text{W/m}\cdot\text{K}$$

$$q = (\bar{h}L)w(T_s - T_\infty) = (2.49)(1.5)(63) = 235\ \text{W}$$

7.32[D] Rework Problem 7.30 if now the wall temperature is fixed—at 193 °F.

▮ With property values from Problem 7.30, we apply (*I-10*), evaluating $F_1(\mathbf{Pr}) \approx 0.50$ from Table I-1:

$$\bar{h} = \frac{4}{3}\left(\frac{k}{L}\right)[F_1(\mathbf{Pr})]\left(\frac{\mathbf{Gr}_L}{4}\right)^{1/4} = \frac{4}{3}\left(\frac{0.0167}{2}\right)(0.50)\left(\frac{1.21 \times 10^9}{4}\right)^{1/4} = 0.734\ \text{Btu/hr}\cdot\text{ft}^2\cdot°\text{F}$$

$$q = \bar{h}A(T_s - T_\infty) = (0.734)(2 \times 5)(193 - 80) = 829\ \text{Btu/hr}$$

7.33[D] Rework Problem 7.31 if now the wall temperature is fixed—at 90 °C.

▮ With property values from Problem 7.31, we apply (*I-10*), evaluating $F_1(\mathbf{Pr}) \approx 0.50$ from Table I-1:

$$\bar{h} = \frac{4}{3}\left(\frac{k}{L}\right)[F_1(\mathbf{Pr})]\left(\frac{\mathbf{Gr}_L}{4}\right)^{1/4} = \frac{4}{3}\left(\frac{0.0289}{0.61}\right)(0.50)\left(\frac{1.21 \times 10^9}{4}\right)^{1/4} = 4.16\ \text{W/m}^2\cdot\text{K}$$

$$q = \bar{h}A(T_s - T_\infty) = (4.16)(0.61 \times 1.5)(90 - 27) = 240\ \text{W}$$

7.34[D] For the vertical wall of Problem 7.17, determine the heat transfer per foot of wall width by the simplest empirical correlation equation for (*a*) laminar flow and (*b*) turbulent flow.

▮ The simplest empirical correlation equation in either case is (*I-17*). From Problem 7.17, $\mathbf{Gr}_L\,\mathbf{Pr} = 10^9$ and, from Table B-4, at $T_f = 100\ °\text{F}$, $k = 0.0156\ \text{Btu/hr}\cdot\text{ft}\cdot°\text{F}$.

(*a*) From Table I-3, $\Lambda = L_1$, $C = 0.59$, and $a = \frac{1}{4}$; therefore,

$$\bar{h}L = Ck(\mathbf{Gr}_L\,\mathbf{Pr})^a = (0.59)(0.0156)(10^{9/4}) = 1.638\ \text{Btu/hr}\cdot\text{ft}\cdot°\text{F}$$

$$\frac{q}{w} = (\bar{h}L)(T_s - T_\infty) = (1.638)(80) = 131\ \text{Btu/hr}\cdot\text{ft}$$

(*b*) With $C = 0.13$ and $a = \frac{1}{3}$,

$$\bar{h}L = Ck(\mathbf{Gr}_L\,\mathbf{Pr})^a = (0.13)(0.0156)(10^3) = 2.03\ \text{Btu/hr}\cdot\text{ft}\cdot°\text{F}$$

$$\frac{q}{w} = (\bar{h}L)(T_s - T_\infty) = (2.03)(80) = 162\ \text{Btu/hr}\cdot\text{ft}$$

7.35[D] For the vertical wall of Problem 7.18, determine the heat transfer per meter of wall width by the simplest empirical correlation equation for (*a*) laminar flow and (*b*) turbulent flow.

▮ The simplest empirical correlation equation in either case is (*I-17*). From Problem 7.18, $\mathbf{Gr}_L\,\mathbf{Pr} = 10^9$ and, from Table B-4, at $T_f = 38\ °\text{C}$, $k = 0.0270\ \text{W/m}\cdot\text{K}$.

(*a*) From Table I-3, $\Lambda = L$, $C = 0.59$, and $a = \frac{1}{4}$; therefore,

$$\bar{h}L = Ck(\mathbf{Gr}_L\,\mathbf{Pr})^a = (0.59)(0.0270)(10^{9/4}) = 2.835\ \text{W/m}\cdot\text{K}$$

$$\frac{q}{w} = (\bar{h}L)(T_s - T_\infty) = (2.835)(60 - 16) = 125\ \text{W/m}$$

(*b*) With $C = 0.13$ and $a = \frac{1}{3}$,

$$\bar{h}L = Ck(\mathbf{Gr}_L\,\mathbf{Pr})^a = (0.13)(0.0270)(10^3) = 3.51\ \text{W/m}\cdot\text{K}$$

$$\frac{q}{w} = (\bar{h}L)(T_s - T_\infty) = (3.51)(60 - 16) = 154\ \text{W/m}$$

7.36[D] Formula (*1*) below results from the integral analysis of laminar free convection along a vertical plate. Tabulate δ (boundary-layer thickness) versus x (height) if the plate is maintained at 213 °F and exposed to air at 100 °F and 1.5 atm pressure. Consider the range $10^4 < \mathbf{Gr}_x\,\mathbf{Pr} < 10^9$.

$$\frac{\delta}{x} = (3.93)\,\mathbf{Pr}^{-1/2}(0.952 + \mathbf{Pr})^{1/4}\,\mathbf{Gr}_x^{-1/4} \qquad (1)$$

▮ The reference temperature for analytical results is

$$T_{\text{ref}} = T_s + (0.38)(T_\infty - T_s) = 213 + (0.38)(100 - 213) = 170\ °\text{F}$$

From Table B-4, at 170 °F and with conversion of the density to 1.5 atm,

$$\rho = (1.5)(0.0623) = 0.0935\ \text{lbm/ft}^3 \qquad \mu g_c = 1.394 \times 10^{-5}\ \text{lbm/ft}\cdot\text{sec} \qquad \mathbf{Pr} = 0.697$$

Also, $\beta = 1/T_\infty = 1/(630\ °\text{R})$; hence

$$\mathbf{Gr}_x = \frac{g\beta\rho^2(T_s - T_\infty)}{(\mu g_c)^2}x^3 = \frac{(32.2\ \text{ft/sec}^2)(0.0935\ \text{lbm/ft}^3)^2[(213-100)\ °\text{R}]}{(630\ °\text{R})(1.394\times10^{-5}\ \text{lbm/ft}\cdot\text{sec})^2}x^3 = (2.60\times10^8\ \text{ft}^{-3})x^3$$

TABLE 7-1

x (ft)	δ (ft)
0.0381	0.0185
0.1	0.0236
0.2	0.0281
0.3	0.0311
0.5	0.0353
0.7	0.0384
1.0	0.0420
1.5	0.0465
1.77	0.0484

Substitution for $\mathbf{Gr}_x$ and $\mathbf{Pr}$ in (*1*) gives

$$\delta = (0.0420\ \text{ft}^{3/4})x^{1/4} \tag{2}$$

for which the extreme values of x are given by

$$(0.697)(2.60 \times 10^8)x_{\text{min}}^3 = 10^4 \quad \text{or} \quad x_{\text{min}} = 0.0381\ \text{ft}$$

and

$$(0.697)(2.60 \times 10^8)x_{\text{max}}^3 = 10^9 \quad \text{or} \quad x_{\text{max}} = 1.77\ \text{ft}$$

See Table 7-1.

7.37 For Problem 7.36, recalculate δ at $x = 1.77$ ft from the similarity solution.

▮ According to Problem 7.12, we have, for $\mathbf{Pr} \approx 0.7$, $\eta = 5.5$. Hence

$$5.5 \approx \frac{\delta}{x}\left(\frac{\mathbf{Gr}_x}{4}\right)^{1/4} \quad \text{or} \quad \delta \approx 5.5x\left(\frac{4\,\mathbf{Pr}}{\mathbf{Gr}_x\,\mathbf{Pr}}\right)^{1/4}$$

Thus, since $x = 1.77$ ft corresponds to $\mathbf{Pr} = 0.697$ and $\mathbf{Gr}_x\,\mathbf{Pr} = 10^9$,

$$\delta \approx (5.5)(1.77)\left[\frac{4(0.697)}{10^9}\right]^{1/4} = 0.071\ \text{ft}$$

which is 46 percent greater than the value in Table 7-1. Poor agreement is due primarily to approximations in the integral analysis.

7.38 Repeat Problem 7.36 for a vertical plate maintained at 101 °C and exposed to air at 38 °C and 1.5 atm pressure. [If the temperatures corresponded *exactly* to those of Problem 7.36, a direct conversion of the result (*2*) from feet to meters would be possible.]

▮ The reference temperature for analytical results is

$$T_{\text{ref}} = T_s + (0.38)(T_\infty - T_s) = 101 + (0.38)(38 - 101) = 77\ ^\circ\text{C}$$

From Table B-4, at 77 °C and with conversion of the density to 1.5 atm,

$$\rho = (1.5)(0.997) = 1.496\ \text{kg/m}^3 \qquad \mu = 2.075 \times 10^{-5}\ \text{Pa}\cdot\text{s} \qquad \mathbf{Pr} = 0.697$$

Also $\beta = 1/T_\infty = 1/(311\ \text{K})$; hence

$$\mathbf{Gr}_x = \frac{g\beta\rho^2(T_s - T_\infty)}{\mu^2}x^3 = \frac{(9.8\ \text{m/s}^2)(1.496\ \text{kg/m}^3)^2[(101 - 38)\ \text{K}]}{(311\ \text{K})(2.075 \times 10^{-5}\ \text{Pa}\cdot\text{s})^2}x^3 = (10.31\ \text{mm}^{-3})x^3$$

with x measured in millimeters. Substitution for $\mathbf{Gr}_x$ and $\mathbf{Pr}$ in (*1*) of Problem 7.36 gives

$$\delta = (2.97\ \text{mm}^{3/4})x^{1/4}$$

for which the extreme values of x are given by

$$(0.697)(10.31)x_{\text{min}}^3 = 10^4 \quad \text{or} \quad x_{\text{min}} = 11.1\ \text{mm}$$

and

$$(0.697)(10.31)x_{\text{max}}^3 = 10^9 \quad \text{or} \quad x_{\text{max}} = 518\ \text{mm}$$

See Table 7-2.

TABLE 7-2

x (mm)	δ (mm)
11.1	5.42
100	9.39
150	10.39
200	11.16
250	11.81
300	12.36
350	12.84
400	13.28
450	13.68
518	14.17

7.39[D] A vertical flat plate maintained at 220 °F is exposed to still air at 90 °F. Determine **(a)** the boundary-layer thickness 1.25 ft from the lower edge; **(b)** h_x at the same height.

(a) From Table B-4, at $T_{ref} = 220 + (0.38)(90 - 220) = 170$ °F,

$$\mathbf{Pr} = 0.697 \qquad \nu = 22.38 \times 10^{-5}\ \text{ft}^2/\text{sec} \qquad k = 0.017\,35\ \text{Btu/hr}\cdot\text{ft}\cdot{}^\circ\text{F}$$

and $\beta = 1/T_\infty = 1/(550\ {}^\circ\text{R})$.

$$\mathbf{Gr}_{1.25} = \frac{(32.2\ \text{ft/sec}^2)[(220-90)\ {}^\circ\text{R}](1.25\ \text{ft})^3}{(550\ {}^\circ\text{R})(2.238 \times 10^{-4}\ \text{ft}^2/\text{sec})^2} = 2.97 \times 10^8$$

$$\mathbf{Gr}_{1.25}\,\mathbf{Pr} = (2.97 \times 10^8)(0.697) < 10^9$$

Thus, the flow at the specified location is laminar, and (*1*) of Problem 7.36 applies.

$$\delta = (3.93)\,\mathbf{Pr}^{-1/2}\,(0.952 + \mathbf{Pr})^{1/4}(\mathbf{Gr}_x)^{-1/4}x = (3.93)(0.697)^{-1/2}(1.649)^{1/4}(2.97 \times 10^8)^{-1/4}(15\ \text{in.})$$
$$= 0.61\ \text{in.}$$

(b) Applying the laminar-flow integral correlation (*I-13*),

$$h_x = \frac{k}{x}(508)\left(\frac{\mathbf{Gr}_x\,\mathbf{Pr}^2}{0.952 + \mathbf{Pr}}\right)^{1/4}$$
$$= \frac{0.017\,35\ \text{Btu/hr}\cdot\text{ft}\cdot{}^\circ\text{F}}{1.25\ \text{ft}}(0.508)\left[\frac{(2.97 \times 10^8)(0.697)^2}{0.952 + 0.697}\right]^{1/4}$$
$$= 0.682\ \text{Btu/hr}\cdot\text{ft}^2\cdot{}^\circ\text{F}$$

7.40[D] A vertical flat plate maintained at 105 °C is exposed to still air at 32 °C. Determine **(a)** the boundary-layer thickness 38 cm from the lower edge; **(b)** h_x at the same height.

(a) From Table B-4, at $T_{ref} = 105 + (0.38)(32 - 105) = 77$ °C,

$$\mathbf{Pr} = 0.697 \qquad \nu = 2.079 \times 10^{-5}\ \text{m}^2/\text{s} \qquad k = 0.0300\ \text{W/m}\cdot\text{K}$$

and $\beta = 1/T_\infty = 1/(305\ \text{K})$.

$$\mathbf{Gr}_{0.38} = \frac{(9.8\ \text{m/s}^2)[(105-32)\ \text{K}](0.38\ \text{m})^3}{(305\ \text{K})(2.079 \times 10^{-5}\ \text{m}^2/\text{s})^2} = 2.98 \times 10^8$$

$$\mathbf{Gr}_{0.38}\,\mathbf{Pr} = (2.98 \times 10^8)(0.697) < 10^9$$

Thus, the flow at the specified location is laminar, and (*1*) of Problem 7.36 applies.

$$\delta = (3.93)\,\mathbf{Pr}^{-1/2}(0.952 + \mathbf{Pr})^{1/4}(\mathbf{Gr}_x)^{-1/4}x = (3.93)(0.697)^{-1/2}(1.649)^{1/4}(2.98 \times 10^8)^{-1/4}(38\ \text{cm})$$
$$= 1.543\ \text{cm}$$

(b) Applying the laminar-flow integral correlation (*I-13*),

$$h_x = \frac{k}{x}(0.508)\left(\frac{\mathbf{Gr}_x\,\mathbf{Pr}^2}{0.952 + \mathbf{Pr}}\right)^{1/4}$$
$$= \frac{0.0300\ \text{W/m}\cdot\text{K}}{0.38\ \text{m}}(0.508)\left[\frac{(2.98 \times 10^8)(0.697)^2}{0.952 + 0.697}\right]^{1/4} = 3.88\ \text{W/m}^2\cdot\text{K}$$

7.41 Derive (*I-10*) and (*I-14*) from (*I-9*) and (*I-13*), respectively.

Both (*I-9*) and (*I-13*) have the form $h_x = Cx^{-1/4}$, where C is independent of x; hence,

$$\bar{h} = \frac{1}{L}\int_0^L Cx^{-1/4}\,dx = \frac{4}{3}CL^{-1/4} = \frac{4}{3}h_L$$

which is tantamount to (*I-10*) or (*I-14*). Note that, as usual, $\overline{\mathbf{Nu}}$ has no significance as an average, but it is a useful and convenient dimensionless term; $\overline{\mathbf{Nu}} \equiv (\bar{h}L/k)$.

7.42[D] A vertical flat plate 4 ft high is maintained at 120 °F and is exposed to still air at 80 °F. Determine the heat transfer per foot of width using the results of the integral analysis.

Assuming turbulent flow, from Table B-4, at $T_f = 100$ °F,

$$\mathbf{Pr} = 0.706 \qquad \nu = 18.10 \times 10^{-5}\ \text{ft}^2/\text{sec} \qquad k = 0.015\,65\ \text{Btu/hr}\cdot\text{ft}\cdot{}^\circ\text{F}$$

Also, $\beta = 1/T_\infty = 1/(540\ °R)$.

$$\mathbf{Gr}_L = \frac{g\beta(T_s - T_\infty)L^3}{\nu^2} = \frac{(32.2\ \text{ft/sec}^2)[(120-80)\ °R](4\ \text{ft})^3}{(540\ °R)(18.10\times10^{-5}\ \text{ft}^2/\text{sec})^2} = 4.66\times10^9$$

$$\mathbf{Gr}_L\,\mathbf{Pr} = (4.66\times10^9)(0.706) = 3.290\times10^9 > 10^9$$

To this turbulent flow (*I-16*) may be applied:

$$\bar{h}L = (0.0210)k(\mathbf{Gr}_L\,\mathbf{Pr})^{2/5} = (0.0210)(0.015\,65)(3.290\times10^9)^{2/5} = 2.107\ \text{Btu/hr}\cdot\text{ft}\cdot°\text{F}$$

$$\frac{q}{w} = (\bar{h}L)(T_s - T_\infty) = (2.107)(40) = 84.3\ \text{Btu/hr}\cdot\text{ft}$$

7.43^D A vertical flat plate 1.22 m high is maintained at 50 °C and is exposed to still air at 27 °C. Determine the heat transfer per meter of width using the results of the integral analysis.

I Assuming turbulent flow, from Table B-4, at $T_f = 38.5\ °C$,

$$\mathbf{Pr} = 0.705 \qquad \nu = 1.69\times10^{-5}\ \text{m}^2/\text{s} \qquad k = 0.0271\ \text{W/m}\cdot\text{K}$$

Also, $\beta = 1/T_\infty = 1/(300\ \text{K})$.

$$\mathbf{Gr}_L = \frac{g\beta(T_s - T_\infty)L^3}{\nu^2} = \frac{(9.8\ \text{m/s}^2)[(50-27)\ \text{K}](1.22\ \text{m})^3}{(300\ \text{K})(1.69\times10^{-5}\ \text{m}^2/\text{s})^2} = 4.77\times10^9$$

$$\mathbf{Gr}_L\,\mathbf{Pr} = (4.77\times10^9)(0.705) = 3.36\times10^9 > 10^9$$

To this turbulent flow (*I-16*) may be applied:

$$\bar{h}L = (0.0210)k(\mathbf{Gr}_L\,\mathbf{Pr})^{2/5} = (0.0210)(0.0271)(3.36\times10^9)^{2/5} = 3.68\ \text{W/m}\cdot\text{K}$$

$$\frac{q}{w} = (\bar{h}L)(T_s - T_\infty) = (3.68)(23) = 84.6\ \text{W/m}$$

7.2 EMPIRICAL CORRELATIONS: EXTERNAL REGIONS

7.44^D A vertical plate 0.6 m wide and 0.3 m high is exposed to gaseous oxygen at 33 °C and 1.0 atm pressure. Calculate the heat loss from the plate (one side) if it is maintained at 44 °C. Use two empirical correlations for $\overline{\mathbf{Nu}}$, and compare with analytical results.

I From Table B-4, at $T_f = (44+33)/2 = 38.5\ °C$,

$$\nu = 1.699\times10^{-5}\ \text{m}^2/\text{s} \qquad k = 0.0276\ \text{W/m}\cdot\text{K} \qquad \mathbf{Pr} = 0.707$$

Also, $\beta = 1/T_\infty = 1/(306\ \text{K})$. Then

$$\mathbf{Gr}_L\,\mathbf{Pr} = \frac{g\beta(T_s - T_\infty)L^3\,\mathbf{Pr}}{\nu^2} = \frac{(9.8)(11)(0.3)^3(0.707)}{(306)(1.699\times10^{-5})^2} = 2.33\times10^7 \quad (\textit{laminar})$$

From (*I-17*), with $\Lambda = L$, $C = 0.59$, and $a = \frac{1}{4}$,

$$\bar{h}L = kC(\mathbf{Gr}_L\,\mathbf{Pr})^a = (0.0276)(0.59)(2.33\times10^7)^{1/4} = 1.13\ \text{W/m}\cdot\text{K}$$

$$q = (\bar{h}L)w(T_s - T_\infty) = (1.13)(0.6)(11) = 7.46\ \text{W}$$

From (*I-19*),

$$\bar{h}L = k\left\{0.68 + \frac{(0.67)(\mathbf{Gr}_L\,\mathbf{Pr})^{1/4}}{[1+(0.492/\mathbf{Pr})^{9/16}]^{4/9}}\right\} = (0.0276)\left\{0.68 + \frac{0.67(2.33\times10^7)^{1/4}}{[1+(0.492/0.707)^{9/16}]^{4/9}}\right\} = 1.00\ \text{W/m}\cdot\text{K}$$

Since $T_{\text{ref}} = 44 + (0.38)(33-44) = 39.8\ °C$ and $T_f = 38.5\ °C$ are nearly the same, the latter temperature may be assumed for (*I-10*), giving

$$\bar{h}L = \frac{4}{3}kF_1(\mathbf{Pr})\left(\frac{\mathbf{Gr}_L}{4}\right)^{1/4} = \frac{4}{3}(0.0276)(0.50)\left[\frac{2.33\times10^7}{4(0.707)}\right]^{1/4} = 0.986\ \text{W/m}\cdot\text{K}$$

which is very nearly 1.00 W/m · K. ($F_1(\mathbf{Pr}) \approx 0.50$ from Table I-1 by interpolation.)

7.45^D A vertical plate 2 ft wide and 1 ft high is exposed to oxygen gas at 90 °F and 1.0 atm pressure. Calculate the heat loss from the plate (one side) if it is maintained at 110 °F. Use two empirical correlations for $\overline{\mathbf{Nu}}$.

▮ At $T_f = (110 + 90)/2 = 100\ °F$, Table B-4 gives

$$\nu = 18.29 \times 10^{-5}\ \text{ft}^2/\text{sec} \qquad k = 0.015\,98\ \text{Btu/hr}\cdot\text{ft}\cdot°\text{F} \qquad \mathbf{Pr} = 0.707$$

Also, $\beta = 1/T_\infty = 1/(550\ °\text{R})$. Then

$$\mathbf{Gr}_L\,\mathbf{Pr} = \frac{g\beta(T_s - T_\infty)L^3\,\mathbf{Pr}}{\nu^2} = \frac{(32.2)(20)(1^3)(0.707)}{(550)(18.29 \times 10^{-5})^2} = 2.47 \times 10^7 \quad (\textit{laminar})$$

From (*I-17*), with $\Lambda = L$, $C = 0.59$, and $a = \frac{1}{4}$,

$$\bar{h}L = kC(\mathbf{Gr}_L\,\mathbf{Pr})^a = (0.015\,98)(0.59)(2.47 \times 10^7)^{1/4} = 0.665\ \text{Btu/hr}\cdot\text{ft}\cdot°\text{F}$$
$$q = (\bar{h}L)w(T_s - T_\infty) = (0.665)(2)(20) = 26.6\ \text{Btu/hr}$$

From (*I-19*),

$$\bar{h}L = k\left\{0.68 + \frac{(0.67)(\mathbf{Gr}_L\,\mathbf{Pr})^{1/4}}{[1 + (0.492/\mathbf{Pr})^{9/16}]^{4/9}}\right\} = (0.015\,98)\left\{0.68 + \frac{0.67(2.47 \times 10^7)^{1/4}}{[1 + (0.492/0.707)^{9/16}]^{4/9}}\right\} = 0.590\ \text{Btu/hr}\cdot\text{ft}\cdot°\text{F}$$

which is about 11 percent below the value obtained with (*I-17*). This lower value is probably more accurate, as (*I-19*) is a more recent correlation than (*I-17*).

7.46[D] Estimate the rate of heat transfer to a cylinder 6 ft in diameter and 6 ft long if it is submerged vertically in engine oil at a temperature of 96 °F. The temperature of the cylinder is maintained at 40 °F.

▮ From Table B-3, at $T_f = 68\ °F$,

$$\nu = 0.0097\ \text{ft}^2/\text{s} \qquad k = 0.084\ \text{Btu/hr}\cdot\text{ft}\cdot°\text{F} \qquad \mathbf{Pr} = 10\,400 \qquad \beta = 0.39 \times 10^{-3}\ °\text{R}^{-1}$$

Hence,

$$\mathbf{Gr}_L\,\mathbf{Pr} = \frac{g\beta(\Delta T)L^3\,\mathbf{Pr}}{\nu^2} = \frac{(32.2\ \text{ft/s}^2)(0.39 \times 10^{-3}\ °\text{R}^{-1})[(96 - 40)\ °\text{R}](6\ \text{ft})^3(10\,400)}{(0.0097\ \text{ft}^2/\text{s})^2}$$
$$= 1.679 \times 10^{10} \quad (\textit{turbulent})$$

Because

$$\frac{35}{\mathbf{Gr}_L^{1/4}} = \frac{35}{[(1.679 \times 10^{10})/10\,400]^{1/4}} = \frac{35}{35.6} < 1 = \frac{D}{L}$$

(*I-18*) is satisfied and the vertical-plate version of (*I-17*) holds ($\Lambda = L$).

$$\bar{h}L = kC(\mathbf{Gr}_L\,\mathbf{Pr})^a = (0.084)(0.13)(1.679 \times 10^{10})^{1/3} = 28.0\ \text{Btu/hr}\cdot\text{ft}\cdot°\text{F}$$
$$q = (\bar{h}L)\pi D(T_s - T_\infty) = (28.0)\pi(6)(40 - 96) = -29\,600\ \text{Btu/hr}$$

By contrast, (*I-20*) yields $q = -36\,900$ Btu/hr, while Fig. I-1 yields $q = -22\,200$ Btu/hr. The spread is due to the very high Prandtl number.

7.47[D] Estimate the rate of heat transfer to a cylinder 1.8 m in diameter and 1.8 m long if it is submerged vertically in engine oil at a temperature of 35 °C. The temperature of the cylinder is maintained at 5 °C.

▮ From Table B-3, at $T_f = 20\ °C$,

$$\nu = 9.01 \times 10^{-4}\ \text{m}^2/\text{s} \qquad k = 0.1453\ \text{W/m}\cdot\text{K} \qquad \mathbf{Pr} = 10\,400 \qquad \beta = 0.702 \times 10^{-3}\ \text{K}^{-1}$$

Hence,

$$\mathbf{Gr}_L\,\mathbf{Pr} = \frac{g\beta(\Delta T)L^3\,\mathbf{Pr}}{\nu^2} = \frac{(9.8\ \text{m/s}^2)(0.702 \times 10^{-3}\ \text{K}^{-1})[(35 - 5)\ \text{K}](1.8\ \text{m})^3(10\,400)}{(9.01 \times 10^{-4}\ \text{m}^2/\text{s})^2}$$
$$= 1.542 \times 10^{10} \quad (\textit{turbulent})$$

Because

$$\frac{35}{\mathbf{Gr}_L^{1/4}} = \frac{35}{[(1.542 \times 10^{10})/(10\,400)]^{1/4}} = \frac{35}{34.9} \approx 1.0 = \frac{D}{L}$$

(*I-18*) is (marginally) satisfied, and the vertical-plate version of (*I-17*) holds ($\Lambda = L$).

$$\bar{h}L = kC(\mathbf{Gr}_L\,\mathbf{Pr})^a = (0.1453)(0.13)(1.542 \times 10^{10})^{1/3} = 47.1\ \text{W/m}\cdot\text{K}$$
$$q = (\bar{h}L)\pi D(T_s - T_\infty) = (47.1\ \text{W/m}\cdot\text{K})\pi(1.8\ \text{m})[(5 - 35)\ \text{K}] = -7.99\ \text{kW}$$

By contrast, (*I-20*) yields $q = -9.96$ kW, while Fig. I-1 yields $q = -6.19$ kW. The spread is due to the very high value of **Pr**.

7.48[D] On a certain summer day the air temperature in New Orleans reaches 105 °F, with the air relatively still. A milkshake in a 4-in.-diam. cup 8 in. tall is exposed to this air but shaded from the sun. Estimate the initial rate of "heat" gained by the milkshake along the cylindrical surface, if its surface temperature is 35 °F; use (*a*) the 'miscellaneous-shapes' approach, (*b*) the 'vertical-plate' approach (if valid).

▌

$$T_f = \frac{T_s + T_\infty}{2} = \frac{35 + 105}{2} = 70\ °\text{F}$$

By interpolation in Table B-4,

$$\nu = 16.14 \times 10^{-5}\ \text{ft}^2/\text{sec} \qquad k = 0.014\,91\ \text{Btu/hr}\cdot\text{ft}\cdot°\text{F} \qquad \mathbf{Pr} = 0.710$$

Also, $\beta = 1/T_\infty = 1/(565\ °\text{R})$. Hence

$$\mathbf{Ra}_L \equiv \mathbf{Gr}_L\,\mathbf{Pr} = \frac{g\beta(\Delta T)L^3\,\mathbf{Pr}}{\nu^2} = \frac{(32.2\ \text{ft/sec}^2)[(105-35)\ °\text{R}]L^3(0.710)}{(565\ °\text{R})(16.14\times 10^{-5}\ \text{ft}^2/\text{sec})^2} = (1.087\times 10^8\ \text{ft}^{-3})L^3$$

(*a*) From the bottom of Table I-3,

$$\Lambda = \frac{L_h L_v}{L_h + L_v} = \frac{(4\ \text{in.})(8\ \text{in.})}{12\ \text{in.}} = \frac{8}{3} = \frac{2}{9}\ \text{ft}$$

whence $\mathbf{Gr}_\Lambda\,\mathbf{Pr} = (1.087\times 10^8)(\frac{2}{9})^3 = 1.197\times 10^6$ (*laminar flow*). Then (*I-17*) gives

$$\bar{h}\Lambda = kC(\mathbf{Gr}_\Lambda\,\mathbf{Pr})^a = (0.014\,91)(0.60)(1.197\times 10^6)^{1/4} = 0.296\ \text{Btu/hr}\cdot\text{ft}\cdot°\text{F}$$

and, since $\Lambda = L/3$,

$$q = 3(\bar{h}\Lambda)\pi D(T_s - T_\infty) = 3(0.295)\pi(\tfrac{4}{12})(35-105) = -65.1\ \text{Btu/hr}$$

(*b*) For $L = 8\ \text{in.} = \frac{2}{3}$ ft, $\mathbf{Gr}_L\,\mathbf{Pr} = (1.087\times 10^8)(\frac{2}{3})^3 = 3.22\times 10^7$ and

$$\frac{35}{\mathbf{Gr}_L^{1/4}} = \frac{35}{[(3.22\times 10^7)/0.710]^{1/4}} = \frac{35}{82.06} < \frac{1}{2} = \frac{D}{L}$$

Thus the top of Table I-3 may be used in conjunction with (*I-17*), to give, since $\Lambda = L$,

$$\bar{h}L = kC(\mathbf{Gr}_L\,\mathbf{Pr})^a = (0.014\,91)(0.59)(3.22\times 10^7)^{1/4} = 0.663\ \text{Btu/hr}\cdot\text{ft}\cdot°\text{F}$$

$$q = (\bar{h}L)\pi D(T_s - T_\infty) = (0.663)\pi(\tfrac{4}{12})(35-105) = -48.6\ \text{Btu/hr}$$

7.49[D] On a certain summer day the air temperature in New Orleans reaches 41 °C, with the air relatively still. A milkshake in a 10-cm-diam. cup 20 cm tall is exposed to this air but shaded from the sun. Estimate the initial rate of "heat" gained by the milkshake along the cylindrical surface, if the surface temperature is 2 °C; use (*a*) the 'miscellaneous-shapes' approach, (*b*) the 'vertical-plate' approach (if valid).

▌

$$T_f = \frac{T_s + T_\infty}{2} = \frac{2 + 41}{2} = 21.5\ °\text{C}$$

Interpolation in Table B-4 gives

$$\nu = 1.499\times 10^{-5}\ \text{m}^2/\text{s} \qquad k = 0.025\,79\ \text{W/m}\cdot\text{K} \qquad \mathbf{Pr} = 0.710$$

Also, $\beta = 1/T_\infty = 1/(314\ \text{K})$; therefore,

$$\mathbf{Gr}_L\,\mathbf{Pr} = \frac{g\beta(\Delta T)L^3\,\mathbf{Pr}}{\nu^2} = \frac{(9.8\ \text{m/s})^2[(41-2)\ \text{K}]L^3(0.710)}{(314\ \text{K})(1.499\times 10^{-5}\ \text{m}^2/\text{s})^2} = (3846\ \text{cm}^{-3})L^3$$

(*a*) From the bottom of Table I-3,

$$\Lambda = \frac{L_h L_v}{L_h + L_v} = \frac{(10\ \text{cm})(20\ \text{cm})}{30\ \text{cm}} = \frac{20}{3}\ \text{cm}$$

whence $\mathbf{Gr}_\Lambda\,\mathbf{Pr} = (3846)(\frac{20}{3})^3 = 1.140\times 10^6$ (*laminar flow*). Then (*I-17*) gives

$$\bar{h}\Lambda = kC(\mathbf{Gr}\,\mathbf{Pr})^a = (0.025\,79)(0.60)(1.140\times 10^6)^{1/4} = 0.5056\ \text{W/m}\cdot\text{K}$$

and, since $\Lambda = L/3$,

$$q = 3(\bar{h}\Lambda)\pi D(T_s - T_\infty) = 3(0.5056)\pi(0.10)(2-41) = -18.58\ \text{W}$$

(*b*) For $L = 20$ cm, $\mathbf{Gr}_L\,\mathbf{Pr} = (3846)(20)^3 = 3.08 \times 10^7$ and

$$\frac{35}{\mathbf{Gr}_L^{1/4}} = \frac{35}{[(3.08 \times 10^7)/0.710]^{1/4}} = \frac{35}{81.2} < \frac{1}{2} = \frac{D}{L}$$

Thus, the top of Table I-3 may be used in conjunction with (*I-17*), to give, since $\Lambda = L$,

$$\bar{h}L = kC(\mathbf{Gr}_L\,\mathbf{Pr})^a = (0.025\,79)(0.59)(3.08 \times 10^7)^{1/4} = 1.134\ \text{W/m}\cdot\text{K}$$

$$q = (\bar{h}L)\pi D(T_s - T_\infty) = (1.132)\pi(0.10)(2 - 41) = -13.89\ \text{W}$$

7.50[D] A vertical 4-in.-diam. cylinder 18 in. tall is maintained at 125 °F in an atmospheric environment of 75 °F. Determine the rate of heat transfer from the cylindrical surface.

▌ From Table B-4, at $T_f = 100$ °F,

$$\nu = 18.10 \times 10^{-5}\ \text{ft}^2/\text{sec} \qquad k = 0.015\,65\ \text{Btu/hr}\cdot\text{ft}\cdot{}^\circ\text{F} \qquad \mathbf{Pr} = 0.706$$

and, at T_∞, $\beta = 1/(535\ {}^\circ\text{R})$. So,

$$\mathbf{Gr}_L = \frac{(32.2\ \text{ft/sec}^2)[(125 - 75)\ {}^\circ\text{R}](1.5\ \text{ft})^3}{(535\ {}^\circ\text{R})(18.10 \times 10^{-5}\ \text{ft}^2/\text{sec})^2} = 3.10 \times 10^8$$

and $\mathbf{Ra}_L \equiv \mathbf{Gr}_L\,\mathbf{Pr} = (3.10 \times 10^8)(0.706) = 2.19 \times 10^8$ (*laminar flow*). For a vertical plate, (*I-17*) would give

$$\bar{h}_{\text{plate}} = \frac{k}{L} C\,\mathbf{Ra}_L^a = \frac{0.015\,65}{\frac{18}{12}}(0.59)(2.19 \times 10^8)^{1/4} = 0.749\ \text{Btu/hr}\cdot\text{ft}^2\cdot{}^\circ\text{F} \qquad (1)$$

However, condition (*I-18*) is not satisfied:

$$\frac{35}{\mathbf{Gr}_L^{1/4}} = \frac{35}{(3.10 \times 10^8)^{1/4}} = 0.264 > \frac{2}{9} = \frac{D}{L}$$

In such a case, the following correction of (*1*) is suggested:

$$\bar{h}_{\text{cyl}} = \bar{h}_{\text{plate}}\left[1 + (1.3)\left(\frac{L/D}{\mathbf{Ra}_L^{1/4}}\right)^{0.9}\right]$$

$$= (0.749)\left\{1 + (1.3)\left[\frac{\frac{9}{2}}{(2.19 \times 10^8)^{1/4}}\right]^{0.9}\right\} = 0.799\ \text{Btu/hr}\cdot\text{ft}^2\cdot{}^\circ\text{F} \qquad (2)$$

from which

$$q = \bar{h}_{\text{cyl}}A(T_s - T_\infty) = (0.799)\pi(\tfrac{4}{12})(\tfrac{18}{12})(125 - 75) = 62.8\ \text{Btu/hr}$$

7.51[D] A vertical 10-cm-diam. cylinder 46 cm tall is maintained at 52 °C in an atmospheric environment of 24 °C. Determine the heat transfer rate from the cylindrical surface.

▌ From Table B-4, at $T_f = 38$ °C,

$$\nu = 1.68 \times 10^{-5}\ \text{m}^2/\text{s} \qquad k = 0.027\,06\ \text{W/m}\cdot\text{K} \qquad \mathbf{Pr} = 0.706$$

and, at T_∞, $\beta = 1/(297\ \text{K})$. So,

$$\mathbf{Gr}_L = \frac{(9.8\ \text{m/s}^2)[(52 - 24)\ \text{K}](0.46\ \text{m})^3}{(297\ \text{K})(1.68 \times 10^{-5}\ \text{m}^2/\text{s})^2} = 3.19 \times 10^8$$

and $\mathbf{Ra}_L \equiv \mathbf{Gr}_L\,\mathbf{Pr} = (3.19 \times 10^8)(0.706) = 2.25 \times 10^8$ (*laminar flow*). For a vertical plate, (*I-17*) *would give*

$$\bar{h}_{\text{plate}} = \frac{k}{L} C\,\mathbf{Ra}_L^a = \frac{0.027\,06}{0.46}(0.59)(2.25 \times 10^8)^{1/4} = 4.25\ \text{W/m}^2\cdot\text{K} \qquad (1)$$

However, condition (*I-18*) is not satisfied:

$$\frac{35}{\mathbf{Gr}_L^{1/4}} = \frac{35}{(3.19 \times 10^8)^{1/4}} = 0.262 > \frac{10}{40} > \frac{10}{46} = \frac{D}{L}$$

In such a case, the following correction of (*1*) is suggested:

$$\bar{h}_{cyl} = \bar{h}_{plate}\left[1 + (1.3)\left(\frac{L/D}{\mathbf{Ra}_L^{1/4}}\right)^{0.9}\right]$$

$$= (4.25)\left\{1 + (1.3)\left[\frac{4.6}{(2.25 \times 10^8)^{1/4}}\right]^{0.9}\right\} = 4.55 \text{ W/m}^2 \cdot \text{K}$$

from which

$$q = \bar{h}_{cyl} A(T_s - T_\infty) = (4.55)\pi(0.10)(0.46)(52 - 24) = 18.4 \text{ W}$$

7.52^{D} A 20-in.-high vertical plate is at a constant surface temperature of 250 °F and exposed to still air at 50 °F. Determine the rate of heat transfer per foot of width by ***(a)*** the exact solution of the boundary-layer equations (Ostrach's results), ***(b)*** the approximate integral solution, and ***(c)*** an empirical correlation.

▌ From Table B-4, at $T_{ref} = 250 + (0.38)(50 - 250) = 174$ °F,

$$\mathbf{Pr} = 0.697 \qquad k = 0.017\,44 \text{ Btu/hr} \cdot \text{ft} \cdot {}^\circ\text{F} \qquad \nu = 22.62 \times 10^{-5} \text{ ft}^2/\text{sec}$$

Moreover, $\beta = 1/T_\infty = 1/(510\ {}^\circ\text{R})$. Hence,

$$\mathbf{Gr}_L = \frac{g\beta(T_s - T_\infty)L^3}{\nu^2} = \frac{(32.2 \text{ ft/sec}^2)[(250 - 50)\ {}^\circ\text{R}](\frac{20}{12} \text{ ft})^3}{(510\ {}^\circ\text{R})(22.62 \times 10^{-5} \text{ ft}^2/\text{sec})^2} = 1.142 \times 10^9$$

and $\mathbf{Gr}_L\,\mathbf{Pr} = (1.142 \times 10^9)(0.697) = 7.96 \times 10^8$ *(laminar flow)*.

(a) By *(I-10)* and interpolation in Table I-1,

$$\bar{h}L = \frac{4}{3} kF_1(\mathbf{Pr})\left(\frac{\mathbf{Gr}_L}{4}\right)^{1/4} = \frac{4}{3}(0.017\,44)(0.4909)\left(\frac{1.142 \times 10^9}{4}\right)^{1/4} = 1.484 \text{ Btu/hr} \cdot \text{ft} \cdot {}^\circ\text{F}$$

$$\frac{q}{w} = (\bar{h}L)(T_s - T_\infty) = (1.484)(200) = 296.8 \text{ Btu/hr} \cdot \text{ft}$$

(b) By *(I-14)*,

$$\bar{h}L = \frac{4}{3} k(0.508)\left(\frac{\mathbf{Gr}_L\,\mathbf{Pr}^2}{0.952 + \mathbf{Pr}}\right)^{1/4} = \frac{4}{3}(0.017\,44)(0.508)\left[\frac{(7.96 \times 10^8)(0.697)}{0.952 + 0.697}\right]^{1/4} = 1.60 \text{ Btu/hr} \cdot \text{ft} \cdot {}^\circ\text{F}$$

$$\frac{q}{w} = (\bar{h}L)(T_s - T_\infty) = (1.60)(200) = 320 \text{ Btu/hr} \cdot \text{ft}$$

(c) The empirical *(I-19)* requires property values, other than β, at $T_f = 150$ °F. We have

$$\nu = 21.16 \times 10^{-5} \text{ ft}^2/\text{sec} \qquad k = 0.016\,86 \text{ Btu/hr} \cdot \text{ft} \cdot {}^\circ\text{F} \qquad \mathbf{Pr} = 0.699$$

and these lead to $\mathbf{Gr}_L\,\mathbf{Pr} = 9.13 \times 10^8$. Then,

$$\bar{h}L = (0.016\,86)\left\{0.68 + \frac{(0.67)(9.13 \times 10^8)^{1/4}}{[1 + (0.492/0.699)^{9/16}]^{4/9}}\right\} = 1.516 \text{ Btu/hr} \cdot \text{ft} \cdot {}^\circ\text{F}$$

$$\frac{q}{w} = (1.516)(200) = 303 \text{ Btu/hr} \cdot \text{ft}$$

7.53^{D} A 50-cm-high vertical plate is at a constant surface temperature of 120 °C and exposed to still air at 10 °C. Determine the heat transfer per meter of width by ***(a)*** the exact solution of the boundary-layer equations (Ostrach's results), ***(b)*** the approximate integral solution, and ***(c)*** an empirical correlation.

▌ From Table B-4, at $T_{ref} = 120 + (0.38)(10 - 120) = 78$ °C,

$$\mathbf{Pr} = 0.697 \qquad k = 0.0300 \text{ W/m} \cdot \text{K} \qquad \nu = 2.089 \times 10^{-5} \text{ m}^2/\text{s}$$

Moreover, $\beta = 1/T_\infty = 1/(283 \text{ K})$. Hence

$$\mathbf{Gr}_L = \frac{g\beta(T_s - T_\infty)L^3}{\nu^2} = \frac{(9.8 \text{ m/s}^2)[(120 - 10) \text{ K}](0.5 \text{ m})^3}{(283 \text{ K})(2.089 \times 10^{-5} \text{ m}^2/\text{s})^2} = 1.090 \times 10^9$$

and $\mathbf{Ra}_L \equiv \mathbf{Gr}_L\,\mathbf{Pr} = (1.090 \times 10^9)(0.697) = 7.597 \times 10^8$ *(laminar flow)*.

(a) By *(I-10)*, with F_1 from *(I-11)*,

$$\bar{h}L = \frac{4}{3} kF_1(\mathbf{Pr})\left(\frac{\mathbf{Gr}_L}{4}\right)^{1/4} = \frac{4}{3}(0.0300)(0.4984)\left(\frac{1.090 \times 10^9}{4}\right)^{1/4} = 2.562 \text{ W/m} \cdot \text{K}$$

$$\frac{q}{w} = (\bar{h}L)(T_s - T_\infty) = (2.562)(120 - 10) = 282 \text{ W/m}$$

(*b*) By (*I-14*)

$$\bar{h}L = \frac{4}{3}k(0.508)\left(\frac{\mathbf{Gr}_L\,\mathbf{Pr}^2}{0.952+\mathbf{Pr}}\right)^{1/4} = \frac{4}{3}(0.0300)(0.508)\left[\frac{(7.597\times 10^8)(0.697)}{0.952+0.697}\right]^{1/4} = 2.72\ \text{W/m}\cdot\text{K}$$

$$\frac{q}{w} = (\bar{h}L)(T_s - T_\infty) = (2.72)(110) = 299\ \text{W/m}$$

(*c*) The empirical (*I-19*) requires property values, excluding β, at $T_f = 65\ °\text{C}$:

$$\nu = 1.956\times 10^{-5}\ \text{m}^2/\text{s} \qquad k = 0.0291\ \text{W/m}\cdot\text{K} \qquad \mathbf{Pr} = 0.700$$

and these lead to $\mathbf{Ra}_L \equiv \mathbf{Gr}_L\,\mathbf{Pr} = 8.71\times 10^8$. Then, since the flow is laminar, (*I-9*) yields

$$\bar{h}L = (0.0291)\left\{0.68 + \frac{(0.67)(8.71\times 10^8)^{1/4}}{[1+(0.492/0.700)^{9/16}]^{4/9}}\right\} = 2.59\ \text{W/m}\cdot\text{K}$$

$$\frac{q}{w} = (2.59)(110) = 284\ \text{W/m}$$

Problems 7.54–7.61 are applications of (*I-21*), (*I-22*), and (*I-23*). For these problem solutions, as given, the characteristic length L is defined as the quarter-perimeter of the plate. Recent work tends to prefer the definition $L \equiv$ area/perimeter; the student may want to repeat these problems, using the latter definition. However, the value of $\bar{h}$ from (*I-22*) is independent of the choice of L.

7.54^D Calculate the heat loss from a horizontal 2-ft by 2-ft plate with an upper surface temperature of 330 °F to air at 70 °F and atmospheric pressure.

▌ Anticipating the use of an empirical correlation, we evaluate properties (except β) at $T_f = 200\ °\text{F}$:

$$\nu = 24.21\times 10^{-5}\ \text{ft}^2/\text{sec} \qquad k = 0.018\,05\ \text{Btu/hr}\cdot\text{ft}\cdot°\text{F} \qquad \mathbf{Pr} = 0.694$$

and $\beta = 1/T_\infty = 1/(530\ °\text{R})$. Then the Rayleigh *number*, $\mathbf{Ra}_L \equiv \mathbf{Gr}_L\mathbf{Pr}$, is

$$\mathbf{Ra}_L = \frac{g\beta(T_s - T_\infty)L^3\,\mathbf{Pr}}{\nu^2} = \frac{(32.2\ \text{ft/sec}^2)[(330-70)\ °\text{R}](2\ \text{ft})^3(0.694)}{(530\ °\text{R})(24.21\times 10^{-5}\ \text{ft}^2/\text{sec})^2} = 1.496\times 10^9 \quad (\textit{turbulent})$$

Correlation (*I-22*) is appropriate:

$$\bar{h}L = k(0.14)(\mathbf{Ra}_L)^{1/3} = (0.018\,05)(0.14)(1.496\times 10^9)^{1/3} = 2.96\ \text{Btu/hr}\cdot\text{ft}\cdot°\text{F}$$

$$q = (\bar{h}L)L(T_s - T_\infty) = (2.96)(2)(260) = 1539\ \text{Btu/hr}$$

7.55^D Calculate the heat loss from a horizontal 0.6-m by 0.6-m plate with an upper surface temperature of 166 °C to air at 21 °C and atmospheric pressure.

▌ Anticipating the use of an empirical correlation, we evaluate properties (except β) at $T_f = 93.5\ °\text{C}$:

$$\nu = 2.249\times 10^{-5}\ \text{m}^2/\text{s} \qquad k = 0.0312\ \text{W/m}\cdot\text{K} \qquad \mathbf{Pr} = 0.694$$

and $\beta = 1/T_\infty = 1/(294\ \text{K})$. Then the Rayleigh number, $\mathbf{Ra}_L \equiv \mathbf{Gr}_L\mathbf{Pr}$, is

$$\mathbf{Ra}_L = \frac{g\beta(T_s - T_\infty)L^3\,\mathbf{Pr}}{\nu^2} = \frac{(9.8\ \text{m/s}^2)[(166-21)\ \text{K}](0.6\ \text{m})^3(0.694)}{(294\ \text{K})(2.249\times 10^{-5}\ \text{m}^2/\text{s})^2} = 1.432\times 10^9 \quad (\textit{turbulent})$$

Correlation (*I-22*) is appropriate:

$$\bar{h}L = k(0.14)(\mathbf{Ra}_L)^{1/3} = (0.0312)(0.14)(1.432\times 10^9)^{1/3} = 4.93\ \text{W/m}\cdot\text{K}$$

$$q = (\bar{h}L)L(T_s - T_\infty) = (4.93)(0.6)(166-21) = 428\ \text{W}$$

7.56^D A horizontal square plate with its lower surface heated to 125 °F is losing heat to still air at a temperature of 75 °F; the length of a side of the plate is 1.5 ft. Determine the rate of heat transfer. (The upper surface is insulated.)

▌ Anticipating the use of an empirical correlation, we evaluate properties (except β) at $T_f = 100\ °\text{F}$:

$$\nu = 18.10\times 10^{-5}\ \text{ft}^2/\text{sec} \qquad k = 0.015\,65\ \text{Btu/hr}\cdot\text{ft}\cdot°\text{F} \qquad \mathbf{Pr} = 0.705$$

and $\beta = 1/T_\infty = 1/(535\ °\text{R})$. Then

$$\mathbf{Gr}_L\,\mathbf{Pr} = \frac{g\beta(\Delta T)L^3\,\mathbf{Pr}}{\nu^2} = \frac{(32.2\ \text{ft/sec}^2)[(125-75)\ °\text{F}](1.5\ \text{ft})^3(0.705)}{(535\ °\text{R})(18.10\times 10^{-5}\ \text{ft}^2/\text{sec})^2} = 2.19\times 10^8$$

Correlation (*I-23*) is appropriate:

$$\bar{h}L = k(0.27)(\mathbf{Gr}_L\,\mathbf{Pr})^{1/4} = (0.015\,65)(0.27)(2.19\times10^8)^{1/4} = 0.514\ \text{Btu/hr}\cdot\text{ft}\cdot{}^\circ\text{F}$$

$$q = (\bar{h}L)L(T_s - T_\infty) = (0.513)(1.5)(50) = 38.5\ \text{Btu/hr}$$

The use of $L = A/P = 0.375$ ft, rather than $L = 1.5$ ft, yields $q = 54.5$ Btu/hr.

7.57^D A horizontal square plate with its lower surface heated to 52 °C is losing heat to still air at a temperature of 24 °C; the plate is 0.46 m on a side. Find the rate of heat transfer. (The upper surface is insulated.)

▌ Anticipating the use of an empirical correlation, we evaluate properties at $T_f = 38$ °C: $\nu = 1.682\times10^{-5}\ \text{m}^2/\text{s}$, $k = 0.027\,06$ W/m·K, $\mathbf{Pr} = 0.705$, and $\beta = 1/T_\infty = 1/(297\ \text{K})$. Then

$$\mathbf{Ra}_L \equiv \mathbf{Gr}_L\,\mathbf{Pr} = \frac{g\beta(\Delta T)L^3\,\mathbf{Pr}}{\nu^2} = \frac{(9.8\ \text{m/s}^2)[(52-24)\ \text{K}](0.46\ \text{m})^3(0.705)}{(297\ \text{K})(1.682\times10^{-5}\ \text{m}^2/\text{s})^2} = 2.24\times10^8$$

Correlation (*I-23*) is appropriate:

$$\bar{h}L = k(0.27)(\mathbf{Ra}_L)^{1/4} = (0.027\,06)(0.27)(2.24\times10^8)^{1/4} = 0.894\ \text{W/m}\cdot\text{K}$$

$$q = (\bar{h}L)L(T_s - T_\infty) = (0.894)(0.46)(52-24) = 11.5\ \text{W}$$

The use of $L = A/P = 0.115$ m, rather than $L = 0.46$ m, yields $q = 16.3$ W.

7.58^D A horizontal 3-ft by 3-ft plate with its upper surface heated to 120 °F is exposed to a large body of gaseous carbon dioxide at 80 °F and a pressure of 1 atm. Estimate the heat transfer rate.

▌ Anticipating the use of an empirical correlation, we evaluate properties at $T_f = 100$ °F:

$$\nu = 9.644\times10^{-5}\ \text{ft}^2/\text{sec} \qquad k = 0.0101\ \text{Btu/hr}\cdot\text{ft}\cdot{}^\circ\text{F} \qquad \mathbf{Pr} = 0.766$$

and $\beta = 1/T_\infty = 1/(540\ {}^\circ\text{R})$. Then

$$\mathbf{Gr}_L\,\mathbf{Pr} = \frac{g\beta(\Delta T)L^3\,\mathbf{Pr}}{\nu^2} = \frac{(32.2\ \text{ft/sec}^2)[(120-80)\ {}^\circ\text{R}](3\ \text{ft})^3(0.766)}{(540\ {}^\circ\text{R})(9.644\times10^{-5}\ \text{ft}^2/\text{sec})^2} = 5.304\times10^9 \quad (\textit{turbulent})$$

Correlation (*I-22*) is appropriate:

$$\bar{h}L = k(0.14)(\mathbf{Gr}_L\,\mathbf{Pr})^{1/3} = (0.0101)(0.14)(5.304\times10^9)^{1/3} = 2.46\ \text{Btu/hr}\cdot\text{ft}\cdot{}^\circ\text{F}$$

$$q = (\bar{h}L)L(T_s - T_\infty) = (2.46)(3)(40) = 295\ \text{Btu/hr}$$

7.59^D A horizontal 1.0-m by 1.0-m plate with its upper surface heated to 49 °C is exposed to a large body of gaseous carbon dioxide at 27 °C and a pressure of 1 atm. Estimate the heat transfer rate.

▌ Anticipating the use of an empirical correlation, we evaluate properties at $T_f = 38$ °C:

$$\nu = 8.959\times10^{-6}\ \text{m}^2/\text{s} \qquad k = 0.017\,46\ \text{W/m}\cdot\text{K} \qquad \mathbf{Pr} = 0.766$$

and $\beta = 1/T_\infty = 1/(300\ \text{K})$. Then

$$\mathbf{Gr}_L\,\mathbf{Pr} = \frac{g\beta(\Delta T)L^3\,\mathbf{Pr}}{\nu^2} = \frac{(9.8\ \text{m/s}^2)[(49-27)\ \text{K}](1.0\ \text{m})^3(0.766)}{(300\ \text{K})(8.959\times10^{-6}\ \text{m}^2/\text{s})^2} = 6.858\times10^9 \quad (\textit{turbulent})$$

Correlation (*I-22*) is appropriate:

$$\bar{h}L = k(0.14)(\mathbf{Gr}_L\,\mathbf{Pr})^{1/3} = (0.017\,46)(0.14)(6.858\times10^9)^{1/3} = 4.64\ \text{W/m}\cdot\text{K}$$

$$q = (\bar{h}L)L(T_s - T_\infty) = (4.64)(1)(49-27) = 102\ \text{W}$$

7.60^D A horizontal 1-ft by 1-ft plate with its upper surface heated is exposed to water at 65 °F. The heat transfer rate from the upper surface is 3200 Btu/hr. Estimate the upper surface temperature. Measured values of β are: at 60 °F, $\beta = 0.085\times10^{-3}\ {}^\circ\text{R}^{-1}$; at 100 °F, $\beta = 0.20\times10^{-3}\ {}^\circ\text{R}^{-1}$.

▌ **First iteration.** Guess $T_s^{(0)} = 95$ °F, making $T_f^{(0)} = \frac{1}{2}(95+65) = 80$ °F. From Table B-3, at 80 °F,

$$\nu = 0.958\times10^{-5}\ \text{ft}^2/\text{sec} \qquad \mathbf{Pr} = 6.12 \qquad k = 0.351\ \text{Btu/hr}\cdot\text{ft}\cdot{}^\circ\text{F}$$

Also, interpolation in the measured liquid data at 80 °F yields $\beta \approx 0.14\times10^{-3}\ {}^\circ\text{R}^{-1}$. Thus,

$$\mathbf{Gr}_L\,\mathbf{Pr} = \frac{g\beta(\Delta T)L^3\,\mathbf{Pr}}{\nu^2} = \frac{(32.2\ \text{ft/sec})(0.14\times10^{-3}\ {}^\circ\text{R}^{-1})[(95-65)\ {}^\circ\text{R}](1\ \text{ft})^3(6.12)}{(0.958\times10^{-5}\ \text{ft}^2/\text{sec})^2} = 9.02\times10^9 \quad (\textit{turbulent})$$

so that $\bar{h}$ may be found from (*I-22*):

$$\bar{h} = \frac{k}{L}(0.14)(\mathbf{Gr}_L\,\mathbf{Pr})^{1/3} = \frac{0.351}{1}(0.14)(9.02\times10^9)^{1/3} = 103\ \text{Btu/hr}\cdot\text{ft}^2\cdot{}^\circ\text{F}$$

$$T_s^{(1)} - T_\infty = \frac{q}{\bar{h}A} = \frac{3200}{(103)(1)(1)} = 31\ {}^\circ\text{F}$$

whence $T_s^{(1)} = 96\ {}^\circ\text{F}$ and $T_f^{(1)} = 80.5\ {}^\circ\text{F} \approx T_f^{(0)}$. Further iteration is clearly unnecessary, and we accept 96 °F as the solution.

7.61^D A horizontal square plate, 0.3 m on a side, with its upper surface heated, is exposed to water at 19 °C. The heat transfer rate from the upper surface is 940 W. Estimate the upper surface temperature. Measured values of β are: at 16 °C, $\beta = 0.153\times10^{-3}$ K; at 38 °C, $\beta = 0.360\times10^{-3}\ \text{K}^{-1}$.

▌ **First iteration.** Guess $T_s^{(0)} = 35\ {}^\circ\text{C}$, making $T_f^{(0)} = \frac{1}{2}(35+19) = 27\ {}^\circ\text{C}$. From Table B-3, at 27 °C,

$$\nu = 8.84\times10^{-7}\ \text{m}^2/\text{s} \qquad \mathbf{Pr} = 6.08 \qquad k = 0.606\ \text{W/m}\cdot\text{K}$$

Also, interpolation in the measured liquid data at 27 °C yields $\beta \approx 2.56\times10^{-4}\ \text{K}^{-1}$. Thus,

$$\mathbf{Gr}_L\,\mathbf{Pr} = \frac{g\beta(\Delta T)L^3\,\mathbf{Pr}}{\nu^2} = \frac{(9.8\ \text{m/s}^2)(2.56\times10^{-4}\ \text{K})^{-1}[(35-19)\ \text{K}](0.3\ \text{m})^3(6.08)}{(8.84\times10^{-7}\ \text{m}^2/\text{s})^2} = 8.43\times10^9 \quad (\textit{turbulent})$$

so that $\bar{h}$ may be found from (*I-22*):

$$\bar{h} = \frac{k}{L}(0.14)(\mathbf{Gr}_L\,\mathbf{Pr})^{1/3} = \frac{0.606}{0.3}(0.14)(8.43\times10^9)^{1/3} = 577\ \text{W/m}^2\cdot\text{K}$$

$$T_s^{(1)} - T_\infty = \frac{q}{\bar{h}A} = \frac{940}{(577)(0.3)(0.3)} = 18.1\ {}^\circ\text{C} \qquad \text{whence} \qquad T_s^{(1)} = 37.1\ {}^\circ\text{C} \ \text{and}\ T_f^{(1)} = 28.0\ {}^\circ\text{C} \approx T_f^{(0)}$$

Further iteration is uncalled for, and we accept 37 °C as the solution.

7.62^D A very long, 6-in.-diam. horizontal cylinder is maintained at 104 °F in a large tank of glycerin, which is at 68 °F. (***a***) Determine the heat transfer rate to the glycerin per foot of cylinder length. (***b***) Repeat part (*a*), assuming vertical-flat-plate theory, with the height L equal to half the circumference.

▌ From Table B-3, at $T_f = 86\ {}^\circ\text{F}$,

$$\nu = 0.0054\ \text{ft}^2/\text{sec} \qquad k = 0.165\ \text{Btu/hr}\cdot\text{ft}\cdot{}^\circ\text{F} \qquad \mathbf{Pr} = 5380 \qquad \beta \approx 0.3\times10^{-3}\ {}^\circ\text{R}^{-1}$$

(***a***)
$$\mathbf{Gr}_D\,\mathbf{Pr} = \frac{g\beta(T_s - T_\infty)D^3\,\mathbf{Pr}}{\nu^2} = \frac{(32.2\ \text{ft/sec}^2)(0.3\times10^{-3}\ {}^\circ\text{R}^{-1})[(104-68)\ {}^\circ\text{R}](\frac{1}{2}\ \text{ft})^3(5380)}{(0.0054\ \text{ft}^2/\text{sec})^2}$$

$$= 8.02\times10^6 \quad (\textit{laminar})$$

By (*I-17*) and Table I-3 (here $\Lambda = D$),

$$\bar{h}D = kC(\mathbf{Gr}_D\,\mathbf{Pr})^a = (0.165)(0.53)(8.02\times10^6)^{1/4} = 4.66\ \text{Btu/hr}\cdot\text{ft}\cdot{}^\circ\text{F}$$

$$\frac{q}{l} = (\bar{h}D)\pi(T_s - T_\infty) = (4.65)\pi(104-68) = 526\ \text{Btu/hr}\cdot\text{ft}$$

(***b***) With $L/D = \pi/2$, we have from (*a*):

$$\mathbf{Gr}_L\,\mathbf{Pr} = (\pi/2)^3(8.02\times10^6) = 3.11\times10^7 \quad (\textit{laminar})$$

Again by (*I-17*) and Table I-3, with $\Lambda = L$,

$$\bar{h}L = kC(\mathbf{Gr}_L\,\mathbf{Pr})^a = (0.165)(0.59)(3.11\times10^7)^{1/4} = 7.3\ \text{Btu/hr}\cdot\text{ft}\cdot{}^\circ\text{F}$$

$$\frac{q}{l} = (\bar{h}L)(T_s - T_\infty) = (7.3)(36) = 263\ \text{Btu/hr}\cdot\text{ft}$$

The heat transfer from both sides of the "plate" is then $2(263) = 526\ \text{Btu/hr}\cdot\text{ft}$, in excellent agreement with (*a*).

7.63^D A very long, 15-cm-diam. horizontal cylinder is maintained at 40 °C in a large tank of glycerin, which is at 20 °C. (***a***) Determine the heat transfer to the glycerin per meter of cylinder length. (***b***) Repeat part (*a*), assuming vertical-flat-plate theory, with the height L equal to half the circumference.

From Table B-3, at $T_f = 30\ °C$,

$$\mathbf{Pr} = 5380 \qquad \nu = 5.01\times10^{-4}\ m^2/s \qquad k = 0.285\ W/m\cdot K \qquad \beta_f \approx 0.54\times10^{-3}\ K^{-1}$$

(*a*)
$$\mathbf{Gr}_D\,\mathbf{Pr} = \frac{g\beta(T_s - T_\infty)D^3\,\mathbf{Pr}}{\nu^2} = \frac{(9.8\ m/s^2)(0.54\times10^{-3}\ K^{-1})[(40-20)\ K](0.15\ m)^3(5380)}{(5.01\times10^{-4}\ m^2/s)^2}$$

$$= 7.65\times10^6 \quad (\textit{laminar})$$

By (*I-17*) and Table I-3, with $\Lambda = D$,

$$\bar{h}D = kC(\mathbf{Gr}_D\,\mathbf{Pr})^a = (0.285)(0.53)(7.65\times10^6)^{1/4} = 7.94\ W/m\cdot K$$

$$\frac{q}{l} = (\bar{h}D)\pi(T_s - T_\infty) = (7.94)\pi(40-20) = 499\ W/m$$

(*b*) With $L/D = \pi/2$, we have from (*a*):

$$\mathbf{Gr}_L\,\mathbf{Pr} = \left(\frac{\pi}{2}\right)^3(7.65\times10^6) = 2.96\times10^7 \quad (\textit{laminar})$$

Again by (*I-17*) and Table I-3, with $\Lambda = L$,

$$\bar{h}L = kC(\mathbf{Gr}_L\,\mathbf{Pr})^a = (0.285)(0.59)(2.96\times10^7)^{1/4} = 12.4\ W/m\cdot K$$

$$\frac{q}{l} = (\bar{h}L)(T_s - T_\infty) = (12.4)(20) = 248\ W/m$$

The heat transfer from both sides of the "plate" is then $2(248) = 496\ W/m$, in excellent agreement with (*a*).

7.64[D] A 4-in.-diam. horizontal cylinder is 12 ft long. Its curved surface (neglect the ends; $L/D = 36$) is at 300 °F, and the surrounding fluid is still air at 100 °F (far from the cylinder). Give three estimates of the "average" Nusselt modulus for free convective heat transfer.

From Table B-4, at $T_f = 200\ °F$, $\mathbf{Pr} = 0.694$ and $\nu = 24.21\times10^{-5}\ ft^2/sec$; also, $\beta = 1/T_\infty = 1/(560\ °R)$.

$$\mathbf{Gr}_D\,\mathbf{Pr} = \frac{g\beta(T_s - T_\infty)D^3\,\mathbf{Pr}}{\nu^2} = \frac{(32.2\ ft/sec^2)[(300-100)\ °R](\frac{4}{12}\ ft)^3(0.694)}{(560\ °R)(24.21\times10^{-5}\ ft^2/sec)^2} = 5.04\times10^6 \quad (\textit{laminar})$$

(*a*) By correlation (*I-24*) and Table I-3, $\overline{\mathbf{Nu}}_D = (0.53)(5.04\times10^6) = 25.1$.

(*b*) By (*I-25*),

$$\overline{\mathbf{Nu}}_D = \left\{0.60 + \frac{(0.387)(5.04\times10^6)^{1/6}}{[1 + (0.559/0.694)^{9/16}]^{8/27}}\right\}^2 = 23.0$$

(*c*) Figure I-3 gives, for $\log(\mathbf{Gr}_D\,\mathbf{Pr}) = 6.70$,

$$\log \overline{\mathbf{Nu}}_D \approx 1.37 \qquad \text{or} \qquad \overline{\mathbf{Nu}}_D \approx 23.4$$

7.65[D] A 10-cm-diam. horizontal cylinder is 3.6 m long. Its curved surface (neglect the ends; $L/D = 36$) is at 150 °C, and the surrounding fluid is still air at 38 °C (far from the cylinder). Give three estimates of the "average" Nusselt modulus for free convective heat transfer.

From Table B-4, at $T_f = 94\ °C$, $\mathbf{Pr} = 0.694$ and $\nu = 2.25\times10^{-5}\ m^2/s$; also, $\beta = 1/T_\infty = 1/(311\ K)$.

$$\mathbf{Gr}_D\,\mathbf{Pr} = \frac{g\beta(T_s - T_\infty)D^3\,\mathbf{Pr}}{\nu^2} = \frac{(9.8\ m/s^2)[(150-38)\ K](0.10\ m)^3(0.694)}{(311\ K)(2.25\times10^{-5}\ m^2/s)^2} = 4.84\times10^6 \quad (\textit{laminar})$$

(*a*) By correlation (*I-24*) and Table I-3, $\overline{\mathbf{Nu}}_D = (0.53)(4.84\times10^6)^{1/4} = 24.9$.

(*b*) By (*I-25*),

$$\overline{\mathbf{Nu}}_D = \left\{0.60 + \frac{(0.387)(4.84\times10^6)^{1/6}}{[1 + (0.559/0.694)^{9/16}]^{8/27}}\right\}^2 = 22.8$$

(*c*) Figure I-3 gives, for $\log(\mathbf{Gr}_D\,\mathbf{Pr}) = 6.68$,

$$\log \overline{\mathbf{Nu}}_D \approx 1.35 \qquad \text{or} \qquad \overline{\mathbf{Nu}}_D \approx 22.4$$

7.66 A 0.1-in.-diam. horizontal wire 10 in. long $(L/D = 100)$ is heated electrically so that the surface temperature is maintained at 122 °F. The wire is immersed in glycerin at 86 °F. Determine the heat lost by the wire. The measured value of β at 86 °F is 0.30×10^{-3} °R^{-1} and at 100 °F is 0.32×10^{-3} °R^{-1}.

From Table B-3, at $T_f = 104$ °F,

$$\nu = 0.0024 \text{ ft}^2/\text{sec} \qquad \mathbf{Pr} = 2450 \qquad k = 0.165 \text{ Btu/hr}\cdot\text{ft}\cdot{}^\circ\text{F}$$

Moreover, by extrapolation of the data to 104 °F, $\beta \approx 0.32 \times 10^{-3}$ °R^{-1}.

$$\mathbf{Gr}_D\,\mathbf{Pr} = \frac{g\beta(T_s - T_\infty)D^3\,\mathbf{Pr}}{\nu^2} = \frac{(32.2 \text{ ft/sec}^2)(0.32 \times 10^{-3}\ {}^\circ\text{R}^{-1})[(122-86)\ {}^\circ\text{R}][(0.1/12)\text{ ft}]^3(2450)}{(0.0024 \text{ ft}^2/\text{sec})^2} = 91.3$$

With such a small parameter, we must use Figure I-3, which gives $(\log \mathbf{Gr}_D\,\mathbf{Pr} = 1.96)$ $\log \overline{\mathbf{Nu}}_D \approx 0.32$, or $\overline{\mathbf{Nu}}_D \approx 2.1$. Then

$$q = (\bar{h}D)\pi L(T_s - T_\infty) = (k\,\overline{\mathbf{Nu}}_D)\pi L(T_s - T_\infty) = (0.165)(2.1)\pi(\tfrac{10}{12})(122-86) = 32.7 \text{ Btu/hr}$$

7.67^D A 2.5-mm-diam. horizontal wire 250 mm long $(L/D = 100)$ is heated electrically so that the surface temperature is maintained at 50 °C. The wire is immersed in glycerin at 30 °C. Determine the heat lost by the wire. The measured value of β at 30 °C is 0.54×10^{-3} K^{-1} and at 38 °C is 0.58×10^{-3} K^{-1}.

From Table B-3, at $T_f = 40$ °C,

$$\nu = 2.23 \times 10^{-4} \text{ m}^2/\text{s} \qquad \mathbf{Pr} = 2450 \qquad k \approx 0.285 \text{ W/m}\cdot\text{K}$$

Moreover, by extrapolation of the data to 40 °C, $\beta \approx 0.58 \times 10^{-3}$ K^{-1}.

$$\mathbf{Gr}_D\,\mathbf{Pr} = \frac{g\beta(T_s - T_\infty)D^3\,\mathbf{Pr}}{\nu^2} = \frac{(9.8 \text{ m/s}^2)(0.58 \times 10^{-3}\text{ K}^{-1})[(50-30)\text{ K}](2.5 \times 10^{-3}\text{ m})^3(2450)}{(2.23 \times 10^{-4}\text{ m}^2/\text{s})^2} = 87.4$$

and $\log(\mathbf{Gr}_D\,\mathbf{Pr}) = 1.94$. From Figure I-3, $\log \overline{\mathbf{Nu}}_D \approx 0.32$, or $\overline{\mathbf{Nu}}_D \approx 2.1$. Thus,

$$q = (\bar{h}D)\pi L(T_s - T_\infty) = (k\,\overline{\mathbf{Nu}}_D)\pi L(T_s - T_\infty) = (0.285)(2.1)\pi(0.250)(20) = 9.39 \text{ W}$$

7.68^D A vertical, heated, 1.0-in.-diam. rod 1.0 ft long is at a uniform temperature of 350 °F and exposed to air at 50 °F and atmospheric pressure. Determine the rate of heat transfer from the rod.

From Table B-4, at $T_f = 200$ °F,

$$\nu = 24.21 \times 10^{-5} \text{ ft}^2/\text{sec} \qquad \mathbf{Pr} = 0.694 \qquad k = 0.0180 \text{ Btu/hr}\cdot\text{ft}\cdot{}^\circ\text{F}$$

Also, $\beta = 1/T_\infty = 1/(510\ {}^\circ\text{R})$. Thus,

$$\mathbf{Gr}_L = \frac{g\beta(T_s - T_\infty)L^3}{\nu^2} = \frac{(32.2 \text{ ft/sec}^2)[(350-50)\ {}^\circ\text{R}](1.0 \text{ ft})^3}{(510\ {}^\circ\text{R})(24.21 \times 10^{-5}\text{ ft}^2/\text{sec})^2} = 3.23 \times 10^8$$

whence $\mathbf{Gr}_L\,\mathbf{Pr} = (3.23 \times 10^8)(0.694) = 2.24 \times 10^8$ (*laminar flow*). Making the test (*I-18*),

$$\frac{35}{\mathbf{Gr}_L^{1/4}} = \frac{35}{(3.23 \times 10^8)^{1/4}} \approx \frac{3}{12} > \frac{1}{12} = \frac{D}{L}$$

we see that flat-plate theory is invalid here; we must use a correlation for vertical cylinders. Because

$$\mathbf{Gr}_D\,\mathbf{Pr}\,\frac{D}{L} = (\mathbf{Gr}_L\,\mathbf{Pr})\left(\frac{D}{L}\right)^4 = (2.24 \times 10^8)(\tfrac{1}{12})^4 \approx 10^4$$

and $\log 10^4 = 4$, use of Figure I-2 is ruled out. Instead, as an approximation we can apply (*I-17*), with constants taken from the bottom of Table I-3:

$$\Lambda = \frac{L_v L_h}{L_v + L_h} = \frac{(1)(\frac{1}{12})}{1 + (\frac{1}{12})} = \frac{1}{13} \text{ ft}$$

$$\mathbf{Gr}_\Lambda\,\mathbf{Pr} = \left(\frac{\frac{1}{13}}{1.0}\right)^3 (2.24 \times 10^8) = 1.02 \times 10^5$$

$$\bar{h} = \frac{kC(\mathbf{Gr}_L\,\mathbf{Pr})^a}{\Lambda} = \frac{(0.0180)(0.60)(1.02 \times 10^5)^{1/4}}{\frac{1}{13}} = 2.51 \text{ Btu/hr}\cdot\text{ft}^2\cdot{}^\circ\text{F}$$

$$q = \bar{h}A(T_s - T_\infty) = (2.51)[\pi(\tfrac{1}{12})(1.0)](350-50) = 197 \text{ Btu/hr}$$

7.69^D A vertical, heated, 25-mm-diam. rod 300 mm long is at a uniform temperature of 177 °C and exposed to air at 10 °C and atmospheric pressure. Determine the rate of heat transfer from the rod.

▮ From Table B-4, at $T_f = 93.5$ °C,

$$\nu = 2.25 \times 10^{-5}\ \mathrm{m^2/s} \qquad \mathbf{Pr} = 0.694 \qquad k = 0.0311\ \mathrm{W/m\cdot K}$$

and $\beta = 1/T_\infty = 1/(283\ \mathrm{K})$. Then

$$\mathbf{Gr}_L = \frac{g\beta(T_s - T_\infty)L^3}{\nu^2} = \frac{(9.8\ \mathrm{m/s^2})[(177-10)\ \mathrm{K}](0.300\ \mathrm{m})^3}{(283\ \mathrm{K})(2.25\times 10^{-5}\ \mathrm{m^2/s})^2} = 3.08 \times 10^8$$

whence $\mathbf{Gr}_L\,\mathbf{Pr} = (3.08\times 10^8)(0.694) = 2.14 \times 10^8$ (*laminar flow*). Making the test (*I-18*),

$$\frac{35}{\mathbf{Gr}_L^{1/4}} = \frac{35}{(3.08\times 10^8)^{1/4}} \approx \frac{3}{12} > \frac{1}{12} = \frac{D}{L}$$

we see that flat-plate theory is invalid here; we must use a correlation for vertical cylinders. Because

$$\mathbf{Gr}_D\,\mathbf{Pr}\,\frac{D}{L} = (\mathbf{Gr}_L\,\mathbf{Pr})\left(\frac{D}{L}\right)^4 = (2.14\times 10^8)\left(\frac{1}{12}\right)^4 \approx 10^4$$

and $\log 10^4 = 4$, use of Figure I-2 is ruled out. Instead, as an approximation we can apply (*I-17*), with constants taken from the bottom of Table I-3:

$$\Lambda = \frac{L_v L_h}{L_v + L_h} = \frac{(0.300)(0.025)}{0.300 + 0.025} = 0.023\ \mathrm{m}$$

$$\mathbf{Gr}_\Lambda\,\mathbf{Pr} = \left(\frac{0.023}{0.300}\right)^3 (2.14\times 10^8) = 9.64 \times 10^4$$

$$\bar{h} = \frac{kC(\mathbf{Gr}_L\,\mathbf{Pr})^a}{\Lambda} = \frac{(0.0311)(0.60)(9.64\times 10^4)^{1/4}}{0.023} = 14.3\ \mathrm{W/m^2\cdot K}$$

$$q = \bar{h}A(T_s - T_\infty) = (14.3)[\pi(0.025)(0.300)](177-10) = 56.3\ \mathrm{W}$$

7.70^D A 0.001-in.-diam. vertical wire at 400 °F is exposed to air at 70 °F and standard atmospheric pressure. Determine the rate of heat transfer for a 12-in. length of the wire.

▮ Interpolation in Table B-4, at $T_f = 235$ °F, with $g = 32.2\ \mathrm{ft/sec^2}$, gives

$$\frac{g\beta}{\nu^2} = 0.875\times 10^6\ \mathrm{°R^{-1}\cdot ft^{-3}} \qquad \mathbf{Pr} = 0.691 \qquad k = 0.0188\ \mathrm{Btu/hr\cdot ft\cdot °F}$$

Hence, using Figure I-2,

$$\mathbf{Gr}_D\,\mathbf{Pr}\,\frac{D}{L} = \left(\frac{g\beta}{\nu^2}\right)(T_s - T_\infty)\,\mathbf{Pr}\,\frac{D^4}{L}$$

$$= (0.875\times 10^6)(400-70)(0.691)\left[\frac{(0.001/12)^4}{1.0}\right] = 9.62\times 10^{-9}$$

$$\log(9.62\times 10^{-9}) = -8.017$$

$$\overline{\mathbf{Nu}}_D \approx 0.32$$

$$q = k\,\overline{\mathbf{Nu}}_D\,\pi L(T_s - T_\infty) \approx (0.0188)(0.32)\pi(1.0)(400-70) = 6.24\ \mathrm{Btu/hr}$$

7.71^D A 25-μm-diam. vertical wire at 205 °C is exposed to air at 21 °C and standard atmospheric pressure. Determine the rate of heat transfer for a 0.30-m length of the wire.

▮ Interpolation in Table B-4, at $T_f = 113$ °C, with $g = 9.8\ \mathrm{m/s^2}$, gives

$$\frac{g\beta}{\nu^2} = 5.55\times 10^7\ \mathrm{K^{-1}\cdot m^{-3}} \qquad \mathbf{Pr} = 0.691 \qquad k = 0.0326\ \mathrm{W/m\cdot K}$$

Hence, using Figure I-2,

$$\mathbf{Gr}_D\,\mathbf{Pr}\,\frac{D}{L} = \left(\frac{g\beta}{\nu^2}\right)(T_s - T_\infty)\,\mathbf{Pr}\,\frac{D^4}{L}$$

$$= (5.55\times 10^7)(205-21)(0.691)\left[\frac{(25\times 10^{-6})^4}{0.30}\right] = 9.19\times 10^{-9}$$

$$\log(9.19 \times 10^{-9}) = -8.04$$
$$\overline{\mathbf{Nu}}_D \approx 0.31$$
$$q = k\,\overline{\mathbf{Nu}}_D\,\pi L(T_s - T_\infty) \approx (0.0326)(0.31)\pi(0.30)(205 - 21) = 1.75\ \text{W}$$

7.72[D] Estimate the rate of heat transfer from a 0.003-in.-diam. wire 1 ft long placed horizontally in light oil at 65 °F. The surface temperature of the wire is maintained at 275 °F and measured parameters at $T_f = 170$ °F are

$$\frac{g\beta}{\nu^2} = 3.36 \times 10^6\ °\text{R}^{-1}\cdot\text{ft}^{-3} \qquad \mathbf{Pr} = 98 \qquad k = 0.075\ \text{Btu/hr}\cdot\text{ft}\cdot°\text{F}$$

▮ $$\mathbf{Gr}_D\,\mathbf{Pr} = \left(\frac{g\beta}{\nu^2}\right)(T_s - T_\infty)D^3\,\mathbf{Pr} = (3.36 \times 10^6)(210)\left(\frac{0.003}{12}\right)^3(98) = 1.08$$

From Table I-3, the guideline is to use Figure I-3; thus,

$$\log 1.08 = 0.0334$$
$$\log \overline{\mathbf{Nu}}_D \approx 0.06$$
$$\overline{\mathbf{Nu}}_D \approx 1.15$$
$$q = k\,\overline{\mathbf{Nu}}_D\,\pi L(T_s - T_\infty) \approx (0.075)(1.15)\pi(1)(210) = 56.9\ \text{Btu/hr}$$

7.73[D] Estimate the rate of heat transfer from a 75-μm-diam. wire 0.3 m long placed horizontally in light oil at 19 °C. The surface temperature of the wire is maintained at 135 °C and measured properties at $T_f = 77$ °C are

$$\frac{g\beta}{\nu^2} = 2.138 \times 10^8\ \text{K}^{-1}\cdot\text{m}^{-3} \qquad \mathbf{Pr} = 98 \qquad k = 0.130\ \text{W/m}\cdot\text{K}$$

▮ $$\mathbf{Gr}_D\,\mathbf{Pr} = \left(\frac{g\beta}{\nu^2}\right)(T_s - T_\infty)D^3\,\mathbf{Pr} = (2.138 \times 10^8)(116)(75 \times 10^{-6})^3(98) = 1.025$$

and $\log 1.025 = 0.0107$. From Figure I-3,

$$\log \overline{\mathbf{Nu}}_D \approx 0.06$$
$$\overline{\mathbf{Nu}}_D \approx 1.15$$
$$q = k\,\overline{\mathbf{Nu}}_D\,\pi L(T_s - T_\infty) \approx (0.130)(1.15)\pi(0.3)(116) = 16.34\ \text{W}$$

7.74[D] The surface temperature of a 6-in.-OD, 10-ft-long pipe carrying steam in a large room is 200 °F. The air in the room is at 14.7 psia and 100 °F. Determine the free convective heat loss from the pipe when it is **(*a*)** vertical and **(*b*)** horizontal.

▮ From Table B-4, by interpolation at $T_f = 150$ °F and assuming standard g,

$$\frac{g\beta}{\nu^2} = \frac{g}{T_\infty \nu^2} = 1.284 \times 10^6\ °\text{R}^{-1}\cdot\text{ft}^{-3} \qquad \mathbf{Pr} = 0.699 \qquad k = 0.0169\ \text{Btu/hr}\cdot\text{ft}\cdot°\text{F}$$

(*a*)

$$\mathbf{Gr}_L\,\mathbf{Pr} = \left(\frac{g\beta}{\nu^2}\right)(T_s - T_\infty)L^3\,\mathbf{Pr} = (1.284 \times 10^6)(200 - 100)(10^3)(0.699)$$
$$= 8.98 \times 10^{10} \quad (\textit{turbulent})$$

By (*I-17*), wherein $\Lambda = L$,

$$\overline{\mathbf{Nu}} = (0.13)(8.98 \times 10^{10})^{1/3} = 582$$
$$q = k\,\overline{\mathbf{Nu}}\,\pi D(T_s - T_\infty) = (0.0169)(582)\pi(\tfrac{6}{12})(200 - 100) = 1540\ \text{Btu/hr}$$

(*b*)

$$\mathbf{Gr}_D\,\mathbf{Pr} = (\tfrac{1}{20})^3(8.98 \times 10^{10}) = 1.12 \times 10^7 \quad (\textit{laminar})$$

By (*I-17*), with $\Lambda = D$,

$$\overline{\mathbf{Nu}} = (0.53)(1.12 \times 10^7)^{1/4} = 30.7$$
$$q = k\,\overline{\mathbf{Nu}}\,\pi L(T_s - T_\infty) = (0.0169)(30.7)\pi(10)(200 - 100) = 1630\ \text{Btu/hr}$$

Surprisingly, this laminar flow transfers more heat than does the turbulent flow of (*a*).

7.75[D] The surface temperature of a 15-cm-OD, 3-m-long pipe carrying steam in a large room is 94 °C. The air in the room is at 101.3 kPa and 38 °C. Determine the free convective heat loss from the pipe when it is **(a)** vertical and **(b)** horizontal.

▮ From Table B-4, by interpolation at $T_f = 66\ °\text{C}$ and assuming standard g,

$$\frac{g\beta}{\nu^2} = \frac{g}{T_\infty \nu^2} = 8.16 \times 10^7\ \text{K}^{-1}\cdot\text{m}^{-3} \qquad \mathbf{Pr} = 0.699 \qquad k = 0.0292\ \text{W/m}\cdot\text{K}$$

(a)

$$\mathbf{Gr}_L\ \mathbf{Pr} = \left(\frac{g\beta}{\nu^2}\right)(T_s - T_\infty)L^3\ \mathbf{Pr} = (8.16 \times 10^7)(56)(3^3)(0.699)$$

$$= 8.62 \times 10^{10} \quad (\textit{turbulent})$$

By (*I-17*), with $\Lambda = L$,

$$\overline{\mathbf{Nu}} = (0.13)(8.62 \times 10^{10})^{1/3} = 574$$

$$q = k\,\overline{\mathbf{Nu}}\,\pi D(T_s - T_\infty) = (0.0292)(574)\pi(0.15)(56) = 442\ \text{W}$$

(b)

$$\mathbf{Gr}_D\ \mathbf{Pr} = \left(\frac{0.15}{3}\right)^3(8.62 \times 10^{10}) = 1.08 \times 10^7 \quad (\textit{laminar})$$

By (*I-17*), with $\Lambda = D$,

$$\overline{\mathbf{Nu}} = (0.53)(1.08 \times 10^7)^{1/4} = 30.4$$

$$q = k\,\overline{\mathbf{Nu}}\,\pi L(T_s - T_\infty) = (0.0292)(30.4)\pi(3)(56) = 469\ \text{W}$$

Surprisingly, this laminar flow transfers more heat than does the turbulent flow of (*a*).

7.76[D] A 1-ft-square plate is inclined from the vertical at an angle of 20°. The surface temperature of the plate is 140 °F, and the temperature of the surrounding air is 60 °F. **(a)** Determine the rate of heat transfer from the plate. **(b)** Repeat part (*a*) with water instead of air; $\beta = 0.085 \times 10^{-3}\ °\text{R}^{-1}$ at 60 °F.

▮ (*a*) From Table B-4, for air at $T_f = 100\ °\text{F}$ and assuming $g = 32.2\ \text{ft/sec}^2$,

$$\frac{g\beta}{\nu^2} = \frac{g}{T_\infty \nu^2} = 1.89 \times 10^6\ °\text{R}^{-1}\cdot\text{ft}^{-3} \qquad \mathbf{Pr} = 0.706 \qquad k = 0.0156\ \text{Btu/hr}\cdot\text{ft}\cdot°\text{F}$$

The Grashof number is modified by the factor $\cos\theta$ (since the acceleration of gravity normal to the plate is $g\cos\theta$):

$$\mathbf{Gr}_L\ \mathbf{Pr} = \left(\frac{g\beta}{\nu^2}\right)(\cos\theta)(T_s - T_\infty)L^3\ \mathbf{Pr}$$

$$= (1.89 \times 10^6)(\cos 20°)(140 - 60)(1^3)(0.706) = 10.0 \times 10^7 \quad (\textit{laminar})$$

Then, by (*I-17*),

$$\overline{\mathbf{Nu}} = (0.59)(10.0 \times 10^7)^{1/4} = 59.0$$

$$q = k\,\overline{\mathbf{Nu}}\,w(T_s - T_\infty) = (0.0156)(59.0)(1)(140 - 60) = 73.6\ \text{Btu/hr}$$

(*b*) From Table B-3, for saturated liquid water at $T_f = 100\ °\text{F}$ and with standard gravity,

$$\frac{g\beta}{\nu^2} = 4.87 \times 10^7\ °\text{R}^{-1}\cdot\text{ft}^{-3} \qquad \mathbf{Pr} = 4.64 \qquad k = 0.361\ \text{Btu/hr}\cdot\text{ft}\cdot°\text{F}$$

Therefore, using (*I-17*),

$$\mathbf{Gr}_L\ \mathbf{Pr} = \left(\frac{g\beta}{\nu^2}\right)(\cos\theta)(T_s - T_\infty)L^3\ \mathbf{Pr}$$

$$= (4.87 \times 10^7)(\cos 20°)(140 - 60)(1^3)(4.64) = 1.70 \times 10^{10} \quad (\textit{turbulent})$$

$$\overline{\mathbf{Nu}} = (0.13)(1.70 \times 10^{10})^{1/3} = 334$$

$$q = k\,\overline{\mathbf{Nu}}\,w(T_s - T_\infty) = (0.361)(334)(1)(140 - 60) = 9646\ \text{Btu/hr}$$

7.77[D] A 30-cm-square plate is inclined from the vertical at an angle of 20°. The surface temperature of the plate is 60 °C, and the temperature of the surrounding air is 16 °C. **(a)** Determine the rate of heat transfer from the plate. **(b)** Repeat part (*a*) with water instead of air; $\beta = 0.153 \times 10^{-3}\ \text{K}^{-1}$ at 16 °C.

▌ (*a*) From Table B-4, for air at $T_f = 38\ °\text{C}$ and assuming $g = 9.8\ \text{m/s}^2$,

$$\frac{g\beta}{\nu^2} = \frac{g}{T_\infty \nu^2} = 1.20 \times 10^8\ \text{K}^{-1} \cdot \text{m}^{-3} \qquad \mathbf{Pr} = 0.695 \qquad k = 0.0271\ \text{W/m} \cdot \text{K}$$

Therefore, using (*I-17*),

$$\mathbf{Gr}_L\ \mathbf{Pr} = \left(\frac{g\beta}{\nu^2}\right)(\cos\theta)(T_s - T_\infty)L^3\ \mathbf{Pr}$$

$$= (1.20 \times 10^8)(\cos 20°)(60 - 16)(0.30)^3(0.695) = 9.32 \times 10^7 \quad (\textit{laminar})$$

$$\overline{\mathbf{Nu}} = (0.59)(9.32 \times 10^7)^{1/4} = 58.0$$

$$q = k\,\overline{\mathbf{Nu}}\,w(T_s - T_\infty) = (0.0271)(58.0)(0.30)(60 - 16) = 20.75\ \text{W}$$

(*b*) From Table B-3, for saturated liquid water at $T_f = 38\ °\text{C}$ and with standard gravity,

$$\frac{g\beta}{\nu^2} = 3.122 \times 10^9\ \text{K}^{-1} \cdot \text{m}^{-3} \qquad \mathbf{Pr} = 4.61 \qquad k = 0.625\ \text{W/m} \cdot \text{K}$$

Therefore, using (*I-17*),

$$\mathbf{Gr}_L\ \mathbf{Pr} = \left(\frac{g\beta}{\nu^2}\right)(\cos\theta)(T_s - T_\infty)L^3\ \mathbf{Pr}$$

$$= (3.122 \times 10^9)(\cos 20°)(60 - 16)(0.30)^3(4.61) = 1.61 \times 10^{10} \quad (\textit{turbulent})$$

$$\overline{\mathbf{Nu}} = (0.13)(1.61 \times 10^{10})^{1/3} = 328$$

$$q = k\,\overline{\mathbf{Nu}}\,w(T_s - T_\infty) = (0.625)(328)(0.30)(60 - 16) = 2706\ \text{W}$$

7.78[D] A 9-in.-long flat plate 1 ft in horizontal width is inclined from the vertical at an angle of 15°. The plate is maintained at a constant temperature of 150 °F and is exposed to gaseous carbon dioxide at 50 °F and 14.7 psia. Determine the rate of heat transfer from the plate, if $g = 32.2\ \text{ft/sec}^2$.

▌ From Table B-4, at $T_f = 100\ °\text{F}$,

$$\nu = 9.64 \times 10^{-5}\ \text{ft}^2/\text{sec} \qquad k = 0.0101\ \text{Btu/hr} \cdot \text{ft} \cdot °\text{F} \qquad \mathbf{Pr} = 0.767 \approx 0.77$$

and, at T_∞, $\beta = 1/(510\ °\text{R})$. Hence,

$$\mathbf{Gr}_L\ \mathbf{Pr} = \frac{(g\cos\theta)\beta(T_s - T_\infty)L^3\ \mathbf{Pr}}{\nu^2} = \frac{(32.2\cos 15°)(150 - 50)(\frac{9}{12})^3(0.77)}{(510)(9.64 \times 10^{-5})^2} = 2.13 \times 10^8 \quad (\textit{laminar})$$

and (*I-17*) gives, with $\Lambda = L$,

$$\overline{\mathbf{Nu}} = (0.59)(2.13 \times 10^8)^{1/4} = 71.3$$

$$q = k\,\overline{\mathbf{Nu}}\,w(T_s - T_\infty) = (0.0101)(71.3)(1)(150 - 50) = 72.0\ \text{Btu/hr}$$

7.79[D] A 0.23-m-long flat plate 0.3 m wide (horizontal dimension) is inclined from the vertical at an angle of 15°. The plate is maintained at a constant temperature of 66 °C and is exposed to gaseous carbon dioxide at 10 °C and 1 atm pressure. Determine the rate of heat transfer from the plate, if $g = 9.8\ \text{m/s}^2$.

▌ From Table B-4, at $T_f = 38\ °\text{C}$,

$$\nu = 8.96 \times 10^{-6}\ \text{m}^2/\text{s} \qquad k = 0.0175\ \text{W/m} \cdot \text{K} \qquad \mathbf{Pr} = 0.77$$

Also, $\beta = 1/T_\infty = 1/(283\ \text{K})$. Then

$$\mathbf{Gr}_L\ \mathbf{Pr} = \frac{(g\cos\theta)\beta(T_s - T_\infty)L^3\ \mathbf{Pr}}{\nu^2} = \frac{(9.8\cos 15°)(66 - 10)(0.23)^3(0.77)}{(283)(8.96 \times 10^{-6})^2} = 2.21 \times 10^8 \quad (\textit{laminar})$$

and (*I-17*) gives, with $\Lambda = L$,

$$\overline{\mathbf{Nu}} = (0.59)(2.21 \times 10^8)^{1/4} = 72.0$$

$$q = k\,\overline{\mathbf{Nu}}\,w(T_s - T_\infty) = (0.0175)(72.0)(0.3)(66 - 10) = 21.2\ \text{W}$$

7.80[D] A smooth rock which approximates a 3-in.-diam. sphere is placed in still water at 60 °F. Using the rather crude method of miscellaneous shapes, estimate the initial rate of heat transfer, if the initial temperature of the rock is 100 °F. Assume standard gravitational acceleration and a value of β at 80 °F (the initial film temperature) of 0.15×10^{-3} °R^{-1}.

From Table B-3, at $T_f = 80$ °F,

$$\nu = 0.958 \times 10^{-5}\ \text{ft}^2/\text{sec} \qquad k = 0.351\ \text{Btu/hr}\cdot\text{ft}\cdot°\text{F} \qquad \mathbf{Pr} = 6.13$$

so that $g\beta/\nu^2 = 5.26 \times 10^7\ °\text{R}^{-1}\cdot\text{ft}^{-3}$. Using the method of miscellaneous shapes (Table I-3),

$$\Lambda = \frac{L_v L_h}{L_v + L_h} = \frac{(3)(3)}{3+3} = 1.5\ \text{in.} \quad (\text{i.e.,}\ \ \Lambda = r)$$

$$\mathbf{Gr}_r\,\mathbf{Pr} = \left(\frac{g\beta}{\nu^2}\right)(T_s - T_\infty)r^3\,\mathbf{Pr} = (5.26 \times 10^7)(100 - 60)\left(\frac{1.5}{12}\right)^3(6.13)$$

$$= 2.52 \times 10^7 \quad (\textit{laminar})$$

$$\overline{\mathbf{Nu}} = (0.60)(2.52 \times 10^7)^{1/4} = 42.5$$

$$q = k\,\overline{\mathbf{Nu}}\,(4\pi r)(T_s - T_\infty) = (0.351)(42.5)(4\pi)\left(\frac{1.5}{12}\right)(100 - 60) = 937\ \text{Btu/hr}$$

7.81[D] Repeat Problem 7.80 using the Churchill correlation, (*I-26*).

From Problem 7.80, $\mathbf{Gr}_D\,\mathbf{Pr} = 2^3\,\mathbf{Gr}_r\,\mathbf{Pr} = 2.02 \times 10^8$; (*I-26*) then gives

$$\overline{\mathbf{Nu}}_D = 2 + \frac{0.589(2.02 \times 10^8)^{1/4}}{[1 + (0.469/6.13)^{9/16}]^{4/9}} = 65.9$$

$$q = k\,\overline{\mathbf{Nu}}_D\,\pi D(T_s - T_\infty) = (0.351)(65.9)\pi(\tfrac{3}{12})(100 - 60) = 727\ \text{Btu/hr}$$

This value is about 22 percent less than the approximate value given by the miscellaneous-shapes approach in Problem 7.80. The present answer is more accurate, since it was obtained with a correlation for spheres.

7.82[D] A smooth rock which approximates a 75-mm-diam. sphere is placed in quiescent water at 16 °C. Using the rather crude method of miscellaneous shapes, estimate the initial rate of heat transfer, if the initial temperature of the rock is 38 °C. Assume standard gravitational acceleration and a value of β at 27 °C (the initial film temperature) of $0.27 \times 10^{-3}\ \text{K}^{-1}$.

From Table B-3, at $T_f = 27$ °C,

$$\nu = 0.89 \times 10^{-6}\ \text{m}^2/\text{s} \qquad k = 0.607\ \text{W/m}\cdot\text{K} \qquad \mathbf{Pr} = 6.13$$

whence $g\beta/\nu^2 = 3.34 \times 10^9\ \text{K}^{-1}\cdot\text{m}^{-3}$. Using the method of miscellaneous shapes (Table I-3),

$$\Lambda = \frac{L_v L_h}{L_v + L_h} = \frac{(75)(75)}{75+75} = 37.5\ \text{mm} \quad (\text{i.e.,}\ \ \Lambda = r)$$

$$\mathbf{Gr}_r\,\mathbf{Pr} = \left(\frac{g\beta}{\nu^2}\right)(T_s - T_\infty)r^3\,\mathbf{Pr} = (3.34 \times 10^9)(38 - 16)(0.0375)^3(6.13)$$

$$= 2.37 \times 10^7 \quad (\textit{laminar})$$

$$\overline{\mathbf{Nu}} = (0.60)(2.37 \times 10^7)^{1/4} = 41.9$$

$$q = k\,\overline{\mathbf{Nu}}\,(4\pi r)(T_s - T_\infty) = (0.607)(41.9)(4\pi)(0.0375)(38 - 16) = 264\ \text{W}$$

7.83[D] Repeat Problem 7.82 using the Churchill correlation, (*I-26*).

From Problem 7.82, $\mathbf{Gr}_D\,\mathbf{Pr} = 2^3\,\mathbf{Gr}_r\,\mathbf{Pr} = 1.90 \times 10^8$; (*I-26*) then gives

$$\overline{\mathbf{Nu}}_D = 2 + \frac{0.589(1.90 \times 10^8)^{1/4}}{[1 + (0.469/6.13)^{9/16}]^{4/9}} = 64.9$$

$$q = k\,\overline{\mathbf{Nu}}_D\,\pi D(T_s - T_\infty) = (0.607)(64.9)\pi(0.075)(38 - 16) = 204\ \text{W}$$

This value is about 23 percent less than the approximate value given by the miscellaneous-shapes approach in Problem 7.82. The present answer is more accurate, since it was obtained with a correlation for spheres.

7.3 EMPIRICAL CORRELATIONS: ENCLOSED REGIONS

7.84[D] Two horizontal surfaces separated by a 5-in. layer of air have upper and lower temperatures of 80 °F and 170 °F, respectively. Determine the rate of heat loss per square foot of the hotter surface.

| From Table B-4, at $T_{avg} = \frac{1}{2}(T_h + T_c) = 125\ °F$,

$$\nu = 19.63 \times 10^{-5}\ ft^2/sec \qquad k = 0.016\,26\ Btu/hr \cdot ft \cdot °F \qquad \mathbf{Pr} = 0.703$$

With $\beta = 1/T_{avg} = 1/(585\ °R)$ and standard g,

$$\mathbf{Gr}_b = \frac{g\beta(T_h - T_c)b^3}{\nu^2} = \frac{(32.2\ ft/sec^2)[(170 - 80)\ °R](\frac{5}{12}\ ft)^3}{(585\ °R)(19.63 \times 10^{-5}\ ft^2/sec)^2} = 9.3 \times 10^6$$

By (*I-29*),

$$\overline{\mathbf{Nu}}_b = (0.068)\ \mathbf{Gr}_b^{1/3} = (0.068)(9.3 \times 10^6)^{1/3} = 14.3$$

$$\bar{h} = \frac{\overline{\mathbf{Nu}}_b\, k}{b} = \frac{(14.3)(0.016\,26)}{\frac{5}{12}} = 0.558\ Btu/hr \cdot ft^2 \cdot °F$$

$$\frac{q}{A} = \bar{h}(T_h - T_c) = (0.558)(170 - 80) = 50.2\ Btu/hr \cdot ft^2$$

7.85[D] Two horizontal surfaces separated by a 125-mm layer of air have upper and lower temperatures of 300 K and 350 K, respectively. Determine the rate of heat exchange per square meter between the two surfaces.

| From Table B-4, at $T_{avg} = \frac{1}{2}(T_h + T_c) = 325\ K = 52\ °C$,

$$\nu = 1.82 \times 10^{-5}\ m^2/s \qquad k = 0.0281\ W/m \cdot K \qquad \mathbf{Pr} = 0.703$$

With $\beta = 1/T_{avg} = 1/(325\ K)$ and standard g,

$$\mathbf{Gr}_b = \frac{g\beta(T_h - T_c)b^3}{\nu^2} = \frac{(9.8\ m/s^2)[(350 - 300)\ K](0.125\ m)^3}{(325\ K)(1.82 \times 10^{-5}\ m^2/s)^2} = 8.89 \times 10^6$$

By (*I-29*),

$$\overline{\mathbf{Nu}}_b = (0.068)\ \mathbf{Gr}_b^{1/3} = (0.068)(8.89 \times 10^6)^{1/3} = 14.1$$

$$\bar{h} = \frac{\overline{\mathbf{Nu}}_b\, k}{b} = \frac{(14.1)(0.0281)}{0.125} = 3.17\ W/m^2 \cdot K$$

$$\frac{q}{A} = \bar{h}(T_h - T_c) = (3.17)(350 - 300) = 158.5\ W/m^2$$

7.86[D] Two horizontal surfaces with air between them are separated by 2 in. Calculate the heat flux ($Btu/hr \cdot ft^2$) if (*a*) the upper surface is at 110 °F and the lower is at 90 °F and (*b*) the upper surface is at 90 °F and the lower is at 110 °F.

| (*a*) From Table B-4, at $T_{avg} = 100\ °F$, $k = 0.0156\ Btu/hr \cdot ft \cdot °F$. Then, by (*I-28*) and (*I-27*),

$$\frac{q}{A} = k\left(\frac{T_h - T_c}{b}\right) = (0.0156)\left(\frac{110 - 90}{\frac{2}{12}}\right) = 1.872\ Btu/hr \cdot ft^2$$

(*b*) From Table B-4, at $T_{avg} = 100\ °F$, $\nu = 18.10 \times 10^{-5}\ ft^2/sec$ and $\mathbf{Pr} = 0.706$; also, $\beta = 1/T_{avg} = 1/(560\ °R)$. Therefore,

$$\mathbf{Gr}_b = \frac{g\beta(T_h - T_c)b^3}{\nu^2} = \frac{(32.2\ ft/sec^2)[(110 - 90)\ °R](\frac{2}{12}\ ft)^3}{(560\ °R)(18.10 \times 10^{-5}\ ft^2/sec)^2} = 1.63 \times 10^5$$

By (*I-29*),

$$\overline{\mathbf{Nu}}_b = (0.195)\ \mathbf{Gr}_b^{1/4} = (0.195)(1.63 \times 10^5)^{1/4} = 3.92$$

$$\bar{h} = \frac{\overline{\mathbf{Nu}}_b\, k}{b} = \frac{(3.92)(0.0156)}{\frac{2}{12}} = 0.367\ Btu/hr \cdot ft^2 \cdot °F$$

$$\frac{q}{A} = \bar{h}(T_h - T_c) = (0.367)(20) = 7.34\ Btu/hr \cdot ft^2$$

7.87^D Two horizontal surfaces with air between them are separated by 5 cm. Calculate the heat flux (W/m^2) if (*a*) the upper surface is at 44 °C and the lower is at 32 °C, and (*b*) the upper surface is at 32 °C and the lower is at 44 °C.

▮ (*a*) From Table B-4, at $T_{avg} = 38$ °C, $k = 0.0270$ W/m·K. Then, by (*I-28*) and (*I-27*),

$$\frac{q}{A} = k\left(\frac{T_h - T_c}{b}\right) = (0.0270)\left(\frac{44 - 32}{0.05}\right) = 6.48 \text{ W/m}^2$$

(*b*) From Table B-4, at $T_{avg} = 38$ °C, $\nu = 1.68 \times 10^{-5}$ m^2/s and $\mathbf{Pr} = 0.706$; also, $\beta = 1/T_{avg} = 1/(311 \text{ K})$. Therefore,

$$\mathbf{Gr}_b = \frac{g\beta(T_h - T_c)b^3}{\nu^2} = \frac{(9.8 \text{ m/s}^2)[(44-32)\text{ K}](0.05 \text{ m})^3}{(311 \text{ K})(1.68 \times 10^{-5} \text{ m}^2/\text{s})^2} = 1.67 \times 10^5$$

By (*I-29*),

$$\overline{\mathbf{Nu}} = (0.195)\,\mathbf{Gr}_b^{1/4} = (0.195)(1.67 \times 10^5)^{1/4} = 3.94$$

$$\bar{h} = \frac{\overline{\mathbf{Nu}}_b\, k}{b} = \frac{(3.94)(0.027)}{0.05} = 2.13 \text{ W/m}^2\cdot\text{K}$$

$$\frac{q}{A} = \bar{h}(T_h - T_c) = (2.13)(44 - 32) = 25.5 \text{ W/m}^2$$

7.88^D In a laboratory experiment, two horizontal surfaces which are separated by a 1-in. layer of water have upper and lower temperatures of 75 °F and 105 °F, respectively. At the location of the laboratory the measured value of g, together with experimentally obtained values of β, ρ, and μ at $T_{avg} = 90$ °F, yields the value

$$\frac{g\beta\rho^2}{(\mu g_c)^2} = 8.5 \times 10^7 \text{ °R}^{-1}\cdot\text{ft}^{-3}$$

Determine the heat flux in Btu/hr·ft^2.

▮
$$\mathbf{Gr}_b = \frac{g\beta}{\nu^2}(T_h - T_c)b^3 = (8.5 \times 10^7 \text{ °R}^{-1}\cdot\text{ft}^{-3})[(105-75)\text{ °R}](\tfrac{1}{12}\text{ ft})^3 = 1.476 \times 10^6$$

From Table B-3, at 90 °F, $k = 0.356$ Btu/hr·ft·°F and $\mathbf{Pr} = 5.38$; so, $\mathbf{Gr}_b\,\mathbf{Pr} = (1.476 \times 10^6)(5.38) = 7.94 \times 10^6$ and (*I-30*) applies.

$$\overline{\mathbf{Nu}}_b = (0.069)(1.476 \times 10^6)^{1/3}(5.38)^{0.407} = 15.58$$

$$\bar{h} = \frac{\overline{\mathbf{Nu}}_b\, k}{b} = \frac{(15.58)(0.356)}{\frac{1}{12}} = 66.6 \text{ Btu/hr}\cdot\text{ft}^2\cdot\text{°F}$$

$$\frac{q}{A} = \bar{h}(T_h - T_c) = (66.6)(105 - 75) = 1998 \text{ Btu/hr}\cdot\text{ft}^2$$

7.89^D In a laboratory experiment, two horizontal surfaces which are separated by a 25-mm layer of water have upper and lower temperatures of 24 °C and 41 °C, respectively. At the location of the laboratory the measured value of g, together with experimentally obtained values of β, ρ, and μ at $T_{avg} = 32.5$ °C, yields the value

$$\frac{g\beta\rho^2}{\mu^2} = 5.4 \times 10^9 \text{ K}^{-1}\cdot\text{m}^{-3}$$

Determine the heat flux in W/m^2.

$$\mathbf{Gr}_b = \frac{g\beta}{\nu^2}(T_h - T_c)b^3 = (5.4 \times 10^9 \text{ K}^{-1}\cdot\text{m}^{-3})[(41-24)\text{ K}](0.025 \text{ m})^3 = 1.434 \times 10^6$$

From Table B-3, at 32.5 °C, $k = 0.616$ W/m·K and $\mathbf{Pr} = 5.38$; thus, $\mathbf{Gr}_b\,\mathbf{Pr} = (1.434 \times 10^6)(5.38) = 7.71 \times 10^6$ and (*I-30*) applies.

$$\overline{\mathbf{Nu}}_b = (0.069)(1.434 \times 10^6)^{1/3}(5.38)^{0.407} = 15.43$$

$$\bar{h} = \frac{\overline{\mathbf{Nu}}_b\, k}{b} = \frac{(15.43)(0.616)}{0.025} = 380 \text{ W/m}^2\cdot\text{K}$$

$$\frac{q}{A} = \bar{h}(T_h - T_c) = (0.380 \text{ kW/m}^2\cdot\text{K})(17 \text{ K}) = 6.46 \text{ kW/m}^2$$

7.90[D] Two parallel vertical walls, each 4 ft high, are separated by a 3-in. air space. Calculate the convective heat flux from the hotter to the cooler wall, if one wall is at 170 °F and the other is at 80 °F.

▌ From Table B-4, at $T_{avg} = 125\ °F$,

$$\nu = 19.63 \times 10^{-5}\ ft^2/sec \qquad k = 0.016\,26\ Btu/hr \cdot ft \cdot °F \qquad \mathbf{Pr} = 0.703$$

Also, $\beta = 1/T_{avg} = 1/(585\ °R)$; hence

$$\mathbf{Gr}_b = \frac{g\beta}{\nu^2}(T_h - T_c)b^3 = \frac{32.2\ ft/sec^2}{(585\ °R)(19.63 \times 10^{-5}\ ft^2/sec)^2}[(170 - 80)\ °R][(\tfrac{3}{12})\ ft]^3 = 2.01 \times 10^6$$

Because $L/b = \frac{48}{3} > 3$, (*I-31*) may be applied to give

$$\overline{\mathbf{Nu}}_b = (0.065)(2.01 \times 10^6)^{1/3}(16)^{-1/9} = 6.03$$

$$\bar{h} = \frac{\overline{\mathbf{Nu}}_b\, k}{b} = \frac{(6.03)(0.016\,26)}{\frac{3}{12}} = 0.392\ Btu/hr \cdot ft^2 \cdot °F$$

$$\frac{q}{A} = \bar{h}(T_h - T_c) = (0.392)(90) = 35.3\ Btu/hr \cdot ft^2$$

7.91[D] Two parallel walls, each 1.25 m high, are separated by a 7.5-cm air space. Calculate the convective heat flux from the hotter to the cooler wall, if one wall is at 77 °C and the other is at 27 °C.

▌ From Table B-4, at $T_{avg} = 52\ °C$,

$$\nu = 1.824 \times 10^{-5}\ m^2/s \qquad k = 0.0281\ W/m \cdot K \qquad \mathbf{Pr} = 0.703$$

and $\beta = 1/T_{avg} = 1/(325\ K)$; therefore

$$\mathbf{Gr}_b = \frac{g\beta}{\nu^2}(T_h - T_c)b^3 = \frac{9.8\ m/s^2}{(325\ K)(1.824 \times 10^{-5}\ m^2/s)^2}[(77 - 27)\ K](0.075\ m)^3 = 1.91 \times 10^6$$

Because $L/b = 1.25/0.075 = \frac{50}{3} > 3$, (*I-31*) may be applied to give

$$\overline{\mathbf{Nu}}_b = (0.065)(1.91 \times 10^6)^{1/3}(\tfrac{50}{3})^{-1/9}$$

$$\bar{h} = \frac{\overline{\mathbf{Nu}}_b\, k}{b} = \frac{(5.90)(0.0281)}{0.075} = 2.21\ W/m^2 \cdot K$$

$$\frac{q}{A} = \bar{h}(T_h - T_c) = (2.21)(50) = 110.5\ W/m^2$$

7.92[D] Estimate the heat transfer rate from a 40-W incandescent bulb at 77 °C to quiescent air at 27 °C. Approximate the bulb as a 50-mm-diam. sphere. What percentage of the electric power is lost by free convection?

▌ From Table B-4 the required parameters, evaluated at $T_f = (T_s + T_\infty)/2 = 52\ °C$, are

$$\nu = (19.63 \times 10^{-5})(0.0929) = 1.824 \times 10^{-5}\ m^2/s \qquad k = (0.016\,26)(1.729) = 0.0281\ W/m \cdot K \qquad \mathbf{Pr} = 0.703$$

Also, $\beta = 1/T_\infty = 1/(300\ K)$. Anticipating use of (*I-26*), we calculate

$$\mathbf{Gr}_D\,\mathbf{Pr} = \frac{g\beta(T_s - T_\infty)D^3\,\mathbf{Pr}}{\nu^2} = \frac{(9.80\ m/s^2)(50\ K)(0.050\ m)^3(0.703)}{(300\ K)(1.824 \times 10^{-5}\ m^2/s)^2} = 4.31 \times 10^5 < 10^{11}$$

$$\overline{\mathbf{Nu}}_D = 2 + \frac{(0.589)(\mathbf{Gr}_D\,\mathbf{Pr})^{1/4}}{[1 + (0.469/\mathbf{Pr})^{9/16}]^{4/9}} = 2 + \frac{(0.589)(4.31 \times 10^5)^{1/4}}{[1 + (0.469/0.703)^{9/16}]^{4/9}} = 13.63$$

$$q = k\,\overline{\mathbf{Nu}}_D\,\pi D(T_s - T_\infty) = (0.0281)(13.63)\pi(0.050)(50) = 3.00\ W$$

The percentage lost by free convection is, therefore,

$$\frac{3.00}{40}(100\%) = 7.51\%$$

This result is of the same order of magnitude as that obtained for the same configuration (but at 127 °C) in forced convection (Problem 5.133). In such cases, both free and forced convection should be considered.

7.93[D] Estimate the heat transfer from a 40-W incandescent bulb at 170 °F to 80 °F quiescent air. Approximate the bulb as a 2.0-in.-diam. sphere. What percentage of the electric power is lost by free convection?

From Table B-4 the required parameters, evaluated at $T_f = (T_s + T_\infty)/2 = 125\ °F$, are

$$\nu = 19.63 \times 10^{-5}\ \text{ft}^2/\text{sec} \qquad k = 0.016\,26\ \text{Btu/hr}\cdot\text{ft}\cdot°\text{F} \qquad \textbf{Pr} = 0.703$$

Also, $\beta = 1/T_\infty = 1/(540\ °R)$. Anticipating use of (*I-26*), we calculate

$$\textbf{Gr}_D\ \textbf{Pr} = \frac{g\beta(T_s - T_\infty)D^3\ \textbf{Pr}}{\nu^2} = \frac{(32.17\ \text{ft/sec}^2)(90\ °\text{R})(\tfrac{2}{12}\ \text{ft})^3(0.703)}{(540\ °\text{R})(19.63 \times 10^{-5}\ \text{ft}^2/\text{sec})^2} = 4.56 \times 10^5 < 10^{11}$$

$$\overline{\textbf{Nu}}_D = 2 + \frac{(0.589)(\textbf{Gr}_D\ \textbf{Pr})^{1/4}}{[1 + (0.469/\textbf{Pr})^{9/16}]^{4/9}} = 2 + \frac{(0.589)(4.56 \times 10^5)^{1/4}}{[1 + (0.469/0.703)^{9/16}]^{4/9}} = 13.80$$

$$q = k\,\overline{\textbf{Nu}}_D\,\pi D(T_s - T_\infty) = (0.016\,26)(13.80)\pi(\tfrac{2}{12})(90) = 10.57\ \text{Btu/hr}$$

Because $10.57\ \text{Btu/hr} \approx 3.10\ \text{W}$, the percentage lost by free convection is

$$\frac{3.10}{40}(100\%) = 7.75\%$$

This result is of the same order of magnitude as that obtained for the same configuration (but at 260 °F) in forced convection (Problem 5.134). In such cases, both free and forced convection should be considered.

7.94[D] What heat load is generated in a restaurant by a 1.0-m by 0.8-m cooking grill which is maintained at 134 °C? The room temperature is 20 °C. Assume still air with the hot grill surface facing upward.

From Table B-4, at $T_f = 77\ °C$,

$$\nu = (22.38 \times 10^{-5})(9.29 \times 10^{-2}) = 2.08 \times 10^{-5}\ \text{m}^2/\text{s} \qquad k = (0.017\,35)(1.729) = 0.0300\ \text{W/m}\cdot\text{K} \qquad \textbf{Pr} = 0.697$$

For air, $\beta = 1/T_\infty = 1/(293\ \text{K})$; hence, using in this case $L = A/P = (0.8\ \text{m}^2)/(3.6\ \text{m}) = 0.222\ \text{m}$,

$$\textbf{Gr}_L\ \textbf{Pr} = \frac{g\beta(T_s - T_\infty)L^3\ \textbf{Pr}}{\nu^2} = \frac{(9.80\ \text{m/s}^2)(114\ \text{K})(0.222\ \text{m})^3(0.697)}{(293\ \text{K})(2.08 \times 10^{-5}\ \text{m}^2/\text{s})^2} = 6.72 \times 10^7$$

and the turbulence correlation (*I-22*) applies. Under our choice of L, $A/L = P$ and

$$q = k\,\overline{\textbf{Nu}}_L\,P(T_s - T_\infty) = (0.0300)[(0.15)(6.72 \times 10^7)^{1/3}](3.6)(114) = 751\ \text{W}$$

7.95[D] What heat load is generated in a restaurant by a 40-in. by 30-in. cooking grill which is maintained at 273 °F? The room temperature is 67 °F. Assume still air with the hot grill surface upward.

From Table B-4, at $T_f = 170\ °F$,

$$\nu = 22.38 \times 10^{-5}\ \text{ft}^2/\text{sec} \qquad k = 0.017\,35\ \text{Btu/hr}\cdot\text{ft}\cdot°\text{F} \qquad \textbf{Pr} = 0.697$$

For air, $\beta = 1/T_\infty = 1/(527\ °R)$; hence, using in this case $L = A/P = (1200\ \text{in}^2)/(140\ \text{in.}) = 8.57\ \text{in.}$,

$$\textbf{Gr}_L\ \textbf{Pr} = \frac{g\beta(T_s - T_\infty)L^3\ \textbf{Pr}}{\nu^2} = \frac{(32.17\ \text{ft/sec}^2)(206\ °\text{R})[(8.57/12)\ \text{ft}]^3(0.697)}{(527\ °\text{R})(22.38 \times 10^{-5}\ \text{ft}^2/\text{sec})^2} = 6.37 \times 10^7$$

and the turbulence correlation (*I-22*) applies. Under our choice of L, $A/L = P$ and

$$q = k\,\overline{\textbf{Nu}}_L\,P(T_s - T_\infty) = (0.017\,35)[(0.15)(6.37 \times 10^7)^{1/3}](\tfrac{140}{12})(206) = 2498\ \text{Btu/hr}$$

7.96[D] What electric power is required to maintain a 0.003-in.-diam., 2-ft-long vertical wire at 260 °F in quiescent air at 80 °F?

At $T_f = (T_s + T_\infty)/2 = 170\ °F$, fluid properties from Table B-4 are

$$\nu = 22.38 \times 10^{-5}\ \text{ft}^2/\text{sec} \qquad k = 0.017\,35\ \text{Btu/hr}\cdot\text{ft}\cdot°\text{F} \qquad \textbf{Pr} = 0.697$$

Also, $\beta = 1/T_\infty = 1/(540\ °R)$; then

$$\textbf{Gr}_D = \frac{g\beta(T_s - T_\infty)D^3}{\nu^2} = \frac{(32.2\ \text{ft/sec}^2)[(260 - 80)\ °\text{R}][(0.003/12)\ \text{ft}]^3}{(540\ °\text{R})(22.38 \times 10^{-5}\ \text{ft}^2/\text{sec})^2} = 3.348 \times 10^{-3}$$

The parameter required for using Figure I-2 is

$$\log\left(\textbf{Gr}_D\ \textbf{Pr}\ \frac{D}{L}\right) = \log\left[(3.348 \times 10^{-3})(0.697)\left(\frac{0.003}{24}\right)\right] = -6.53$$

From Figure I-2, $\overline{\mathbf{Nu}}_D \approx 0.37$, and so

$$q = k\,\overline{\mathbf{Nu}}_D\,\pi L(T_s - T_\infty) = (0.017\,35)(0.37)\pi(2)(180) = 7.26 \text{ Btu/hr}$$

Thus an electric power input of 7.26 Btu/hr = 2.13 W is required.

7.97^D What electric power is required to maintain a 75-μm-diam., 0.60-m-long vertical wire at 127 °C in quiescent air at 27 °C?

▌ At $T_f = (T_s + T_\infty)/2 = 77$ °C, fluid properties from Table B-4 are

$$\nu = (22.38 \times 10^{-5})(0.092\,903) = 2.079 \times 10^{-5} \text{ m}^2/\text{s} \qquad k = (0.017\,35)(1.7295) = 0.0300 \text{ W/m}\cdot\text{K} \qquad \mathbf{Pr} = 0.697$$

Also, $\beta = 1/T_\infty = 1(300 \text{ K})$; then

$$\mathbf{Gr}_D = \frac{g\beta(T_s - T_\infty)D^3}{\nu^2} = \frac{(9.80 \text{ m/s}^2)[(127 - 27) \text{ K}](75 \times 10^{-6} \text{ m})^3}{(300 \text{ °K})(2.079 \times 10^{-5} \text{ m}^2/\text{s})^2} = 3.185 \times 10^{-3}$$

The parameter required for using Figure I-2 is

$$\log\left(\mathbf{Gr}_D \, \mathbf{Pr} \, \frac{D}{L}\right) = \log\left[(3.185 \times 10^{-3})(0.697)\left(\frac{75 \times 10^{-6}}{0.60}\right)\right] = -6.56$$

From Figure I-2, $\overline{\mathbf{Nu}}_D \approx 0.37$, and so

$$q = k\,\overline{\mathbf{Nu}}_D\,\pi L(T_s - T_\infty) = (0.0300)(0.37)\pi(0.60)(100) = 2.09 \text{ W}$$

which is the required electric power.

7.98^D Atmospheric air is contained between two parallel, vertical plates separated by a distance of 1 in. The plates, which are 6 ft high and 4 ft wide, are at temperatures of 120 °F and 40 °F. Estimate the rate of heat transfer across the air space.

▌ From Table B-4, at $T_{\text{avg}} = 80$ °F, $\nu = 16.88 \times 10^{-5}$ ft²/sec and $k = 0.015\,16$ Btu/hr · ft · °F; also, $\beta = 1/T_{\text{avg}} = 1/(540 \text{ °R})$. Applying (*I-31*) $(L/b = \frac{72}{1})$,

$$\mathbf{Gr}_b = \frac{g\beta(T_h - T_c)b^3}{\nu^2} = \frac{(32.2 \text{ ft/sec}^2)[(120 - 40) \text{ °R}](\frac{1}{12} \text{ ft})^3}{(540 \text{ °R})(16.88 \times 10^{-5} \text{ ft}^2/\text{sec})^2} = 9.69 \times 10^4$$

$$\bar{h} = \frac{k}{b}(0.18)\,\mathbf{Gr}_b^{1/4}\left(\frac{L}{b}\right)^{-1/9} = \frac{0.015\,16}{\frac{1}{12}}(0.18)(9.69 \times 10^4)^{1/4}(72)^{-1/9} = 0.359 \text{ Btu/hr}\cdot\text{ft}^2\cdot\text{°F}$$

$$q = \bar{h}A(T_h - T_c) = (0.359 \text{ Btu/hr}\cdot\text{ft}^2\cdot\text{°F})(24 \text{ ft}^2)[(120 - 40) \text{ °F}] = 689.3 \text{ Btu/hr}$$

7.99^D Atmospheric air is contained between two parallel, vertical plates separated by a distance of 25 mm. The plates, which are 1.8 m high and 1.2 m wide, are at temperatures of 49 °C and 5 °C. Estimate the rate of heat transfer across the air space.

▌ From Table B-4, at $T_{\text{avg}} = 27$ °C, $\nu = 1.568 \times 10^{-5}$ m²/s and $k = 0.026\,22$ W/m · K; also, $\beta = 1/T_{\text{avg}} = 1/(300 \text{ K})$. Applying (*I-31*) $(L/b = 1.8/0.025 = 72)$,

$$\mathbf{Gr}_b = \frac{g\beta(T_h - T_c)b^3}{\nu^2} = \frac{(9.80 \text{ m/s}^2)[(49 - 5) \text{ K}](0.025 \text{ m})^3}{(300 \text{ K})(1.568 \times 10^{-5} \text{ m}^2/\text{s})^2} = 9.13 \times 10^4$$

$$\bar{h} = \frac{k}{b}(0.18)\,\mathbf{Gr}_b^{1/4}\left(\frac{L}{b}\right)^{-1/9} = \frac{0.026\,22}{0.025}(0.18)(9.13 \times 10^4)^{1/4}(72)^{-1/9} = 2.040 \text{ W/m}^2\cdot\text{K}$$

$$q = \bar{h}A(T_h - T_c) = (2.04 \text{ W/m}^2\cdot\text{K})(2.16 \text{ m}^2)[(49 - 5) \text{ K}] = 194 \text{ W}$$

7.100^D Air at 50 psia is contained between two concentric spheres of diameters 4 in. and 3 in. Estimate the heat transfer rate when the inner sphere is maintained at 120 °F and the outer sphere is held at 40 °F.

▌ Equations (*I-32*) through (*I-35*) may be used here. At $T_{\text{avg}} = 80$ °F, Table B-4 gives

$$\mu g_c = 1.241 \times 10^{-5} \text{ lbm/sec}\cdot\text{ft} \qquad k = 0.015\,16 \text{ Btu/hr}\cdot\text{ft}\cdot\text{°F} \qquad \mathbf{Pr} = 0.708$$

Since the air is above atmospheric pressure, the density is given by

$$\rho = \frac{p}{RT} = \frac{50(144)\ \text{lbf/ft}^2}{(53.3\ \text{ft}\cdot\text{lbf/lbm}\cdot{}^\circ\text{R})(540\ {}^\circ\text{R})} = 0.250\ \text{lbm/ft}^3$$

which gives a kinematic viscosity of

$$\nu = \frac{\mu g_c}{\rho} = \frac{1.241\times10^{-5}\ \text{lbm/sec}\cdot\text{ft}}{0.250\ \text{lbm/ft}^3} = 4.96\times10^{-5}\ \text{ft}^2/\text{sec}$$

Then, with $\beta = 1/T_{\text{avg}} = 1/(540\ {}^\circ\text{R})$ and $b = \frac{1}{24}$ ft,

$$\mathbf{Gr}_b = \frac{g\beta(T_h - T_c)b^3}{\nu^2} = \frac{(32.2\ \text{ft/sec}^2)[(120-40)\ {}^\circ\text{R}](\frac{1}{24}\ \text{ft})^3}{(540\ {}^\circ\text{R})(4.96\times10^{-5}\ \text{ft}^2/\text{sec})^2} = 1.41\times10^5$$

Noting that $\mathbf{Gr}_s/\mathbf{Gr}_b$ must be dimensionless, we have from (*I-35*):

$$\mathbf{Gr}_s = \frac{b}{(D_oD_i)^4}\frac{\mathbf{Gr}_b}{(D_i^{-7/5}+D_o^{-7/5})^5} = \frac{0.5}{(4\times3)^4}\left[\frac{1.41\times10^5}{(3^{-7/5}+4^{-7/5})^5}\right] = 575$$

$\mathbf{Gr}_s\,\mathbf{Pr} = (575)(0.708) = 407$ (OK)

$$k_{\text{eff}} = k(0.74)\left(\frac{\mathbf{Pr}}{0.861+\mathbf{Pr}}\right)^{1/4}(\mathbf{Gr}_s\,\mathbf{Pr})^{1/4} = (0.015\,16)(0.74)\left(\frac{0.708}{0.861+0.708}\right)^{1/4}(407)^{1/4} = 0.0413\ \text{Btu/hr}\cdot\text{ft}\cdot{}^\circ\text{F}$$

$$q = k_{\text{eff}}\pi\left(\frac{D_iD_o}{b}\right)(T_h - T_c) = \frac{(0.0413\ \text{Btu/hr}\cdot\text{ft}\cdot{}^\circ\text{F})\pi(\frac{3}{12}\ \text{ft})(\frac{4}{12}\ \text{ft})}{\frac{1}{24}\ \text{ft}}[(120-40)\ {}^\circ\text{F}] = 20.76\ \text{Btu/hr}$$

7.101[D] Air at 344.7 kPa (3.403 atm) is contained between two concentric spheres of diameters 100 and 75 mm. Estimate the rate of heat transfer when the inner sphere is maintained at 50 °C and the outer sphere is held at 5 °C.

▌ Equations (*I-32*) through (*I-35*) may be used here. At $T_{\text{avg}} = 27$ °C, Table B-4 gives

$\mu = 1.847\times10^{-5}$ Pa·s $k = 0.026\,22$ W/m·K $\mathbf{Pr} = 0.708$ $\rho_{1\ \text{atm}} = 1.177\ \text{kg/m}^3$

The corrected density, $\rho = (1.177)(3.403/1) = 4.005\ \text{kg/m}^3$, gives a kinematic viscosity of

$$\nu = \frac{\mu}{\rho} = \frac{1.847\times10^{-5}\ \text{kg/m}\cdot\text{s}}{4.005\ \text{kg/m}^3} = 4.61\times10^{-6}\ \text{m}^2/\text{s}$$

Then, with $\beta = 1/T_{\text{avg}} = 1/(300\ \text{K})$ and $b = 0.0125$ m,

$$\mathbf{Gr}_b = \frac{g\beta(T_h - T_c)b^3}{\nu^2} = \frac{(9.8\ \text{m/s}^2)[(50-5)\ \text{K}](0.0125\ \text{m})^3}{(300\ \text{K})(4.61\times10^{-6}\ \text{m}^2/\text{s})^2} = 1.35\times10^5$$

Noting that $\mathbf{Gr}_s/\mathbf{Gr}_b$ must be dimensionless, we have from (*I-35*):

$$\mathbf{Gr}_s = \frac{b}{(D_oD_i)^4}\left[\frac{\mathbf{Gr}_b}{(D_i^{-7/5}+D_o^{-7/5})^5}\right] = \frac{12.5}{[(100)(75)]^4}\left[\frac{1.35\times10^5}{(75^{-7/5}+100^{-7/5})^5}\right] = 550$$

$\mathbf{Gr}_s\,\mathbf{Pr} = (550)(0.708) = 389$ (OK)

$$k_{\text{eff}} = k(0.74)\left(\frac{\mathbf{Pr}}{0.861+\mathbf{Pr}}\right)^{1/4}(\mathbf{Gr}_s\,\mathbf{Pr})^{1/4} = (0.026\,22)(0.74)\left(\frac{0.708}{0.861+0.708}\right)^{1/4}(389)^{1/4} = 0.070\,62\ \text{W/m}\cdot\text{K}$$

$$q = k_{\text{eff}}\pi\left(\frac{D_iD_o}{b}\right)(T_h - T_c) = \frac{(0.070\,62\ \text{W/m}\cdot\text{K})\pi(0.075\ \text{m})(0.100\ \text{m})}{0.0125\ \text{m}}[(50-5)\ \text{K}] = 5.99\ \text{W}$$

CHAPTER 8
Boiling and Condensation

Unless otherwise specified, the following nomenclature will be used throughout this chapter:

$c_l \equiv$ specific heat of saturated liquid, J/kg · K or Btu/lbm · °F
$C_{sf} \equiv$ surface-fluid constant (Table F-1)
$g \equiv$ local gravitational acceleration m/s^2 or ft/sec^2
$g_c = 32.17$ lbm · ft/lbf · sec^2 (constant in the British Engineering System)
$h_{fg} \equiv$ enthalpy of vaporization, J/kg or Btu/lbm
$\mathbf{Pr}_l \equiv$ Prandtl number of saturated liquid
$q/A \equiv$ heat flux, W/m^2 or Btu/hr · ft^2
$T_s - T_{sat} \equiv$ excess temperature, K or °F
$\mu_l \equiv$ liquid viscosity, Pa · s ($\mu_l g_c \equiv$ liquid viscosity, lbm/ft · hr)
$\sigma \equiv$ surface tension, N/m or lbf/ft
$\rho_l \equiv$ density of saturated liquid, kg/m^3 or lbm/ft^3
$\rho_v \equiv$ density of saturated vapor, kg/m^3 or lbm/ft^3

The surface-fluid constant, for which some values are given in Table F-1, is a function of the surface roughness (number of nucleating sites) and the angle of contact between the bubble and the heating surface. Values of surface tension σ for water is given as a function of temperature by

$$\sigma = (0.005\,28)(1 - 0.0013T)$$

where T is in °F and σ is in lbf/ft.

8.1 NUCLEATE BOILING

8.1 Heat transfer in phase-change processes—vaporization and its inverse, condensation—is determined by an equation of the same type as in simple convection, i.e.

$$q = hA(T_s - T_{sat}) \tag{1}$$

where the temperature differential, $T_s - T_{sat}$, is that between the heating (cooling) surface and the saturated liquid. Discuss the factors affecting the heat transfer coefficient h.

▌ Since phase-change processes involve changes in density, viscosity, specific heat, and thermal conductivity of the fluid, while the fluid's latent heat is either liberated (condensation) or absorbed (vaporization), the heat transfer coefficient for boiling and condensation is much more complicated than that for single-phase convective processes. Because of this, most engineering calculations involving boiling and condensation are made from empirical correlations.

As boiling is predominantly a local phenomenon, the heat transfer coefficient h is normally given without the overbar, as in (*1*). To be conservative in regard to burnout of heating elements, one often uses the maximum local h in (*1*).

8.2 The behavior of a fluid during boiling is highly dependent upon the excess temperature, $\Delta T \equiv T_s - T_{sat}$, measured from the boiling point of the fluid. Figure 8-1 indicates six different regimes for typical *pool boiling* (stationary fluid); the heat-flux curve is commonly called the *boiling curve*. Outline the behaviors expected in the different regimes.

▌ *Regime I.* Heat is transferred by free convection (Chapter 7).

Regime II. Bubbles begin to appear at the heating surface and rise to the free surface individually.

Regime III. The boiling action becomes so vigorous that individual bubbles combine with others very rapidly to form a vapor bubble column reaching to the free surface.

Regime IV. Bubbles form so rapidly that they blanket the heating surface, preventing fresh fluid from moving in to take their place. The increased resistance of this film reduces the heat flux, and the heat transfer decreases with increasing temperature differential. Because the film intermittently collapses and reappears, the regime is very unstable.

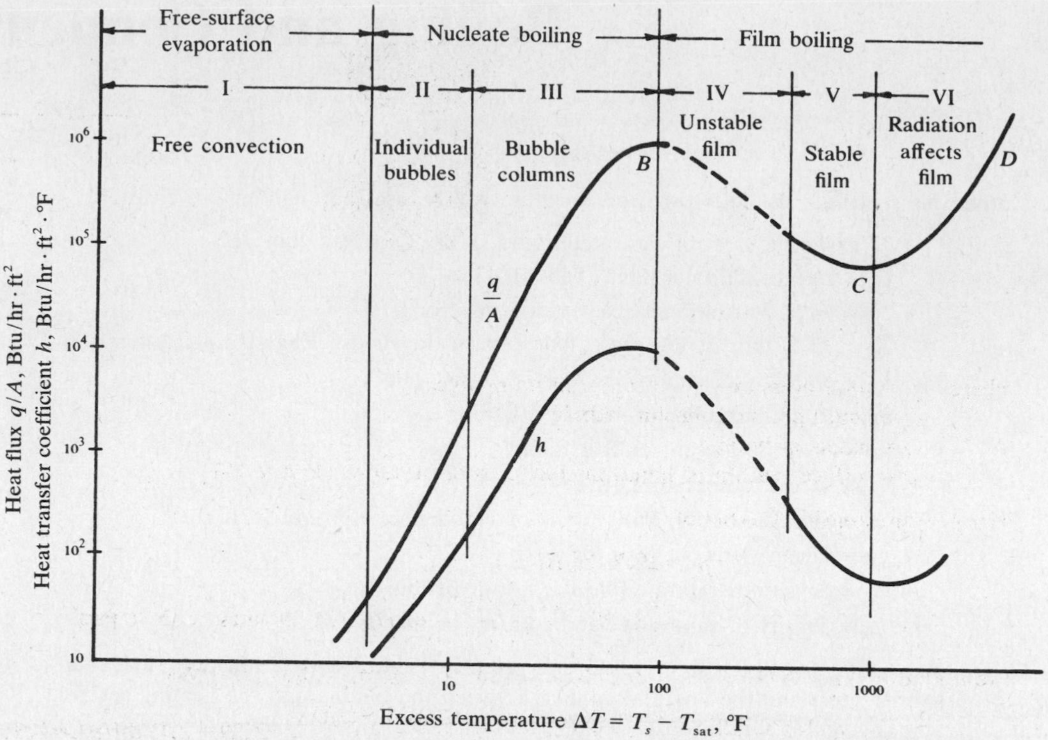

Fig. 8-1. Horizontal, Chromel C, 0.040-in.-diameter heating wire in water at 1 atm.

Regime V. The film on the heater surface becomes stable.

Regime VI. As ΔT reaches about 1000 °F, radiant heat transfer comes into play—in fact, becomes predominant—and the heat flux again rises with increasing ΔT.

The point of peak heat flux, point B, is variously referred to as the *burnout point*, the *boiling crisis*, and the *departure from nucleate boiling*.

8.3 How is it that the excess temperature $T_s - T_{sat}$, and not the difference between surface temperature and fluid bulk temperature, drives the heat transfer in boiling?

▌ The heat transfer from a plate of given size at, say 105 °C, to a pool of liquid water at atmospheric pressure is essentially the same whether the water is at a bulk temperature of 80 °C, 95 °C, or 100 °C. (In reality, there is a slight effect due to subcooling, but this is omitted in engineering calculations.) It is, therefore, not possible to develop meaningful, repeatable correlations at any temperature except the saturation temperature.

8.4 Show that the expansion velocity of a spherical bubble is inversely proportional to the square of its radius.

▌ For a purely radial displacement of the surrounding fluid (assumed nonviscous), the continuity equation in spherical coordinates becomes

$$\frac{1}{r^2}\frac{d}{dr}(r^2V_r) = 0 \qquad \text{or} \qquad r^2V_r = C \qquad \text{or} \qquad V_r = \frac{C}{r^2}$$

8.5 In an isothermal laboratory experiment utilizing water at 60 °F in an open container, spherical vapor bubbles of approximately $\frac{1}{10}$-in. diameter are observed at a fluid depth of 1 ft. The laboratory barometer reading is 29.25 ± 0.01 cmHg. Determine the vapor pressure inside the bubble, assuming static conditions and no substance other than water vapor in the bubble. Is the effect of the surface tension negligible? Is the fluid head of 1 ft negligible? (Consider the accuracy of the barometer.)

▌ Using the nomenclature of Fig. 8-2, a simple force balance gives

$$p_i = p_o = \frac{2\sigma}{r} \tag{1}$$

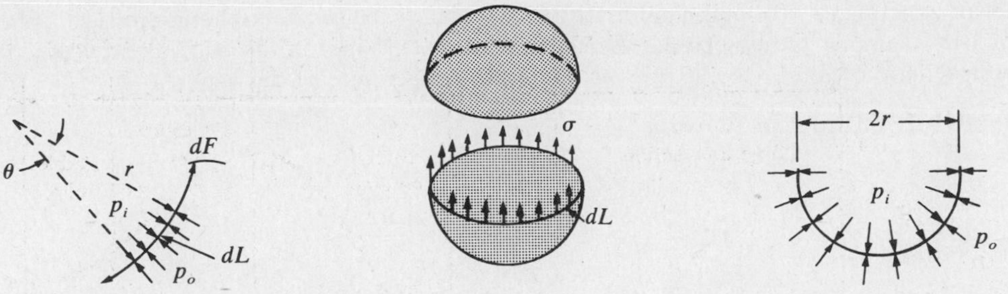

Fig. 8-2

where p_i = total pressure $(p_v + p_g)$ inside bubble
p_o = liquid pressure outside bubble
r = radius of bubble
σ = surface tension of liquid in contact with its vapor (Figure F-1)

The pressure outside the bubble is

$$p_o = 29.25 \text{ cmHg} + 1 \text{ ftH}_2\text{O}$$
$$= \left[29.25 + \frac{(2.54)(12)}{13.59}\right] \text{cmHg} = 31.49 \text{ cmHg} = 15.48 \text{ psia}$$

From Figure F-1 at 60 °F, $\sigma \approx 0.005$ lbf/ft; also $r = \frac{1}{240}$ ft. Thus (*1*) gives

$$p_i = p_v = p_o + \frac{2\sigma}{r} = (15.48 \times 144)\frac{\text{lbf}}{\text{ft}^2} + \frac{2(0.005 \text{ lbf/ft})}{\frac{1}{240} \text{ ft}} = (2229 + 2.4) \text{ lbf/ft}^2 \approx 2231.4 \frac{\text{lbf}}{\text{ft}^2}$$

Since we can detect pressure change within ± 0.01 cmHg $= \pm 0.708$ lbf/ft^2 and since the surface tension contributes 2.4 lbf/ft^2, it is not negligible. Also, liquid head is clearly not negligible.

8.6 For a spherical bubble containing a noncondensable gas, infer the equilibrium relation between bubble size and the amount of superheat, $T_v - T_{sat}$, from the Clausius–Clapeyron equation,

$$\frac{dp}{dT} = \frac{h_{fg}}{v_v T_v} = \frac{p_v h_{fg}}{R_v T_v^2} \tag{1}$$

where R_v is the effective "gas" constant of the vapor.

▌ By (*1*) of Problem 8.5,

$$p_v - p_l = \frac{2\sigma}{r} - p_g \tag{2}$$

Making the approximation

$$\frac{dp}{dT} \approx \frac{p_v - p_l}{T_v - T_{sat}}$$

in (*1*), one finds

$$T_v - T_{sat} \approx \frac{R_v T_v^2}{p_v h_{fg}}(p_v - p_l) \tag{3}$$

Together (*2*) and (*3*) give

$$T_v - T_{sat} \approx \frac{R_v T_v^2}{p_v h_{fg}}\left(\frac{2\sigma}{r} - p_g\right) \tag{4}$$

which is the equilibrium relationship between bubble radius and amount of superheat.

8.7 Referring to Problem 8.6, discuss bubble growth and collapse in terms of the amount of superheat.

▌ If the liquid temperature surrounding the bubble results in $T_l - T_{sat}$ greater than $T_v - T_{sat}$ as given by (*4*) of Problem 8.6, a bubble of radius r containing a noncondensable gas exerting pressure p_g will grow; if $T_l - T_{sat}$ is less than this $T_v - T_{sat}$, the bubble will collapse. It should also be evident that the effect of the noncondensable gas is to reduce the amount of superheat necessary for bubble growth.

8.8 In an experiment with bulk-fluid boiling of water at standard atmospheric pressure, spherical vapor bubbles of $\frac{1}{8}$-in. diameter are observed. Assuming pure water vapor in the bubble and vapor pressure equal to atmospheric pressure, 14.696 psia, calculate the temperature of the vapor.

▮ By (4) of Problem 8.6, with $p_g = 0$,

$$\Delta T = \frac{R_v T_v^2}{p_v h_{fg}} \left(\frac{2\sigma}{r}\right)$$

Substitute

$$r = \frac{1}{(16)(12)} \text{ ft}$$

$$\sigma = 0.004 \text{ lbf/ft}$$

$$R_v = 85.8 \text{ ft} \cdot \text{lbf/lbm} \cdot {}^\circ\text{R}$$

$$p_v = (14.7)(144) = 2117 \text{ lbf/ft}^2$$

$$h_{fg} = (970.3 \text{ Btu/lbm})(778 \text{ ft} \cdot \text{lbf/lbm} \cdot \text{Btu})$$

$$T_v = 212\ {}^\circ\text{F} = 672\ {}^\circ\text{R}$$

to obtain

$$\Delta T = \frac{(85.8)(672)^2}{(2117)[(970.3)(778)]}\left[\frac{2(0.004)}{1/(16)(12)}\right] \approx 0.374\ {}^\circ\text{R} = 0.374\ {}^\circ\text{F}$$

$$T_v = \Delta T + T_{\text{sat}} \approx 212.374\ {}^\circ\text{F}$$

8.9[D] Using Fig. 8-1, estimate the excess temperature for a 0.040-in.-diam., horizontal, 6-in.-long Chromel C wire submerged in water at atmospheric pressure. The voltage drop in the wire is 14.7 V and the current is 42.8 A.

▮ An energy balance gives

$$q = EI = hA\,\Delta T$$

Since $1 \text{ W} = 1 \text{ V} \cdot \text{A}$,

$$EI = [(14.7)(42.8) \text{ W}](3.413 \text{ Btu/hr} \cdot \text{W}) = 2147.32 \text{ Btu/hr}$$

The surface area of the wire is

$$A = \pi DL = \pi\left(\frac{0.040}{12} \text{ ft}\right)\left(\frac{6}{12} \text{ ft}\right) = 0.005\,236 \text{ ft}^2$$

Therefore

$$\frac{q}{A} = \frac{2147.32}{0.005\,236} = 4.1 \times 10^5 \text{ Btu/hr} \cdot \text{ft}^2$$

From Fig. 8-1, $\Delta T \approx 50\ {}^\circ\text{F}$.

8.10[D] A 1.0-mm-diam. Chromel C wire, 150 mm long, is submerged horizontally in water at atmospheric pressure. The wire has a steady-state applied voltage drop of 10.1 V and a current of 52.3 A. Determine the heat flux and the approximate wire temperature.

▮ The electric energy input rate is

$$q = EI = (10.1)(52.3) = 528.23 \text{ W}$$

The wire surface area is

$$A = \pi DL = \pi(1.0 \times 10^{-3} \text{ m})(150 \times 10^{-3} \text{ m}) = 4.7124 \times 10^{-4} \text{ m}^2$$

The boiling energy flux is

$$\frac{q}{A} = \frac{528.23 \text{ W}}{4.7124 \times 10^{-4} \text{ m}^2} = 1.121 \text{ MW/m}^2$$

To approximate the excess temperature using Fig. 8-1, convert q/A to British Engineering units.

$$\frac{q}{A} = (1.121 \times 10^6 \text{ W/m}^2)\left(\frac{1 \text{ Btu/hr} \cdot \text{ft}^2}{3.152\,48 \text{ W/m}^2}\right) = 355\,573 \text{ Btu/hr} \cdot \text{ft}^2$$

From Fig. 8-1,

$$\Delta T \approx (40\ ^\circ\text{F})\left(\frac{5\ ^\circ\text{C}}{9\ ^\circ\text{F}}\right) = 22\ ^\circ\text{C}$$

and

$$T_s = 100 + 22 = 122\ ^\circ\text{C}$$

8.11[D] A 6-in.-long, 0.040-in.-diam. nickel wire submerged horizontally in water at 100 psig requires 131.8 A at 2.18 V to maintain the wire at 350.08 °F. Calculate the heat transfer coefficient.

▮ From the steam tables, the saturation temperature at $p = 100 + 14.7 = 114.7$ psia is 337.92 °F; therefore $\Delta T = 350.08 - 337.92 = 12.16$ °F. A heat balance on the wire gives

$$q = EI = hA\,\Delta T$$

whence

$$h = \frac{EI}{A\,\Delta T} = \frac{[(2.18)(131.8)\ \text{W}](3.413\ \text{Btu/hr}\cdot\text{W})}{\pi[(0.040/12)\ \text{ft}](\frac{6}{12}\ \text{ft})(12.16\ ^\circ\text{F})}\left(\frac{1\ \text{W/m}^2\cdot\text{K}}{0.176\,12\ \text{Btu/hr}\cdot\text{ft}^2\cdot{}^\circ\text{F}}\right) = 87.4\ \text{kW/m}^2\cdot\text{K}$$

8.12 For liquid flow over a single sphere, the following empirical relation gives the average heat transfer coefficient:

$$\frac{hD}{k_\infty} = (1.2 + 0.53\,\mathbf{Re}^{0.54})\,\mathbf{Pr}^{0.3}\left(\frac{\mu_\infty}{\mu_s}\right)^{0.25} \tag{1}$$

valid when $1 < \mathbf{Re} < 2\times 10^5$ and all properties except μ_s are evaluated at the free-stream temperature. Estimate $\bar{h}$ if the vapor bubble rise velocity in a pool of saturated water at 1 atm pressure is 3 m/s and the 3.2-mm bubble diameter remains constant.

▮ At $T_\infty = 100$ °C

$$k_\infty = (0.393)(1.7296) = 0.6797\ \text{W/m}\cdot\text{K}$$
$$\nu_\infty = (0.316\times 10^{-5})(0.0929) = 2.94\times 10^{-7}\ \text{m}^2\text{/s}$$
$$\rho = (59.97)(16.018) = 960.60\ \text{kg/m}^3$$
$$\mu_\infty = \mu_s = \nu\rho = 2.824\times 10^{-4}\ \text{Pa}\cdot\text{s}$$

Then, since $\mathbf{Re} = V_\infty D/\nu_\infty$,

$$\bar{h} = \frac{0.6797\ \text{W/m}\cdot\text{K}}{0.0032\ \text{m}}\left\{1.2 + 0.53\left[\frac{(3\ \text{m/s})(0.0032\ \text{m})}{2.94\times 10^{-7}\ \text{m}^2\text{/s}}\right]^{0.54}\right\}(1.74)^{0.3}(1)^{0.25} = 36.40\ \text{kW/m}^2\cdot\text{K}$$

8.13 In Problem 8.12, how much will the vapor temperature drop as the bubble moves 0.3 m?

▮ By an energy balance and assuming no phase change

$$\bar{h}A(T_v - T_{\text{sat}})\,\Delta t = (\rho c_p V)_v\,\Delta T$$

where $T_v - T_{\text{sat}} = 0.374$ °F $= 0.208$ °C (see Problem 8.8; $\frac{1}{8}$ in. ≈ 3.2 mm), $c_p = (1.0070)(4.184) = 4.2133$ kJ/kg·K, and

$$\Delta t = \frac{\text{distance}}{\text{velocity}} = \frac{0.3\ \text{m}}{3\ \text{m/s}} = 0.1\ \text{s}$$

Therefore,

$$\Delta T = \frac{\bar{h}A(T_v - T_{\text{sat}})\,\Delta t}{(\rho c_p V)_v} = \frac{(36.40\ \text{kW/m}^2\cdot\text{K})(4\pi)(0.0016\ \text{m})^2(0.208\ \text{K})(0.1\ \text{s})}{(960.60\ \text{kg/m}^3)(4.2133\ \text{kJ/kg}\cdot\text{K})\frac{4}{3}\pi(0.0016\ \text{m})^3} = 0.3503\ ^\circ\text{C}$$

8.14[D] A tungsten wire submerged horizontally in water at atmospheric pressure is electrically heated, maintaining the surface at 225 °F. The wire is 0.040 in. in diameter and 12 in. long; $E = 2$ V and $I = 20$ A. Determine the heat transfer coefficient.

▮ An energy balance gives

$$q = hA\,\Delta T = EI = (2\ \text{V})(20\ \text{A})\left(\frac{1\ \text{W}}{1\ \text{V}\cdot\text{A}}\right)(3.413\ \text{Btu/hr}\cdot\text{W}) = 136.52\ \text{Btu/hr}$$

$$A = \pi DL = \pi\left(\frac{0.040}{12}\ \text{ft}\right)(1\ \text{ft}) = 0.0105\ \text{ft}^2$$

$$\frac{q}{A} = \frac{136.52}{0.0105} = 13\,002\ \text{Btu/hr}\cdot\text{ft}^2$$

Then, since $\Delta T = 225 - 212 = 13\ °\mathrm{F}$,

$$h = \frac{q/A}{\Delta T} = \frac{13\,002\ \mathrm{Btu/hr \cdot ft^2}}{13\ °\mathrm{F}} = 1000\ \mathrm{Btu/hr \cdot ft^2 \cdot °F}$$

8.15 What is the effect of liquid pressure on boiling heat transfer?

Numerous investigations indicate that the heat flux increases with increasing pressure. This is in keeping with the equilibrium requirement for excess temperature, (*4*) of Problem 8.6. For a given cavity size, a larger surrounding pressure (and hence a larger vapor pressure) requires a smaller value of $T_v - T_{\mathrm{sat}}$. Thus, a bubble can exist at a lower excess temperature, and the heat flux for a high pressure is greater than that for a low pressure at the same excess temperature.

8.16 The most commonly accepted general correlation for heat transfer in the nucleate boiling regimes (II and III of Fig. 8-1) is that due to W. M. Rohsenow.

$$\left(\frac{q}{A}\right)_b = (\mu_l g_c) h_{fg} \left[\frac{(g/g_c)(\rho_l - \rho_v)}{\sigma}\right]^{1/2} \left[\frac{c_l (T_s - T_{\mathrm{sat}})}{h_{fg}\, \mathbf{Pr}_l^{1.7}\, C_{\mathrm{sf}}}\right]^3 \qquad (1)$$

(When using SI units, replace g_c by 1.) Using this correlation, estimate the heat flux which would occur in nucleate boiling of saturated water at 400 °F with a platinum heater at 425 °F.

$c_l = 1.08\ \mathrm{Btu/lbm \cdot °F}$ $\qquad h_{fg} = 825.9\ \mathrm{Btu/lbm}$ $\qquad \mathbf{Pr}_l = 0.927$ $\qquad T_s - T_{\mathrm{sat}} = 25\ °\mathrm{F}$

$$\frac{g}{g_c} = 1.0\ \mathrm{lbf/lbm} \quad \text{(standard gravity)}$$

$$\mu_l g_c = (0.09 \times 10^{-3}\ \mathrm{lbm/ft \cdot sec})(3600\ \mathrm{sec/hr}) = 0.328\ \mathrm{lbm/ft \cdot hr} \qquad \sigma \approx 0.0024\ \mathrm{lbf/ft}$$

$$\rho_l = \frac{1}{0.018\,64}\ \mathrm{lbm/ft^3} = 53.6\ \mathrm{lbm/ft^3} \qquad \rho_v = \frac{1}{1.8630}\ \mathrm{lbm/ft^3} = 0.536\ \mathrm{lbm/ft^3} \qquad C_{\mathrm{sf}} = 0.013$$

Substituting in (*1*),

$$\left(\frac{q}{A}\right)_b = (0.328\ \mathrm{lbm/ft \cdot hr})(825.9\ \mathrm{Btu/lbm}) \left[\frac{(1.0\ \mathrm{lbf/lbm})(53.06\ \mathrm{lbm/ft^3})}{0.0024\ \mathrm{lbf/ft}}\right]^{1/2}$$

$$\times \left[\frac{(1.08\ \mathrm{Btu/lbm \cdot °F})(25\ °\mathrm{F})}{(825.9\ \mathrm{Btu/lbm})(0.927)^{1.7}(0.013)}\right]^3 = 9.37 \times 10^5\ \mathrm{Btu/ft^2 \cdot hr}$$

It is important to notice from the Rohsenow correlation that the heat flux in nucleate boiling varies as the cube of the excess temperature.

8.17 Estimate the surface temperature of a brass heater required to maintain a heat flux of 650 000 $\mathrm{Btu/hr \cdot ft^2}$ in nucleate boiling of water at 700 psia.

The saturation temperature corresponding to a pressure of 700 psia for steam is 503 °F.

$$c_l = 1.19\ \mathrm{Btu/lbm \cdot °F} \qquad C_{\mathrm{sf}} = 0.006$$

$$\frac{g}{g_c} = 1.0\ \mathrm{lbf/lbm} \quad \text{(standard gravity)}$$

$$h_{fg} = 710.2\ \mathrm{Btu/lbm} \qquad \mathbf{Pr}_l = 0.87$$

$$\mu_l g_c = (0.071 \times 10^{-3}\ \mathrm{lbm/ft \cdot sec})(3600\ \mathrm{sec/hr}) = 0.256\ \mathrm{lbm/ft \cdot hr} \qquad \sigma = 0.0017\ \mathrm{lbf/ft}$$

$$\rho_l = \frac{1}{0.020\,50}\ \mathrm{lbm/ft^3} = 48.8\ \mathrm{lbm/ft^3} \qquad \rho_v = \frac{1}{0.6556}\ \mathrm{lbm/ft^3} = 1.52\ \mathrm{lbm/ft^3}$$

Substitute in (*1*) of Problem 8.16 and solve for T_s:

$$6.5 \times 10^5\ \mathrm{Btu/hr \cdot ft^2} = (0.256\ \mathrm{lbm/ft \cdot hr})(710.2\ \mathrm{Btu/lbm}) \left[\frac{(1.0\ \mathrm{lbf/lbm})(47.3\ \mathrm{lbm/ft^3})}{0.0017\ \mathrm{lbf/ft}}\right]^{1/2}$$

$$\times \left[\frac{(1.19\ \mathrm{Btu/lbm \cdot °F})(T_s - 503\ °\mathrm{F})}{(710.2\ \mathrm{Btu/lbm})(0.87)^{1.7}(0.006)}\right]^3$$

$$6.5 \times 10^5 = (182)(2.78 \times 10^4)^{1/2}[0.354(T_s - 503\ °\mathrm{F})]^3$$

$$[0.354(T_s - 503\ °\mathrm{F})]^3 = 21.4$$

$$0.354(T_s - 503\ °\mathrm{F}) = 2.778$$

$$T_s = 511\ °\mathrm{F}$$

8.18 Estimate the heat flux to water undergoing nucleate pool boiling on a very small copper wire. The conditions are atmospheric pressure and an excess temperature of 40 °F.

▮

$$c_l = 1.0 \text{ Btu/lbm} \cdot \text{°F} \qquad C_{sf} = 0.013$$

$$\frac{g}{g_c} = 1.0 \text{ lbf/lbm} \quad \text{(standard gravity)}$$

$$h_{fg} = 970.3 \text{ Btu/lbm} \qquad \mathbf{Pr}_l = 1.78 \qquad T_s - T_{sat} = 40 \text{ °F}$$

$$\mu_l g_c = (0.194 \times 10^{-3} \text{ lbm/ft} \cdot \text{sec})(3600 \text{ sec/hr}) = 0.699 \text{ lbm/ft} \cdot \text{hr}$$

$$\sigma = 0.004 \text{ lbf/ft}$$

$$\rho_l = \frac{1}{0.016\,72} \text{ lbm/ft}^3 = 59.7 \text{ lbm/ft}^3 \qquad \rho_v = \frac{1}{26.799} \text{ lbm/ft}^3 = 0.0373 \text{ lbm/ft}^3$$

By (*1*) of Problem 8.16,

$$\left(\frac{q}{A}\right)_b = (0.699 \text{ lbm/ft} \cdot \text{hr})(970.3 \text{ Btu/lbm})\left[\frac{(1.0 \text{ lbf/lbm})(59.56 \text{ lbm/ft}^3)}{0.004 \text{ lbf/ft}}\right]^{1/2}$$

$$\times \left[\frac{(1.0 \text{ Btu/lbm} \cdot \text{°F})(40 \text{ °F})}{(970.3 \text{ Btu/lbm})(1.78)^{1.7}(0.013)}\right]^3 = 142\,000 \text{ Btu/hr} \cdot \text{ft}^2$$

8.19 Water is boiling on a 0.005-in.-diam. nickel wire at 1 atm pressure and 20 °F excess temperature. Estimate the heat flux.

▮

$$C_{sf} = 0.006 \qquad \sigma = 0.004 \text{ lbf/ft} \qquad c_l = 1 \text{ Btu/lbm} \cdot \text{°F} \qquad \mathbf{Pr}_l = 1.78$$

$$\rho_l = 59.85 \text{ lbm/ft}^3 \qquad \rho_v = 0.0372 \text{ lbm/ft}^3 \qquad h_{fg} = 970.3 \text{ Btu/lbm}$$

$$\frac{g}{g_c} = 1.0 \text{ lbf/lbm} \quad \text{(standard gravity)}$$

$$\mu_l g_c = (0.194 \times 10^{-3} \text{ lbm/ft} \cdot \text{sec})(3600 \text{ sec/hr}) = 0.699 \text{ lbm/ft} \cdot \text{hr}$$

Using (*1*) of Problem 8.16,

$$\left(\frac{q}{A}\right)_b = (0.699 \text{ lbm/hr})(970.3 \text{ Btu/lbm})\left[\frac{(1.0 \text{ lbf/lbm})(59.81 \text{ lbm/ft}^3)}{0.004 \text{ lbf/ft}}\right]^{1/2}$$

$$\times \left[\frac{(1 \text{ Btu/lbm} \cdot \text{°F})(20 \text{ °F})}{(970.3 \text{ Btu/lbm})(1.78)^{1.7}(0.006)}\right]^3 = 177\,500 \text{ Btu/hr} \cdot \text{ft}^2$$

8.20 A heated brass plate is submerged vertically in water at atmospheric pressure. The plate is maintained at 232 °F. What is the heat transfer rate per unit area?

$$C_{sf} = 0.006 \quad \text{(Table F-1)} \qquad \sigma \approx 0.004 \quad \text{(Figure F-1)}$$

$$c_l = 1.01 \text{ Btu/lbm} \cdot \text{°F} \quad \text{(Table B-3)} \qquad \mathbf{Pr}_l = 1.58 \quad \text{(Table B-3)}$$

$$\rho_l = 59.44 \quad \text{(Table B-3)} \qquad \rho_v = 0.0362 \text{ lbm/ft}^3 \quad \text{(Table B-4)}$$

$$h_{fg} = 957.48 \text{ Btu/lbm} \quad \text{(steam tables)} \qquad \frac{g}{g_c} = 1.0 \text{ lbf/lbm} \quad \text{(standard gravity)}$$

$$\mu_l g_c = \rho_l \nu_l = (59.44)(0.288 \times 10^{-5}) = 0.171 \times 10^{-3} \text{ lbm/ft} \cdot \text{sec} = 0.617 \text{ lbm/ft} \cdot \text{hr}$$

Substituting in (*1*) of Problem 8.16,

$$\left(\frac{q}{A}\right)_f = (0.617 \text{ lbm/ft} \cdot \text{hr})(957.48 \text{ Btu/lbm})\left[\frac{(1.0 \text{ lbf/lbm})(59.40 \text{ lbm/ft}^3)}{0.004 \text{ lbf/ft}}\right]^{1/2}$$

$$\times \left\{\frac{(1 \text{ Btu/lbm} \cdot \text{°F})[(232 - 212) \text{ °F}]}{(957.48 \text{ Btu/lbm})(1.58)^{1.7}(0.006)}\right\}^3 = 2.94 \times 10^5 \text{ Btu/hr} \cdot \text{ft}^2$$

8.21 Estimate the heat flux for water boiling on a 10-mil-diam. platinum wire at 1 atm pressure, 30 °F excess temperature, and standard gravity.

▮ Since Table F-1 contains a value of C_{sf} for this fluid–surface combination, that is, $C_{sf} = 0.013$, we can use the Rohsenow equation. (Note from Fig. 8-1 that at the stated surface excess temperature, the regime is nucleate boiling.) The appropriate fluid properties are

$c_l = 1.0$ Btu/lbm · °F $h_{fg} = 970.3$ Btu/lbm $\mathrm{Pr}_l = 1.78$ $\mu_l g_c = 0.699$ lbm/ft · hr
$\rho_l = 59.85$ lbm/ft³ $\rho_v = 0.0373$ lbm/ft³

Then, by (*1*) of Problem 8.16,

$$\left(\frac{q}{A}\right)_b = (0.699\ \mathrm{lbm/ft\cdot hr})(970.3\ \mathrm{Btu/lbm})\left[\frac{(1.0\ \mathrm{lbf/lbm})(59.81\ \mathrm{lbm/ft^3})}{0.004\,03\ \mathrm{lbf/ft}}\right]^{1/2}$$

$$\times\left[\frac{(1.0\ \mathrm{Btu/lbm\cdot{}^\circ F})(30\ {}^\circ\mathrm{F})}{(970.3\ \mathrm{Btu/lbm})(1.78)^{1.7}(0.013)}\right]^3 = 58\,800\ \mathrm{Btu/hr\cdot ft^2}$$

8.22 At the point of maximum heat transfer (point *B* of Fig. 8-1), the recommended correlation, due to Zuber, is

$$\left(\frac{q}{A}\right)_{\max} = 0.18\rho_v h_{fg}\left[\frac{\sigma(\rho_l - \rho_v)gg_c}{\rho_v^2}\right]^{1/4}\left(\frac{\rho_l}{\rho_l + \rho_v}\right)^{1/2} \qquad (1)$$

(When using SI units, replace g_c by 1.) Observe that the peak heat flux is independent of the heating element. Using this correlation, determine the peak heat flux for nucleate boiling of water at 1 standard atm pressure and standard gravity.

$$\rho_l = \frac{1}{0.016\,719}\ \mathrm{lbm/ft^3} = 59.81\ \mathrm{lbm/ft^3} \qquad \rho_v = \frac{1}{26.799}\ \mathrm{lbm/ft^3} = 0.0373\ \mathrm{lbm/ft^3}$$

$$h_{fg} = 970.3\ \mathrm{Btu/lbm} \qquad \sigma \approx 0.004\ \mathrm{lbf/ft}$$

Then

$$\left(\frac{q}{A}\right)_{\max} = (0.18)(0.0373\ \mathrm{lbm/ft^3})(970.3\ \mathrm{Btu/lbm})$$

$$\times\left[\frac{(0.004\ \mathrm{lbf/ft})(59.77\ \mathrm{lbm/ft^3})(32.17\ \mathrm{ft/sec^2})(32.17\ \mathrm{lbm\cdot ft/lbf\cdot sec^2})}{(0.0373\ \mathrm{lbm/ft^3})^2}\right]^{1/4}\left(\frac{59.81}{59.85}\right)^{1/2}$$

$$= 133\ \mathrm{Btu/sec\cdot ft^2} = 4.81\times 10^5\ \mathrm{Btu/hr\cdot ft^2}$$

8.23 In a laboratory experiment, a current of 193 A suffices to burn out a 12-in.-long, 0.040-in.-diam. nickel wire which is submerged horizontally in water at atmospheric pressure. What was the voltage at burnout?

By Problem 8.22, $(q/A)_{\max} = 4.81\times 10^5$ Btu/hr · ft². The burnout voltage, E_b, must satisfy $E_b I = q_{\max}$. Thus, recalling that 1 W = 1 V · A,

$$E_b = \frac{q_{\max}}{I} = \frac{A}{I}\left(\frac{q}{A}\right)_{\max} = \frac{\pi[(0.040/12)\ \mathrm{ft}](1\ \mathrm{ft})}{193\ \mathrm{A}}(4.81\times 10^5\ \mathrm{Btu/hr\cdot ft^2})\left(\frac{1\ \mathrm{W}}{3.413\ \mathrm{Btu/hr}}\right) = 7.65\ \mathrm{V}$$

8.24 Estimate the peak heat flux, in MW/m², for nucleate boiling of water at normal atmospheric pressure and standard gravity.

Use (*1*) of Problem 8.22 in SI units. As data,

$$\rho_l = (59.81)(16.02) = 958.2\ \mathrm{kg/m^3} \qquad \rho_v = (0.0373)(16.02) = 0.60\ \mathrm{kg/m^3}$$

$$h_{fg} = (970.3\ \mathrm{Btu/lbm})\left(\frac{1\ \mathrm{lbm}}{0.454\ \mathrm{kg}}\right)(1054.8\ \mathrm{J/Btu}) = 2.25\ \mathrm{MJ/kg}$$

$$\sigma \approx (0.004\ \mathrm{lbf/ft})(4.448\ \mathrm{N/lbf})\left(\frac{1}{0.3048}\ \frac{\mathrm{ft}}{\mathrm{m}}\right) = 0.0584\ \mathrm{N/m}$$

Thus, remembering that 1 N = 1 kg · m/s² and 1 J = 1 W · s,

$$\left(\frac{q}{A}\right)_{\max} = (0.18)(0.60\ \mathrm{kg/m^3})(2.25\ \mathrm{MJ/kg})\left[\frac{(0.0584\ \mathrm{N/m})(957.6\ \mathrm{kg/m^3})(9.8\ \mathrm{m/s^2})}{(0.60\ \mathrm{kg/m^3})^2}\right]^{1/4}\left(\frac{958.2}{958.8}\right)^{1/2}$$

$$= 1.517\ \mathrm{MW/m^2}$$

8.25 A heated nickel plate at 222 °F is submerged horizontally in water at atmospheric pressure. What is the heat transfer per unit area?

For an excess temperature $T_s - T_{sat} = 222 - 212 = 10\ °F$, Fig. 8-1 indicates that the boiling is most likely nucleate, with (*1*) of Problem 8.16 being valid. The required parameters, from Table B-3 except where noted, are

$$h_{fg} = 970.4\ \text{Btu/lbm}\quad \text{(steam tables)}\qquad \sigma = 0.004\ \text{lbf/ft}\quad \text{(Figure F-1)}$$

$$\rho_l = 59.97\ \text{lbm/ft}^3\qquad c_l = 1.007\ \text{Btu/lbm}\cdot°\text{F}\qquad \rho_v = 0.0373\ \text{lbm/ft}^3\quad \text{(steam tables)}\qquad \text{Pr}_l = 1.74$$

$$\frac{g}{g_c} = 1.0\ \text{lbf/lbm}\quad \text{(standard gravity)}\qquad \mu_l g_c = 0.682\ \text{lbm/ft}\cdot\text{hr}\qquad C_{sf} = 0.006\quad \text{(Table F-1)}$$

The heat transfer rate is

$$\left(\frac{q}{A}\right)_b = (0.682\ \text{lbm/ft}\cdot\text{hr})(970.4\ \text{Btu/lbm})\left\{\frac{(1.0\ \text{lbf/lbm})[(59.97 - 0.0373)\ \text{lbm/ft}^3]}{0.004\ \text{lbf/ft}}\right\}^{1/2}$$
$$\times\left[\frac{(1.007\ \text{Btu/lbm}\cdot°\text{F})(10\ °\text{F})}{(970.4\ \text{Btu/lbm})(1.74)^{1.7}(0.006)}\right]^3 = 2.48\times10^4\ \text{Btu/hr}\cdot\text{ft}^2$$

The peak heat flux for water at 1 atm was found in Problem 8.23 to be

$$\left.\frac{q}{A}\right|_{max} = 4.81\times10^5\ \text{Btu/hr}\cdot\text{ft}^2$$

By Problem 8.23, $(q/A)_b < (q/A)_{max}$; so nucleate boiling does in fact occur.

8.26 A second correlation, of Rohsenow and Griffith, which is widely used for determining the peak heat flux in nucleate boiling, is

$$\left(\frac{q}{A}\right)_{max} = V_o \rho_v h_{fg}\left(\frac{g}{g_o}\right)^{1/4}\left(\frac{\rho_l - \rho_v}{\rho_v}\right)^{0.6} \tag{1}$$

with $V_0 = 143\ \text{ft/hr} = 0.0121\ \text{m/s}$ and $g_o = 32.17\ \text{ft/sec}^2 = 9.8\ \text{m/s}^2$. Using this equation (which is based on the energy transport by the vapor bubbles), repeat Problem 8.24.

$$\left(\frac{q}{A}\right)_{max} = (0.0121\ \text{m/s})(0.60\ \text{kg/m}^3)(2.25\ \text{MJ/kg})(1.0)^{1/4}\left(\frac{957.6}{0.60}\right)^{0.6} = 1.364\ \text{MW/m}^2$$

8.27 A third correlation for peak heat flux in nucleate boiling, determined by Kutateladze through dimensional analysis, is

$$\left(\frac{q}{A}\right)_{max} = (0.14)(\rho_v)^{1/2} h_{fg}[\sigma(\rho_l - \rho_v)gg_c]^{1/4} \tag{1}$$

(When using SI units, replace g_c by 1.) Repeat Problem 8.24 using this correlation.

$$\left(\frac{q}{A}\right)_{max} = (0.14)(0.60\ \text{kg/m}^3)^{1/2}(2.25\ \text{MJ/kg})[(0.0584\ \text{N/m})(957.6\ \text{kg/m}^3)(9.8\ \text{m/s}^2)]^{1/4} = 1.181\ \text{MW/m}^2$$

8.28 A simplifying assumption for the boiling of water is that $h \propto p^{0.4}$ (see Problem 8.15); i.e.,

$$h = h_a\left(\frac{p}{p_a}\right)^{0.4} \tag{1}$$

Some simple correlations for h_a, the heat transfer coefficient at standard atmospheric pressure p_a, are given in Table 8-1. Estimate the heat transfer per unit area from a horizontal flat plate submerged in water at atmospheric pressure, if the plate is held at 106 °C.

TABLE 8-1

configuration	h_a, Btu/hr·ft²·°F	range of validity, Btu/hr·ft²
Horizontal surface	$h_a = 151(\Delta T)^{1/3}$	$q/A < 5000$
(in wide vessel)	$h_a = (0.168)(\Delta T)^3$	$5000 < q/A < 75\,000$
Vertical surface	$h_a = 87(\Delta T)^{1/7}$	$q/A < 1000$
(in wide vessel)	$h_a = (0.24)(\Delta T)^3$	$1000 < q/A < 20\,000$
Vertical tube	$h_a \approx 189(\Delta T)^{1/3}$	$q/A < 5000$
(interior)	$h_a \approx (0.21)(\Delta T)^3$	$5000 < q/A < 75\,000$

Table 8-1 requires all data in British Engineering units; thus,

$$\Delta T = [(106 - 100)\ ^\circ\text{C}]\left(\frac{9\ ^\circ\text{F}}{5\ ^\circ\text{C}}\right) = 10.8\ ^\circ\text{F}$$

$$h = h_a = 151(\Delta T)^{1/3} = 151(10.8)^{1/3} = 333.77\ \text{Btu/hr}\cdot\text{ft}^2\cdot{}^\circ\text{F}$$

$$\frac{q}{A} = h\,\Delta T = (333.77\ \text{Btu/hr}\cdot\text{ft}^2\cdot{}^\circ\text{F})(10.8\ ^\circ\text{F}) = 3604.75\ \text{Btu/hr}\cdot\text{ft}^2 < 5000\ \text{Btu/hr}\cdot\text{ft}^2$$

or, converting back to SI,

$$\frac{q}{A} = (3604.75\ \text{Btu/hr}\cdot\text{ft}^2)\left(\frac{1\ \text{kW}}{3413\ \text{Btu/hr}}\right)\frac{1\ \text{ft}^2}{(0.3048\ \text{m})^2} = 11.37\ \text{kW/m}^2$$

8.29 If the plate of Problem 8.25 were copper, what would be the heat flux, in kW/m^2?

We note that in the Rohsenow equation, (*1*) of Problem 8.16, all parameters are identical to those in Problem 8.25 except the surface–fluid constant C_{sf}; therefore, if the boiling remains nucleate,

$$\left(\frac{q}{A}\right)_{\text{copper}} = \left(\frac{C_{\text{sf nickel}}}{C_{\text{sf copper}}}\right)^3\left(\frac{q}{A}\right)_{\text{nickel}}$$

From Table F-1, $C_{\text{sf copper}} = 0.013$; hence

$$\left(\frac{q}{A}\right)_{\text{copper}} = \left(\frac{0.006}{0.013}\right)^3(2.48\times10^4\ \text{Btu/hr}\cdot\text{ft}^2)\left(\frac{1\ \text{kW/m}^2}{317.1\ \text{Btu/hr}\cdot\text{ft}^2}\right) = 7.688\ \text{kW/m}^3$$

The boiling is certainly nucleate, since the peak heat flux, which is independent of the heater material, has the same value as in Problem 8.22.

8.30 How does the result of Problem 8.25 compare with that obtained from Table 8-1?

From Table 8-1, $h = h_a = 151(\Delta T)^{1/3}$, so that

$$\frac{q}{A} = h\,\Delta T = 151(\Delta T)^{4/3} = 151(10)^{4/3} = 3.25\times10^3\ \text{Btu/hr}\cdot\text{ft}^2$$

which deviates by a great amount. In this case an engineer would favor the result from the Rohsenow correlation.

8.31 A brass plate which is submerged horizontally in water at atmospheric pressure is heated at the rate of $0.7\ \text{MW/m}^2$. At what temperature, in °C, must the plate be held?

Assume that nucleate boiling occurs. Except for the excess temperature ΔT and the surface–fluid constant C_{sf}, the parameters are identical to those in Problem 8.25. We may, then, write

$$\frac{(q/A)_{\text{brass}}}{(q/A)_{\text{nickel}}} = \left[\frac{(\Delta T)_{\text{brass}}}{(\Delta T)_{\text{nickel}}}\,\frac{C_{\text{sf nickel}}}{C_{\text{sf brass}}}\right]^3$$

From Table F-1, $C_{\text{sf brass}} = 0.006$, and from Problem 8.25,

$$\left(\frac{q}{A}\right)_{\text{nickel}} = (2.48\times10^4\ \text{Btu/hr}\cdot\text{ft}^2)\left(\frac{3.1537\ \text{W/m}^2}{\text{Btu/hr}\cdot\text{ft}^2}\right) = 0.0782\ \text{MW/m}^2$$

$$(\Delta T)_{\text{nickel}} = 10\ ^\circ\text{F} = 5.56\ ^\circ\text{C}$$

Using these values in the above relation, we get

$$\frac{0.7}{0.0782} = \left[\frac{(\Delta T)_{\text{brass}}}{5.56\ ^\circ\text{C}}\left(\frac{0.006}{0.006}\right)\right]^3$$

Solving,

$$(\Delta T)_{\text{brass}} = (5.56\ ^\circ\text{C})\left(\frac{0.7}{0.0782}\right)^{1/3} = 11.54\ ^\circ\text{C}$$

and

$$(T_s)_{\text{brass}} = (\Delta T)_{\text{brass}} + T_{\text{sat}} = 11.54 + 100 = 111.54\ ^\circ\text{C}$$

Since the excess temperature $\Delta T = 11.54\ ^\circ\text{C} = 20.78\ ^\circ\text{F}$, the assumption of nucleate boiling appears to be reasonable from Fig. 8-1.

8.32 Determine the dimensional constant for use in the first equation of Table 8-1 when SI units are used.

▮ The constant 151 has units

$$\frac{[h]}{[\Delta T]^{1/3}} = \frac{\text{Btu/hr}\cdot\text{ft}^2\cdot{}^\circ\text{F}}{{}^\circ\text{F}^{1/3}}$$

Thus

$$h_a = \left(\frac{151\ \text{Btu/hr}\cdot\text{ft}^2\cdot{}^\circ\text{F}}{{}^\circ\text{F}^{1/3}}\right)(\Delta T)^{1/3} = \left[\left(\frac{151\ \text{Btu/hr}\cdot\text{ft}^2\cdot{}^\circ\text{F}}{{}^\circ\text{F}^{1/3}}\right)\left(\frac{5.6783\ \text{W/m}^2\cdot\text{K}}{\text{Btu/hr}\cdot\text{ft}^2\cdot{}^\circ\text{F}}\right)\left(\frac{9\ {}^\circ\text{F}}{5\ \text{K}}\right)^{1/3}\right](\Delta T)^{1/3}$$

$$= (1.043\ \text{kW/m}^2\cdot\text{K}^{4/3})(\Delta T)^{1/3}$$

where now h_a is in $\text{kW/m}^2\cdot\text{K}$ and ΔT is in K. This relation is valid for

$$\frac{q}{A} < (5000\ \text{Btu/hr}\cdot\text{ft}^2)\left(\frac{3.1537\ \text{W/m}^2}{\text{Btu/hr}\cdot\text{ft}^2}\right) = 15\,769\ \text{W/m}^2 \approx 16\ \text{kW/m}^2$$

8.33 Rework Problem 8.28, using the result of Problem 8.32.

▮
$$\frac{q}{A} = h_a\,\Delta T = (1.043\ \text{kW/m}^2\cdot\text{K}^{4/3})(6\ \text{K})^{4/3} = 11.37\ \text{kW/m}^2$$

8.34 Estimate the maximum nucleate boiling heat flux for water in a pressure cooker at 30 psia.

▮ From the steam tables, Table B-4, and Figure F-1,

$$\rho_v = \left(\frac{1}{13.746}\right)(16.02) = 1.1647\ \text{kg/m}^3 \qquad \rho_l = \left(\frac{1}{0.017\,01}\right)(16.02) = 941.82\ \text{kg/m}^3$$

$$h_{fg} = (945.3\ \text{Btu/lbm})\left(\frac{1\ \text{lbm}}{0.454\ \text{kg}}\right)(1054.8\ \text{J/Btu}) = 2.196\ \text{MJ/kg}$$

$$\sigma = (0.004\ \text{lbf/ft})(4.448\ \text{N/lbf})\left(\frac{1\ \text{ft}}{0.3048\ \text{m}}\right) = 0.0584\ \text{N/m}$$

Using (*1*) of Problem 8.27,

$$\left(\frac{q}{A}\right)_{\max} = (0.14)(\rho_v)^{1/2}h_{fg}[\sigma(\rho_l - \rho_v)gg_c]^{1/4}$$

$$= (0.14)(1.1647)^{1/2}(2.196\ \text{MJ/kg})[(0.0584)(941.82 - 1.1647)(9.8)(1.0)]^{1/4} = 1.598\ \text{MW/m}^2$$

8.2 FILM BOILING

8.35 Approximate the heat flux in film boiling from a 0.40-in.-diam. horizontal tube in saturated water at atmospheric pressure. The surface temperature of the tube is 488 °F. Use Bromley's equation, (*F-2*), neglecting radiation.

▮ The saturation temperature of water at atmospheric pressure is 212 °F; hence the temperature for property evaluations is $T_f = \frac{1}{2}(488 + 212) = 350\ {}^\circ\text{F}$.

$$\Delta T = T_s - T_{\text{sat}} = 488 - 212 = 276\ {}^\circ\text{F} \qquad k_{vf} = 0.0173\ \text{Btu/hr}\cdot\text{ft}\cdot{}^\circ\text{F}$$

$$\rho_{vf} = 0.0306\ \text{lbm/ft}^3 \qquad \rho_l = \frac{1}{0.016\,72}\ \text{lbm/ft}^3 = 59.8\ \text{lbm/ft}^3$$

$$g = 32.17\ \text{ft/sec}^2 \qquad h_{fg} = 970.3\ \text{Btu/lbm}\ \ (\text{at } 212\ {}^\circ\text{F}) \qquad c_{pvf} = 0.473\ \text{Btu/lbm}\cdot{}^\circ\text{F}$$

$$D = 0.40\ \text{in.} = 0.033\ \text{ft} \qquad \mu_{vf}g_c = 1.025\times10^{-5}\ \text{lbm/ft}\cdot\text{sec}$$

Assuming $h_r = 0$, and making the time units consistent,

$$h = h_c = (0.62)\Bigg[(0.0173\ \text{Btu/hr}\cdot\text{ft}\cdot{}^\circ\text{F})^3(0.0306\ \text{lbm/ft}^3)(59.77\ \text{lbm/ft}^3)(32.17\ \text{ft/sec}^2)$$

$$\times\frac{970.3\ \text{Btu/lbm} + 0.4(0.473\ \text{Btu/lbm}\cdot{}^\circ\text{F})(276\ {}^\circ\text{F})}{(0.033\ \text{ft})(1.025\times10^{-5}\ \text{lbm/ft}\cdot\text{sec})(276\ {}^\circ\text{F})}\left(\frac{3600\ \text{sec}}{1\ \text{hr}}\right)\Bigg]^{1/4} = 36.50\ \text{Btu/hr}\cdot\text{ft}^2\cdot{}^\circ\text{F}$$

$$\frac{q}{A} = h\,\Delta T = (36.50)(276) = 10\,070\ \text{Btu/hr}\cdot\text{ft}^2$$

8.36 A 2-in.-diam. polished copper bar is submerged horizontally in a pool of water at atmospheric pressure and 68 °F. The bar is maintained at 300 °F. Estimate the heat transfer rate per foot of the bar.

▌ The excess temperature is $\Delta T = T_s - T_{sat} = 300 - 212 = 88$ °F, which may lie in the nucleate boiling regime III of Fig. 8-1; therefore (*1*) of Problem 8.16 will be tried and compared with the peak heat flux, from Problem 8.22.

The required parameters taken from Table B-3, except where noted, are given below. Note that the liquid parameters are evaluated at the saturation condition at 212 °F, since the temperature of the pool of water has little effect on the heat transfer.

$$h_{fg} = 970.4 \text{ Btu/lbm (steam tables)} \qquad \rho_l = 59.97 \text{ lbm/ft}^3 \qquad \rho_v = 0.0373 \text{ lbm/ft}^3 \text{ (steam tables)}$$

$$\frac{g}{g_c} = 1.0 \text{ lbf/lbm (standard gravity)}$$

$$\mu_l g_c = \nu_l \rho_l = (0.316 \times 10^{-5} \text{ ft}^2/\text{sec})(59.97 \text{ lbm/ft}^3) = 1.895 \times 10^{-4} \text{ lbm/ft} \cdot \text{sec} = 0.682 \text{ lbm/ft} \cdot \text{hr}$$

$$\sigma = 0.004 \text{ lbf/ft (Figure F-1)} \qquad c_l = 1.007 \text{ Btu/lbm} \cdot \text{°F} \qquad \mathbf{Pr}_l = 1.74 \qquad C_{sf} = 0.013 \text{ (Table F-1)}$$

Thus,

$$\frac{q}{A} = (0.682 \text{ lbm/ft} \cdot \text{hr})(970.4 \text{ Btu/lbm})\left\{\frac{(1.0 \text{ lbf/lbm})[(59.97 - 0.0373) \text{ lbm/ft}^3]}{0.004 \text{ lbf/ft}}\right\}^{1/2}$$

$$\times \left[\frac{(1.007 \text{ Btu/lbm} \cdot \text{°F})(88 \text{ °F})}{(970.4 \text{ Btu/lbm})(1.74)^{1.7}(0.013)}\right]^3 = 1.666 \times 10^6 \text{ Btu/hr} \cdot \text{ft}^2$$

But this exceeds $(q/A)_{max} = 4.81 \times 10^5$ Btu/hr · ft²; therefore, film boiling exists, and (*F-1*), (*F-2*), and (*F-3*) apply. Four additional vapor properties are required, evaluated at the mean film temperature,

$$T_f = \frac{T_s + T_{sat}}{2} = \frac{300 + 212}{2} = 256 \text{ °F}$$

hence, from Table B-4:

$$k_{vf} = 0.0150 \text{ Btu/hr} \cdot \text{ft} \cdot \text{°F} \qquad c_{pvf} = 0.482 \text{ Btu/lbm} \cdot \text{°F} \qquad \rho_{vf} = 0.0348 \text{ lbm/ft}^3$$

$$\mu_{vf} g_c = 8.98 \times 10^{-6} \text{ lbm/ft} \cdot \text{sec} = 0.0323 \text{ lbm/ft} \cdot \text{hr}$$

and

$$\epsilon = 0.023 \text{ (Table B-6)} \qquad \sigma = 0.1714 \times 10^{-8} \text{ Btu/hr} \cdot \text{ft}^2 \cdot \text{°R}^4 \text{ (Stefan–Boltzmann constant)}$$

By (*F-2*),

$$h_c = (0.62)\left[\frac{k_{vf}^3 \rho_{vf}(\rho_l - \rho_{vf}) g (h_{fg} + 0.4 c_{pvf} \Delta T)}{D(\mu_{vf} g_c) \Delta T}\right]^{1/4} = (0.62)$$

$$\times \left\{\frac{(0.0150 \text{ Btu/hr} \cdot \text{ft} \cdot \text{°F})^3 (0.0348 \text{ lbm/ft}^3)[(59.97 - 0.0348) \text{ lbm/ft}^3][(32.2 \text{ ft/sec}^2)(3600 \text{ sec/hr})^2]}{(\frac{2}{12} \text{ ft})(0.0323 \text{ lbm/ft} \cdot \text{hr})(88 \text{ °F})}\right\}^{1/4}$$

$$\times [970.4 \text{ Btu/lbm} + 0.4(0.482 \text{ Btu/lbm} \cdot \text{°F})(88 \text{ °F})]^{1/4} = 49.74 \text{ Btu/hr} \cdot \text{ft}^2 \cdot \text{°F}$$

By (*F-3*),

$$h_r = \frac{\sigma\epsilon(T_s^4 - T_{sat}^4)}{T_s - T_{sat}} = \frac{(0.1714 \times 10^{-8})(0.023)[(760)^4 - (672)^4]}{88} = 0.0581 \text{ Btu/hr} \cdot \text{ft}^2 \cdot \text{°F}$$

The radiation heat transfer coefficient is negligible, which could have been guessed from the outset since the surface temperature is relatively low. Therefore, (*F-1*) reduces to $h = h_c$, and

$$\frac{q}{A} = h_c \Delta T = (49.74)(88) = 4377 \text{ Btu/hr} \cdot \text{ft}^2$$

Since

$$A = \pi D L = \pi(\tfrac{2}{12} \text{ ft})L = 0.524L$$

the heat transfer per length of bar is

$$\frac{q}{L} = (0.524)(4377) = 2294 \text{ Btu/hr} \cdot \text{ft}$$

8.37 Determine the excess temperature at minimum heat flux for Problem 8.36.

▌ Substitute the parameters of Problem 8.36 in (*F-6*), using consistent time units, to find

$$\Delta T_C = (0.127)\left\{\frac{(0.0348\ \text{lbm/ft}^2)(970.4\ \text{Btu/lbm})}{(0.0150\ \text{Btu/hr}\cdot\text{ft}\cdot{}^\circ\text{F})[(1\ \text{hr}/3600\ \text{sec})]}\right\}\left\{\frac{(32.2\ \text{ft/sec}^2)[(59.97-0.0373)\ \text{lbm/ft}^3]}{(59.97+0.0373)\ \text{lbm/ft}^3}\right\}^{2/3}$$

$$\times\left\{\frac{0.004\ \text{lbf/ft}}{(1.0\ \text{lbf/lbm})[(59.97-0.0373)\ \text{lbm/ft}^3]}\right\}^{1/2}\left\{\frac{(0.555\ \text{lbm/ft}\cdot\text{hr})[(1\ \text{hr}/3600\ \text{sec})]}{(32.2\ \text{ft/sec}^2)[(59.97-0.0373)\ \text{lbm/ft}^3]}\right\}^{1/3} = 366\ {}^\circ\text{F}$$

8.38 Obtain the approximate solution (*F-5*) of (*F-1*). [Note that (*F-4*) is just a refinement of (*F-5*).]

▌ Multiply (*F-1*) through by $h^{1/3}$, obtaining

$$h^{4/3} = h_c^{4/3} + h_r h^{1/3} = h_c^{4/3}\left[1 + \frac{h_r}{h_c}\left(\frac{h}{h_c}\right)^{1/3}\right]$$

or

$$h = h_c\left[1 + \frac{h_r}{h_c}\left(\frac{h}{h_c}\right)^{1/3}\right]^{3/4} \qquad (1)$$

Now, for $h_r \ll h_c$, one may replace $(h/h_c)^{1/3}$ in (*1*) by $(1)^{1/3} = 1$ and may retain only the first two terms of the resulting binomial expansion:

$$h \approx h_c\left[1 + \frac{3}{4}\frac{h_r}{h_c}\right] = h_c + \frac{3}{4}h_r$$

8.3 FLOW BOILING

8.39 Figure 8-3 shows a typical *flow boiling* curve. A simplified approach for determination of the heat transfer in flow boiling is to sum the convective effect, either forced or natural convection without boiling, and the boiling effect:

$$\frac{q}{A} = \frac{q}{A}\bigg|_{\text{conv}} + \frac{q}{A}\bigg|_{\text{boil}} \qquad (1)$$

Here $(q/A)_{\text{conv}} = \bar{h}(T_s - T_b)$, where T_b is the bulk temperature, and $\bar{h}$ is given by the appropriate relation from Chapter 5 or 6. What change in heat flux would occur in Problem 8.9 if the water were flowing normal to the wire at 10 ft/sec? Assume the water temperature to be 212 °F.

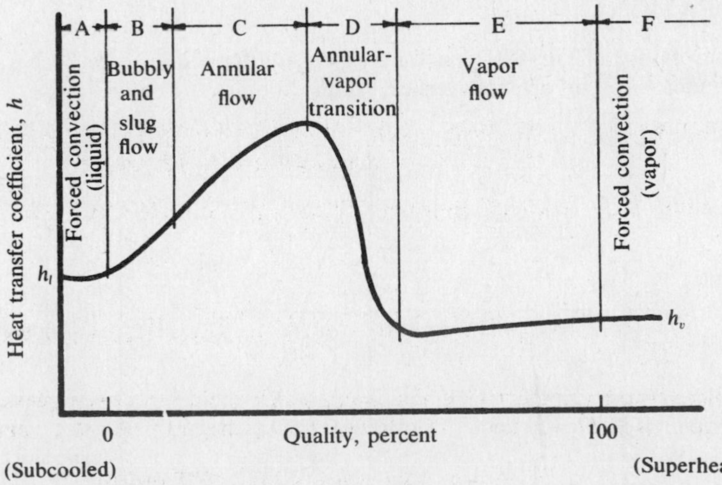

Fig. 8-3

▌ The change is just $(q/A)_{\text{conv}}$. Evaluating the fluid (vapor) properties at the film temperature

$$T_f = \frac{T_\infty + T_s}{2} = T_{\text{sat}} + \frac{\Delta T}{2} = 212 + \frac{50}{2} = 237\ {}^\circ\text{F}$$

we get the following values from Table B-4:

$k = 0.0145\ \text{Btu/hr}\cdot\text{ft}\cdot{}^\circ\text{F}$ $\qquad$ $\mathbf{Pr} = 1.0528$ $\qquad$ $\nu = 2.431\times10^{-4}\ \text{ft}^2/\text{sec}$

Therefore

$$\mathbf{Re}_D = \frac{V_\infty D}{\nu} = \frac{(10\ \text{ft/sec})[(0.040/12)\ \text{ft}]}{2.431 \times 10^{-4}\ \text{ft}^2/\text{sec}} = 137.12$$

At this Reynolds number, (*G-10*) and Table G-1 yield

$$\bar{h} = \frac{0.0145\ \text{Btu/hr} \cdot \text{ft} \cdot {}^\circ\text{F}}{(0.040/12)\ \text{ft}} (0.683)(1.0528)^{1/3}(137.12)^{0.466} = 29.94\ \text{Btu/hr} \cdot \text{ft}^2 \cdot {}^\circ\text{F}$$

and so

$$\left.\frac{q}{A}\right|_{\text{conv}} = \bar{h}(T_s - T_\infty) = (29.94\ \text{Btu/hr} \cdot \text{ft}^2 \cdot {}^\circ\text{F})[(262 - 212)\ {}^\circ\text{F}] = 1497\ \text{Btu/hr} \cdot \text{ft}^2$$

8.40 Film boiling occurs when water flows normal to a polished 15-mm-diam. copper tube at the rate of 3 m/s. Determine the boiling heat transfer coefficient when the tube is maintained at 114 °C.

▌ The tube size and fluid velocity are such that (*F-7*) gives h_c, while (*F-3*) gives h_r. Except where noted, the required fluid properties, evaluated at

$$T_f = \frac{T_s + T_{\text{sat}}}{2} = \frac{114 + 100}{2} = 107\ {}^\circ\text{C}$$

are taken from Table B-4:

$$k_{vf} = (0.0142)(1.7296) = 0.024\,56\ \text{W/m} \cdot \text{K} \qquad c_{pf} = (0.492)(4184) = 2058.5\ \text{J/kg} \cdot \text{K}$$
$$\rho_{vf} = (0.0366)(16.02) = 0.5863\ \text{kg/m}^3 \qquad h_{fg} = 2.25 \times 10^6\ \text{J/kg} \quad \text{(steam tables)}$$

Also, $\sigma = 5.6697 \times 10^{-8}\ \text{W/m}^2 \cdot \text{K}^4$ (Stefan–Boltzmann constant) and $\epsilon = 0.023$ (from Table B-6). Then

$$h_c = (2.7)\left\{\frac{(3\ \text{m/s})(0.024\,56\ \text{W/m} \cdot \text{K})(0.5863\ \text{kg/m}^3)[(2.25 \times 10^6\ \text{J/kg}) + 0.4(2058.5\ \text{J/kg} \cdot \text{K})(14\ \text{K})]}{(0.015\ \text{m})(14\ \text{K})}\right\}^{1/2}$$
$$= 1841.58\ \text{W/m}^2 \cdot \text{K}$$

$$h_r = \frac{(5.6697 \times 10^{-8}\ \text{W/m}^2 \cdot \text{K}^4)(0.023)[(387)^4 - (373)^4]\ \text{K}^4}{(387 - 373)\ \text{K}} = 0.2863\ \text{W/m}^2 \cdot \text{K}$$

Since $h_r \ll h_c$, the total heat transfer coefficient is given by (*F-5*) as

$$h \approx 1841.58 + \tfrac{3}{4}(0.2863) = 1841.79\ \text{W/m}^2 \cdot \text{K}$$

8.4 FILM CONDENSATION

8.41 Using the nomenclature of Fig. 8-4, derive an expression for the average heat transfer coefficient $\bar{h}$ for laminar film condensation on a (cool) vertical plate.

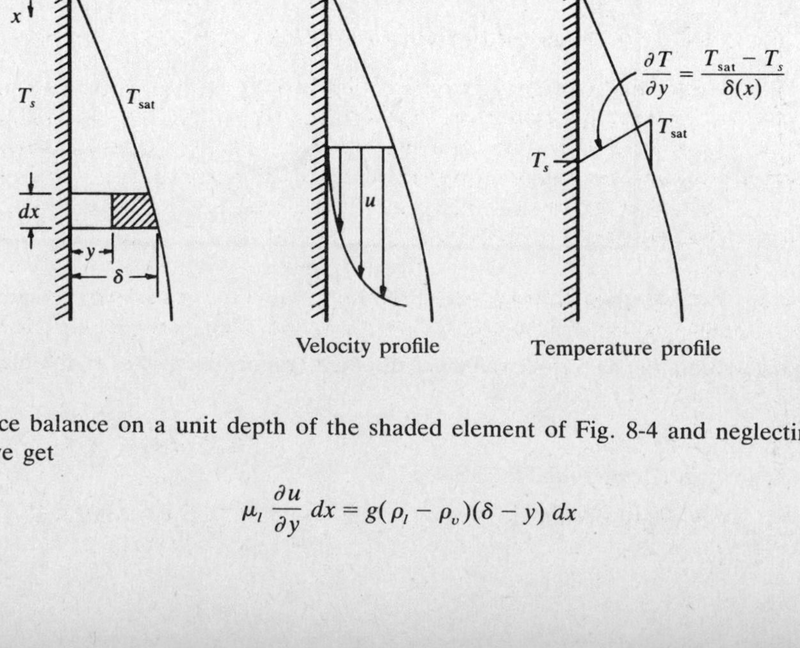

Fig. 8-4

▌ Making a force balance on a unit depth of the shaded element of Fig. 8-4 and neglecting inertia terms (low velocity), we get

$$\mu_l \frac{\partial u}{\partial y}\, dx = g(\rho_l - \rho_v)(\delta - y)\, dx \tag{1}$$

[To adapt (*1*) to British Engineering units, replace g by g/g_c (or μ_l by $\mu_l g_c$).] The term on the left is the viscous shear force at y, and the term on the right is the difference between the weight and buoyancy forces. The underlying assumptions are (1) linear temperature gradient in the film; (2) uniform surface temperature T_s; (3) pure vapor at the saturation temperature T_{sat}; and (4) negligible shear at the liquid–vapor interface, i.e., low velocity. For the no-slip boundary condition, $u=0$ at $y=0$, (*1*) integrates to give

$$u = \frac{g(\rho_l - \rho_v)}{\mu_l}\left(y\delta - \frac{y^2}{2}\right) \tag{2}$$

The condensate mass-flow rate per unit depth, $\dot{m}'$, at any downward distance x is given by

$$\dot{m}' = \int_0^{\delta} \rho_l u \, dy = \frac{\rho_l g(\rho_l - \rho_v)\delta^3}{3\mu_l} \tag{3}$$

from which the rate of change of mass flow with respect to condensate thickness is

$$\frac{d\dot{m}'}{d\delta} = \frac{\rho_l g(\rho_l - \rho_v)\delta^2}{\mu_l} \tag{4}$$

The excess mass-flow rate $d\dot{m}'$ must come from the condensation at the interface which is given by

$$d\dot{m}' = \frac{dq'}{h_{fg}} \tag{5}$$

where h_{fg} is the enthalpy of vaporization. Moreover, since the liberated heat is conducted through the film,

$$dq' = \frac{k_l(T_{sat} - T_s)}{\delta}\, dx \tag{6}$$

Combining (*4*), (*5*), and (*6*), we obtain, upon integration, the functional relation between δ and x:

$$\delta(x) = \left[\frac{4\mu_l k_l x(T_{sat} - T_s)}{\rho_l g(\rho_l - \rho_v)h_{fg}}\right]^{1/4} \tag{7}$$

Since the heat convected into the film is conducted through it to the plate

$$h_x(T_{sat} - T_s) = \frac{k_l(T_{sat} - T_s)}{\delta(x)} \qquad \text{or} \qquad h_x = \frac{k_l}{\delta(x)} \tag{8}$$

By (*8*), $h_x \propto x^{-1/4}$; hence, averaging over the height L of the plate, one finds

$$\bar{h} = \frac{4}{3}h_L = \frac{4}{3}\frac{k_l}{\delta(L)} = (0.943)\left[\frac{\rho_l g(\rho_l - \rho_v)h_{fg}k_l^3}{\mu_l L(T_{sat} - T_s)}\right]^{1/4} \tag{9}$$

With g replaced with $g \sin\phi$, (*9*) holds for a plate inclined at angle ϕ with the horizontal. For better agreement with experiment, the numerical coefficient in (*9*) is increased by 20 percent, yielding (in British Engineering units) the formula (*F-8*) given in Appendix F. In addition, (*F-8*) involves, instead of h_{fg} (evaluated at T_{sat}), the corrected enthalpy difference h^*_{fg}; see Problem 8.42.

8.42 Explain why the result of Problem 8.41 is improved by replacing h_{fg} by h^*_{fg} as given by (*F-9*).

▌ Equation (*5*) of Problem 8.41 is tantamount to the assumption that, in film condensation, the enthalpy change (cooling) occurs entirely in the phase transition, at T_{sat}. But there is additional, transfilm cooling of the liquid, from T_{sat} to T_s; this may be accounted for by an additive correction to h_{fg}: $h^*_{fg} = h_{fg} + \Delta$. On the assumption of a linear temperature profile (Fig. 8-4), a straightforward calculation yields $\Delta = \frac{3}{8}c_{pl}(T_{sat} - T_s)$. To include advection in the film, $\frac{3}{8}$ is changed to 0.68 by some authors. Even then, the Δ correction to h_{fg} is usually small. We will use $\frac{3}{8}$.

8.43 A vertical plate, 18 in. high and maintained at 88 °F, is exposed to saturated steam at atmospheric pressure. Determine the rate of heat transfer and the condensate mass flow per foot of plate width in film condensation.

▌ At $T_f = (212 + 88)/2 = 150$ °F,

$$\rho_l = 61.2 \text{ lbm/ft}^3 \qquad c_{pl} = 1.00 \text{ Btu/lbm}\cdot\text{°F}$$
$$\mu_l g_c = 0.292\times10^{-3} \text{ lbm/ft}\cdot\text{sec} \qquad \rho_v = 0.0372 \text{ lbm/ft}^3 \quad (\text{at } T_{sat})$$
$$k_l = 0.384 \text{ Btu/hr}\cdot\text{ft}\cdot\text{°F} \qquad h_{fg} = 970.3 \text{ Btu/lbm} \quad (\text{at } T_{sat})$$

Assuming laminar condensate flow, (*F-9*) and (*F-8*) give

$$\rho_l(\rho_l-\rho_v)\approx\rho_l^2=(61.2)^2=3750\ (\text{lbm/ft}^3)^2$$

$$h_{fg}^*=970.3+\tfrac{3}{8}(1.00)(212-88)=1016.8\ \text{Btu/lbm}$$

$$\bar{h}=1.13\left\{\frac{(0.384\ \text{Btu/hr}\cdot\text{ft}\cdot{}^\circ\text{F})^3(32.2\ \text{ft/sec}^2)[3750\ (\text{lbm/ft}^3)]^2(1016.8\ \text{Btu/lbm})}{(0.292\times10^{-3}\ \text{lbm/ft}\cdot\text{sec})(1.5\ \text{ft})(124\ {}^\circ\text{F})[(1\ \text{hr/3600 sec})]}\right\}^{1/4}=930\ \text{Btu/hr}\cdot\text{ft}^2\cdot{}^\circ\text{F}$$

Then $$q'=\bar{h}L(T_{sat}-T_s)=(930)(1.5)(212-88)=1.73\times10^5\ \text{Btu/hr}\cdot\text{ft}$$

and, by (*5*) of Problem 8.41,

$$\dot{m}'=\frac{q'}{h_{fg}^*}=\frac{1.73\times10^5}{1016.8}=170\ \text{lbm/hr}\cdot\text{ft}$$

Checking the Reynolds number via (*F-12*):

$$\mathbf{Re}_f=\frac{4\dot{m}'}{\mu_l g_c}=\frac{4(170\ \text{lbm/hr}\cdot\text{ft})}{(0.292\times10^{-3}\ \text{lbm/ft}\cdot\text{sec})(3600\ \text{sec/hr})}=647<1800$$

and the film flow is indeed laminar.

8.44 For the situation of Problem 8.41, estimate the height of plate necessary for condensate flow just to become turbulent ($\mathbf{Re}_f=1800$).

By (*F-12*) and (*F-8*), $\mathbf{Re}_f\propto L^{3/4}$, or $L\propto\mathbf{Re}_f^{4/3}$, in the laminar regime. Thus,

$$L_{trans}=(1.5\ \text{ft})(\tfrac{1800}{647})^{4/3}=5.87\ \text{ft}$$

8.45 Saturated ammonia at 10 °F condenses on a plate 14 in. long inclined 75° to the horizontal and maintained at −10 °F. Determine the rate of heat transfer to the plate and the rate of condensate per foot of plate width. The following properties of ammonia at $T_f=0$ °F (h_{fg} and v_v are at T_{sat}) may be used:

$h_{fg}=611.8$ Btu/lbm $v_l=0.024\,19$ ft³/lbm $c_{pl}=1.085$ Btu/lbm·°F
$\mu_l g_c=17.1\times10^{-5}$ lbm/ft·sec $v_v=9.116$ ft³/lbm $k_l=0.316$ Btu/hr·ft·°F

Assume the condensate flow to be laminar and use (*F-8*).

$$\rho_l(\rho_l-\rho_v)=\left(\frac{1}{0.024\,19}\right)\left(\frac{1}{0.024\,19}-\frac{1}{9.116}\right)\frac{\text{lbm}^2}{\text{ft}^6}\approx1704\ (\text{lbm/ft}^3)^2$$

$$h_{fg}^*=611.8+\tfrac{3}{8}(1.085)[10-(-10)]\approx620\ \text{Btu/lbm}$$

$$\sin75^\circ\approx0.966$$

$$\bar{h}=(1.13)\left[\frac{(0.316\ \text{Btu/hr}\cdot\text{ft}\cdot{}^\circ\text{F})^3(32.17\ \text{ft/sec}^2)(1704\ \text{lbm}^2/\text{ft}^6)(620\ \text{Btu/lbm})(0.966)}{(17.1\times10^{-5}\ \text{lbm/ft}\cdot\text{sec})(\tfrac{14}{12}\ \text{ft})(20\ {}^\circ\text{F})}(3600\ \text{sec/hr})\right]^{1/4}$$

$$=1110\ \text{Btu/hr}\cdot\text{ft}^2\cdot{}^\circ\text{F}$$

Check $\mathbf{Re}_f$ to see if laminar assumption was valid. By (*F-12*),

$$\mathbf{Re}_f=\frac{4\bar{h}(A/P)(T_{sat}-T_s)}{(\mu_l g_c)h_{fg}^*}=\frac{4(1110\ \text{Btu/hr}\cdot\text{ft}^2\cdot{}^\circ\text{F})(\tfrac{14}{12}\ \text{ft})(20\ {}^\circ\text{F})}{(17.1\times10^{-5}\ \text{lbm/ft}\cdot\text{sec})(3600\ \text{sec/hr})(620\ \text{Btu/lbm})}=271\quad(\textit{laminar})$$

Heat transfer per unit width is

$$q'=\bar{h}L(T_{sat}-T_s)=(1110\ \text{Btu/hr}\cdot\text{ft}^2\cdot{}^\circ\text{F})(\tfrac{14}{12}\ \text{ft})(20\ {}^\circ\text{F})=25\,900\ \text{Btu/hr}\cdot\text{ft}$$

Rate of condensation per unit width is

$$\dot{m}'=\frac{q'}{h_{fg}^*}=\frac{25\,900\ \text{Btu/hr}\cdot\text{ft}}{620\ \text{Btu/lbm}}=41.8\ \text{lbm/hr}\cdot\text{ft}$$

8.46 Determine the effective heat transfer coefficient for saturated Freon-12 at 140 °F experiencing film condensation on a horizontal $\tfrac{1}{2}$-in.-OD tube held at 80 °F by an internal flow of cool air. The Freon properties at the film temperature (h_{fg} and v_v are at T_{sat}) are

$h_{fg}=54.313$ Btu/lbm $v_l=0.012\,924$ ft³/lbm $c_{pl}=0.2456$ Btu/lbm·°F
$\mu_l g_c=15.75\times10^{-5}$ lbm/ft·sec $v_v=0.267\,69$ ft³/lbm $k_l=0.0395$ Btu/hr·ft·°F

$$\rho_l(\rho_l - \rho_v) = \left(\frac{1}{0.012\,924}\right)\left(\frac{1}{0.012\,924} - \frac{1}{0.267\,69}\right) = 5698\ \text{lbm}^2/\text{ft}^6$$

$$h^*_{fg} = (54.313) + \tfrac{3}{8}(0.2456)(140 - 80) = 59.839\ \text{Btu/lbm}$$

Then, by (*F-10*),

$$\bar{h} = (0.725)\left[\frac{k_l^3 g \rho_l(\rho_l - \rho_v)h^*_{fg}}{D(\mu_l g_c)(T_{\text{sat}} - T_s)}\right]^{1/4}$$

$$= (0.725)\left[\frac{(0.0395\ \text{Btu/hr}\cdot\text{ft}\cdot°\text{F})^3(32.17\ \text{ft/sec}^2)(5698\ \text{lbm}^2/\text{ft}^6)(59.839\ \text{Btu/lbm})}{(\frac{1}{24}\ \text{ft})(15.75 \times 10^{-5}\ \text{lbm/ft}\cdot\text{sec})(60\ °\text{F})}(3600\ \text{sec/hr})\right]^{1/4}$$

$$= 203.3\ \text{Btu/hr}\cdot\text{ft}^2\cdot°\text{F}$$

8.47 Repeat Problem 8.46 for saturated Freon-12 vapor condensing on a bank of $\frac{1}{2}$-in.-OD tubes, 10 in each vertical column and 12 in each horizontal row. If the tubes are 2 ft long, estimate the total rate of condensation.

By comparison of (*F-10*) and (*F-10*) with D replaced by $10D$ it is obvious that

$$\bar{h}_{10} = \bar{h}_1(\tfrac{1}{10})^{1/4}$$

where h_1 is the coefficient for a single tube as calculated in Problem 8.46. Thus,

$$\bar{h}_{10} = (203.3\ \text{Btu/hr}\cdot\text{ft}^2\cdot°\text{F})(\tfrac{1}{10})^{1/4} = 114.3\ \text{Btu/hr}\cdot\text{ft}^2\cdot°\text{F}$$

This applies to each tube (no correction necessary for 12 tubes in each row). Total area:

$$A = N(\pi DL) = (120)(\pi)(\tfrac{1}{24}\ \text{ft})(2\ \text{ft}) = 10\pi\ \text{ft}^2$$

$$q = \bar{h}A(T_{\text{sat}} - T_s) = (114.3\ \text{Btu/hr}\cdot\text{ft}^2\cdot°\text{F})(10\pi\ \text{ft}^2)(60\ °\text{F}) = 215\,450\ \text{Btu/hr}$$

$$\dot{m} = \frac{q}{h^*_{fg}} = \frac{215\,450\ \text{Btu/hr}}{59.839\ \text{Btu/lbm}} = 3600\ \text{lbm/hr}$$

8.48^D A vertical plate 1 ft high is maintained at 188 °F and exposed to saturated steam at atmospheric pressure. Calculate the heat transfer and condensation rate per hour per foot of plate width.

The liquid properties are evaluated at the film temperature, $T_f = (212 + 188)/2 = 200\ °\text{F}$. Thus,

$$\rho_l = 60.1\ \text{lbm/ft}^3 \qquad k_l = 0.394\ \text{Btu/hr}\cdot\text{ft}\cdot°\text{F}$$

$$\mu_l g_c = 0.205 \times 10^{-3}\ \text{lbm/ft}\cdot\text{sec} \qquad c_{pl} = 1.00\ \text{Btu/lbm}\cdot°\text{F}$$

The vapor density is $\rho_v = 0.0372\ \text{lbm/ft}^3$ at saturation conditions, and $h_{fg} = 970.3\ \text{Btu/lbm}$. We shall assume the condensate flow to be laminar and use (*F-8*):

$$\rho_l(\rho_l - \rho_v) \approx \rho_l^2 = (60.1)^2 = 3612\ (\text{lbm/ft}^3)^2$$

$$h^*_{fg} = 970.3 + \tfrac{3}{8}(1.00)(212 - 188) = 979.3\ \text{Btu/lbm}$$

$$\bar{h} = (1.13)\left[\frac{(0.394\ \text{Btu/hr}\cdot\text{ft}\cdot°\text{F})^3(32.17\ \text{ft/sec}^2)[3612\ (\text{lbm/ft}^3)^2](979.3\ \text{Btu/lbm})}{(0.205 \times 10^{-3}\ \text{lbm/ft}\cdot\text{sec})(1\ \text{ft})(24\ °\text{F})[(1\ \text{hr}/3600\ \text{sec})]}\right]^{1/4} = 1697\ \text{Btu/hr}\cdot\text{ft}^2\cdot°\text{F}$$

$$q' = \bar{h}L(T_{\text{sat}} - T_s) = (1697\ \text{Btu/hr}\cdot\text{ft}^2\cdot°\text{F}^2)(1\ \text{ft})(24\ °\text{F}) = 40\,728\ \text{Btu/hr}\cdot\text{ft}$$

$$\dot{m}' = \frac{q'}{h^*_{fg}} = \frac{40\,728\ \text{Btu/hr}\cdot\text{ft}}{979.3\ \text{Btu/lbm}} = 41.6\ \text{lbm/hr}\cdot\text{ft}$$

Checking the Reynolds number by (*F-12*):

$$\text{Re}_f = \frac{4\dot{m}'}{\mu_l g_c} = \frac{4(41.6\ \text{lbm/hr}\cdot\text{ft})}{(0.205 \times 10^{-3}\ \text{lbm/ft}\cdot\text{sec})(3600\ \text{sec/hr})} = 225 \quad (\text{OK})$$

8.49 For laminar film condensation, what is the ratio of heat transfer of a horizontal tube of large diameter to that of a vertical tube of the same size for the same temperature difference?

Dividing (*F-10*) by (*F-8*) with $\sin\phi = 1$, we get

$$\frac{\bar{h}_{\text{hor}}}{\bar{h}_{\text{ver}}} = (0.64)\left(\frac{L}{D}\right)^{1/4} \qquad (1)$$

8.50 What L/D ratio will produce the same laminar-film-condensation-controlled heat transfer rate to a tube in both the vertical and horizontal orientations? Assume the tube diameter is large compared with the condensate thickness.

▮ By (*1*) of Problem 8.49,

$$1 = (0.64)\left(\frac{L}{D}\right)^{1/4} \qquad \text{or} \qquad \frac{L}{D} = 5.96$$

8.51^D A wide vertical cooling fin, approximating a flat plate 0.3 m high, is exposed to steam at atmospheric pressure. The fin is maintained at 90 °C by cooling water. Determine the heat transfer rate and also the condensate mass flow rate per unit width.

▮ At

$$T_f = \frac{T_{sat} + T_s}{2} = \frac{100 + 90}{2} = 95\ \text{°C}$$

$$\rho_l = \left(\frac{1}{0.016\,654}\right)(16.02) = 961.9\ \text{kg/m}^3 \quad \text{(steam tables)}$$

$$\rho_v = \left(\frac{1}{26.8}\right)(16.02) = 0.598\ \text{kg/m}^3 \quad \text{(steam tables at } T_{sat}\text{)}$$

$$h_{fg} = 970.4\ \text{Btu/lbm} = 2.26\ \text{MJ/kg} \quad \text{(steam tables at } T_{sat}\text{)}$$

$$k_l = (0.3913)(1.7296) = 0.6767\ \text{W/m}\cdot\text{K} \qquad \mu_l = 300\ \mu\text{Pa}\cdot\text{s}$$

Thus, for laminar film flow, (*F-8*), with $h_{fg} \approx h^*_{fg}$, gives

$$\bar{h} = (1.13)\left\{\frac{(961.9\ \text{kg/m}^3)(9.8\ \text{m/s}^2)[(961.9 - 0.598)\ \text{kg/m}^3](2.26 \times 10^6\ \text{J/kg})(0.6767\ \text{W/m}\cdot\text{K})^3}{(300 \times 10^{-6}\ \text{Pa}\cdot\text{s})(0.3\ \text{m})(10\ \text{K})}\right\}^{1/4}$$

$$= 10.4\ \text{kW/m}^2\cdot\text{K}$$

$$q' = \bar{h}L(T_{sat} - T_s) = (10.4\ \text{kW/m}^2\cdot\text{K})(0.3\ \text{m})(10\ \text{K}) = 31.2\ \text{kW/m}$$

$$\dot{m}' = \frac{q'}{h_{fg}} = \frac{0.0312\ \text{MW/m}}{2.26\ \text{MJ/kg}} = 0.0138\ \text{kg/s}\cdot\text{m}$$

Checking the Reynolds number by (*F-12*):

$$\mathbf{Re}_f = \frac{4\dot{m}'}{\mu_l} = \frac{4(0.0138\ \text{kg/s}\cdot\text{m})}{300 \times 10^{-6}\ \text{Pa}\cdot\text{s}} = 184 \quad \text{(OK)}$$

8.52 A horizontal, 2-in.-OD tube is surrounded by saturated steam at 2.0 psia. The tube is maintained at 90 °F. What is the average heat transfer coefficient?

▮ The average heat transfer coefficient is given by (*F-10*), which requires the following property data, taken from Table B-3 and the steam tables. The liquid properties are evaluated at the mean film temperature, $T_f = (T_{sat} + T_s)/2 = 108$ °F.

$$\rho_l = 62.03\ \text{lbm/ft}^3 \qquad k_l = 0.364\ \text{Btu/hr}\cdot\text{ft}\cdot\text{°F}$$

$$\rho_v = 0.005\,76\ \text{lbm/ft}^3 \quad (\text{at } T_{sat}) \qquad \mu_l g_c = 4.26 \times 10^{-4}\ \text{lbm/ft}\cdot\text{sec}$$

$$h_{fg} = 1022.1\ \text{Btu/lbm} \quad (\text{at } T_{sat})$$

Using these data and writing $h^*_{fg} \approx h_{fg}$,

$$\bar{h} = (0.725) \times \left\{\frac{(62.03\ \text{lbm/ft}^3)(32.2\ \text{ft/sec}^2)[(62.03 - 0.005\,76)\ \text{lbm/ft}^3](1022.1\ \text{Btu/lbm})(0.364\ \text{Btu/hr}\cdot\text{ft}\cdot\text{°F})^3}{(4.26 \times 10^{-4}\ \text{lbm/ft}\cdot\text{sec})(\frac{2}{12}\ \text{ft})[(126 - 90)\ \text{°F}][(1\ \text{hr}/3600\ \text{sec})]}\right\}^{1/4}$$

$$= 1241.6\ \text{Btu/hr}\cdot\text{ft}^2\cdot\text{°F}$$

8.53 Rework Problem 8.52, using the corrected enthalpy function (*F-9*).

▮ At $T_f = 108$ °F the specific heat of water is 0.998 Btu/lbm · °F; therefore,

$$h^*_{fg} = 1022.1\ \text{Btu/lbm} + \tfrac{3}{8}(0.998\ \text{Btu/lbm}\cdot\text{°F})[(126 - 90)\ \text{°F}] = 1035.6\ \text{Btu/lbm}$$

Hence, a more accurate result for Problem 8.52 is

$$\bar{h} = (1241.6)\left(\frac{1035.6}{1022.1}\right)^{1/4} = 1245.7 \text{ Btu/hr}\cdot\text{ft}^2\cdot{}^\circ\text{F}$$

giving an error of 0.33 percent, negligible in most engineering calculations.

8.54 A vertical plate 4 ft high is maintained at 140 °F in the presence of saturated steam at atmospheric pressure. Estimate the heat transfer rate per unit width and the condensation rate per unit width.

▮ At

$$T_f = \frac{T_{\text{sat}} + T_s}{2} = \frac{212 + 140}{2} = 176\ {}^\circ\text{F}$$

$$\rho_l = 60.81 \text{ lbm/ft}^3 \qquad k_l = 0.386 \text{ Btu/hr}\cdot\text{ft}\cdot{}^\circ\text{F}$$

$$c_{pl} = 1.0023 \text{ Btu/lbm}\cdot{}^\circ\text{F} \qquad h^*_{fg} \approx h_{fg} = 970.4 \text{ Btu/lbm} \quad (\text{steam tables at } T_{\text{sat}})$$

$$\mu_l g_c = 0.238 \times 10^{-3} \text{ lbm/ft}\cdot\text{sec} \qquad \rho_v = 0.0373 \text{ lbm/ft}^3 \quad (\text{steam tables at } T_{\text{sat}})$$

Temporarily assuming laminar flow, we eliminate $\bar{h}$ between (*F-8*) and (*F-12*) to obtain, since $D_h = 4L$ and $\phi = 90°$,

$$\mathbf{Re}_f = (4.52)\left[\frac{\rho_l g(\rho_l - \rho_v)k_l^3(T_{\text{sat}} - T_s)^3}{(\mu_l g_c)^5 h_{fg}^3} L^3\right]^{1/4} = (4.52)$$

$$\times \left\{\frac{(60.81 \text{ lbm/ft}^3)(32.2 \text{ ft/sec}^2)[(60.81 - 0.0373) \text{ lbm/ft}^3](0.386 \text{ Btu/hr}\cdot\text{ft}\cdot{}^\circ\text{F})^3[(212-140)\ {}^\circ\text{F}]^3}{(0.238\times 10^{-3} \text{ lbm/ft}\cdot\text{sec})^5(970.4 \text{ Btu/lbm})^3(3600 \text{ sec/hr})^3}(4 \text{ ft})^3\right\}^{1/4}$$

$$= 1203 < 1800$$

Thus, the flow really is laminar, and (*F-12*) and (*5*) of Problem 8.41 yield

$$\dot{m}' = \frac{\mathbf{Re}_f(\mu_l g_c)}{4} = \frac{1203(0.238\times 10^{-3} \text{ lbm/ft}\cdot\text{sec})}{4} = 0.0716 \text{ lbm/sec}\cdot\text{ft} = 258 \text{ lbm/hr}\cdot\text{ft}$$

$$q' = \dot{m}'h^*_{fg} = (258 \text{ lbm/hr}\cdot\text{ft})(970.4 \text{ Btu/lbm}) = 250\,363 \text{ Btu/hr}\cdot\text{ft}$$

8.55 How would doubling the plate height of Problem 8.54 affect the heat transfer rate?

▮ The fluid parameters are identical to those of Problem 8.54. The flow will likely be turbulent, however. Indeed, (*F-13*) and (*F-12*) give

$$\mathbf{Re}_f = (0.002\,96)\left[\frac{\rho_l g(\rho_l - \rho_v)k_l^3(T_{\text{sat}} - T_s)^3}{(\mu_l g_c)^5 h_{fg}^3} L^3\right]^{5/9} = (0.002\,96)$$

$$\times \left\{\frac{(60.81 \text{ lbm/ft}^3)(32.2 \text{ ft/sec}^2)[(60.81 - 0.0373) \text{ lbm/ft}^3](0.386 \text{ Btu/hr}\cdot\text{ft}\cdot{}^\circ\text{F})^3[(212-140)\ {}^\circ\text{F}]^3}{(0.238\times 10^{-3} \text{ lbm/ft}\cdot\text{sec})^5(970.4 \text{ Btu/lbm})^3(3600 \text{ sec/hr})^3}(8 \text{ ft})^3\right\}^{5/9}$$

$$= 2303 > 1800$$

Since $q' \propto \mathbf{Re}_f$, the ratio of the heat transfer rates is

$$\frac{q_8'}{q_4'} = \frac{2302}{1203} = 1.91$$

and the heat transfer rate is approximately doubled by doubling the plate height.

The heat transfer rate increases more rapidly, however, as the plate is heightened further. For example, increasing the plate height by a factor of 4 (to 16 ft) gives more than a sixfold increase in heat transfer. This illustrates the effect which eddy diffusion has on the heat transfer rate.

CHAPTER 9
Heat Exchangers

9.1 The temperature profiles for a parallel-flow, flat-plate-type heat exchanger are shown in Figure J-5*a*, Appendix J. Knowing the surface area and the value of the overall heat transfer coefficient U where

$$q = UA\,\overline{\Delta T} \tag{J-1}$$

calculate $\overline{\Delta T}$ in terms of the four temperatures T_{hi}; T_{ho}; T_{ci}; T_{co}?

I We shall assume that

1. U is constant throughout the exchanger.
2. The system is adiabatic; heat exchange takes place only between the two fluids.
3. The temperatures of both fluids are constant over a given cross section and can be represented by bulk temperatures.
4. The specific heats of the fluids are constant (evaluated at the appropriate mean temperature).

Based upon these assumptions, the heat transfer between the hot and cold fluids for a differential length dx is

$$dq = U(T_h - T_c)\,dA \tag{1}$$

The energy gained by the cold fluid is equal to that given up by the hot fluid, i.e.,

$$dq = \dot{m}_c c_c\,dT_c = -\dot{m}_h c_h\,dT_h \tag{2}$$

where $\dot{m}$ is the mass flow rate and c is the specifc heat. Solving for the temperature differentials from equation (2) and subtracting, we get

$$d(T_h - T_c) = -\left(\frac{1}{\dot{m}_h c_h} + \frac{1}{\dot{m}_c c_c}\right) dq \tag{3}$$

Eliminating dq between (1) and (3) yields

$$\frac{d(T_h - T_c)}{T_h - T_c} = -U\left(\frac{1}{\dot{m}_h c_h} + \frac{1}{\dot{m}_c c_c}\right) dA \tag{4}$$

which integrates to give

$$\ln\frac{\Delta T_2}{\Delta T_1} = -UA\left(\frac{1}{\dot{m}_h c_h} + \frac{1}{\dot{m}_c c_c}\right) \tag{5}$$

where the ΔT terms are as shown in Figure J-5*a*.

From an energy balance on each fluid,

$$\dot{m}_h c_h = \frac{q}{T_{hi} - T_{ho}} \qquad \dot{m}_c c_c = \frac{q}{T_{co} - T_{ci}}$$

and substitution of these expressions into (5) gives

$$\ln\frac{\Delta T_2}{\Delta T_1} = -UA\,\frac{(T_{hi} - T_{ho}) + (T_{co} - T_{ci})}{q}$$

or, in terms of the differences in end temperatures,

$$q = UA\,\frac{\Delta T_2 - \Delta T_1}{\ln(\Delta T_2/\Delta T_1)} \tag{J-2}$$

Upon comparing this result with equation (J-1), we see that

$$\overline{\Delta T} = \frac{\Delta T_2 - \Delta T_1}{\ln(\Delta T_2/\Delta T_1)} \equiv (\Delta T)_{lm}$$

This average effective temperature difference is called the log-mean temperature difference (LMTD). It can easily be shown that the subscripts 1 and 2 may be interchanged without changing the value of $(\Delta T)_{lm}$; hence, the designation of ends for use in (J-2) is arbitrary.

9.2[D] Table J-1 lists overall U values for various fluid-to-fluid heat exchanger applications. For water flowing inside and in crossflow over tubes (as in shell-and-tube heat exchangers), a set of h values is $\bar{h}_i = 480\ \text{Btu/hr}\cdot\text{ft}^2\cdot{}^\circ\text{F}$, $\bar{h}_o = 370\ \text{Btu/hr}\cdot\text{ft}^2\cdot{}^\circ\text{F}$. If the tube is mild steel, OD = 1.18 in. and ID = 1.00 in., compare the U_o value with that listed in Table J-1. The average tube wall (metal) temperature is 212 °F.

▮ The overall U_o formulation for this problem neglecting "fouling" is, by (*5*) of Problem 1.75,

$$U_o = \frac{1}{A_o \sum R_{th}} = \frac{1}{A_o\left[\dfrac{1}{A_i\bar{h}_i} + \dfrac{\ln(r_o/r_i)}{(A_o/r_o)k_{i-o}} + \dfrac{1}{A_o\bar{h}_o}\right]} \tag{J-3}$$

where A_o and A_i are the outside tube surface area and inside tube surface area, respectively, and the term containing k_{i-o} is the resistance due to the tube wall based on outside surface area. This equation allows "extended surface" design on the outside of the tube. For "straight" tubing where $A_o = 2\pi r_o L$, the equation simplifies to [see (*1*) of Problem 1.76]

$$U_o = \frac{1}{\dfrac{r_o}{r_i\bar{h}_i} + r_o\dfrac{\ln(r_o/r_i)}{k_{i-o}} + \dfrac{1}{\bar{h}_o}} \tag{J-4}$$

In this problem,

$$r_o = \frac{1.18}{2} = 0.59\ \text{in.} \qquad r_i = \frac{1.00}{2} = 0.50\ \text{in.}$$

From Appendix B-1, k for mild steel at 212 °F is

$$k_{i-o} = 26\ \text{Btu/hr}\cdot\text{ft}\cdot{}^\circ\text{F}$$

Then

$$U_o = \frac{1}{\dfrac{0.59}{(0.50)(480\ \text{Btu/hr}\cdot\text{ft}^2\cdot{}^\circ\text{F})} + \dfrac{(0.59\ \text{in.})\ln(0.59/0.50)}{26\ \text{Btu/hr}\cdot\text{ft}\cdot{}^\circ\text{F}}\left(\dfrac{1\ \text{ft}}{12\ \text{in.}}\right) + \dfrac{1}{370\ \text{Btu/hr}\cdot\text{ft}^2\cdot{}^\circ\text{F}}}$$

$$= 182.6\ \text{Btu/hr}\cdot\text{ft}^2\cdot{}^\circ\text{F}$$

The range of values given in Table J-1 for water-to-water heat exchangers is 150–300 $\text{Btu/hr}\cdot\text{ft}^2\cdot{}^\circ\text{F}$. In this problem the effects of "fouling" of the heat exchanger surfaces have been omitted. See Problem 9.20.

9.3[D] In Problem 9-2, determine the error in U_o introduced by neglecting the conductive term in (*J-4*).

▮ $$U_o \approx \frac{1}{(r_o/r_i\bar{h}_i) + (1/\bar{h}_o)} = \left(\frac{1}{\dfrac{0.59}{0.50(480)} + \dfrac{1}{370}}\right)\frac{\text{Btu}}{\text{hr}\cdot\text{ft}^2\cdot{}^\circ\text{F}} = \frac{1}{0.002\,46 + 0.002\,70} = 193.8\ \text{Btu/hr}\cdot\text{ft}^2\cdot{}^\circ\text{F}$$

$$\%\ \text{error} = \frac{182.6 - 193.8}{182.6} = -6.1\%$$

This percentage error is often smaller than the inaccuracy of the correlation equations used to determine $\bar{h}_i$ and $\bar{h}_o$ (Chapters 5 and 6), and the conductive resistance is often neglected in calculating original (clean surface) values of U_o.

9.4[D] Table J-1 lists overall U values for various fluid-to-fluid heat exchanger applications. For water flowing inside and in crossflow over tubes (as in shell-and-tube heat exchangers), a set of h values is $\bar{h}_i = 2725\ \text{W/m}^2\cdot{}^\circ\text{C}$, $\bar{h}_o = 2100\ \text{W/m}^2\cdot{}^\circ\text{C}$. If the tube is mild steel, OD = 3.00 cm and ID = 2.50 cm, (*a*) compare the U_o value with that listed in Table J-1 and (*b*) examine the effect of omitting the conductive resistance of the tube wall. The average tube wall (metal) temperature is 100 °C.

▮ (*a*) The value of U_o (based on outside tube surface area and for no "extended" tube surface) is

$$U_o = \frac{1}{\dfrac{r_o}{r_i\bar{h}_i} + r_o\dfrac{\ln(r_o/r_i)}{k_{i-o}} + \dfrac{1}{\bar{h}_o}} \tag{J-4}$$

In this problem,

$$r_o = \frac{3.00}{2} = 1.50\ \text{cm} \qquad r_i = \frac{2.50}{2} = 1.25\ \text{cm}$$

From Appendix B-1, k for mild steel at 100 °C is $k_{i-o} = 45.0$ W/m · °C. Then

$$U_o = \cfrac{1}{\cfrac{1.50\text{ cm}}{(1.25\text{ cm})(2725\text{ W/m}^2\cdot{}^\circ\text{C})} + \cfrac{(1.50\text{ cm})\ln(1.50/1.25)}{45.0\text{ W/m}\cdot{}^\circ\text{C}}\left(\cfrac{1\text{ m}}{100\text{ cm}}\right) + \cfrac{1}{2100\text{ W/m}^2\cdot{}^\circ\text{C}}}$$

$$= 1024\text{ W/m}^2\cdot{}^\circ\text{C}$$

(*b*) Omitting the resistance of the tube wall.

$$U_o \approx \frac{1}{0.000\,440 + 0.000\,476} = 1091\text{ W/m}^2\cdot{}^\circ\text{C}$$

The error in omitting the conductive resistance is

$$\%\text{ error} = \frac{1024 - 1091}{1024} = -6.5\%$$

In this problem the effects of "fouling" the heat transfer surfaces have been neglected. For inclusion of fouling, see Problem 9.21.

9.5[D] Water at a flow rate of 20 lbm/min is cooled in a double-pipe heat exchanger (see Figure J-1) from 90 °F to 70 °F by a cold brine solution which enters at 8 °F and exits at 22 °F. If the overall heat transfer coefficient is 180 Btu/hr · ft² · °F, what heat exchanger area is required for (*a*) parallel flow and (*b*) counterflow?

▮ The heat transfer from the water is given by

$$q = \dot{m}c_p\,\Delta T$$

where $c_p = 0.9985$ Btu/lbm · °F is taken from Table B-3 at the mean temperature 80 °F; therefore,

$$q = (20\text{ lbm/min})(60\text{ min/hr})(0.9985\text{ Btu/lbm}\cdot{}^\circ\text{F})[(90-70)\ {}^\circ\text{F}] = 23\,964\text{ Btu/hr}$$

(*a*) Figure J-5*a* is a qualitative representation of the temperature distribution for the parallel-flow case. The log-mean temperature difference is given by equation (*J-2*).

$$(\Delta T)_{lm} = \frac{\Delta T_2 - \Delta T_1}{\ln(\Delta T_2/\Delta T_1)}$$

for which $\Delta T_1 = 90 - 8 = 82$ °F and $\Delta T_2 = 70 - 22 = 48$ °F; hence,

$$(\Delta T)_{lm} = \frac{48 - 82}{\ln \frac{48}{82}} = 63.49\ {}^\circ\text{F}$$

and

$$A = \frac{q}{U(\Delta T)_{lm}} = \frac{23\,964\text{ Btu/hr}}{(180\text{ Btu/hr}\cdot\text{ft}^2\cdot{}^\circ\text{F})(63.49\ {}^\circ\text{F})} = 2.10\text{ ft}^2$$

(*b*) The temperature distribution for the counterflow case is shown qualitatively in Figure J-5*b*, from which

$$\Delta T_1 = 90 - 22 = 68\ {}^\circ\text{F} \qquad \text{and} \qquad \Delta T_2 = 70 - 8 = 62\ {}^\circ\text{F}$$

Equation (*J-2*) yields

$$(\Delta T)_{lm} = \frac{62 - 68}{\ln \frac{62}{68}} = 64.95\ {}^\circ\text{F}$$

and

$$A = \frac{q}{U(\Delta T)_{lm}} = \frac{23\,964\text{ Btu/hr}}{(180\text{ Btu/hr}\cdot\text{ft}^2\cdot{}^\circ\text{F})(64.95\ {}^\circ\text{F})} = 2.05\text{ ft}^2$$

9.6[D] Water flowing at the rate of 0.15 kg/s is heated from 40 °C to 80 °C in a counterflow double-pipe heat exchanger (see Figures J-1 and J-5*b*). The hot fluid is oil and the overall heat transfer coefficient is 250 W/m² · K. Find the heat transfer area, if the oil enters at 105 °C and leaves at 70 °C.

▮ The heat gained by the water is given by $q = \dot{m}_w c_w (T_{wo} - T_{wi})$. From Table B-3, the specific heat of water at $(40 + 80)/2 = 60$ °C is

$$c_w = (0.994)(4.184 \times 10^3) = 4.181\text{ kJ/kg}\cdot\text{K}$$

This gives a heat transfer rate

$$q = (0.15\ \text{kg/s})(4.181\ \text{kJ/kg}\cdot\text{K})[(80 - 40)\ \text{K}] = 25.1\ \text{kW}$$

Using Figure J-5*b*, where $\Delta T_2 = 70 - 40 = 30\ °\text{C}$ and $\Delta T_1 = 105 - 80 = 25\ °\text{C}$, we get with equation (*J-2*) (see Problem 9.1)

$$A = \frac{q}{U}\frac{\ln(\Delta T_2/\Delta T_1)}{\Delta T_2 - \Delta T_1} = \frac{2.51\times 10^4\ \text{W}}{250\ \text{W/m}^2\cdot\text{K}}\left[\frac{\ln\frac{30}{25}}{(30-25)\ \text{K}}\right] = 3.66\ \text{m}^2$$

9.7[D] Using a double-pipe counterflow heat exchanger, water at a flow rate of 50 lbm/min is heated from 55 °F to 95 °F by oil which enters at 200 °F and leaves at 140 °F. Determine the heat exchanger area for an overall heat transfer coefficient of 42 Btu/hr · ft² · °F.

▌ With $c_p = 1.001$ from Table B-3, at 75 °F, the total heat transfer to the water is

$$q = \dot{m}c_p\,\Delta T = (50\ \text{lbm/min})(60\ \text{min/hr})(1.001\ \text{Btu/lbm}\cdot°\text{F})[(95 - 55)\ °\text{F}] = 120\,120\ \text{Btu/hr}$$

Referring to Figure J-5*b*, the log-mean temperature difference is given by

$$(\Delta T)_{lm} = \frac{\Delta T_2 - \Delta T_1}{\ln(\Delta T_2/\Delta T_1)} \qquad (J\text{-}2)$$

where $\Delta T_2 = 140 - 55 = 85\ °\text{F}$ and $\Delta T_1 = 200 - 95 = 105\ °\text{F}$; hence,

$$(\Delta T)_{lm} = \frac{85 - 105}{\ln\frac{85}{105}} = \frac{-20}{-0.211} = 94.6\ °\text{F}$$

and the area is

$$A = \frac{q}{U(\Delta T)_{lm}} = \frac{120\,120\ \text{Btu/hr}}{(42\ \text{Btu/hr}\cdot\text{ft}^2\cdot°\text{F})(94.6\ °\text{F})} = 30.2\ \text{ft}^2$$

9.8[D] For the same parameters as in Problem 9.7, what area is required when using a shell-and-tube heat exchanger with the water making one shell pass and the oil making two tube passes?

▌ In this case we need a correction factor F from Figure J-6 to use in the relation

$$q = UAF(\Delta T)_{lm} \qquad (J\text{-}5)$$

where $(\Delta T)_{lm}$ is that for the concentric double-pipe counterflow heat exchanger, Figures J-1 and J-5*b*. From the nomenclature of Figure J-6: $T_i = 55\ °\text{F}$, $T_o = 95\ °\text{F}$, $t_i = 200\ °\text{F}$, and $t_o = 140\ °\text{F}$; therefore, the dimensionless parameters required to get the correction factor are

$$P = \frac{t_o - t_i}{T_i - t_i} = \frac{140 - 200}{55 - 200} = 0.414 \qquad Z = \frac{T_i - T_o}{t_o - t_i} = \frac{55 - 95}{140 - 200} = 0.666$$

From Figure J-6, $F \approx 0.95$ and the area is (see Problem 9.7)

$$A = \frac{q}{UF(\Delta T)_{lm}} \approx \frac{30.2\ \text{ft}^2}{0.95} = 31.8\ \text{ft}^2$$

9.9 Repeat Problem 9.7 for a heat exchanger having two shell passes and four tube passes.

▌ The dimensionless parameters for this case are the same as those of Problem 9.8, but Figure J-7 must be used to obtain F, which approaches unity. Therefore, $A \approx 30.2\ \text{ft}^2$.

9.10[D] Using a double-pipe counterflow heat exchanger (see Figure J-1), water at a flow rate of 23 kg/min is heated from 13 °C to 35 °C by oil which enters at 94 °C and leaves at 60 °C. Determine the heat exchanger area for an overall heat transfer coefficient of 238 W/m² · K.

▌ With $c_p = 4.188\ \text{kJ/kg}\cdot\text{K}$ (at 24 °C), the total heat transfer rate to the water is

$$q = \dot{m}c_p\,\Delta T = (23\ \text{kg/min})(1\ \text{min}/60\ \text{s})(4.188\ \text{kJ/kg}\cdot\text{K})[(35 - 13)\ \text{K}] = 35.319\ \text{kW}$$

Referring to Figure J-5*b*, the log-mean temperature difference is given by

$$(\Delta T)_{lm} = \frac{\Delta T_2 - \Delta T_1}{\ln(\Delta T_2/\Delta T_1)} \qquad (J\text{-}2)$$

where $\Delta T_2 = 60 - 13 = 47\ °\text{C}$ and $\Delta T_1 = 94 - 35 = 59\ °\text{C}$; hence,

$$(\Delta T)_{lm} = \frac{47 - 59}{\ln\frac{47}{59}} = \frac{-12}{-0.227} = 52.77\ °\text{C}$$

and the area is

$$A = \frac{q}{U(\Delta T)_{lm}} = \frac{35\,319\ \text{W}}{(238\ \text{W/m}^2\cdot\text{K})(52.77\ \text{K})} = 2.81\ \text{m}^2$$

9.11[D] For the same parameters as in Problem 9.10, what area is required when using a shell-and-tube heat exchanger with the water making one shell pass and the oil making two tube passes?

▌ In this case we need a correction factor F from Figure J-6 to use in the relation

$$q = UAF(\Delta F)_{lm} \tag{J-5}$$

From Problem 9.10 and the nomenclature of Figure J-6: $T_i = 13\ ^\circ\text{C}$, $T_o = 35\ ^\circ\text{C}$, $t_i = 94\ ^\circ\text{C}$, and $t_o = 60\ ^\circ\text{C}$; therefore, the dimensionless parameters required to get the correction factor are

$$P = \frac{t_o - t_i}{T_i - t_i} = \frac{60-94}{13-94} = 0.420 \qquad Z = \frac{T_i - T_o}{t_o - t_i} = \frac{13-35}{60-94} = 0.647$$

From Figure J-6, $F \approx 0.95$ and the area is (see Problem 9.10)

$$A = \frac{q}{UF(\Delta T)_{lm}} \approx \frac{2.81\ \text{m}^2}{0.95} = 2.96\ \text{m}^2$$

9.12 Design (determine the heat exchange area) a counterflow double-pipe heat exchanger (see Figure J-1) to cool hot oil with a specific heat $c_p = 2.05\ \text{kJ/kg}\cdot\text{K}$ from 195 °C to 65 °C. The hot oil flow rate is 0.7 kg/s. Cold oil $(c_p = 1.84\ \text{kJ/kg}\cdot\text{K})$ exits at 149 °C at the rate of 1.0 kg/s.

▌ The unknown inlet temperature of the cold oil may be found from an energy balance on the two fluids, i.e.,

$$\dot{m}_c c_c (T_{co} - T_{ci}) = \dot{m}_h c_h (T_{hi} - T_{ho})$$

$$T_{ci} = T_{co} - \frac{\dot{m}_h c_h}{\dot{m}_c c_c}(T_{hi} - T_{ho}) = 149\ ^\circ\text{C} - \frac{(0.70\ \text{kg/s})(2.05\ \text{kJ/kg}\cdot\text{K})}{(1.0\ \text{kg/s})(1.84\ \text{kJ/kg}\cdot\text{K})}[(195-65)\ ^\circ\text{C}] = 47.6\ ^\circ\text{C}$$

Knowing the inlet and exit temperature for both fluids, we may now use equation (*J-2*) in conjunction with an energy balance on one fluid to determine the area. From an energy balance on the hot fluid, the heat transfer is given by

$$q = \dot{m}_h c_h (T_{hi} - T_{ho}) = (0.70\ \text{kg/s})(2.05\ \text{kJ/kg}\cdot\text{K})[(195-65)\ \text{K}] = 186.6\ \text{kW}$$

Referring to Figure J-5*b* and using (*J-2*), we have

$$A = \frac{q}{U}\left[\frac{\ln(\Delta T_2/\Delta T_1)}{\Delta T_2 - \Delta T_1}\right]$$

where

$$\Delta T_2 = T_{ho} - T_{ci} = 65 - 47.6 = 17.4\ ^\circ\text{C} \qquad \Delta T_1 = T_{hi} - T_{co} = 195 - 149 = 46\ ^\circ\text{C}$$

The "design" requires choosing an appropriate U value. From Table J-1, suitable values of U for oil-to-oil heat exchangers are from 170–312 W/m²·K. (The exact value depends upon all the factors involved in determining h values for the center pipe flow and the annular flow, specifically velocity, diameters, fluid properties, etc.) Choosing $U \approx 300\ \text{W/m}^2\cdot\text{K}$, we obtain

$$A \approx \frac{186.6\ \text{kW}}{0.300\ \text{kW/m}^2\cdot\text{K}}\left[\frac{\ln(17.4/46)}{(17.4-46)\ \text{K}}\right] = 21.1\ \text{m}^2$$

9.13 What area would be required for the conditions of Problem 9.6 if a shell-and-tube heat exchanger were substituted for the double-pipe heat exchanger? The water makes one shell pass, and the oil makes two tube passes.

▌ Assuming that the overall heat transfer coefficient remains at 250 W/m²·K, we must get a correction factor F from Figure J-6 to use in

$$q = UAF(\Delta T)_{lm} \tag{J-5}$$

The temperatures for use in the figure are

$$T_i = 40\ ^\circ\text{C} \qquad T_o = 80\ ^\circ\text{C} \qquad t_i = 105\ ^\circ\text{C} \qquad t_o = 70\ ^\circ\text{C}$$

The dimensionless ratios are

$$P = \frac{t_o - t_i}{T_i - t_i} = \frac{70 - 105}{40 - 105} = 0.54 \qquad Z = \frac{T_i - T_o}{t_o - t_i} = \frac{40 - 80}{70 - 105} = 1.14$$

which with Figure J-6 gives a correction factor $F \approx 0.62$; therefore (see Problem 9.6)

$$A = \frac{q}{UF(\Delta T)_{lm}} \approx \frac{3.66 \text{ m}^2}{0.62} = 5.90 \text{ m}^2$$

Clearly, the shell-and-tube heat exchanger is less area efficient than the concentric double-pipe, counterflow unit.

9.14 Hot engine oil ($c_p = 1.9$ kJ/kg · K) enters a crossflow heat exchanger, with both fluids unmixed, at 100 °C and a flow rate of 0.13 kg/s. The cooling fluid is water which enters at 40 °C and exits at 80 °C. The water flow rate is 0.05 kg/s and $c_p = 4.181$ kJ/kg · K. The heat transfer area is 3.0 m². (*a*) What is the oil exit temperature? (*b*) What is the overall heat transfer coefficient?

I (*a*) Using subscripts *e* for engine oil and *w* for water, we have from an energy balance

$$T_{eo} = T_{ei} - \frac{\dot{m}_e c_e}{\dot{m}_w c_w}(T_{wo} - T_{wi}) = 100\ ^\circ\text{C} - \frac{(0.13 \text{ kg/s})(1.9 \text{ kJ/kg}\cdot\text{K})}{(0.05 \text{ kg/s})(4.181 \text{ kJ/kg}\cdot\text{K})}[(80 - 40)\ ^\circ\text{C}] = 52.74\ ^\circ\text{C}$$

(*b*) In the nomenclature of Figure J-9,

$$T_i = 100\ ^\circ\text{C} \qquad T_o = 52.74\ ^\circ\text{C} \qquad t_i = 40\ ^\circ\text{C} \qquad t_o = 80\ ^\circ\text{C}$$

Evaluating the dimensionless parameters, we get

$$P = \frac{t_o - t_i}{T_i - t_i} = \frac{80 - 40}{100 - 40} = 0.67 \qquad Z = \frac{T_i - T_o}{t_o - t_i} = \frac{100 - 52.74}{80 - 40} = 1.18$$

and the correction factor is $F \approx 0.67$.

The heat transfer rate is given by an energy balance on (say) the water and equation (*J-5*):

$$q = \dot{m}_w c_w (T_{wo} - T_{wi}) = UAF(\Delta T)_{lm} = UAF\frac{\Delta T_2 - \Delta T_1}{\ln(\Delta T_2/\Delta T_1)} \tag{1}$$

where $(\Delta T)_{lm}$ is for the concentric pipe counterflow heat exchanger, Figure J-5*b*. Thus,

$$\Delta T_2 = T_{eo} - T_{wi} = 52.74 - 40 = 12.74\ ^\circ\text{C}$$

$$\Delta T_1 = T_{ei} - T_{wo} = 100 - 80 = 20\ ^\circ\text{C}$$

From the above heat transfer equation (*1*),

$$U = \frac{\dot{m}_w c_w (T_{wo} - T_{wi}) \ln(\Delta T_2/\Delta T_1)}{AF(\Delta T_2 - \Delta T_1)} \approx \frac{(0.05 \text{ kg/s})(4.181 \text{ kJ/kg}\cdot\text{K})[(80 - 40)\text{ K}]\ln(12.74/20)}{(3.0 \text{ m}^2)(0.67)[(12.74 - 20)\text{ K}]}$$
$$= 0.26 \text{ kW/m}^2\cdot\text{K}$$

This value of *U*, 260 W/m² · K, is within the range suggested in Table J-1 for water to lubricating oil.

9.15[D] Cooling water at 60 °F and a flow rate of 57.5 tons/hr is available to cool a 25-ton-per-hour air flow from 120 °F to 100 °F. Using a crossflow heat exchanger with both fluids unmixed and assuming the overall heat transfer coefficient to be 25 Btu/hr · ft² · °F, estimate the required heat exchange surface area.

I The total heat transfer rate from the air is

$$q = (\dot{m}c_p\,\Delta T)_a \approx (50\,000 \text{ lbm/hr})(0.24 \text{ Btu/lbm}\cdot{}^\circ\text{F})[(120 - 100)\ ^\circ\text{F}] = 240\,000 \text{ Btu/hr}$$

where $c_{p,a} \approx 0.24$ Btu/lbm · °F.

From an energy balance we can get the exit water temperature, i.e.,

$$\dot{m}_a c_{pa}\,\Delta T_a = \dot{m}_w c_{pw}\,\Delta T_w = 240\,000 \text{ Btu/hr}$$

so that with $c_{p,w} \approx 1.0$ Btu/lbm · °F

$$\Delta T_w \approx \frac{240\,000 \text{ Btu/hr}}{(115\,000 \text{ lbm/hr})(1.0 \text{ Btu/lbm}\cdot{}^\circ\text{F})} = 2.09\ ^\circ\text{F}$$

and

$$T_{wo} = T_{wi} + \Delta T_w \approx 60 + 2.09 = 62.09\ ^\circ\text{F}$$

The surface area is given by (J-5) as

$$A = \frac{q}{UF(\Delta T)_{lm}}$$

where F is taken from Figure J-9 and ΔT_{lm} is the log-mean temperature difference for a counterflow, double-pipe heat exchanger having the same fluid inlet and outlet temperatures. Referring to Figure J-5(b), $\Delta T_1 = 120 - 62.09 = 57.91$ °F and $\Delta T_2 = 100 - 60 = 40$ °F. Equation (J-2) gives the log-mean temperature difference as

$$(\Delta T)_{lm} = \frac{\Delta T_2 - \Delta T_1}{\ln(\Delta T_2/\Delta T_1)} = \frac{40 - 57.91}{\ln(40/57.91)} = 48.40\ ^\circ\text{F}$$

Letting the lowercase temperature nomenclature of Figure J-9 represent the air, the required dimensionless parameters are

$$P = \frac{t_o - t_i}{T_i - t_i} = \frac{100 - 120}{60 - 120} = 0.33 \qquad Z = \frac{T_i - T_o}{t_o - t_i} = \frac{60 - 62.09}{100 - 120} = 0.10$$

The correction factor, from Figure J-9, is approximately unity; therefore

$$A \approx \frac{240\,000\ \text{Btu/hr}}{(25\ \text{Btu/hr}\cdot\text{ft}^2\cdot{}^\circ\text{F})(1.0)(48.40\ ^\circ\text{F})} = 198.3\ \text{ft}^2$$

9.16 What error would have been introduced in Problem 9.15 if the arithmetic-mean temperature difference, defined by $(\Delta T)_{am} \equiv (\Delta T_2 + \Delta T_1)/2$, had been used rather than the log-mean temperature difference, $(\Delta T)_{lm}$?

▌ The arithmetic-mean temperature difference is

$$(\Delta T)_{am} = \frac{40 + 57.91}{2} = 48.96\ ^\circ\text{F}$$

This gives an area of

$$A_{am} = \left(\frac{48.40}{48.96}\right)(198.3) = 196.0\ \text{ft}^2$$

which underspecifies the area by

$$\text{Error} = \frac{198.3 - 196.0}{198.3} \times 100\% = 1.16\%$$

9.17[D] Estimate the surface area required in a crossflow heat exchanger with both fluids unmixed to cool 22.68 metric tons per hour of compressed air from 49 °C to 38 °C with water entering at 16 °C flowing at the rate of 52.16 tons/hr. Assume that the average value of the overall heat transfer coefficient is 142 W/m$^2\cdot$K.

▌ The total rate of heat transfer from the air is

$$q = (\dot{m}c_p\,\Delta T)_a = (22\,680\ \text{kg/hr})(1.004\ \text{kJ/kg}\cdot\text{K})[(49 - 38)\ \text{K}]\,\frac{1\ \text{hr}}{3600\ \text{s}} = 69.577\ \text{kW}$$

From an energy balance we can get the exit water temperature, i.e.,

$$\dot{m}_a c_{pa}\,\Delta T_a = \dot{m}_w c_{pw}\,\Delta T_w = 69.577\ \text{kW}$$

$$\Delta T_w = \frac{(69.577\ \text{kW})(3600\ \text{s/hr})}{(52\,160\ \text{kg/hr})(4.184\ \text{kJ/kg}\cdot\text{K})} = 1.15\ ^\circ\text{C}$$

and

$$T_{wo} = T_{wi} + \Delta T_w = 16 + 1.15 = 17.15\ ^\circ\text{C}$$

The surface area is given by (J-5) as

$$A = \frac{q}{UF(\Delta T)_{lm}}$$

where F is taken from Figure J-9 and $(\Delta T)_{lm}$ is the log-mean temperature difference for a counterflow double-pipe heat exchanger having the same fluid inlet and outlet temperatures. Referring to Figure J-5b, $\Delta T_1 = 49.0 - 17.15 = 31.85$ °C and $\Delta T_2 = 38 - 16 = 22$ °C. Equation (J-2) gives the log-mean temperature difference as

$$(\Delta T)_{lm} = \frac{\Delta T_2 - \Delta T_1}{\ln(\Delta T_2/\Delta T_1)} = \frac{22 - 31.85}{\ln(22/31.85)} = 26.62\ ^\circ\text{C}$$

Letting the lowercase temperature nomenclature of Figure J-9 represent the air, the required dimensionless parameters are

$$P = \frac{t_o - t_i}{T_i - t_i} = \frac{38 - 49}{16 - 49} = 0.33 \qquad Z = \frac{T_i - T_o}{t_o - t_i} = \frac{16 - 17.15}{38 - 49} = 0.105$$

The correction factor, from Figure J-9, is approximately unity; therefore

$$A \approx \frac{69.577 \text{ kW}}{(0.142 \text{ kW/m}^2 \cdot \text{K})(1.0)(26.62 \text{ K})} = 18.41 \text{ m}^2$$

9.18 For what value of $\Delta T_2/\Delta T_1$ is the arithmetic-mean temperature difference,

$$(\Delta T)_{am} \equiv \frac{\Delta T_2 + \Delta T_1}{2}$$

5 percent larger than the log-mean temperature difference, $(\Delta T)_{lm}$?

▮ We have

$$\frac{(\Delta T)_{am}}{(\Delta T)_{lm}} = \frac{\frac{1}{2}(\Delta T_2 + \Delta T_1)}{(\Delta T_2 - \Delta T_1)/\ln(\Delta T_2/\Delta T_1)} = \frac{1}{2}\,\frac{(\Delta T_2/\Delta T_1) + 1}{(\Delta T_2/\Delta T_1) - 1}\ln\frac{\Delta T_2}{\Delta T_1}$$

For $(\Delta T)_{am}/(\Delta T)_{lm} = 1.05$,

$$\frac{(\Delta T_2/\Delta T_1) + 1}{(\Delta T_2/\Delta T_1) - 1}\ln\frac{\Delta T_2}{\Delta T_1} = 2.10$$

Solving by trial, $\Delta T_2/\Delta T_1 = 2.2$.

It can be shown analytically that $(\Delta T)_{am}/(\Delta T)_{lm}$ is a strictly increasing function of $\Delta T_2/\Delta T_1$ for $\Delta T_2/\Delta T_1 \geq 1$. Consequently, the simple arithmetic-mean temperature difference gives results to within 5 percent when the end temperature differences vary by no more than a factor of 2.2.

9.19 When new, a heat exchanger transfers 10 percent more heat than it does after being in service for 6 months. Assuming that it operates between the same temperature differentials and that there is insufficient scale buildup to change the effective surface area, determine the effective overall fouling factor in terms of its clean (new) overall heat transfer coefficient.

▮ The heat exchange ratio may be writen as

$$\frac{q_{\text{clean}}}{q_{\text{dirty}}} = \frac{U_{\text{clean}} A\,\overline{\Delta T}}{U_{\text{dirty}} A\,\overline{\Delta T}} = 1.10 \qquad \text{or} \qquad U_{\text{dirty}} = \frac{U_{\text{clean}}}{1.10} \tag{1}$$

By definition the effective overall fouling factor R_f is

$$R_f \equiv \frac{1}{U_{\text{dirty}}} - \frac{1}{U_{\text{clean}}} \tag{J-6}$$

Thus, substituting expression (1) into equation (J-6)

$$R_f = \frac{1}{10U_{\text{clean}}}$$

9.20[D] Reconsider the shell-and-tube heat exchanger of Problem 9.2. The water inside the tubes is treated boiler feedwater above 125 °F, and the water outside the tubes is seawater below 125 °F. Determine U_o (based on outside tube area), including fouling effects.

▮ In this case, the "dirty" value of U_o is given by equation (J-4) modified to include $R_{f,i}$ and $R_{f,o}$; this is

$$U_{o,d} = \frac{1}{\dfrac{r_o}{r_i\bar{h}_i} + \dfrac{r_o R_{f,i}}{r_i} + \dfrac{r_o \ln(r_o/r_i)}{k_{i-o}} + R_{f,o} + \dfrac{1}{\bar{h}_o}} \tag{J-7}$$

From Table J-3 the fouling factors are

$$R_{f,i} = 0.001 \text{ hr} \cdot \text{ft}^2 \cdot {}^\circ\text{F/Btu} \qquad R_{f,o} = 0.0005 \text{ hr} \cdot \text{ft}^2 \cdot {}^\circ\text{F/Btu}$$

and from Problem 9.2

$$\frac{r_o}{r_i\bar{h}_i} = 0.00246 \text{ hr} \cdot \text{ft}^2 \cdot {}^\circ\text{F/Btu} \qquad \frac{r_o \ln(r_o/r_i)}{k_{i-o}} = 0.000\,313 \text{ hr} \cdot \text{ft}^2 \cdot {}^\circ\text{F/Btu} \qquad \frac{1}{\bar{h}_o} = 0.002\,70 \text{ hr} \cdot \text{ft}^2 \cdot {}^\circ\text{F/Btu}$$

Substituting into equation (*J-7*)

$$U_{o,d} = \frac{1 \text{ Btu/hr}\cdot\text{ft}^2\cdot°\text{F}}{0.002\,46 + (0.59/0.50)(0.001) + 0.000\,313 + 0.0005 + 0.002\,70} = 139.8 \text{ Btu/hr}\cdot\text{ft}^2\cdot°\text{F}$$

Fouling has reduced the U_o value by 23.4 percent from the clean (new) condition. Clearly, fouling can be important in heat exchanger design.

9.21 Reconsider Problem 9.4, where now the shellside fluid is seawater which enters at 28 °C and exits at <52 °C (125 °F), and the fluid inside the tube is treated boiler feedwater with a minimum temperature of 150 °C. Determine the U_o for normal service.

▌ In normal service, the tubes will experience inside and outside fouling, and the "dirty" U_o expression is

$$U_{o,d} = \frac{1}{\dfrac{r_o}{r_i\bar{h}_i} + \dfrac{r_o}{r_i}R_{f,i} + \dfrac{r_o \ln(r_o/r_i)}{k_{i-o}} + R_{f,o} + \dfrac{1}{\bar{h}_o}} \qquad (J\text{-}7)$$

From Table J-3 the fouling factors are

$$R_{f,i} = 0.0002 \text{ m}^2\cdot\text{K/W} \qquad R_{f,o} = 0.000\,09 \text{ m}^2\cdot\text{K/W}$$

From Problem 9.4

$$\frac{r_o}{r_i\bar{h}_i} = 0.000\,440 \qquad r_o\frac{\ln(r_o/r_i)}{k_{i-o}} = 0.000\,060\,8 \qquad \frac{1}{\bar{h}_o} = 0.000\,476$$

Substituting into equation (*J-9*) yields

$$U_{o,d} = \frac{1}{0.000\,440 + (1.5/1.25)(0.0002) + 0.000\,060\,8 + 0.000\,09 + 0.000\,476} = 765 \text{ W/m}^2\cdot\text{K}$$

This value is 25 percent lower than the "clean" value of U_o calculated in Problem 9.4.

9.22 A water flow rate of 250 lbm/hr is heated from 60 °F to 110 °F in a concentric, double-pipe, parallel-flow heat exchanger. The hot fluid is oil flowing at 300 lbm/hr which enters at 400 °F. The oil c_p is 0.45 Btu/lbm·°F. ***(a)*** What heat exchanger area is required for an overall heat transfer coefficient $U_o = 45$ Btu/hr·ft²·°F? ***(b)*** Determine the number of transfer units (NTU). ***(c)*** Calculate the effectiveness of the heat exchanger.

▌ ***(a)*** At an average water temperature of 85 °F, the specific heat is $c_{pw} = 0.998$ Btu/lbm·°F; therefore, an energy balance gives

$$(\dot{m}c_p\,\Delta T)_{\text{oil}} = (\dot{m}c_p\,\Delta T)_{\text{water}}$$

$$(300 \text{ lbm/hr})(0.45 \text{ Btu/lbm}\cdot°\text{F})[(400 - T_o)\,°\text{F}] = (250 \text{ lbm/hr})(0.998 \text{ Btu/lbm}\cdot°\text{F})[(110 - 60)\,°\text{F}]$$

$$T_{\text{oil}}\Big|_o = 400 - \frac{250(0.998)(50)}{300(0.45)} = 307.59\,°\text{F}$$

This gives end temperature differences: $\Delta T_1 = 400 - 60 = 340$ °F and $\Delta T_2 = 307.59 - 110 = 197.59$ °F, and the log-mean temperature difference is

$$(\Delta T)_{lm} = \frac{\Delta T_2 - \Delta T_1}{\ln(\Delta T_2/\Delta T_1)} = \frac{197.59 - 340}{\ln(197.59/340)} = 262.39\,°\text{F}$$

The total heat transfer is $q = (\dot{m}c_p\,\Delta T)_{\text{water}} = 12\,475$ Btu/hr, and the area is given by

$$A_o = \frac{q}{U_o(\Delta T)_{lm}} = \frac{12\,475 \text{ Btu/hr}}{(45 \text{ Btu/hr}\cdot\text{ft}^2\cdot°\text{F})(262.39\,°\text{F})} = 1.06 \text{ ft}^2$$

(b) The number of transfer units is given by $\text{NTU} \equiv U_oA_o/C_{\min}$ where $C_{\min}$ is the minimum value of $\dot{m}c_p$ over the fluids. Now

$$(\dot{m}c_p)_{\text{water}} = (250 \text{ lbm/hr})(0.998 \text{ Btu/lbm}\cdot°\text{F}) = 249.5 \text{ Btu/hr}\cdot°\text{F}$$

$$(\dot{m}c_p)_{\text{oil}} = (300 \text{ lbm/hr})(0.45 \text{ Btu/lbm}\cdot°\text{F}) = 135 \text{ Btu/hr}\cdot°\text{F}$$

Therefore, $C_{\min} = 135$ Btu/hr·°F and

$$\text{NTU} = \frac{(45 \text{ Btu/hr}\cdot\text{ft}^2\cdot°\text{F})(1.06 \text{ ft}^2)}{135 \text{ Btu/hr}\cdot°\text{F}} = 0.35$$

(*c*) The effectiveness expression given in Table J-2 for this configuration is

$$\epsilon = \frac{1 - \exp[-\mathbf{NTU}(1 + C_{min}/C_{max})]}{1 + C_{min}/C_{max}} = \frac{1 - \exp\{-(0.35)[1 + (135/249.5)]\}}{1 + (135/249.5)} = 27.0\%$$

which agrees very well with that shown graphically in Figure J-10 with

$$C \equiv \frac{C_{min}}{C_{max}} = \frac{135}{249.5} = 0.54$$

Note the low value of **NTU** is caused by the low value of U_o, but this value of U_o is mid to upper range of typical values of water-to-oil heat exchangers as shown in Table J-1.

9.23[D] Water enters a counterflow double-pipe heat exchanger at 38 °C, flowing at the rate of 0.076 kg/s. It is heated by oil ($c_p = 1.88$ kJ/kg · K) flowing at the rate of 0.152 kg/s from an inlet at a temperature of 116 °C. For an area of 1.30 m² and an overall heat transfer coefficient of 340 W/m² · K, determine the total heat transfer rate.

▮ Since the inlet temperatures are known and the $(\dot{m}c_p)$ products can be calculated, equation (*J-9*) may be used to determine the heat transfer after finding the heat exchanger effectiveness, ϵ. Since the specific heat of water is approximately 4.186 kJ/kg · K,

$$(\dot{m}c_p)_{water} = (0.076\ \text{kg/s})(4.186\ \text{kJ/kg}\cdot\text{K}) = 0.318\ \text{kW/K} = C_{max}$$

$$(\dot{m}c_p)_{oil} = (0.152\ \text{kg/s})(1.88\ \text{kJ/kg}\cdot\text{K}) = 0.286\ \text{kW/K} = C_{min}$$

Hence

$$C = \frac{C_{min}}{C_{max}} = \frac{0.286}{0.318} = 0.90$$

and

$$\mathbf{NTU} = \frac{UA}{C_{min}} = \frac{(340\ \text{W/m}^2\cdot\text{K})(1.30\ \text{m}^2)}{(286\ \text{W/m}^2\cdot\text{K})} = 1.55$$

Using these parameters with Figure J-11, we get $\epsilon \approx 0.62$; hence

$$q = \epsilon C_{min}(T_{hi} - T_{ci}) \approx (0.62)(0.286\ \text{kW/K})[(116 - 38)\ \text{K}] = 13.83\ \text{kW}$$

9.24[D] Water enters a counterflow double-pipe heat exchanger at 100 °F, flowing at the rate of 10.0 lbm/min. It is heated by oil ($c_p = 0.45$ Btu/lbm · °F) flowing at the rate of 20.0 lbm/min from an inlet at a temperature of 240 °F. For an area of 14.0 ft² and an overall heat transfer coefficient of 60 Btu/hr · ft² · °F, determine the total rate of heat transfer.

▮ Since the inlet temperatures are known and the $(\dot{m}c_p)$ products can be calculated, equation (*J-9*) may be used to determine the heat transfer after finding the heat exchanger effectiveness, ϵ. Since the specific heat of water is approximately 1.0 Btu/lbm · °F,

$$(\dot{m}c_p)_{water} = (10.0\ \text{lbm/min})(1.0\ \text{Btu/lbm}\cdot{}^\circ\text{F}) = 10.0\ \text{Btu/min}\cdot{}^\circ\text{F} = C_{max}$$

$$(\dot{m}c_p)_{oil} = (20.0\ \text{lbm/min})(0.45\ \text{Btu/lbm}\cdot{}^\circ\text{F}) = 9.0\ \text{Btu/min}\cdot{}^\circ\text{F} = C_{min}$$

Therefore

$$C = \frac{C_{min}}{C_{max}} = \frac{9.0}{10.0} = 0.90$$

and

$$\mathbf{NTU} = \frac{UA}{C_{min}} = \frac{(60\ \text{Btu/hr}\cdot\text{ft}^2\cdot{}^\circ\text{F})(14.0\ \text{ft}^2)}{(9.0\ \text{Btu/min}\cdot{}^\circ\text{F})(60\ \text{min/hr})} = 1.56$$

Using these parameters with Figure J-11, we get $\epsilon \approx 0.62$; hence

$$q = \epsilon C_{min}(T_{hi} - T_{ci}) \approx (0.62)(9.0\ \text{Btu/min}\cdot{}^\circ\text{F})(60\ \text{min/hr})[(240 - 100)\ {}^\circ\text{F}] = 46\,872\ \text{Btu/hr}$$

9.25[D] Water enters a crossflow heat exchanger (both fluids unmixed) at 16 °C and flows at the rate of 7.68 kg/s to cool 8.0 kg/s of air from 120 °C. For an overall heat transfer coefficient of 225 W/m² · K and an exchanger surface area of 143 m², what is the exit air temperature?

▮ Taking the specific heats of the air and water to be constant at 1.005 and 4.187 kJ/kg · K, respectively, we get

$$C_{air} = (\dot{m}c_p)_{air} = (8.0\ \text{kg/s})(1.005\ \text{kJ/kg}\cdot\text{K}) = 8.04\ \text{kW/K}$$

$$C_{water} = (\dot{m}c_p)_{water} = (7.68\ \text{kg/s})(4.187\ \text{kJ/kg}\cdot\text{K}) = 32.16\ \text{kW/K}$$

which gives

$$C = \frac{C_{min}}{C_{max}} = \frac{8.04}{32.16} = 0.25$$

Also,

$$\text{NTU} = \frac{AU}{C_{min}} = \frac{(143\ \text{m}^2)(225\ \text{W/m}^2\cdot\text{K})}{8040\ \text{W/K}} = 4.00$$

and from Figure J-14, $\epsilon \approx 0.93$.

The heat transfer rate is given by (*J-9*) as

$$q = \epsilon C_{min}(T_{hi} - T_{ci}) \approx (0.93)(8.04\ \text{kW/K})[(120 - 16)\ \text{K}] = 778\ \text{kW}$$

An energy balance on the air then gives the exit temperature:

$$q = \dot{m}c_p(T_i - T_o)$$

$$T_o = -\frac{q}{\dot{m}c_p} + T_i = -\frac{778\ \text{kW}}{8.04\ \text{kW/°C}} + 120\ \text{°C} = 23.3\ \text{°C}$$

9.26[D] Water enters a crossflow heat exchanger (both fluids unmixed) at 60 °F and flows at the rate of 60 000 lbm/hr to cool 80 000 lbm/hr of air from 250 °F. For an overall heat transfer coefficient of 40 Btu/hr · ft² · °F and an exchanger surface area of 2600 ft², what is the exit air temperature?

▮ Taking the specific heats of the air and water to be constant at 0.24 and 1.0 Btu/lbm · °F, respectively, we get

$$C_{air} = (\dot{m}c_p)_{air} = (80\,000\ \text{lbm/hr})(0.24\ \text{Btu/lbm}\cdot\text{°F}) = 19\,200\ \text{Btu/hr}\cdot\text{°F}$$

$$C_{water} = (\dot{m}c_p)_{water} = (60\,000\ \text{lbm/hr})(1.0\ \text{Btu/lbm}\cdot\text{°F}) = 60\,000\ \text{Btu/hr}\cdot\text{°F}$$

which gives

$$C = \frac{C_{min}}{C_{max}} = \frac{19\,200}{60\,000} = 0.32$$

Also,

$$\text{NTU} = \frac{AU}{C_{min}} = \frac{(2600\ \text{ft}^2)(40\ \text{Btu/hr}\cdot\text{ft}^2\cdot\text{°F})}{19\,000\ \text{Btu/hr}\cdot\text{°F}} = 5.42$$

From Figure J-14, $\epsilon \approx 0.94$; hence, by (*J-9*),

$$q = \epsilon C_{min}(T_{hi} - T_{ci}) \approx (0.94)(19\,200\ \text{Btu/hr}\cdot\text{°F})[(250 - 60)\ \text{°F}] = 3.43 \times 10^6\ \text{Btu/hr}$$

An energy balance on the air then gives the exit temperature:

$$q = \dot{m}c_p(T_i - T_o)$$

$$T_o = -\frac{q}{\dot{m}c_p} + T_i = -\frac{3.43 \times 10^6\ \text{Btu/hr}}{19\,200\ \text{Btu/hr}\cdot\text{°F}} + 250\ \text{°F} = -178.6\ \text{F} + 250\ \text{°F} = 71.4\ \text{°F}$$

9.27 Solve Problem 9.26 for a 1-shell-pass, 10-tube-pass heat exchanger.

▮ The same dimensionless parameters hold, but the heat exchanger effectiveness is lower. From Figure J-12, $\epsilon \approx 0.83$; therefore, by proportion,

$$T_o \approx \frac{0.83}{0.94}(-178.6\ \text{°F}) + 250\ \text{°F} = 92.3\ \text{°F}$$

9.28 Give the two possible expressions of ϵ in terms of the four inlet and outlet temperatures, T_{hi}; T_{ho}; T_{ci}; T_{co}.

▮ By definition

$$\epsilon = \frac{\text{actual heat transfer}}{\text{maximum possible heat transfer}} = \frac{\text{actual heat transfer}}{C_{min}(T_{hi} - T_{ci})}$$

There are two possibilities:

1. $C_{min} = C_h$:

$$\text{Actual} \quad q = C_h(T_{hi} - T_{ho})$$

$$\epsilon_h = \frac{C_h(T_{hi} - T_{ho})}{C_{min}(T_{hi} - T_{ci})} = \frac{T_{hi} - T_{ho}}{T_{hi} - T_{ci}} \tag{J-10}$$

2. $C_{min} = C_c$:

$$\text{Actual} \quad q = C_c(T_{co} - T_{ci})$$

$$\epsilon_c = \frac{C_c(T_{co} - T_{ci})}{C_{min}(T_{hi} - T_{ci})} = \frac{T_{co} - T_{ci}}{T_{hi} - T_{ci}} \tag{J-11}$$

9.29 Hot water at 180 °F enters the tubes of a 2-shell-pass, 8-tube-pass heat exchanger at the rate of 80 lbm/min, heating helium from 80 °F. The overall heat transfer coefficient is 20 Btu/hr · ft^2 · F, and the exchanger area is 240 ft^2. If the water exits at 120 °F, determine the exit temperature of the helium and its mass-flow rate.

▌ Since the flow rate of the helium is unknown, there is no way to determine a priori the miniumum heat-capacity rate, i.e., the fluid having the smaller value of $\dot{m}c_p$. Assuming that the minimum fluid is the water,

$$C_{water} = (\dot{m}c_p)_{water} = (80 \text{ lbm/min})(1.0 \text{ Btu/lbm} \cdot °\text{F}) = 80 \text{ Btu/min} \cdot °\text{F}$$

Based upon this assumption, the number of transfer units is

$$\textbf{NTU} = \frac{UA}{C_{min}} = \frac{(20 \text{ Btu/hr} \cdot \text{ft}^2 \cdot °\text{F})(240 \text{ ft}^2)}{(80 \text{ Btu/min} \cdot °\text{F})(60 \text{ min/hr})} = 1.00$$

Since $C_h < C_c$ by the starting assumption, equation (*J-10*) holds, giving

$$\epsilon_h = \frac{T_{hi} - T_{ho}}{T_{hi} - T_{ci}} = \frac{180 - 120}{180 - 80} = 0.60$$

From Figure J-13 these parameters give $C = C_{min}/C_{max} \approx 0.25$, which validates the initial assumption of the water as the minimum fluid. Hence,

$$C_{He} \approx \frac{C_{water}}{0.25} = 320 \text{ Btu/min} \cdot °\text{F} = (\dot{m}c_p)_{He}$$

In the temperature range being considered, the specific heat of helium is 1.242 Btu/lbm · °F; therefore,

$$\dot{m}_{He} = \frac{C_{He}}{(c_p)_{He}} = \frac{(320 \text{ Btu/min} \cdot °\text{F})}{1.242 \text{ Btu/lbm} \cdot °\text{F}} = 258 \text{ lbm/min}$$

An energy balance gives the helium exit temperature, i.e.,

$$[\dot{m}c_p(T_i - T_o)]_{water} = [\dot{m}c_p(T_o - T_i)]_{He}$$

$$(80 \text{ Btu/min} \cdot °\text{F})[(180 - 120)\ °\text{F}] = (320 \text{ Btu/min} \cdot °\text{F})[(T_o - 80)\ °\text{F}]$$

$$T_o\Big|_{He} = 15 + 80 = 95\ °\text{F}$$

9.30[D] Hot commercial heat transfer oil is used in a crossflow heat exchanger to heat a dye solution in a manufacturing plant. The mixed-flow dye solution ($c_p = 1.12$ Btu/lbm · °F) enters at 60 °F and exits at 130 °F at a flow rate of 3000 lbm/hr. The unmixed-flow oil ($c_p = 0.46$ Btu/lbm · °F) enters at 400 °F, and the flow system produces an overall heat transfer coefficient of 50 Btu/hr · ft^2 · °F. For an exchanger surface area of 80 ft^2, what mass flow rate of oil is required, and what is its exit temperature?

▌ The given conditions do not permit the direct determination of the fluid with the smaller $\dot{m}c_p$. If the minimum fluid is the dye solution, we can solve for ϵ and **NTU** and use Figure J-15 to get $C_{mixed}/C_{unmixed}$, which would permit the oil flow rate to be determined. If the minimum fluid is the oil, a solution requires trial and error. Assuming that the minimum fluid is the dye solution, equation (*J-11*) gives

$$\epsilon_{dye} = \epsilon_c = \frac{T_{co} - T_{ci}}{T_{hi} - T_{ci}} = \frac{130 - 60}{400 - 60} = 0.21$$

Under the same assumption,

$$C_{mixed} = C_{dye} = (\dot{m}c_p)_{dye} = (3000 \text{ lbm/hr})(1.12 \text{ Btu/lbm} \cdot °\text{F}) = 3360 \text{ Btu/hr} \cdot °\text{F}$$

and

$$\text{NTU} = \frac{UA}{C_{\text{mixed}}} = \frac{(50\ \text{Btu/hr}\cdot\text{ft}^2\cdot{}^\circ\text{F})(80\ \text{ft}^2)}{3360\ \text{Btu/hr}\cdot{}^\circ\text{F}} = 1.19$$

With these parameters, it is apparent from Figure J-15 that the ratio $C_{\text{mixed}}/C_{\text{unmixed}}$ does not exist; therefore, the minimum fluid is the oil.

We must now assume a value of C_{unmixed} (which amounts to assuming $\dot{m}_{\text{oil}}$), determine the pertinent parameters, and compare calculated results with Figure J-15 until a solution is found. A first trial in this procedure is outlined below; subsequent assumptions and results are presented in Table 9-1. Assume

$$C_{\text{unmixed}} = C_{\text{oil}} = C_{\text{min}} = 1000\ \text{Btu/hr}\cdot{}^\circ\text{F}$$

Then

$$\frac{C_{\text{mixed}}}{C_{\text{unmixed}}} = \frac{3360}{1000} = 3.36 \qquad \text{NTU} = \frac{UA}{C_{\text{min}}} = \frac{50(80)}{1000} = 4.00$$

From Figure J-15, $\epsilon \approx 0.84$; an energy balance gives

$$\Delta T_h = \Delta T_{\text{oil}} = \frac{(\dot{m}c_p\,\Delta T)_{\text{dye}}}{(\dot{m}c_p)_{\text{oil}}} = \frac{C_{\text{mixed}}(\Delta T)_{\text{dye}}}{C_{\text{unmixed}}} = (3.36)(130-60) = 235.2\ {}^\circ\text{F}$$

$$\epsilon_h = \epsilon_{\text{oil}} = \frac{(\Delta T)_{\text{oil}}}{T_{hi} - T_{ci}} = \frac{235.2}{400-60} = 0.69$$

and the assumed C_{unmixed} is incorrect.

TABLE 9-1

trial no.	C_{unmixed} (assumed)	$\frac{C_{\text{mixed}}}{C_{\text{unmixed}}}$	NTU	ΔT_h	ϵ calculated	ϵ from Fig. J.15
1	1000 Btu/hr · °F	3.36	4.00	235.2 °F	0.69	0.84
2	2000	1.68	2.00	117.6	0.35	0.67
3	1500	2.24	2.67	156.8	0.46	0.77
4	880	3.82	4.55	267.3	0.79	0.87
5	750	4.48	5.33	313.6	0.92	0.92

The required parameters may now be determined.

$$\dot{m}_{\text{oil}} = \frac{C_{\text{oil}}}{(c_p)_{\text{oil}}} = \frac{750\ \text{Btu/hr}\cdot{}^\circ\text{F}}{0.46\ \text{Btu/lbm}\cdot{}^\circ\text{F}} = 1630\ \text{lbm/hr}$$

$$(\Delta T)_{\text{oil}} = 313.6 = T_{hi} - T_{ho} = 400 - T_{ho}$$

$$T_{ho} = 400 - 313.6 = 86.4\ {}^\circ\text{F}$$

CHAPTER 10

Radiation

Problems of this chapter employ the following notations for radiant power (heat):

$$\begin{aligned} q_i &\equiv \text{(net power from surface } i) \\ &\equiv \text{(power emitted by surface } i) - \text{(power absorbed by surface } i) \\ q_{i\to j} &\equiv \text{(power emitted by surface } i \text{ and } \textit{falling on} \text{ surface } j) \\ q'_{i-j} &\equiv \text{(power emitted by surface } i \text{ and } \textit{absorbed by} \text{ surface } j) \\ q_{i-j} &\equiv \text{(net heat transfer from surface } i \text{ to surface } j) \end{aligned}$$

Note the identity $q_{i-j} = q'_{i-j} - q'_{j-i} = -q_{j-i}$, which finds frequent use.

10.1 BASIC CONCEPTS

10.1 What is the photon energy E at the longest wavelength in the thermal band of the electromagnetic radiation spectrum?

▌ The propagation velocity c, which is the speed of light, is given by

$$c = \lambda\nu = 3 \times 10^8 \text{ m/s}$$

where λ is the wavelength and ν is the frequency. The energy E is given by

$$E = h\nu = \frac{hc}{\lambda}$$

where h is Planck's constant, 6.626×10^{-34} J · s. Referring to Fig. 10-1, the longest wavelength in the thermal band is $\lambda = 10^{-4}$ m; therefore,

$$E = \frac{hc}{\lambda} = \frac{(6.626 \times 10^{-34} \text{ J}\cdot\text{s})(3 \times 10^8 \text{ m/s})}{10^{-4} \text{ m}} = 1.988 \times 10^{-22} \text{ J}$$

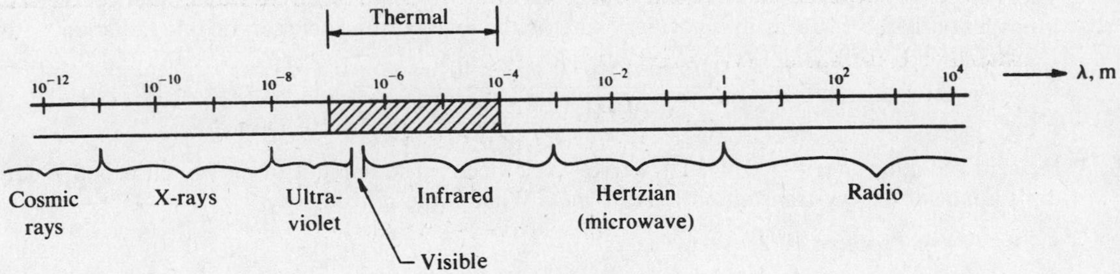

Fig. 10-1

10.2 Referring to Fig. 10-2, where

$\alpha \equiv$ fraction of incident energy absorbed $\equiv$ absorptivity
$\rho \equiv$ fraction of incident energy reflected $\equiv$ reflectivity
$\tau \equiv$ fraction of incident energy transmitted $\equiv$ transmissivity

derive the relationship among α, ρ, and τ.

▌ From an energy balance, we get

$$G = \rho G + \alpha G + \tau G$$

Therefore,

$$\rho + \alpha + \tau = 1 \qquad (1)$$

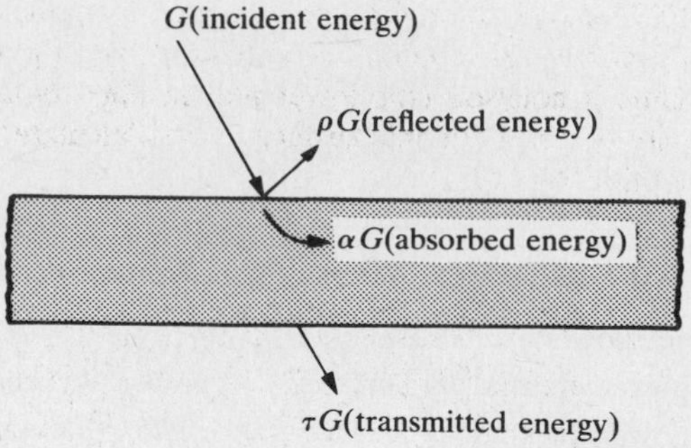

Fig. 10-2

10.3 The incident solar radiant flux at the earth's mean orbital radius from the sun (93 000 000 miles) is $G_E \approx$ 444.7 Btu/hr · ft². Using this fact, determine the solar flux in the vicinity of the planet Mercury, which has a mean orbital radius of 36 000 000 miles.

▮ Considering the sun as a point source, we have

$$G_M = G_E\left(\frac{r_E}{r_M}\right)^2 = (444.7)(\tfrac{93}{36})^2 = 2968 \text{ Btu/hr} \cdot \text{ft}^2$$

10.4 The total incident radiant flux upon a body is 2200 W/m². Of this amount, 450 W/m² is reflected and 900 W/m² is absorbed by the body. Find the transmissivity τ.

▮ By (*1*) of Problem 10.2,

$$\tau = 1 - \rho - \alpha = 1 - \tfrac{450}{2200} - \tfrac{900}{2200} = 0.386$$

10.5 A photon has an energy of 0.23 pJ. Calculate its frequency (in hertz) and its wavelength in vacuum (in picometers).

▮

$$\nu = \frac{E}{h} = \frac{0.23 \times 10^{-12} \text{ J}}{6.626 \times 10^{-34} \text{ J} \cdot \text{s}} = 3.48 \times 10^{20} \text{ s}^{-1} = 3.48 \times 10^{20} \text{ Hz}$$

$$\lambda = \frac{c}{\nu} = \frac{3 \times 10^8 \text{ m/s}}{3.48 \times 10^{20} \text{ Hz}} = 8.64 \times 10^{-13} \text{ m} = 0.864 \text{ pm}$$

10.6 The wavelength of a radiating system is 0.037 pm (0.037×10^{-12} m). Assuming that the velocity of propagation is the same as in a vacuum, what is the energy of a quantum of this radiation?

▮

$$E = \frac{hc}{\lambda} = \frac{(6.626 \times 10^{-34} \text{ J} \cdot \text{s})(3 \times 10^8 \text{ m/s})}{0.037 \times 10^{-12} \text{ m}} = 5.37 \text{ pJ} = 5.37 \times 10^{-5} \text{ erg}$$

10.7 Incident radiation ($G = 500$ W/m²) strikes an object. The amount of energy absorbed is 150 W/m², and the amount of energy transmitted is 25 W/m². What is the reflectivity ρ?

▮ By (*1*) of Problem 10.2,

$$\rho = 1 - \frac{150 + 25}{500} = 0.65$$

10.8 Radiation of 370 W/m² is incident upon a shallow layer of liquid which has an absorptivity of 0.3 and a transmissivity of 0.12. Find the rate of energy reflected.

▮ By (*1*) of Problem 10.2,

$$\rho G = (1 - \alpha - \tau)G = (1 - 0.3 - 0.12)(370 \text{ W/m}^2) = 215 \text{ W/m}^2$$

10.9 Radiation strikes an object with transmissivity of 0.03 and reflectivity of 0.50; the absorbed flux is indirectly measured (by the heating effect) as 94 W/m². Compute the incident flux.

▮ By (*1*) of Problem 10.2,

$$\frac{94 \text{ W/m}^2}{G} + 0.03 + 0.50 = 1 \qquad \text{or} \qquad G = 200 \text{ W/m}^2$$

10.10 The upper plot in Fig. 10-3*a* shows the approximate spectral characteristics of the reflectivity of an opaque surface on the Apollo XII command module. Irradiation from the sun can be approximated as shown in the graph of monochromatic irradiation (the lower plot in Fig. 10-3*a*) $G_\lambda \equiv dG/d\lambda$. **(*a*)** Plot the monochromatic absorptivity α_λ vs. λ for this surface. **(*b*)** Calculate the total irradiation $G_{\text{total}} = \int_0^\infty G_\lambda \, d\lambda$. **(*c*)** Find the energy absorption rate G_A.

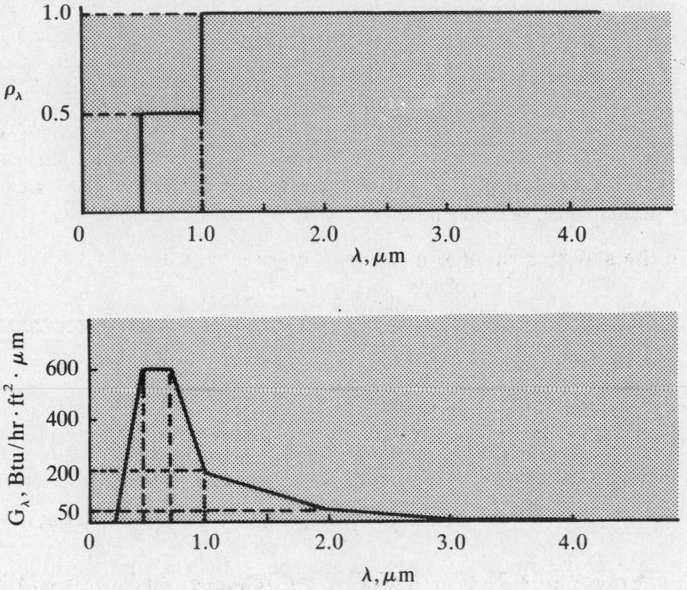

Fig. 10-3*a*

▮ (*a*) Since the surface is opaque, $\tau_\lambda = 0$ for all λ, whence, for every λ,

$$\alpha_\lambda + \rho_\lambda = 1$$

or $\alpha_\lambda = 1 - \rho_\lambda$ (Fig. 10-3*b*).

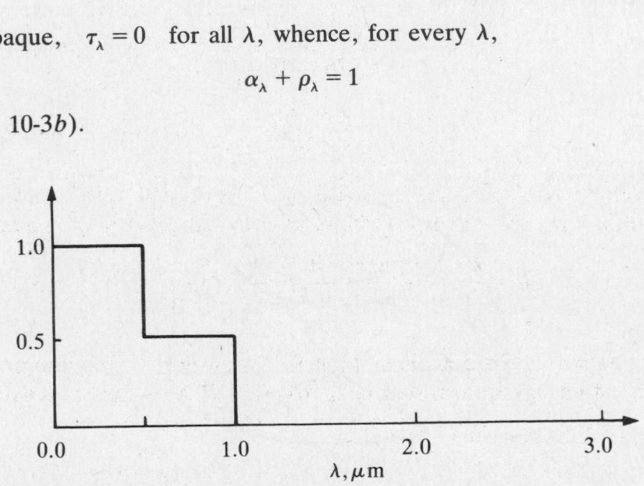

Fig. 10-3*b*

(*b*) Total irradiation is the area under the G_λ-curve; this is the sum of seven areas $\Delta_i G$:

$i = 1$	$0 < \lambda < 0.25$	$\Delta_1 G = 0$ Btu/hr · ft²
$i = 2$	$0.25 < \lambda < 0.50$	$\Delta_2 G = \frac{1}{2}(0.25)(600) = 75$ Btu/hr · ft²
$i = 3$	$0.50 < \lambda < 0.75$	$\Delta_3 G = (0.25)(600) = 150$ Btu/hr · ft²
$i = 4$	$0.75 < \lambda < 1.0$	$\Delta_4 G = \frac{1}{2}(0.25)(400) + 200(0.25) = 100$ Btu/hr · ft²
$i = 5$	$1.0 < \lambda < 2.0$	$\Delta_5 G = (1.0)(50) + \frac{1}{2}(1.0)(150) = 125$ Btu/hr · ft²
$i = 6$	$2.0 < \lambda < 3.0$	$\Delta_6 G = \frac{1}{2}(1.0)(50) = 25$ Btu/hr · ft²
$i = 7$	$\lambda > 3.0$	$\Delta_7 G = 0$

$$G_{\text{total}} = 0 + 75 + 150 + 100 + 125 + 25 + 0 = 475 \text{ Btu/hr} \cdot \text{ft}^2$$

(*c*) By definition of α_λ,

$$G_A = \int_0^\infty \alpha_\lambda G_\lambda \, d\lambda = \int_0^\infty \alpha_\lambda \, dG = \sum_{i=1}^{7} (\alpha_\lambda)_i \, \Delta_i G$$

$$= (1.0)(0) + (1.0)(75 \text{ Btu/hr} \cdot \text{ft}^2) + (0.5)(150 \text{ Btu/hr} \cdot \text{ft}^2) + (0.5)(100 \text{ Btu/ hr} \cdot \text{ft}^2) + 0 + 0 + 0$$

$$= 200 \text{ Btu/hr} \cdot \text{ft}^2$$

10.11 What physical ratio determines whether a real surface is an almost-specular reflector [Fig. 10-4(*a*)] or an almost-diffuse reflector [Fig. 10-4(*b*)]?

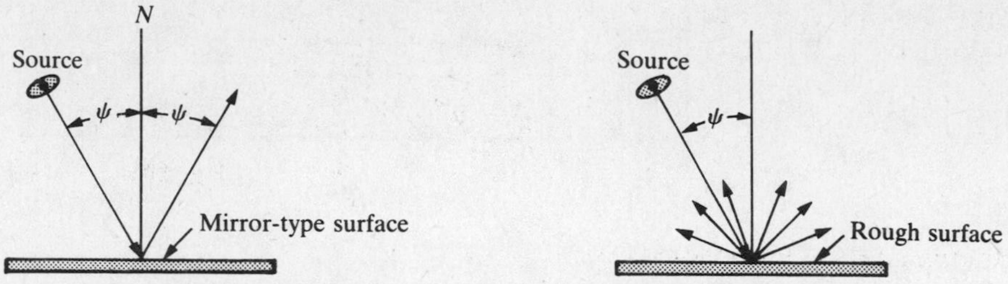

(*a*) Specular reflector (*b*) Diffuse reflector **Fig. 10-4**

I If the roughness dimension (e.g., mean pit depth) for a real surface is considerably smaller than the wavelength of incident irradiation, the surface behaves as a specular reflector; if the roughness dimension is large with respect to wavelength, the surface reflects diffusely.

10.12 Radiation intensity I is the radiant power (energy per unit time) per unit solid angle per unit area of the emitter projected normal to the line of view of the receiver from the radiating element. For the geometry depicted in Fig. 10-5, find the relation between total emissive power and radiant intensity for the small patch dA_1.

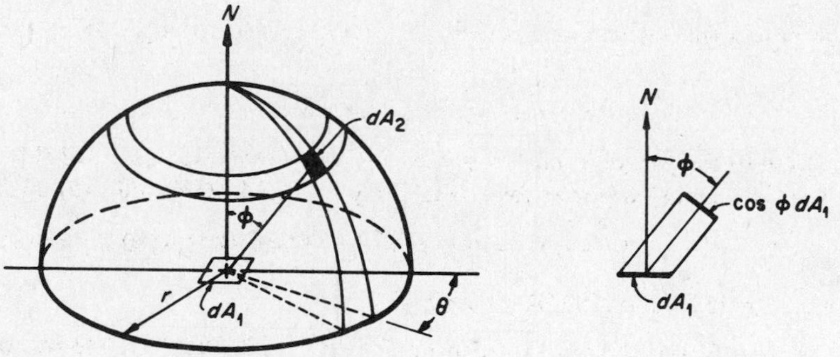

Fig. 10-5

I The energy radiated from dA_1 to dA_2 is given by

$$dq_{1\to 2} = I(\cos\phi \, dA_1) \, d\omega \tag{1}$$

Here

$$d\omega \equiv \frac{dA_2}{r^2} = \sin\phi \, d\theta \, d\phi \tag{2}$$

is the solid angle subtended by dA_2, and $\cos\phi \, dA_1$ is the area of the emitting surface projected normal to the line of view to the receiving surface. Substituting (*2*) into (*1*) and integrating over the hemispherical surface yields the desired relationship:

$$E \equiv \frac{q_{1\to 2}}{dA_1} = \int_0^{2\pi} \int_0^{\pi/2} I \cos\phi \sin\phi \, d\phi \, d\theta \tag{3}$$

If the emitting surface is diffuse and $I = \text{constant}$, (*3*) integrates to

$$E = \pi I \tag{4}$$

10.2 BLACKBODY RADIATION; THE STEFAN–BOLTZMANN LAW

10.13 In Fig. 10-6, a single bundle (ray) of energy is shown entering a small opening into a cavity (Hohlraum) with resulting specular reflections from the inner surfaces. If the entering energy flux is G_0, and if the material has absorptivity $\alpha_m < 1$, express the flux after each successive reflection.

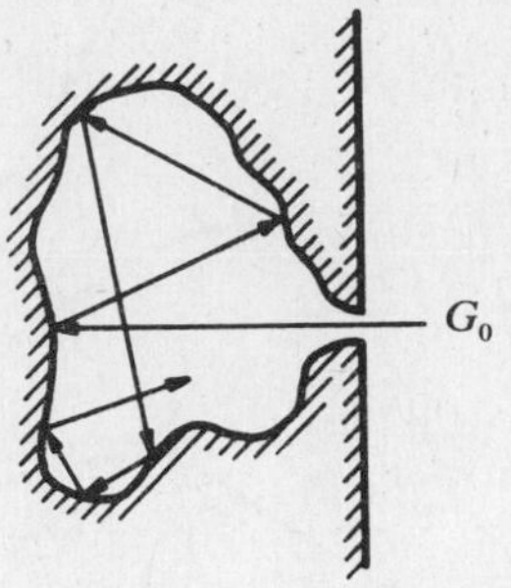

Fig. 10-6

▌ The flux after the first reflection is

$$G_1 = (1 - \alpha_m)G_0$$

after the second,

$$G_2 = (1 - \alpha_m)G_1 = (1 - \alpha_m)^2 G_0$$

. . . ; after the nth,

$$G_n = (1 - \alpha_m)^n G_0$$

10.14 Show that a Hohlraum approximates a blackbody (an ideal body for which $\alpha = 1.0$).

▌ Refer to Problem 10.13. Any ray emerging from the Hohlraum will have suffered a very large number of internal reflections. The overall absorptivity will then be

$$\alpha = \frac{G_0 - G_m}{G_0} = 1 - (1 - \alpha_m)^n \to 1$$

10.15 The emissive power (power radiated per unit area) of a blackbody is given by the Stefan–Boltzmann equation:

$$E_b = \sigma T^4 \qquad (1)$$

where T is absolute temperature and $\sigma = 5.6697 \times 10^{-8}\ \text{W/m}^2 \cdot \text{K}^4$ or $0.1714 \times 10^{-8}\ \text{Btu/hr} \cdot \text{ft}^2 \cdot {}^\circ\text{R}^4$. Determine the total emissive power for a blackbody at (*a*) 1100 K and (*b*) 1520 °F.

▌ (*a*)

$$E_b = \sigma T^4 = (5.6697 \times 10^{-8}\ \text{W/m}^2 \cdot \text{K}^4)(1100\ \text{K})^4 = 83.0\ \text{kW/m}^2$$

(*b*) With $T = 1520\ {}^\circ\text{F} = 1980\ {}^\circ\text{R}\ (= 1100\ \text{K})$

$$E_b = \sigma T^4 = (0.1714 \times 10^{-8}\ \text{Btu/hr} \cdot \text{ft}^2 \cdot {}^\circ\text{R}^4)(1980\ {}^\circ\text{R})^4 = 26\,300\ \text{Btu/hr} \cdot \text{ft}^2 \quad (= 83.0\ \text{kW/m}^2)$$

10.16 A blackbody has an emissive power of 2540 W/m². Determine its temperature in °C.

▌

$$T = \left(\frac{E_b}{\sigma}\right)^{1/4} = \left(\frac{2540\ \text{W/m}^2}{5.6697 \times 10^{-8}\ \text{W/m}^2 \cdot \text{K}^4}\right)^{1/4} = 460\ \text{K} = 187\ {}^\circ\text{C}$$

10.17 Figure 10-7 is a plot of the monochromatic emissive power of a blackbody at several absolute temperatures. The first accurate expression for $E_{b\lambda}$ was determined by Max Planck; it is

$$E_{b\lambda} = \frac{C_1 \lambda^{-5}}{\exp(C_2/\lambda T) - 1} \qquad (1)$$

in which

$$C_1 = 3.742 \times 10^8\ \text{W} \cdot \mu\text{m}^4/\text{m}^2 = 1.187 \times 10^8\ \text{Btu} \cdot \mu\text{m}^4/\text{hr} \cdot \text{ft}^2$$

$$C_2 = 1.4387 \times 10^4\ \mu\text{m} \cdot \text{K} = 2.5896 \times 10^4\ \mu\text{m} \cdot {}^\circ\text{R}$$

Show that (*1*) is consistent with the Stefan–Boltzmann equation, (*1*) of Problem 10.15.

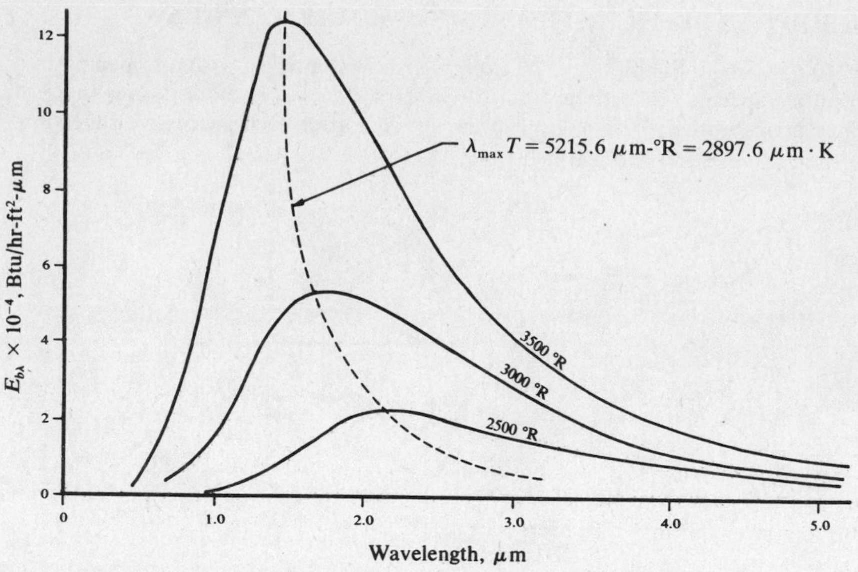

Fig. 10-7

▮ For the total emissive power we have

$$E_b = \int_0^\infty E_{b\lambda}\, d\lambda = \int_0^\infty \frac{C_1 \lambda^{-5}\, d\lambda}{\exp(C_2/\lambda T) - 1}$$

Choose the new integration variable $\omega = C_2/\lambda T$ to obtain

$$E_b = \left(\frac{C_1}{C_2^4} \int_0^\infty \frac{\omega^3\, d\omega}{e^\omega - 1}\right) T^4 = \text{constant} \times T^4 \tag{2}$$

10.18 Find the value of Planck's constant σ implied by Problem 10.17.

▮ According to (2) of Problem 10.17,

$$\sigma = \frac{C_1}{C_2^4} \int_0^\infty \frac{\omega^3\, d\omega}{e^\omega - 1}$$

Now
$$\frac{1}{e^\omega - 1} = \frac{e^{-\omega}}{1 - e^{-\omega}} = \sum_{k=1}^{\infty} e^{-k\omega}$$

so that, interchanging summation and integration, and consulting standard mathematical handbooks,

$$\sigma = \frac{C_1}{C_2^4} \sum_{k=1}^{\infty} \int_0^\infty \omega^3 e^{-k\omega}\, d\omega = \frac{C_1}{C_2^4} (3!) \sum_{k=1}^{\infty} \frac{1}{k^4} = \frac{C_1}{C_2^4} (3!)\left(\frac{\pi^4}{90}\right) = \frac{\pi^4 C_1}{15 C_2^4}$$

Substituting the numerical values of C_1 and C_2 given in Problem 10.17,

$$\sigma = \frac{\pi^4 (3.742 \times 10^8\ \text{W} \cdot \mu\text{m}^4/\text{m}^2)}{15(1.4387 \times 10^4\ \mu\text{m} \cdot \text{K})^4} = 5.672 \times 10^{-8}\ \text{W/m}^2 \cdot \text{K}^4$$

10.19 Referring to Fig. 10-7, the shift in location of the maximum value of the monochromatic emissive power to shorter wavelengths with increasing temperature is evident. This wavelength shift is described by *Wien's displacement law*. Using Planck's formula (1) of Problem 10.17, determine the equation for Wien's displacement law.

▮ Differentiating Planck's formula at constant T,

$$\frac{\partial E_{b\lambda}}{\partial \lambda} = \frac{(e^{C_2/\lambda T} - 1)(-5C_1\lambda^{-6}) - (C_1\lambda^{-5})(e^{C_2/\lambda T})(-C_2/\lambda^2 T)}{(e^{C_2/\lambda T} - 1)^2}$$

For a maximum, the numerator on the right must vanish. This gives, after cancellation of common factors,

$$x - \frac{C_2}{5(1 - e^{-C_2/x})} = 0 \tag{1}$$

where $x \equiv \lambda_{max} T$. For $\lambda \approx 1\ \mu m$, $T \approx 10^3$ K, we have $C_2 \approx 10\lambda T$. Thus, as a first approximation, we neglect $e^{-C_2/x}$ in comparison to 1, and (*1*) gives

$$x = \frac{C_2}{5} = 2877\ \mu m \cdot K$$

If this value is used as the first approximation, x_0, in Newton's iterative method, a single iteration gives

$$x_1 = \frac{1 - 6e^{-5}}{1 - 7e^{-5} + e^{-10}} \left(\frac{C_2}{5}\right) = \frac{0.9596}{0.9528}(2877) = 2897.6\ \mu m \cdot K$$

which we accept as the final value.

Therefore,

$$\lambda_{max} T = 2897.6\ \mu m \cdot K \tag{2}$$

which plots as the dashed curve through the peak values of emissive power in Fig. 10-7.

10.20^D For a blackbody maintained at 390 K, determine **(*a*)** the total emissive power, **(*b*)** the wavelength at which the maximum monochromatic emissive power occurs, **(*c*)** the maximum monochromatic emissive power.

▮ (*a*)

$$E_b = \sigma T^4 = (10^8 \sigma)\left(\frac{T}{100}\right)^4 = (5.6697)(3.9)^4 = 1312\ W/m^2$$

(*b*)

$$\lambda_{max} T = 2897.6\ \mu m \cdot K \quad (\textit{Wien's law})$$

$$\lambda_{max} = \frac{2897.6\ \mu m \cdot K}{390\ K} = 7.43\ \mu m$$

(*c*) By (*1*) of Problem 10.17,

$$E_{b\lambda_{max}} = \frac{3.742 \times 10^8}{(7.43)^5[\exp(14\,387/2897.6) - 1]} = 116.1\ W/m^2 \cdot \mu m$$

10.21 Determine the monochromatic emissive power at 2.30 μm of a blackbody at a temperature of 1640 K.

▮ By (*1*) of Problem 10.17,

$$E_{b\lambda} = \frac{(3.742 \times 10^8\ W \cdot \mu m^4/m^2)(2.30\ \mu m)^{-5}}{\exp[14\,387\ \mu m \cdot K/(2.30\ \mu m)(1640\ K)] - 1} = 0.1311\ MW/m^2 \cdot \mu m$$

10.22^D Determine λ_{max} and the maximum value of the monochromatic emissive power of a blackbody at 3500 °R.

▮ From (*2*) of Problem 10.19 and (*1*) of Problem 10.17,

$$\lambda_{max} = \frac{5215.6\ \mu m \cdot °R}{3500\ °R} = 1.49\ \mu m$$

$$E_{b\lambda} = \frac{(1.187 \times 10^8\ Btu \cdot \mu m^4/hr \cdot ft^2)(1.49\ \mu m)^{-5}}{\exp[(2.5896 \times 10^4\ \mu m \cdot °R)/(1.49\ \mu m)(3500\ °R)] - 1} = 1.135 \times 10^5\ Btu/hr \cdot ft^2 \cdot \mu m$$

10.23 Show how to use Table E-1 to calculate, at given temperature T, the fraction of total power radiated in the band (λ_1, λ_2) of the blackbody spectrum.

▮ The fraction in the band $(0, \lambda)$ is given by (*1*) of Problem 10.17 as

$$\frac{1}{\sigma T^4}\int_0^{\lambda} E_{b\gamma}\, d\gamma = \frac{C_1}{\sigma}\int_0^{\lambda T} \frac{dx}{x^5[\exp(C_2/x) - 1]} \equiv \varphi(\lambda T)$$

a dimensionless function of λT which is tabulated in Table E-1. Thus, for a blackbody at absolute temperature T,

$$\text{Fraction of power in } (\lambda_1, \lambda_2) = \varphi(\lambda_2 T) - \varphi(\lambda_1 T) \tag{1}$$

$$\text{Power in } (\lambda_1, \lambda_2) = [\varphi(\lambda_2 T) - \varphi(\lambda_1 T)]\sigma T^4 \tag{2}$$

In most engineering texts, $\varphi(\lambda T)$ is given as $E_{b(0-\lambda T)}/(\sigma T^4)$.

10.24 A real surface has a total emissive power E less than that of a blackbody. The ratio of the total emissive power of a body to that of a blackbody at the same temperature is the total emissivity (or total hemispherical emissivity), ϵ:

$$\epsilon \equiv \frac{E}{E_b} \tag{1}$$

Some numerical values of total emissivity are presented in Table B-6 of Appendix B; general trends are shown in Fig. 10-8. Determine the radiant heat flux from fireclay brick at 1273 K.

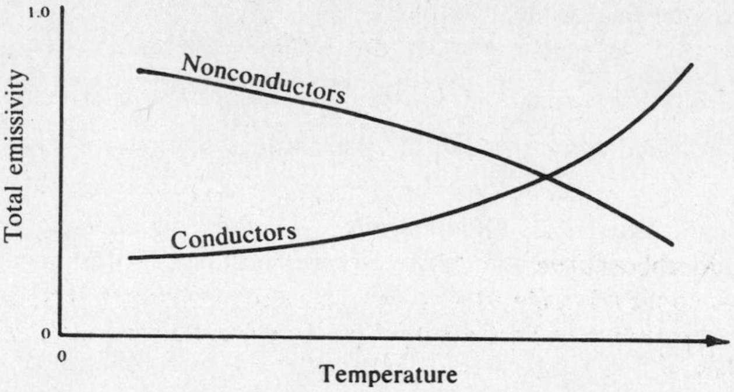

Fig. 10-8

▌ From Table B-6, $\epsilon = 0.75$, whence

$$E = \epsilon(10^8\sigma)\left(\frac{T}{100}\right)^4 = (0.75)(5.6697)(12.73)^4 = 111\,670\ \text{W/m}^2 = 0.111\,67\ \text{MW/m}^2$$

10.25 Determine the radiant heat loss per square centimeter from a mild-steel plate being formed at 1145 K.

▌ The emittance of a blackbody at 1145 K is given by the Stefan–Boltzmann equation

$$E_b = (10^8\sigma)\left(\frac{T}{100}\right)^4 (5.6697)(11.45^4) = 97\,450\ \text{W/m}^2$$

Applying a linear interpolation to the data of Table B-6 yields

$$\epsilon_{1600} \approx 0.29$$

and thus,

$$E_b = (10^8\sigma)\left(\frac{T}{100}\right)^4 (5.6697)(11.45^4) = 97\,450\ \text{W/m}^2$$

10.26 Assuming the sun to be a sphere of radius $R = 433\,000$ miles that radiates as a blackbody, calculate its surface temperature.

▌ Conservation of energy gives, as in Problem 10.3, $G_S = G_E(r_E/R)^2$. But, by assumption, $G_S = E_b = \sigma T_S^4$; hence

$$T_S = \left(\frac{G_E}{\sigma}\right)^{1/4}\left(\frac{r_E}{R}\right)^{1/2} = \left(\frac{444.7\ \text{Btu/hr}\cdot\text{ft}^2}{0.1714\times 10^{-8}\ \text{Btu/hr}\cdot\text{ft}^2\cdot{}^\circ\text{R}^4}\right)^{1/4}\left(\frac{93\,000}{433}\right)^{1/2} = 10\,460\ {}^\circ\text{R}$$

10.27 The radiant heat transfer from a plate of 2 cm^2 area at 1100 K to a very cold enclosure is 3.0 W. What is the emissivity of the plate at this temperature?

▌
$$\epsilon = \frac{E}{E_b} = \frac{(3.0\ \text{W})/(2\times 10^{-4}\ \text{m}^2)}{(5.67\times 10^{-8}\ \text{W/m}^2\cdot\text{K}^4)(1100\ \text{K})^4} = 0.181$$

10.28 By means of a "thought experiment" find the relationship between absorptivity and emissivity.

▌ Consider a black enclosure as shown in Fig. 10-9. It contains a small body, say body 1, which is also a blackbody. Under equilibrium conditions, energy is absorbed by the small body at the same rate as it is emitted, A_1E_b. Imagine body 1 replaced by body 2, which is not a blackbody but has the same size, shape, geometric position, and orientation as body 1. The energy impinging upon this second body is the same as before, $A_1E_b = A_2E_b$. An energy balance on the second body under steady-state conditions is

$$\alpha_2 A_2 E_b = A_2 E$$

and consequently, for any arbitrary body,

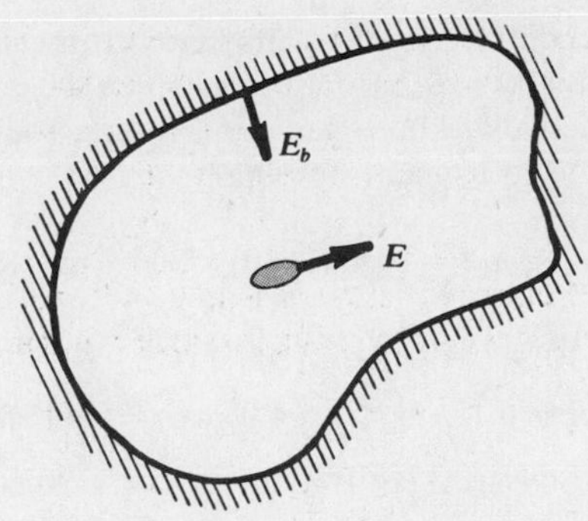

Fig. 10-9

$$\alpha = \frac{E}{E_b} \equiv \epsilon \tag{1}$$

This is known as Kirchhoff's law.

Note that (*1*) is applicable only to a surface receiving blackbody irradiation from surroundings at the same temperature as itself (see Problem 10.31). For monochromatic irradiation, however, it can be shown that Kirchhoff's law is

$$\alpha_\lambda(T) = \epsilon_\lambda(T) \equiv \frac{E_\lambda}{E_{b\lambda}} \tag{2}$$

That is, for specified wavelength, the monochromatic absorptivity and the monochromatic emissivity of a surface at a given surface temperature are equal, regardless of the temperature of the source of the irradiation.

10.29 Reflectivity measurements, which are relatively easy to make, are often used to obtain other surface radiation properties. A set of reflectivity measurements for a certain nontransmitting solid surface at 1000 °R is roughly graphed in Fig. 10-10. Estimate the total emissive power at this temperature.

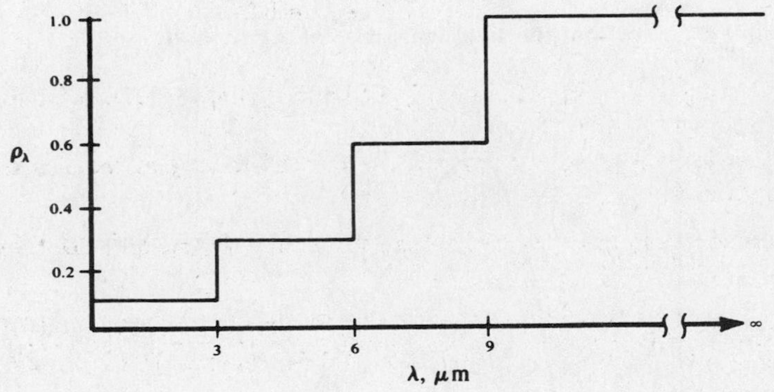

Fig. 10-10

▮ By Kirchhoff's law,

$$E = \int_0^\infty \epsilon_\lambda E_{b\lambda}\, d\lambda = \int_0^\infty \alpha_\lambda E_{b\lambda}\, d\lambda$$

But for this solid surface, $\alpha_\lambda = 1 - \rho_\lambda$; so, using Problem 10.23 and Table E-1 for values of $\varphi(\lambda T)$,

$$\begin{aligned}\epsilon = \frac{E}{\sigma T^4} &= \int_0^\infty \frac{(1-\rho_\lambda)E_{b\lambda}}{\sigma T^4}\, d\lambda \\ &= (1-0.1)\varphi(3000\ \mu\text{m}\cdot{}^\circ\text{R}) + (1-0.3)[\varphi(6000) - \varphi(3000)] + (1-0.6)[\varphi(9000) - \varphi(6000)] \\ &= (0.2)\varphi(3000) + (0.3)\varphi(6000) + (0.4)\varphi(9000) = 0.363\end{aligned}$$

and so

$$E = \epsilon\sigma T^4 = (0.363)(0.1714 \times 10^{-8}\ \text{Btu/hr}\cdot\text{ft}^2\cdot{}^\circ\text{R}^4)(1000\ {}^\circ\text{R})^4 = 622.2\ \text{Btu/hr}\cdot\text{ft}^2$$

10.30 An instrument box housing a control system is subjected to thermal radiation from molten iron in a smelting operation. The housing emissivity is 0.2 from 0.1 to 4.9 μm and is approximately zero for all other wavelengths. Assuming incident blackbody radiation from a source at 2600 °F, how much power is absorbed per square foot of the instrument housing?

▮ $T = 2600 + 460 = 3060\ °\text{R}$ $\qquad \lambda_1 T = (0.1)(3060) = 306\ \mu\text{m}\cdot°\text{R}$ $\qquad \lambda_2 T = (4.9)(3060) = 15\,000\ \mu\text{m}\cdot°\text{R}$

By (2) of Problem 10.23 and Table E-1,

$$(\text{Incident power in } 0.1\ \mu\text{m} < \lambda < 4.9\ \mu\text{m}) = [\varphi(15\,000) - \varphi(306)]\sigma T^4$$
$$\approx (0.8689)(0.1714\ \text{Btu/hr}\cdot\text{ft}^2\cdot°\text{R}^4)(30.60\ °\text{R})^4 = 131\,000\ \text{Btu/hr}\cdot\text{ft}^2$$

Hence, by Kirchhoff's law,

$$\text{Absorbed power} = \alpha \times (\text{incident power}) = \epsilon \times (\text{incident power}) = (0.2)(131\,000) = 26\,000\ \text{Btu/hr}\cdot\text{ft}^2$$

10.31 Show that ϵ and α are not necessarily equal for irradiation of a surface at temperature T by a source at a different temperature, T^*.

▮ The total emissivity is, by (2) of Problem 10.28,

$$\epsilon = \frac{E}{E_b} = \frac{\int_0^\infty \epsilon_\lambda(T)E_{b\lambda}(T)\,d\lambda}{\sigma T^4} = \epsilon(T)$$

The total absorptivity is, by definition,

$$\alpha = \frac{\text{energy absorbed}}{\text{energy incident}} = \frac{\int_0^\infty \alpha_\lambda(T)G_\lambda(T^*)\,d\lambda}{G(T^*)} = \alpha(T, T^*)$$

Since α depends on T^*, while ϵ does not, the two will generally be unequal.

10.32 As seen in Fig. 10-11, the monochromatic emissive power of a real surface is not a constant fraction of that of a black surface. A very useful idealization is that of a *gray body*, defined by

$$\epsilon_\lambda(T) = \epsilon_g = \text{constant}$$

for all λ and all T. Prove that the total emissivity of a gray body is ϵ_g.

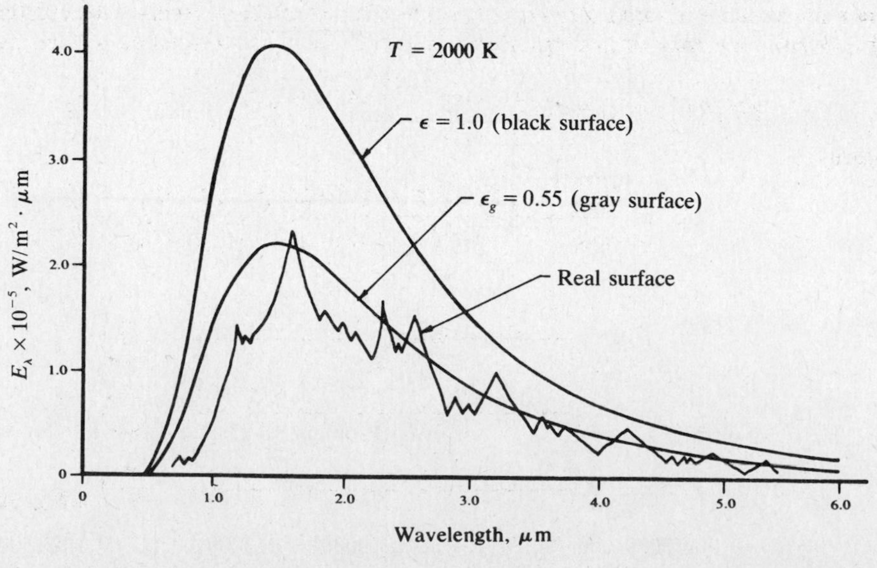

Fig. 10-11

▮
$$E = \int_0^\infty E_\lambda\,d\lambda = \int_0^\infty \epsilon_\lambda E_{b\lambda}\,d\lambda = \epsilon_g \int_0^\infty E_{b\lambda}\,d\lambda = \epsilon_g E_b$$

Hence $\epsilon_g = E/E_b \equiv \epsilon$.

10.33 Develop an expression for the net radiant heat transfer between two parallel infinite black planes at T_1 and T_2, respectively, shown in Fig. 10-12.

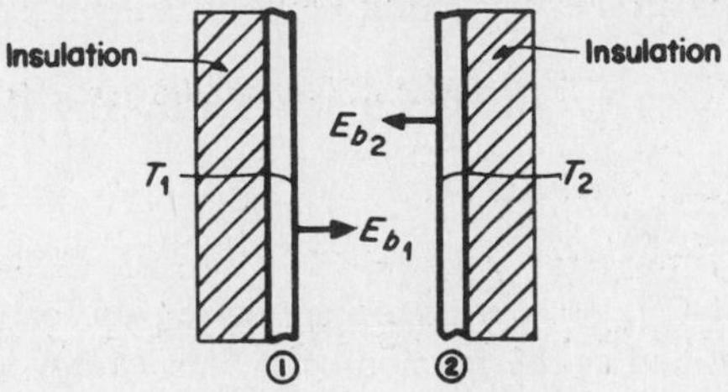

Fig. 10-12

All radiant energy leaving surface 1 impinges upon surface 2, and vice versa. Since both surfaces are black, all incident energy is absorbed, i.e., $q'_{i-j} = q_{i\to j}$. Thus

$$q_{1-2} = q_{1\to 2} - q_{2\to 1} = E_{b1}A_1 - E_{b2}A_2$$

or, per unit surface area (since $A_1 = A_2 = A$),

$$\frac{q_{1-2}}{A} = E_{b1} - E_{b2} = \sigma(T_1^4 - T_2^4) \qquad (1)$$

10.34 A large plane, perfectly insulated on one face and maintained at a fixed temperature T_1 on the bare face, which has an emissivity of 0.90, loses 200 W/m^2 when exposed to surroundings at nearly 0 K. A second plane the same size as the first is also perfectly insulated on one face, but its bare face has an emissivity of 0.45. When the bare face of the second plane is maintained at a fixed temperature T_2 and exposed to surroundings at nearly 0 K, it loses 100 W/m^2. These two planes are brought together so that the parallel bare faces are only 1 cm apart, and the heat supply to each is so adjusted that their respective temperatures remain unchanged. What will be the net heat flux between the planes?

From $E = \epsilon\sigma T^4$,

$$\left(\frac{T_1}{T_2}\right)^4 = \left(\frac{E_1}{E_2}\right)\left(\frac{\epsilon_2}{\epsilon_1}\right) = (2)\left(\frac{1}{2}\right) = 1$$

Thus $T_1 = T_2$, and $q_{1-2}/A = 0$, by (1) of Problem 10.33.

10.35 Two large parallel black planes are maintained at temperatures T_1 and T_2. Later a third black plane, which reaches a temperature T_3 after steady-state conditions have been reached, is placed between these two (Fig. 10-13). What is the ratio of heat transferred with this third plane installed to that transferred initially?

$$q = \sigma(T_1^4 - T_3^4) = \sigma(T_3^4 - T_2^4) \qquad T_1^4 - T_3^4 = T_3^4 - T_2^4 \qquad T_3^4 = \frac{(T_1^4 + T_2^4)}{2}$$

Therefore

$$\frac{q_{1-3}}{q_{1-2}} = \frac{\sigma\{T_1^4 - [(T_1^4 + T_2^4)/2]\}}{\sigma(T_1^4 - T_2^4)} = \frac{T_1^4 - T_2^4}{2(T_1^4 - T_2^4)} = \frac{1}{2}$$

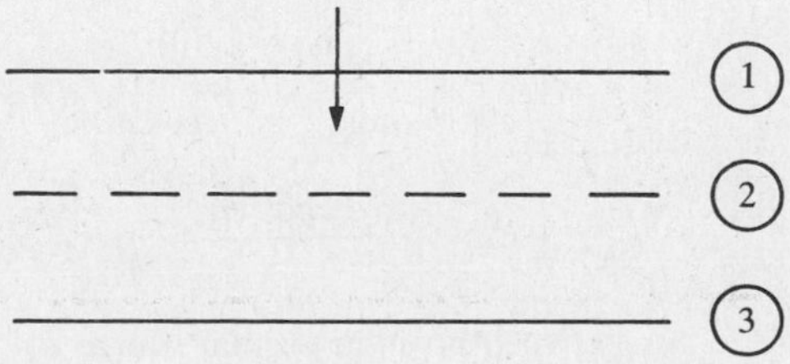

Fig. 10-13

10.36 A window in an experimental aircraft has an area of 1 ft^2 and transmits 3 percent of the incident thermal radiation between 1.5 and 5.0 μm. It is opaque to thermal radiation at all other wavelengths. Determine the heat flux transmitted through the window from a blackbody radiation source at 2340 °F.

$T = 2340\ °F = 2800\ °R \qquad \lambda_1 T = (1.5)(2800) = 4200\ \mu m\cdot °R \qquad \lambda_2 T = (5.0)(2800) = 14\,000\ \mu m\cdot °R$

By (2) of Problem 10.23 and Table E-1,

$$P_b \equiv (\text{incident power in } 1.5 < \lambda < 5.0\ \mu\text{m})$$
$$= [\varphi(14\,000) - \varphi(4200)]\sigma T^4 = (0.7202)(0.1714\ \text{Btu/hr}\cdot\text{ft}^2\cdot{}^\circ\text{R}^4)(28.00\ {}^\circ\text{R})^4 = 7.65\times10^4\ \text{Btu/hr}\cdot\text{ft}^2$$

whence $$q = \tau A P_b = (0.03)(1\ \text{ft}^2)(7.65\times10^4\ \text{Btu/hr}\cdot\text{ft}^2) = 2290\ \text{Btu/hr}$$

10.3 CONFIGURATION FACTORS

10.37 Using the nomenclature of Fig. 10-14, develop an expression for the *configuration factor* between patches dA_1 and dA_2, which is defined as the fraction of radiant energy leaving dA_1 that strikes dA_2 directly, where dA_1 is assumed to be emitting energy diffusely. The term "directly" means that none of the energy is transferred by reflection or reradiation from other surfaces.

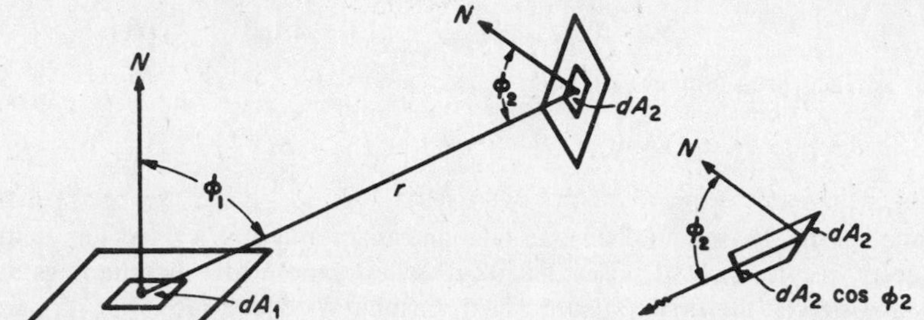

Fig. 10-14

▮ Because intensity decreases as the square of distance, the power radiated from dA_1 and incident upon dA_2 is

$$dq_{1\to2} = \frac{I_1(dA_1\cos\phi_1)}{r^2}(dA_2\cos\phi_2) = I_1\cos\phi_1\,dA_1\,d\omega_{1\to2} \tag{1}$$

[cf. (1) of Problem 10.12, where I_1 is independent of ϕ_1 and where

$$d\omega_{1\to2} = \frac{dA_2\cos\phi_2}{r^2}$$

is the solid angle subtended at dA_1 by dA_2].

Now the total energy radiated from dA_1 is, by (4) of Problem 10.12,

$$dq_1 = E\,dA_1 = \pi I_1\,dA_1 \tag{2}$$

Together (1) and (2) give the configuration factor as

$$F_{dA_1\to dA_2} \equiv \frac{dq_{1\to2}}{dq_1} = \frac{1}{\pi}\cos\phi_1\,d\omega_{1\to2} \tag{3}$$

10.38 Refer to Problem 10.37. Show that if the two patches have the same area, the same configuration factor results when the roles of diffuse source and receiver are interchanged.

▮ By (3) of Problem 10.37,

$$F_{dA_2\to dA_1} = \frac{1}{\pi}\cos\phi_2\,d\omega_{2\to1} = \frac{1}{\pi}\cos\phi_2\,\frac{dA_1\cos\phi_1}{r^2}$$
$$= \frac{1}{\pi}\cos\phi_1\,\frac{dA_2\cos\phi_2}{r^2} = \frac{1}{\pi}\cos\phi_1\,d\omega_{1\to2} = F_{dA_1\to dA_2}$$

10.39 Express the configuration factor between an infinitesimal source dA_1 and a finite receiving surface A_2.

▮ From (3) of Problem 10.37,

$$F_{dA_1\to A_2} = \int_{A_2} F_{dA_1\to dA_2} = \frac{1}{\pi}\int_{A_2}\cos\phi_1\,d\omega_{1\to2} \tag{1}$$

The results of the integration for a particular configuration are presented in Figure E-1. See also Problem 10.40. The first equality in (1) expresses the theorem of *subdivision of the receiving surface*.

10.40 Evaluate $F_{dA_1-A_2}$ for the configuration of Fig. 10-15 (a planar point source above the center of a parallel circular receiver).

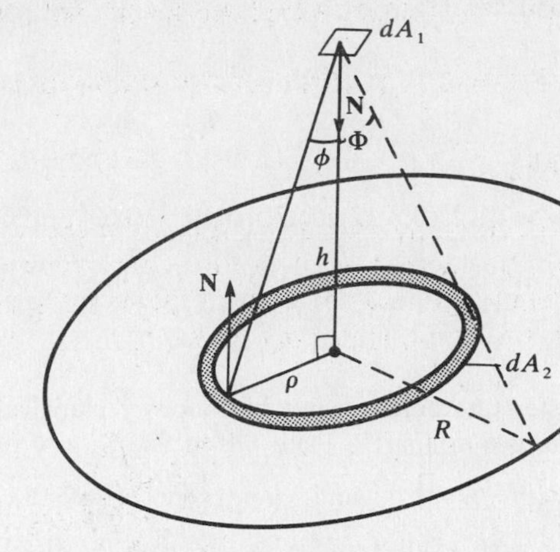

Fig. 10-15

In this case $\phi_1 = \phi_2 = \phi$; for convenience, choose $\rho = h \tan \phi$ as the independent variable. Then

$$d\omega_{1\to2} = \frac{(2\pi\rho\ d\rho)\cos\phi}{\rho^2 + h^2}$$

which, together with

$$\cos^2\phi = \frac{1}{1+\tan^2\phi} = \frac{h^2}{\rho^2+h^2}$$

gives, by (*1*) of Problem 10.39,

$$F_{dA_1\to A_2} = h^2 \int_0^R \frac{2\rho\ d\rho}{(\rho^2+h^2)^2} = \frac{R^2}{R^2+h^2} = \sin^2\Phi \qquad (1)$$

10.41 Find the configuration factor between a planar point source and an infinite parallel plane.

Letting $R\to\infty$ or $\Phi\to\pi/2$ in (*1*) of Problem 10.40, $F\to1$ (as it obviously must), for all h.

10.42 Derive an expression for the configuration factor between two finite diffuse surfaces.

In the notation of Problem 10.37 and for constant I_1,

$$F_{A_1\to A_2} \equiv F_{1\to2} = \frac{q_{1\to2}}{q_1} = \frac{\int_{A_2}\int_{A_1} \cos\phi_1\ dA_1\ d\omega_{1\to2}}{\pi A_1} \qquad (1)$$

10.43 Which curve of Fig. 10-16 represents the configuration factor F for parallel black planes of finite size?

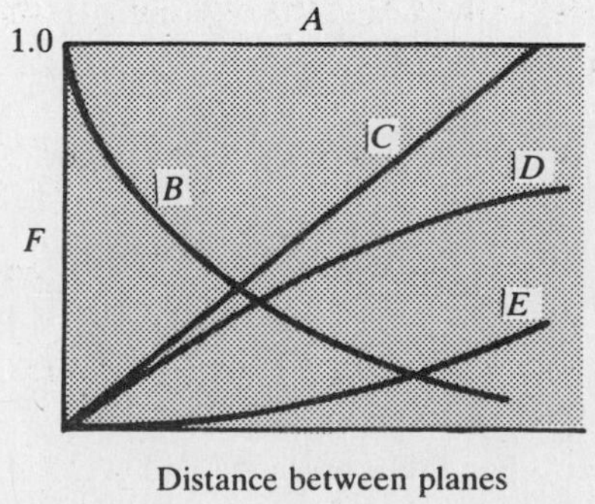

Fig. 10-16

From Figure E-2, assuming that the planes are of identical dimensions the configuration factor is seen to be unity when the distance c is zero and decreases as the distance c increases, which implies that the ratios γ and β decrease. Thus curve B is the solution.

10.44[D] Two blackbody rectangles, 6 ft by 12 ft, are parallel and directly opposed; they are 12 ft apart. If surface 1 is at $T_1 = 200\ ^\circ\text{F}$ and surface 2 is at $T_2 = 600\ ^\circ\text{F}$, determine **(a)** the net rate of heat transfer q_{1-2}; **(b)** the net energy loss rate from the 200 °F surface (side facing surface 2 only) if the surroundings other than surface 2 behave as a blackbody at 0 °R.

▮ **(a)** Since $A_1 = A_2 = A$ and $F_{1\to2} = F_{2\to1} = F$,

$$q_{1-2} = q'_{i-j} - q'_{j-i} = q_{i\to j} - q_{j\to i} = FA\sigma(T_1^4 - T_2^4) \qquad (1)$$

From Figure E-2, with $\beta = 1.0$ and $\gamma = 0.5$, $F \approx 0.12$; hence

$$q_{1-2} \approx (0.12)(6\times12)(0.1714\times10^{-8})(660^4 - 1060^4) = -15\,886\ \text{Btu/hr}$$

(b) $$q_1 = AE_{b1} - AFE_{b2} = A\sigma(T_1^4 - FT_2^4) \approx (72)(0.1714\times10^{-8})[(660)^4 - (0.12)(1060)^4] = 4720\ \text{Btu/hr}$$

10.45[D] Determine the net heat transfer rate between two blackbody rectangles parallel and directly opposed. The plates are 4 by 6 cm and spaced 6 m apart. Plate 1 is at 920 K, and plate 2 is at 810 K.

▮ Here, from Figure E-2, with $\beta = 1.0$ and $\gamma = 0.666$, $F \approx 0.15$, and (*1*) of Problem 10.44 gives

$$q_{1-2} \approx (0.15)(4\times6)(5.67\times10^{-8})(920^4 - 810^4) = 58\,360\ W$$

or 58.36 kW.

10.46 A large black enclosure consists of a box as shown in Fig. 10-17. Surface 1 (bottom) is at 530 K, surface 2 (top) is at 450 K, and all vertical surfaces (including the back wall, 3) are at 480 K. Find the net heat transfer rate q_{1-2}.

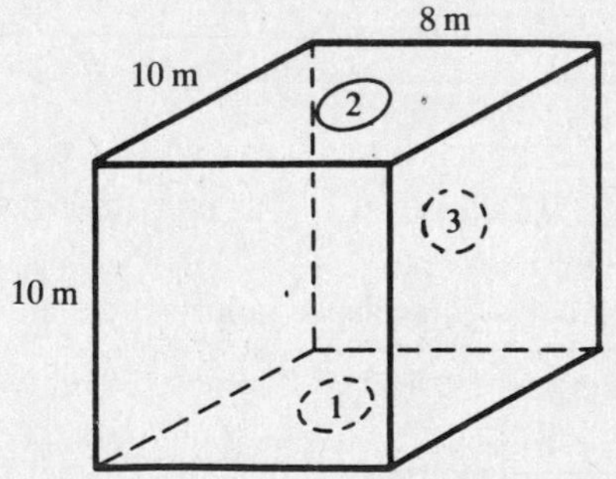

Fig. 10-17

▮ From Figure E-2 with $\beta = 1.0$ and $\gamma = 0.8$, $F_{1-2} \approx 0.168$; (*1*) of Problem 10.44 yields

$$q_{1-2} \approx (0.168)(80)(5.6697)(5.3^4 - 4.5^4) = 28\,880\ \text{W} = 28.88\ \text{kW}$$

10.47 Referring to Problem 10.39, express the analogous theorem of *subdivision of the emitting surface*.

▮ By (*1*), (*2*), and (*3*) of Problem 10.37,

$$F_{A_1\to dA_2} = \frac{\int_{A_1} dq_{1\to2}}{q_1} = \frac{\int_{A_1} (I_1 \cos\phi_1\, d\omega_{1\to2})\, dA_1}{\int_{A_1} \pi I_1\, dA_1}$$

$$= \frac{\pi I_1 \int_{A_1} F_{dA_1\to dA_2}\, dA_1}{\pi I_1 A_1} = \frac{1}{A_1}\int_{A_1} F_{dA_1\to dA_2}\, dA_1$$

$$\equiv (\text{average value of } F_{dA_1\to dA_2} \text{ over } A_1)$$

10.48 Prove the *reciprocity theorem*: For finite areas A_1 and A_2, $A_1F_{1\to2} = A_2F_{2\to1}$.

▮ By (*1*) of Problem 10.42,

$$A_1F_{1\to2} = \frac{1}{\pi}\int_{A_2}\int_{A_1} \frac{\cos\phi_1 \cos\phi_2}{r^2}\, dA_1\, dA_2$$

Clearly, we can write the similar expression

$$A_2F_{2\to 1} = \frac{1}{\pi}\int_{A_1}\int_{A_2} \frac{\cos\phi_2 \cos\phi_1}{r^2}\, dA_2\, dA_1$$

since the assignment of subscripts 1 and 2 is arbitrary. The integrand of both expressions is the same, and it is a continuous function; hence the order of integration is immaterial and the theorem follows. (Problem 10.38 is now seen as a special case of the reciprocity theorem.)

10.49 Establish the reciprocity theorem through a physical argument.

▌ Let two black surfaces, A_1 and A_2, come to thermal equilibrium (common temperature T) in an environment at 0 K. Then

$$0 = q_{1-2} = F_{1\to 2}A_1\sigma T^4 - F_{2\to 1}A_2\sigma T^4 = (A_1F_{1\to 2} - A_2F_{2\to 1})\sigma T^4$$

whence the theorem since $\sigma T^4 \neq 0$.

10.50 Assuming the two plane surfaces of Fig. 10-18 at right angles are black, determine the radiant heat transfer from surface 1 to surface 2.

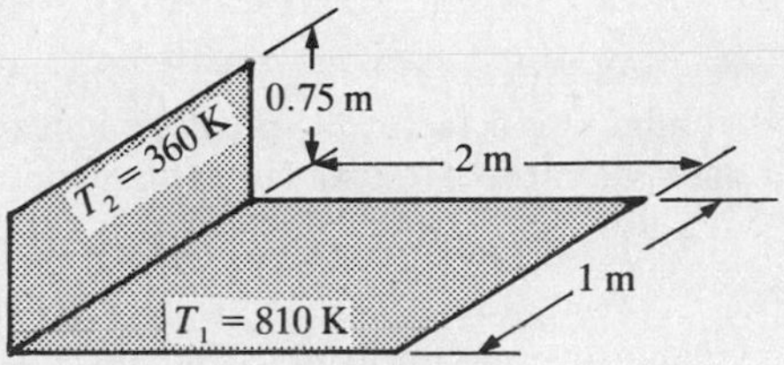

Fig. 10-18

▌ By the reciprocity theorem,

$$q_{1-2} = F_{1\to 2}A_1\sigma T_1^4 - F_{2\to 1}A_2\sigma T_2^4 = F_{1\to 2}A_1\sigma(T_1^4 - T_2^4) \tag{1}$$

From Figure E-3, with $\beta = \frac{2}{1} = 2.0$ and $\gamma = 0.75$, $F_{1\to 2} \approx 0.1$; thus

$$q_{1-2} = (0.1)(2\times 1)(5.6697)(8.1^4 - 3.6^4) = 4691 \text{ W}$$

10.51 Rework Problem 10.44(*b*) if the (black) surroundings are at 70 °F.

▌ From the loss, 4720 Btu/hr, corresponding to zero radiation by the surroundings, one must subtract the power actually received from the surroundings:

$$F_{\text{surr}\to A_1}A_{\text{surr}}\sigma T_{\text{surr}}^4 = F_{A_1\to\text{surr}}A_1\sigma T_{\text{surr}}^4 = (1 - F_{1-2})A_1\sigma T_{\text{surr}}^4$$

$$\approx (1 - 0.12)(72)(0.1714\times 10^{-8})(530^4) = 8569 \text{ Btu/hr}$$

Thus surface 1 loses $4720 - 8569 = -3849$ Btu/hr (a net power gain).

10.52 For the configuration illustrated in Figure E-5, with the flat annular area between the two cylinders at one end designated as A_3, obtain expressions for $F_{1\to 3}$, $F_{3\to 1}$, $F_{3\to 2}$, and $F_{3\to 3}$ in terms of $F_{1\to 1}$, $F_{1\to 2}$, and the three areas A_1, A_2, and A_3. ($F_{3\to 3}$ is the configuration factor between the two annular areas at opposite ends.)

▌ An evident property of the configuration factor is stated in the *enclosure theorem*: $\Sigma_j F_{i\to j} = 1$, where A_j ranges over the complete environment of A_i, including A_i itself. Thus

$$2F_{1\to 3} + F_{1\to 2} + F_{1\to 1} = 1 \quad\text{or}\quad F_{1\to 3} = \tfrac{1}{2}(1 - F_{1\to 2} - F_{1\to 1})$$

and by reciprocity

$$F_{3\to 1} = \frac{A_1}{2A_3}(1 - F_{1\to 2} - F_{1\to 1})$$

Also, since $F_{2\to 2} = 0$,

$$2F_{2\to 3} + F_{2\to 1} = 1 \quad\text{or}\quad F_{2\to 3} = \tfrac{1}{2}(1 - F_{2\to 1})$$

and reciprocity gives

$$F_{2\to3} = \frac{1}{2}\left(1 - \frac{A_1}{A_2} F_{1\to2}\right) \qquad \text{and} \qquad F_{3\to2} = \frac{A_2}{2A_3}\left(1 - \frac{A_1}{A_2} F_{1\to2}\right)$$

Again by the enclosure theorem,

$$F_{3\to3} + F_{3\to1} + F_{3\to2} = 1$$

or

$$F_{3\to3} = 1 - F_{3\to1} - F_{3\to2} = 1 - \frac{A_1}{2A_3}(1 - F_{1\to2} - F_{1\to1}) - \frac{A_2}{2A_3}\left(1 - \frac{A_1}{A_2} F_{1\to2}\right)$$

$$= 1 - \frac{A_1 + A_2}{2A_3} + \frac{A_1}{2A_3}(2F_{1\to2} + F_{1\to1})$$

10.53 A small aircraft component, originally at 290 K, is placed in a large oven at 560 K. Approximate the initial net instantaneous heat transfer to the component by radiation if all surfaces are blackbodies, the component is approximately spherical in shape with a diameter of 4 cm, and the oven is cubical with 4-m sides.

▌ The net radiative flux from the oven to the component is, by (*1*) of Problem 10.50,

$$q_{1-2} = A_1 F_{1\to2}\sigma(T_1^4 - T_2^4)$$

where subscript 1 denotes oven and 2 denotes component. $F_{1\to2}$ would be quite difficult to evaluate geometrically; however, the configuration factor from component to oven, $F_{2\to1}$, is unity since all radiation leaving the component impinges upon the oven walls. Thus, applying the reciprocity theorem yields

$$q_{1-2} = 4\pi(0.02)^2(1)(5.6697)(5.6^4 - 2.9^4) = 26.0\ \text{W}$$

10.54 A hemispherical shell and a plane form an enclosure (Fig. 10-19). What is the configuration factor $F_{1\to2}$?

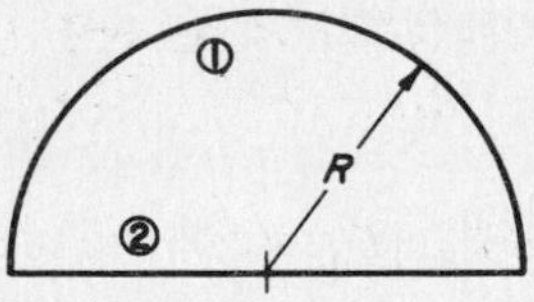

Fig. 10-19

▌
$$A_1 F_{1\to2} = A_2 F_{2\to1} \qquad \text{or} \qquad F_{1\to2} = \left(\frac{A_2}{A_1}\right) F_{2\to1}$$

But $F_{2\to2} + F_{2\to1} = 1$ and $F_{2\to2} = 0$; therefore

$$F_{1\to2} = \frac{A_2}{A_1}(1) = \frac{\pi R^2}{4\pi R^2/2} = \frac{1}{2}$$

10.55 Two parallel rectangular planes, 1 and 2, of dimensions a by $2a$ are joined on their long edge by a third plane 3 perpendicular to them and of height $1.5a$ (Fig. 10-20). Determine the configuration factors $F_{1\to2}$ and $F_{1\to3}$.

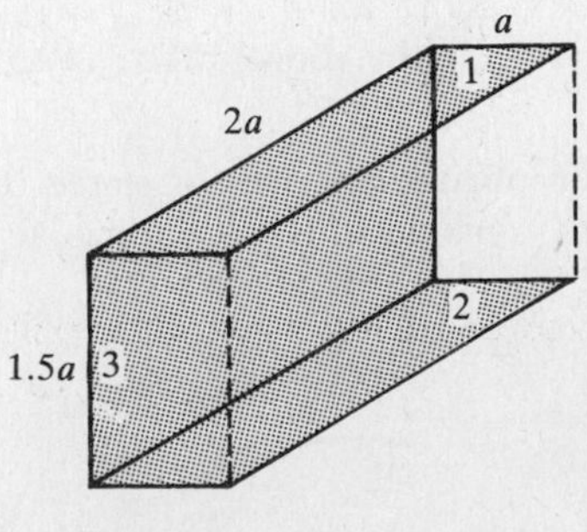

Fig. 10-20

▌ For $F_{1\to2}$, use Figure E-2.

$$\beta = \frac{2a}{1.5a} = 1.333 \qquad \gamma = \frac{a}{1.5a} = 0.667$$

Thus,

$$F_{1\to 2} = 0.17$$

For $F_{1\to 3}$, use Figure E-3.

$$\beta = \frac{a}{2a} = 0.50 \qquad \gamma = \frac{1.5a}{2a} = 0.75$$

Thus,

$$F_{1\to 3} = 0.28$$

10.56 Assuming the surfaces in Fig. 10-21 to be black, determine the radiant heat transfer from surface 2 to surface 1.

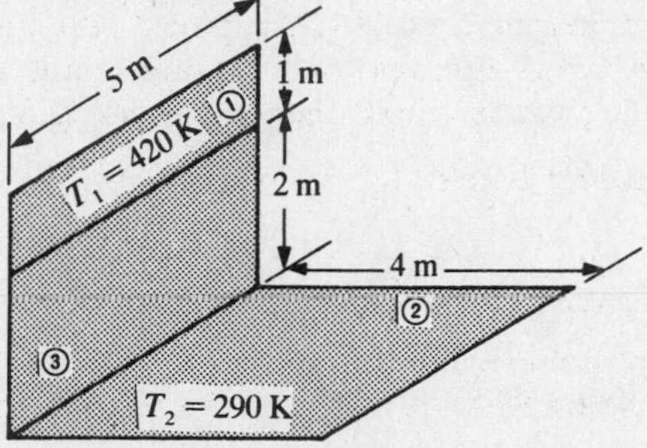

Fig. 10-21

By subdivision of the receiving surface (Problem 10.39),

$$F_{2\to 1} = F_{2\to (1,3)} - F_{2\to 3}$$

For $F_{2\to(1,3)}$, $\beta = \frac{4}{5} = 0.8$; $\gamma = \frac{3}{5} = 0.6$ and Figure E-3 gives $F_{2\to(1,3)} = 0.20$. For F_{2-3}, $\beta = \frac{4}{5} = 0.8$; $\gamma = \frac{2}{5} = 0.4$ and Figure E-3 gives $F_{2\to 3} = 0.16$. Therefore, with $F_{2\to 1} = 0.20 - 0.16 = 0.04$, (*1*) of Problem 10.50 gives

$$q_{2-1} = A_2F_{2-1}\sigma(T_2^4 - T_1^4) = (4\times 5)(0.04)(5.6697)(2.9^4 - 4.2^4) = -1091 \text{ W}$$

10.57 Approximate the radiant energy leaving a 30-mm-diam. sphere at 1200 K and impinging upon a 1-m by 1.5-m wall 1 m away from the sphere (Fig. 10-22). Assume all surfaces to be blackbodies.

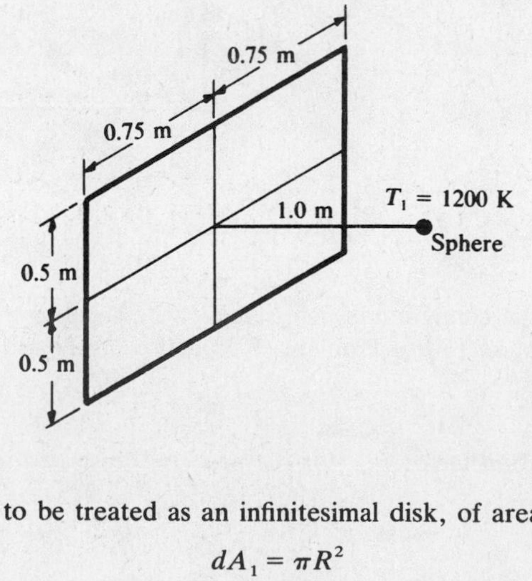

Fig. 10-22

The sphere is small enough to be treated as an infinitesimal disk, of area

$$dA_1 = \pi R^2$$

From Figure E-1, with

$$\beta = 0.75 \qquad \text{and} \qquad \gamma = 0.5$$

the configuration factor to one-fourth the wall is approximately 0.021. Thus, for the entire wall A_2,

$$F_{dA_1\to A_2} \approx 4(0.021) = 0.084$$

and

$$q_{dA_1\to A_2} = F_{dA_1\to A_2}(\sigma T_1^4)(dA_1) = (0.084)(5.6697\times10^{-8}\ \mathrm{W/m^2\cdot K^4})(1200\ \mathrm{K})^4\pi(15\times10^{-3}\ \mathrm{m})^2 = 6.98\ \mathrm{W}$$

10.58 Find the configuration factor $F_{1\to2}$ for the plane wall surfaces shown in Fig. 10-23.

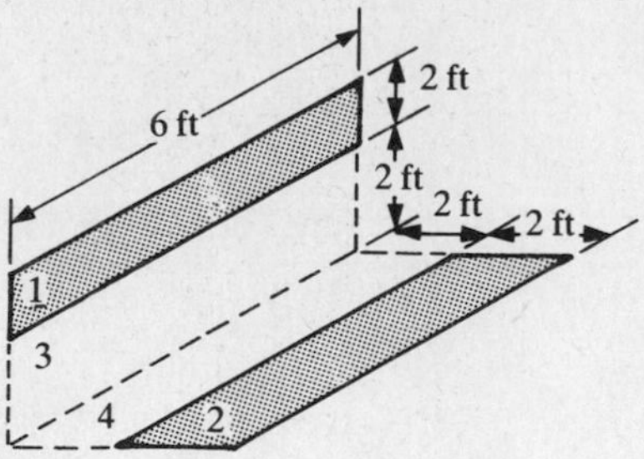

Fig. 10-23

By Problem 10.39, with $A_1 = A_3$,

$$F_{(1,3)\to(4,2)} = F_{(1,3)\to4} + F_{(1,3)\to2} = F_{(1,3)\to4} + \frac{F_{1\to2}+F_{3\to2}}{2} \tag{1}$$

Also,

$$F_{3\to(4,2)} = F_{3\to4} + F_{3\to2} \tag{2}$$

Elimination of $F_{3\to2}$ between (*1*) and (*2*) leads to

$$F_{1\to2} = 2F_{(1,3)\to(4,2)} - 2F_{(1,3)\to4} - F_{3\to(4,2)} + F_{3\to4} \tag{3}$$

Observe that the *only* symmetry exploited [in (*1*)] is $A_1 = A_3$. Each term on the right of (*3*) can be evaluated from Figure E-3; thus

$$F_{1\to2} \approx 2(0.23) - 2(0.16) - (0.33) + (0.27) = 0.08$$

10.59 Find the configuration factor for the surfaces shown in Fig. 10-24.

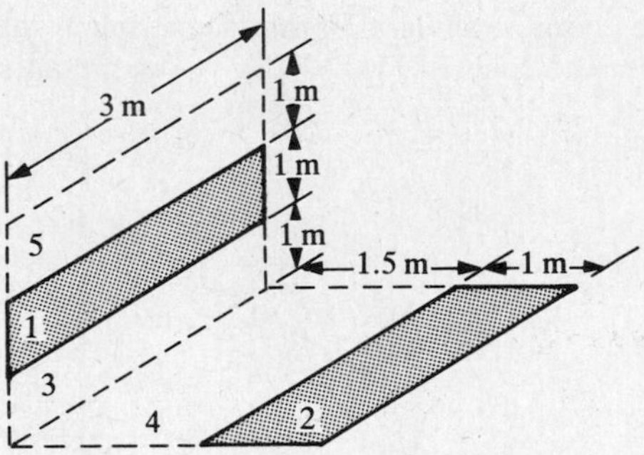

Fig. 10-24

Disregard A_5; then the configuration is essentially the same as that treated in Problem 10.58 (with $A_1 = A_3$). Thus, by (*3*) of Problem 10.58 and using Figure E-3,

$$F_{1\to2} \approx 2(0.24) - 2(0.21) - 0.33 + 0.32 = 0.05$$

10.60 Determine the configuration factor F_{1-2} for the problem illustrated in Fig. 10-25.

$$F_{1\to(2,3)} = F_{1\to2} + F_{1\to3} = 2F_{1\to2} \quad \textit{(by symmetry)}$$

From Figure E-2, $F_{1\to(2,3)} = 0.1285$; hence $F_{1\to2} = 0.1285/2 = 0.0642$.

10.61 For two concentric cylinders as shown in Figure E-5, having $r_1 = 2.0$ ft, $r_2 = 1.0$ ft, and $L = 2.0$ ft, determine $F_{1\to1}$, $F_{1\to2}$, and $F_{3\to3}$, where the end annular plane area is A_3. Use the results of Problem 10.52.

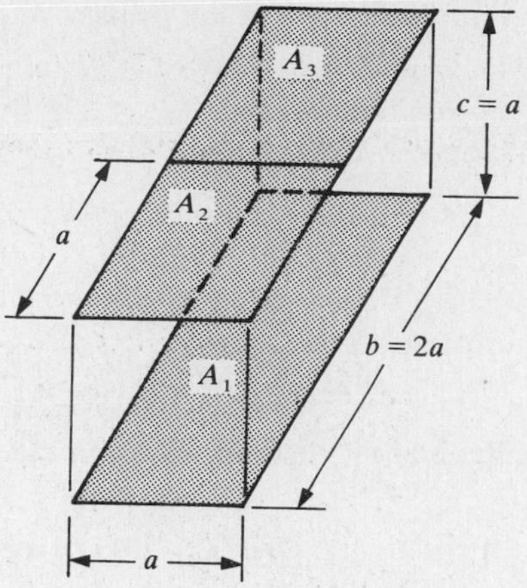

Fig. 10-25

▌ From Figure E-5, with $\gamma = 2.0$ and $\beta = 2.0$, $F_{1\to1} = 0.23$; $F_{1\to2} = 0.335$. The areas are

$$A_1 = 2\pi(2.0)(2.0) = 25.133\ \text{ft}^2 \qquad A_2 = 2\pi(1.0)(2.0) = 12.566\ \text{ft}^2 \qquad A_3 = \pi(2.0^2 - 1.0^2) = 9.425\ \text{ft}^2$$

From Problem 10.52,

$$F_{3\to3} = 1 - \frac{25.133 + 12.566}{2(9.425)} + \frac{25.133}{2(9.425)}[2(0.335) + 0.23] = 0.200$$

10.62 Determine the configuration factor $F_{1\to2}$ for Fig. 10-26.

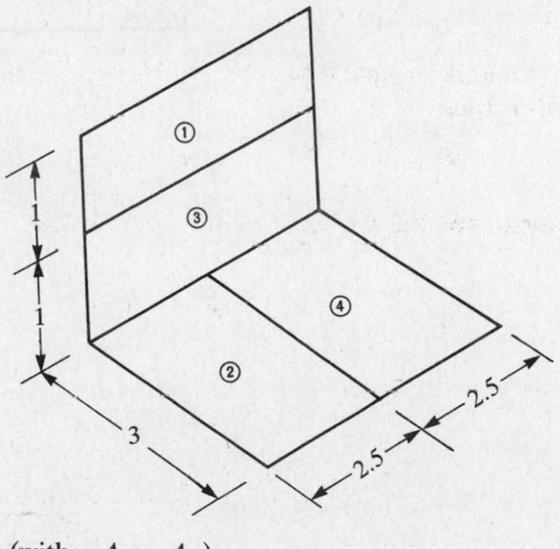

Fig. 10-26

▌ By subdivision and symmetry (with $A_1 = A_3$),

$$F_{(1,3)\to(2,4)} = \tfrac{1}{2}[F_{1\to(2,4)} + F_{3\to(2,4)}]$$
$$= \tfrac{1}{2}[2F_{1\to2} + F_{3\to(2,4)}]$$

or

$$F_{1\to2} = F_{(1,3)\to(2,4)} - \tfrac{1}{2}F_{3\to(2,4)} = 0.285 - \tfrac{1}{2}(0.37) = 0.10$$

10.63 A plate with an emissivity of 0.4 is attached to the side of a spaceship so that it is perfectly insulated from the inside of the ship. Assuming that outer space is a blackbody at absolute zero, determine the equilibrium temperature of the plate if the radiant heat flux from the sun is 3154 W/m².

▌ An energy balance for steady state is

$$\left(\frac{q}{A}\right)_{\text{from sun}} = \left(\frac{q}{A}\right)_{\text{to space}}$$

Thus,

$$3154\ \text{W/m}^2 = \epsilon\sigma(T_p^4 - T_{\text{surr}}^4)$$

$$T_p = \left[\frac{3154\ \text{W/m}^2}{(0.4)(5.6697 \times 10^{-8}\ \text{W/m}^2\cdot\text{K}^4)}\right]^{1/4} = 611\ \text{K}$$

10.64 Refer to Figure E-4. Is the figure still usable if the point of interest (γ, β) falls within the geometrical inset?

▮ Yes, it is. The inset covers part of the region $\gamma > 1$, $\beta > 1$. But, applying the reciprocity theorem to the two disks, we have

$$\pi r_1^2 F_{1\to 2}(\gamma, \beta) = \pi r_2^2 F_{1\to 2}\left(\frac{1}{\beta}, \frac{1}{\gamma}\right)$$

or, since $\gamma\beta = r_2/r_1$,

$$F_{1\to 2}(\gamma, \beta) = (\gamma\beta)^2 F_{1\to 2}\left(\frac{1}{\beta}, \frac{1}{\gamma}\right)$$

If the point (γ, β) is hidden, the point $(1/\beta, 1/\gamma)$ will be in the clear.

10.4 GRAY BODY RADIATION; THE ELECTRICAL ANALOGY

10.65 For the gray surface (Problem 10.32) indicated in Fig. 10-27, the *radiosity* J is the total radiant energy leaving the surface per unit area per unit time and G is the irradiation from all other surfaces. Derive an expression for the heat transfer from the surface. Assume the body absorbs αG and that no energy is transmitted, i.e., $\tau = 0$.

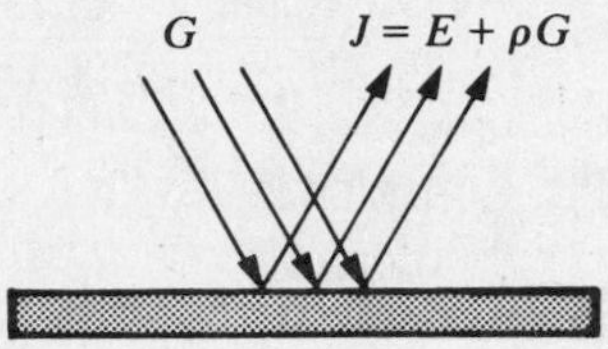

Fig. 10-27

▮ An energy balance at the surface gives

$$\frac{q}{A} = J - G \qquad (1)$$

But, by Problem 10.32, $E = \epsilon E_b$, so that

$$J = \epsilon E_b + \rho G = \epsilon E_b + (1-\epsilon)G \qquad (2)$$

Eliminating G between (*1*) and (*2*) gives

$$q = \frac{E_b - J}{(1-\epsilon)/\epsilon A} \qquad (3)$$

10.66 Give an electrical analog to (*3*) of Problem 10.65.

▮ See Fig. 10-28.

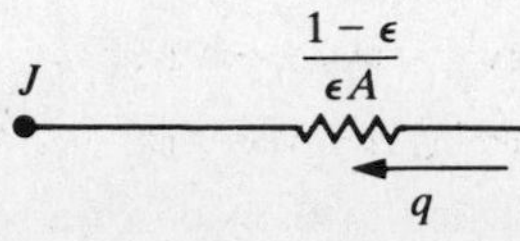

Fig. 10-28

10.67 The configuration factor $F_{1\to 2}$ between two (gray) surfaces can be treated as though the measure of "spacial conductivity" whereby only part of the radiation from surface 1 reaches surface 2. Derive an expression for the heat transfer q_{1-2}, accounting for only this spacial resistance.

▮ See Fig. 10-29. The heat transfer from surface 1 to "space" is given by

$$q_{1\text{-space}} = \frac{J_1}{R_{\text{space}}} = \frac{J_1}{1/A_1F_{1\to 2}} = A_1F_{1\to 2}J_1$$

Similarly,

$$q_{2\text{-space}} = A_2F_{2\to 1}J_2$$

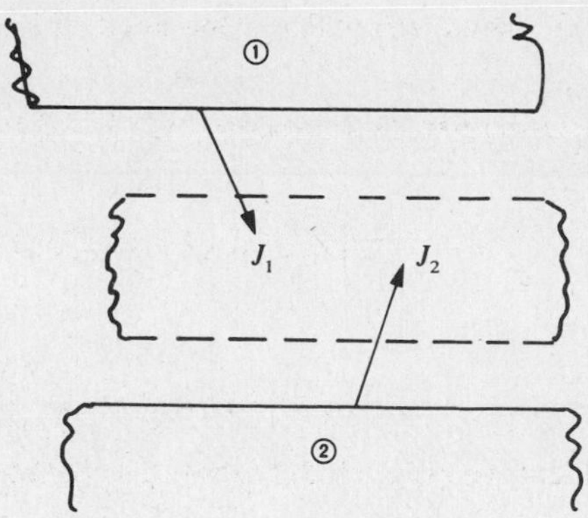

Fig. 10-29

Making an energy balance on "space" gives

$$q_{1-2} = A_1F_{1\to2}J_1 - A_2F_{2\to1}J_2$$

But $A_1F_{1\to2} = A_2F_{2\to1}$ from the reciprocity theorem; therefore,

$$q_{1-2} = A_1F_{1\to2}(J_1 - J_2) \qquad (1)$$

This expression generalizes the blackbody formula (*1*) of Problem 10.50.

10.68 Devise an electrical analog for (*1*) of Problem 10.67.

▮ See Fig. 10-30.

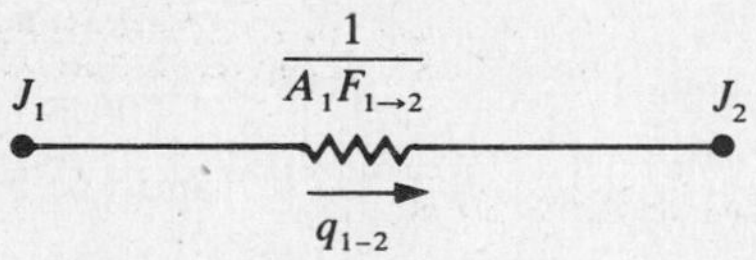

Fig. 10-30

10.69 Using Problems 10.65 through 10.68, determine an expression for radiant energy exchange between two gray bodies, and give the electrical analog.

▮ The electrical analog may be obtained by combining circuit components; see Fig. 10-31. The expression of Ohm's law is now

$$q_{1-2} = \frac{E_{b1} - E_{b2}}{\Sigma R} = \frac{E_{b1} - E_{b2}}{(1-\epsilon_1)/\epsilon_1A_1 + 1/A_1F_{1\to2} + (1-\epsilon_2)/\epsilon_2A_2} \qquad (1)$$

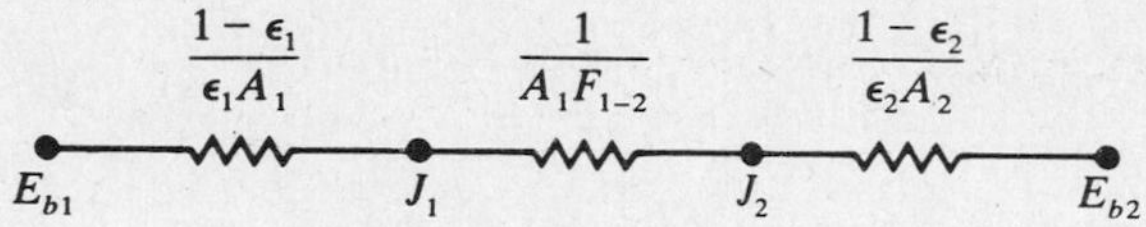

Fig. 10-31

10.70 Determine the net heat flux between two infinite parallel gray planes.

▮ Since this is a two-body system exchanging heat only between the two planes, (*1*) of Problem 10.69 is applicable:

$$q_{1-2} = \frac{E_{b1} - E_{b2}}{(1-\epsilon_1)/\epsilon_1A_1 + 1/A_1F_{1\to2} + (1-\epsilon_2)/\epsilon_2A_2}$$

But since the planes are identical, $A_1 = A_2 = A$ and since they are infinite in extent, $F_{1\to2} = F_{2\to1} = 1.0$. Then

$$\frac{q_{1-2}}{A} = \frac{\sigma(T_1^4 - T_2^4)}{(1-\epsilon_1)/\epsilon_1 + 1 + (1-\epsilon_2)/\epsilon_2} = \frac{\sigma(T_1^4 - T_2^4)}{1/\epsilon_1 + 1/\epsilon_2 - 1} \qquad (1)$$

10.71 Derive (*1*) of Problem 10.70 by tracing the energy rays.

▮ See Fig. 10-32. For a ray which leaves surface 1, a fraction is absorbed by surface 2, and the remainder is reflected back toward surface 1. At surface 1 a second partial absorption and re-reflection occurs, etc.

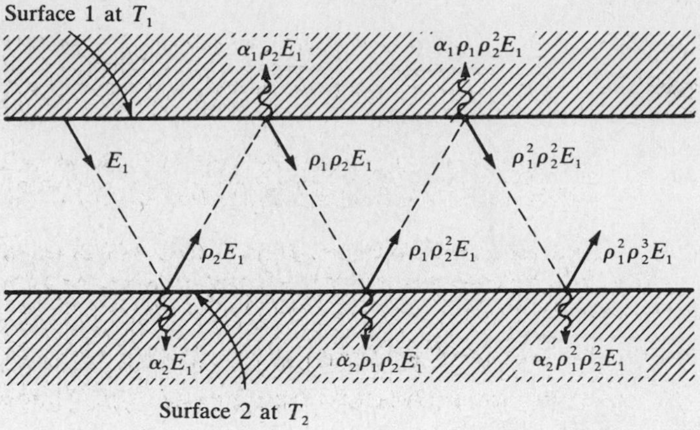

Fig. 10-32

The radiant transfer per unit area from plane 1 to plane 2 is

$$\frac{q'_{1-2}}{A} = \alpha_2E_1 + \alpha_2\rho_1\rho_2E_1 + \alpha_2\rho_1^2\rho_2^2E_1 + \cdots = \alpha_2E_1[1 + \rho_1\rho_2 + (\rho_1\rho_2)^2 + \cdots] = \frac{\alpha_2E_1}{1-\rho_1\rho_2}$$

Likewise

$$\frac{q'_{2-1}}{A} = \frac{\alpha_1E_2}{1-\rho_1\rho_2}$$

Thus, the net heat transfer from plane 1 to plane 2 is

$$\frac{q_{1-2}}{A} = \frac{q'_{1-2}}{A} - \frac{q'_{2-1}}{A} = \frac{\alpha_2E_1 - \alpha_1E_2}{1-\rho_1\rho_2} \qquad (1)$$

But for gray bodies, $E = \epsilon\sigma T^4$, $\epsilon = \alpha$, and $\rho = 1 - \epsilon$, and (1) becomes

$$\frac{q_{1-2}}{A} = \frac{\epsilon_2\epsilon_1\sigma T_1^4 - \epsilon_1\epsilon_2\sigma T_2^4}{1-(1-\epsilon_1)(1-\epsilon_2)} = \frac{\sigma(T_1^4 - T_2^4)}{1/\epsilon_1 + 1/\epsilon_2 - 1}$$

10.72 Generalizing Problem 10.69, show the electrical analogs for (*a*) three-gray-surface and (*b*) four-gray-surface systems.

▌ (*a*) See Fig. 10-33. (*b*) See Fig. 10-34.

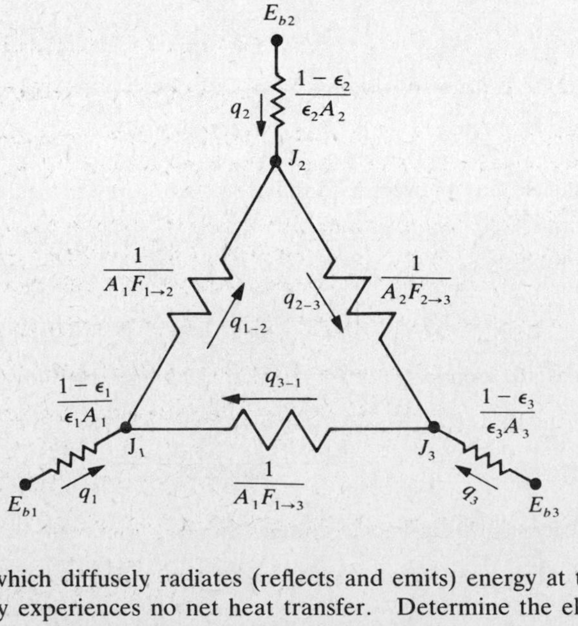

Fig. 10-33

10.73 A *reradiating surface* is one which diffusely radiates (reflects and emits) energy at the same rate it receives incident radiation and thereby experiences no net heat transfer. Determine the electrical analog and the heat transfer equation for a closed system consisting of two gray surfaces and one reradiating surface.

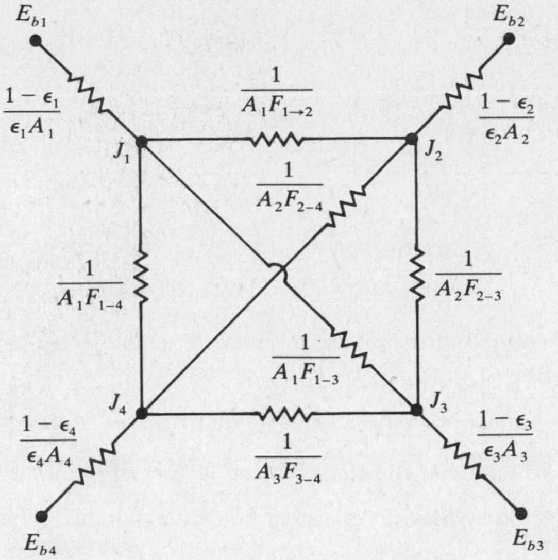

Fig. 10-34

▮ The network of Fig. 10-31 (Problem 10.69) is modified to include a node representing the reradiating surface r; the result is Fig. 10-35. The net heat transfer from body 1 to body 2 is found by dividing the potential difference by the sum of the series resistances. The center series resistance is for the two parallel branches from J_1 to J_2, and thus

$$q_{1-2} = \frac{\sigma(T_1^4 - T_2^4)}{\dfrac{1-\epsilon_1}{A_1\epsilon_1} + \dfrac{1}{A_1F_{1\to2} + 1/(1/A_1F_{1\to r} + 1/A_2F_{2\to r})} + \dfrac{1-\epsilon_2}{A_2\epsilon_2}} \qquad (1)$$

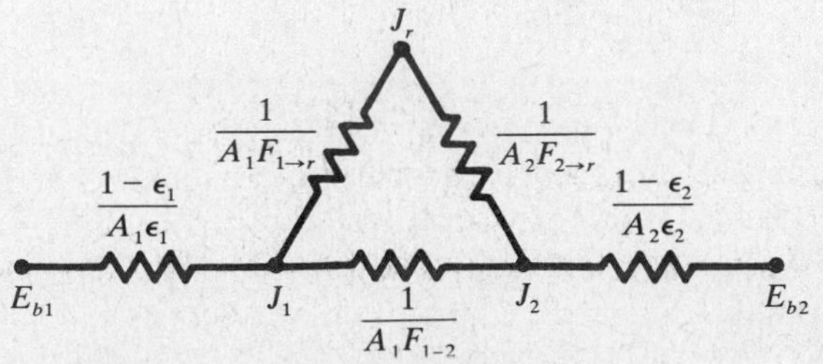

Fig. 10-35

10.74 In an experimental kitchen a beef roast wrapped in aluminum foil is placed in the center of a gas oven to cook. The oven is preheated to 400 °F, and heating is accomplished by the gas flame underneath the bottom surface of the oven. The oven side walls and top are well insulated (reradiating), and the physical dimensions are 2 by 2 by 2 ft. The roast can be approximated as an 8-in.-diam. sphere. Assuming oven-wall emissivity of 0.8, aluminum-foil emissivity of 0.1, and an initial meat temperature of 70 °F, calculate the total initial radiant heat flux to the roast.

▮ For this problem Fig. 10-35 and (*1*) of Problem 10.73 are applicable. Denote the heater surface (bottom of oven) by subscript 1, the roast by subscript 2, and the other five oven walls by r:

$$T_1 = 400 + 460 = 860\ \text{°R} \qquad T_2 = 70 + 460 = 530\ \text{°R} \qquad \epsilon_1 = 0.8 \qquad \epsilon_2 = 0.1$$

The configuration factors are best determined from the reciprocity theorem. Since all radiant energy from the roast impinges upon the oven walls and each wall receives an equal amount, we have

$$F_{2\to1} = \tfrac{1}{6} \quad \text{and} \quad F_{2\to r} = \tfrac{5}{6}$$

By reciprocity

$$F_{1\to2} = \frac{A_2}{A_1} F_{2\to1} = \frac{\pi(\frac{8}{12})^2}{2\times2}\,\frac{1}{6} = 0.058$$

Finally, by the enclosure theorem,

$$F_{1\to r} = 1.0 - F_{1\to1} - F_{1\to2} = 1.0 - 0 - 0.058 = 0.942$$

Calculating the three thermal resistances, we have $A_1 = 4\ \text{ft}^2$ and $A_2 = (4\pi/9)\ \text{ft}^2$; thus

$$\frac{1-\epsilon_1}{A_1\epsilon_1} = \frac{1-0.8}{4(0.8)} = 0.0625\ \text{ft}^{-2}$$

$$\frac{1}{A_1F_{1-2} + 1/(1/A_1F_{1\to r} + 1/A_2F_{2\to r})} = \frac{1}{4(0.058) + 1/[1/(4)(0.942) + 1/(4\pi/9)(\frac{5}{6})]} = 0.89\ \text{ft}^{-2}$$

$$\frac{1-\epsilon_2}{A_2\epsilon_2} = \frac{1-0.1}{(4\pi/9)(0.1)} = 6.45\ \text{ft}^{-2}$$

Then
$$q_{1-2} = \frac{(0.1714\ \text{Btu/hr}\cdot\text{ft}^2\cdot{}^\circ\text{R}^4)[(8.60\ {}^\circ\text{R})^4 - (5.30\ {}^\circ\text{R})^4]}{(0.0625 + 0.89 + 6.45)\ \text{ft}^{-2}} = 108\ \text{Btu/hr}$$

10.75 For a closed system composed of two *black* surfaces, A_1 and A_2, and a reradiating surface r, define *modified configuration factors* $\bar{F}_{i\to j}$ $(i, j = 1, 2)$, where

$$\bar{F}_{i\to j} \equiv (\text{fraction of power from } A_i \text{ that reaches } A_j \text{ directly or via reradiation})$$

Express $\bar{F}_{1\to 2}$ in terms of (unmodified) configuration factors of the system.

▮ It can be shown that the reciprocity theorem holds for modified configuration factors, and hence that

$$q_{1-2} = A_1\bar{F}_{1\to 2}\sigma(T_1^4 - T_2^4) \tag{1}$$

Comparing this with (*1*) of Problem 10.73—in which we set $\epsilon_1 = \epsilon_2 = 1$ to convert the gray surfaces to black surfaces—we have at once:

$$\bar{F}_{1\to 2} = F_{1\to 2} + \frac{1}{1/F_{1\to r} + (A_1/A_2)(1/F_{2\to r})} \tag{2}$$

10.76 Refer to Problem 10.75. Show that, despite the form of (*2*), $\bar{F}_{1\to 2}$—and hence q_{1-2}—is completely independent of the geometry of the reradiating surface.

▮ By the enclosure theorem,

$$F_{1\to r} = 1 - F_{1\to 1} - F_{1\to 2} \qquad \text{and} \qquad F_{2\to r} = 1 - F_{2\to 1} - F_{2\to 2}$$

and substitution into (*2*) of Problem 10.75 shows $\bar{F}_{1\to 2}$ to be independent of the geometry of r.

10.77 With reference to Problems 10.75 and 10.76, express $\bar{F}_{1\to 2}$ in terms of $F_{1\to 2}$, assuming planar or convex black surfaces $(F_{1\to 1} = F_{2\to 2} = 0)$.

▮ Substitution of $F_{1\to r} = 1 - F_{1\to 2}$ and $F_{2\to r} = 1 - F_{2\to 1} = 1 - (A_1/A_2)F_{1\to 2}$ in (*2*) of Problem 10.75 yields, after some algebra,

$$\bar{F}_{1\to 2} = \frac{(A_2/A_1) - (F_{1\to 2})^2}{1 - 2F_{1\to 2} + (A_2/A_1)} \tag{1}$$

In the special case $A_1 = A_2$, this becomes

$$\bar{F}_{1\to 2} = \frac{1 + F_{1\to 2}}{2} \tag{2}$$

10.78 Draw a diagram that makes result (*2*) of Problem 10.77 intuitive.

▮ See Fig. 10-36.

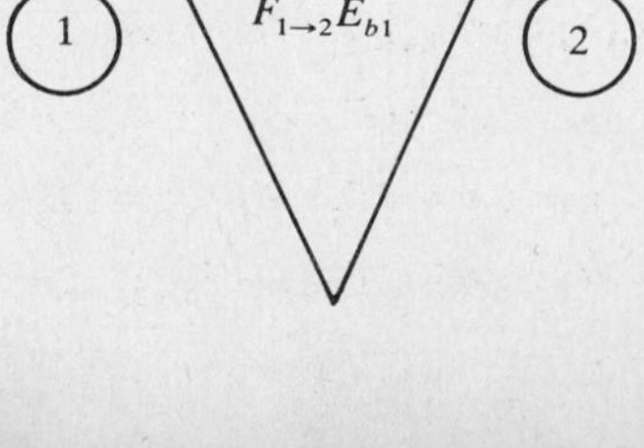

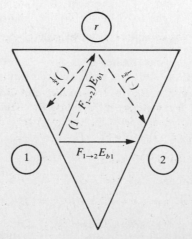

Fig. 10-36

10.79 Two concentric parallel disks are 3 ft apart, with disk 1 (1-ft radius) being at 200 °F and disk 2 (1.5-ft radius) being at 400 °F. Both disks are blackbodies, with a single reradiating surface in the form of a right frustum of a cone enclosing them. Calculate the heat transfer q_{2-1}.

▮ (*1*) of Problem 10.75 and (*1*) of Problem 10.77 apply here. Instead of Figure E-4 we use the exact formula

$$F_{1\to 2} = \tfrac{1}{2}\{x - [x^2 - 4(\beta\gamma)^2]^{1/2}\}$$

where $x \equiv 1 + (1+\beta^2)\gamma^2$, and β and γ are as defined in Figure E-4. Here, $\beta = 1.5/3 = 0.5$ and $\gamma = \frac{3}{1} = 3$; hence,

$$x = 1 + (1 + 0.25)9 = 12.25$$

and

$$F_{1\to 2} = \tfrac{1}{2}\{12.25 - [(12.25)^2 - 4(1.5)^2]^{1/2}\} = 0.1865$$

From this, and the values $A_1 = \pi$, $A_2 = 2.25\pi$, we have

$$\bar{F}_{1\to 2} = \frac{(A_2/A_1) - (F_{1\to 2})^2}{1 - 2F_{1\to 2} + (A_2/A_1)} = \frac{2.25 - (0.1865)^2}{1 - 2(0.1865) + 2.25} = 0.770$$

$$q_{1-2} = A_1\bar{F}_{1\to 2}\sigma(T_1^4 - T_2^4) = \pi(0.770)(0.1714\times 10^{-8})[(660)^4 - (860)^4] = -1481.2 \text{ Btu/hr} = -q_{2-1}$$

10.80 Two parallel infinite black planes are maintained at 200 °C and 300 °C. (***a***) Determine the net rate of heat transfer per unit area. (***b***) Repeat for the case where both temperatures are lowered by 100 °C and determine the ratio of the reduced heat transfer to the original value.

▮ (*a*) Denoting as plane 1 the hotter plane,

$$\frac{q_{1-2}}{A} = \sigma(T_1^4 - T_2^4) = (5.6697\times 10^{-8})[(573.15)^4 - (473.15)^4] = 3276.78 \text{ W/m}^2$$

(*b*)

$$\frac{q_{1-2}}{A} = (5.6697\times 10^{-8})[(473.15)^4 - (373.15)^4] = 1742.31 \text{ W/m}^2$$

$$\frac{(q_{1-2}/A)_{200-100}}{(q_{1-2}/A)_{300-200}} = \frac{1742.31}{3276.78} = 0.5317$$

A reduction in temperature of 100 °C reduces the net heat transfer rate approximately 47 percent!

10.81 For a closed system of n black surfaces, write a general expression for q_i, the net power from surface i.

▮

$$\begin{aligned} q_i &\equiv (\text{power out}) - (\text{power in}) \\ &= A_i\sigma T_i^4 - \sum_{j=1}^{n} F_{j\to i}A_j\sigma T_j^4 \\ &= A_i\sigma\left(T_i^4 - \sum_{j=1}^{n} F_{i\to j}T_j^4\right) \\ &= A_i\sigma\sum_{j=1}^{n} F_{i\to j}(T_i^4 - T_j^4) \qquad (1) \end{aligned}$$

where the reciprocity and enclosure theorems were successively applied. Note that the coefficient of $F_{i\to i}$ in (*1*) is zero.

10.82 Two blackbody rectangles, 0.6 m by 1.2 m, are parallel and directly opposed. The bottom rectangle is at $T_1 = 500$ K and the top rectangle is at $T_2 = 900$ K. The two rectangles are 1.2 m apart. Determine the rate at which the bottom rectangle is losing energy if the surroundings (other than the top rectangle) are considered to be a blackbody at $T_3 = 300$ K.

▮ For this case, (*1*) of Problem 10.81 yields

$$q_1 = A_1\sigma[F_{1\to 2}(T_1^4 - T_2^4) + F_{1\to 3}(T_1^4 - T_3^4)]$$

From Figure E-2, with $\beta = 1$ and $\gamma = 0.5$, $F_{1\to 2} = 0.12$; and

$$F_{1\to 3} = 1 - F_{1\to 1} - F_{1\to 2} = 1 - 0 - 0.12 = 0.88$$

Thus

$$q_1 = (0.72)(5.67\times10^{-8})[(0.12)(500^4-900^4)+(0.88)(500^4-300^4)] = -954\ \text{W}$$

i.e., a gain of 954 W.

10.83 *Gray body configuration factors* $\mathscr{F}_{i\to j}$ are defined by analogy to blackbody configuration factors $F_{i\to j}$. They too obey reciprocity and give heat transfer via the usual formula

$$q_{i-j} = A_i\mathscr{F}_{i\to j}(E_{bi}-E_{bj}) \tag{1}$$

Calculate $\mathscr{F}_{1\to2}$ for a closed system of two gray surfaces.

Comparing (*1*) with (*1*) of Problem 10.69 yields at once

two-gray-surface enclosure

$$\frac{1}{\mathscr{F}_{1\to2}} = \frac{1}{F_{1\to2}} + \frac{A_1}{A_2}\left(\frac{1}{\epsilon_2}-1\right)+\left(\frac{1}{\epsilon_1}-1\right) \tag{2}$$

For reference, some allied results are given below.

one gray surface (2) enclosing a convex or plane gray body (1)

$$\frac{1}{\mathscr{F}_{1\to2}} = \frac{1}{\epsilon_1} + \frac{A_1}{A_2}\left(\frac{1}{\epsilon_2}-1\right) \tag{3}$$

two convex gray surfaces with no other radiation present

$$\mathscr{F}_{1\to2} = \frac{F_{1\to2}}{\dfrac{1}{\epsilon_1\epsilon_2}-\left(\dfrac{1}{\epsilon_1}-1\right)\left(\dfrac{1}{\epsilon_2}-1\right)\dfrac{A_1}{A_2}(F_{1\to2})^2} \tag{4}$$

one reradiating zone enclosing two active gray surfaces

$$\frac{1}{\mathscr{F}_{1\to2}} = \frac{1}{\bar{F}_{1\to2}} + \frac{A_1}{A_2}\left(\frac{1}{\epsilon_2}-1\right)+\left(\frac{1}{\epsilon_1}-1\right) \tag{5}$$

In (*5*), $\bar{F}_{1\to2}$ is given by (*1*) of Problem 10.77 (assuming plane or convex surfaces).

10.84 Repeat Problem 10.79 for both disks gray with $\epsilon_1=\epsilon_2=0.7$.

By (*5*) of Problem 10.83, using areas and $\bar{F}_{1\to2}$ from Problem 10.79,

$$\frac{1}{\mathscr{F}_{1\to2}} = \frac{1}{0.770} + \left(\frac{\pi}{2.25\pi}\right)\left(\frac{1}{0.7}-1\right)+\left(\frac{1}{0.7}-1\right) = 1.9177 \qquad \text{or} \qquad \mathscr{F}_{1\to2} = 0.521\,43$$

Then, by (*1*) of Problem 10.83,

$$q_{1-2} = \pi(0.521\,43)(0.1714\times10^{-8})[(660)^4-(860)^4] = -1003.10\ \text{Btu/hr} = -q_{2-1}$$

10.85 Show how the three-gray-surface enclosure can be "solved" by use of the electrical analogy (Problem 10.72).

At node 1 of Fig. 10-33 Kirchhoff's current law gives $q_1 - q_{1-2} + q_{3-1} = 0$. But, by Ohm's law,

$$q_1 = \frac{E_{b1}-J_1}{(1-\epsilon_1)/\epsilon_1A_1} \qquad q_{1-2} = \frac{J_1-J_2}{1/A_1F_{1\to2}} \qquad q_{3-1} = \frac{J_3-J_1}{1/A_1F_{1\to3}}$$

so that

$$\left(\frac{\epsilon_1}{1-\epsilon_1}+F_{1\to2}+F_{1\to3}\right)J_1 - F_{1\to2}J_2 - F_{1\to3}J_3 = \frac{\epsilon_1E_{b1}}{1-\epsilon_1} \tag{1}$$

Nodes 2 and 3 provide two similar equations. The set of three simultaneous linear equations determines the three unknown potentials J_1, J_2, J_3; and from these, all pertinent heat transfer quantities can be computed.

10.86 A convex gray body having a surface area of 4 m^2 has $\epsilon_1=0.35$ and $T_1=680$ K. This is completely enclosed by a gray surface having an area of 36 m^2, $\epsilon_2=0.75$, and $T_2=310$ K. Find the net rate of heat transfer q_{1-2} between the two surfaces.

▌ By (3) of Problem 10.83,

$$\frac{1}{\mathscr{F}_{1\to2}} = \frac{1}{\epsilon_1} + \frac{A_1}{A_2}\left(\frac{1}{\epsilon_2} - 1\right) = \frac{1}{0.35} + \frac{4}{36}\left(\frac{1}{0.75} - 1\right) = 2.89$$

$$\mathscr{F}_{1\to2} = 0.3455$$

Thus, by (1) of Problem 10.83,

$$q_{1-2} = A_1\mathscr{F}_{1\to2}\sigma(T_1^4 - T_2^4) = 4(0.3455)(5.6697)(6.8^4 - 3.1^4) = 16 \text{ kW}$$

10.87 *Radiation shielding* is often used to reduce the radiative heat transfer. Determine an expression for the heat flux with shielding, using the nomenclature of Fig. 10-37. Treat the three elements as infinite planes.

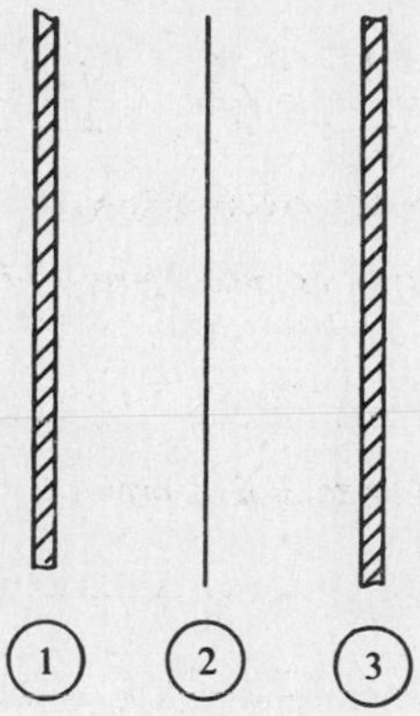

Fig. 10-37

▌ The flux between 1 and 3, without the shields, is given by (1) of Problem 10.70:

$$\left(\frac{q_{1-3}}{A}\right)_0 = \frac{\sigma(T_1^4 - T_3^4)}{1/\epsilon_1 + 1/\epsilon_3 - 1} \tag{1}$$

With the shield, in the steady state,

$$\frac{q_{1-2}}{A} = \frac{q_{2-3}}{A} \equiv \frac{q}{A}$$

or

$$\frac{q}{A} = \frac{\sigma(T_1^4 - T_2^4)}{1/\epsilon_1 + 1/\epsilon_2 - 1} = \frac{\sigma(T_2^4 - T_3^4)}{1/\epsilon_2 + 1/\epsilon_3 - 1} \tag{2}$$

For the case where $\epsilon_1 = \epsilon_3$, (1) gives

$$T_2^4 = \tfrac{1}{2}(T_1^4 + T_3^4) \tag{3}$$

If also $\epsilon_1 = \epsilon_2 = \epsilon_3 \equiv \epsilon$, then

$$\frac{q}{A} = \frac{1}{2}\left[\frac{\sigma(T_1^4 - T_3^4)}{2/\epsilon - 1}\right] = \frac{1}{2}\left(\frac{q_{1-3}}{A}\right)_0 \tag{4}$$

In general, for n radiation shields each having the same emissivity as the two active walls,

$$\frac{q}{A} = \frac{1}{n+1}\left(\frac{q_{1-3}}{A}\right)_0 \tag{5}$$

10.88 Two parallel metal walls of a kitchen oven have temperatures $T_1 = 506$ K and $T_3 = 300$ K, and emissivity $\epsilon_1 = \epsilon_3 = 0.30$, where subscripts 1 and 3 denote the inner and outer walls, respectively. The space between the walls is filled with a rock-wool-type insulation. Assuming this insulation material to be transparent to thermal radiation, calculate the radiant heat transfer rate per unit area between the two walls (*a*) with no radiation shield and (*b*) for one radiation shield of aluminum foil having $\epsilon_2 = 0.09$.

▌ (*a*) By (1) of Problem 10.87,

$$\frac{q_{1-3}}{A} = \frac{\sigma(T_1^4 - T_3^4)}{1/\epsilon_1 + 1/\epsilon_3 - 1} = \frac{5.67(5.06^4 - 3^4)}{1/0.3 + 1/0.3 - 1} = 574.9 \text{ W/m}^2$$

(*b*) By (*2*) and (*3*) of Problem 10.87,

$$\frac{q}{A} = \frac{1}{2}\,\frac{\sigma(T_1^4 - T_3^4)}{1/\epsilon_1 + 1/\epsilon_2 - 1} = \frac{1}{2}\,\frac{(5.67)(5.06^4 - 3.00^4)}{1/0.30 + 1/0.09 - 1} = 121\ \text{W/m}^2$$

10.89 In a manufacturing operation, a 3-m by 5-m plate of rough carbon at 500 K is placed in a large room where the average temperature is 320 K. Estimate the heat loss from the plate due to radiation.

▮ For a plate area A_1 very small compared to the area of the enclosing room A_2, (*3*) of Problem 10.83 gives

$$\frac{\epsilon_1}{\mathscr{F}_{1\to 2}} = 1 + \frac{A_1}{A_2}\left(\frac{\epsilon_1}{\epsilon_2} - \epsilon_1\right) \approx 1 \qquad \text{or} \qquad \mathscr{F}_{1\to 2} \approx \epsilon_1$$

From Table B-6, $\epsilon_1 = 0.77$; therefore,

$$q_{1-2} = A_1\mathscr{F}_{1\to 2}\sigma(T_1^4 - T_2^4) \approx (15\ \text{m}^2)(0.77)(5.67\ \text{W/m}^2\cdot\text{K}^4)[(5.00\ \text{K})^4 - (3.20\ \text{K})^4] = 34.0\ \text{kW}$$

10.90 A 2-m by 3-m shallow pan is filled with pure water. It is placed on a well-insulated stand on top of a high building. It is known that the heat loss by radiation to the sky (assumed to be a blackbody at absolute zero) can cause the water to freeze on a calm night. Estimate the heat flux to the sky assuming that the water is at 274 K and that it is a gray body with $\epsilon = 0.6$. Assume further that the water "sees" only the sky.

▮ Since the water surface (1) is "enclosed" by the sky (2), we have from Problem 10.89, $\mathscr{F}_{1\to 2} \approx \epsilon_1 = 0.6$, and

$$q_{1-2} \approx A_1\epsilon_1\sigma(T_1^4 - T_2^4) = (6\ \text{m}^2)(0.6)(5.6697\ \text{W/m}^2\cdot\text{K}^4)(2.74^4 - 0) = 1150\ \text{W}$$

10.91 Two perpendicular walls have a common edge. Each wall is 4 by 8 ft, the 4-ft edge being common (Fig. 10-38). Wall 1 is vertical and has an emissivity of 0.7 and a temperature of 1200 °F. Wall 2 is horizontal and has an emissivity of 0.4 and a temperature of 800 °F. Calculate the net heat transfer from wall 1 to wall 2. Assume no radiation or reflection from other surfaces or surroundings.

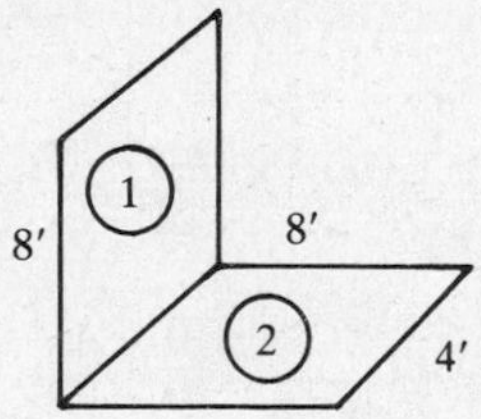

Fig. 10-38

▮ From Figure E-3, with $\beta = \frac{8}{4} = 2$ and $\gamma = \frac{8}{4} = 2$, $F_{1\to 2} = 0.14$. Then, by (*1*) of Problem 10.69, with

$$E_{b1} - E_{b2} = (0.1714)(16.6^4 - 12.6^4) = 8.7 \times 10^3\ \text{Btu/hr}\cdot\text{ft}^2$$

$$\frac{1-\epsilon_1}{\epsilon_1 A_1} = \frac{1-0.7}{(0.7)(32)} = 0.0134\ \text{ft}^{-2} \qquad \frac{1}{A_1F_{1\to 2}} = \frac{1}{(0.14)(32)} = 0.223\ \text{ft}^{-2} \qquad \frac{1-\epsilon_2}{\epsilon_2 A_2} = \frac{1-0.4}{(0.4)(32)} = 0.0469\ \text{ft}^{-2}$$

$$q_{1-2} = \frac{8.7\times 10^3}{1.34\times 10^{-2} + 22.3\times 10^{-2} + 4.69\times 10^{-2}} = 3.08\times 10^4\ \text{Btu/hr}$$

10.92[D] Parallel and opposite rectangles, each 8 by 16 m, are 8 m apart and are connected by reradiating refractory walls. The emissivity and temperature of one are 0.90 and 1200 °C; the corresponding values for the other are 0.80 and 800 °C. Determine the net radiant heat transfer between the plates.

▮ Applying (*1*) of Problem 10.73, with $A_1 = A_2 = 128\ \text{m}^2$, $F_{1\to 2} \approx 0.28$ (from Figure E-2),

$$F_{1\to r} = 1 - F_{1\to 2} \approx 0.72 = F_{2\to r}$$

we have

$$q_{1-2} = \frac{(5.67)(14.73^4 - 10.73^4)}{\dfrac{1-0.90}{(128)(0.90)} + \dfrac{1}{(128)(0.28) + 1/[1/(128)(0.72) + 1/(128)(0.72)]} + \dfrac{1-0.80}{(128)(0.80)}} = 12.76\ \text{MW}$$

10.93[D] Parallel and opposite 3-ft by 6-ft plates, having temperatures of 1000 and 300 °F, are located 10 ft apart. The plates have emissivities of 0.8 and 0.3, respectively. What is the net radiant-energy interchange if the plates are enclosed by reradiating walls?

▌ Applying (*1*) of Problem 10.73, with $A_1 = A_2 = 18\ \text{ft}^2$, $F_{1\to 2} \approx 0.052$ (from Figure E-2),

$$F_{1\to r} = 1 - F_{1\to 2} \approx 0.948 = F_{2\to r}$$

we have

$$q_{1-2} = \frac{(0.1714)(14.60^4 - 7.60^4)}{\dfrac{1-0.8}{(18)(0.8)} + \dfrac{1}{(18)(0.052) + 1/[1/(18)(0.948) + 1/(18)(0.948)]} + \dfrac{1-0.3}{(18)(0.3)}} = 29\,000\ \text{Btu/hr}$$

10.94 In an industrial operation, a wall (1) of fireclay brick is maintained at 1940 °F. The wall is 10 ft high and 15 ft wide. The floor (2) of concrete tiles is 15 by 20 ft and is at a temperature of 100 °F. Calculate the radiant heat flux from the wall to the floor (*a*) assuming no radiation or reflection from other surfaces or surroundings; (*b*) if the surroundings are reradiating surfaces.

▌ From Table B-6, $\epsilon_1 = 0.75$, $\epsilon_2 = 0.63$. From Figure E-3,

$$\beta = \tfrac{10}{15} = \tfrac{2}{3} \qquad \gamma = \tfrac{20}{15} = \tfrac{4}{3} \qquad F_{1\to 2} = 0.26$$

(*a*) (*1*) of Problem 10.69 gives

$$q_{1-2} = \frac{(0.1714)(24.00^4 - 5.60^4)}{\dfrac{1-0.75}{(0.75)(150)} + \dfrac{1}{(150)(0.26)} + \dfrac{1-0.63}{(0.63)(300)}} = 1.9 \times 10^6\ \text{Btu/hr}$$

(*b*) Applying (*1*) of Problem 10.73, with

$$F_{1\to r} = 1 - F_{1\to 2} = 0.74$$

$$F_{2\to r} = 1 - F_{2-1} = 1 - (A_1/A_2)F_{1\to 2} = 1 - \left(\frac{1}{2}\right)(0.26) = 0.87$$

we have

$$q_{1-2} = \frac{(0.1714)(24.00^4 - 5.60^4)}{\dfrac{1-0.75}{(150)(0.75)} + \dfrac{1}{(150)(0.26) + 1/[1/(150)(0.74) + 1/(300)(0.87)]} + \dfrac{1-0.63}{(300)(0.63)}} = 4.45 \times 10^6\ \text{Btu/hr}$$

10.95 The first and third layers of a three-layer foil insulation are at 80 and 0 °F, respectively. The emissivity of the foil is 0.05. Neglecting conduction and convection, determine the heat flux.

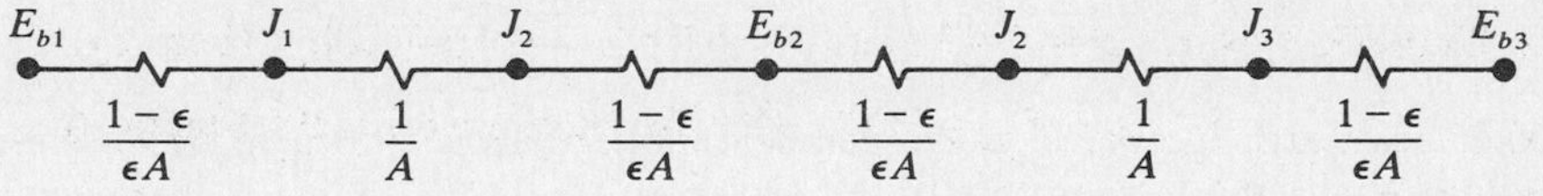

Fig. 10-39

▌ Since $F_{1\to 2} = F_{2\to 3} = 1.0$, and assuming that surface areas are the same and surfaces "see" only each other, the electrical analogy is given by Fig. 10-39. Thus, with $T_1 = 80\ °\text{F} = 540\ °\text{R}$, $T_3 = 0\ °\text{F} = 460\ °\text{R}$,

$$E_{b1} - E_{b3} = \sigma(T_1^4 - T_3^4) = (0.1714)(5.4^4 - 4.6^4) = 69.0\ \text{Btu/hr}\cdot\text{ft}^2$$

$$\Sigma R = \frac{1-\epsilon}{\epsilon A} + \frac{1}{A} + \frac{1-\epsilon}{\epsilon A} + \frac{1-\epsilon}{\epsilon A} + \frac{1}{A} + \frac{1-\epsilon}{\epsilon A} = \frac{1}{A}\left[4\left(\frac{1-\epsilon}{\epsilon}\right) + 2\right] = \frac{1}{A}\left[4\left(\frac{0.95}{0.05}\right) + 2\right] = \frac{78}{A}$$

$$q_{1-3} = \frac{69.0\ \text{Btu/hr}\cdot\text{ft}^2}{78/A} \qquad \text{or} \qquad \frac{q_{1-3}}{A} = 0.885\ \text{Btu/hr}\cdot\text{ft}^2$$

10.96 Two flat parallel square plates have equal areas of 25 ft² and are separated by a distance of 1 ft. The net radiant heat exchange is 3000 Btu/hr, and the temperature and emissivity of the cooler plate are 140 °F and 0.4, respectively. If the emissivity of the other plate is 0.7, determine its temperature.

▌ Assume that the plates exchange energy only with each other. Figure E-2, for $\beta = \gamma = 5$, gives $F_{1\to 2} = 0.7$. We have

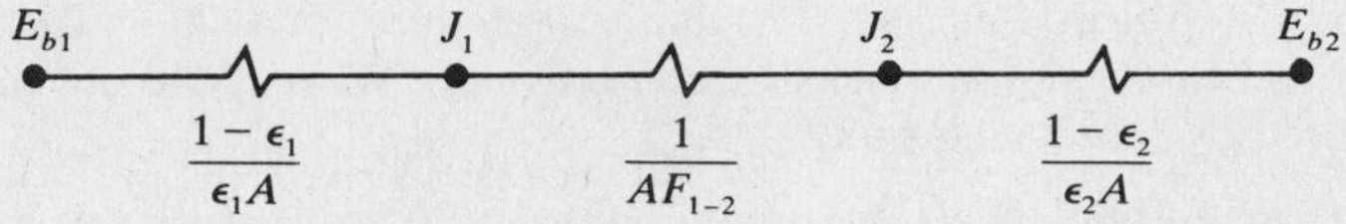

Fig. 10-40

$$q_{1-2} = \frac{E_{b1} - E_{b2}}{\Sigma R} = 3000 \text{ Btu/hr}$$

where

$$E_{b1} - E_{b2} = (0.1714)\left[\left(\frac{T_1}{100}\right)^4 - (6.00)^4\right] = (0.1714 \times 10^{-8} T_1^4 \ \text{Btu/hr} \cdot \text{ft}^2 \cdot {}^\circ\text{R}^4) - (222 \text{ Btu/hr} \cdot \text{ft}^2)$$

and, from Fig. 10-40,

$$\Sigma R = \frac{1-\epsilon_1}{\epsilon_1 A} + \frac{1}{AF_{1\to 2}} + \frac{1-\epsilon_2}{\epsilon_2 A} = \frac{1}{25 \text{ ft}^2}\left(\frac{0.3}{0.7} + \frac{1}{0.7} + \frac{0.6}{0.4}\right) = 0.134 \text{ ft}^{-2}$$

Therefore

$$(0.1714 \times 10^{-8} T_1^4 \ \text{Btu/hr} \cdot \text{ft}^2 \cdot {}^\circ\text{R}^4) - (222 \text{ Btu/hr} \cdot \text{ft}^2) = (3000 \text{ Btu/hr})(0.134 \text{ ft}^{-2})$$

Solving, $T_1 = 777\ {}^\circ\text{R} = 317\ {}^\circ\text{F}$.

10.97 The wall of a building consists of two large parallel planes. The emissivity and temperature of the inner wall are 0.5 and 20 °C, respectively. A radiation shield with an emissivity of 0.09 is placed between the walls. If the radiant heat loss is 2 W/m², determine the equilibrium temperature of the shield.

I Apply (*2*) of Problem 10.87:

$$\frac{q}{A} = \frac{\sigma(T_1^4 - T_2^4)}{1/\epsilon_1 + 1/\epsilon_2 - 1}$$

$$2 \text{ W/m}^2 = \frac{(5.67 \times 10^{-8} \text{ W/m}^2 \cdot \text{K}^4)[(293 \text{ K})^4 - T_2^4]}{1/0.5 + 1/0.09 - 1}$$

$$T_2 = 289 \text{ K} = 16\ {}^\circ\text{C}$$

10.98 A long, rusty 2-in.-OD pipe is in a concrete tunnel 2 ft square. The tunnel wall is at 80 °F and has an emissivity of 0.9; the corresponding values for the pipe surface are 300 °F and 0.73. Determine the radiant heat transfer per linear foot of pipe.

I By (*1*) and (*3*) of Problem 10.83,

$$\frac{q_{1-2}}{L} = \frac{\pi D \sigma (T_1^4 - T_2^4)}{\dfrac{1}{\epsilon_1} + \dfrac{\pi D}{4w}\left(\dfrac{1}{\epsilon_2} - 1\right)} = \frac{\pi(\frac{2}{12})(0.1714)(7.60^4 - 5.40^4)}{\dfrac{1}{0.73} + \dfrac{\pi(\frac{2}{12})}{4(2)}\left(\dfrac{1}{0.9} - 1\right)} = 162 \text{ Btu/hr} \cdot \text{ft}$$

10.99 The radiant heat exchange between two parallel infinite walls is 25 Btu/hr · ft². The emissivities of the walls are 0.3 and 0.2, and the temperature of the cooler wall is 70 °F. What is the temperature of the warmer wall?

I Solve (*1*) of Problem 10.70 for the unknown temperature:

$$25 \text{ Btu/hr} \cdot \text{ft}^2 = \frac{(0.1714 \text{ Btu/hr} \cdot \text{ft}^2 \cdot {}^\circ\text{R}^4)[(T_1/100)^4 - (530\ {}^\circ\text{R}/100)^4]}{(1/0.3) + (1/0.2) - 1}$$

$$25(3.33 + 5 - 1) = (0.1714)\left(\frac{T_1}{100}\right)^4 {}^\circ\text{R}^{-4} - 0.1714(5.3)^4$$

$$T_1^4 = \left(\frac{318.6}{0.1714}\right) \times 10^8\ {}^\circ\text{R}^4$$

$$T_1 = 657\ {}^\circ\text{R} = 197\ {}^\circ\text{F}$$

10.100 A cryogen at −221 °F is to be stored in the inner of two eccentric polished brass spheres with diameters of 9 and 12 in., respectively. If the emissivity of the polished brass is 0.03, calculate the total radiant heat exchange between the spheres. Neglect the temperature drop through the metal and assume that the outer-sphere temperature is 70 °F.

▌ By (*1*) and (*3*) of Problem 10.83,

$$q_{1-2}=\frac{A_1\sigma(T_1^4-T_2^4)}{\dfrac{1}{\epsilon_1}+\dfrac{A_1}{A_2}\left(\dfrac{1}{\epsilon_2}-1\right)}=\frac{\pi(\frac{9}{12})^2(0.1714)(2.39^4-5.30^4)}{(1/0.03)+(\frac{9}{12})^2[(1/0.03)-1]}=-4.45\ \text{Btu/hr}=-q_{2-1}$$

10.101 Three thin sheets of polished copper are placed parallel and very close to each other. If the temperatures of the outside sheets are 720 °F and 100 °F, calculate the temperature of the middle sheet.

▌ Treating the middle sheet as a radiation shield between the outer sheets, with $\epsilon_1=\epsilon_3$, we have from (*3*) of Problem 10.87

$$T_2=\left(\frac{T_1^4+T_3^4}{2}\right)^{1/4}=\left(\frac{1180^4+560^4}{2}\right)^{1/4}\ \text{°R}=1005\ \text{°R}=545\ \text{°F}$$

10.102 What is the net heat flux in Problem 10.101?

▌ From Table B-5, $\epsilon_1=\epsilon_2=\epsilon=0.023$. Then by (*1*) of Problem 10.70

$$\frac{q_{1-2}}{A}=\frac{\sigma(T_1^4-T_2^4)}{1/\epsilon_1+1/\epsilon_2-1}=\frac{(0.1714)(11.80^4-10.05^4)}{2/0.023-1}=18.3\ \text{Btu/hr}\cdot\text{ft}^2=q_{2-3}=q$$

10.103 Two 4-ft-square parallel flat plates are 2 ft apart. Plate 1 is maintained at a temperature of 1740 °F and plate 2 at 720 °F. The emissivities are 0.4 and 0.8, respectively. Considering the surroundings black at 0 °R and including multiple interreflections, determine **(a)** the net radiant exchange and **(b)** the heat input required to maintain the temperature of plate 1.

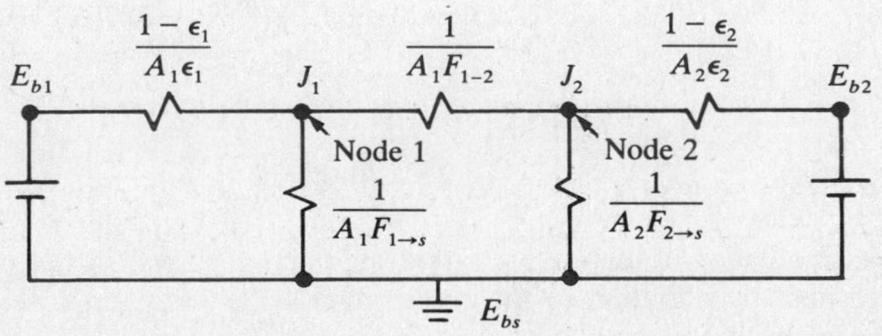

Fig. 10-41

▌ The equivalent network is shown in Fig. 10-41.

$$A_1=A_2=16\ \text{ft}^2\qquad T_1=1740\ \text{°F}=2200\ \text{°R}$$
$$\epsilon_1=0.4\qquad T_2=720\ \text{°F}=1180\ \text{°R}\qquad \epsilon_2=0.8$$

From Figure E-2, with $\beta=2$ and $\gamma=2$, $F_{1\to2}=0.41=F_{2\to1}$. Then

$$F_{1\to s}=1-F_{1\to2}=0.59=1-F_{2\to1}=F_{2\to s}$$

Kirchhoff's current law at node 1 yields

$$\frac{E_{b1}-J_1}{(1-\epsilon_1)/\epsilon_1A_1}=\frac{J_1-J_2}{1/A_1F_{1-2}}+\frac{J_1-E_{bs}}{1/A_1F_{1-s}}$$

$$\left[\frac{(0.4)(16)}{0.6}\right](E_{b1}-J_1)=(0.41)(16)(J_1-J_2)+(0.59)(16)(J_1-E_{bs})$$

$$1.67J_1-0.41J_2=0.67E_{b1}\tag{1}$$

Similarly, at node 2,

$$0.082J_1-1.00J_2=-0.8E_{b2}\tag{2}$$

Solving simultaneously,

$$J_1=0.406E_{b1}+0.199E_{b2}=0.1714[(0.406)(22.00)^4+(0.199)(11.80)^4]=1.7\times10^4\ \text{Btu/hr}\cdot\text{ft}^2$$

$$J_2=0.082J_1+0.8E_{b2}=0.082(1.7\times10^4)+0.8(0.1714)(11.80)^4=4.052\times10^3\ \text{Btu/hr}\cdot\text{ft}^2$$

(*a*) The net radiant exchange from plate 1 to plate 2 is

$$q_{1-2} = \frac{J_1 - J_2}{1/A_1 F_{1\to 2}} = (16)(0.41)[(17 - 4.055) \times 10^3 \text{ Btu/hr}] = 8.49 \times 10^4 \text{ Btu/hr}$$

(*b*)

$$\text{Heat required} = q_1 = \frac{E_{b1} - J_1}{(1-\epsilon_1)/A_1\epsilon_1} = (0.4/0.6)(16)[(0.1714)(22.00^4) - 1.7 \times 10^4]$$

$$= 2.47 \times 10^5 \text{ Btu/hr}$$

10.104 The wall of a house consists of two large parallel planes. The outer and inner planes have emissivities of 0.9 and 0.4, respectively. If the respective outer and inner temperatures on a summer day are 120 °F and 70 °F, what will be the radiant heat flux per unit area (*a*) without and (*b*) with an aluminum radiation shield having $\epsilon = 0.06$?

I Let subscript 1 denote the outer plane, 2 denote the shield, and 3 denote the inner plane.

(*a*) By (*1*) of Problem 10.87,

$$\frac{q}{A} = \frac{\sigma(T_1^4 - T_3^4)}{1/\epsilon_1 + 1/\epsilon_3 - 1} = \frac{0.1714[(\frac{580}{100})^4 - (\frac{530}{100})^4]}{1/0.9 + 1/0.4 - 1} = 22.5 \text{ Btu/hr} \cdot \text{ft}^2$$

(*b*) By (*2*) of Problem 10.87,

$$\frac{q}{A} = \frac{\sigma(T_1^4 - T_2^4)}{1/\epsilon_1 + 1/\epsilon_2 - 1} = \frac{\sigma(T_2^4 - T_3^4)}{1/\epsilon_2 + 1/\epsilon_3 - 1}$$

Solving for T_2 gives

$$\frac{(\frac{580}{100})^4 - (T_2/100)^4}{1/0.9 + 1/0.06 - 1} = \frac{(T_2/100)^4 - (\frac{530}{100})^4}{1/0.06 + 1/0.4 - 1}$$

$$\left(\frac{T_2}{100}\right)^4 = 967$$

and thus

$$\frac{q}{A} = \frac{0.1714[(\frac{580}{100})^4 - 967]}{1/0.9 + 1/0.06 - 1} = 1.68 \text{ Btu/hr} \cdot \text{ft}^2$$

10.105 A thermocouple is shielded in a high-temperature air duct as shown in the cross section of Fig. 10-42. The thermocouple loses heat by radiation to the cooler duct wall, this energy being obtained by convection from the hot air. Determine an approximate expression for the heat transfer without the shield.

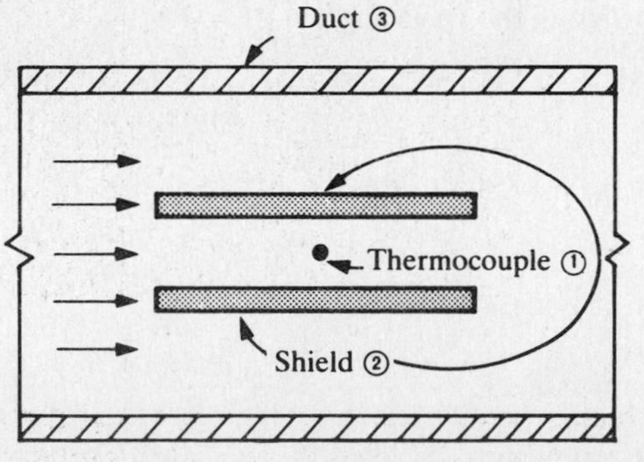

Fig. 10-42

I

$$q_{1-3} \approx \epsilon_1 A_1 \sigma(T_1^4 - T_3^4) \quad [\textit{see Problem 10.89}]$$

10.106 Refer to Problem 10.105. Give an expression for the heat transfer for the shielded thermocouple.

I Although the shield "sees" itself, it exchanges no net flux with itself; thus it may be treated as a single convex body, and (*2*) of Problem 10.83 applies. Since $F_{1\to 2} \approx 1$ (to the extent that the thermocouple "sees" only the shield),

$$\frac{1}{\mathscr{F}_{1\to 2}} \approx 1 + \frac{A_1}{A_2}\left(\frac{1}{\epsilon_2} - 1\right) + \left(\frac{1}{\epsilon_1} - 1\right) = \frac{1}{\epsilon_1} + \frac{A_1}{A_2}\left(\frac{1}{\epsilon_2} - 1\right)$$

and so

$$q_{1-2} = A_1\mathcal{F}_{1\to2}(E_{b1} - E_{b2}) \approx \frac{A_1\sigma(T_1^4 - T_2^4)}{1/\epsilon_1 + (A_1/A_2)[(1/\epsilon_2) - 1]} \approx \epsilon_1 A_1 \sigma(T_1^4 - T_2^4)$$

where the last step presumes $A_1/A_2 \ll 1$.

10.107 An unshielded thermocouple $(\epsilon_1 = 0.06)$ reaches 2500 °F in a combustion duct. The oxidized duct $(\epsilon_3 = 0.3)$ reaches 1000 °F. How much heat is transferred by radiation if the duct is 4 in. in diameter by 3 ft long and the diameter of the thermocouple is 0.06 in.?

▌ Since the thermocouple is small compared with the duct, Problem 10.105 gives

$$q_{1-3} \approx (0.06)\left[\pi\left(\frac{0.06}{12}\right)^2\right](0.1714 \times 10^{-8})(2960^4 - 1460^4) = 0.58 \text{ Btu/hr}$$

10.108 If a 1-ft-long, 2-in.-diam. aluminum shield $(\epsilon_2 = 0.09)$ is placed around the thermocouple of Problem 10.107, what is the heat transfer from the thermocouple to the duct, and what is the temperature of the shield?

▌ Because all heat impinging on the shield (2) is given up to the duct (3), the heat transfer is as computed in Problem 10.107: $q_{1-3} = 0.58$ Btu/hr. Now, since

$$\frac{A_1}{A_2} = \frac{\pi(0.06)^2}{\pi(2)(12)} = 1.5 \times 10^{-4} \ll 1$$

Problem 10.106 gives $q_{1-2} \approx \epsilon_1 A_1 \sigma(T_1^4 - T_2^4)$. But

$$q_{1-2} = q_{1-3} \qquad \text{or} \qquad \epsilon_1 A_1 \sigma(T_1^4 - T_2^4) \approx \epsilon_1 A_1 \sigma(T_1^4 - T_3^4)$$

Thus, to our order of approximation, $T_2 = T_3 = 1000$ °F; i.e., the shield becomes in effect the duct.

10.5 GASEOUS EMISSIONS

10.109 Discuss the approach to solving radiation problems involving gases and vapors with nonsymmetrical molecules, which emit or absorb radiant energy only within certain wavelength bands (Fig. 10-43).

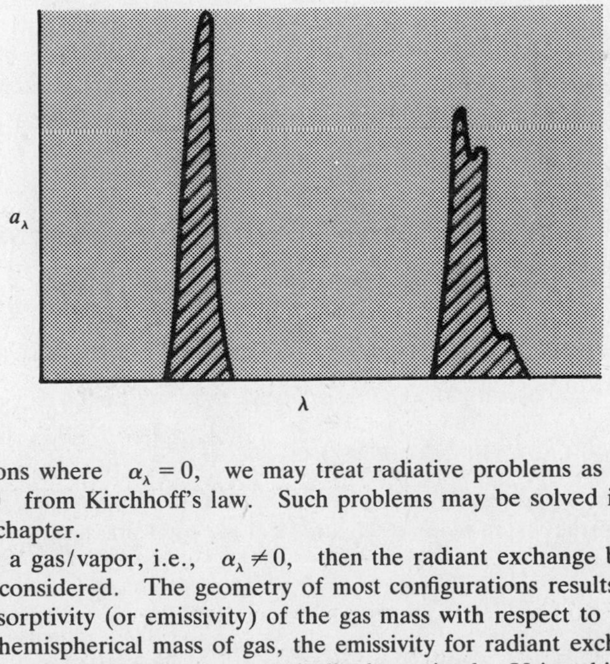

Fig. 10-43

▌ In the wavelength regions where $\alpha_\lambda = 0$, we may treat radiative problems as if the gas/vapor is not present, since $\alpha_\lambda = \epsilon_\lambda = 0$ from Kirchhoff's law. Such problems may be solved in the manner of all previous problems in this chapter.

If there is absorption by a gas/vapor, i.e., $\alpha_\lambda \neq 0$, then the radiant exchange between the gas/vapor and other bodies must be considered. The geometry of most configurations results in a rather complicated integration to yield the absorptivity (or emissivity) of the gas mass with respect to a boundary element.

In the special case of a hemispherical mass of gas, the emissivity for radiant exchange from the gas to the center of the hemispherical base can readily be analytically determined. Using this approach, H. C. Hottel and R. B. Egbert determined the effective emissivity of a hemispherical gas system of radius L at a partial pressure p_i radiating to a black surface element located at the center of the hemispherical base. Their results for carbon dioxide and water vapor are given in Figures E-6 through E-9. These results are also applicable to other shapes of practical interest by use of the equivalent beam lengths given in Table E-2.

For a mixture containing two gases/vapors, carbon dioxide and water vapor, for example, an approximate emissivity can be obtained by adding the individually determined emissivities.

10.110 Determine the emissivity of water vapor at 3000 °F in a spherical container of 1.5-in. diameter. The partial pressure of the vapor is 0.3 atm, and the total pressure is 1.7 atm.

The equivalent hemispherical beam length is

$$L = \tfrac{2}{3}(1.5) = 1.0 \text{ ft} \qquad p_{wL} = 0.3(1.0) = 0.3 \qquad \frac{p_w + p_T}{2} = \frac{0.3 + 1.7}{2} = 1.0$$

From Figure E-9, $C_{pw} = 1.5$ and from Figure E-8 $(\epsilon_w)_1 = 0.055$. Thus,

$$(\epsilon_w)_{\text{actual}} = (1.5)(0.055) = 0.0825$$

10.111 Determine the emissivity of water vapor at 3540 °F between two large parallel walls which are 8 in. apart. The partial pressure of the water vapor is 0.2 atm, and the total pressure is 1.4 atm.

$$T = 3540 \text{ °F} = 4000 \text{ °R}$$

From Table E-2,

$$L = 1.8 \times 8/12 \text{ ft} = 1.2 \text{ ft} \qquad p_w L = 0.2(1.2) = 0.24 \text{ ft} \cdot \text{atm} \qquad \frac{p_w + p_T}{2} = \frac{0.2 + 1.4}{2} = 0.8 \text{ atm}$$

From Figure E-9,

$$C_{pw} = 1.35$$

From Figure E-8, $(\epsilon_w)_1 = 0.035$; thus

$$(\epsilon_w)_{\text{actual}} = (1.35)(0.035) \approx 0.047$$

10.112 Determine the emissivity of carbon dioxide at 2040 °F in a 3-ft cubical container. The partial pressure of the carbon dioxide is 0.06 atm, and the total pressure is 0.2 atm.

From Table E-2,

$$L = \tfrac{2}{3}(3 \text{ ft}) = 2 \text{ ft} \qquad \text{and} \qquad p_{cd}L = (0.06)(2) = 0.12 \text{ ft} \cdot \text{atm}$$

From Figure E-6, at $T = 2040 \text{ °F} = 2500 \text{ °R}$,

$$(\epsilon_{cd})_1 = 0.066 \quad (\textit{at 1 atm})$$

From Figure E-7, at $p_T = 0.2$ atm,

$$C_{cd} = 0.66$$

Thus,

$$(\epsilon_{cd})_{\text{actual}} = (0.63)(0.066) = 0.042$$

10.113 Determine the emissivity of carbon dioxide at 3040 °F in a very long 1.5-ft-diam. cylindrical passage. The partial pressure of the carbon dioxide is 0.04 atm, and the total pressure is 0.3 atm.

From Table E-2, $L = (1)(1.5) = 1.5$ ft.

$$T = 3040 \text{ °F} = 3500 \text{ °R} \qquad p_c = 0.04 \text{ atm} \qquad p_c L = (0.04)(1.5) = 0.06 \text{ ft} \cdot \text{atm}$$

From Figure E-6, $(\epsilon_{cd})_1 = 0.031$ and from Figure E-7, $C_{cd} = 0.69$.

$$(\epsilon_{cd})_{\text{actual}} = (0.69)(0.031) = 0.021$$

10.114 Calculate the emissive power of water vapor at 1040 °F in a spherical container of 3 ft diameter. The partial pressure of the water vapor is 0.8 atm, and the total pressure is 1.0 atm.

$$T = 1040 \text{ °F} = 1500 \text{ °R}$$

From Table E-2, $L = \tfrac{2}{3}(3) = 2.0$ ft.

$$p_w L = 0.8(2) = 1.6 \text{ ft} \cdot \text{atm} \qquad \frac{p_w + p_T}{2} = \frac{0.8 + 1}{2} = 0.9 \text{ atm}$$

From Figure E-9, $C_{pw} = 1.3$ and from Figure E-8, $(\epsilon_w)_1 = 0.3$.

$$\epsilon_{\text{actual}} = (1.3)(0.3) = 0.39$$

$$\text{Emissive power} \equiv E = \epsilon \sigma T^4 = (0.39)(0.1714)(15.00)^4 = 3.38 \times 10^3 \text{ Btu/hr} \cdot \text{ft}^2$$

10.115 Gaseous products of combustion at 2000 °F leave a furnace through a very long cylindrical flue, which is 2.5 ft in diameter. The partial pressure of CO_2 in the mixture is 0.09 atm, and the total pressure is 1.1 atm. Determine the radiative heat transfer per linear foot from the carbon dioxide to the flue wall if the wall emissivity is unity and its temperature is 400 °F.

Since the gas is enclosed by the flue, the use of an effective emissivity for the gas permits it to be treated as a single body enclosed by a larger body. Thus, by (*1*) and (*3*) of Problem 10.83,

$$q_{c-f} = \frac{A_c \sigma (T_c^4 - T_f^4)}{1/\epsilon_c + (A_c/A_f)[(1/\epsilon_f) - 1]}$$

where subscript c denotes CO_2
subscript f denotes flue

Since $\epsilon_f = 1.0$, this reduces to

$$q_{c-f} = A\epsilon_c \sigma (T_c^4 - T_f^4) \tag{1}$$

where A is the effective area of the CO_2 (same as inner flue area).

Find ϵ_c:

$$T_c = 2000\ °\text{F} = 2460\ °\text{R} \qquad p_c = 0.3\ \text{atm} \qquad L = \text{diam.} = 2.5\ \text{ft} \quad (\textit{from Table E-2})$$

Thus

$$p_c L = (0.09)(2.5) = 0.225\ \text{ft} \cdot \text{atm}$$

From Figure E-6, $\epsilon_c \approx 0.085$. From Figure E-7 at $p_T = 1.1$ atm, $C_{pc} \approx 1.0$. Therefore,

$$(\epsilon_c)_{p_T} = 1(\epsilon_c)_1 = 0.085$$

Now

$$A = \pi\, dl$$

$$\frac{q_{c-f}}{l} = \pi\, d\epsilon_c\, \sigma(T_c^4 - T_f^4) = \pi(2.5\ \text{ft})(0.085)(0.1714\ \text{Btu/hr} \cdot \text{ft}^2 \cdot °\text{R}^4)[(24.60^4 - 8.60^4)\ °\text{R}^4]$$

$$= 4.13 \times 10^4\ \text{Btu/hr} \cdot \text{ft}$$

10.116 Suppose that the products of combustion in Problem 10.115 resulted from burning natural gas with an excess of air, and that the partial pressures of the other major constituents are: water vapor = 0.15 atm; oxygen, $O_2 = 0.05$ atm; nitrogen, $N_2 = 0.81$ atm. What is the total rate of heat transfer from the hot gases to the flue per lineal foot?

N_2 and O_2 have nonpolar, symmetrical molecular structures. They do not emit or absorb significant amounts of energy. Thus, the radiant flux from the products to the flue is that due to CO_2 and H_2O. The contribution due to CO_2 was found in Problem 10.115. For H_2O (water vapor), we have, analogous to (*1*) of Problem 10.115,

$$q_{w-f} = A\epsilon_w \sigma (T_w^4 - T_f^4)$$

Find ϵ_w:

$$T_w = 2000\ °\text{F} = 2460\ °\text{R} \qquad p_w = 0.15\ \text{atm} \qquad L = \text{diam.} = 2.5\ \text{ft} \quad (\textit{from Table E-2})$$

Thus,

$$p_w L = (0.15)(2.5) = 0.375\ \text{ft} \cdot \text{atm}$$

From Figure E-8, $(\epsilon_w)_1 \approx 0.101$

$$\frac{p_w + p_T}{2} = \frac{0.15 + 1.1}{2} = 0.625$$

From Figure E-9,

$$C_{pw} \approx 1.15$$

Hence,

$$(\epsilon_w)_{\text{actual}} = (0.101)(1.15) = 0.116$$

Thus,

$$\frac{q_{w-f}}{l} = \pi(2.5 \text{ ft})(0.116)(0.1714 \text{ Btu/hr}\cdot\text{ft}^2\cdot{}^\circ\text{R}^4)[(24.60^4 - 8.60^4)^\circ\text{R}^4] = 5.63\times10^4 \text{ Btu/hr}\cdot\text{ft}$$

The total radiant flux from the products of combustion to the flue is

$$\frac{q}{l} = \frac{q_{c-f}}{l} + \frac{q_{w-f}}{l} = 4.13\times10^4 + 5.63\times10^4 = 9.76\times10^4 \text{ Btu/hr}\cdot\text{ft}$$

10.117 Determine the effective emissivity (for radiation from the gas to the surface) of CO_2 gas at 2500 °R in a very long cylinder which is 2 ft in diameter. The partial pressure of the CO_2 is 0.2 atm and the gas system total pressure is 0.3 atm.

▌ From Table E-2, $L = 1\times D = 2$ ft. From Figure E-6 at $p_{cd}L = 0.2\times2 = 0.4$ atm · ft and $T =$ 2500 °R, $(\epsilon_{cd})_1 \approx 0.103$. From Figure E-7 at $p = 0.3$ atm and $p_{cd}L = 0.4$ atm · ft, $C_{cd} \approx 0.78$. Thus

$$(\epsilon_{cd})_p \approx (0.78)(0.103) = 0.08$$

10.118 A combustion exhaust gas at 2500 °R has a CO_2 partial pressure of 0.08 atm, a water vapor partial pressure of 0.16 atm, and a total gas system pressure of 2.0 atm. Estimate the effective gas mixture emissivity in a long cylindrical flue 3 ft in diameter. The other major gas constituents are O_2 and N_2.

▌ The O_2 and N_2 constituents do not absorb or emit radiant energy in the temperature range of this problem. We may approximate the gas system emissivity by linear addition of the individual emissivities of the CO_2 and the water vapor; thus,

CO_2:

$$L = 1\times D = 3 \text{ ft} \qquad p_{cd}L = 0.08\times3 = 0.24 \text{ atm}\cdot\text{ft}$$

From Figure E-6,

$$(\epsilon_{cd})_1 \approx 0.085$$

From Figure E-7,

$$C_{cd} \approx 1.2 \quad \text{and} \quad (\epsilon_{cd})_p \approx (0.085)(1.2) = 0.102$$

H_2O:

$$L = 1\times D = 3 \text{ ft} \qquad p_wL = 0.16\times3 = 0.48 \text{ atm}\cdot\text{ft}$$

From Figure E-8,

$$(\epsilon_w)_1 \approx 0.115$$

From Figure E-9, at $(p_w + p)/2 = 2.16/2 = 1.08$ atm,

$$C_w \approx 1.5 \quad \text{and} \quad (\epsilon_w)_p \approx (0.115)(1.5) = 0.173$$

It follows that $\epsilon_{\text{total}} \approx 0.102 + 0.173 = 0.275$.

APPENDIX A

TABLE A-1 Conversion Factors for Single Terms

to convert from	to	multiply by
	Energy	
Btu (thermochemical)	joule	1054.350 264 48
calorie (thermochemical)	joule	4.184
foot lbf	joule	1.355 817 9
foot poundal	joule	0.042 140 110
kilowatt hour	joule	3.60×10^{6}
watt hour	joule	3600
	Force	
dyne	newton	1.00×10^{-5}
ounce force (avoirdupois)	newton	0.278 013 85
pound force, lbf (avoirdupois)	newton	4.448 221 615 26
poundal	newton	0.138 254 954 3
	Length	
angstrom	meter	1.00×10^{-10}
foot	meter	0.3048
inch	meter	0.0254
micron	meter	1.00×10^{-6}
mil	meter	2.54×10^{-5}
mile (U.S. statute)	meter	1609.344
yard	meter	0.9144
	Mass	
gram	kilogram	1.00×10^{-3}
lbm (avoirdupois)	kilogram	0.453 592 37
ounce mass (avoirdupois)	kilogram	0.028 349 523
ton (long)	kilogram	1016.0469
ton (metric)	kilogram	1000
ton (short, 2000 lbm)	kilogram	907.184 74
	Temperature	
Celsius	Kelvin	$K = C + 273.15$
Fahrenheit	Celsius	$C = \frac{5}{9}(F - 32)$
Fahrenheit	Kelvin	$K = \frac{5}{9}(F + 459.67)$
Kelvin	Celsius	$C = K - 273.15$
Rankine	Kelvin	$K = \frac{5}{9}R$

TABLE A-2 Conversion Factors for Compound Terms

to convert from	to	multiply by
	Acceleration	
foot/second^2	meter/second^2	0.3048
inch/second^2	meter/second^2	0.0254
	Density	
gram/centimeter^3	kilogram/meter^3	1000
lbm/foot^3	kilogram/meter^3	16.018 463
slug/foot^3	kilogram/meter^3	515.379
	Energy/Area · Time	
*Btu/foot^2 · hour	watt/meter^2	3.152 480 8
*calorie/cm^2 · minute	watt/meter^2	697.333 33
	Power	
Btu/second	watt	1054.350 264 4
calorie/second	watt	4.184
foot · lbf/second	watt	1.355 817 9
horsepower (550 ft · lbf/second)	watt	745.699 87
horsepower (electric)	watt	746.000 00
	Pressure	
atmosphere	pascal	$1.013\,25 \times 10^5$
bar	pascal	1.00×10^5
millimeter of mercury (0 °C)	pascal	133.322
centimeter of water (4 °C)	pascal	98.0638
dyne/centimeter^2	pascal	0.100
lbf/inch^2 (psi)	pascal	6894.7572
newton/meter^2	pascal	1.00
torr (0 °C)	pascal	133.322
	Speed	
foot/second	meter/second	0.3048
kilometer/hour	meter/second	0.277 777 78
knot (international)	meter/second	0.514 444 44
mile/hour (U.S. statute)	meter/second	0.447 04
	Thermal Conductivity	
Btu · inch/foot^2 · second · °F	watt/meter · kelvin	518.873 15
Btu/foot · hour · °F	watt/meter · kelvin	1.729 577 1

*All Btu and calorie terms in Table A-2 are thermochemical values.

TABLE A-2 (continued)

to convert from	to	multiply by
	Viscosity	
centipoise	pascal · second	1.00×10^{-3}
centistoke	meter²/second	1.00×10^{-6}
foot²/second	meter²/second	0.092 903 04
kilogram/meter · second	pascal · second	1.00
lbm/foot · second	pascal · second	1.488 163 9
lbf · second/foot²	pascal · second	47.880 258
poise	pascal · second	0.10
poundal · second/ft²	pascal · second	1.488 163 9
slug/foot · second	pascal · second	47.880 258
stoke	meter²/second	1.00×10^{-4}
	Volume	
fluid ounce (U.S.)	meter³	$2.957\,352\,95 \times 10^{-5}$
foot³	meter³	0.028 316 846 5
gallon (British)	meter³	$4.546\,087 \times 10^{-3}$
gallon (U.S. dry)	meter³	$4.404\,883\,77 \times 10^{-3}$
gallon (U.S. liquid)	meter³	$3.785\,411\,78 \times 10^{-3}$
liter (H_2O at 4 °C)	meter³	$1.000\,028 \times 10^{-3}$
liter (SI)	meter³	1.00×10^{-3}
pint (U.S. liquid)	meter³	$4.731\,764\,73 \times 10^{-4}$
quart (U.S. liquid)	meter³	$9.463\,529\,5 \times 10^{-4}$
yard³	meter³	0.764 554 857

APPENDIX B

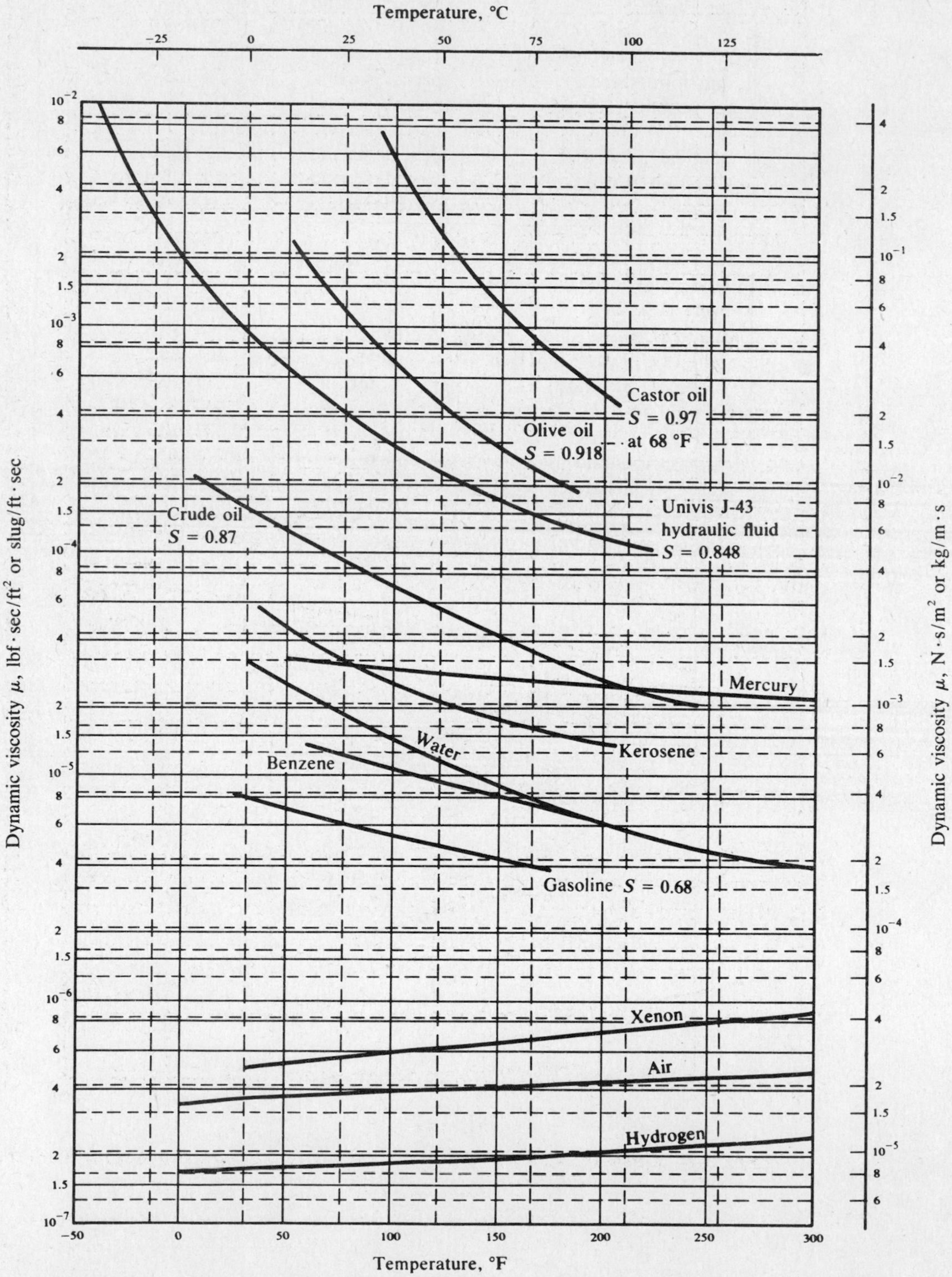

Figure B-1 Dynamic (absolute) viscosity of fluids. Specific gravity (S) values apply at 70 °F.

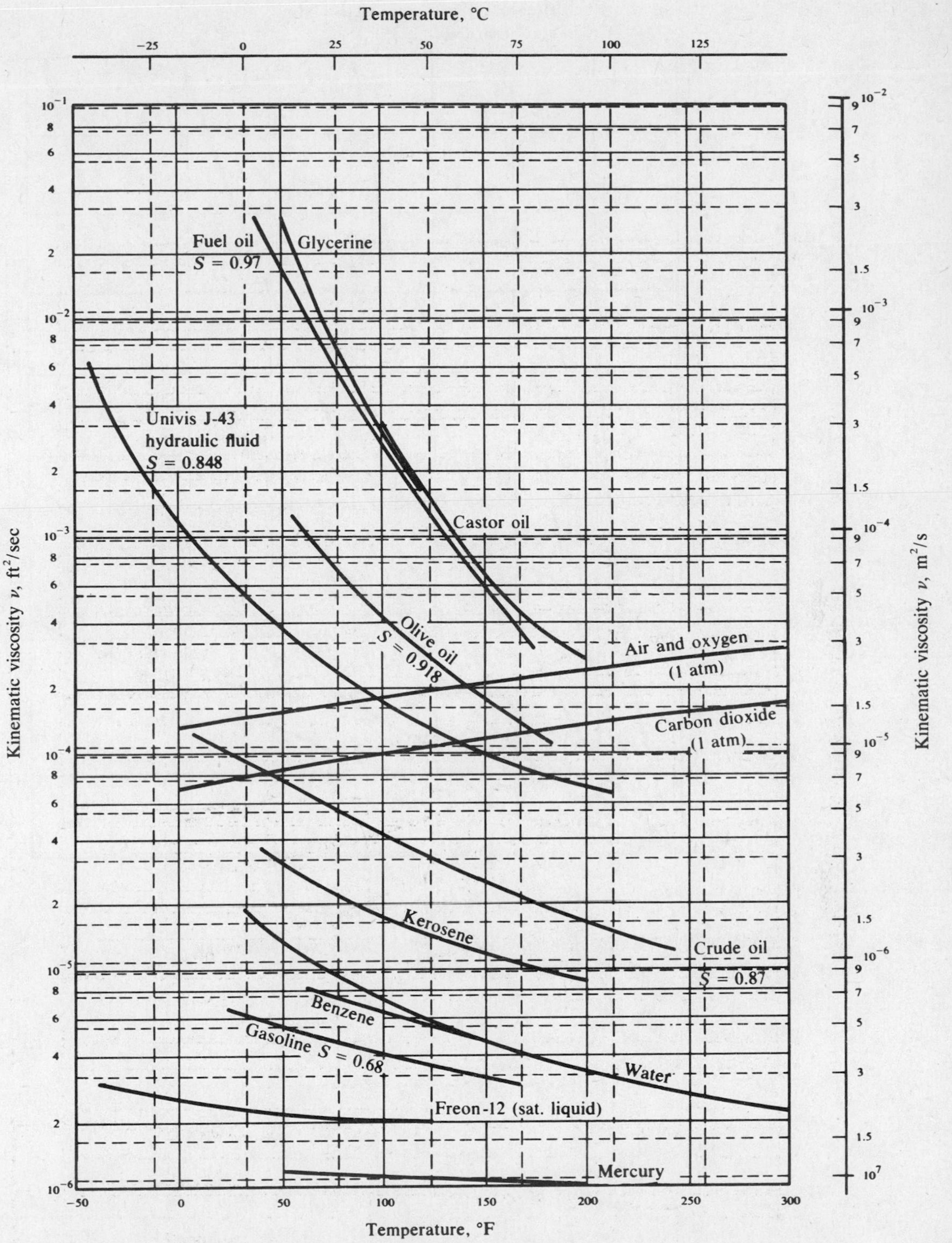

Figure B-2 Kinematic viscosity of fluids. Specific gravity (S) values apply at 70 °F.

p_c = critical pressure (3208 psia = 218.3 atm)
T_c = critical temperature (1165.3 °R)

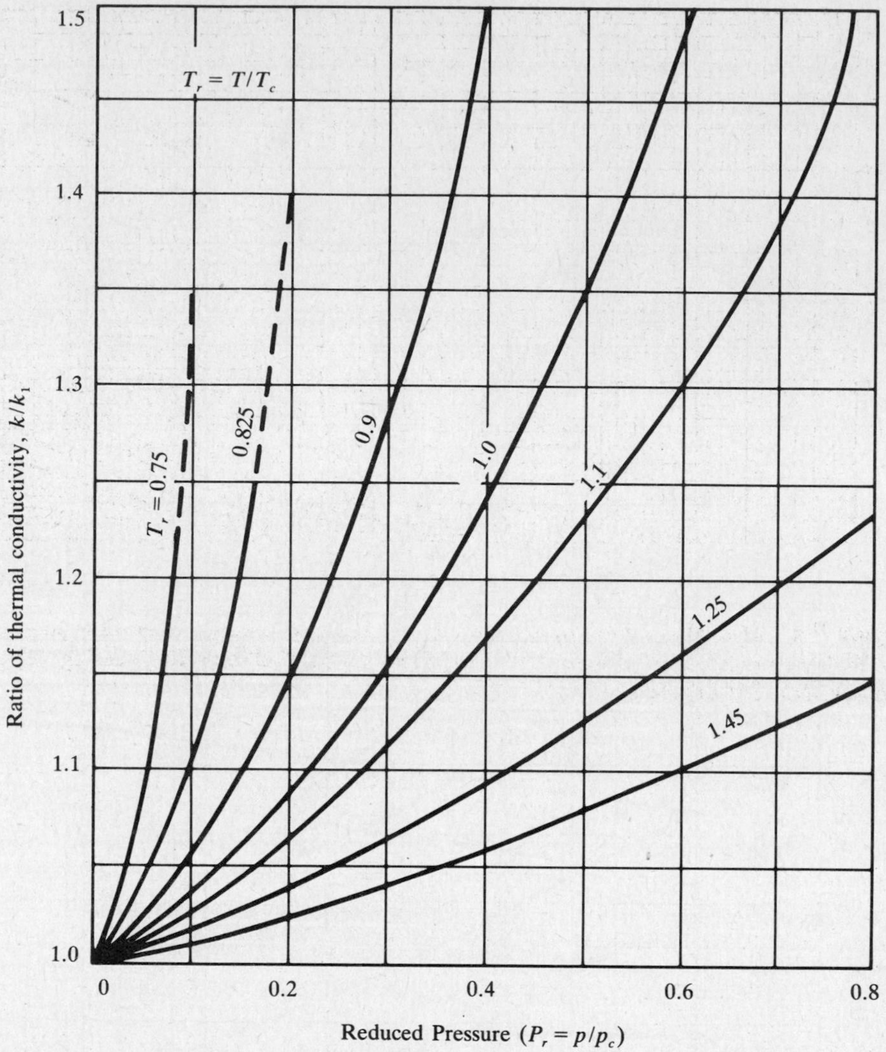

Figure B-3 Ratio of steam thermal conductivity k to the value k_1, at one atmosphere and the same temperature

μ_1 = dynamic viscosity at 1 atm and same temperature
p_c = critical pressure

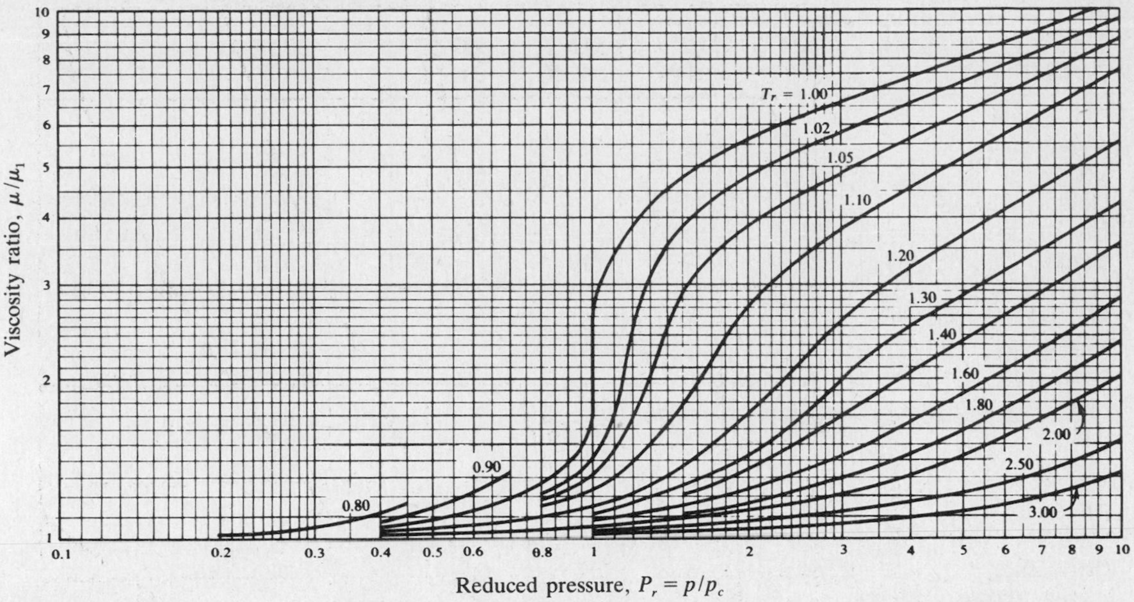

Figure B-4 Generalized correlation chart of the dynamic viscosity of gases at high pressures. [From E. W. Comings, B. J. Mayland, and R. S. Egly, Univ. of Illinois Engineering Experiment Station Bulletin No. 354 (1944).]

TABLE B-1 Properties of Metals

material	k, Btu/hr·ft·°F 32 °F 0 °C	212 °F 100 °C	572 °F 300 °C	932 °F 500 °C	c_p, Btu/lbm·°F 32 °F 0 °C	ρ, lbm/ft^3 32 °F 0 °C	α, ft^2/hr 32 °F 0 °C
Metals—pure							
Aluminum	117	119	133	156	0.208	169	3.33
Copper	224	218	212	207	0.091	558	4.42
Gold	169	170	—	—	0.030	1203	4.68
Iron	35.8	36.6	—	—	0.104	491	0.70
Lead	20.1	19	18	—	0.030	705	0.95
Magnesium	91	92	—	—	0.232	109	3.60
Molybdenum	72	68	64	62	0.060	638	1.88
Nickel	54	48	37	—	0.106	556	0.92
Silver	241	240	—	—	0.056	655	6.57
Tin	38	34	—	—	0.054	456	1.54
Zinc	65.1	63	58	—	0.091	446	1.60
Alloys							
Admiralty metal	65	64					
Brass, 70% Cu, 30% Zn	61.5	74	85	—	0.092	532	1.26
Bronze, 75% Cu, 25% Sn	15	—	—	—	0.082	541	0.34
Cast iron, plain	33	31.8	27.7	24.8	0.11	474	0.63
alloy	30	28.3	27	—	0.10	455	0.66
Constantan, 60% Cu, 40% Ni	12.4	12.8	—	—	0.10	557	0.22
18-8 stainless steel, Type 304	8.0	9.4	10.9	12.4	0.11	488	0.15
Type 347	8.0	9.3	11.0	12.8	0.11	488	0.15
Steel, mild, 1% C	26.5	26	25	22	0.11	490	0.49
SI units	**W/m·K**				**J/kg·K**	**kg/m^3**	**m^2/s**
To convert to SI units multiply tabulated values by	1.729 577				4.184 $\times 10^3$	1.601 846 $\times 10^1$	2.580 640 $\times 10^{-5}$

TABLE B-2 Properties of Nonmetals*

substance	T °F	T °C	c_p, Btu/lbm·°F	ρ, lbm/ft^3	k, Btu/hr·ft·°F	α, ft^2/hr
Structural						
Asphalt	68	20			0.43	
Bakelite	68	20	0.38	79.5	0.134	0.0044
Bricks						
Common	68	20	0.20	100	0.40	0.02
Face	68	20		128	0.76	
Carborundum brick	1110	600			10.7	
	2550	1400			6.4	
Chrome brick	392	200			1.34	0.036
	1022	550	0.20	188	1.43	0.038
	1652	900			1.15	0.031
Diatomaceous earth (fired)	400	205			0.14	
	1600	870			0.18	
Fireclay brick (burnt 2426 °F, 1330 °C)	932	500			0.60	0.020
	1472	800	0.23	128	0.62	0.021
	2012	1100			0.63	0.021
Fireclay brick (burnt 2642 °F, 1450 °C)	932	500			0.74	0.022
	1472	800	0.23	145	0.79	0.024
	2012	1100			0.81	0.024
Fireclay brick (Missouri)	392	200			0.58	0.015
	1112	600	0.23	165	0.85	0.022
	2552	1400			1.02	0.027
Magnesite	400	205			2.2	
	1200	650	0.27		1.6	
	2200	1205			1.1	
Cement, Portland				94	0.17	
Cement, mortar	75	24			0.67	
Concrete	68	20	0.21	119–144	0.47–0.81	0.019–0.027
Concrete, cinder	75	24			0.44	
Glass, plate	68	20	0.2	169	0.44	0.013
Glass, borosilicate	86	30		139	0.63	
Plaster, gypsum	70	21	0.2	90	0.28	0.016
Plaster, metal lath	70	21			0.27	
Plaster, wood lath	70	21			0.16	
Stone						
Granite			0.195	165	1.0–2.3	0.031–0.071
Limestone	210–570	100–300	0.217	155	0.73–0.77	0.022–0.023
Marble	68	20	0.193	156–169	1.6	0.054
Sandstone	68	20	0.17	135–144	0.94–1.2	0.041–0.049
SI units			**J/kg·K**	**kg/m^3**	**W/m·K**	**m^2/s**
To convert to SI units multiply tabulated values by			4.184 $\times 10^3$	1.601 846 $\times 10^1$	1.729 577	2.580 640 $\times 10^{-5}$

*Adapted with permission of Macmillan Publishing Company from A. Chapman, *Heat Transfer*, 2d ed., Macmillan, London, 1967. Copyright 1967 by Alan J. Chapman.

TABLE B-2 (continued)

substance	T °F	T °C	c_p, Btu/lbm·°F	ρ, lbm/ft^3	k, Btu/hr·ft·°F	α, ft^2/hr
Structural (cont.)						
Wood, cross grain:						
Balsa	86	30		8.8	0.032	
Cypress	86	30		29	0.056	
Fir	75	24	0.65	26.0	0.063	0.0037
Oak	86	30	0.57	38–30	0.096	0.0049
Yellow pine	75	24	0.67	40	0.085	0.0032
White pine	86	30		27	0.065	
Wood, radial:						
Oak	68	20	0.57	38–30	0.10–0.12	0.0043–0.0047
Fir	68	20	0.65	26.0–26.3	0.08	0.0048
Insulating						
Asbestos	−328	−200		29.3	0.043	
	32	0			0.090	
Asbestos	32	0		36.0	0.087	
	212	100			0.111	
	392	200			0.120	
	752	400			0.129	
Asbestos	−328	−200		43.5	0.09	
	32	0			0.135	
Asbestos cement					1.2	
Asbestos sheet	124	51			0.096	
Asbestos felt (40 laminations per inch)	100	38			0.033	
	300	149			0.040	
	500	260			0.048	
Asbestos felt (20 laminations per inch)	100	38			0.045	
	300	149			0.055	
	500	260			0.065	
Balsam wool	90	32		2.2	0.023	
Cardboard, corrugated					0.037	
Celotex	90	32			0.028	
Corkboard	86	30		10	0.025	
Cork, ground	86	30		9.4	0.025	
SI units			**J/kg·K**	**kg/m^3**	**W/m·K**	**m^2/s**
To convert to SI units multiply tabulated values by			4.184 $\times 10^3$	1.601 846 $\times 10^1$	1.729 577	2.580 640 $\times 10^{-5}$

TABLE B-2 (continued)

substance	T °F	T °C	c_p, Btu/lbm · °F	ρ, lbm/ft^3	k, Btu/hr · ft · °F	α, ft^2/hr
Insulating (cont.)						
Diatomaceous earth (powdered)	200	93		14	0.033	
	400	204			0.039	
	600	316			0.046	
Felt, hair	20	−7		11.4	0.0212	
	100	38			0.0254	
	200	93			0.0299	
Fiber insulating board	70	21		14.8	0.028	
Glass wool	20	−7		1.5	0.0217	
	100	38			0.0313	
	200	93			0.0435	
Glass wool	20	−7		4.0	0.0179	
	100	38			0.0239	
	200	93			0.0317	
Glass wool	20	−7		6.0	0.0163	
	100	38			0.0218	
	200	93			0.0288	
Kapok	86	30			0.020	
Magnesia, 85%	100	38		16.9	0.039	
	200	93			0.041	
	300	149			0.043	
	400	204			0.046	
Rock wool	20	−7		4.0	0.0150	
	100	38			0.0224	
	200	93			0.0317	
Rock wool	20	−7		8.0	0.0171	
	100	38			0.0228	
	200	93			0.0299	
Rock wool	20	−7		12.0	0.0183	
	100	38			0.0226	
	200	93			0.0281	
Miscellaneous						
Aerogel, silica	248	120		8.5	0.013	
Clay	68	20	0.21	91.0	0.739	0.039
Coal, anthracite	68	20	0.30	75–94	0.15	0.005–0.006
Coal, powdered	86	30	0.31	46	0.067	0.005
Cotton	68	20	0.31	5	0.034	0.075
Earth, coarse	68	20	0.44	128	0.30	0.0054
Ice	32	0	0.46	57	1.28	0.048
Rubber, hard	32	0		74.8	0.087	
Sawdust	75	24			0.034	
Silk	68	20	0.33	3.6	0.021	0.017
SI units			**J/kg · K**	**kg/m^3**	**W/m · K**	**m^2/s**
To convert to SI units multiply tabulated values by			4.184 $\times 10^3$	1.601 846 $\times 10^1$	1.729 577	2.580 640 $\times 10^{-5}$

TABLE B-3 Properties of Liquids in Saturated State*

T °F	T °C	ρ, lbm/ft^3	c_p, Btu/lbm·°F	ν, ft^2/sec	k, Btu/hr·ft·°F	α, ft^2/hr	Pr	β, 1/°R
Water (H_2O)								
32	0	62.57	1.0074	1.925×10^{-5}	0.319	5.07×10^{-3}	13.6	
68	20	62.46	0.9988	1.083	0.345	5.54	7.02	0.10×10^{-3}
104	40	62.09	0.9980	0.708	0.363	5.86	4.34	
140	60	61.52	0.9994	0.514	0.376	6.02	3.02	
176	80	60.81	1.0023	0.392	0.386	6.34	2.22	
212	100	59.97	1.0070	0.316	0.393	6.51	1.74	
248	120	59.01	1.015	0.266	0.396	6.62	1.446	
284	140	57.95	1.023	0.230	0.395	6.68	1.241	
320	160	56.79	1.037	0.204	0.393	6.70	1.099	
356	180	55.50	1.055	0.186	0.390	6.68	1.004	
392	200	54.11	1.076	0.172	0.384	6.61	0.937	
428	220	52.59	1.101	0.161	0.377	6.51	0.891	
464	240	50.92	1.136	0.154	0.367	6.35	0.871	
500	260	49.06	1.182	0.148	0.353	6.11	0.874	
537	280	46.98	1.244	0.145	0.335	5.74	0.910	
572	300	44.59	1.368	0.145	0.312	5.13	1.019	
Ammonia (NH_3)								
−58	−50	43.93	1.066	0.468×10^{-5}	0.316	6.75×10^{-3}	2.60	
−40	−40	43.18	1.067	0.437	0.316	6.88	2.28	
−22	−30	42.41	1.069	0.417	0.317	6.98	2.15	
−4	−20	41.62	1.077	0.410	0.316	7.05	2.09	
14	−10	40.80	1.090	0.407	0.314	7.07	2.07	
32	0	39.96	1.107	0.402	0.312	7.05	2.05	
50	10	39.09	1.126	0.396	0.307	6.98	2.04	
68	20	38.19	1.146	0.386	0.301	6.88	2.02	1.36×10^{-3}
86	30	37.23	1.168	0.376	0.293	6.75	2.01	
104	40	36.27	1.194	0.366	0.285	6.59	2.00	
122	50	35.23	1.222	0.355	0.275	6.41	1.99	
Carbon dioxide (CO_2)								
−58	−50	72.19	0.44	0.128×10^{-5}	0.0494	1.558×10^{-3}	2.96	
−40	−40	69.78	0.45	0.127	0.0584	1.864	2.46	
−22	−30	67.22	0.47	0.126	0.0645	2.043	2.22	
−4	−20	64.45	0.49	0.124	0.0665	2.110	2.12	
14	−10	61.39	0.52	0.122	0.0635	1.989	2.20	
SI Units		kg/m^3	J/kg·K	m^2/s	W/m·K	m^2/s	—	1/K
To convert to SI units multiply tabulated values by		$1.601\,846 \times 10^1$	4.184×10^3	$9.290\,304 \times 10^{-2}$	1.729 577	$2.580\,640 \times 10^{-5}$	—	1.80

*Adapted by permission from E. R. G. Eckert and R. M. Drake, Jr., *Heat and Mass Transfer*, 2d ed., McGraw-Hill Book Company, New York, 1959.

TABLE B-3 (continued)

T		ρ, lbm/ft^3	c_p, Btu/lbm · °F	ν, ft^2/sec	k, Btu/hr · ft · °F	α, ft^2/hr	Pr	β, 1/°R
°F	°C							
Carbon dioxide (CO_2) (cont.)								
32	0	57.87	0.59	0.117	0.0604	1.774	2.38	
50	10	53.69	0.75	0.109	0.0561	1.398	2.80	
68	20	48.23	1.2	0.098	0.0504	0.860	4.10	7.78×10^{-3}
86	30	37.32	8.7	0.086	0.0406	0.108	28.7	
Sulfur dioxide (SO_2)								
−58	−50	97.44	0.3247	0.521×10^{-5}	0.140	4.42×10^{-3}	4.24	
−40	−40	95.94	0.3250	0.456	0.136	4.38	3.74	
−22	−30	94.43	0.3252	0.399	0.133	4.33	3.31	
−4	−20	92.93	0.3254	0.349	0.130	4.29	2.93	
14	−10	91.37	0.3255	0.310	0.126	4.25	2.62	
32	0	89.80	0.3257	0.277	0.122	4.19	2.38	
50	10	88.18	0.3259	0.250	0.118	4.13	2.18	
68	20	86.55	0.3261	0.226	0.115	4.07	2.00	1.08×10^{-3}
86	30	84.86	0.3263	0.204	0.111	4.01	1.83	
104	40	82.98	0.3266	0.186	0.107	3.95	1.70	
122	50	81.10	0.3268	0.174	0.102	3.87	1.61	
Methylchloride (CH_3Cl)								
−58	−50	65.71	0.3525	0.344×10^{-5}	0.124	5.38×10^{-3}	2.31	
−40	−40	64.51	0.3541	0.342	0.121	5.30	2.32	
−22	−30	63.46	0.3564	0.338	0.117	5.18	2.35	
−4	−20	62.39	0.3593	0.333	0.113	5.04	2.38	
14	−10	61.27	0.3629	0.329	0.108	4.87	2.43	
32	0	60.08	0.3673	0.325	0.103	4.70	2.49	
50	10	58.83	0.3726	0.320	0.099	4.52	2.55	
68	20	57.64	0.3788	0.315	0.094	4.31	2.63	
86	30	56.38	0.3860	0.310	0.089	4.10	2.72	
104	40	55.13	0.3942	0.303	0.083	3.86	2.83	
122	50	53.76	0.4034	0.295	0.077	3.57	2.97	
Dichlorodifluoromethane (Freon-12)(CCl_2F_2)								
−58	−50	96.56	0.2090	0.334×10^{-5}	0.039	1.94×10^{-3}	6.2	1.46×10^{-3}
−40	−40	94.81	0.2113	0.300	0.040	1.99	5.4	
−22	−30	92.99	0.2139	0.272	0.040	2.04	4.8	
−4	−20	91.18	0.2167	0.253	0.041	2.09	4.4	
14	−10	89.24	0.2198	0.238	0.042	2.13	4.0	
SI Units		**kg/m^3**	**J/kg · K**	**m^2/s**	**W/m · K**	**m^2/s**	—	**1/K**
To convert to SI units multiply tabulated values by		1.601 846 $\times 10^1$	4.184 $\times 10^3$	9.290 304 $\times 10^{-2}$	1.729 577	2.580 640 $\times 10^{-5}$	—	1.80

TABLE B-3 (continued)

T								
°F	°C	ρ, lbm/ft^3	c_p, Btu/lbm · °F	ν, ft^2/sec	k, Btu/hr · ft · °F	α, ft^2/hr	Pr	β, 1/°R
		Dichlorodifluoromethane (Freon-12)(CCl_2F_2) (cont.)						
32	0	87.24	0.2232	0.230	0.042	2.16	3.8	
50	10	85.17	0.2268	0.219	0.042	2.17	3.6	
68	20	83.04	0.2307	0.213	0.042	2.17	3.5	
86	30	80.85	0.2349	0.209	0.041	2.17	3.5	
104	40	78.48	0.2393	0.206	0.040	2.15	3.5	
122	50	75.91	0.2440	0.204	0.039	2.11	3.5	
		Eutectic calcium chloride solution (29.9% $CaCl_2$)						
−58	−50	82.39	0.623	39.13×10^{-5}	0.232	4.52×10^{-3}	312	
−40	−40	82.09	0.6295	26.88	0.240	4.65	208	
−22	−30	81.79	0.6356	18.49	0.248	4.78	139	
−4	−20	81.50	0.642	11.88	0.257	4.91	87.1	
14	−10	81.20	0.648	7.49	0.265	5.04	53.6	
32	0	80.91	0.654	4.73	0.273	5.16	33.0	
50	10	80.62	0.660	3.61	0.280	5.28	24.6	
68	20	80.32	0.666	2.93	0.288	5.40	19.6	
86	30	80.03	0.672	2.44	0.295	5.50	16.0	
104	40	79.73	0.678	2.07	0.302	5.60	13.3	
122	50	79.44	0.685	1.78	0.309	5.69	11.3	
		Glycerin [$C_3H_5(OH)_3$]						
32	0	79.66	0.540	0.0895	0.163	3.81×10^{-3}	84.7×10^3	
50	10	79.29	0.554	0.0323	0.164	3.74	31.0	
68	20	78.91	0.570	0.0127	0.165	3.67	12.5	0.28×10^{-3}
86	30	78.54	0.584	0.0054	0.165	3.60	5.38	
104	40	78.16	0.600	0.0024	0.165	3.54	2.45	
122	50	77.72	0.617	0.0016	0.166	3.46	1.63	
		Ethylene glycol [$C_2H_4(OH_2)$]						
32	0	70.59	0.548	61.92×10^{-5}	0.140	3.62×10^{-3}	615	
68	20	69.71	0.569	20.64	0.144	3.64	204	0.36×10^{-3}
104	40	68.76	0.591	9.35	0.148	3.64	93	
140	60	67.90	0.612	5.11	0.150	3.61	51	
176	80	67.27	0.633	3.21	0.151	3.57	32.4	—
212	100	66.08	0.655	2.18	0.152	3.52	22.4	
SI units		**kg/m^3**	**J/kg · K**	**m^2/s**	**W/m · K**	**m^2/s**	—	**1/K**
To convert to SI units multiply tabulated values by		1.601 846 $\times 10^1$	4.184 $\times 10^3$	9.290 304 $\times 10^{-2}$	1.729 577	2.580 640 $\times 10^{-5}$	—	1.80

TABLE B-3 (continued)

T °F	*T* °C	ρ, lbm/ft^3	c_p, Btu/lbm · °F	ν, ft^2/sec	*k*, Btu/hr · ft · °F	α, ft^2/hr	Pr	β, 1/°R
				Engine oil (unused)				
32	0	56.13	0.429	0.0461	0.085	3.53×10^{-3}	47 100	
68	20	55.45	0.449	0.0097	0.084	3.38	10 400	0.39×10^{-3}
104	40	54.69	0.469	0.0026	0.083	3.23	2 870	
140	60	53.94	0.489	0.903×10^{-3}	0.081	3.10	1 050	
176	80	53.19	0.509	0.404	0.080	2.98	490	
212	100	52.44	0.530	0.219	0.079	2.86	276	
248	120	51.75	0.551	0.133	0.078	2.75	175	
284	140	51.00	0.572	0.086	0.077	2.66	116	
320	160	50.31	0.593	0.060	0.076	2.57	84	
				Mercury (Hg)				
32	0	850.78	0.0335	0.133×10^{-5}	4.74	166.6×10^{-3}	0.0288	
68	20	847.71	0.0333	0.123	5.02	178.5	0.0249	1.01×10^{-4}
122	50	843.14	0.0331	0.112	5.43	194.6	0.0207	
212	100	835.57	0.0328	0.0999	6.07	221.5	0.0162	
302	150	828.06	0.0326	0.0918	6.64	246.2	0.0134	
392	200	820.61	0.0375	0.0863	7.13	267.7	0.0116	
482	250	813.16	0.0324	0.0823	7.55	287.0	0.0103	
600	316	802	0.032	0.0724	8.10	316	0.0083	
SI units		**kg/m^3**	**J/kg · K**	**m^2/s**	**W/m · K**	**m^2/s**	—	**1/K**
To convert to SI units multiply tabulated values by		$1.601\,846 \times 10^1$	4.184×10^3	$9.290\,304 \times 10^{-2}$	1.729 577	$2.580\,640 \times 10^{-5}$	—	1.80

TABLE B-4 Properties of Gases at Atmospheric Pressure*

T °F	*T* °C	ρ, lbm/ft^3	c_p, Btu/lbm · °F	μg_c, lbm/ft · sec	ν, ft^2/sec	k, Btu/hr · ft · °F	α, ft^2/hr	Pr
				Air				
−280	−173	0.2248	0.2452	0.4653×10^{-5}	2.070×10^{-5}	0.005 342	0.096 91	0.770
−190	−123	0.1478	0.2412	0.6910	4.675	0.007 936	0.2226	0.753
−100	−73	0.1104	0.2403	0.8930	8.062	0.010 45	0.3939	0.739
−10	−23	0.0882	0.2401	1.074	10.22	0.012 87	0.5100	0.722
80	27	0.0735	0.2402	1.241	16.88	0.015 16	0.8587	0.708
170	77	0.0623	0.2410	1.394	22.38	0.017 35	1.156	0.697
260	127	0.0551	0.2422	1.536	27.88	0.019 44	1.457	0.689
350	177	0.0489	0.2438	1.669	31.06	0.021 42	1.636	0.683
440	227	0.0440	0.2459	1.795	40.80	0.023 33	2.156	0.680
530	277	0.0401	0.2482	1.914	47.73	0.025 19	2.531	0.680
620	327	0.0367	0.2520	2.028	55.26	0.026 92	2.911	0.680
710	377	0.0339	0.2540	2.135	62.98	0.028 62	3.324	0.682
800	427	0.0314	0.2568	2.239	71.31	0.030 22	3.748	0.684
890	477	0.0294	0.2593	2.339	79.56	0.031 83	4.175	0.686
980	527	0.0275	0.2622	2.436	88.58	0.033 39	4.631	0.689
1070	577	0.0259	0.2650	2.530	97.68	0.034 83	5.075	0.692
1160	627	0.0245	0.2678	2.620	106.9	0.036 28	5.530	0.696
1250	677	0.0232	0.2704	2.703	116.5	0.037 70	6.010	0.699
1340	727	0.0220	0.2727	2.790	126.8	0.039 01	6.502	0.702
1520	827	0.0200	0.2772	2.955	147.8	0.041 78	7.536	0.706
1700	927	0.0184	0.2815	3.109	169.0	0.044 10	8.514	0.714
1880	1027	0.0169	0.2860	3.258	192.8	0.046 41	9.602	0.722
2060	1127	0.0157	0.2900	3.398	216.4	0.048 80	10.72	0.726
2240	1227	0.0147	0.2939	3.533	240.3	0.050 98	11.80	0.734
2420	1327	0.0138	0.2982	3.668	265.8	0.053 48	12.88	0.741
2600	1427	0.0130	0.3028	3.792	291.7	0.055 50	14.00	0.749
2780	1527	0.0123	0.3075	3.915	318.3	0.057 50	15.09	0.759
2960	1627	0.0116	0.3128	4.029	347.1	0.0591	16.40	0.767
3140	1727	0.0110	0.3196	4.168	378.8	0.0612	17.41	0.783
3320	1827	0.0105	0.3278	4.301	409.9	0.0632	18.36	0.803
3500	1927	0.0100	0.3390	4.398	439.8	0.0646	19.05	0.831
3680	2027	0.0096	0.3541	4.513	470.1	0.0663	19.61	0.863
3860	2127	0.0091	0.3759	4.611	506.9	0.0681	19.92	0.916
4160	2293	0.0087	0.4031	4.750	546.0	0.0709	20.21	0.972
				Helium				
−456	−271		1.242	5.66×10^{-7}		0.0061		
−400	−240	0.0915	1.242	33.7	3.68×10^{-5}	0.0204	0.1792	0.74
−200	−129	0.211	1.242	84.3	39.95	0.0536	2.044	0.70
−100	−73	0.0152	1.242	105.2	69.30	0.0680	3.599	0.694
0	−18	0.0119	1.242	122.1	102.8	0.0784	5.299	0.70
200	93	0.008 29	1.242	154.9	186.9	0.0977	9.490	0.71
SI units		**kg/m^3**	**J/kg · K**	**Pa · s**	**m^2/s**	**W/m · K**	**m^2/s**	
To convert to SI units multiply tabulated values by		$1.601\,846 \times 10^1$	4.184×10^3	1.488 164	$9.290\,304 \times 10^{-2}$	1.729 577	$2.580\,640 \times 10^{-5}$	

*Adapted by permission from E. R. G. Eckert and R. M. Drake, Jr., *Heat and Mass Transfer*, 2d ed., McGraw-Hill Book Company, New York, 1959.

TABLE B-4 (continued)

T °F	*T* °C	ρ, lbm/ft^3	c_p, Btu/lbm·°F	μg_c, lbm/ft·sec	ν, ft^2/sec	*k*, Btu/hr·ft·°F	α, ft^2/hr	Pr
				Helium (cont.)				
400	204	0.006 37	1.242	184.8	289.9	0.114	14.40	0.72
600	316	0.005 17	1.242	209.2	404.5	0.130	20.21	0.72
800	427	0.004 39	1.242	233.5	531.9	0.145	25.81	0.72
1000	538	0.003 76	1.242	256.5	682.5	0.159	34.00	0.72
1200	649	0.003 30	1.242	277.9	841.0	0.172	41.98	0.72
				Hydrogen				
−406	−243	0.052 89	2.589	1.079×10^{-6}	2.040×10^{-5}	0.0132	0.0966	0.759
−370	−223	0.031 81	2.508	1.691	5.253	0.0209	0.262	0.721
−280	−173	0.015 34	2.682	2.830	18.45	0.0384	0.933	0.712
−190	−123	0.010 22	3.010	3.760	36.79	0.0567	1.84	0.718
−100	−73	0.007 66	3.234	4.578	59.77	0.0741	2.99	0.719
−10	−23	0.006 13	3.358	5.321	86.80	0.0902	4.38	0.713
80	27	0.005 11	3.419	6.023	117.9	0.105	6.02	0.706
170	77	0.004 38	3.448	6.689	152.7	0.119	7.87	0.697
260	127	0.003 83	3.461	7.300	190.6	0.132	9.95	0.690
350	177	0.003 41	3.463	7.915	232.1	0.145	12.26	0.682
440	227	0.003 07	3.465	8.491	276.6	0.157	14.79	0.675
530	277	0.002 79	3.471	9.055	324.6	0.169	17.50	0.668
620	327	0.002 55	3.472	9.599	376.4	0.182	20.56	0.664
800	427	0.002 18	3.481	10.68	489.9	0.203	26.75	0.659
980	527	0.001 91	3.505	11.69	612	0.222	33.18	0.664
1160	627	0.001 70	3.540	12.62	743	0.238	39.59	0.676
1340	727	0.001 53	3.575	13.55	885	0.254	46.49	0.686
1520	827	0.001 39	3.622	14.42	1039	0.268	53.19	0.703
1700	927	0.001 28	3.670	15.29	1192	0.282	60.00	0.715
1880	1027	0.001 18	3.720	16.18	1370	0.296	67.40	0.733
1940	1060	0.001 15	3.735	16.42	1429	0.300	69.80	0.736
				Oxygen				
−280	−173	0.2492	0.2264	5.220×10^{-6}	2.095×10^{-5}	0.005 22	0.092 52	0.815
−190	−123	0.1635	0.2192	7.721	4.722	0.007 90	0.2204	0.773
−100	−73	0.1221	0.2181	9.979	8.173	0.010 54	0.3958	0.745
−10	−23	0.0975	0.2187	12.01	12.32	0.013 05	0.6120	0.725
80	27	0.0812	0.2198	13.86	17.07	0.015 46	0.8662	0.709
170	77	0.0695	0.2219	15.56	22.39	0.017 74	1.150	0.702
260	127	0.0609	0.2250	17.16	28.18	0.020 00	1.460	0.695
350	177	0.0542	0.2285	18.66	34.43	0.022 12	1.786	0.694
440	227	0.0487	0.2322	20.10	41.27	0.024 11	2.132	0.697
530	277	0.0443	0.2360	21.48	48.49	0.026 10	2.496	0.700
620	327	0.0406	0.2399	22.79	56.13	0.027 92	2.867	0.704
SI units		kg/m^3	J/kg·K	Pa·s	m^2/s	W/m·K	m^2/s	
To convert to SI units multiply tabulated values by		1.601 846 $\times 10^1$	4.184 $\times 10^3$	1.488 164	9.290 304 $\times 10^{-2}$	1.729 577	2.580 640 $\times 10^{-5}$	

TABLE B-4 (continued)

T °F	T °C	ρ, lbm/ft^3	c_p, Btu/lbm · °F	μg_c, lbm/ft · sec	ν, ft^2/sec	k, Btu/hr · ft · °F	α, ft^2/hr	Pr
				Nitrogen				
−280	−173	0.2173	0.2561	4.611×10^{-6}	2.122×10^{-5}	0.005 460	0.098 11	0.786
−100	−73	0.1068	0.2491	8.700	8.146	0.010 54	0.3962	0.747
80	27	0.0713	0.2486	11.99	16.82	0.015 14	0.8542	0.713
260	127	0.0533	0.2498	14.77	27.71	0.019 27	1.447	0.691
440	227	0.0426	0.2521	17.27	40.54	0.023 02	2.143	0.684
620	327	0.0355	0.2569	19.56	55.10	0.026 46	2.901	0.686
800	427	0.0308	0.2620	21.59	70.10	0.029 60	3.668	0.691
980	527	0.0267	0.2681	23.41	87.68	0.032 41	4.528	0.700
1160	627	0.0237	0.2738	25.19	98.02	0.035 07	5.404	0.711
1340	727	0.0213	0.2789	26.88	126.2	0.037 41	6.297	0.724
1520	827	0.0194	0.2832	28.41	146.4	0.039 58	7.204	0.736
1700	927	0.0178	0.2875	29.90	168.0	0.041 51	8.111	0.748
				Carbon dioxide				
−64	−53	0.1544	0.187	7.462×10^{-6}	4.833×10^{-5}	0.006 243	0.2294	0.818
−10	−23	0.1352	0.192	8.460	6.257	0.007 444	0.2868	0.793
80	27	0.1122	0.208	10.051	8.957	0.009 575	0.4103	0.770
170	77	0.0959	0.215	11.561	12.05	0.011 83	0.5738	0.755
260	127	0.0838	0.225	12.98	15.49	0.014 22	0.7542	0.738
350	177	0.0744	0.234	14.34	19.27	0.016 74	0.9615	0.721
440	227	0.0670	0.242	15.63	23.33	0.019 37	1.195	0.702
530	277	0.0608	0.250	16.85	27.71	0.022 08	1.453	0.685
620	327	0.0558	0.257	18.03	32.31	0.024 91	1.737	0.668
				Carbon monoxide				
−64	−53	0.096 99	0.2491	9.295×10^{-6}	9.583×10^{-5}	0.011 01	0.4557	0.758
−10	−23	0.0525	0.2490	10.35	12.14	0.012 39	0.5837	0.750
80	27	0.071 09	0.2489	11.990	16.87	0.014 59	0.8246	0.737
170	77	0.060 82	0.2492	13.50	22.20	0.016 66	1.099	0.728
260	127	0.053 29	0.2504	14.91	27.98	0.018 64	1.397	0.722
350	177	0.047 35	0.2520	16.25	34.32	0.0252	1.720	0.718
440	227	0.042 59	0.2540	17.51	41.11	0.022 32	2.063	0.718
530	277	0.038 72	0.2569	18.74	48.40	0.024 05	2.418	0.721
620	327	0.035 49	0.2598	19.89	56.04	0.025 69	2.786	0.724
				Ammonia (NH_3)				
−58	−50	0.0239	0.525	4.875×10^{-6}	2.04×10^{-4}	0.0099	0.796	0.93
32	0	0.0495	0.520	6.285	1.27	0.0127	0.507	0.90
122	50	0.0405	0.520	7.415	1.83	0.0156	0.744	0.88
212	100	0.0349	0.534	8.659	2.48	0.0189	1.015	0.87
302	150	0.0308	0.553	9.859	3.20	0.0226	1.330	0.87
392	200	0.0275	0.572	11.08	4.03	0.0270	1.713	0.84
SI units		**kg/m^3**	**J/kg · K**	**Pa · s**	**m^2/s**	**W/m · K**	**m^2/s**	
To convert to SI units multiply tabulated values by		$1.601\,846 \times 10^1$	4.184×10^3	1.488 164	$9.290\,304 \times 10^{-2}$	1.729 577	$2.580\,640 \times 10^{-5}$	

TABLE B-4 (continued)

T °F	*T* °C	ρ, lbm/ft^3	c_p, Btu/lbm · °F	μg_c, lbm/ft · sec	ν, ft^2/sec	*k*, Btu/hr · ft · °F	α, ft^2/hr	Pr
Steam (H_2O vapor)								
224	107	0.0366	0.492	8.54×10^{-6}	2.33×10^{-4}	0.0142	0.789	1.060
260	127	0.0346	0.481	9.03	2.61	0.0151	0.906	1.040
350	177	0.0306	0.473	10.25	3.35	0.0173	1.19	1.010
440	227	0.0275	0.474	11.45	4.16	0.0196	1.50	0.996
530	277	0.0250	0.477	12.66	5.06	0.0219	1.84	0.991
620	327	0.0228	0.484	13.89	6.09	0.0244	2.22	0.986
710	377	0.0211	0.491	15.10	7.15	0.0268	2.58	0.995
800	427	0.0196	0.498	16.30	8.31	0.0292	2.99	1.000
890	477	0.0183	0.506	17.50	9.56	0.0317	3.42	1.005
980	527	0.0171	0.514	18.72	10.98	0.0342	3.88	1.010
1070	577	0.0161	0.522	19.95	12.40	0.0368	4.38	1.019
SI units		**kg/m^3**	**J/kg · K**	**Pa · s**	**m^2/s**	**W/m · K**	**m^2/s**	
To convert to SI units multiply tabulated values by		1.601 846 $\times 10^1$	4.184 $\times 10^3$	1.488 164	9.290 304 $\times 10^{-2}$	1.729 577	2.580 640 $\times 10^{-5}$	

TABLE B-5 Critical Constants and Molecular Weights of Gases

gas	molecular weight	critical constants* p_c, atm	critical constants* T_c, °R
Air	28.95	37.2	238.4
Argon	39.944	48.0	272.0
Oxygen	32	49.7	277.8
Nitrogen	28.02	33.5	226.9
Carbon dioxide	44.01	73.0	547.7
Carbon monoxide	28.01	35.0	241.5
Hydrogen	2.02	12.8	59.9
Ethyl alcohol	46.0	63.1	929.3
Benzene	78.0	47.7	1011
Freon-12	120.92	39.6	692.4
Ammonia	17.03	111.5	730.0
Helium	4.00	2.26	9.47
Mercury vapor	200.61	>200	>3281.7
Methane	16.03	45.8	343.2
Propane	44.09	42.0	665.9
Water	18.016	218.3	1165.3
Xenon	131.3	58.0	521.55

TABLE B-6 Normal Total Emissivity of Various Surfaces*

surface	*T*, °F	emissivity ϵ
Metals and their oxides		
Aluminum		
Highly polished plate, 98.3% pure	440–1070	0.039–0.057
Commercial sheet	212	0.09
Heavily oxidized	299–940	0.20–0.31
Brass		
Highly polished		
73.2% Cu, 26.7% Zn	476–674	0.028–0.031
62.4% Cu, 36.8% Zn, 0.4% Pb, 0.3% Al	494–710	0.033–0.037
82.9% Cu, 17.0% Zn	530	0.030
Hard-rolled, polished, but direction of polishing visible	70	0.038
Dull plate	120–660	0.22
Copper		
Polished	242	0.023
	212	0.052
Plate, heated long time, covered with thick oxide layer	77	0.78
Gold, pure, highly polished	440–1160	0.018–0.035
Iron and steel (not including stainless)		
Steel, polished	212	0.066
Iron, polished	800–1880	0.14–0.38
Cast iron, newly turned	72	0.44
Cast iron, turned and heated	1620–1810	0.60–0.70
Mild steel	450–1950	0.20–0.32
Oxidized surfaces		
Iron plate, pickled, then rusted red	68	0.61
Iron, dark-gray surface	212	0.31
Rough ingot iron	1700–2040	0.87–0.95
Sheet steel with strong, rough oxide layer	75	0.80
Lead		
Unoxidized, 99.96% pure	260–440	0.057–0.075
Gray oxidized	75	0.28
Oxidized at 300 °F	390	0.63
Magnesium, magnesium oxide	530–1520	0.55–0.20
Molybdenum		
Filament	1340–4700	0.096–0.202
Massive, polished	212	0.071
Monel metal, oxidized at 1110 °F	390–1110	0.41–0.46
Nickel		
Polished	212	0.072
Nickel oxide	1200–2290	0.59–0.86
Nickel alloys		
Copper nickel, polished	212	0.059
Nichrome wire, bright	120–1830	0.65–0.79
Nichrome wire, oxidized	120–930	0.95–0.98

*Abstracted by permission from H. C. Hottel in W. H. McAdams (ed.), *Heat Transmission*, 3d ed., pp. 472–478. Copyright 1954, McGraw-Hill Book Company.

TABLE B-6 (continued)

surface	*T*, °F	emissivity ϵ
Platinum; polished plate, pure	440–1160	0.054–0.104
Silver		
Polished, pure	440–1160	0.020–0.032
Polished	100–700	0.022–0.031
Stainless steels		
Polished	212	0.074
Type 301	450–1725	0.54–0.63
Tin, bright tinned iron	76	0.043 and 0.064
Tungsten, filament	6000	0.39
Zinc, galvanized sheet iron, fairly bright	82	0.23
Refractories, building materials, paints, and miscellaneous		
Alumina (85–99.5% Al_2O_3, 0–12% SiO_2, 0–1% Ge_2O_3); effect of mean grain size, μm		
10 μm		0.30–0.18
50 μm		0.39–0.28
100 μm		0.50–0.40
Asbestos, board	74	0.96
Brick		
Red, rough, but no gross irregularities	70	0.93
Fireclay	1832	0.75
Carbon		
T-carbon (Gebruder Siemens) 0.9% ash, started with emissivity of 0.72 at 260 °F but on heating changed to values given	260–1160	0.81–0.79
Filament	1900–2560	0.526
Rough plate	212–608	0.77
Lampblack, rough deposit	212–932	0.84–0.78
Concrete tiles	1832	0.63
Enamel, white fused, on iron	66	0.90
Glass		
Smooth	72	0.94
Pyrex, lead, and soda	500–1000	0.95–0.85
Paints, lacquers, varnishes		
Snow-white enamel varnish on rough iron plate	73	0.906
Black shiny lacquer, sprayed on iron	76	0.875
Black shiny shellac on tinned iron sheet	70	0.821
Black matte shellac	170–295	0.91
Black or white lacquer	100–200	0.80–0.95
Flat black lacquer	100–200	0.96–0.98
Porcelain, glazed	72	0.92
Quartz, rough, fused	70	0.93
Roofing paper	69	0.91
Rubber, hard, glossy plate	74	0.94
Water	32–212	0.95–0.963

APPENDIX C

Conduction Shape Factors*

physical system	schematic	shape factor	restrictions
Isothermal sphere of radius r buried in **semi-infinite** medium having isothermal surface	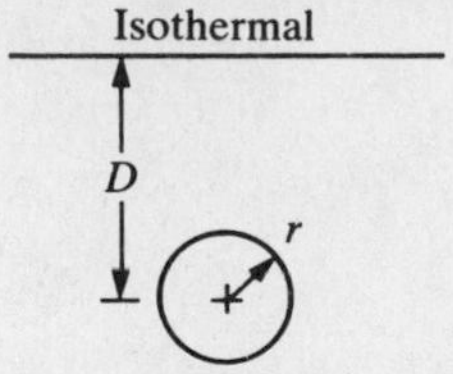	$\dfrac{4\pi r}{1 + r/2D}$	
Isothermal sphere buried in **semi-infinite** medium with insulated surface	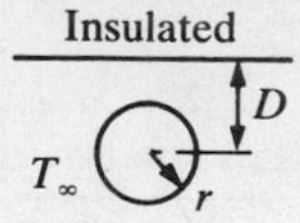	$\dfrac{4\pi r}{1 + r/2D}$	
Two isothermal spheres buried in infinite medium	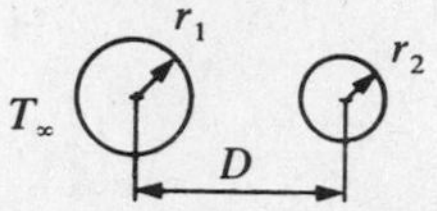	$\dfrac{4\pi}{\dfrac{r_2}{r_1}\left[1 - \dfrac{(r_1/D)^4}{1-(r_2/D)^2}\right] - \dfrac{2r_2}{D}}$	$D > 5r_{max}$
Thin rectangular plate of length L, buried in semi-infinite medium having isothermal surface	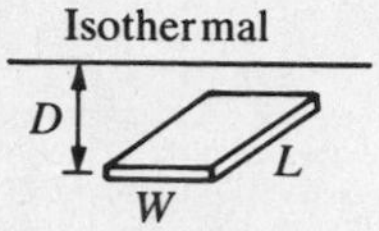	$\dfrac{\pi W}{\ln(4W/L)}$ $\dfrac{2\pi W}{\ln(4W/L)}$	$D = 0$ $D \gg W$
Eccentric cylinders of length L	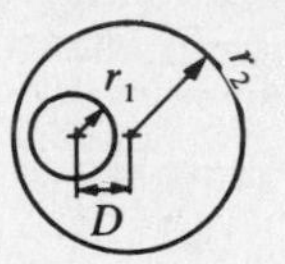	$\dfrac{2\pi L}{\cosh^{-1}\left(\dfrac{r_1^2 + r_2^2 - D^2}{2r_1r_2}\right)}$	$L \gg r_2$
Cylinder centered in a square of length L	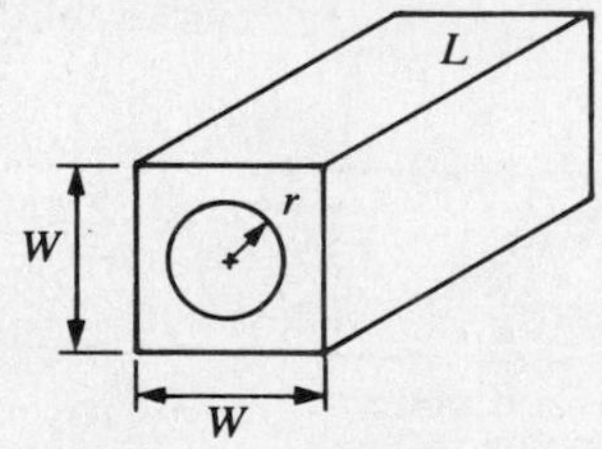	$\dfrac{2\pi L}{\ln(0.54W/r)}$	$L \gg W$
Isothermal cylinder of radius r buried in semi-infinite medium having isothermal surface	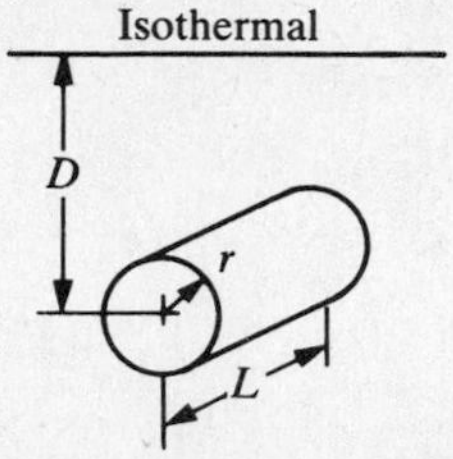	$\dfrac{2\pi L}{\cosh^{-1}(D/r)}$ $\dfrac{2\pi L}{\ln(2D/r)}$ $\dfrac{2\pi L}{\ln\dfrac{L}{r}\left\{\dfrac{\ln[L/(2D)]}{1-\ln(L/r)}\right\}}$	$L \gg r$ $L \gg r$, $D > 3r$ $D \gg r$, $L \gg D$
Isothermal sphere of radius r buried in **infinite** medium		$4\pi r$	

*Summarized from:

(1) R. V. Andrews, "Solving Conductive Heat Transfer Problems with Electrical-Analogue Shape Factors," *Chem. Eng. Prog.*, **51**(2):67 (1955).

(2) J. E. Sunderland and K. R. Johnson, "Shape Factors for Heat Conduction through Bodies with Isothermal or Convective Boundary Conditions," *Trans. ASHRAE*, **70**:237–241 (1964).

Conduction Shape Factors (continued)

physical system	schematic	shape factor	restrictions
Conduction between two isothermal cylinders buried in infinite medium	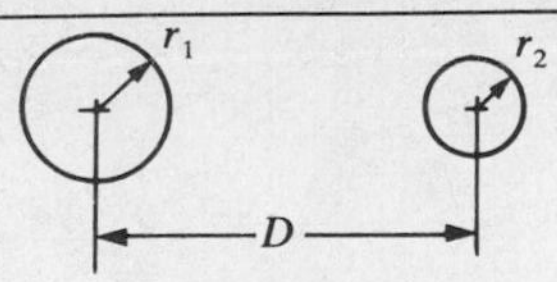	$\dfrac{2\pi L}{\cosh^{-1}\left(\dfrac{D^2 - r_1^2 - r_2^2}{2r_1r_2}\right)}$	$L \gg r$ $L \gg D$
Isothermal cylinder of radius r placed in semi-infinite medium as shown	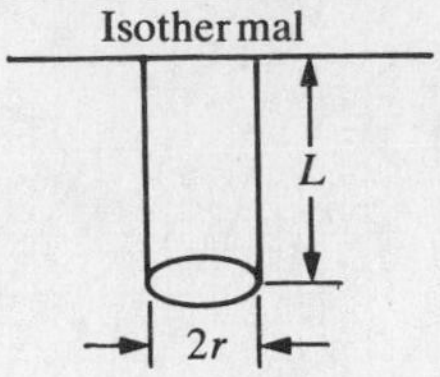	$\dfrac{2\pi L}{\ln(2L/r)}$	$L \gg 2r$
Isothermal rectangular parallelepiped buried in semi-infinite medium having isothermal surface	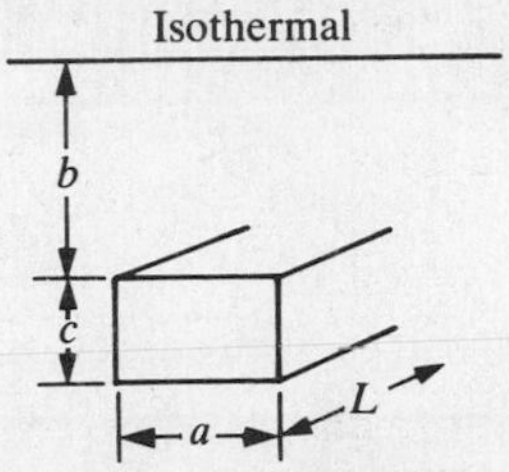	$1.685L\left[\log\left(1+\dfrac{b}{a}\right)\right]^{-0.59}$ $\times\left(\dfrac{b}{c}\right)^{-0.078}$	See Andrews
Plane wall	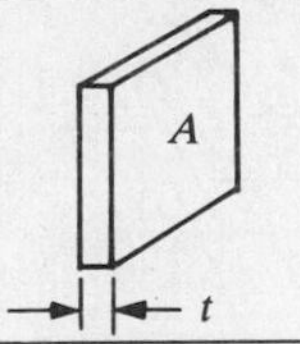	$\dfrac{A}{t}$	One-dimensional heat flow
Hollow cylinder, length L	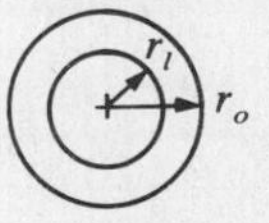	$\dfrac{2\pi L}{\ln(r_o/r_i)}$	$L \gg r$
Hollow sphere	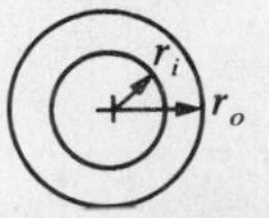	$\dfrac{4\pi r_o r_i}{r_o - r_i}$	
Conduction through an edge formed by intersection of two plane walls with inner wall temperature T_1 and outer wall temperature T_2	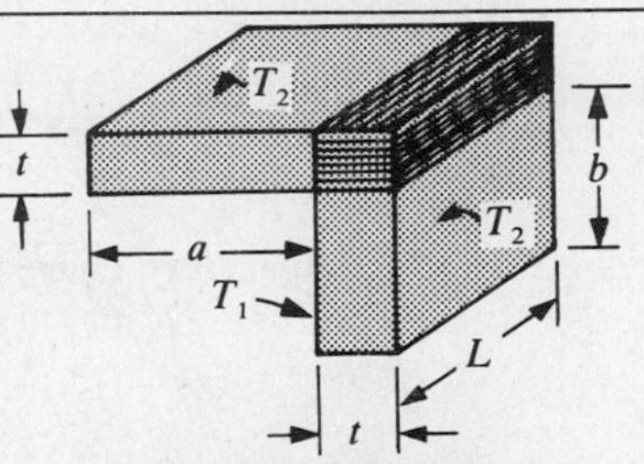	$S = 0.54L$	$a > t/5$ $b > t/5$
Conduction through a corner at intersection of three plane walls, each of thickness t, with uniform inner temperature T_1 and outer temperature T_2	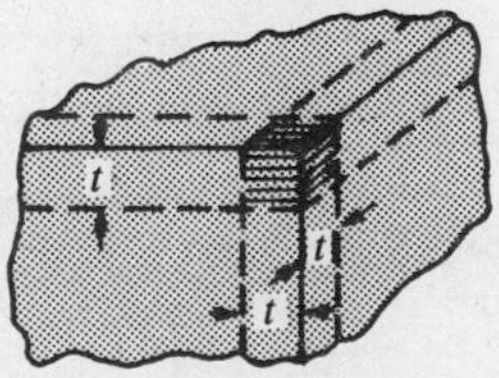	$S = 0.15t$	Inside dimensions $> t/5$

APPENDIX D

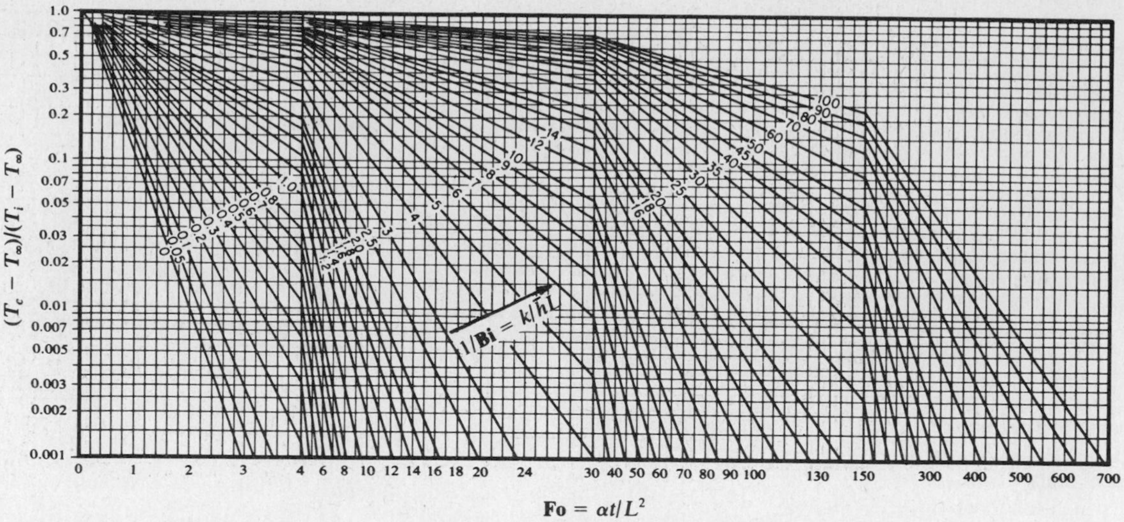

Figure D-1 Centerline temperature for a slab. [From M. P. Heisler, *Trans. ASME*, **69**:227 (1947).]

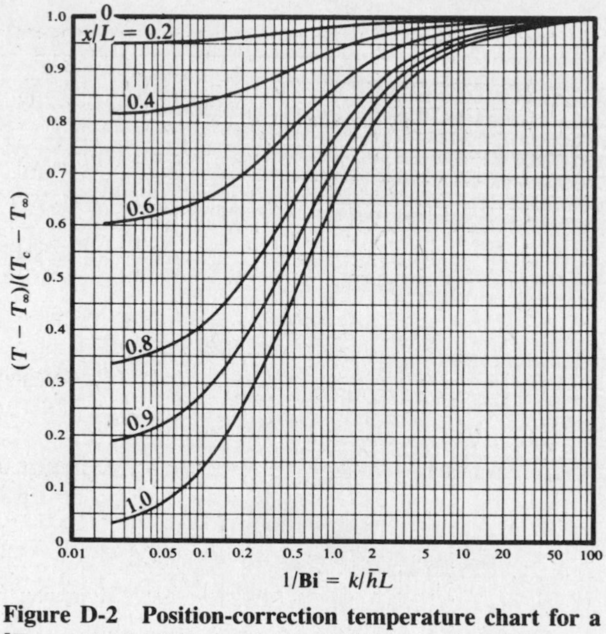

Figure D-2 Position-correction temperature chart for a slab. [From M. P. Heisler, *Trans. ASME*, **69**:227 (1947).]

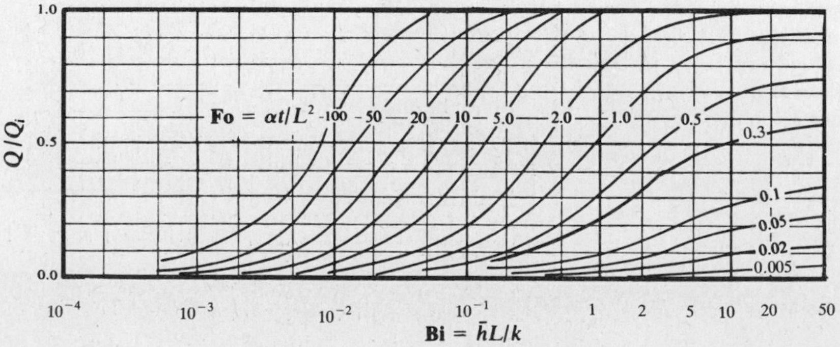

Figure D-3 Total heat flow, slab.

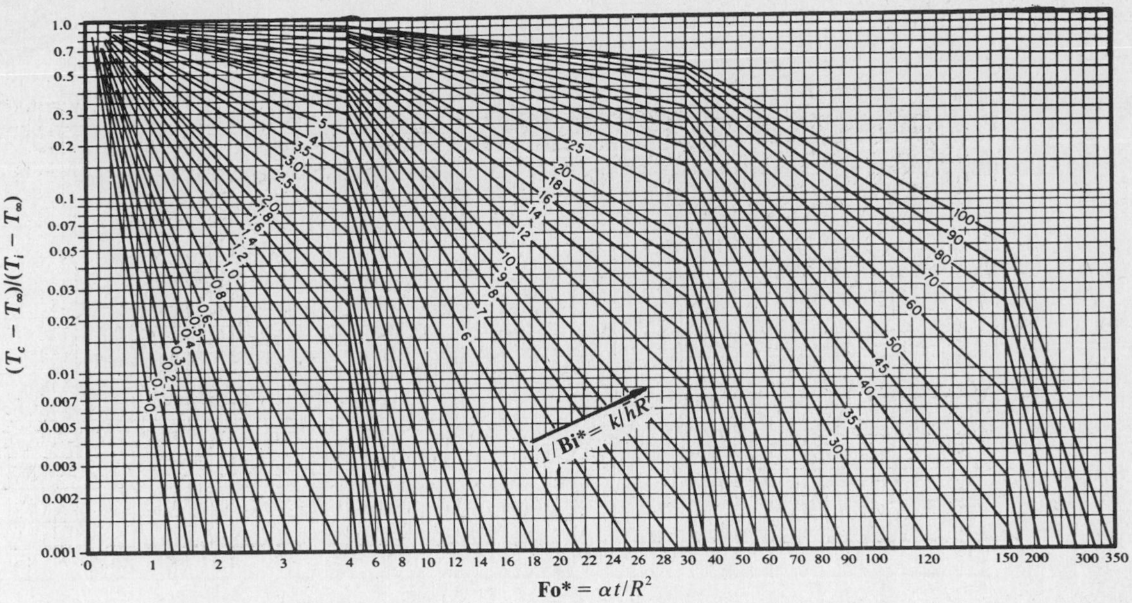

Figure D-4 Centerline temperature for a solid cylinder. [From M. P. Heisler, *Trans. ASME*, **69**:227 (1947).]

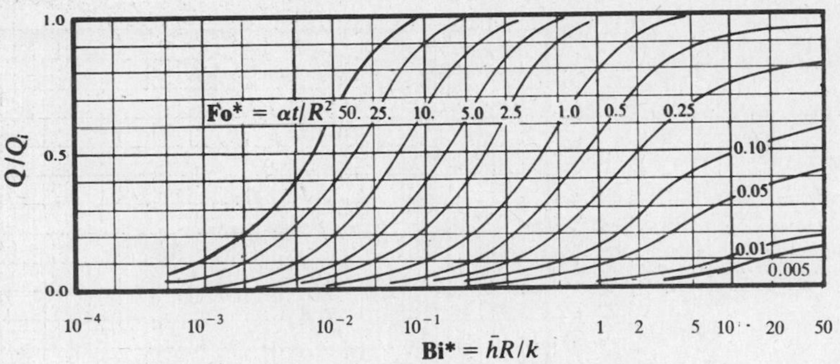

Figure D-5 Total heat flow, solid cylinder.

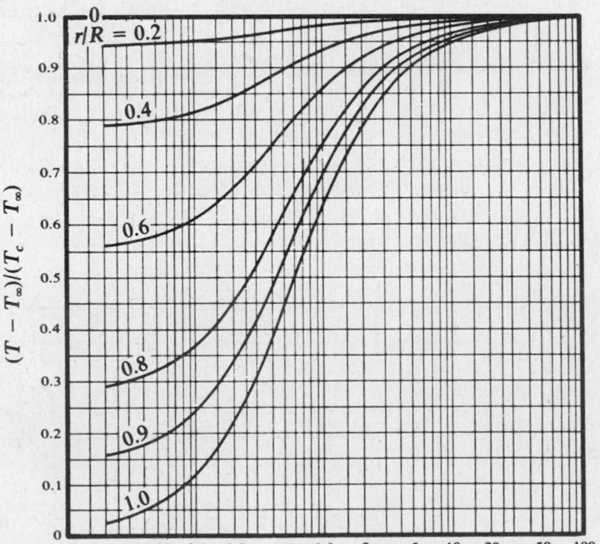

Figure D-6 Position-correction temperature chart for an infinitely long cylinder. [From M. P. Heisler, *Trans. ASME*, **69**:227 (1947).]

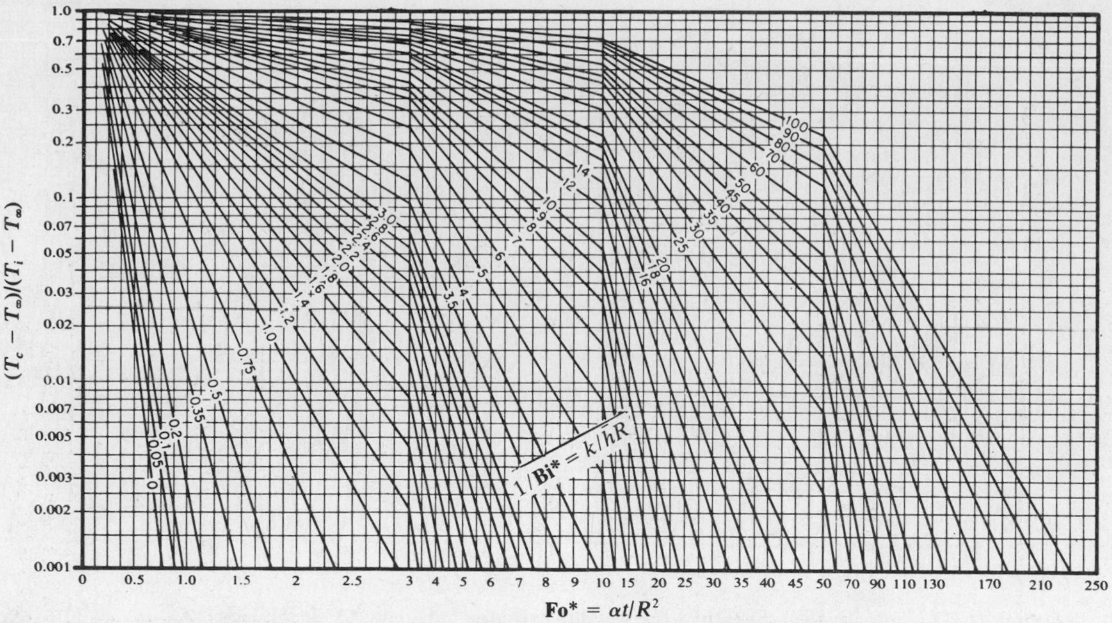

Figure D-7 Center temperature of a solid sphere. [From M. P. Heisler, *Trans. ASME*, **69**:227 (1947).]

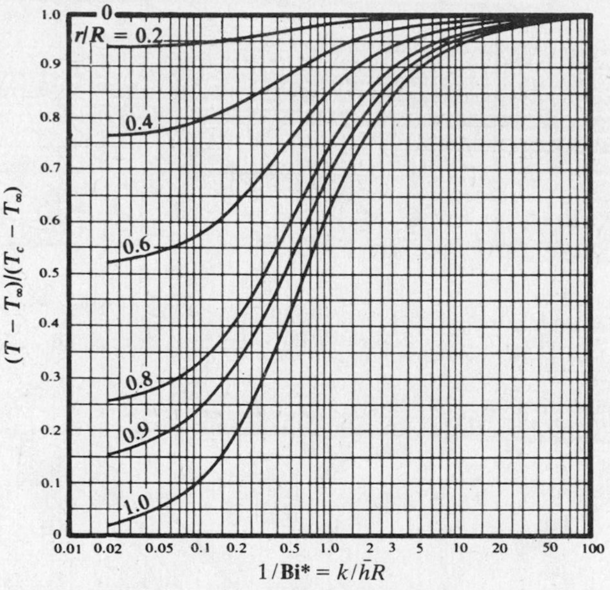

Figure D-8 Position-correction temperature chart for a solid sphere. [From M. P. Heisler, *Trans. ASME*, **69**:227 (1947).]

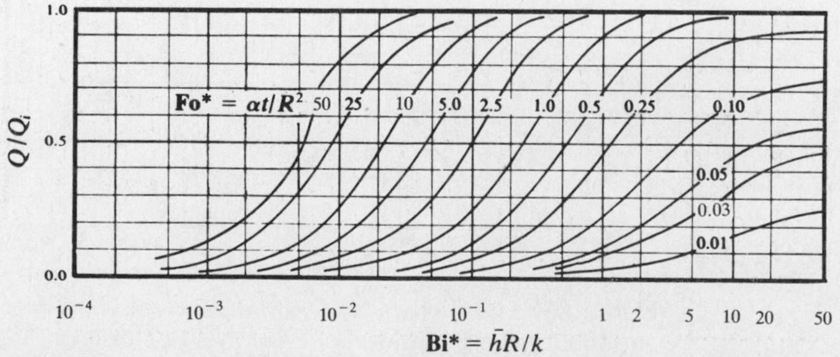

Figure D-9 Total heat flow, solid sphere.

TABLE D-1 Summary of Transient, Two-Dimensional Finite-Difference Equations ($\Delta x = \Delta y$)

	explicit method		
configuration	finite-difference equation	stability criterion	implicit method
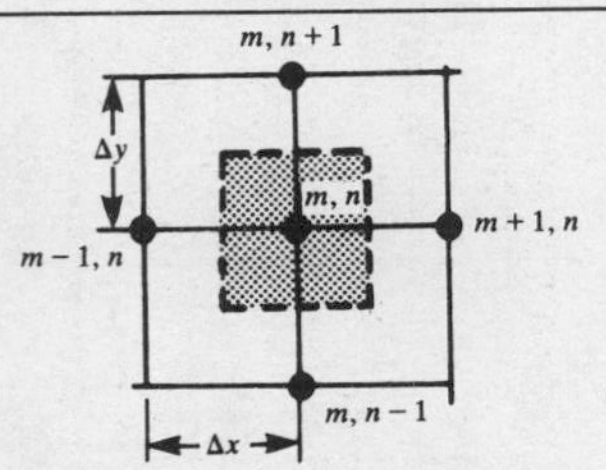	$T^{p+1}_{m,n} = \mathbf{Fo}\,(T^p_{m+1,n} + T^p_{m-1,n} + T^p_{m,n+1} + T^p_{m,n-1}) + (1 - 4\,\mathbf{Fo})T^p_{m,n}$ 1. Interior node	$\mathbf{Fo} \le \frac{1}{4}$	$(1 + 4\,\mathbf{Fo})T^{p+1}_{m,n} - \mathbf{Fo}\,(T^{p+1}_{m+1,n} + T^{p+1}_{m-1,n} + T^{p+1}_{m,n+1} + T^{p+1}_{m,n-1}) = T^p_{m,n}$
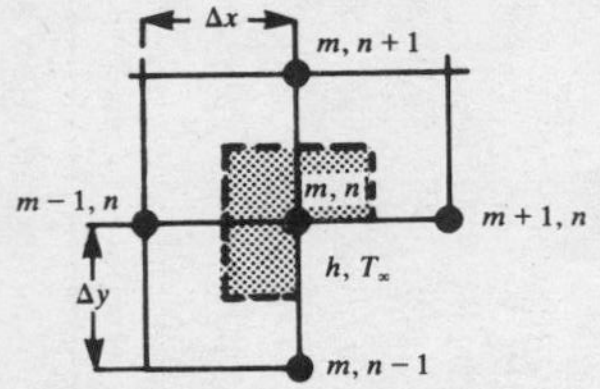	$T^{p+1}_{m,n} = \frac{2}{3}\,\mathbf{Fo}\,(T^p_{m+1,n} + 2T^p_{m-1,n} + 2T^p_{m,n+1} + T^p_{m,n-1} + 2\,\mathbf{Bi}\,T_\infty) + (1 - 4\,\mathbf{Fo} - \frac{4}{3}\,\mathbf{Bi\,Fo})T^p_{m,n}$ 2. Node at interior corner with convection	$\mathbf{Fo}\,(3 + \mathbf{Bi}) \le \frac{3}{4}$	$[1 + 4\,\mathbf{Fo}\,(1 + \frac{1}{3}\,\mathbf{Bi})]T^{p+1}_{m,n} - \frac{2}{3}\,\mathbf{Fo}\,(T^{p+1}_{m+1,n} + 2T^{p+1}_{m-1,n} + 2T^{p+1}_{m,n+1} + T^{p+1}_{m,n-1}) = T^p_{m,n} + \frac{4}{3}\,\mathbf{Bi\,Fo}\,T_\infty$
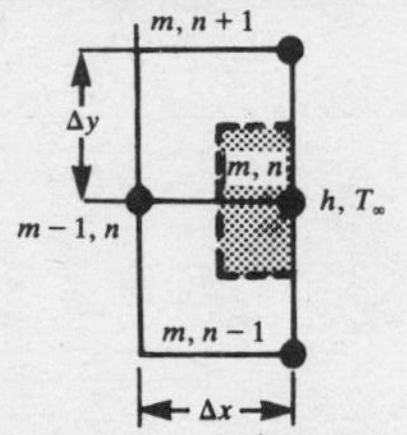	$T^{p+1}_{m,n} = \mathbf{Fo}\,(2T^p_{m-1,n} + T^p_{m,n+1} + T^p_{m,n-1} + 2\,\mathbf{Bi}\,T_\infty) + (1 - 4\,\mathbf{Fo} - 2\,\mathbf{Bi\,Fo})T^p_{m,n}$ 3. Node at plane surface with convection*	$\mathbf{Fo}\,(2 + \mathbf{Bi}) \le \frac{1}{2}$	$[1 + 2\,\mathbf{Fo}\,(2 + \mathbf{Bi})]T^{p+1}_{m,n} - \mathbf{Fo}\,(2T^{p+1}_{m-1,n} + T^{p+1}_{m,n+1} + T^{p+1}_{m,n-1}) = T^p_{m,n} + 2\,\mathbf{Bi\,Fo}\,T_\infty$
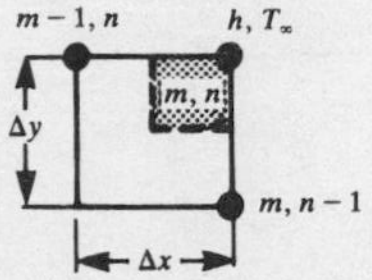	$T^{p+1}_{m,n} = 2\,\mathbf{Fo}\,(T^p_{m-1,n} + T^p_{m,n-1} + 2\,\mathbf{Bi}\,T_\infty) + (1 - 4\,\mathbf{Fo} - 4\,\mathbf{Bi\,Fo})T^p_{m,n}$ 4. Node at exterior corner with convection	$\mathbf{Fo}\,(1 + \mathbf{Bi}) \le \frac{1}{4}$	$[1 + 4\,\mathbf{Fo}\,(1 + \mathbf{Bi})]T^{p+1}_{m,n} - 2\,\mathbf{Fo}\,(T^{p+1}_{m-1,n} + T^{p+1}_{m,n-1}) = T^p_{m,n} + 4\,\mathbf{Bi\,Fo}\,T_\infty$

*To obtain the finite-difference equation and/or stability criterion for an adiabatic surface (or surface of symmetry), simply set **Bi** equal to zero. Adapted by permission from Frank P. Incropera and David P. DeWitt, *Fundamentals of Heat and Mass Transfer*, 2d ed., Copyright 1985 by John Wiley & Sons, Inc., New York.

APPENDIX E

TABLE E-1 Planck Radiation Functions

λT, μm·°R	$\varphi(\lambda T)$	λT, μm·°R	$\varphi(\lambda T)$
1 000	0.0000	10 400	0.7183
1 200	0.0000	10 600	0.7284
1 400	0.0000	10 800	0.7380
1 600	0.0000	11 000	0.7472
1 800	0.0003	11 200	0.7561
2 000	0.0010	11 400	0.7645
2 200	0.0025	11 600	0.7726
2 400	0.0053	11 800	0.7803
2 600	0.0098	12 000	0.7878
2 800	0.0164	12 200	0.7949
3 000	0.0254	12 400	0.8017
3 200	0.0368	12 600	0.8082
3 400	0.0507	12 800	0.8145
3 600	0.0668	13 000	0.8205
3 800	0.0851	13 200	0.8263
4 000	0.1052	13 400	0.8318
4 200	0.1269	13 600	0.8371
4 400	0.1498	13 800	0.8422
4 600	0.1736	14 000	0.8471
4 800	0.1982	14 200	0.8518
5 000	0.2232	14 400	0.8564
5 200	0.2483	14 600	0.8607
5 400	0.2735	14 800	0.8649
5 600	0.2986	15 000	0.8689
5 800	0.3234	16 000	0.8869
6 000	0.3477	17 000	0.9018
6 200	0.3715	18 000	0.9142
6 400	0.3948	19 000	0.9247
6 600	0.4174	20 000	0.9335
6 800	0.4394	21 000	0.9411
7 000	0.4607	22 000	0.9475
7 200	0.4812	23 000	0.9531
7 400	0.5010	24 000	0.9579
7 600	0.5201	25 000	0.9621
7 800	0.5384	26 000	0.9657
8 000	0.5561	27 000	0.9689
8 200	0.5730	28 000	0.9717
8 400	0.5892	29 000	0.9742
8 600	0.6048	30 000	0.9764
8 800	0.6197	40 000	0.9891
9 000	0.6340	50 000	0.9941
9 200	0.6477	60 000	0.9965
9 400	0.6608	70 000	0.9977
9 600	0.6733	80 000	0.9984
9 800	0.6853	90 000	0.9989
10 000	0.6968	100 000	0.9991
10 200	0.7078		

Adapted from R. V. Dunkle, "Thermoradiation Tables and Applications," *Trans. ASME*, **76**:549–552 (1954).

TABLE E-2

configuration	equivalent beam length, L
Space between infinite planes	1.8 × distance between planes
Sphere	$\frac{2}{3}$ × diameter
Infinitely long cylinder	1 × diameter
Cube	$\frac{2}{3}$ × side
Arbitrary surface	≈3.6 × (volume/surface area)

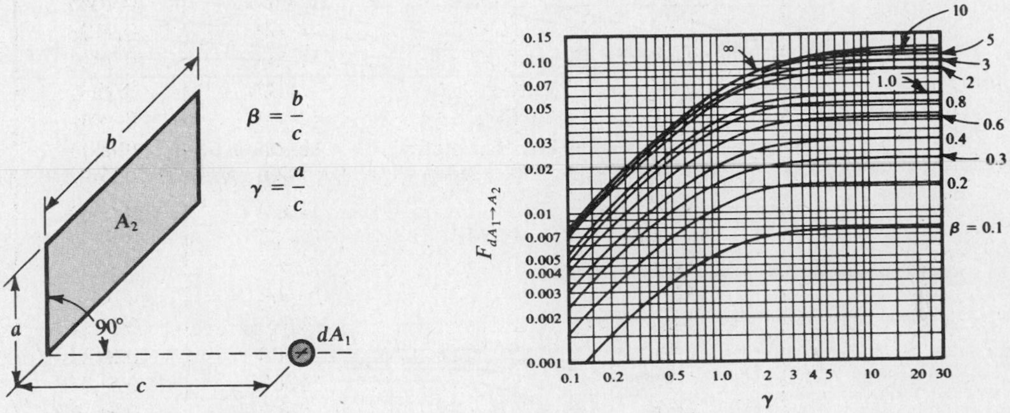

Figure E-1 Spherical point source to a plane rectangle. (Adapted from D. C. Hamilton and W. R. Morgan, *NACA Tech. Note TN-2836*, 1952.)

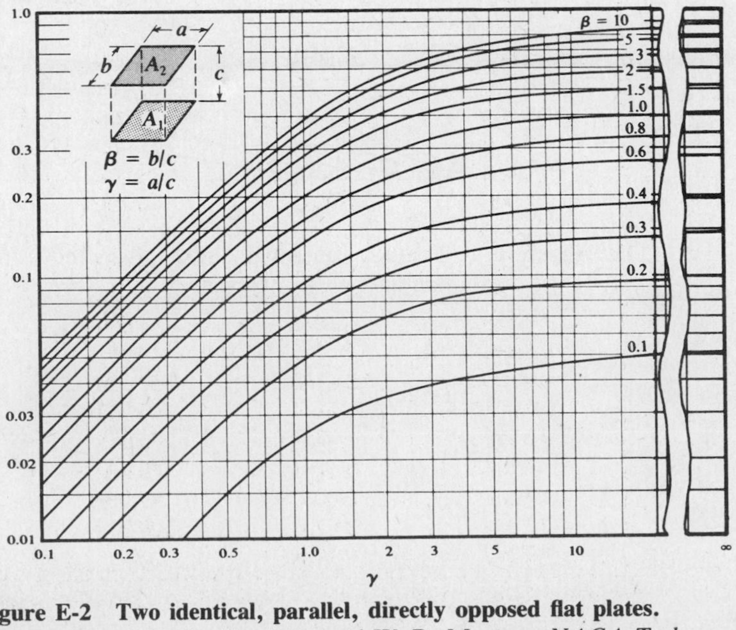

Figure E-2 Two identical, parallel, directly opposed flat plates. (Adapted from D. C. Hamilton and W. R. Morgan, *NACA Tech. Note TN-2836*, 1952.)

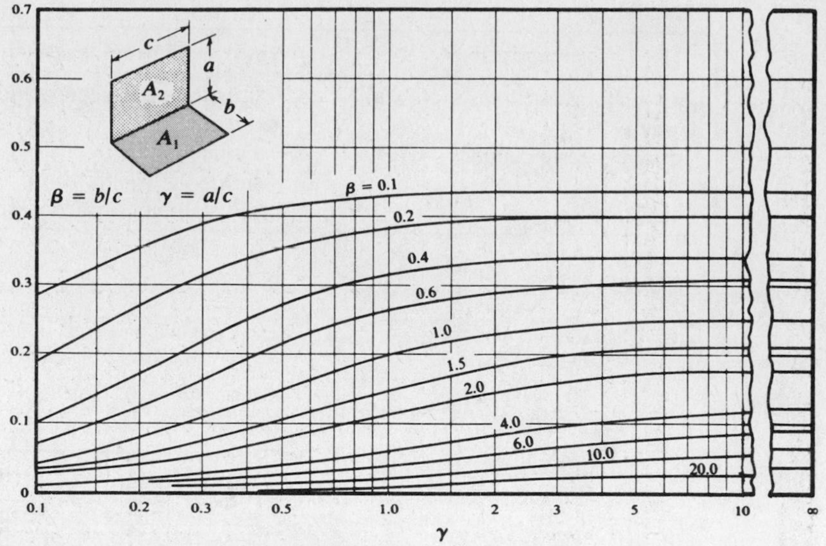

Figure E-3 Two perpendicular flat plates with a common edge. (Adapted from D. C. Hamilton and W. R. Morgan, *NACA Tech. Note TN-2836*, 1952.)

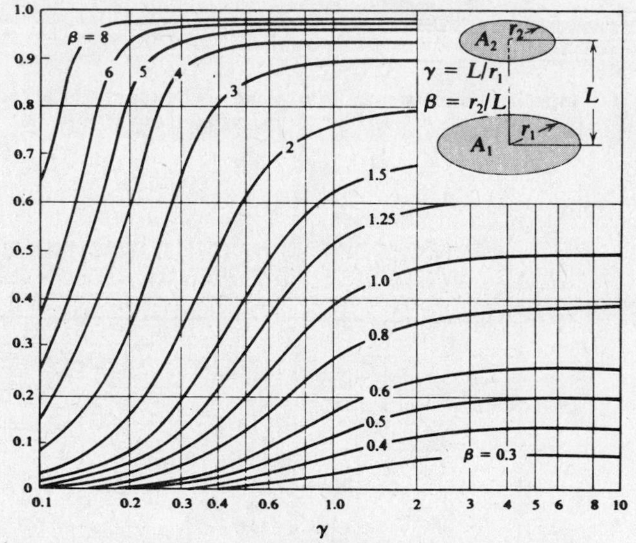

Figure E-4 Parallel, concentric disks. (Adapted from D. C. Hamilton and W. R. Morgan, *NACA Tech. Note TN-2836*, 1952.)

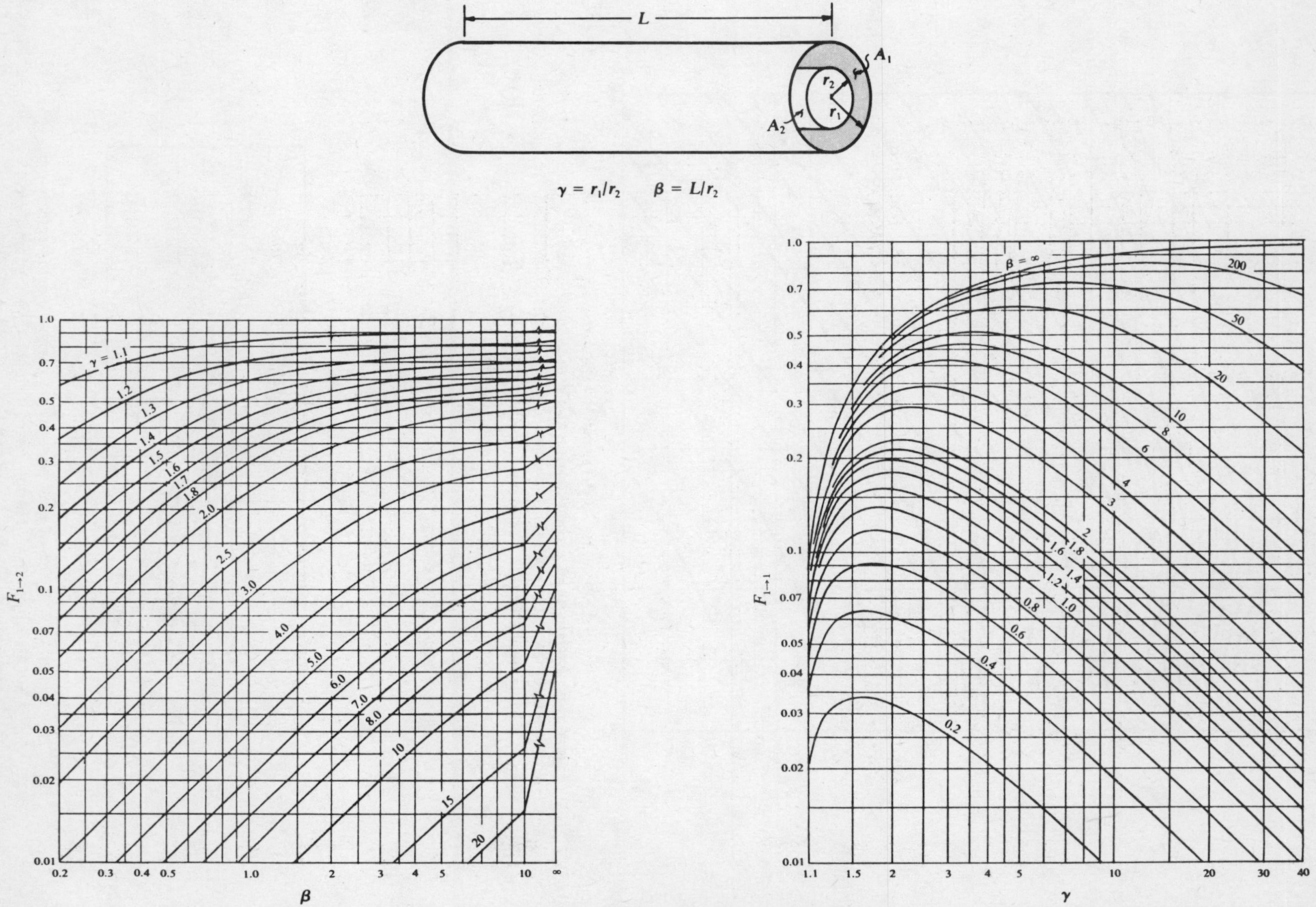

Figure E-5 Concentric cylinders of finite length. (Adapted from D. C. Hamilton and W. R. Morgan, *NACA Tech. Note TN-2836*, 1952.)

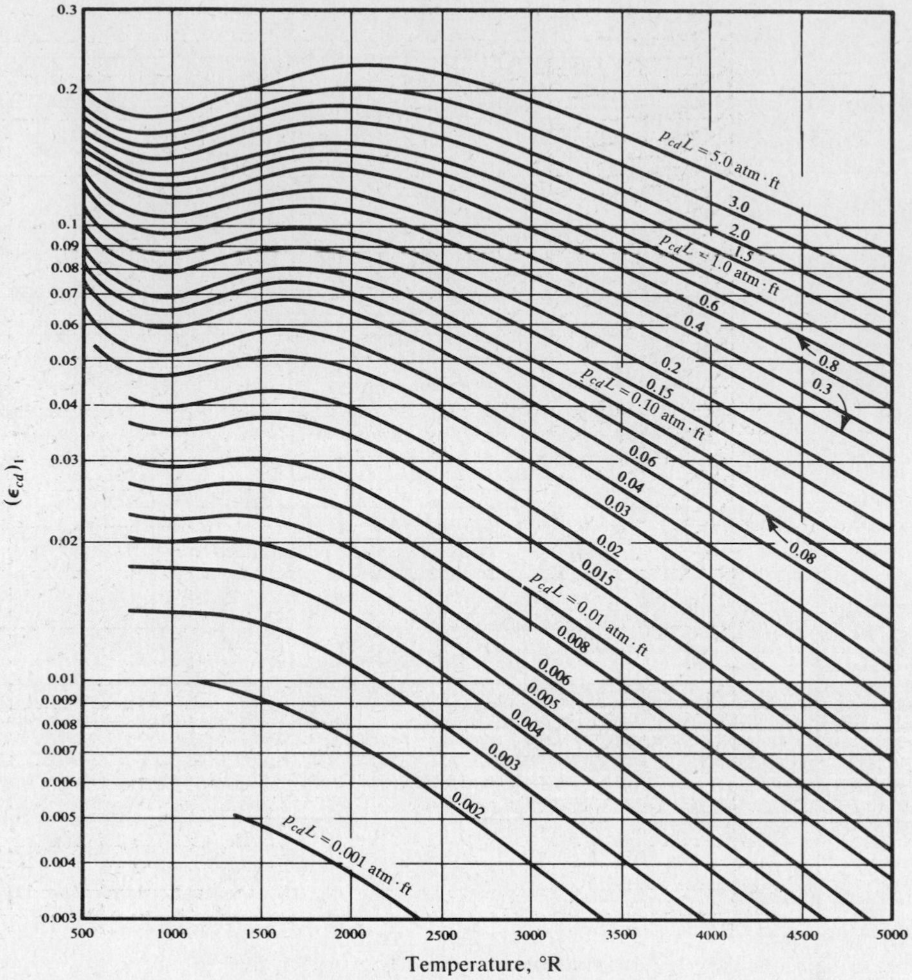

Figure E-6 Emissivity of carbon dioxide at 1 atm total pressure. (Adapted from H. C. Hottel, chap. 4 in W. C. McAdams, *Heat Transmission*, 3d ed. Copyright 1954. McGraw-Hill Book Company. Used by permission.)

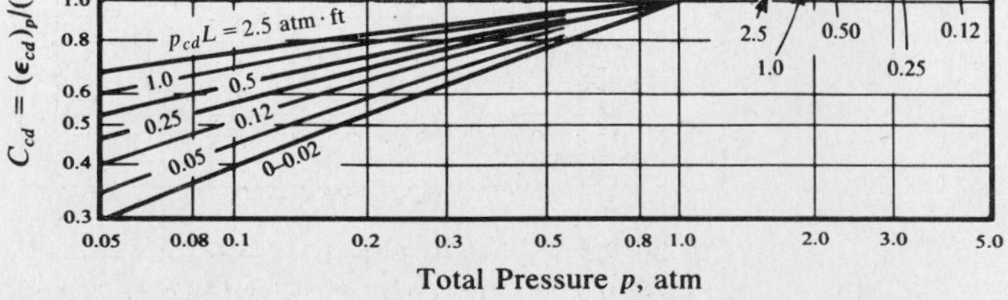

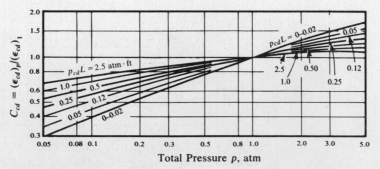

Figure E-7 Effect of total pressure on carbon dioxide emissivity. (Adapted from H.C. Hottel, chap. 4 in W. C. McAdams, *Heat Transmission*, 3d ed. Copyright 1954. McGraw-Hill Book Company. Used by permission.)

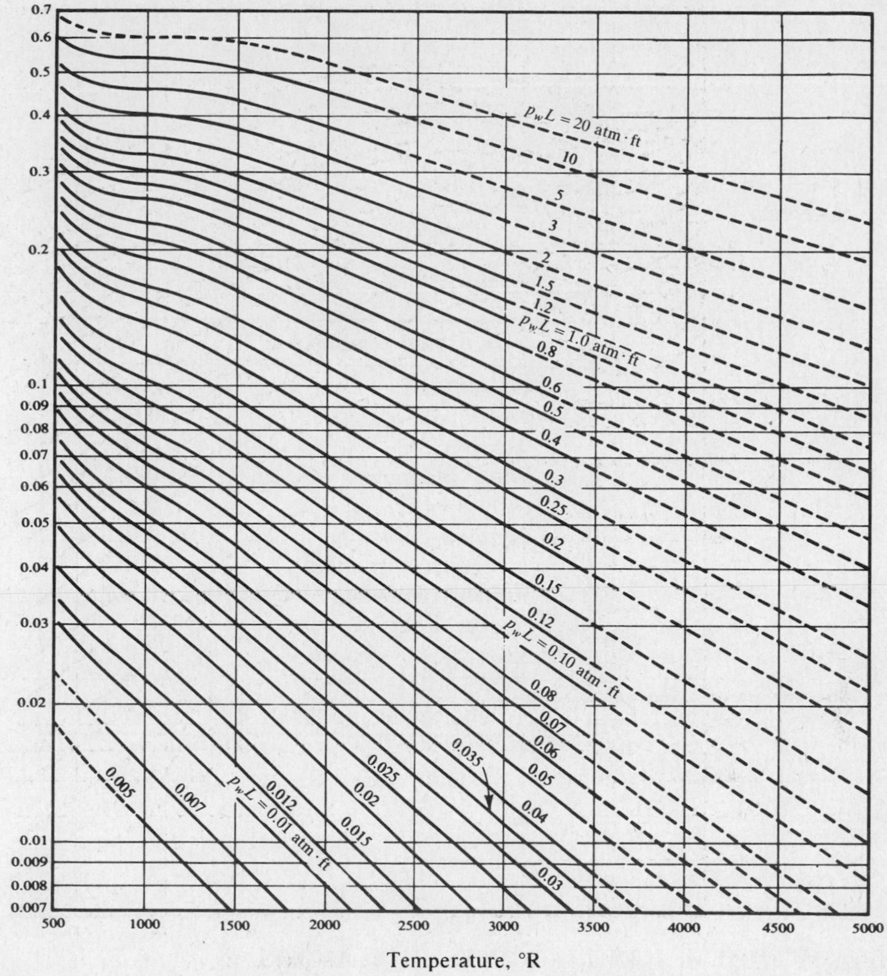

Figure E-8 Emissivity of hypothetical water vapor system at 1 atm total pressure. (Adapted from H. C. Hottel, chap. 4 in W. C. McAdams, *Heat Transmission*, 3d ed. Copyright 1954. McGraw-Hill Book Company. Used by permission.)

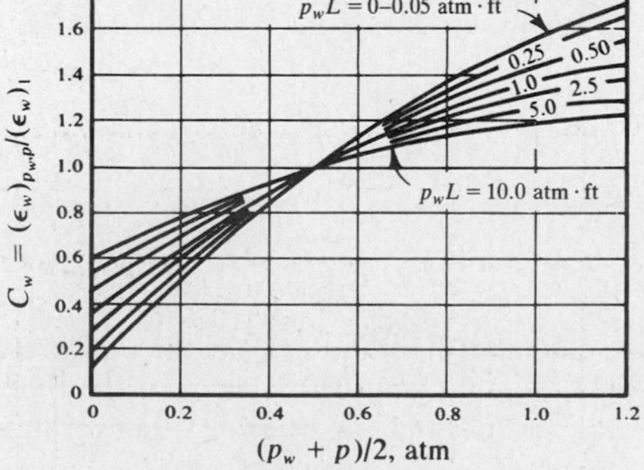

Figure E-9 Effect of partial and total pressures on emissivity of water vapor. (Adapted from H. C. Hottel, chap. 4 in W. C. McAdams, *Heat Transmission*, 3d ed. Copyright 1954. McGraw-Hill Book Company. Used by permission.)

APPENDIX F

(In any of the stated formulas that involves g_c, British units are understood. The formula becomes valid in the SI if g_c is replaced by 1.)

TABLE F-1

surface–fluid combination	C_{sf}
Water–brass	0.006
Water–copper	0.013
Water–nickel	0.006
Water–platinum	0.013
CCl_4–copper	0.013
Benzene–chromium	0.010
n-Pentane–chromium	0.015
Ethyl alcohol–chromium	0.0027
Isopropyl alcohol–copper	0.0025
35% K_2CO_3–copper	0.0054
50% K_2CO_3–copper	0.0027
n-Butyl alcohol–copper	0.0030

F.1 NONFLOW FILM BOILING (REGIMES IV, V, AND VI)

Horizontal Tube

$$h = h_c\left(\frac{h_c}{h}\right)^{1/3} + h_r \quad \text{(an implicit equation for } h\text{)} \tag{F-1}$$

$$h_c = (0.62)\left[\frac{k_{vf}^3 \rho_{vf}(\rho_l - \rho_{vf}) g (h_{fg} + 0.4 c_{pvf}\,\Delta T)}{D(\mu_{vf} g_c)\,\Delta T}\right]^{1/4} \tag{F-2}$$

$$h_r = \frac{\sigma\epsilon(T_s^4 - T_{sat}^4)}{T_s - T_{sat}} \tag{F-3}$$

±0.3 percent:
$$h \approx h_c + h_r\left[\frac{3}{4} + \frac{1}{4}\left(\frac{h_r}{h_c}\right)\left(\frac{1}{2.62 + h_r/h_c}\right)\right] \quad \left(0 < \frac{h_r}{h_c} < 10\right) \tag{F-4}$$

±5 percent:
$$h \approx h_c + \frac{3}{4}h_r \quad \left(\frac{h_r}{h_c} < 1\right) \tag{F-5}$$

In (*F-3*) σ is the Stefan–Boltzmann constant and ϵ is the emissivity of the surface; T_s and T_{sat} must be on an absolute scale. In (*F-2*), D is the outside diameter of the tube, and, as indicated by the additional subscript f, vapor properties are taken at the mean film temperature, $T_f = (T_s + T_{sat})/2$.

Excess Temperature at Point of Minimum Heat Flux (Point C, Fig. 8-1)

$$\Delta T_C = (0.127)\left(\frac{\rho_{vf} h_{fg}}{k_{vf}}\right)\left[\frac{g(\rho_l - \rho_v)}{\rho_l + \rho_v}\right]^{2/3}\left[\frac{\sigma}{(g/g_c)(\rho_l - \rho_v)}\right]^{1/2}\left[\frac{\mu_f g_c}{g_0(\rho_l - \rho_v)}\right]^{1/3} \tag{F-6}$$

Properties designated by a subscript f in (*F-6*) are evaluated at the mean film temperature, $T_f = (T_s + T_{sat})/2$; g_0 is the earth's standard gravitation, 32.17 ft/sec^2 or 9.81 m/s^2.

F.2 FLOW FILM BOILING

Horizontal Tube

$$h_c = (2.7)\left[\frac{V_\infty k_{vf}\rho_{vf}(h_{fg} + 0.4c_{pf}\,\Delta T)}{D\,\Delta T}\right]^{1/2} \qquad V_\infty > 2\sqrt{gD} \tag{F-7}$$

Use (*F-1*), (*F-2*), and (*F-3*) when $V_\infty < 2\sqrt{gD}$.

F.3 LAMINAR FILM CONDENSATION

Vertical and Inclined Surfaces (ϕ = angle with horizontal)

$$\bar{h} = (1.13)\left[\frac{\rho_l g(\rho_l - \rho_v)h_{fg}^* k_l^3}{(\mu_l g_c)L(T_{sat} - T_s)}\sin\phi\right]^{1/4} \tag{F-8}$$

where

$$h_{fg}^* = h_{fg} + \tfrac{3}{8}c_{pl}(T_{sat} - T_s) \tag{F-9}$$

Vertical tubes. Equation (*F-8*), with $\sin\phi = 1$, is also valid for the inside and outside surfaces of vertical tubes if the tubes are large in diameter D compared to the film thickness δ. However, (*F-8*) is not valid for inclined tubes, since the film flow would not be parallel to the axis of the tube.

Horizontal tubes

$$\bar{h} = (0.725)\left[\frac{\rho_l g(\rho_l - \rho_v)h_{fg}^* k_l^3}{(\mu_l g_c)D(T_{sat} - T_s)}\right]^{1/4} \tag{F-10}$$

When condensation takes place in a bank of n horizontal tubes arranged in a vertical tier, the condensate from an upper tube flows onto lower tubes, affecting the heat transfer rate. In this case, an estimate can be made of the heat transfer, in the absence of empirical relations which account for splashing and other effects, by replacing D in (*F-10*) by nD.

When a liquid film is vigorous enough, heat is transferred not only by conduction but also by eddy diffusion, a characteristic of turbulence. This may occur on tall vertical surfaces or in banks of horizontal tubes. When such behavior occurs the laminar relations are no longer valid. This change occurs when the film Reynolds number, $\mathbf{Re}_f$, defined by

$$\mathbf{Re}_f \equiv \frac{VD_h\rho_l}{\mu_l g_c} = \frac{4\rho_l AV}{P(\mu_l g_c)} \tag{F-11}$$

is approximately 1800; here the hydraulic diameter, $D_h \equiv 4A/P$, is the characteristic length. In this relation, A is the area over which the condensate flows and P is the wetted perimeter. For inclined surfaces of width W, $A/P = LW/W = L$; for vertical tubes, $A/P = \pi DL/\pi D = L$; and, for horizontal tubes, $A/P = \pi DL/L = \pi D$.

Noting that $\dot{m} = \rho_l AV$ and that $\dot{m}' = \dot{m}/P$, the film Reynolds number may be expressed as

$$\mathbf{Re}_f = \frac{4\dot{m}'}{\mu_l g_c} = \frac{\bar{h}D_h(T_{sat} - T_s)}{h_{fg}^*(\mu_l g_c)} \tag{F-12}$$

where $\dot{m}'$ is the condensate mass flow per unit width for surfaces or per unit length for tubes. Its maximum value occurs at the lower edge of the surface.

Turbulent film condensation on vertical surfaces

$$\bar{h} = (0.0076)\,\mathbf{Re}_f^{0.4}\left[\frac{\rho_l g(\rho_l - \rho_v)k_l^3}{(\mu_l g_c)^2}\right]^{1/3} \tag{F-13}$$

which is valid for $\mathbf{Re}_f > 1800$.

APPENDIX G
Equations Used in Chapter 5

G.1 FLAT PLATE IN LAMINAR FLOW

A. Local h_x: $$\mathbf{Nu}_x = (0.332)\,\mathbf{Re}_x^{1/2}\,\mathbf{Pr}^{1/3} \qquad (G\text{-}1)$$

Valid for: $\mathbf{Re}_x < \mathbf{Re}_{\text{crit}}$ gases or liquids

$\mathbf{Pr} > 0.6$ $\quad T_s = \text{const.}$

Evaluate properties at $T_f = (T_s + T_\infty)/2$.

B. Average $\bar{h}$: $$\overline{\mathbf{Nu}} \equiv \frac{\bar{h}L}{k} = (0.664)\,\mathbf{Re}_L^{1/2}\,\mathbf{Pr}^{1/3} \qquad (G\text{-}2)$$

Valid for: $\mathbf{Re}_x < \mathbf{Re}_{\text{crit}}$ gases or liquids

$\mathbf{Pr} > 0.6$ $\quad T_s = \text{const.}$

Evaluate properties at $T_f = (T_s + T_\infty)/2$.

C. Local h_x: $$\mathbf{Nu}_x = (0.332)\,\mathbf{Re}_x^{1/2}\,\mathbf{Pr}^{1/3}\left[1 - \left(\frac{x_i}{x}\right)^{3/4}\right]^{-1/3} \qquad (G\text{-}3)$$

Valid for: Unheated starting length of surface, x_i.

$\mathbf{Re}_x < \mathbf{Re}_{\text{crit}}$ gases or liquids

$\mathbf{Pr} > 0.6$ $\quad T_s = \text{const.}$

Evaluate properties at $T_f = (T_s + T_\infty)/2$.

D. Local h_x: $$\mathbf{Nu}_x = (0.453)\,\mathbf{Re}_x^{1/2}\,\mathbf{Pr}^{1/3} \qquad (G\text{-}4)$$

Valid for: $\mathbf{Re}_x < \mathbf{Re}_{\text{crit}}$ gases or liquids

$0.6 < \mathbf{Pr} < 50$ $\quad q_s = \text{const.}$

Evaluate properties at $T_f = (T_s + T_\infty)/2$.

G.2 FLAT PLATE IN MIXED FLOW

A. Local h_x: $$\mathbf{Nu}_x = (0.0296)\,\mathbf{Re}_x^{4/5}\,\mathbf{Pr}^{1/3} \qquad (G\text{-}5)$$

Valid for: $\mathbf{Re}_x < 10^7$ gases or liquids, turbulent

$0.6 < \mathbf{Pr} < 60$ $\quad T_s = \text{const.}$

Evaluate properties at $T_f = (T_s + T_\infty)/2$.

B. Local h_x:

$$\mathbf{Nu}_x = \frac{(0.0296)\,\mathbf{Re}_x^{4/5}\,\mathbf{Pr}}{1 + (0.860)\,\mathbf{Re}_x^{-1/10}\left[(\mathbf{Pr} - 1) + \ln\left(\frac{5\,\mathbf{Pr} + 1}{6}\right)\right]} \qquad (G\text{-}6)$$

Valid for: $\mathbf{Re}_x < 10^7$ gases or liquids, turbulent

$0.6 < \mathbf{Pr} < 60$ $\quad T_s = \text{const.}$

Evaluate properties at $T_f = (T_s + T_\infty)/2$.

C. Average $\bar{h}$: $$\overline{\mathbf{Nu}} \equiv \frac{\bar{h}L}{k} = \mathbf{Pr}^{1/3}\,(0.037\,\mathbf{Re}_L^{4/5} - \mathscr{A}) \qquad (G\text{-}7)$$

Valid for: $\mathbf{Re}_L < 10^7$ gases or liquids, mixed laminar/turbulent

$0.6 < \mathbf{Pr} < 60$ $\quad T_s = \text{const.}$ $\quad \mathscr{A}$-values from Problem 5.71

Evaluate properties at $T_f = (T_s + T_\infty)/2$.

D. Average $\bar{h}$:

$$\overline{\mathbf{Nu}} \equiv \frac{\bar{h}L}{k} = (0.036)\,\mathbf{Pr}^{0.43}\,(\mathbf{Re}_L^{0.8} - 9200)\left(\frac{\mu_\infty}{\mu}\right)^{1/4} \qquad (G\text{-}8)$$

Valid for: $10^5 < \mathbf{Re}_L < 5.5 \times 10^6$ mixed laminar/turbulent

$0.7 < \mathbf{Pr} < 380$, $T_s = \text{const.}$ liquids

$$0.26 < \frac{\mu_\infty}{\mu_s} < 3.5$$

Evaluate all properties except μ_s at T_∞.

E. Local h_x:

$$\mathbf{Nu}_x = 1.04(\mathbf{Nu}_x)_{T_s=\text{const.}} \qquad (G\text{-}9)$$

where $(\mathbf{Nu}_x)_{T_s=\text{const.}}$ is by equation (*G-5*) or (*G-6*).

Valid for: $\mathbf{Re}_x < 10^7$ gases or liquids, turbulent

$0.6 < \mathbf{Pr} < 60$ $q_s = \text{const.}$

Evaluate properties at $T_f = (T_s + T_\infty)/2$.

G.3 CROSS FLOW OVER A CYLINDER

A. Average $\bar{h}$:

$$\overline{\mathbf{Nu}} \equiv \frac{\bar{h}D}{k} = C\,\mathbf{Pr}^{1/3}\,\mathbf{Re}_D^n \qquad \textit{Hilpert equation} \qquad (G\text{-}10)$$

Valid for: $0.4 < \mathbf{Re}_D < 4 \times 10^5$ gas flow

C and *n* from Table G-1 (for liquid flow also—circular cylinders only).
Evaluate all properties at $T_f = (T_s + T_\infty)/2$.

TABLE G-1

configuration	$\mathbf{Re}_D$	C	n
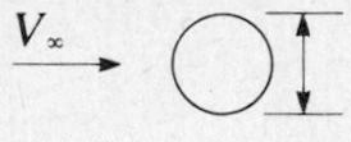 D	0.4–4 4–40 40–4000 4000–40 000 40 000–400 000	0.989 0.911 0.683 0.193 0.0266	0.330 0.385 0.466 0.618 0.805
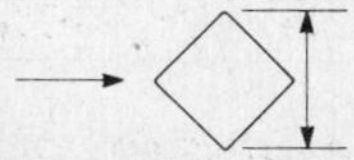 D	5000–100 000	0.246	0.588
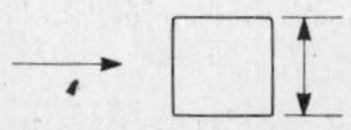 D	5000–100 000	0.102	0.675
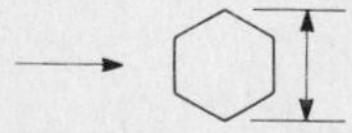 D	5000–19 500 19 500–100 000	0.160 0.0385	0.638 0.782
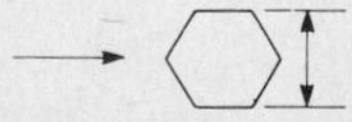 D	5000–100 000	0.153	0.638
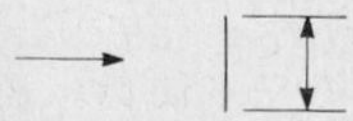 D	4000–15 000	0.228	0.731

B. Average $\bar{h}$:

$$\overline{\mathbf{Nu}}_\infty \equiv \frac{\bar{h}D}{k_\infty} = C\,\mathbf{Re}^n_{D,\infty}\,\mathbf{Pr}^m_\infty\left(\frac{\mathbf{Pr}_\infty}{\mathbf{Pr}_s}\right)^{1/4} \qquad \textit{Zhukauskas equation} \qquad (G\text{-}11)$$

Valid for: $1.0 < \mathbf{Re}_{D,\infty} < 10^6$ gases and liquids

$0.7 < \mathbf{Pr}_\infty < 500$

C and n from Table G-2.

$\mathbf{Pr}_\infty < 10$: $m = 0.37$ $\qquad$ $\mathbf{Pr}_\infty > 10$: $m = 0.36$

Evaluate all properties at T_∞, except $\mathbf{Pr}_s$ at T_s.

TABLE G-2

$\mathbf{Re}_{D,\infty}$	C	n
1 to 40	0.75	0.4
40 to 1000	0.51	0.5
10^3 to 2×10^5	0.26	0.6
2×10^5 to 10^6	0.076	0.7

C. Average $\bar{h}$:

$$\overline{\mathbf{Nu}} \equiv \frac{\bar{h}D}{k} = 0.3 + \frac{(0.62)\,\mathbf{Re}_D^{1/2}\,\mathbf{Pr}^{1/3}}{[1 + (0.4/\mathbf{Pr})^{2/3}]^{1/4}}\left[1 + \left(\frac{\mathbf{Re}_D}{28\,200}\right)^{5/8}\right]^{4/5} \qquad \textit{Churchill–Bernstein equation} \qquad (G\text{-}12)$$

Valid for: $\mathbf{Re}_D\,\mathbf{Pr} > 0.2$ gases and liquids

Evaluate all properties at T_f.

G.4 FLOW OVER A SPHERE

A. Average $\bar{h}$:

$$\overline{\mathbf{Nu}}_\infty \equiv \frac{\bar{h}D}{k_\infty} = 2 + (0.4\,\mathbf{Re}^{1/2}_{D,\infty} + 0.06\,\mathbf{Re}^{2/3}_{D,\infty})\,\mathbf{Pr}^{0.4}_\infty\left(\frac{\mu_\infty}{\mu_s}\right)^{1/4} \qquad \textit{Whitaker equation} \qquad (G\text{-}13)$$

Valid for: $0.71 < \mathbf{Pr}_\infty < 380$

$3.5 < \mathbf{Re}_{D,\infty} < 7.6 \times 10^4$

$1.0 < \dfrac{\mu_\infty}{\mu_s} < 3.2$

Evaluate all properties except μ_s at T_∞.

B. Average $\bar{h}$ (*falling drop*):

$$\overline{\mathbf{Nu}}_\infty \equiv \frac{\bar{h}D}{k_\infty} = 2 + (0.6)\,\mathbf{Re}^{1/2}_{D,\infty}\,\mathbf{Pr}^{1/3}_\infty\left[25\left(\frac{D}{x}\right)^{0.7}\right] \qquad (G\text{-}14)$$

where x is distance measured from start of fall. Properties are evaluated at T_∞.
Assumption: For each instantaneous location x, a quasisteady state exists, with $V^2_\infty = 2gx$ (free-fall velocity).
When x is unknown, average value during fall can be estimated by omitting the bracketed factor, which accounts for droplet oscillations.

G.5 FLOW THROUGH TUBE BUNDLES

A. Average $\bar{h}$:

$$\overline{\mathbf{Nu}} \equiv \frac{\bar{h}D}{k} = C_1\,\mathbf{Re}^n_{D,\max} \qquad \textit{Grimison equation} \qquad (G\text{-}15)$$

Valid for: Air only, with $\mathbf{Pr} = 0.7$.

$$2000 < \mathbf{Re}_{D,\max} < 40\,000$$

$$N_L \equiv \text{no. tubes in flow direction} \geq 10$$

C_1 and n from Figure G-1 and Table G-3.
Evaluate all properties at $T_f = (T_\infty + T_s)/2$.

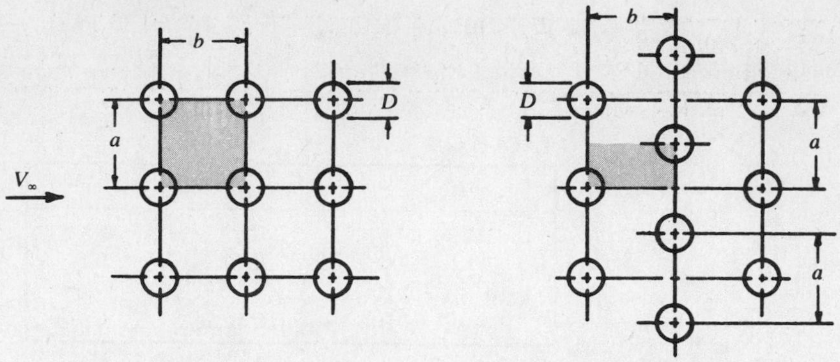

(*a*) In-Line (*b*) Staggered

Figure G-1

TABLE G-3*

	a/D							
	1.25		1.5		2		3	
b/D	C_1	n	C_1	n	C_1	n	C_1	n
In-line tubes:								
1.25	0.348	0.592	0.275	0.608	0.100	0.704	0.0633	0.752
1.5	0.367	0.586	0.250	0.620	0.101	0.702	0.0678	0.744
2	0.418	0.570	0.299	0.602	0.229	0.632	0.198	0.648
3	0.290	0.601	0.357	0.584	0.374	0.581	0.286	0.608
Staggered tubes:								
0.6							0.213	0.636
0.9					0.446	0.571	0.401	0.581
1			0.497	0.558				
1.125					0.478	0.565	0.518	0.560
1.25	0.518	0.556	0.505	0.554	0.519	0.556	0.522	0.562
1.5	0.451	0.568	0.460	0.562	0.452	0.568	0.488	0.568
2	0.404	0.572	0.416	0.568	0.482	0.556	0.449	0.570
3	0.310	0.592	0.356	0.580	0.440	0.562	0.421	0.574

*From E. D. Grimison, *Trans. ASME*, **59**: 583–594 (1937).

B. Average $\bar{h}$:

$$\overline{\mathbf{Nu}} \equiv \frac{\bar{h}D}{k} = (1.13)C_1(\mathbf{Re}_{D,\max})^n\,\mathbf{Pr}^{1/3} \qquad (G\text{-}16)$$

Valid for:

$$\mathbf{Pr} > 0.7$$

$$2000 < \mathbf{Re}_{D,\max} < 40\,000$$

$$N_L \geq 10$$

C_1 and n from Figure G-1 and Table G-3.

C. Average $\bar{h}$: For $N_L < 10$, use either (*G-15*) (air) or (*G-16*) (**Pr**>0.7) to obtain $\bar{h}_{10}$. Modify by the ratio $\bar{h}/\bar{h}_{10}$ given in Table G-4.

TABLE G-4*

	number of tubes								
	1	**2**	**3**	**4**	**5**	**6**	**7**	**8**	**9**
Staggered	0.68	0.75	0.83	0.89	0.92	0.95	0.97	0.98	0.99
In-line	0.64	0.80	0.87	0.90	0.92	0.94	0.96	0.98	0.99

*From W. M. Kays and R. K. Lo, Stanford University *Tech. Rpt. 15*, 1952.

APPENDIX H

Equations for Chapter 6

H.1 STEADY PIPE FLOW

$$\textit{all conditions} \quad \frac{dT_b}{dx} = \frac{q_s''P}{\dot{m}c_{p,b}} = \frac{P}{\dot{m}c_{p,b}} h_x(T_s - T_b) \tag{H-1}$$

$$\textit{constant } q_s'' \quad T_b(x) = T_{b,i} + \frac{q_s''P}{\dot{m}c_{p,b}} x \tag{H-2}$$

$$\textit{constant } T_s \quad T_b(x) = T_s - (T_s - T_{b,i}) \exp\left[\frac{-Px\bar{h}(0,x)}{\dot{m}c_{p,b}}\right] \tag{H-3}$$

$$\left.\begin{matrix}\textit{constant } q_s'' \\ \textit{thermally developed}\end{matrix}\right\} \quad q_{\text{conv}} = \bar{h}\, PL(T_s - T_b) \qquad (T_s - T_b = \text{const.}) \tag{H-4}$$

$$\textit{constant } T_s \quad q_{\text{conv}} = \bar{h}PL(\Delta T)_{\text{l.m.}} \qquad \left[(\Delta T)_{\text{l.m.}} \equiv \frac{T_{b,e} - T_{b,i}}{\ln \dfrac{T_s - T_{b,i}}{T_s - T_{b,e}}}\right] \tag{H-5}$$

$$\left.\begin{matrix}\textit{constant } q_s'' \\ \textit{laminar} \\ \textit{developed profiles}\end{matrix}\right\} \quad \mathbf{Nu}_{D,\infty} \equiv \frac{h_\infty D}{k_b} \approx 4.36 \tag{H-6}$$

where $x = \infty$ stands for any location beyond the entry region.

$$\left.\begin{matrix}\textit{constant } T_s \\ \textit{laminar} \\ \textit{developed profiles}\end{matrix}\right\} \quad \mathbf{Nu}_{D,\infty} \equiv \frac{h_\infty D}{k_b} \approx 3.66 \tag{H-7}$$

where $x = \infty$ stands for any location beyond the entry region.

Sieder–Tate Equation (laminar flow, developing profiles, constant T_s)

$$\overline{\mathbf{Nu}} \equiv \frac{\bar{h}D}{k_b} = (1.86)\left(\frac{\mathbf{Re}_{D,b}\,\mathbf{Pr}_b}{L/D}\right)^{1/3}\left(\frac{\mu_b}{\mu_s}\right)^{0.14} \tag{H-8}$$

valid for $0.48 < \mathbf{Pr}_b < 16\,700$
$0.0044 < \mu_b/\mu_s < 9.75$
$\overline{\mathbf{Nu}} \geq 3.72$

Evaluate properties other than μ_s at $\bar{T}_b \equiv (T_{b,i} + T_{b,e})/2$.

H.2 HAUSEN EQUATION (laminar flow, developing temperature profile)

Define

$$\mathbf{Nu}\,(x) \equiv \begin{cases} h_x D/k_b & \text{for constant } q_s'' \\ \bar{h}(0,x)D/k_b & \text{for constant } T_s \end{cases}$$

Then

$$\mathbf{Nu}\,(x) = \mathbf{Nu}_{D,\infty} + \frac{K_1[(D/x)\,\mathbf{Re}_D\,\mathbf{Pr}]}{1 + K_2[(D/x)\,\mathbf{Re}_D\,\mathbf{Pr}]^n} \tag{H-9}$$

Parameters from Table H-1.

TABLE H-1

wall condition	inlet velocity	Pr	$Nu_{D,\infty}$	K_1	K_2	n
Constant q_s''	Parabolic*	Any	4.36	0.023	0.0012	1.0
Constant q_s''	Developing	0.7	4.36	0.036	0.0011	1.0
Constant T_s	Parabolic*	Any	3.66	0.0668	0.04	$\frac{2}{3}$
Constant T_s	Developing	0.7	3.66	0.104	0.016	0.8

*See Problem 4.65.

H.3 FULLY DEVELOPED TURBULENT FLOW AT CONSTANT T_s OR q_s

For the three correlations that follow, define

$$\mathbf{Nu}_{D,\infty} \equiv \begin{cases} h_x D/k_b & \text{for } 10 < L/D < 60, \quad \text{properties at } T_b(x) \\ \bar{h}_\infty D/k_b & \text{for } L/D > 60, \quad \text{properties at } \bar{T}_b \end{cases}$$

$$\mathbf{Nu}_{D,\infty} = (0.023)\,\mathbf{Re}_{D,b}^{0.8}\,\mathbf{Pr}_b^{1/3} \qquad \textit{Chilton–Colburn} \tag{H-10}$$

valid for $\mathbf{Re}_{D,b} > 10^4$
$0.7 < \mathbf{Pr}_b < 50$
$|T_s - T_b| < 60$ °C (100 °F) gases
< 6 °C (10 °F) liquids

$$\mathbf{Nu}_{D,\infty} = (0.023)\,\mathbf{Re}_{D,b}^{0.8}\,\mathbf{Pr}_b^{n} \qquad \textit{Dittus–Boelter} \tag{H-11}$$

valid for $\mathbf{Re}_{D,b} > 10^4$
$0.7 < \mathbf{Pr}_b < 160$
$n = 0.4$ (fluid heated); $n = 0.3$ (fluid cooled)
$|T_s - T_b| < 6$ °C (10 °F) liquids
< 60 °C (100 °F) gases

$$\mathbf{Nu}_{D,\infty} = (0.027)\,\mathbf{Re}_{D,b}^{0.8}\,\mathbf{Pr}_b^{1/3}\left(\frac{\mu_b}{\mu_s}\right)^{0.14} \qquad \textit{Sieder–Tate} \tag{H-12}$$

valid for $\mathbf{Re}_{D,b} > 10^4$
$0.7 < \mathbf{Pr}_b < 1.67 \times 10^4$
μ_s evaluated at T_s
$|T_s - T_b| < 60$ °C but larger than with (*H-10*) or (*II-11*)

H.4 ENTRY REGION CORRECTIONS

$$\frac{\bar{h}_{\text{ent}}}{\bar{h}_\infty} = (1.11)\left[\frac{\mathbf{Re}_D^{1/5}}{(L/D)^{4/5}}\right]^{0.275} \tag{H-13}$$

valid for $L/D < (L/D)_{\text{crit}} \equiv (0.623)\,\mathbf{Re}_D^{1/4}$
All entry conditions

$$\frac{\bar{h}_{\text{ent}}}{\bar{h}_\infty} = 1 + \frac{C}{L/D} \tag{H-14}$$

valid for $(L/D)_{\text{crit}} < (L/D) < 60$
Entry conditions as specified in Table H-2

TABLE H-2

inlet configuration	C
Bell-mouthed with screen	1.4
Calming section, $L/D = 11.2$	1.4
$L/D = 2.8$	3.0
45° bend	5.0
90° bend	7.0

$$\overline{\mathbf{Nu}}_D \equiv \frac{\bar{h}_{\text{ent}} D}{k_b} = (0.036)\,\mathbf{Re}_{D,b}^{0.8}\,\mathbf{Pr}_b^{1/3}\left(\frac{D}{L}\right)^{0.055} \qquad \textit{Nusselt equation} \qquad (H\text{-}15)$$

valid for $10 < L/D < 400$
All entry conditions

$$\left.\begin{array}{l}\textit{abrupt contraction}\\ \textit{from reservoir}\end{array}\right\} \quad \frac{\bar{h}_{\text{ent}}}{\bar{h}_\infty} = \begin{cases} 1 + (D/L)^{0.7} & 2 < L/D < 20 \\ 1 + (6D/L) & 20 < L/D < 60 \end{cases} \qquad (H\text{-}16)$$

$$\textit{rough tubes} \quad \frac{\bar{h}_{\text{rough}}}{\bar{h}_{\text{smooth}}} = \left(\frac{f_{\text{rough}}}{f_{\text{smooth}}}\right)^n \qquad (H\text{-}17)$$

where $f_{\text{smooth}} = (0.184)\,\mathbf{Re}_{D,b}^{-0.2}$ and $n \equiv (0.68)\,\mathbf{Pr}_b^{0.215}$

valid for $10^4 < \mathbf{Re}_{D,b} < 10^5$ (turbulent flow)
$1 < f_{\text{rough}}/f_{\text{smooth}} < 3$

$$\overline{\mathbf{Nu}}_{D,\infty} \equiv \frac{\bar{h}_\infty D}{k_b} = \frac{\phi^2\,\mathbf{Re}_{D,b}\,\mathbf{Pr}_b}{1.07 + (12.7)\phi(\mathbf{Pr}_b^{2/3} - 1)}\left(\frac{\mu_b}{\mu_s}\right)^n \qquad \textit{Petukhov equation} \qquad (H\text{-}18)$$

where $\phi \equiv (5.15 \log \mathbf{Re}_{D,b} - 4.64)^{-1}$

valid for developed velocity profile
$0.8 < \mu_b/\mu_s < 40$
$5000 < \mathbf{Re}_{D,b} < 1.25 \times 10^5$
$2 < \mathbf{Pr}_b < 140$
$n = 0$ (gases)
$n = 0.11$ (heating liquid)
$n = 0.25$ (cooling liquid)

APPENDIX I

Grashof Number

$$\mathbf{Gr}_x \equiv \frac{g\beta(T_s - T_\infty)x^3}{\nu^2} \propto \frac{\text{buoyancy force}}{\text{viscous force}} \qquad (I\text{-}1)$$

Coefficient of Volumetric Expansion

$$\beta \equiv \frac{1}{v}\left(\frac{\partial v}{\partial T}\right)_p = -\frac{1}{\rho}\left(\frac{\partial \rho}{\partial T}\right)_p \qquad (I\text{-}2)$$

where $v = \rho^{-1}$ is specific volume.

Boundary-Layer Differential Equations (x positive upward)

$$\textbf{\textit{x-momentum}} \quad u\frac{\partial u}{\partial x} + v\frac{\partial u}{\partial y} = g\beta(T - T_\infty) + \nu\frac{\partial^2 u}{\partial y^2} \qquad (I\text{-}3)$$

$$\textbf{\textit{energy}} \quad u\frac{\partial T}{\partial x} + v\frac{\partial T}{\partial y} = \alpha\frac{\partial^2 T}{\partial y^2} \qquad (I\text{-}4)$$

$$\textbf{\textit{continuity}} \quad \frac{\partial u}{\partial x} + \frac{\partial v}{\partial y} = 0 \qquad (I\text{-}5)$$

Pohlhausen's Similarity Parameter

$$\eta \equiv \frac{y}{x}\left(\frac{\mathbf{Gr}_x}{4}\right)^{1/4} \qquad (I\text{-}6)$$

Dimensionless Temperature

$$\theta \equiv \frac{T - T_\infty}{T_s - T_\infty} \qquad (I\text{-}7)$$

Reference Temperature for Property Evaluation (Except β for Gases)

$$T_{\text{ref}} \equiv T_s + C(T_\infty - T_s) \qquad (I\text{-}8)$$

where $C = 0.38$ [preferable for laminar equations from analyses]
or $C = 0.50$ [yields $T_{\text{ref}} = T_f$; frequently used in textbooks for all properties, including β for gases]

Vertical Wall at Constant Temperature: Ostrach's Similarity Solutions for h_x and $\bar{h}$

$$\mathbf{Nu}_x \equiv \frac{h_x x}{k} = F_1(\mathbf{Pr})\left(\frac{\mathbf{Gr}_x}{4}\right)^{1/4} \quad \text{where} \quad F_1(\mathbf{Pr}) \equiv -\left.\frac{\partial\theta}{\partial\eta}\right|_{\eta=0} \qquad (I\text{-}9)$$

$$\overline{\mathbf{Nu}} \equiv \frac{\bar{h}L}{k} = \frac{4}{3}F_1(\mathbf{Pr})\left(\frac{\mathbf{Gr}_L}{4}\right)^{1/4} \qquad (I\text{-}10)$$

Valid for $10^4 < \mathbf{Gr}\,\mathbf{Pr} < 10^9$ (*laminar regime*)

Evaluate properties at T_{ref}.

for gases $\beta = \dfrac{1}{T_\infty \text{ in absolute units}}$

F_1-values from Table I-1 or from (*I-11*).

TABLE I-1

Pr	0.01	0.72	0.733	1.0	2.0	10.0	100.0	1000.0
F_1**(Pr)**	0.0812	0.5046	0.5080	0.5671	0.7165	1.1694	2.191	3.966

$$F_1(\mathbf{Pr}) \approx \frac{0.75\,\mathbf{Pr}^{1/2}}{(0.609 + 1.221\,\mathbf{Pr}^{1/2} + 1.238\,\mathbf{Pr})^{1/4}} \qquad (I\text{-}11)$$

Vertical Wall with Uniform Heat Flux: Sparrow & Gregg Similarity Solution for $\bar{h}$

$$\overline{\mathbf{Nu}} \equiv \frac{\bar{h}L}{k} = F_2(\mathbf{Pr})\left(\frac{\mathbf{Gr}_L}{4}\right)^{1/4} \quad \text{[parameters as with } (I\text{-}9)] \qquad (I\text{-}12)$$

F_2-values from Table I-2.

TABLE I-2

Pr	0.1	1.0	10.0	100.0
F_2(**Pr**)	0.335	0.811	1.656	3.083

Vertical Wall at Constant Temperature: Integral Solutions for h_x and $\bar{h}$

laminar

$$\mathbf{Nu}_x \equiv \frac{h_x x}{k} = (0.508)\left(\frac{\mathbf{Gr}_x\,\mathbf{Pr}^2}{0.952 + \mathbf{Pr}}\right)^{1/4} \qquad (I\text{-}13)$$

$$\overline{\mathbf{Nu}} \equiv \frac{\bar{h}L}{k} = \frac{4}{3}(0.508)\left(\frac{\mathbf{Gr}_L\,\mathbf{Pr}^2}{0.952 + \mathbf{Pr}}\right)^{1/4} \qquad (I\text{-}14)$$

[parameters as with (*I-9*)].

turbulent

$$\mathbf{Nu}_x \equiv \frac{h_x x}{k} = (0.0295)\left[\frac{\mathbf{Gr}_x\,\mathbf{Pr}^{7/6}}{1 + (0.494)\,\mathbf{Pr}^{2/3}}\right]^{2/5} \qquad (I\text{-}15)$$

$$\overline{\mathbf{Nu}} \equiv \frac{\bar{h}L}{k} = (0.0210)(\mathbf{Gr}_L\,\mathbf{Pr})^{2/5} \qquad (I\text{-}16)$$

Valid for $10^9 < \mathbf{Gr}\,\mathbf{Pr} < 10^{12}$

Evaluate properties at T_{ref}.

for gases

$$\beta = \frac{1}{T_\infty \text{ in absolute units}}$$

TABLE I-3

configuration	characteristic length, Λ	$\mathbf{Gr}_\Lambda\,\mathbf{Pr}$	C	a
Vertical Plates and Large Cylinders				
Laminar	L_v	10^{-1} to 10^4	See Figure I-1	
Laminar	L_v	10^4 to 10^9	0.59	$\frac{1}{4}$
Turbulent	L_v	10^9 to 10^{12}	0.13	$\frac{1}{3}$
Small Vertical Cylinders (Wires)				
	D	10^{-10} to 10^{-4}	See Figure I-2	
Inclined Plates (small θ)				
Multiply Grashof number by cos θ, where θ is the angle of inclination from the vertical, and use vertical plate constants				
Long Horizontal Cylinders (0.002 in. $< D <$ 12 in.)				
Laminar	D	$<10^4$	See Figure I-3	
Laminar	D	10^4 to 10^9	0.53	$\frac{1}{4}$
Turbulent	D	10^9 to 10^{12}	0.13	$\frac{1}{3}$
Fine Horizontal Wires ($D <$ 0.002 in.)				
Laminar	D		0.4	0
Miscellaneous Solid Shapes (spheres, short cylinders, blocks)				
Laminar	$\frac{1}{\Lambda} = \frac{1}{L_v} + \frac{1}{L_h}$	10^{-4} to 10^4	See Figure I-4	
Laminar		10^4 to 10^9	0.60	$\frac{1}{4}$

Empirical Correlation (Laminar or Turbulent Flow, Surface Heated or Cooled)

$$\overline{\mathbf{Nu}} \equiv \frac{\bar{h}\Lambda}{k} = C(\mathbf{Gr}_\Lambda\,\mathbf{Pr})^a \tag{I-17}$$

Evaluate configuration parameters Λ, C, and a from Table I-3.
Evaluate properties at $T_f \equiv (T_s + T_\infty)/2$.

for gases $$\beta = \frac{1}{T_\infty \text{ in absolute units}}$$

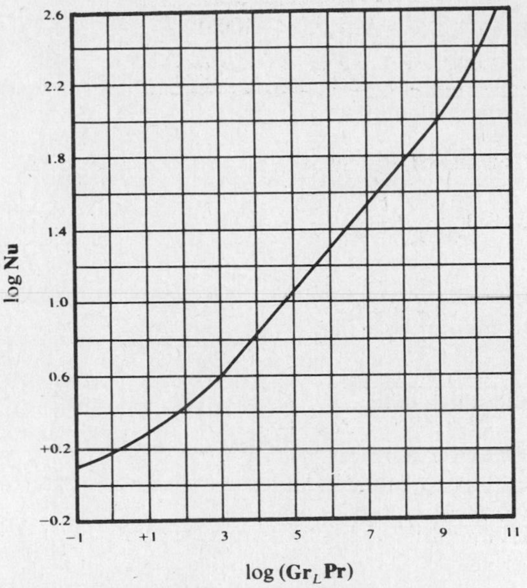

Figure I-1 Correlation for heated vertical plates. (Adapted from W. H. McAdams, *Heat Transmission*, 3d ed., p. 173. Copyright 1954. McGraw-Hill Book Company. Used by permission.)

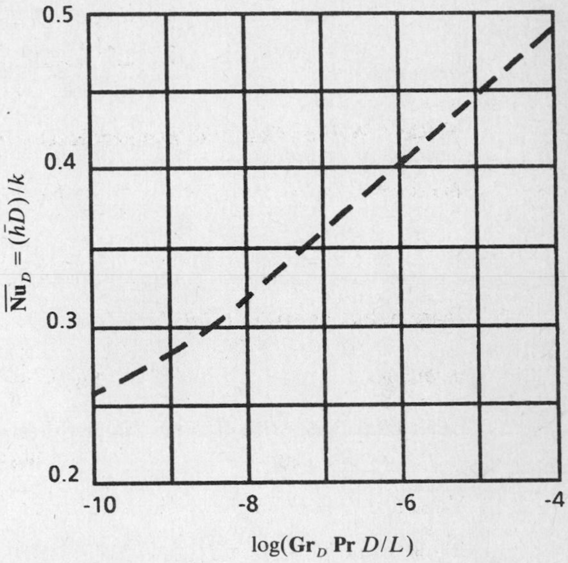

Figure I-2 Approximate correlation for small vertical cylinders.

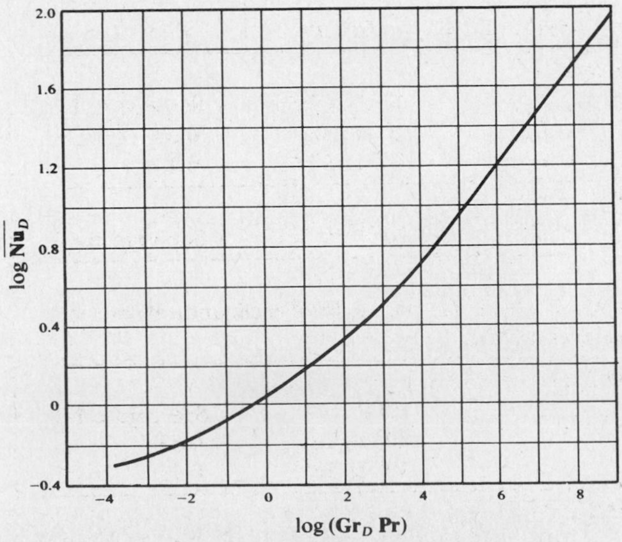

Figure I-3 Correlation for horizontal cylinders. (Adapted from W. H. McAdams, *Heat Transmission*, 3d ed. Copyright 1954. McGraw-Hill Book Company. Used by permission.)

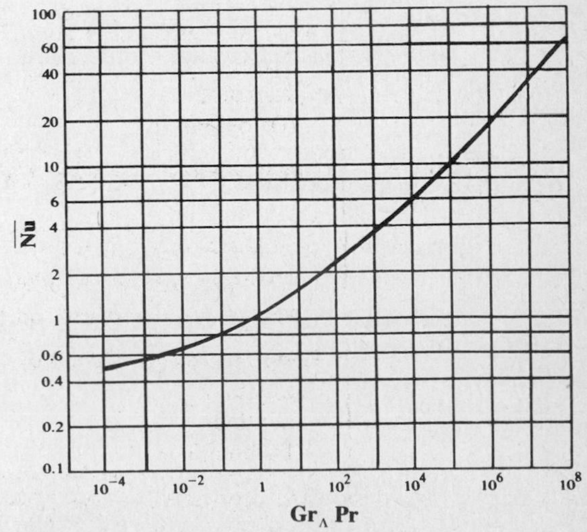

Figure I-4 Correlation for miscellaneous solid shapes.

Condition for Treating Vertical Cylinder as Vertical Plate

$$\frac{D}{L} > \frac{35}{\mathbf{Gr}_L^{1/4}} \tag{I-18}$$

Empirical Correlation for Vertical Wall at Constant Temperature or Uniform Heat Flux

laminar

$$\overline{\mathbf{Nu}}_L \equiv \frac{\bar{h}L}{k} = 0.68 + \frac{(0.67)(\mathbf{Gr}_L\ \mathbf{Pr})^{1/4}}{[1+(0.492/\mathbf{Pr})^{9/16}]^{4/9}} \tag{I-19}$$

Valid for

$$10^4 < \mathbf{Ra}_L < 10^9$$

Evaluate properties at T_f.

$$T_s = T_{L/2} \quad \text{for uniform heat flux}$$

for gases

$$\beta = \frac{1}{T_\infty \text{ in absolute units}}$$

laminar or turbulent

$$\overline{\mathbf{Nu}}_L \equiv \frac{\bar{h}L}{k} = \left\{0.825 + \frac{0.387(\mathbf{Gr}_L\ \mathbf{Pr})^{1/6}}{[1+(0.492/\mathbf{Pr})^{9/16}]^{8/27}}\right\}^2 \tag{I-20}$$

Valid for

$$10^4 < \mathbf{Gr}_L\ \mathbf{Pr} < 10^{12}$$

Evaluate properties at T_f.

$$T_s = T_{L/2} \quad \text{for uniform heat flux}$$

for gases

$$\beta = \frac{1}{T_\infty \text{ in absolute units}}$$

Horizontal Plate with Heated Upper Surface or Cooled Lower Surface

laminar

$$\overline{\mathbf{Nu}}_L \equiv \frac{\bar{h}L}{k} = (0.54)(\mathbf{Gr}_L\ \mathbf{Pr})^{1/4} \tag{I-21}$$

Valid for constant surface temperature

$$\begin{pmatrix} L = \frac{1}{2}(L_1 + L_2) \\ 10^5 < \mathbf{Gr}_L\ \mathbf{Pr} < 2\times 10^7 \end{pmatrix} \quad \text{or} \quad \begin{pmatrix} L = A_s/P \\ 10^4 < \mathbf{Gr}_L\ \mathbf{Pr} < 10^7 \end{pmatrix}$$

Evaluate properties at T_f.

for gases

$$\beta = \frac{1}{T_\infty \text{ in absolute units}}$$

turbulent

$$\overline{\mathbf{Nu}}_L \equiv \frac{\bar{h}L}{k} = C(\mathbf{Gr}_L\ \mathbf{Pr})^{1/3} \tag{I-22}$$

Valid for constant surface temperature

$$\begin{pmatrix} L = \frac{1}{2}(L_1 + L_2),\ C = 0.14 \\ 2\times 10^7 < \mathbf{Gr}_L\ \mathbf{Pr} < 3\times 10^{10} \end{pmatrix} \quad \text{or} \quad \begin{pmatrix} L = A_s/P,\ C = 0.15 \\ 10^7 < \mathbf{Gr}_L\ \mathbf{Pr} < 10^{11} \end{pmatrix}$$

Evaluate properties at T_f.

for gases

$$\beta = \frac{1}{T_\infty \text{ in absolute units}}$$

Horizontal Plate with Heated Lower Surface or Cooled Upper Surface

$$\overline{\mathbf{Nu}}_L \equiv \frac{\bar{h}L}{k} = (0.27)(\mathbf{Gr}_L\ \mathbf{Pr})^{1/4} \tag{I-23}$$

Valid for constant surface temperature

$$\begin{pmatrix} L = \frac{1}{2}(L_1 + L_2) \\ 3\times 10^5 < \mathbf{Gr}_L\ \mathbf{Pr} < 3\times 10^{10} \end{pmatrix} \quad \text{or} \quad \begin{pmatrix} L = A_s/P \\ 10^5 < \mathbf{Gr}_L\ \mathbf{Pr} < 10^{10} \end{pmatrix}$$

Evaluate properties at T_f.

for gases

$$\beta = \frac{1}{T_\infty \text{ in absolute units}}$$

Long Horizontal Cylinder at Constant Temperature

standard correlation

$$\overline{\mathbf{Nu}}_D \equiv \frac{\bar{h}D}{k} = C(\mathbf{Gr}_D\,\mathbf{Pr})^a \qquad (I\text{-}24)$$

Evaluate properties at T_f.

for gases

$$\beta = \frac{1}{T_\infty \text{ in absolute units}}$$

Evaluate C and a from Table I-3.

Churchill–Chu correlation

$$\overline{\mathbf{Nu}}_D \equiv \frac{\bar{h}D}{k} = \left\{0.60 + \frac{(0.387)(\mathbf{Gr}_D\,\mathbf{Pr})^{1/6}}{[1 + (0.559/\mathbf{Pr})^{9/16}]^{8/27}}\right\}^2 \qquad (I\text{-}25)$$

Valid for $10^{-5} < \mathbf{Gr}_D\,\mathbf{Pr} < 10^{12}$

Evaluate properties at T_f.

for gases

$$\beta = \frac{1}{T_\infty \text{ in absolute units}}$$

Region Outside a Sphere

Churchill correlation

$$\overline{\mathbf{Nu}}_D \equiv \frac{\bar{h}D}{k} = 2 + \frac{(0.589)(\mathbf{Gr}_D\,\mathbf{Pr})^{1/4}}{[1 + (0.469/\mathbf{Pr})^{9/16}]^{4/9}} \qquad (I\text{-}26)$$

Valid for $\mathbf{Gr}_D\,\mathbf{Pr} < 10^{11}$ and $\mathbf{Pr} > 0.7$

Evaluate properties at T_f.

for gases

$$\beta = \frac{1}{T_\infty \text{ in absolute units}}$$

Notation for Rectangular Enclosures (Figure I-5)

$b \equiv$ width of region
$L \equiv$ length of region
$T_h \equiv$ heated-surface temperature
$T_c \equiv$ cooled-surface temperature

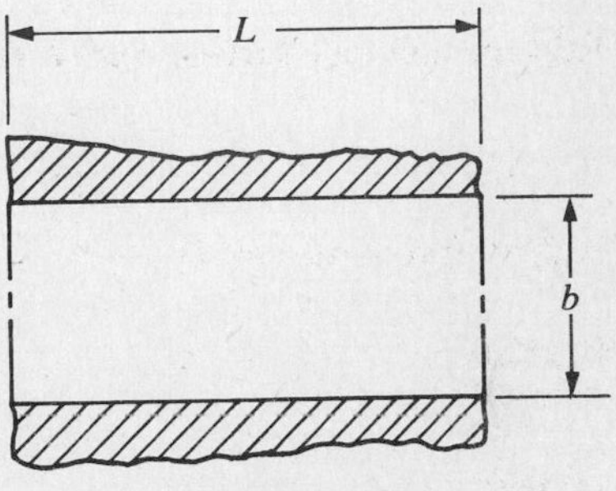

Figure I-5

net heat influx

$$q = \bar{h}A(T_h - T_c) \qquad (I\text{-}27)$$

temperature for property evaluation

$$T_{\text{avg}} \equiv \tfrac{1}{2}(T_h + T_c)$$

Horizontal Rectangular Enclosure: Upper Plate at T_h

$$\overline{\mathbf{Nu}}_b \equiv \frac{\bar{h}b}{k} = 1.0 \quad \text{(i.e., conductive transfer)} \qquad (I\text{-}28)$$

Horizontal Rectangular Enclosure: Lower Plate at T_h

correlation for air

$$\overline{\mathbf{Nu}}_b \equiv \frac{\bar{h}b}{k} = \begin{cases} (0.195)\,\mathbf{Gr}_b^{1/4} & 10^4 < \mathbf{Gr}_b < 4 \times 10^5 \\ (0.068)\,\mathbf{Gr}_b^{1/3} & 4 \times 10^5 < \mathbf{Gr}_b \end{cases} \qquad (I\text{-}29)$$

Globe & Dropkin correlation for liquids

$$\overline{\mathbf{Nu}}_b \equiv \frac{\bar{h}b}{k} = (0.069)\,\mathbf{Gr}_b^{1/3}\,\mathbf{Pr}^{0.407} \qquad (I\text{-}30)$$

Valid for $3 \times 10^5 < \mathbf{Gr}_b\,\mathbf{Pr} < 7 \times 10^9$

Vertical Rectangular Enclosure (Air): $L/b > 3$

For $L/b < 3$, use vertical plate in free-air surroundings.

$$\overline{\mathbf{Nu}}_b \equiv \frac{\bar{h}b}{k} = \begin{cases} 1 & 0 < \mathbf{Gr}_b < 2 \times 10^3 \\ (0.18)\,\mathbf{Gr}_b^{1/4}\,(L/b)^{-1/9} & 2 \times 10^3 < \mathbf{Gr}_b < 2 \times 10^5 \\ (0.065)\,\mathbf{Gr}_b^{1/3}\,(L/b)^{-1/9} & 2 \times 10^5 < \mathbf{Gr}_b < 1.1 \times 10^7 \end{cases} \qquad (I\text{-}31)$$

Region Between Concentric Spheres

$$q = k_{\text{eff}}\,\pi\left(\frac{D_i D_o}{b}\right)(T_h - T_c) \qquad (I\text{-}32)$$

in which

$$b \equiv \frac{D_o - D_i}{2} \qquad (I\text{-}33)$$

$$\frac{k_{\text{eff}}}{k} \equiv (0.74)\left(\frac{\mathbf{Pr}}{0.861 + \mathbf{Pr}}\right)^{1/4} (\mathbf{Gr}_s\,\mathbf{Pr})^{1/4} \qquad (I\text{-}34)$$

$$\mathbf{Gr}_s \equiv \frac{b}{(D_o D_i)^4}\left[\frac{\mathbf{Gr}_b}{(D_i^{-7/5} + D_o^{-7/5})^5}\right] \qquad (I\text{-}35)$$

Valid for $10^2 < \mathbf{Gr}_s\,\mathbf{Pr} < 10^4$

Evaluate properties at $T_{\text{avg}} = (T_h + T_c)/2$.

APPENDIX J

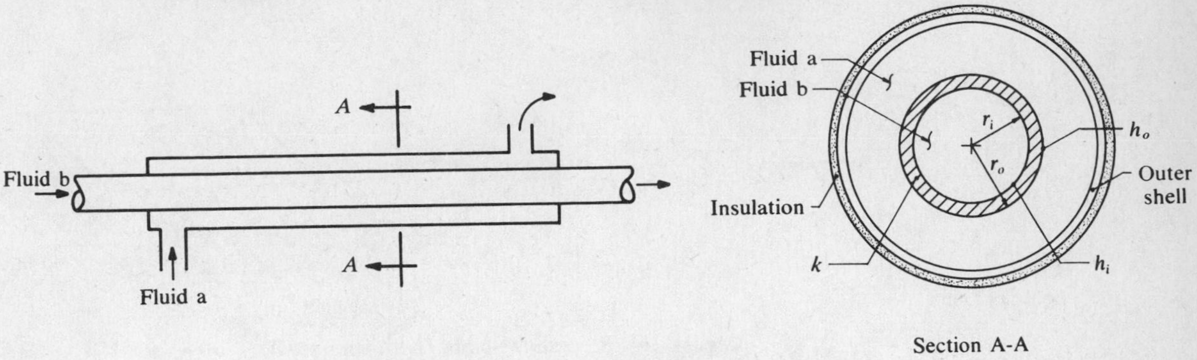

Figure J-1 Double-pipe heat exchanger.

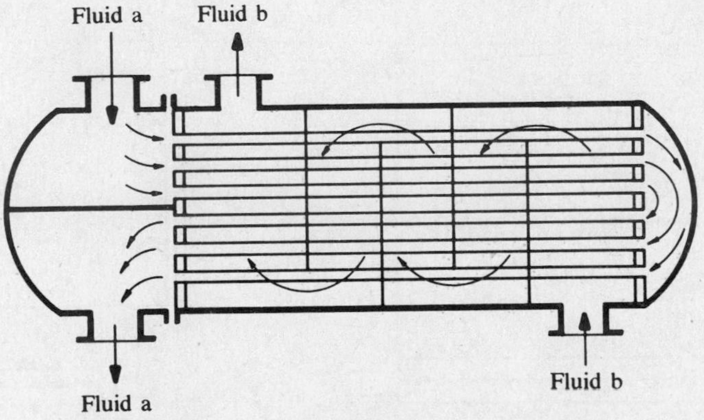

Figure J-2 Shell-and-tube heat exchanger.

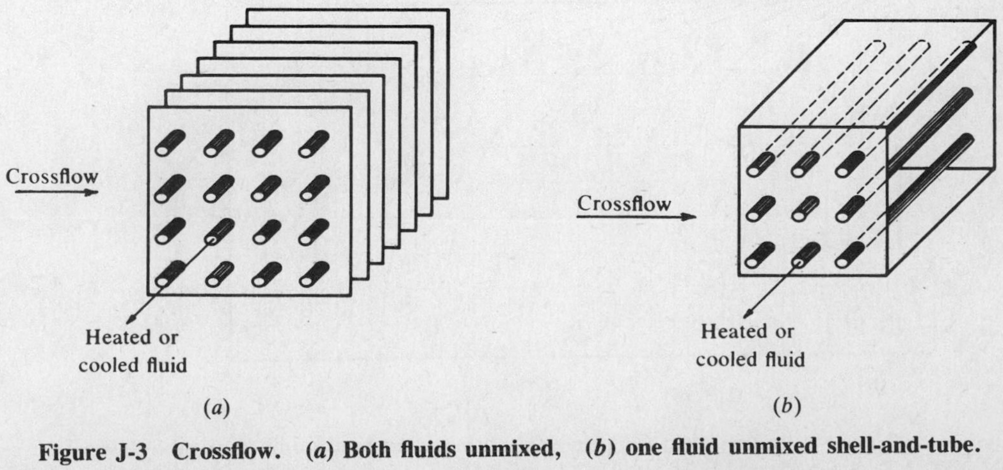

Figure J-3 Crossflow. (*a*) Both fluids unmixed, (*b*) one fluid unmixed shell-and-tube.

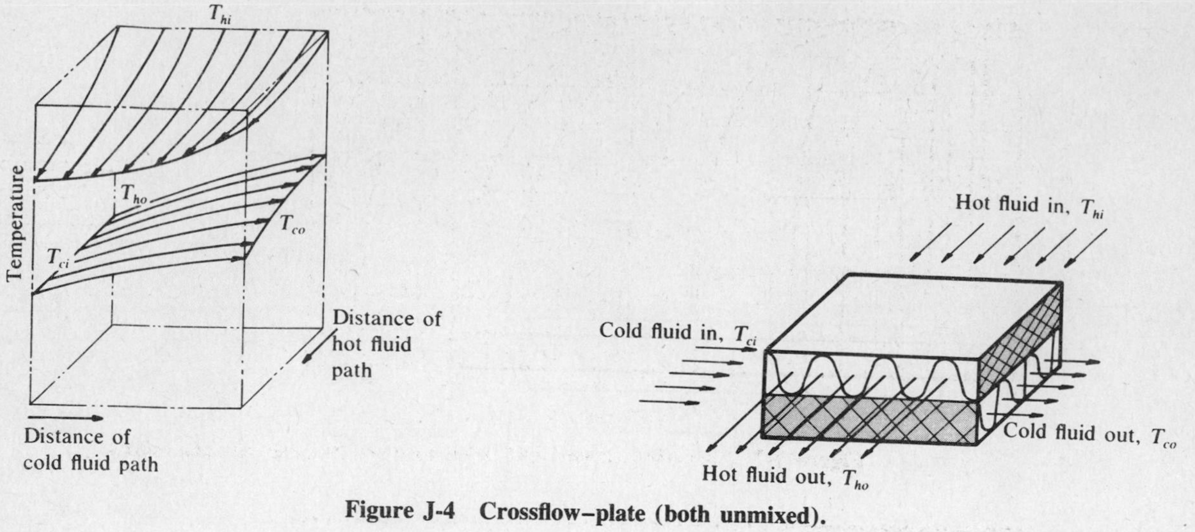

Figure J-4 Crossflow–plate (both unmixed).

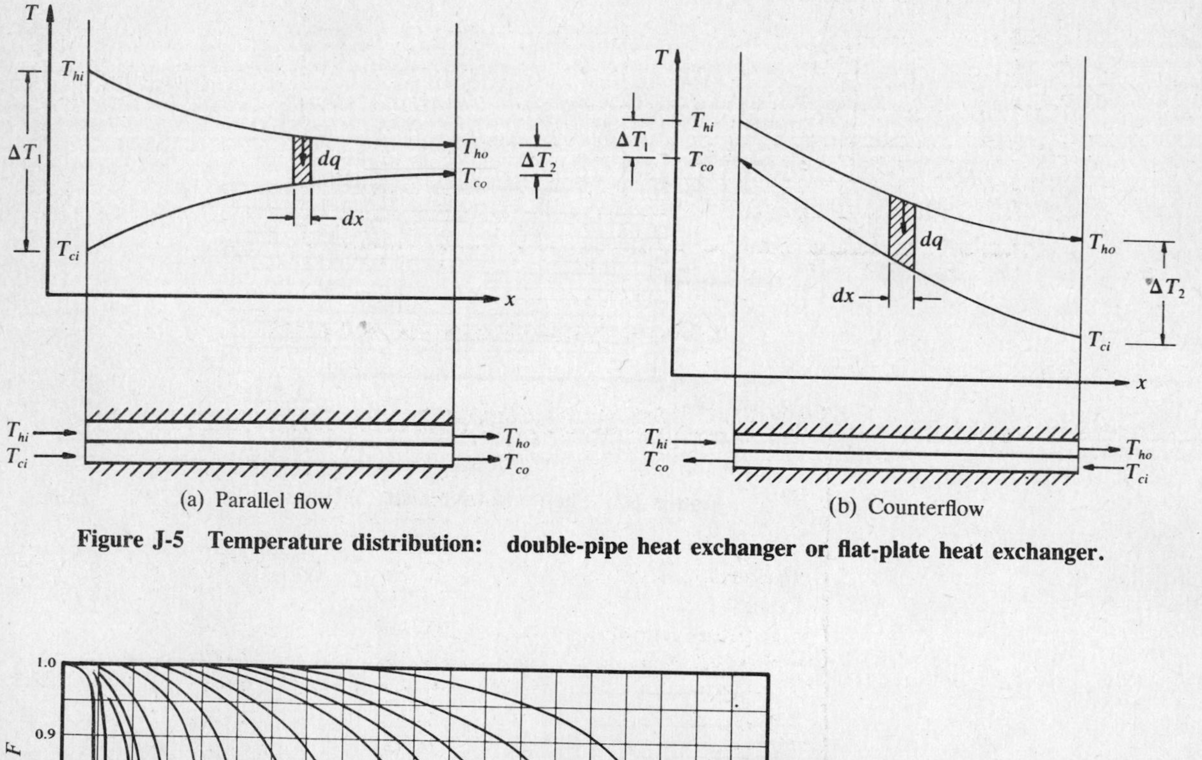

(a) Parallel flow (b) Counterflow

Figure J-5 Temperature distribution: double-pipe heat exchanger or flat-plate heat exchanger.

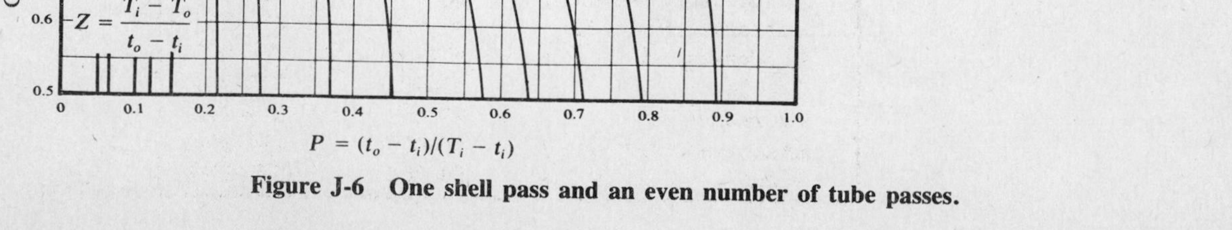

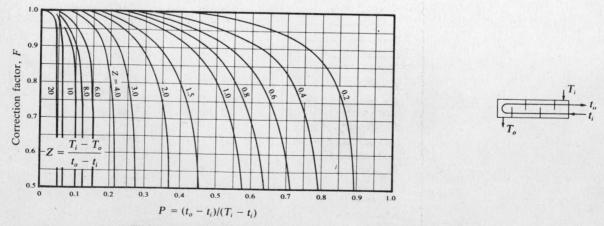

Figure J-6 One shell pass and an even number of tube passes.

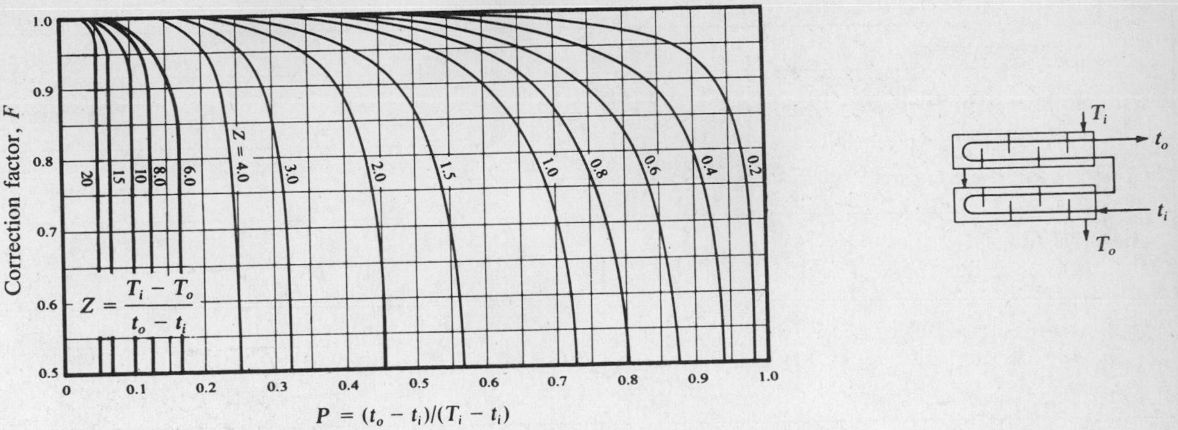

Figure J-7 Two shell passes and twice an even number of tube passes.

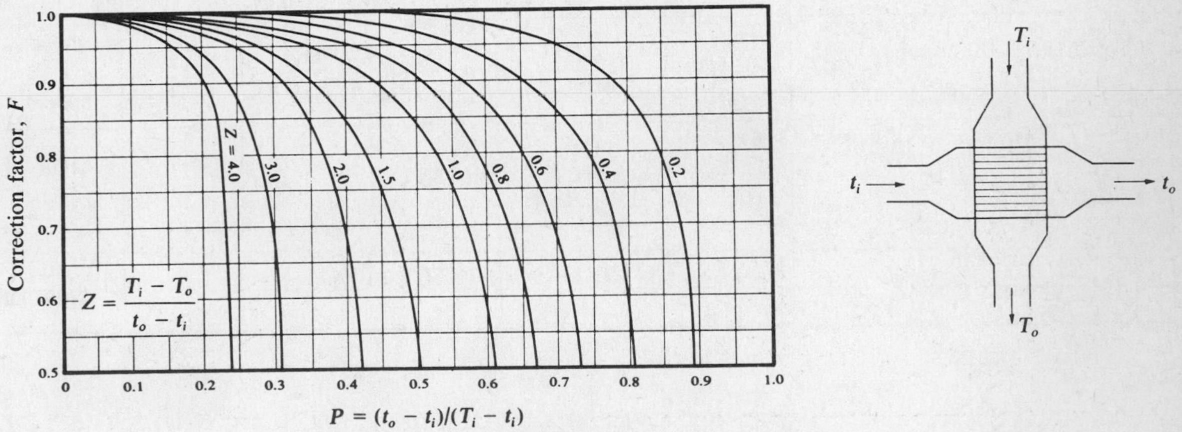

Figure J-8 Crossflow with one fluid mixed.

TABLE J-1 Approximate U Values

fluid	U Btu/hr · ft^2 · °F	U W/m^2 · K
Oil to oil	30–55	170–312
Organics to organics	10–60	57–340
Steam to		
Aqueous solutions	100–600	567–3400
Fuel oil, heavy	10–30	57–170
Light	30–60	170–340
Gases	5–50	28–284
Water	175–600	993–3400
Water to		
Alcohol	50–150	284–850
Brine	100–200	567–1135
Compressed air	10–30	57–170
Condensing alcohol	45–120	255–680
Condensing ammonia	150–250	850–1420
Condensing Freon-12	80–150	454–850
Condensing oil	40–100	227–567
Gasoline	60–90	340–510
Lubricating oil	20–60	113–340
Organic solvents	50–150	284–850
Water	150–300	850–1700

TABLE J-2 Heat-Exchanger Effectiveness

exchanger type	effectiveness	see graph in
Parallel-flow: single-pass	$\epsilon = \dfrac{1-\exp[-\mathbf{NTU}(1+C)]}{1+C}$	Figure J-10
Counterflow: single-pass	$\epsilon = \dfrac{1-\exp[-\mathbf{NTU}(1-C)]}{1-C\exp[-\mathbf{NTU}(1-C)]}$	Figure J-11
Shell-and-tube (one shell pass; 2, 4, 6, etc., tube passes)	$\epsilon_1 = 2\left\{1+C+\dfrac{1+\exp[-\mathbf{NTU}(1+C^2)^{1/2}]}{1-\exp[-\mathbf{NTU}(1+C^2)^{1/2}]}(1+C^2)^{1/2}\right\}^{-1}$	Figure J-12
Shell-and-tube (n shell passes; $2n$, $4n$, $6n$, etc., tube passes)	$\epsilon_n = \left[\left(\dfrac{1-\epsilon_1 C}{1-\epsilon_1}\right)^n - 1\right]\left[\left(\dfrac{1-\epsilon_1 C}{1-\epsilon_1}\right)^n - C\right]^{-1}$	Figure J-13 for $n=2$
Crossflow (both streams unmixed)	$\epsilon \approx 1-\exp\{C(\mathbf{NTU})^{0.22}[\exp[-C(\mathbf{NTU})^{0.78}]-1]\}$	Figure J-14
Crossflow (both streams mixed)	$\epsilon = \mathbf{NTU}\left[\dfrac{\mathbf{NTU}}{1-\exp(-\mathbf{NTU})}+\dfrac{(\mathbf{NTU})(C)}{1-\exp[-(\mathbf{NTU})(C)]}-1\right]^{-1}$	
Crossflow (stream C_{min} unmixed)	$\epsilon = C\{1-\exp[-C[1-\exp(-\mathbf{NTU})]]\}$	Figure J-15 (dashed curves)
Crossflow (stream C_{max} unmixed)	$\epsilon = 1-\exp\{-C[1-\exp[-(\mathbf{NTU})(C)]]\}$	Figure J-15 (solid curves)

$C = C_{min}/C_{max}$

TABLE J-3 Fouling Factors

fluid	R_f	
	hr · ft² · °F/Btu	m² · K/W
Seawater below 125 °F	0.0005	0.000 09
Seawater above 125 °F	0.001	0.0002
Treated boiler feedwater above 125 °F	0.001	0.0002
Fuel oil	0.005	0.0009
Quenching oil	0.004	0.0007
Alcohol vapors	0.0005	0.000 09
Steam, non-oil-bearing	0.0005	0.000 09
Industrial air	0.002	0.0004
Refrigerating liquid	0.001	0.0002

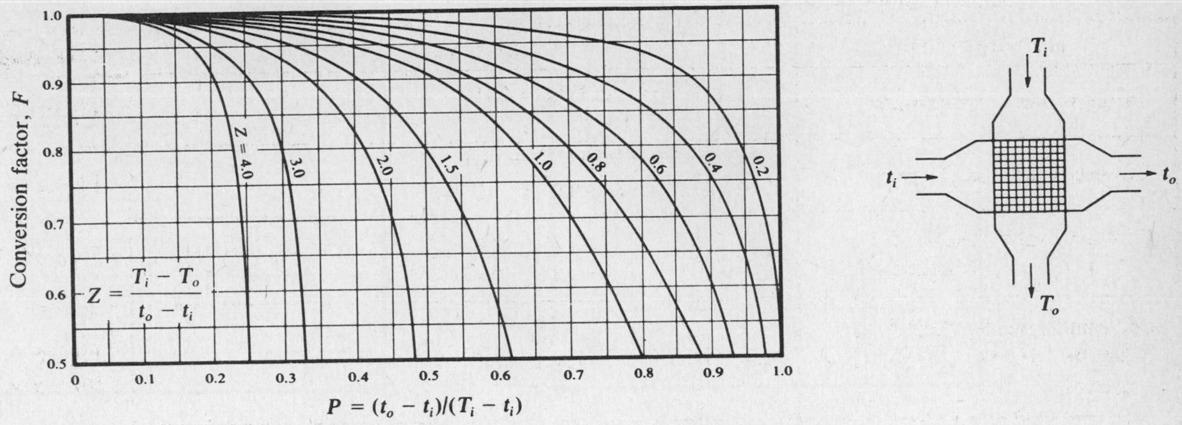

Figure J-9 Crossflow with both fluids unmixed.

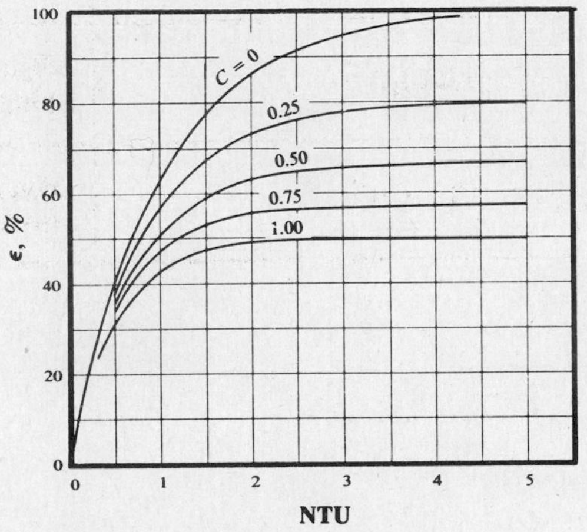

Figure J-10

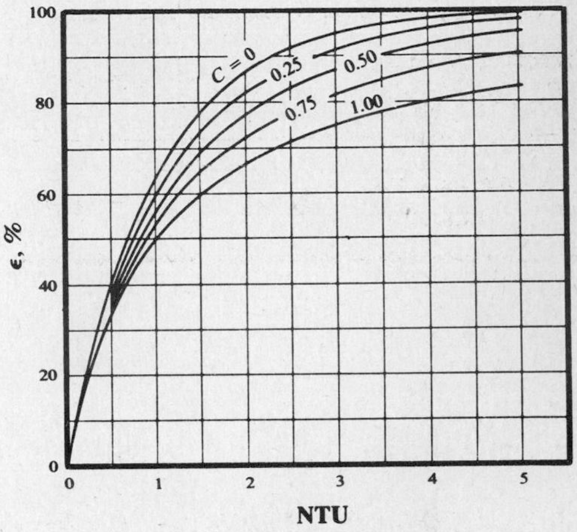

Figure J-11

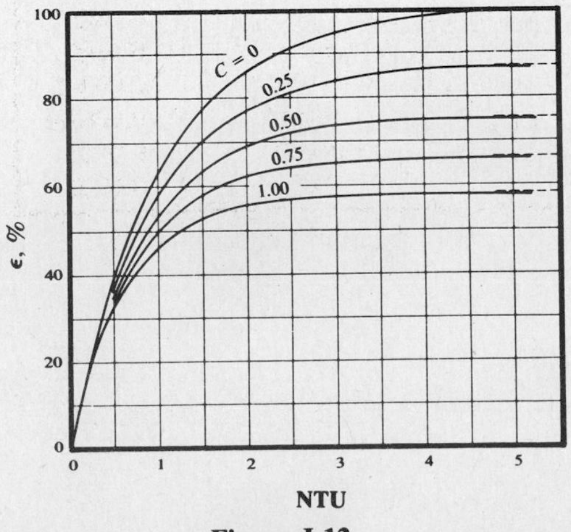

Figure J-12

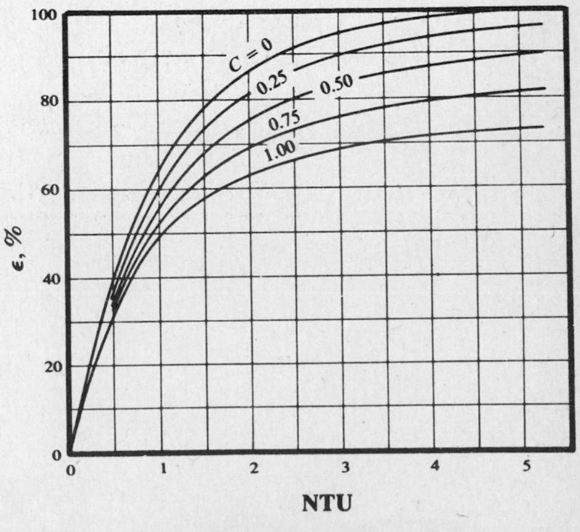

Figure J-13

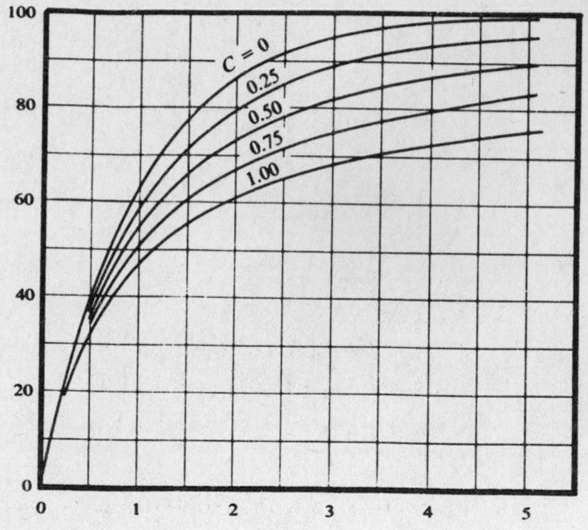

Figure J-14

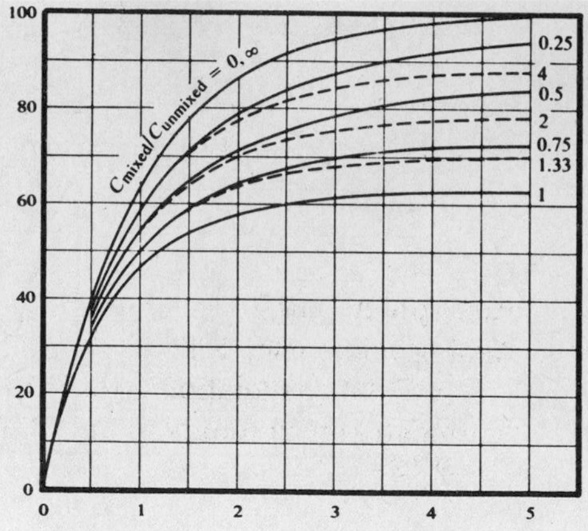

Figure J-15

Definitions and Equations

$$U \equiv \frac{q}{A\,\overline{\Delta T}} = \frac{1}{A \sum R_{\text{th}}} \tag{J-1}$$

$$q = UA\,\frac{\Delta T_2 - \Delta T_1}{\ln(\Delta T_2/\Delta T_1)} \equiv UA(\Delta T)_{lm} \tag{J-2}$$

$$U_o = \frac{1}{A_o \sum R_{\text{th}}} = \frac{1}{A_o\left[\dfrac{1}{A_i\bar{h}_i} + \dfrac{\ln(r_o/r_i)}{(A_o/r_o)k_{i-o}} + \dfrac{1}{A_o\bar{h}_o}\right]} \tag{J-3}$$

$$U_o = \frac{1}{\dfrac{r_o}{r_i\bar{h}_i} + r_o\left[\dfrac{\ln(r_o/r_i)}{k_{i-o}}\right] + \dfrac{1}{\bar{h}_o}} \tag{J-4}$$

$$q = UAF(\Delta T)_{lm} \tag{J-5}$$

$$R_f \equiv \frac{1}{U_{\text{dirty}}} - \frac{1}{U_{\text{clean}}} \tag{J-6}$$

$$U_{o,d} = \frac{1}{\dfrac{r_o}{r_i\bar{h}_i} + \dfrac{r_o}{r_i}R_{f,i} + \dfrac{r_o\ln(r_o/r_i)}{k_{i-o}} + R_{f,o} + \dfrac{1}{\bar{h}_o}} \tag{J-7}$$

$$\epsilon \equiv \frac{\text{actual heat transfer}}{\text{maximum possible heat transfer}} \tag{J-8}$$

$$q = \epsilon(\dot{m}c)_{\min}(T_{hi} - T_{ci}) \tag{J-9}$$

$$\epsilon_h = \frac{C_h(T_{hi} - T_{ho})}{C_{\min}(T_{hi} - T_{ci})} = \frac{T_{hi} - T_{ho}}{T_{hi} - T_{ci}} \tag{J-10}$$

$$\epsilon_c = \frac{C_c(T_{co} - T_{ci})}{C_{\min}(T_{hi} - T_{ci})} = \frac{T_{co} - T_{ci}}{T_{hi} - T_{ci}} \tag{J-11}$$

Index